Principles of PHYSICAL CHEMISTRY
With Applications to the Biological Sciences
Second Edition

Principles of PHYSICAL CHEMISTRY
With Applications to the Biological Sciences
Second Edition

David Freifelder
University of California, San Diego
formerly Brandeis University

Jones and Bartlett Publishers, Inc.
Boston · Portola Valley

Copyright © 1985 by Jones and Bartlett Publishers, Inc., 20 Park Plaza, Boston, MA 02116. Original © 1982 by Science Books International, Inc. All rights reserved. No part of the material protected by this copyright notice may be reproduced or utilized in any form, electronic or mechanical, including photocopying, recording, or by any information storage and retrieval system, without written permission from the copyright owner.

Editorial offices: Jones and Bartlett Publishers, Inc., 30 Granada Court, Portola Valley, CA 94025.
Sales and customer service offices: 20 Park Plaza, Boston, MA 02210.

Library of Congress Cataloging in Publication Data
Freifelder, David Michael
    Principles of physical chemistry, with applications to the biological sciences.

    Rev. ed. of: Physical chemistry for students of biology and chemistry. © 1982.
    Includes bibliographies and index.
    1. Chemistry, Physical and theoretical.   I. Freifelder, David Michael. Physical chemistry for students of biology and chemistry. II. Title. III. Title: Physical chemistry, with applications to the biological sciences.
QD453.2.F73   1985   541.3   83-19999
ISBN 0-86720-046-4

Printed in the United States of America

PREFACE TO THE SECOND EDITION

From its conception *Principles of Physical Chemistry with Applications to the Biological Sciences* was written with the belief that it should be possible to make physical chemistry more meaningful and accessible to students in the life sciences. As explained in the Preface to the first edition, my *modus operandi* was to limit coverage to subjects of immediate value to such students, to minimize the mathematical derivations, and to include subjects that are of particular interest to biologists and not usually found in traditional physical chemistry textbooks.

In an effort to discover whether the goals of the first edition had been achieved, my publishers and I solicited information from instructors who used the book during the first years of publication. We asked for lists of typographical and textual errors, suggestions about topics to be eliminated or added, suggestions for improving clarity, and opinions about the appropriateness of the level of the mathematics. Many instructors responded to our request. Some offered the same suggestions; of course, there were also marked differences in opinion. Every response was noted, compared with others, and considered carefully. This new edition has been revised in accord with these comments.

Apart from corrections of numerous errors in the mathematics, the most significant change in *Principles of Physical Chemistry with Applications to the Biological Sciences* is the addition of a new chapter (Chapter 19) concerned with the structure of crystals, liquids, and liquid crystals, and the use of X-ray crystallography in determining molecular structure. X-ray diffraction as applied to macromolecules is, of course, a mathematically complex subject; a comprehensive treatment would be inconsistent with the

PREFACE

level of presentation chosen for this book. Thus, I have merely attempted to explain some of the more elementary features of X-ray crystallography and provide a basic vocabulary; this knowledge should enable a student to study more advanced texts, if desired. Furthermore, a section on calorimetry has been added to supplement the treatment of thermodynamics, so the student will have some knowledge of the origin of the numbers found in thermodynamic tables. The topic of adiabatic expansion, omitted from the first edition, has also been added. For completeness, the concept of fugacity is now included in the treatment of real systems. A section on free energy and the significance of standard states has also been expanded and the relation between free energy changes and concentration has been clarified with several examples. The quantum mechanics section has been expanded slightly with a section describing tunneling. The phenomenon of independently folded domains in proteins is also an important addition to the chapter on macromolecules. Finally, summaries within chapters (*Comparison of Isothermal, Adiabatic, Reversible, and Irreversible Expansions*) and several appendices (*Exact and Inexact Differentials; Some Useful Thermodynamic Relations; Physical Properties of Water*) have been added.

In an attempt to reduce the mathematical complexity of discussions, derivations have been moved from within chapters and placed in appendices at the ends of those chapters. The student should now be able to understand the meaning and applications of the Debye-Hückel theory without resorting to advanced mathematics and physics that may not be part of his or her background. The section on C_0t analysis and several sections containing serious typographical errors have been corrected.

I hope that all users—professors and students alike—will find the second edition to be an improvement.

I wish to thank the numerous instructors, students, and reviewers who responded to my request for feedback. Special commendation must go to Arthur Bartlett and Donald Jones, my publishers, who were willing to have a second edition follow shortly after the first edition. Their faith and support reflect their commitment to maintaining the highest standards in their publications.

San Diego, Calif.
April, 1984

David Freifelder

PREFACE TO THE FIRST EDITION

Several years ago I was asked to revise a successful physical chemistry textbook so that it would be more useful to biology students and to chemistry students aiming for careers in the life sciences. Since I had been especially concerned about the deficient training in basic physical chemistry possessed by many students and researchers in the life sciences, I enthusiastically began the project. It did not take long for me to discover two great difficulties.

First, the needs of the researcher in the life sciences are, indeed, quite different from those of the chemist or the physicist—the traditional textbooks in physical chemistry include a great deal of material that will never be used by the life scientist yet omit subjects of considerable relevance.

Second, many students in the life sciences do not have the background in mathematics or the commitment to learn the mathematics needed to follow the derivations encountered in traditional physical chemistry textbooks.

Keeping the above two points in mind as I attempted my revision, I observed that by the time I had pared away the topics not useful to the life scientist, there was little left of the original manuscript. I then realized that if the book that had been requested were to exist, it would have to be written from scratch.

I decided to take this new task upon myself; so began a project that was to take me about 1500 hours and to consume reams and reams of paper. Through it all, I determined never to forget my own experiences in studying the material I was now writing about.

As a student of physical chemistry in 1956 I sat in class month after month watching derivations being developed on the blackboard and spent hundreds of hours solving problems that began with the words "Show that..." or "Derive...." After having completed a year of physical chemistry, despite having received a good grade I found myself wondering what I had *truly* learned. This question seemed far more difficult to answer than any of the problems I had worked in the course; the course had seemed inordinately difficult whereas other courses in chemistry and physics had not seemed so. What was the reason for this? The answer to the latter question was simply that at no point had the practical usefulness of physical chemistry in the life sciences been made manifest. Physical chemistry has traditionally been taught as an abstract subject and is often unrelated to the kind of phenomena encountered in the laboratory and in real life, in part, I believe, because physical chemistry textbooks are usually written by persons whose expertise enables them not only to see the overall picture and the logical harmony of the subject easily but also to believe that these are obvious to all students, as well. Such authors most likely learned physical chemistry easily and spent little time probing the connection between physical chemistry and common experience—the very connection that is not at all obvious to the average student.

Many new books have appeared recently that have been designed to teach physical chemistry to potential biologists or to prepare chemistry students for subsequent courses related to the life sciences. The publication of these books might have suggested to me that I should not expend any time in writing my own, but reading each has reinforced my feeling that today's student, completing a course designed by a physical chemist possessing traditional expertise, will still be left with the feeling of confusion that confronted me 25 years ago. I continue to believe that the student heading for a career in the life sciences needs a physical chemistry textbook written by an experimental biologist who did not find the learning of physical chemistry to be a trivial task. This is the book that has resulted from that belief.

The book is not exclusively for biology majors—this should be emphasized. At present, a student planning to enter a graduate program in the life sciences or to apply to medical school may, as an undergraduate, major in a biological field, in chemistry, or in physics (or even in the classics, economics, or history, which is true of some of my own students). This book is for all of these people. Furthermore, the strong practical orientation of the book makes it one that should not be ignored for chemistry programs designed to train experimental chemists.

The reader who compares this book with other books about physical chemistry and also with those specifically meant for the student in the life sciences will discover that this book differs in several ways.

First, this book relies less on mathematics than traditional texts do. I appreciate the fact that physical chemistry has a firm mathematical basis, yet this is often beyond the training of the student and can become intimidating. Mathematics is useful if it provides deeper understanding of the material. If an author's major requirement is that the student understand

PREFACE

how equation 6 mathematically becomes equation 7, then the emphasis on mathematics may impede learning. In this book, no mathematics more advanced than elementary calculus and simple probability theory is used and even these subjects are reviewed in appendices given at the end of the book.

A second important difference is that thermodynamics is given a molecular foundation. Thermodynamics is a marvelously logical subject that does not require molecular explanation. If the two basic postulates, the First and Second Laws of Thermodynamics, are accepted, all else follows logically; it is unnecessary to think about the fact that matter consists of atoms and molecules. To approach thermodynamics as a logical system enables one to use thermodynamics as an elegant tool for deriving numerous valuable equations. Yet such an approach ignores the fact that the systems being discussed do indeed consist of molecules. It has been my experience that to ignore this fact is not satisfying to experimental scientists or to biochemists and molecular biologists, who spend their professional lives trying to unravel the molecular basis of living phenomena and who prefer to know how molecules behave in a given system. Thus, I have presented thermodynamics by combining its postulates with the physical notions of molecular motion and probability. Thermodynamics loses some of its logical beauty in this treatment but the student will benefit by knowing what one actually means by such concepts as internal energy, free energy, entropy, and so forth. Furthermore, the molecular viewpoint used in this book also makes the theory of solutions clearer because, with this viewpoint, one is not hesitant to discuss cohesive forces early in the theory.

A third difference that should be noticeable in this book is the extensive discussion of macromolecules, such as proteins and nucleic acids, in the examples that are part of each chapter. The reason for this emphasis is simple. Most biologists devote more time to the properties of macromolecules than to smaller molecules. Furthermore, in the life sciences physical chemistry can be applied more directly to the study of macromolecules than to smaller molecules.

A fourth difference is the inclusion of lengthy discussions — a chapter for each — of properties of macromolecules and photochemistry, subjects of importance to all endeavoring to study the life sciences, whether approaching from backgrounds in chemistry or biology. Furthermore, quantum mechanics is introduced primarily as a means of understanding spectroscopy, which reflects the practical needs of many experimental scientists.

A fifth difference appears in the problems listed at the end of each chapter. Many problems of a variety of types are provided. Most of these tend to be either very experimental in nature or to require molecular explanations and are designed to give practice in solving the problems that will be encountered when doing laboratory work. Derivations and theoretical notions, which are the mainstay of problem lists in traditional books, have only occasionally been included in this book. Answers to all problems are given, often in detail, at the end of the book.

PREFACE

Once the manuscript for this book was finished, many people helped me to doublecheck the accuracy and soundness of every aspect of it. I received expert advice from Professors Robert Abeles, Irving Epstein, Tom Hollocher, William Jencks, Chris Miller, and Serge Timasheff, all of Brandeis University. In order to obtain the viewpoints of students who had already successfully learned physical chemistry within the past few years, I enlisted the aid of Mike White, who read the entire manuscript, and of Bruce Breit, Alec Cheung, Bruce Gomes, Jon Greene, Les Lang, and Rob Rottapel. To be useful, a textbook must be understandable to the novice. Thus various chapters were read by several undergraduates; Gil Drozdow, Jeff Friedman, and Drew Weissman explained to me any parts that gave them difficulty, and these were reexamined and revised when this was necessary. Finally, when the revised manuscript was completed, another beginning biochemist, Gregg Bannett, read most of it, and further adjustments were made where these were needed. For help in convincing a publisher that the finished manuscript should be published, I thank Peter von Hippel of the University of Oregon, who read early drafts and gave them his stamp of approval. My final thanks must go to Mildred Kravitz, who hates typing but who faithfully joined me in typing about 3000 pages of manuscript. And to my children, Rachel and Joshua, who hope some day to write their own books and who tolerated my typewriter on the dining room table for about a year.

Lexington, Massachusetts   David Freifelder

CONTENTS

Preface to the Second Edition,   iii

Preface to the First Edition,   ix

Chapter 1   UNITS OF MEASUREMENT AND THE GAS LAWS,   1

- 1–1   Units of Measurement,   1
- 1–2   Gases,   6
  References,   15
  Problems,   15

Chapter 2   KINETIC THEORY,   17

- 2–1   The Kinetic Theory of Gases,   18
- 2–2   Energy,   20
- 2–3   Motion of Gas Molecules,   24
- 2–4   Distribution of Kinetic Energies and Molecular Velocities,   30
- 2–5   Brownian Motion,   36
  References,   38
  Problems,   39

Chapter 3 INTERMOLECULAR FORCES, 41

- 3–1 Failure of the Ideal Gas Law: Evidence for an Attractive Force, 41
- 3–2 The Van Der Waals Gas Law and Its Consequences, 44
- 3–3 The Nature of the Attractive Force, 48
- 3–4 Relation Between Intermolecular Forces and Physical Properties, 62
- 3–5 Chemical Bonds, 66
- 3–6 Weak Bonds in Biological Systems, 69
- 3–7 Summary of Various Molecular Interactions, 76
- References, 77
- Problems, 77

Chapter 4 THERMODYNAMICS—THE FIRST LAW, 80

- 4–1 Thermodynamic Laws, 81
- 4–2 The First Law of Thermodynamics, 82
- 4–3 Enthalpy or Heat Contents, 91
- 4–4 Thermochemistry—The Application of the Enthalpy Concept, 96
- References, 114
- Problems, 115

Chapter 5 THE SECOND LAW OF THERMODYNAMICS, 118

- 5–1 The Second Law, 118
- 5–2 The Entropy Concept, 129
- 5–3 Evaluation of the Entropy of a System, 140
- 5–4 The Entropy of a Solution in Water: An Effect in Biological Systems, 149
- References, 153
- Problems, 154

Chapter 6 FREE ENERGY, 156

- 6–1 The Gibbs Free Energy, G, 157
- 6–2 Free Energy and the Equilibrium Constant, 170

6-3 Summary of Methods for Evaluation Thermodynamic Functions, 175
6-4 Application of Free Energy and Biological Processes, 177
References, 190
Problems, 191

Chapter 7 SOLUTIONS OF UNCHARGED MOLECULES, 194

7-1 Methods of Describing Concentration, 195
7-2 The Chemical Potential, 196
7-3 Ideal Solutions, 199
7-4 Colligative Properties, 203
7-5 Solution of Gases in Liquids, 216
7-6 Factors That Determine the Solubility of a Solid, 219
7-7 Partitioning—The Distribution of a Solute Between Two Immiscible Solvents, 221
7-8 Solutions at Temperatures for which the Components Are Near the Melting Points—The Phase Rule, 224
7-9 Nonideal Solutions, 227
References, 237
Problems, 237

Chapter 8 SOLUTIONS OF ELECTROLYTES, 242

8-1 The Ionic Theory, 242
8-2 Need for a Theory of Solutions of Electrolytes, 246
8-3 The Concept of Activity for Electrolytes, 250
8-4 The Debye—Hückel Theory, 256
8-5 Other Considerations about Ionic Strength, 266
8-6 Hydration of Ions, 268
8-7 Salting-Out Proteins, 272
References, 274
Problems, 274

Chapter 9 ACID-BASE EQUILIBRIA, 277

9-1 Dissociating Systems, 277
9-2 Acid-Base Mixtures, 286

9–3 Amino Acids and Proteins, 301
9–4 Metal Ion (Coordination) Complexes, 312
References, 317
Problems, 317

Chapter 10 ELECTROCHEMISTRY AND OXIDATION-REDUCTION REACTIONS, 320

10–1 Review of Basic Electrochemical Concepts, 321
10–2 Electrochemical Reactions and Electrochemical Cells, 321
10–3 Reference Electrodes, 333
10–4 Standard Electrode Potentials and Thermodynamic Quantities, 335
10–5 Concentration Cells and Ion-Selective Electrodes, 341
10–6 Use of Reduction Potentials to Calculate Concentration at Equilibrium, 345
10–7 Potentials of Biochemical Reactions, 347
10–8 The Terminal Oxidation Chain in Living Cells, 350
References, 351
Problems, 352

Chapter 11 CHEMICAL KINETICS, 354

11–1 The Concept of Reaction Mechanisms, 355
11–2 Rate Laws, Order, and Molecularity, 356
11–3 Rate Laws for Reactions of Different Order, and the Determination of Their Rate Constants, 359
11–4 Rate Equations for More Complex Reactions, 366
11–5 Fast Reactions, 372
11–6 Chain Reactions, 378
11–7 Effect of Temperature on the Reaction Rate, 379
11–8 Transition-State Theory, 381
11–9 Diffusion-Controlled Reactions, 384
11–10 Reactions Between Ions in Solution, 387
References, 389
Problems, 390

Chapter 12 CATALYSIS AND ENZYME KINETICS, 396

12–1 Catalysis, 396
12–2 Enzymatic Reactions, 404
References, 416
Problems, 416

Chapter 13 SURFACES, 418

13–1 Surface Tension of Liquids, 418
13–2 Surface Films and Monolayers, 426
13–3 Adsorption, 439
13–4 Practical Aspects of Adsorption, 453
13–5 Micelles and the Bilayer Model of Membrane Structure, 456
13–6 Detergents, 460
13–7 Flotation, 462
References, 464
Problems, 465

Chapter 14 ELECTRICAL AND TRANSPORT PROPERTIES OF SURFACES AND MEMBRANES, 468

14–1 A Precaution about Electrical Units, 468
14–2 Electrical Potentials across Surfaces, 469
14–3 Membranes, 476
14–4 Effect of Ions on the Conductivity of Charged Membranes, 488
References, 493
Problems, 493

Chapter 15 THE PROPERTIES OF MACROMOLECULES, 495

15–1 Chemical and Physical Properties of Macromolecules, 495
15–2 Changes in the Structure of Macromolecules, 513
15–3 Movement of Macromolecules, 531
15–4 Determination of Molecular Weights, 546
References, 565
Problems, 566

Chapter 16 LIGAND BINDING, 570

- 16–1 Molecules with One Binding Site, 571
- 16–2 Macromolecules with Several Binding Sites, 574
- 16–3 Cooperative Binding, 578
- 16–4 Allostery, 584
- 16–5 Experimental Methods to Measure Binding, 595
- References, 600
- Problems, 600

Chapter 17 QUANTUM MECHANICS AND SPECTROSCOPY, 604

- 17–1 Origins of the Quantum Theory, 604
- 17–2 The New Quantum Mechanics, 612
- 17–3 Molecular Spectroscopy, 628
- 17–4 Fluorescence, 639
- 17–5 Proteins and Nucleic Acids, 646
- References, 660
- Problems, 660

Chapter 18 PHOTOCHEMISTRY, RADIATION CHEMISTRY, AND RADIOBIOLOGY, 663

- 18–1 Light, 663
- 18–2 Absorption of Light, 665
- 18–3 Scattering of Light, 673
- 18–4 Photochemical Mechanisms, 677
- 18–5 Photochemical Damage to Biological Molecules and to Living Organisms, 685
- 18–6 Ionizing Radiation, 689
- 18–7 Hit Theory, 707
- 18–8 Effect of the Environment on Radiosensitivity, 715
- References, 717
- Problems, 718

Chapter 19   **SOLIDS AND LIQUIDS,**   723

    19–1   Crystals,   723
    19–2   Liquids,   741
    19–3   Liquid Crystals,   742
            References,   748
            Problems,   749

Appendix I   **SOME NECESSARY MATHEMATICAL RELATIONS,**   750

Appendix II   **SOME STATISTICAL RELATIONS,**   757

Appendix III   **SOME USEFUL THERMODYNAMIC RELATIONS,**   760

Appendix IV   **A STATISTICAL DERIVATION OF RAOULT'S LAW,**   761

Appendix V   **CALCULATION OF SOLUTE ACTIVITY FROM THE SOLVENT ACTIVITY,**   763

Appendix VI   **PHYSICAL PROPERTIES OF WATER,**   765

Appendix VII   **DERIVATION OF THE DEBYE-HÜCKEL LIMITING LAW,**   766

               ANSWERS TO SELECTED PROBLEMS,   770

               INDEX,   801

Principles of
PHYSICAL CHEMISTRY
With Applications to the Biological Sciences

Second Edition

CHAPTER 1

UNITS OF MEASUREMENT AND THE GAS LAWS

Physical chemistry includes a large number of subjects, most of which are related to some degree. These topics can be described in simple and elegant mathematical terms. However, in order to integrate these subjects, one must have a common language; for example, there must be some way to equate the electrical energy of a battery with the kinetic energy of a collection of gas molecules. Hence, it is appropriate that we begin our investigation of physical chemistry by examining the units by which the relevant parameters are expressed. This is the subject of the first part of this chapter.

Many of the basic concepts of physical chemistry and thermodynamics can be developed by discussing the properties of gases. This is not the only approach but is the simplest and the traditional one. The properties of gases are described in most beginning textbooks in chemistry. However, a quick review, as presented in the second part of this chapter, is valuable, because we can then make use of gases to develop the subject of intermolecular forces (Chapter 3) and the laws of thermodynamics (Chapters 4 and 5).

1-1 UNITS OF MEASUREMENT

In the course of learning physical science, students often encounter several different systems of units of measurement and this has probably resulted in confusion and been at the root of many arithmetic errors. The two basic systems of mechanical units are the centimeter-gram-second (cgs) and the meter-kilogram-second (mks) systems. In the study of electrostatics, electrical charge can be expressed in three ways—in electrostatic units (esu), electromagnetic units (emu), and coulombs. Usually mechanical and electrical units are combined; combination forms the cgs-esu (Gaussian) and the mks-coulomb systems.

To some extent, the units that are in use have varied from one branch of science to another, either for historical reasons or because one system seems more applicable than another. On numerous occasions committees have discussed the standardization of units but usually without agreement. However, in 1960 an international commission defined a system of units for universal use by all scientists. This system is called the International System of Units, or SI units, for *système internationale.* Since 1960 there has not been agreement among laboratory scientists about whether this system should replace the more familiar systems. The principal problem has been that in the United States these units represent a departure from the units that have been used for decades in chemistry, physics, or biology, and throughout the world the SI units differ from those used by biochemists.

There are two major problems in the transition from Gaussian to SI units. One of these is that the values of the fundamental constants such as the Planck constant h, the Boltzmann constant k, the gas constant R, and the charge of the electron e differ, and most of the Gaussian values were committed to memory long ago. The second problem is that equations often have different forms in the two systems. For example, a factor of 4π appears in certain equations describing electrical phenomena if the charge of the electron is expressed as an SI unit or as a coulomb unit but is absent if the Gaussian electrostatic unit is used.

There is no doubt that the SI units will some day be universally adopted and that modern books should contribute to this tendency. However, for the present, the use of SI units presents numerous problems.

SI units call for meters (m) or fractions such as micrometers (1 μm = 10^{-6} m) and nanometers (1 nm = 10^{-9} m) to measure distances, and the cubic decimeter, which equals one liter, to measure volume. The units μm and nm are useful and sould replace the micron, ($\mu = \mu$m), and the ångström, Å (10 Å = 1 nm). However, enormous resistance is met when scientists are asked to give up the familiar liter, cubic centimeter, and milliliter. In this book we will express distance in the units that seem most appropriate for the size and will carefully indicate for each equation what unit of length is to be used. Molecular distances will usually be expressed in ångströms or microns, and wavelengths will be given in nanometers, both in accord with current usage. Small volumes will be given in cm^3 and large volumes in liters.

Pressure is an especially troublesome item. The SI unit of pressure is the pascal, yet it is rare to find any measuring instrument calibrated in pascals. It is far more common to see the millimeter of mercury (mm Hg), the torr, or the atmosphere used as a measure of pressure. It is exceedingly important to be aware of the units of pressure in a problem or an experimental system. This is because the gas constant R is frequently a factor in equations in which pressure is measured. The units of R are volume-pressure per degree, so that R must always be expressed in the appropriate units for pressure and volume. It is important that students be able to work in the various systems of expressing pressure, because each appears repeatedly in the laboratory. Thus, in order to provide practice in these units, examples and problems will be given in the various systems.

The two major units of energy are the erg (cgs) and the joule, J (an SI unit). One $J = 10^7$ erg. The unit of heat has traditionally been the calorie, cal—1 cal = 4.184 J = 4.184×10^7 erg. However, chemists and biochemists have customarily used the calorie as a measure of both heat and energy. To change this practice meets with resistance because most chemists have a "feeling" for the size of one kilocalorie. That is, if a chemical reaction has a free energy change of -50 kilocalorie per mole, a chemist knows whether this is an efficient or an inefficient reaction. This is just like the American who knows that an air temperature of 90° Fahrenheit is quite hot, but is not sure whether 32° Celsius is comfortable or not. The disadvantage in using the calorie as a unit of energy is the necessity to divide the value in joules by 4.184, which is not a particularly easy number to remember. Clearly, if students initially learn to use the joule as an energy unit, they will have the same feeling for the joule that older chemists have for the kilocalorie. Thus, in this book, both energy and heat will be expressed in joules, as required in the SI system. In some cases, the value will also be given in kilocalories (in parentheses) when it is a commonly known number. The reader should remember to use the value of the gas constant R of 8.314 J K^{-1} mol^{-1}. (The symbol K is used to symbolize the unit kelvin, formerly called the degree Kelvin.)

The use of molecular versus molar units also can lead to confusion. We have adopted the following convention in this book. Mass in grams is denoted by a lower case m; molecular weight in grams per mole is designated by a capital M. The relation between m and M is of course $M = \mathcal{N}m$, in which $\mathcal{N}$ is Avogadro's number (6×10^{23}), which we denote by a script $\mathcal{N}$. The gas constant R also refers to one mole; the Boltzmann constant k refers to one molecule and is related to R by $R = \mathcal{N}k$. The value of k is 1.38×10^{-23} J K^{-1} or, in cgs units, 1.38×10^{-16} erg K^{-1}.

Electrical units produce considerable complexity. However, the universal use of the volt, V (an SI unit), as a measure of electrical potential clearly indicates how to proceed. The volt is defined as 1 V = 1 J per coulomb, which necessitates the use of the coulomb, C (an SI unit), as the unit of charge. The volt and the coulomb are used exclusively in this book. In this system the charge of the electron, e, is $e = 1.6 \times 10^{-19}$ C, and the electrical potential ϕ of a charge e at a distance r from the charge is $\phi = e/4\pi\epsilon_0 r$, in which ϵ_0, the permittivity of vacuum, has the value $\epsilon_0 = 8.85 \times 10^{-12}$, and r is measured in meters: it is important to notice that r is in meters because the electric field strength, which appears repeatedly in surface chemistry and in electrophoresis, is defined as voltage per unit distance; therefore, since the volt is an SI unit, the distance should be in meters. In laboratory practice, however, the electric field is invariably expressed as volts per centimeter. In consulting other textbooks, especially older ones, electrostatic units may be found. Expressed in these units, the charge of the electron is 4.8×10^{-10} esu, or, 4.8×10^{-10} statcoulombs. In the electrostatic unit system, electrical potential must be expressed in statvolts: 300 V = 1 statvolt. The esu-statvolt system is confusing and is generally not recommended. Sometimes it is difficult, when obtaining an equation of electrostatics from a reference book, to know what units have been used. The presence of ϵ_0 is a clear indication that volts and coulombs should be used. If ϵ_0 is not to be seen, one cannot be sure; however, if an even multiple of π is found, it is likely that the SI volt-coulomb system is being employed.

In the following, several tables are provided that compare SI units with cgs units. Table 1-1 lists the base units, those units from which all others are derived. Tables 1-2 and 1-3 list the derived units for the SI and cgs systems. Table 1-4 lists conversion factors between various units. Finally, Table 1-5 presents the values of the fundamental constants in SI and cgs units.

TABLE 1-1 SI and non-SI base units.

Physical variable	SI Name	SI Symbol	Non-SI Name	Non-SI Symbol
Mass	Kilogram	kg	Gram	g
Length	Meter	m	Centimeter	cm
Time	Second	s	Second	s or sec

TABLE 1-2 The derived SI units.

Physical variable	Name	Symbol	Definition
Force	Newton	N	$kg\ m\ sec^{-2}$
Pressure	Pascal	Pa	$kg\ m^{-1}\ sec^{-2}$ or $N\ m^{-2}$
Energy	Joule	J	$kg\ m^2\ sec^{-2}$
Volume	Cubic decimeter	dm^3	$10^{-3}\ m^{-3}$

SEC. 1–1   UNITS OF MEASUREMENT

TABLE 1-3   The derived non-SI units.

Physical variable	Name	Symbol	Definition
Force	Dyne	dyn	g cm sec^{-1} or 10^{-5} N
Pressure	Atmosphere	atm	1.013×10^6 gm cm^{-1} sec^{-2}
Pressure	Torr	torr	1/760 atm
Pressure	Millimeter of mercury	mm Hg	1/760 atm
Energy	Erg	erg	g cm^2 sec^{-2} or 10^{-7} J
Energy	Calorie	cal	4.184 J
Volume	Cubic centimeter	cm^3	cm^3
Volume	Liter	l	10^3 cm^3

TABLE 1-4   Conversion factors between various units.

1 liter (l) = 1000 cm^3 = 1000 cubic centimeter
1 meter (m) = 100 cm = 10^6 μm = 10^9 nm
1 ångström (Å) = 10 nm = 10^{-8} cm = 10^{-10} m
1 joule (J) = 10^7 erg
1 calorie (cal) = 4.184 J
1 electron volt (eV) = 1.6×10^{-19} J
1 newton (N) = 1 kg m sec^{-2} = 10^5 dyne (dyn)
1 pascal (Pa) = 1 N m^{-2} sec^{-2}
1 atmosphere (atm) = 1.013×10^6 dyn cm^{-2} = 1.013×10^5 N cm^{-2} = 1.013×10^5 Pa
1 Pa = 9.87×10^{-6} atm = 10 dyn cm^{-2}
1 torr = 1 mm Hg = 1/760 atm = 133.3 Pa
1 electrostatic unit (esu) = 3.33×10^{-10} coulomb (C)
1 C = 3×10^9 esu
1 statvolt = 300 volt (V)

TABLE 1-5   Values of fundamental constants in SI and Gaussian units.

Constant	Symbol	SI value	Gaussian value
Avogadro's number	$\mathcal{N}$	6.022×10^{23} mol^{-1}	6.022×10^{23} mol^{-1}
Electronic charge	e	1.602×10^{-19} C	4.803×10^{-10} esu
Planck constant	h	6.626×10^{-34} J sec	6.626×10^{-27} erg sec
Boltzmann constant	k	1.381×10^{-23} J K^{-1}	1.381×10^{-16} erg K^{-1}
Gas constant*	R	8.314 J K^{-1} mol^{-1}	8.314×10^7 erg K^{-1} mol^{-1}
Faraday constant	$\mathcal{F}$	96,485 C mol^{-1}	2.9×10^{14} esu mol^{-1}
Speed of light	c	2.998×10^8 m sec^{-1}	2.998×10^{10} cm sec^{-1}
Permittivity of vacuum	ϵ_0	8.854×10^{-12} C^2 N^{-1} m^{-2}	Not used
Acceleration of gravity	g	9.80 m sec^{-2}	980 cm sec^{-2}

*Also, the value of 0.08206 l atm K^{-1} mol^{-1} is useful.

1-2 GASES

An analysis of the behavior of gases is a common way to begin the study of physical chemistry even though the modern laboratory worker does not usually study gases. This starting point is selected mainly because gases are easier to understand and simpler to describe mathematically than are solids and liquids, although the latter are of greater interest to the life scientist. Furthermore, many of the factors that determine the basic properties of matter are uncovered by a thorough examination of gases. It is curious, though, that most of the principles we have come to understand are derived from an analysis of what is called the ideal gas.

To behave ideally, a gas must have the following three properties: (1) the individual molecules neither attract nor repel one another; (2) the volume of each molecule is zero; and (3) the gas does not liquefy at any pressure or temperature. Although no real gas has these properties, many gases have certain qualities of an ideal gas when the temperature is far above the boiling point and when the pressure is low (as long as one does not examine the gas too closely).

The Basic Properties of Gases

A gas may be defined as that form of matter that expands without limit to fill the available space of a container. Thus, if one ignores the random fluctuations that occur within microscopic volumes in any container, all volume elements have the same number of molecules. These molecules collide with the walls of the container and thereby exert pressure. (For the reader who has forgotten, pressure is defined as the force per unit area.) The pressure of a gas is uniform on all walls of any container regardless of its shape or size.

Intuition tells us that if a gas were contained in a particular volume and if the volume were increased (without a temperature change), the gas would expand, thereby reducing the number of molecules per unit volume, and the gas pressure would decrease. Robert Boyle, in the 17th century, studied this carefully and, using the simple apparatus shown in Figure 1-1, discovered the fact commonly known as Boyle's Law:

> *for a particular mass of gas at a single temperature the product of the volume* V *and the pressure* P *is constant,*

or

$$PV = \text{a constant, at constant temperature} . \tag{1}$$

We will see in Chapter 3 that careful measurements show that this equation is not strictly obeyed.

FIGURE 1-1 Robert Boyle's J-tube. A gas is trapped above the mercury at the sealed end of the J-tube. The pressure on the gas equals the weight of the excess height of Hg in the large branch of the J-tube divided by the cross-sectional area of the tube. Addition or removal of Hg allowed Boyle to vary the pressure. Since the tube has a constant cross-section, the volume of the gas is proportional to the length of the portion of the tube containing the gas, and the pressure of the gas is proportional to the height of the Hg column, as indicated.

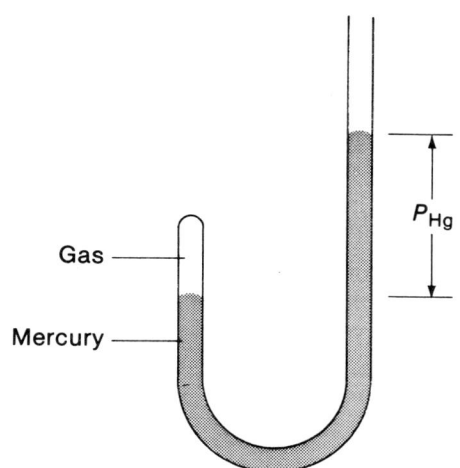

Boyle's Law is easy to rationalize if one realizes that the pressure of a gas is a result of collisions of the gas molecules with the walls of the container. Thus, if the volume of a container were to increase, the number of collisions per unit time per unit area would decrease in proportion to the volume increase. This will be explained more fully in Chapter 2.

EXAMPLE 1–A The pressure change during expansion of a gas in a closed cylinder (calculated in different units of pressure).

A gas is enclosed in a vertical cylinder into which has been fitted a weighted cylindrical piston having a mass of 10 kg and a radius of 5 cm. The volume of the gas is 6 l at 25°C. It is possible to reduce the mass of the piston by removing some of the auxiliary masses. When this is done, the gas expands to 25 l. The cylinder is contained in a constant temperature bath, so the temperature remains at 25°C. What will the final pressure be and what will the mass of the piston be after the expansion?

The area of the piston remains constant, so that the pressure (that is, the force per unit area) is proportional to the mass of the piston. To convert mass to force, the mass must be multiplied by g, the acceleration of gravity. Thus, the initial force acting on the mass is $(10^4 \text{ g})(980 \text{ cm sec}^{-2}) = 9.8 \times 10^6$ dyn. The volume increases by a factor of 25/6, so that the force, according to Boyle's Law, must decrease by the same factor. Therefore the force of the piston after expansion is $(6/25)(9.8 \times 10^6) = 2.35 \times 10^6$ dyn and the mass is $(6/25)(10^4) = 2400$ g. The final pressure is $(2.35 \times 10^6)/\pi 5^2 \text{cm}^2 = 3 \times 10^4$ dyn cm^{-2}.

In SI units, mass is expressed in kg, distance is expressed in m, and g = 9.8 m sec^{-1}. Thus the initial force is $(10)(9.8) = 98$ N and the force of the piston after expansion is $(6/25)(98) = 23.5$ N. The pressure is then $23.5/\pi(0.05)^2 = 3 \times 10^3$ N m^{-2} sec^{-2} = 3×10^3 Pa. The final mass is $(6/25)(10) = 2.4$ kg.

If we wish to express the pressure in atmospheres, we note that a pressure of 1 atm supports a column of mercury 76 cm high. The force exerted by such a column that is 1 cm^2 in area is $(76)(13.7)(980) = 1.013 \times 10^6$ dyn, in which 13.7 is the density of mercury in g cm^{-3}. Therefore the final pressure is $3 \times 10^4/(1.013 \times 10^6) = 0.03$ atm.

Boyle's Law is a limiting law in that it is valid only at fairly low pressures and high temperatures. In Chapter 3 we discuss the reasons for its failure for most real gases. A hypothetical gas that would obey Boyle's Law at all pressures and temperatures is called an *ideal gas;* at low pressure and high temperature one generally assumes that all gases behave ideally.

Boyle's Law can be described in graphical form, as shown in Figure 1-2, in which the pressure P is plotted against the volume V. The resulting curves for various temperatures are hyperbolic; a curve obtained for a particular temperature is called an *isotherm*.

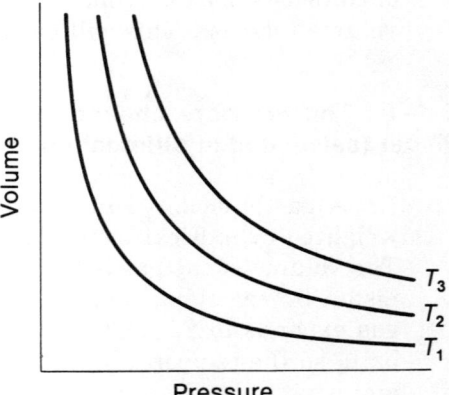

FIGURE 1-2 A plot of Boyle's Law at various temperatures, $T_1 < T_2 < T_3$, showing the hyperbolic relation between P and V. Since each curve represents a single temperature, the curve is known as an *isotherm*.

Note that Equation 1 refers to a particular mass of gas—that is, the constant in the equation depends upon weight. Amadeo Avogadro observed that for a large number of gases examined at relatively low pressure, the volume of one mole of gas is approximately the same when measured at the same temperature. This volume, known as the *gram molecular volume*, V_m, is 22.4 l at 0°C and a pressure of 1 atm. From this finding, in 1811 Avogadro put forth the hypothesis (proved many years later by others) that

equal volumes of gases at the same pressure and temperature contain the same number of molecules,

a statement known as Avogadro's Law.

For hundreds of years it has been recognized that gases expand and pressure increases when a gas is heated, which can be seen in Figure 1-3.

FIGURE 1-3 Effect of temperature on the volume and pressure of an ideal gas. The volume increases linearly with rising temperature, if the pressure is constant. The constant-pressure lines in panel (a) are called *isobars*. Panel (b) shows the *isometrics*, the curves that for a particular molar volume show the linear increase of pressure with rising temperature.

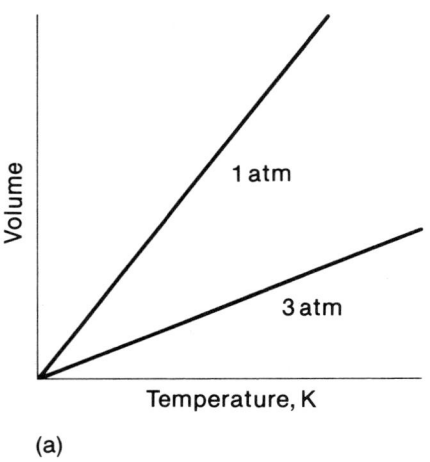
(a)

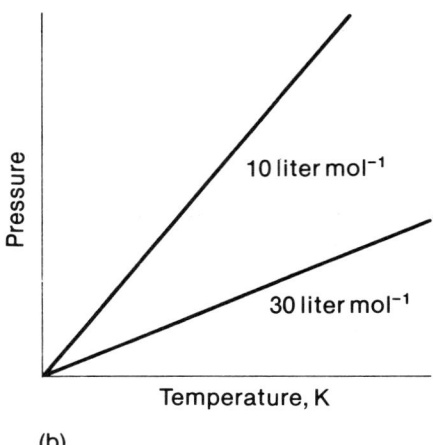
(b)

This effect of temperature was first studied systematically in 1702 by Guillaume Amontons, who observed that for any particular volume of air there is a linear decrease in gas pressure with a linear decrease in temperature. Of greater significance was his observation that the pressure would be zero at some very low temperature, which his crude data suggested was −240°C. Since negative pressure seemed to be an impossible concept, Amontons introduced the important idea of *absolute zero*, a lower limit to temperature. A century later this phenomenon was examined more carefully by Joseph Gay-Lussac, who obtained the important relation

$$P = P_0(1 + \alpha t) , \qquad (2)$$

in which P_0 is the pressure at 0°C, t is the temperature in degrees Celsius at which P is measured, and α is a constant whose measured value he found to be 1/273. Thus $P = 0$ at $t = -273°C$. This temperature is called *absolute zero* and it is designated 0 kelvin or 0 K. More precise measurement has showed that absolute zero is −273.15°C. Using this new temperature scale, the absolute temperature T is $t + 273.15$, so that Equation 2 can be stated as

$$\frac{P}{T} = \frac{P_0}{T_0} = \text{constant} . \qquad (3)$$

Jacques A. C. Charles also studied the dependence on temperature of the volume V of a gas at constant pressure and observed that at relatively low pressures

$$V = V_0(1 + \alpha t), \quad (4)$$

in which t is again the temperature in °C and V_0 is the volume at 0°C and α has the same value found by Gay-Lussac. This equation is called Charles's Law and is commonly written in terms of the absolute temperature T as

$$\frac{V}{T} = \text{constant, at a particular value of } P. \quad (5)$$

EXAMPLE 1–B The stress limit of a gas tank.

A 10,000-liter outdoor storage tank is filled with a gas at a pressure of 50 atm at 20°C. The stress limit of the tank is 60 atm. Is the tank safe?
From Gay-Lussac's Law, $P_1/T_1 = P_2/T_2$, so that the temperature at which the stress limit would be reached is $(293/50) \times 60 = 351.6$ K $= 78.6$°C. This temperature would not be one that would be reached in nature, so the tank is probably safe.

Note that Equations 2 and 4 both indicate that P and V are zero at absolute zero. This is, of course, an unwarranted conclusion, because the equations fail to consider the fact that at very low temperatures, but well above 0 K, the gases liquefy and then solidify. We will see in Chapter 2 that in fact $P = 0$ at 0 K because of a complete cessation of molecular motion, but that $V \neq 0$.

The Ideal Gas Law

Using the observations that PV equals a constant proportional to the number of moles and that both P and V are linear functions of temperature T, one can easily derive the ideal gas law as follows.
We consider a gas initially at a pressure, volume, and temperature P_1, V_1, and T_1, and change the state of the gas to the new values P_2, V_2, and T_2. We perform this change in two stages—first raising the pressure at constant temperature (T_1), which results in a new volume V', and then raising the temperature at the second pressure (P_2). The first change is described by Boyle's Law, $P_1V_1 = P_2V'$ and the second by $V'/T_1 = V_2/T_2$. Combining these equations yields

$$\frac{P_1V_1}{T_1} = \frac{P_2V_2}{T_2} \qquad \text{or} \qquad \frac{PV}{T} = \text{constant}.$$

SEC. 1-2 GASES

From Avogadro's Law, at constant P and T the volume is proportional to the number n of moles of the gas, so the constant is proportional to n. The constant of proportionality is called R, the *ideal gas constant*. Thus,

$$PV = nRT. \qquad (6)$$

This equation is called the *ideal gas law* or the equation of state of an ideal gas. The values of R in various units are listed below:

$$
\begin{array}{ll}
8.314 \times 10^7 & \text{erg K}^{-1}\,\text{mol}^{-1} \\
8.314 & \text{J K}^{-1}\,\text{mol}^{-1}\,(\text{SI units}) \\
1.987 & \text{cal K}^{-1}\,\text{mol}^{-1} \\
1.987 \times 10^{-3} & \text{kcal K}^{-1}\,\text{mol}^{-1} \\
0.0821 & \text{l atm K}^{-1}\,\text{mol}^{-1}
\end{array}
$$

It is important to realize that (1) the ideal gas law is a statement derived from experimental observations and (2) it is valid only at relatively low pressures and above the boiling point of each gas. This is not to say that it has no value, because at atmospheric pressure, V for real gases usually differs by less than 1 percent from that predicted by Equation 6.

EXAMPLE 1-C Measurement of the molecular weight of an ideal gas.

An evacuated bulb weighs 82.3 g when it is empty and 89.6 g when filled with an unnamed gas at a pressure of one atmosphere and a temperature of 25°C. When the bulb is filled with O_2, it weighs 84.7 g at the same temperature and pressure. What is the molecular weight of the unnamed gas?

The O_2 weighs $84.7 - 82.3 = 2.4$ g; the molecular weight of oxygen gives a value of $2.4/32 = 0.075$ mol. If the unnamed gas behaves as an ideal gas, one mole of it will occupy the same volume as one mole of the oxygen (assuming that O_2 is an ideal gas). Therefore since 0.075 mol of the gas weighs $89.6 - 82.3 = 7.3$ g, its molecular weight is $7.3/0.075 = 97.3$.

EXAMPLE 1-D Determine the formula of a substance whose vapor does not behave strictly as an ideal gas.

A 200-ml evacuated bulb is filled with 2.95 g of a liquid and then sealed. The formula of the liquid has been found to be $(CH)_x$ but x is unknown. The bulb is heated to 75°C, at which temperature the liquid is completely vaporized; the pressure in the bulb is 5.3 atm. We assume that although the vapor is not an ideal gas, its behavior does not differ greatly from that of an ideal gas. What is the formula of the liquid?

The number of moles in the bulb is $PV/RT = (5.3\ \text{atm})(0.2\ \text{l})/(0.082)(348)$ or 0.0371 mol. If the vapor were ideal, its molecular weight would be $2.95/0.0371 = 79.5$. The molecular weight of substances having the basic formula $(CH)_x$ is $13x$. Therefore $x = 79.5/13$ or 6.11. However, x must be an integer; we can safely conclude that the discrepancy is a result of nonideality of the vapor and that $x = 6$. Thus the formula must be C_6H_6.

Mixtures of Gases

An important observation was made in 1801 by John Dalton. He found that in a mixture of gases the total pressure of the mixture is the sum of the pressures each gas would exert if it were the only gas present, at the same volume and temperature. This is called Dalton's Law of Partial Pressures and is usually stated as follows:

in a mixture of gases the total pressure is equal to the sum of the partial pressures of its components.

Thus

$$P = p_1 + p_2 + \cdots + p_n, \tag{7}$$

which, according to the ideal gas law, becomes

$$P = \frac{n_1 RT}{V} + \frac{n_2 RT}{V} + \cdots + \frac{n_n RT}{V}$$

$$= (n_1 + n_2 + \cdots + n_n)\frac{RT}{V} = \sum n_i \frac{RT}{V}. \tag{8}$$

These equations are especially useful in the analysis of experiments in which a gas is collected over a liquid, since the total pressure of the system then equals the pressure of the gas plus the pressure of the liquid vapor in equilibrium with the liquid (the vapor pressure).

EXAMPLE 1–E Determination of the amount of H_2 generated in a chemical reaction.

A chemical reaction is being studied in which H_2 is generated. To avoid the danger of explosion, the H_2 is collected in a vessel completely filled with an organic liquid, in which H_2 is insoluble. The liquid is in contact with the atmosphere, which is at a pressure of 760 mm Hg while the experiment is being conducted. As H_2 is produced, liquid is displaced. When 500 ml of H_2 has been collected, the reaction is terminated. The vapor pressure of the liquid is 30.4 mm Hg at 25°C. How much H_2 was generated in the reaction?

The final pressure of the H_2 is $760 - 30.4 = 729.6$ mm Hg or 0.96 atm. Since $V = 0.5$ l, $T = 298$ K, and $R = 0.082$ l atm K^{-1} mol^{-1}, the number of moles of H_2 is $PV/RT = 0.02$ mol.

EXAMPLE 1–F Calculation of the partial pressure of the components of air entering the lungs.

For dry air the mole fractions of N_2, O_2, and CO_2 are 0.79, 0.2095, and 0.0005, respectively. When air at atmospheric pressure enters the lungs, it

SEC. 1-2 GASES

becomes saturated with water vapor, the pressure of which is 48 mm Hg at 37°C, so its partial pressure is 760 − 48 = 712 mm Hg. From Equations 7 and 8 the pressure fraction equals the mole fraction. Thus, the partial pressures of the gases are: N_2, 0.79 × 712 = 562.48 mm Hg; O_2, 0.2095 × 712 = 149.16 mm Hg; and CO_2, 0.0005 × 712 = 0.36 mm Hg.

Let us now consider what partial pressure actually means. According to Avogadro's Law, at a particular temperature and total pressure, the number of molecules per mole is a constant for all gases. This is equivalent to stating that there is only enough space in a given volume for a given

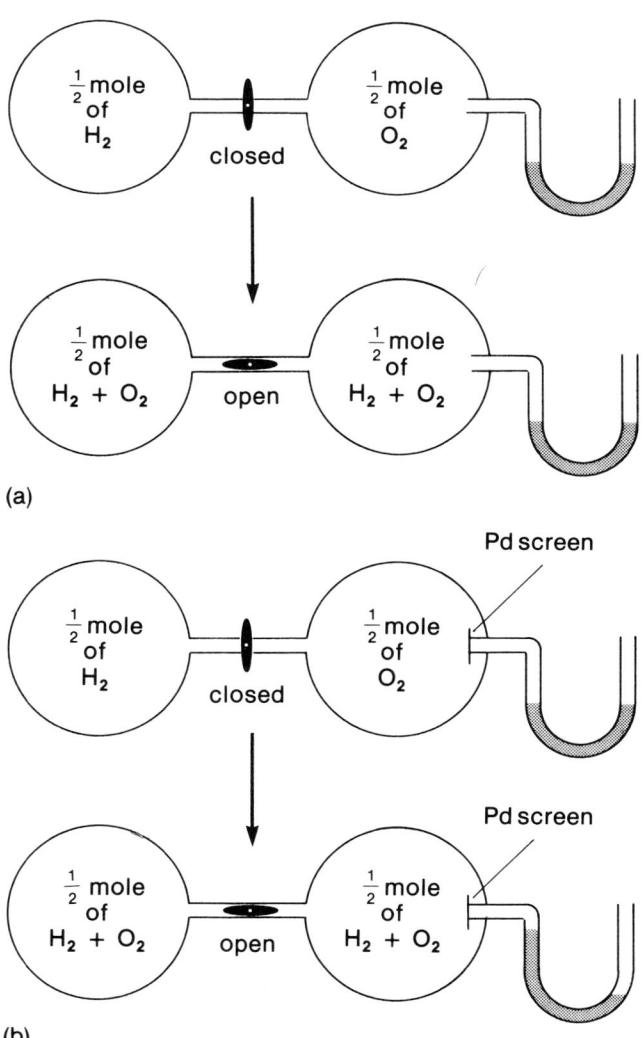

FIGURE 1-4 Demonstration of the partial pressure of H_2, using palladium foil, which is permeable only to H_2. The upper assemblage in both panels (a) and (b) shows the initial state of the system in which the stopcock is closed. The presence or absence of the Pd foil does not affect the state at this point. The lower assemblage in panel (a) shows the state of the system after the stopcock is opened. The lower assemblage in panel (b) shows the state of a system with a Pd foil, after the stopcock is opened.

number of molecules. As long as we assume that the sum of the volumes of the molecules themselves (that is, not counting the space between the molecules) is very small compared to the size of a container, then the available space should be independent of the identity of the molecules. Therefore if $\frac{1}{2}$ mole of ideal gas A is added to $\frac{1}{2}$ mole of ideal gas B, the total number of molecules (that is, those in one mole) occupies the same volume as would either one mole of A or one mole of B: if the total pressure on the walls of the container is one atmosphere, then half of this pressure must come from A and half from B; and if we could measure the pressure only of A, we should obtain a value of $\frac{1}{2}$ atmosphere. The following example, in which $\frac{1}{2}$ mole of H_2 and $\frac{1}{2}$ mole of O_2 are mixed, shows this.

Consider the apparatus shown in Figure 1-4, in which each bulb contains $\frac{1}{2}$ mole of a gas and the pressure is one atmosphere. If the stopcock is opened, thus connecting the two bulbs, the gases will mix and the pressure detected by the manometer will remain at one atmosphere—that is, the levels of Hg in the two arms of the manometer remain the same. Suppose a thin film of palladium metal is placed over the tube leading to the manometer before the stopcock is opened, as shown in panel (b). Palladium is permeable to H_2 but not to O_2. The manometer would still indicate a pressure of one atmosphere, because the palladium would not affect the pressure of the H_2 molecules on the surface of Hg in the manometer. However, when the stopcock is opened, the gases mix in the two spheres but since only H_2 can pass through the palladium screen, only H_2 pressure is measured in the left arm of the manometer; since the H_2 occupies twice its original volume (both spheres rather than only one), it has half its original pressure and the mercury level in the manometer changes to balance the reduced H_2 pressure within the spheres against the unchanged pressure beyond the closed system.

Real Gases

No gas obeys the ideal gas law at high pressures or near its boiling point. This is for two reasons: (1) Between the molecules there are attractive forces, the strength of which increases as the molecules become nearer to one another. This force accounts for such phenomena as condensation and the fact that the volume above certain high pressures is smaller than expected for an ideal gas. (2) The molecules have finite size, so that there is a limit to the minimum distance that can exist between their centers. This accounts for the incompressibility of liquids and solids. To account for these discrepancies, equations have been derived to replace the ideal gas law. The most important equation is called the van der Waals equation. This equation and related phenomena will be discussed in Chapter 3.

For all practical purposes the behavior of a typical gas can be described adequately by the ideal gas law if the pressure is on the order of one atmosphere and the temperature is far above the boiling point of the gas.

REFERENCES

A more detailed discussion of the gas laws can be found in standard textbooks of physical chemistry, such as the following:

Atkins, P.W. 1978. *Physical Chemistry.* W.H. Freeman and Co. San Francisco.
Barrow, G.M. 1979. *Physical Chemistry.* McGraw-Hill. New York.
Castellan, G.W. 1971. *Physical Chemistry.* Addison-Wesley. Reading, Massachusetts.
Daniels, F., and R.A. Alberty. 1979. *Physical Chemistry.* John Wiley. New York.
Levine, I.N. 1978. *Physical Chemistry.* McGraw-Hill. New York.
Moore, W.J. 1972. *Physical Chemistry.* Prentice-Hall. Englewood Cliffs, New Jersey.

Other readings are the following:

Neville, R.G. 1962. "The Discovery of Boyle's Law." *Journal of Chemical Education, 39,* 356.
Thomson, G.W., and D.R. Doustin. 1971. "Determination of Pressure and Volume." In *Techniques of Chemistry,* edited by A. Weissberger and B.W. Rossiter, Vol. 23. Wiley-Interscience. New York.

PROBLEMS

1. A 1-l evacuated bulb is allowed to fill with O_2 at 35°C and a pressure of 1 atm.

   (a) Assuming that O_2 is an ideal gas, what is the difference between the mass of the filled flask and the evacuated flask?

   (b) At what temperature will filling the flask with CO_2 produce the same weight-increase that filling the flask with O_2 at 35°C will?

2. One gram of CO_2 and one gram of ethane are put in a 1-l flask at 30°C. What is the partial pressure of each gas, the total pressure of the gas, and the mole percent of each gas?

3. One liter of N_2 at a pressure of 2 atm is to be placed in a 100 cm^3 container. If the temperature is to remain constant, what will the pressure be in the container?

4. An automobile tire is filled to a pressure of 32 pounds per square inch when the air temperature is 10°C. After driving for three hours, the tire pressure is found to be 35 pounds per square inch. Explain why.

5. A bulb having a volume of 125 cm^3 contains 0.5262 g of a gas at 742.3 mm Hg at 80°C.

   (a) What is the molecular weight of the gas, if it is an ideal gas?

   (b) Chemical analysis of the gas shows that it contains slightly more than two moles of hydrogen per mole of carbon and no other elements. What is the likely formula of the gas?

6. The boiling point of acetic acid is 118.5°C. If the vapor is an ideal gas, what is the density in g cm^{-3} of the vapor at the boiling point?

7. The coefficient of thermal expansion of a gas is defined as

$$\frac{1}{V}\left(\frac{\partial V}{\partial T}\right)_P.$$

   What is its value for an ideal gas?

8. An evacuated flask weighs 36.8362 g. When filled with dry O_2 at 1 atm and 25°C, it weighs 36.9736 g. When filled with a mixture of methane and ethane, it

weighs 36.9529 g. What fraction, in mole percent, of the gas in the bulb is methane?

9. What pressure is exerted by 90 g of CO_2 in a 2-l vessel at 20°C?

10. A good vacuum pumping system can reduce the pressure of a gas to 10^{-6} torr. At a temperature of 20°C what is the number of molecules per cm^3 of a gas at this pressure?

11. Why does a hot air balloon rise?

12. In the summer in seaside communities it is common that if it has been sunny all day, in the afternoon a cool breeze develops that blows from the sea onto the land. This is called a sea breeze.

(a) What is the cause of the sea breeze?

(b) Explain why a sea breeze is rare in the winter, especially if there is snow on the ground.

CHAPTER 2

KINETIC THEORY

The bulk properties of matter, and transformations such as phase changes and chemical reactions, can be elegantly analyzed by thermodynamics, as we will see in Chapters 4 and 5. Thermodynamics is concerned with pressures, volumes, temperatures, and various forms of energy, and is based upon two empirical observations—that energy is neither created nor destroyed, and that in the absence of an external supply of energy all functioning systems gradually lose the ability to perform work. These two observations are formulated as the First and Second Laws of Thermodynamics (Chapters 4 and 5) and from these laws many properties of matter can be deduced logically and simply. In thermodynamics, matter is thought of as a continuous substance, and energy is viewed as flowing in and out of systems as if it were a continuous fluid; no recourse is made to the facts that matter consists of particles and that energy is quantized into indivisible units. Ultimately we wish to understand nature in molecular, atomic, and subatomic terms; here classical thermodynamics fails and thus to the student it sometimes seems arbitrary and irrelevant. This quality may be especially troublesome for students of the life sciences, whose training emphasizes explanations in molecular terms and whose exposure to the mathematical logic of physics is often minimal. However, thermodynamics is important and extraordinarily valuable because of its predictive ability and its capacity to treat complex phenomena with extreme mathematical simplicity. In order to strengthen the application of thermodynamics to other topics in this book, we will study kinetic theory and molecular statistics before examining thermodynamics; then, after a thermodynamic argument is presented, we can restate it in molecular terms, thus providing greater meaning.

In kinetic theory, we shall see that temperature, pressure, and energy are manifestations of molecular motion, and that the laws of thermodynamics follow from the fact that the macroscopic properties of a system are an expression of the average behavior of the enormous number of molecules contained in the system.

2-1 THE KINETIC THEORY OF GASES

Assumptions in the Theory

It is convenient to begin our molecular explanation of the properties of matter with the kinetic theory of gases, because a gas is easier to understand than a solid or a liquid. In the initial exposition of this theory we make three simplifying assumptions: (1) the molecules of a gas occupy no space; that is, they are points having zero volume; (2) they collide elastically and infrequently; and (3) except during collision, there are no forces of attraction or repulsion between them. A gas having such properties is called an *ideal gas*. All three assumptions are incorrect, but nonetheless, the simple theory neatly explains certain properties of gases, at least over a small range of pressure and temperature. Once the ideal gas is understood, these assumptions can be replaced by the properties of real molecules and many of the properties of real gases will then be explainable.

The Kinetic Theory of Gas Pressure

Consider an ideal gas consisting of molecules of mass m contained in a cube having side l and volume $V = l^3$. For any molecule its velocity c (which is a vector quantity) can be expressed as three mutually perpendicular components u, v, and w parallel to the x, y, and z axes of the cube. That is,

$$c^2 = u^2 + v^2 + w^2 . \tag{1}$$

Collisions between the molecules and the wall of the cube occur; since all collisions are perfectly elastic, the velocity of a colliding molecule changes direction but not magnitude. In an elastic collision the angle of incidence of the molecule equals the angle of reflection so that the velocity component u changes sign from $+u$ to $-u$ (or $-u$ to $+u$) and hence the momentum component changes from $+mu$ to $-mu$ (or vice versa). Thus, the momentum change is $2mu$.

The time it takes for the u component to move between the two walls perpendicular to the x axis is l/u, so that the change in the x component of momentum in this time is $(2mu)(u/l)$ or $2mu^2/l$. If the box contains N molecules, each will have a velocity that can be resolved into u, v, and w

components, but these components may differ from one molecule to the next. Thus, it is necessary to consider an *average velocity;* because the momentum term contains the square of the velocity, we define $\overline{u^2}$ as the average value of the square of the velocity component u, so the change per unit time in the x component of momentum is $2Nm\overline{u^2}/l$. The rate of change of momentum is a force; hence $2Nm\overline{u^2}/l$ is the force exerted on the two parallel walls against which the u components collide. The area of each parallel wall is l^2; hence the *pressure, P,* which is nothing other than the force perpendicular to a unit area, is

$$P = \left(\frac{2Nm\overline{u^2}}{l}\right)\Big/ 2l^2 = \frac{Nm\overline{u^2}}{l^3} = \frac{Nm\overline{u^2}}{V}, \qquad (2)$$

in which V is the volume of the cube. Since motion of the molecule in each of the three mutually perpendicular directions is equally probable, $\overline{u^2} = \overline{v^2} = \overline{w^2}$, and thus $\overline{c^2} = 3\overline{u^2}$. Therefore

$$P = \frac{Nm\overline{c^2}}{3V}. \qquad (3)$$

The quantity $\overline{c^2}$ is called the *mean square speed,* and $(\overline{c^2})^{1/2}$, which will appear in other derivations, is the *root mean square speed.* Although this derivation has been simplified by choosing a cubic container, Equation 3 is valid for any shape, because any three-dimensional solid can be constructed from an infinite number of infinitesimal cubes.

The kinetic energy ϵ_K of a molecule with velocity c is $\frac{1}{2}mc^2$, so that the total kinetic energy E of a volume V of gas containing N molecules is $\frac{1}{2}Nm\overline{c^2}$. Thus, Equation 3 may be written as

$$PV = \frac{1}{3}Nm\overline{c^2} = \frac{2}{3}N\epsilon_K = \frac{2}{3}E. \qquad (4)$$

This equation states that PV is a constant because the energy of the gas is constant and is the kinetic expression of Boyle's Law (which Boyle determined empirically).

Another way of thinking about the constancy of PV is the following. An increase in the volume of a container does not alter the velocities of the particles within the container (that is, if the temperature is constant). However, a longer time is required for a particle to traverse the box, so there are fewer collisions with the wall per unit time and hence the pressure is lower.

Partial Pressure

We can now derive *Dalton's Law of Partial Pressures.* Consider a gas mixture consisting of i different classes of molecules, each class having a particular total kinetic energy E_i (for instance, because each class differs in mass). We again assume that the molecules do not interact; the total energy E is the sum of the energies of the i classes, $E = \sum (E_i)$. We may then write Equation 4 as $P_i V = \frac{2}{3} E_i$ or $V \sum (P_i) = \frac{2}{3} \sum (E_i) = \frac{2}{3} E$. Since $PV = \frac{2}{3} E$, $P = \sum (P_i)$; that is, *the total pressure of a mixture of gases equals the sum of the partial pressures of each gas*. We can look at the law of partial pressures in another way. We assume that for an ideal gas the molecules do not interact with one another. In a mixture, molecules of one gas (type 1) also do not interact with molecules of a second gas (type 2). Thus, every molecule (of both gases) behaves as if it has the whole container to itself, and the pressure resulting from the molecules of one type is just proportional to the number of molecules of that type. Since at a single temperature the average energy of a single molecule of type 1 is the same as that of a single molecule of type 2, the average energy of all of the type 1 molecules is proportional to the number of molecules of type 1; and the fraction of the total energy attributable to type 1 molecules is the same as the fraction of all molecules that are type 1 molecules.

2–2 ENERGY

Relation between Kinetic Energy and Temperature

Everyday experience provides evidence for a relation between temperature and motion. Hands are warmed by rubbing them together, a fire can be produced by friction between two sticks, and a metal wire becomes hot merely by being bent back and forth rapidly. Therefore, we expect to find a relation between temperature and kinetic energy (the energy of a particle due to its motion). This relation may be demonstrated simply by referring to the ideal gas law, $PV = nRT$, in which n is the number of moles of the gas, R is the gas constant, and T is the absolute temperature. Thus, combining the ideal gas law and Equation 4,* we obtain

$$\frac{2}{3} N \epsilon_K = nRT . \tag{5}$$

Since $n = N/\mathcal{N}$, in which $\mathcal{N}$ is Avogadro's number,

$$\epsilon_K = \frac{3RT}{2\mathcal{N}} = \frac{3}{2} kT , \tag{6}$$

*The careful reader might notice that Equation 5 is a result of combining an empirical law (the ideal gas law) and a theoretical law (Equation 4) so that Equation 5 and all subsequent equations derived from it are empirical. This is true of a great deal of the kinetic theory of gases, although many of the empirical laws can be derived from the theory of statistical mechanics. In actuality, few of the equations in the kinetic theory are correct because the requirements of quantum statistics are overlooked. We shall not worry about this because, except for particular cases that we will mention, the kinetic theory is an adequate and useful theory.

SEC. 2-2   ENERGY

and $k = R/\mathcal{N}$ is the Boltzmann constant; or, if $E_K = \mathcal{N} \epsilon_K$ is the kinetic energy of one mole of gas,

$$E_K = \frac{3}{2}RT . \tag{7}$$

In the following section we go one step further in the kinetic theory and relax the assumption that molecules have no volume and consider the energy of a diatomic molecule. This does not alter the constancy of the product PV or its relation to the energy of the system. What it does accomplish, though, is to include in the total energy not only the translational kinetic energy but also the energy of other motions, namely, rotation and internal vibration.

Equipartition of Energy: Polyatomic Molecules

In the previous analysis of a gas in a box, each molecule was considered to be undergoing three independent translational motions parallel to the x, y, and z axes. In this way the kinetic energy of one mole of a gas can also be thought of as consisting of three components, each having a value of $\frac{1}{3}(\frac{3}{2}RT) = \frac{1}{2}RT$. Since the position of each molecule is specified in three dimensions by three coordinates, the three components are $\frac{1}{2}mu^2$, $\frac{1}{2}mv^2$, and $\frac{1}{2}mw^2$, in which u, v, and w are the velocities in the x, y, and z directions, respectively, and m is the mass. The expression for translational motion of a polyatomic molecule is no different, because all of the component atoms are linked to one another and the translational motion of a polyatomic molecule can be conveniently described by the motion of the center of mass of its constituent atoms.

A polyatomic molecule is capable of rotational as well as translational motion, and this rotational motion can be resolved into three rotational components about the x, y, and z axes; the rotational kinetic energy is described by a relation similar to that for translational motion—namely,

$$\frac{1}{2}I_x\omega_x^2 + \frac{1}{2}I_y\omega_y^2 + \frac{1}{2}I_z\omega_z^2 , \tag{8}$$

in which I_x, I_y, and I_z are the moments of inertia about the x, y, and z axes, and ω_x, ω_y, and ω_z are the angular velocities of rotation. The number of modes of rotation is independent of the number of atoms in the molecule but does depend upon its shape. For example, if a polyatomic molecule is linear, it can be thought of as existing in two-dimensional space, and only two perpendicular axes are needed to describe its rotation. That is, one motion is rotation in a plane about the center of mass of the molecule and the second motion is rotation in a plane perpendicular to the first plane. All other rotations can be resolved into these two components.

A diatomic molecule is also capable of vibration. Only one mode of vibration is possible, namely the stretching and contraction of the bond connecting the atoms. The number of modes of vibration increases with the number of atoms per molecule, inasmuch as each bond can be stretched or bent.* The number of vibrational modes can be counted as follows. A three-dimensional mechanical system containing N associated particles (in this case, N is the number of atoms per polyatomic molecule) is described by three coordinates per particle or by a total of $3N$ coordinates, so that there are $3N$ independent components of motion.

For molecules consisting of many atoms it is difficult to count the number of vibrational modes (and errors are often made in the attempt). Thus it is more convenient to obtain this number by subtracting the number of translational components (3) and rotational components (2 or 3) from $3N$, because these numbers are small and independent of N. Thus, there are $3N - 6$ and $3N - 5$ modes for the vibrational motion of a nonlinear and a linear molecule, respectively.

The energy of a vibrational mode is described by the sum of a kinetic and a potential energy, and is usually written as

$$\frac{1}{2}m\left(\frac{dr}{dt}\right)^2 + \frac{1}{2}a(\Delta r)^2 \; , \tag{9}$$

in which m is a mass, dr/dt is a linear velocity, Δr is the distance of the vibrating particle from its mean position (that is, its location if it were at rest) and a is a constant that measures the restoring force (Figure 2-1).

Let us now examine the expressions for the energy of translation, rotation, and vibration and note in each the number of coordinates and

FIGURE 2-1 A vibrating sphere connected to a rigid wall by a spring. The sphere moves a maximum distance Δr from its position at rest. The arrows indicate the direction of motion of the sphere.

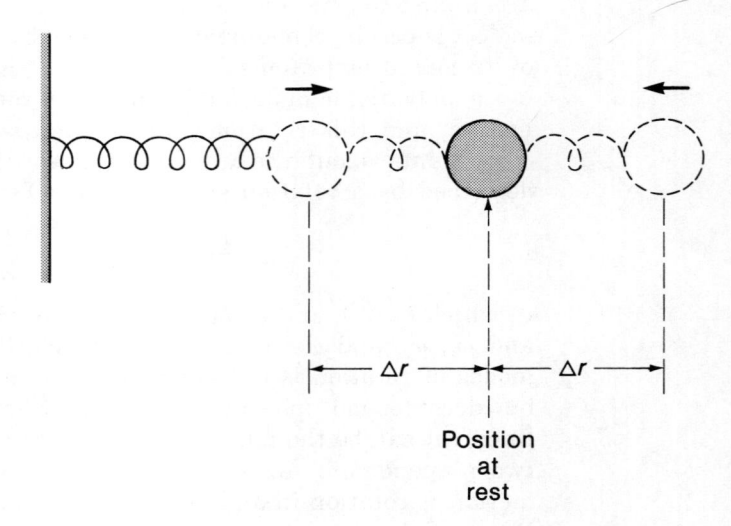

---

*Students have often asked if the bond of a diatomic molecule can be bent. For such a molecule, "bending" is just the combination of contraction with translational motion of the entire molecule.

SEC. 2–2 ENERGY

velocities that appear as a second-power term. For translation there are three (x, y, and z), for rotation there are two and three for linear and nonlinear molecules, respectively, and for vibration there are two (the kinetic and potential energy terms). Counting the number of such terms yields the *principle of equipartition,* which can be stated as follows: if the energy of a single molecule is written as a sum of terms, each of which contains either a coordinate (for example, x, y, angle, and so forth) or a velocity component raised to the second power, then the average energy ϵ_K consists of $\frac{1}{2}kT$ multiplied by the number of such terms (see Equation 6 for the origin of the $\frac{1}{2}kT$ factor). Thus we can make a list for a molecule containing N atoms (see Table 2-1).

TABLE 2-1 Comparison of molecular motion and energy, as obtained with the principle of equipartition.

Kind of motion	Monatomic molecule	Polyatomic molecule Linear	Polyatomic molecule Nonlinear
Translational			
Number of modes	3	3	3
Energy per mole	$\frac{3}{2}RT$	$\frac{3}{2}RT$	$\frac{3}{2}RT$
Rotational			
Number of modes	0	2	3
Energy per mole	0	RT	$\frac{3}{2}RT$
Vibrational			
Number of modes	0	$3N-5$	$3N-6$
Energy per mole	0	$(3N-5)RT$	$(3N-6)RT$
Total energy	$\frac{3}{2}RT$	$(3N-\frac{5}{2})RT$	$(3N-3)RT$

The principle of equipartition provides a simple way to estimate the energy of an ideal gas consisting of any type of molecule.* This is of particular value because this energy is equivalent to the internal energy, U, which is described in the First Law of Thermodynamics in Chapter 4.

EXAMPLE 2–A What is the total energy of CO_2 and of NH_3?

CO_2 is a linear molecule consisting of three atoms. Thus the total energy is $[(3 \times 3) - \frac{5}{2}]RT = (\frac{13}{2})RT$. NH_3 is nonlinear, and thus its total energy is $[(3 \times 4) - 3]RT = 9RT$. Note that we do not need to know any details of the structure of these molecules other than that they are either linear or nonlinear.

---

*We have said "estimate" because the values obtained are always slightly lower than the experimental values and a quantum mechanical treatment is necessary to obtain a more precise value.

Apart from the possible predictive value, what is the law of equipartition telling us? According to this principle, if a gas is warmed, the energy provided by the increase in temperature should be taken up by each of the $3N$ components. The importance of this concept lies in the fact that *each of the $3N$ components gets the same amount of energy*. Thus, in a polyatomic molecule an increase in temperature primarily increases the total vibrational energy rather than the kinetic energy. For example, butane (C_4H_9) has three translational modes, three rotational modes, and 33 vibrational modes, and if butane is heated, $(3N - 6)/(3N - 3) = \frac{33}{36}$ or 92 percent of the added energy is vibrational energy. It is also the case that when a slowly moving polyatomic molecule collides with a fast-moving monatomic molecule, the polyatomic molecule can absorb more energy from a monatomic molecule than can a diatomic molecule.

The equipartition principle also tells us something we have not suspected. It says that if energy is added to a gas, the temperature rise should depend upon how the energy is distributed. One measure of the energy distribution is the heat capacity at constant volume C_V, which we will discuss in detail in Chapter 4. This is defined as $(\partial E/\partial T)_V$ and should be calculable from the expressions in Table 2-1. That is, C_V should equal $\frac{3}{2}R$, $(3N - \frac{5}{2})R$, and $(3N - 3)R$ for a monatomic, linear polyatomic, and nonlinear polyatomic gas, respectively. The value of $\frac{3}{2}R$ is observed experimentally for monatomic gases. However, for diatomic gases the observed value is about 30 percent too low and, furthermore, its value depends upon temperature. For polyatomic molecules the deviation is even greater. The equipartition principle fails to explain these discrepancies. Since it is correct for monatomic gases, in which all energy is translational, its error must be in how it treats rotation and vibration. We should not be too surprised at this, since the theory applies to an ideal gas and the measurements are made with real gases. Yet this cannot be the whole story, because the data for real monatomic gases agree with the theory. Thus we must conclude that there is something wrong with the way that we have treated energy. The real explanation is found in quantum mechanical considerations. In the quantum theory it is shown that the energy distribution of rotations and vibrations is not continuous; only discrete values are possible. When this is taken into account, values of C_V can be calculated that agree more closely with the experimental values.

2–3 MOTION OF GAS MOLECULES

Effusion of Gases

Equation 3 can be combined with the ideal gas law to show that the root mean square speed is inversely proportional to the square root of the mass of a particle. That is,

$$(\overline{c^2})^{1/2} = \left(\frac{3RT}{\mathcal{N}m}\right)^{1/2}. \tag{10}$$

A similar but slightly different equation can be derived from the average speed $\langle c \rangle$:

$$\langle c \rangle = \left(\frac{8RT}{\pi \mathcal{N}m}\right)^{1/2}. \tag{11}$$

This equation has been confirmed experimentally by an analysis of effusion. Consider a container filled with a gas and having a very small orifice from which gas molecules may escape. The volume of the container and the diameter of the orifice are assumed to be of such a size that there is no appreciable flow of gas toward the orifice; thus the number of molecules passing through equals the number that by random motion happen to enter the orifice. This number is proportional to the average speed. A simple argument (which we will not carry out) yields

$$\frac{dn'}{dt} = \frac{1}{4} n \langle c \rangle, \tag{12}$$

in which n is the number of molecules per unit volume and dn'/dt is the number of molecules striking the unit area per unit time. Substituting Equation 11 and using ρ, the gas density, one obtains, for the weight W or volume V of gas effusing in unit time per unit area of the orifice,

$$\frac{dW}{dt} = \rho \left(\frac{RT}{2\pi \mathcal{N}m}\right)^{1/2} \quad \text{or} \quad \frac{dV}{dt} = \left(\frac{RT}{2\pi \mathcal{N}m}\right)^{1/2}. \tag{13}$$

Thomas Graham confirmed the velocity equation experimentally in 1848, and it is sometimes called Graham's Law of Effusion.

Effusion is interesting because it can be used to separate gases having different molecular weights. For example, it has been used to separate isotopes of a single element. Large containers having permeable barriers consisting of very fine pores are filled with a gas. The gas passing through the barrier and the contents remaining in the vessel are enriched for isotopes of lower and higher m, respectively. Repeated passage of separated fractions through the system increases the percent enrichment. Note that this method can be used for elements which are solids and liquids at room temperature merely by heating above the boiling point.

Collisions

In the preceding sections it was assumed for mathematical simplicity that molecules do not collide. For the calculations that have been made, the fact that these assumptions are incorrect does not significantly alter the validity of the results. In this section we will discard these assumptions and examine those properties of gases that are a consequence of both the finite size of the atoms and the high collision rate.

Consider a collection of spherical molecules having diameter d moving at random without intermolecular attraction or repulsion. When the distance between the centers of two molecules equals d, a collision results. Another way of thinking about this is to assume that all but one molecule are stationary and that one molecule, A, is moving through a cylinder of diameter $2d$ (Figure 2-2); then a collision occurs if the center of any other molecule is inside the tube, or on the surface of the tube (a grazing hit). If A moves with average speed $\langle c \rangle$, it sweeps out a volume $\pi d^2 \langle c \rangle$ in one second. If this volume contains n molecules per cm^3, A experiences $n\pi d^2 \langle c \rangle$ collisions per second. To extend the calculation to a collection of molecules all of which are in motion, one must consider the relative velocity between the molecules. Since the molecules are moving in all directions, the average collision occurs between molecules moving perpendicular to one another, and this introduces a factor of $\sqrt{2}$ (Figure 2-3) into the term; then the collision frequency z made by a single molecule is

$$z = \sqrt{2}n\pi d^2 \langle c \rangle . \tag{14}$$

This is often written with σ, the collision cross-section, replacing πd^2. The total number Z of collisions per unit volume per unit time is obtained by multiplying by n and by $\frac{1}{2}$ since, when A collides with B, B collides with A,

FIGURE 2-2 Molecular collisions. Molecule A is moving leftward and collides with B at grazing incidence because their centers are precisely d units apart.

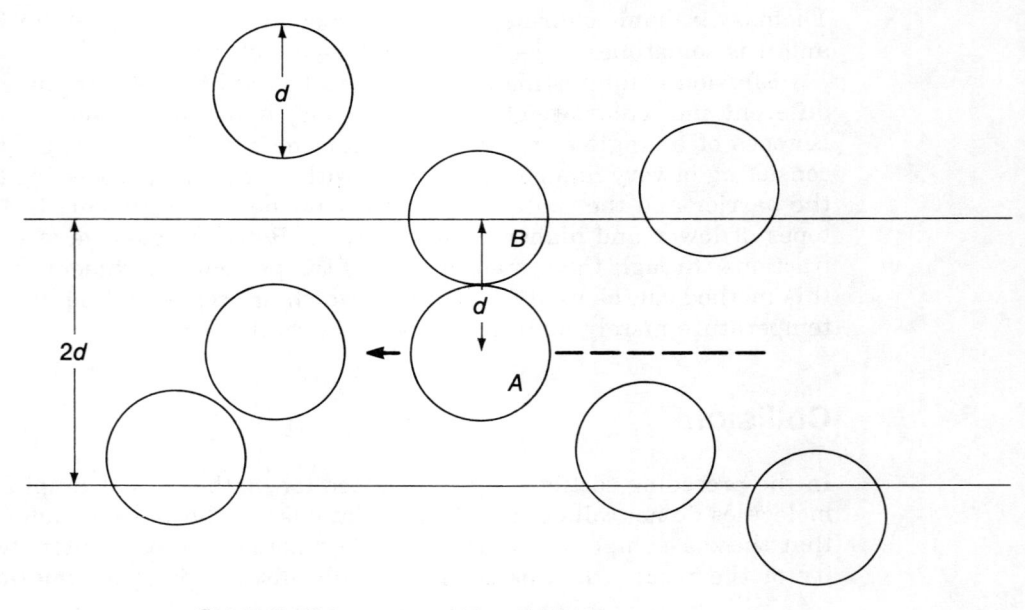

SEC. 2-3   MOTION OF GAS MOLECULES

FIGURE 2-3 Relative speed $\sqrt{2}\langle c\rangle$ resulting from the centers of two molecules approaching one another at right angles at the average speed.

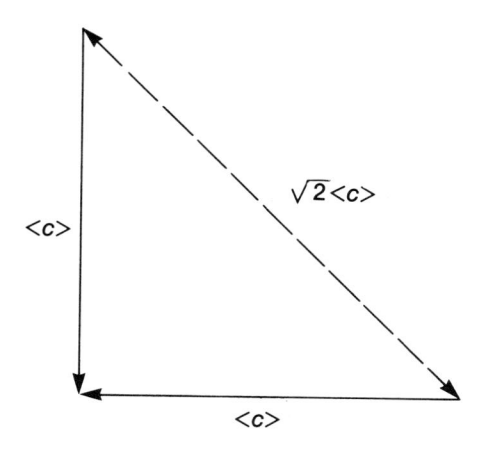

and we do not want to count this as two collisions. Thus

$$Z = \tfrac{1}{2}\sqrt{2}\, n^2 \pi d^2 \langle c\rangle \;, \tag{15}$$

which, by substituting Equation 11, can be written

$$Z = \pi d^2 n^2 \left(\frac{4RT}{\pi \mathcal{N} m}\right)^{1/2} \tag{16}$$

The number of collisions per unit time is very large. For instance, for molecules like N_2 and O_2, the value is approximately 5×10^{28} collisions per second per cm^3 at room temperature and a pressure of 1 atm.

In the study of transport processes (for example, diffusion) one needs to know the number of collisions Z with a particular area A. This is denoted Z_A, and is

$$Z_A = n\left(\frac{RT}{2\pi \mathcal{N} m}\right)^{1/2} . \tag{17}$$

This can be written in terms of pressure P instead of n since the number of molecules per unit volume $= P/RT$. Thus

$$Z_A = P/(2\pi m k T)^{1/2} . \tag{18}$$

Equation 18 is used to derive another useful expression describing molecular effusion into a vacuum. Thus if an orifice has an area A_o, the number of molecules emerging per unit time is simply the number $Z_A A_o$ colliding with an imaginary plane across the orifice, or

$$Z_A A_o = PA_o/(2\pi m k T)^{1/2} . \tag{19}$$

This expression is the basis of measuring the vapor pressure of a solid, a quantity often difficult to measure in traditional ways because it is so small. Thus, a solid is placed in a container having a small orifice of area A_o. The rate of loss of mass is measured and this equals the right-hand term of Equation 19. If m is known, P can be calculated.

Knowing the number of collisions is sometimes of practical importance, as we see in the following example.

EXAMPLE 2–B The requirement for a vacuum in metal evaporation and in electron microscopy.

The large number of collisions experienced by a particle passing through air necessitates the use of vacuum in some experimental situations. One example is vacuum evaporation or the depositing of very thin layers of metal onto samples being prepared for electron microscopy. Most biological samples are almost completely transparent to electrons and therefore are invisible in the electron microscope. To produce absorbance, the sample is coated with a thin layer of metal atoms. This is accomplished by boiling a heavy metal (for example, platinum or gold) and allowing the vapor to condense onto the specimen. A common experimental setup is shown in Figure 2-4. We make the simplifying assumption that a metal atom will reach the sample only if it makes no collisions with a molecule of air (this is of course not strictly true). A significant number of the Pt atoms leave the surface of the metal only when the boiling point T_b is reached. If we consider motion only in the direction of the sample, the kinetic energy of a particle $\frac{1}{2}mc^2 = \frac{1}{2}kT_b$, so that $c = (kT_b/m)^{1/2}$. The Pt atom sweeps out a cylinder of volume $\pi(r_1 + r_2)^2 c$ per second, in which r_1 and r_2 are the radii of the air molecules and Pt atoms. Introducing $\sqrt{2}$ to account for the motion

FIGURE 2-4 Apparatus for vacuum evaporation. The bell jar is evacuated by vacuum and diffusion pumps. The tungsten filament, around which a metal wire to be evaporated is wrapped, is heated to the boiling point of the metal. In about 15 seconds the metal is boiled away and forms a metallic film on the sample. The thickness of the film is proportional to the amount of metal wire wrapped around the filament and inversely proportional to the square of the distance from the filament to the sample. [SOURCE: After *Physical Biochemistry, First Edition,* by David Freifelder. W. H. Freeman and Company. Copyright © 1976.]

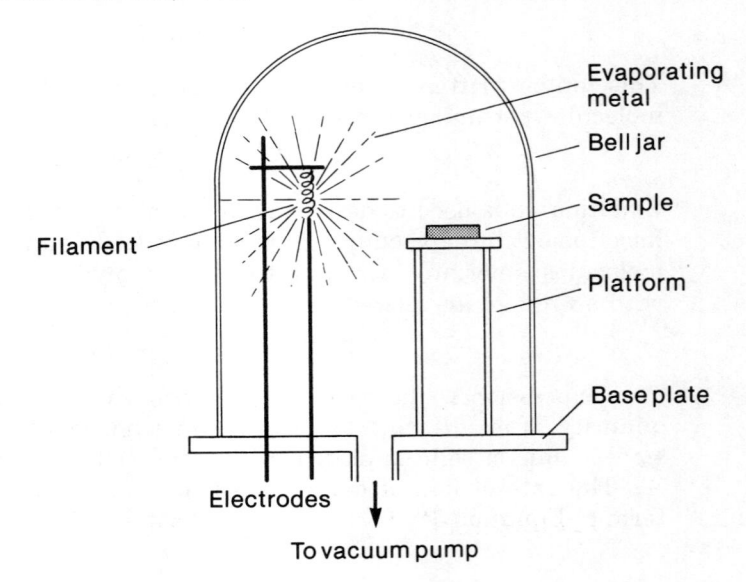

of the air molecules, we see that the collision frequency is $\sqrt{2}n\pi(r_1 + r_2)^2(kT_b/m)^{1/2}$ per second. A typical distance between the source and the sample is 10 cm; thus the time to traverse that distance is $10/(kT_b/m)^{1/2}$, and the number of collisions made in covering that distance is $10\sqrt{2}n\pi(r_1 + r_2)^2$. If this is to be less than 1, the number n of air molecules per unit volume must be less than $1/[10\sqrt{2}(r_1 + r_2)^2] \approx 3 \times 10^{13}$ molecules per cm^3, which is equivalent to 10^{-6} atm. In practice, vacuum evaporators are operated at a pressure of about 5×10^{-7} atm.

This reasoning also enables one to understand the very coarse granular appearance of the background in electron micrographs when a poor vacuum is achieved in the evaporation apparatus. If n is large, many Pt atoms do not reach the samples because of accumulated deflections. However, sometimes clusters of atoms boil from the metal. These clusters have a large cross-section but their larger mass (which increases more rapidly than the radius as the cluster size increases) compensates for the higher collision frequency by reducing the angle of deflection. Thus, with a poor vacuum, the number of Pt atoms reaching the sample per unit time decreases but the ratio of clusters to single atoms increases, thus producing a grainy metal layer.

In an electron microscope each electron must traverse about 800 cm before reaching the viewing screen. A calculation similar to that above explains why the microscope column must maintain a great vacuum (of 10^{-8} atm or greater).

Mean Free Path

For a single molecule the average number of collisions per second is $z = \sqrt{2}n\pi d^2 \langle c \rangle$, and in one second, the molecule travels a distance equal to $\langle c \rangle$. Therefore, the average distance a molecule travels between collisions—a distance denoted by λ—is $\langle c \rangle/z$, or

$$\lambda = \frac{1}{\sqrt{2}n\pi d^2} ; \qquad (20)$$

λ is called the *mean free path*. A typical value of λ for a gas at room temperature and atmospheric pressure is 10^{-5} cm. Knowledge of the mean free path is sometimes of practical value. For instance, in Example 2–B, the calculation could have been carried out in terms of λ, requiring that λ be greater than the distance having to be traversed by the Pt atom or the electron. Another example is the following.

EXAMPLE 2–C The design of incandescent bulbs.

In an ordinary light bulb, electric current passes through a tungsten filament, making the metal white hot. To prevent immediate combustion of the filament, the bulb is evacuated, thus removing O_2. In addition to the emitted light, tungsten atoms are also thrown off at the high temperature because the kinetic energy of some of the atoms allows them to escape from the attractive force that operates between atoms in a solid. At very low pressures the tungsten atoms strike the glass wall of the bulb and form a black film that will rapidly darken the bulb. More important, the loss of tungsten by the filament causes weakening and rapid breakage of the filament. This loss can be avoided if argon is present at a pressure such that the mean free path of the tungsten atoms is less than the diameter of the filament. Under these conditions the emitted tungsten atoms collide with the argon atoms and many of them bounce back onto the filament. Thus, filling an incandescent bulb with argon gas lengthens the lifetime of the bulb considerably and prevents blackening of the glass.

2–4 DISTRIBUTION OF KINETIC ENERGIES AND MOLECULAR VELOCITIES

The Boltzmann Distribution

When particles collide with one another, momentum is exchanged. Thus, even if a gas could be created all of whose molecules had the identical momentum at the instant of creation, the high collision frequency would soon create a population of molecules having a distribution of momenta. This distribution has an average value, and at equilibrium most molecules have a velocity near the average value—that is, the distribution is narrow.

The following simple argument suggests the expected form of the distribution. Consider a particle in a gas having the average kinetic energy. The particle will experience many collisions, some of which will increase its energy (a *plus* collision) and some of which will result in a decrease of its energy (a *minus* collision). To increase its energy to a value E units above the average value, the particle must have a series of plus collisions. If each plus collision provides one energy unit, there must be E plus collisions. If the probability of having a plus collision is $1/q$, in which q is some constant, the probability $P(E)$ of having E plus collisions in a row is $P(E) = (1/q)^E$. Since $\ln P(E) = E \ln(1/q) = -E \ln q = -Ea$, in which a is a constant, then $P(E) = e^{-aE}$. This means that the probability of a particle having energy E above the average kinetic energy is a negative exponential function of E. By using the methodology of statistical mechanics, one can derive the very general expression known as the Boltzmann distribution:

$$n = n_0 e^{-\epsilon/kT} , \qquad (21)$$

in which n_0 is the number of molecules in any given state, n is the number whose potential energy ϵ per molecule is above that of the particular state, and k is the Boltzmann constant. This equation is often written as a ratio of the number of molecules in two states 1 and 2:

$$\frac{n_1}{n_2} = e^{-(\epsilon_1 - \epsilon_2)/kT} . \qquad (22)$$

Note that the units of ϵ measure energy per molecule. If one wishes to use energy per mole, k must be replaced by R.

EXAMPLE 2–D The barometric formula or the variation of air pressure with altitude.

The Boltzmann distribution can be shown to apply to the variation of air pressure above the Earth. Consider a cylinder of air of 1-cm² cross-section whose axis is perpendicular to the Earth. The weight of gas in a horizontal disc of thickness dx is mg, in which m is the mass of a gas molecule and g is the acceleration of gravity, or $\rho g\, dx$, in which ρ is the density of the gas in g cm^{-3}. If the variation in pressure P with height h above the Earth is dP/dx, the difference in pressure between the top and the bottom of the disc is $(dP/dx)dx$. Since pressure is force per unit area A or, in this case, weight per cm², then

$$\left(\frac{dP}{dx}\right) dx = \rho g A\, dx \ (A = 1 \text{ cm}^2) . \qquad (23)$$

For an ideal gas, $\rho = Pm/kT$, in which m is the mass in grams,* so that

$$\frac{dP}{P} = -\frac{mg}{kT} dx . \qquad (24)$$

If the pressure is P_0 at $h = 0$, we can integrate from those values to the pressure P at $x = h$, to obtain

$$\ln(P/P_0) = -mgh/kT , \qquad (25)$$

or

$$P = P_0 e^{-mgh/kT} . \qquad (26)$$

This is equivalent in form to the Boltzmann distribution, in which mgx is the potential energy at a point x due to gravity. This equation is sometimes called the *barometric formula*.

EXAMPLE 2–E Determination of molecular weights from Perrin's formula.

An equation similar to Equation 26 and called Perrin's formula can be used to determine the mass of a small particle or a macromolecule. If a

---

*If mass is expressed as molecular weight $M = \mathcal{N}m$, then k must be replaced by R in the subsequent equations.

suspension of identical small particles or a solution of macromolecules is placed in a gravitational field, collisions between the particles or the macromolecules and the solvent molecules will yield a distribution of the larger particles described by the Boltzmann equation, with $\epsilon = m'gx$, in which m' is an effective mass equal to the mass of the particle corrected for the buoyancy of the solvent. Thus, if v is the volume of the molecule, $m' = m - v\rho_l$, in which ρ_l is the density of the liquid. A proper analysis shows that for dissolved macromolecules, v should not be the volume but the partial specific volume $\bar{v}$ (see Chapter 7). Thus, in the correct analysis, m' is replaced by $m(1 - \bar{v}\rho)$. The Boltzmann equation then becomes

$$n = n_0 e^{-m(1-\bar{v}\rho_l)gh/kT} \tag{27}$$

or

$$m = \ln\left(\frac{n_0}{n}\right)\frac{kT}{(1 - \bar{v}\rho_l)gh}, \tag{28}$$

in which h is the vertical distance separating the points at which n and n_0 are measured. Since n_0/n can be replaced by c_0/c, in which c_0 and c are the concentrations at two points separated by the distance h, m can often be evaluated by determining optical absorbance (which is usually proportional to concentration) at various heights along a cylinder containing a solution.*

Normally this type of measurement is not carried out in a gravitational field but in a centrifugal field, in which case mgh is replaced by $m\omega^2\Delta r$ in which ω is the rotational speed and Δr is a radial distance from the axis of rotation. This will be discussed in more detail in Chapter 15.

The Maxwell-Boltzmann Distribution

The Boltzmann distribution can be extended to define the distribution of velocities (and therefore of kinetic energies) in a population. The derivation, which will be given in outline only, begins by assuming that the *fraction dn/n_0* of molecules having a velocity between u and $u + du$ is represented by an equation of the form

$$\frac{dn}{n_0} = Ae^{-mu^2/2kT}du, \tag{29}$$

in which u is the velocity component parallel to the x axis and A is a constant that is determined by the requirement that the sum of all fractions of molecules having velocities from $-\infty$ to ∞ must be 1. That is,

---

*Many years ago in the author's laboratory a test tube containing a very concentrated and turbid suspension of bacteriophages stood undisturbed for about a year at a constant temperature. It was observed that the turbidity decreased continually from the bottom to the top of the tube. The phages had come to equilibrium with gravity, and from a known relation between turbidity the number of phages per cm³, the molecular weight of the phage was measured.

SEC. 2-4 DISTRIBUTION OF KINETIC ENERGIES

$$A = \int_{-\infty}^{\infty} e^{-mu^2/2kT} du = 1,$$

from which $A = (m/2\pi kT)^{1/2}$, so that

$$\frac{dn}{n_0} = \left(\frac{m}{2\pi kT}\right)^{1/2} e^{-mu^2/2kT} du. \tag{30}$$

To extend this to three dimensions we note that the fraction of molecules having a velocity between $u + du$, $v + dv$, and $w + dw$, in the x, y, and z directions, is the product of the fractions for all directions; that is,

$$\frac{dn}{n_0} = \left(\frac{m}{2\pi kT}\right)^{3/2} e^{-(m/2kT)(u^2+v^2+w^2)} du\, dv\, dw. \tag{31}$$

This can be rewritten in terms of a single velocity c (in any direction) defined by $c^2 = u^2 + v^2 + w^2$, or

$$\frac{dn}{n_0} = 4\pi \left(\frac{m}{2\pi kT}\right)^{3/2} e^{-mc^2/2kT} c^2\, dc. \tag{32}$$

This is called the Maxwell-Boltzmann distribution. A plot of this distribution is shown in Figure 2-5 for two different temperatures. In the second panel of the figure the most probable speed c_m (the peak of the distribution), the average speed $\langle c \rangle$ (the arithmetical mean), and the root mean square speed $(\overline{c^2})^{1/2}$ are indicated. Expressions for these speeds are

$$c_m = \sqrt{\frac{2\pi kT}{m}}, \qquad \langle c \rangle = \sqrt{\frac{8kT}{\pi m}}, \qquad (\overline{c^2})^{1/2} = \sqrt{\frac{3kT}{m}}, \tag{33}$$

whose ratios are $c_m : \langle c \rangle : (\overline{c^2})^{1/2} = 1 : 1.128 : 1.224$.

Several features of the velocity distribution should be noted.

1. The distribution shifts to higher speeds as T increases, so that the average speed increases with rising T.
2. The breadth of the distribution increases with T; that is, the fraction of the molecules having very high velocities increases.
3. The areas of all curves are identical at all temperatures.

The velocity distribution can be converted to an energy distribution. Since the kinetic energy ϵ per molecule is $\frac{1}{2}mc^2$, then

$$c = \sqrt{\frac{2}{m}}\,\epsilon^{1/2} \qquad \text{and} \qquad dc = \frac{1}{\sqrt{2m}}\,\epsilon^{-1/2} d\epsilon. \tag{34}$$

By substituting into the velocity distribution (Equation 32), we obtain for dn_ϵ the number of molecules having kinetic energies between ϵ and $\epsilon + d\epsilon$,

$$dn_\epsilon = 2\pi n_0 \left(\frac{1}{\pi kT}\right)^{3/2} \epsilon^{1/2}\, e^{-\epsilon/kT} d\epsilon. \tag{35}$$

This distribution is plotted in Figure 2-6(a).

34   CHAP. 2   **KINETIC THEORY**

FIGURE 2-5 (a) The Maxwell-Boltzmann distribution at two temperatures. (b) Position of the most probable speed c_m, the average speed $\langle c \rangle$, and the root mean square speed $(\overline{c^2})^{1/2}$.

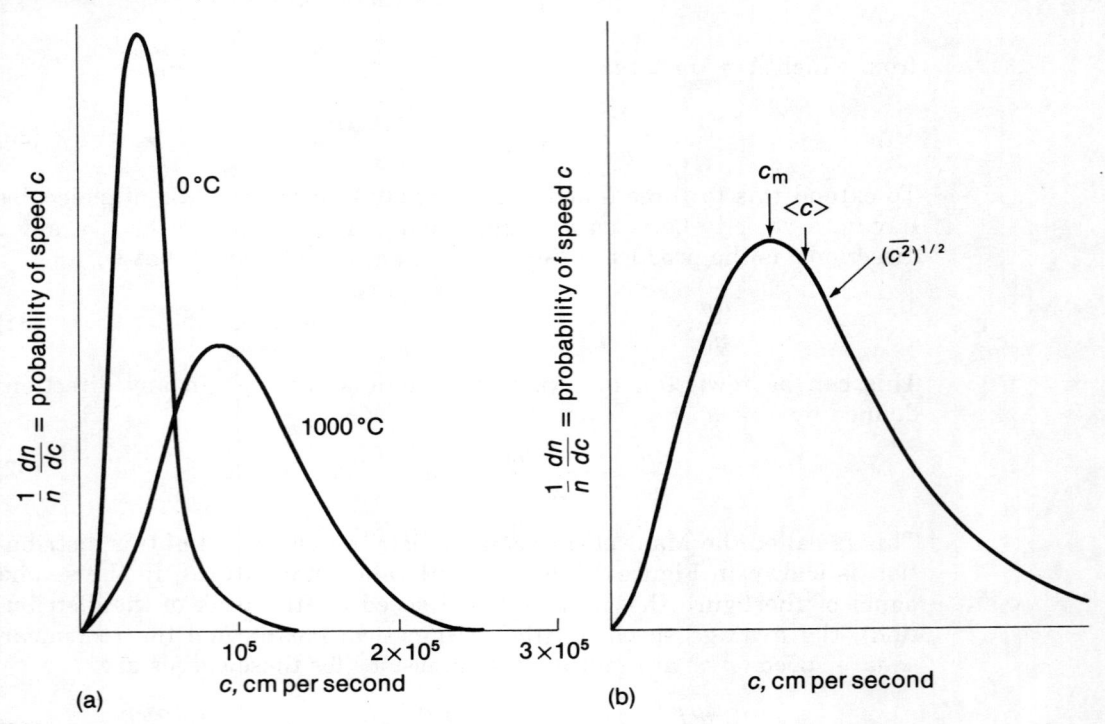

FIGURE 2-6 (a) Energy distribution at 25°C. (b) Fraction of molecules having energies greater than ϵ' at 25°C and 125°C.

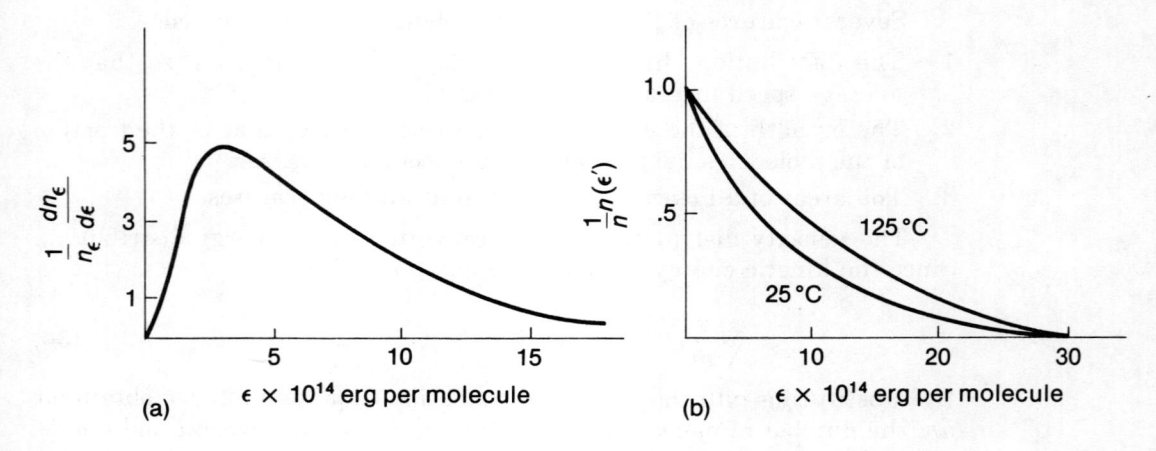

A connection between the energy distribution and chemical phenomena becomes apparent if one calculates the fraction of the molecules having kinetic energies greater than a particular value ϵ'. This calculation is beyond the scope of this book but the result is plotted in Figure 2-6(b). The most significant point is that the fraction of molecules having energies greater than ϵ' increases markedly with temperature, a property which is true also of liquids. This effect is a major cause of the increase with temperature of the rates of most chemical reactions, because in most reactions only molecules having an energy greater than a critical value can react chemically. Therefore, as T rises, the fraction of molecules whose energies exceed this critical value increases. This can also be viewed in terms of increasing the concentration of reactive molecules, which increases the probability of a favorable collision that results in reaction. Thus, the reaction rate rises with temperature. This will be examined in greater detail in Chapter 11.

Relation between the Boltzmann Distribution and Vapor Pressure

If a liquid is placed in an evacuated container, a portion of the liquid evaporates. The pressure exerted on the walls of the container is called the *vapor pressure*. For a particular temperature the value of the vapor pressure varies from one substance to the next, increasing as the attractive force between the molecules of the liquid decreases. As the temperature rises, the ability of a molecule to escape from the liquid increases, so that the vapor pressure increases. With increasing temperature this continues until a temperature is reached at which the vapor pressure equals the external pressure; at this temperature the liquid boils. If the external pressure is 1 atm the temperature is defined as the normal boiling point T_b. If the external pressure is increased, for example, by pushing a piston into the container, a higher temperature will need to be attained before the vapor pressure equals the external pressure; the boiling point is then greater than T_b. A solid also has a vapor pressure, although it is in general much smaller than that of a liquid since the attractive force between molecules is greater in solids than in liquids. For a few solids there exists a temperature below the melting temperature at which the vapor pressure of the solid is 1 atm and sublimation occurs; this temperature is called the *normal sublimation temperature*.

The existence of a vapor pressure and its increase with T are both consequences of the energy distribution just described. The energy distributions of both liquids and solids are also Maxwell-Boltzmann distributions with the slight change that the energy in Equation 35 must be a sum of both the kinetic energy and the potential energies of intermolecular interactions. Thus, for a liquid, even at temperatures far below the boiling point, a fraction of the molecules will have energies greater than that of the attractive force between molecules of the liquid and these molecules will escape to the gaseous phase. Equation 35 and Figure 2-6(b) show that this fraction is greater at higher temperatures, thus explaining the increase in vapor pressure with temperature.

Another consequence of the distribution is that molecules not only leave but also rejoin the liquid—that is, there is a continual exchange of molecules between the liquid and vapor phases. This is for the following reason. A molecule that has entered the vapor phase can lose energy in collisions with other molecules in the vapor. If its energy drops below the cohesive energy and if the molecule by chance collides with the surface of the liquid, it will be recaptured by the liquid. This exchange of molecules between liquid and vapor is easily observed if a container of D_2O (deuterium oxide) is exposed to the atmosphere (which contains H_2O vapor). A molecule of D_2O can leave the liquid and be replaced by a molecule of H_2O from the vapor. Even though there is a small difference in the cohesive energies and the energy distribution of H_2O and D_2O, the great excess of H_2O in the atmosphere results in a gradual conversion of the liquid D_2O to nearly pure liquid H_2O.*

2-5 BROWNIAN MOTION

Small particles suspended in a liquid are in constant motion because the particles are incessantly bombarded with molecules of the liquid. This motion is called Brownian motion after the Scottish botanist Robert Brown (1773–1858), who observed this phenomenon with a microscope while viewing pollen grains suspended in water. In the first years of the 20th century, the French physicist Jean Perrin recognized that if Brownian motion is a result of molecular collisions, small particles at equilibrium should be distributed in a gravitational field according to Equation 27. Thus, Perrin placed an aqueous suspension of microscopic particles in a 0.1-mm-deep well in a glass microscope slide and observed with a microscope the number of particles in a 1-μm-thick layer at various depths in the well. Perrin found that the distribution obeyed Equation 27. This experiment was instrumental in establishing the molecular theory of matter and, in part for this work, Perrin received the Nobel Prize for physics in 1926.

Once the molecular basis of Brownian motion was realized, a quantitative theory was derived by Albert Einstein. Realizing that there is little point in evaluating the motion imparted by a single collision when a huge number of collisions occur in the time it takes to make a single measurement, Einstein calculated the *net* displacement occurring in a particular

---

*This occurs to the chagrin of laboratory workers who leave bottles of expensive D_2O unstoppered.

SEC. 2-4   DISTRIBUTION OF KINETIC ENERGIES

time interval. The moving particle is viewed as ploughing through a viscous liquid so that the motion of the particle is impeded by (1) a friction force proportional to its velocity, with a proportionality constant f that is somehow related to the shape of the particle, and (2) a force resulting from collisions with solvent molecules or arising from any other forces imposed by the fluid. The derivation of the relevant equation is complex and beyond the scope of this book. The result of Einstein's calculation is the equation

$$\langle (\Delta x)^2 \rangle = 2kTt/f . \tag{36}$$

The quantity $(\Delta x)^2$ has the following meaning. A particle is observed first at time 0 and then at time t. During this time interval it undergoes a displacement whose projection on the x axis is Δx. The same particle is observed at later times, always separated by the same interval t, that is, at times $2t, 3t, \ldots$, and Δx is determined for each interval. These values are squared and their mean is computed, thus giving $\langle (\Delta x)^2 \rangle$. This process is shown in Figure 2-7.

FIGURE 2-7 A hypothetical path of a particle undergoing Brownian motion. The particle moves in the direction indicated by the arrows; Δx_1, Δx_2, Δx_3, and Δx_4 are the horizontal displacements after time intervals t, $2t$, $3t$, and $4t$, respectively.

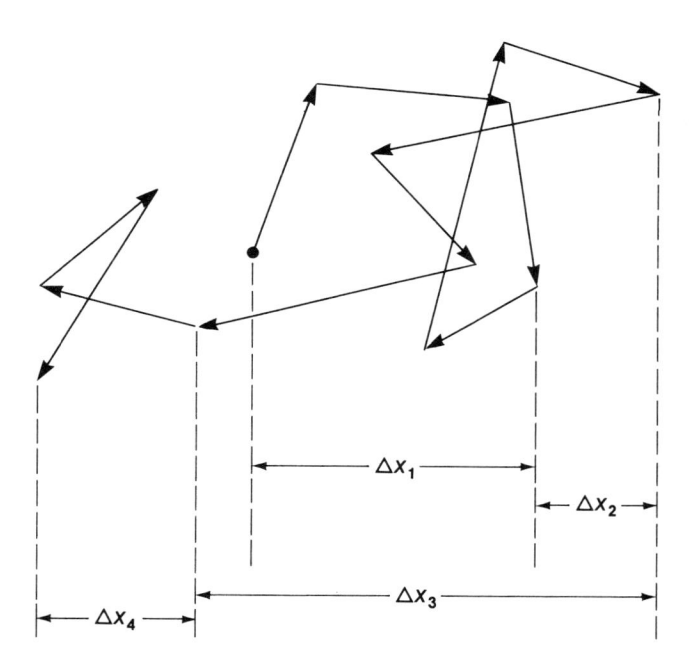

We now assume that the particle can be approximated as a sphere of radius a, in which case, by Stokes' Law,

$$f = 6\pi\eta a . \tag{37}$$

Thus

$$\langle (\Delta x)^2 \rangle = \frac{kTt}{3\pi\eta a} . \tag{38}$$

This equation is known as the Einstein-Smoluchowski equation. The significance of this equation is simply that if it is satisfied by experimental measurement, it will support the idea that Brownian motion is a result of collisions with solvent molecules. Perrin did in fact show that $(1/t)\langle(\Delta x)^2\rangle$ is a constant for a particular value of η and of a and that it is proportional to $1/\eta$ and to $1/a$. A surprising fact obtained from Equation 38 is that $\langle(\Delta x)^2\rangle$ is independent of the mass of the particles, which was confirmed by Perrin. There is also the implication that $(\Delta x)^2$ varies linearly with kT; however, this is not the case, because f decreases markedly with increasing T. This equation proved to be very important because Perrin used it to measure the Boltzmann constant k. A use of Equation 38 is shown in the following example.

EXAMPLE 2–F Measurement of the viscosity of the cytoplasm within living cells.

Most living cells contain small particles that move around by Brownian motion. If photographs of a cell are taken at distinct time intervals and the positions of individual particles are noted, the displacement Δx can be measured as a function of time. Since the radii of larger particles can be measured quite accurately, the internal viscosity of the cell can be calculated from Equation 38. In studies such as these, it was observed by Robert Uretz and Robert Perry at the University of Chicago that the internal viscosity varied with the stage of the life cycle of newt fibroblasts. A related study is given in Example 2–G.

EXAMPLE 2–G Identification of solid or semisolid regions within living cells.

In living eukaryotic cells the mitotic spindle cannot be seen with an ordinary microscope or a phase contrast system. However, the growth of the spindle was once measured by time-lapse photography, by observing the region of the cytoplasm in which intracellular particles do not undergo visible Brownian motion. The mitotic spindle is semisolid (that is, η is very high); therefore the displacements of particles trapped within the spindle region are so small that the particles appear to be stationary. Thus the area of stationary particles was noted and measurement of this area at various times allowed the measurement of the growth of the spindle.

REFERENCES

A more detailed discussion of these topics can be found in the following books.

Carpenter, D.K. 1966. "Kinetic Theory, Temperature, and Equilibrium." *J. Chem. Educ.* 43, 332.

Cowling, T.G. 1950. *Molecules in Motion.* University Press. London.

Dence, J.B. 1972. "Heat Capacity and the Equipartition Theorem." *J. Chem. Educ.* 49, 798.

Hildebrand, J.H. 1963. *An Introduction to Molecular Kinetic Theory.* Reinhold. New York.
Jeans, J.H. 1954. *Introduction to the Kinetic Theory of Gases.* McGraw-Hill. New York.
Kauzmann, W. 1966. *Thermal Properties of Gases.* Vol. I. W.A. Benjamin. Menlo Park, Calif.
Loeb, L.B. 1961. *Kinetic Theory of Gases.* Dover. New York.
Tabor, D. 1969. *Gases, Liquids, and Solids.* Penguin. Baltimore.

PROBLEMS

1. What is the mean free path of neon at 20°C and a pressure of 1 atm? Of 10^{-4} atm? The radius of neon is 1.12 Å.

2. What fraction of the molecules in a gas has kinetic energy between $0.9\,kT$ and $1.1\,kT$ at 20°C?

3. What is the average distance separating the molecules of one mole of a gas at a pressure of 10 atm if its density is 0.00278 g cm^{-3} at 20°C? The molecular weight of the gas is 35.

4. What is the root mean square speed of O_2 at 25°C at a pressure of 1 atm?

5. What is the kinetic energy in joules of one mole of a gas at 25°C?

6. Assuming a pressure of 1 atm and at 0°C, calculate for N_2 the number of collisions per second per molecule and the total number of collisions per cm^3 per second. The radius of N_2 is 1.87 Å.

7. It is claimed that the density of intergalactic space is 1 H atom per 100 liters. The collision diameter of a hydrogen atom is 2 Å. What is the mean free path of an H atom in intergalactic space? (Use light years as a measure of distance.)

8. A plate whose mass is 1 kg is suspended by strings so that the plane of the plate is parallel to the surface of the earth. How many balls having a mass of 0.5 g and moving at a velocity of 500 cm sec^{-1} would have to be shot upward against the plate to keep it suspended in the air when the string is cut?

9. While on the moon a lunar lander whose volume is 20 m^3 is struck by a micrometeorite that makes a hole of 0.1 mm radius. The pressure in the ship is 0.6 atm and the temperature is 25°C. If the return trip to Earth takes 27 hours, can the astronauts return before the pressure drops to one-fourth the initial value?

10. What is the difference in pressure between the top and the bottom of a container of N_2 at sea level and 1 atm total pressure and 20°C, if the container is 200 cm high?

11. What mass of CO_2 collides every second with a leafy plant (in dry air) whose surface area is 200 cm^2? The temperature is 25°C and the pressure is 1 atm. Dry air is 0.033 percent CO_2 by volume.

12. A 100-liter vessel containing H_2 that is one mole percent deuterium (D_2) has a small opening 0.2 mm in diameter that allows the gas to escape into vacuum. The pressure in the vessel is 0.1 atm and the temperature is 20°C. If the gas is allowed to escape for 30 minutes, what will be the mole percent of D_2 in the vessel after this time?

13. If a beam of light passes through a narrow slit some distance away from the light source and then falls on a distant screen, the light pattern on the screen has the dimensions of the slit, except for a slight symmetric spreading and fuzziness resulting from diffraction. The situation is quite different for a beam of atoms, as will be explored in this problem. An oven containing metallic cesium heated above the boiling point has a small opening from which Cs atoms emerge. In order to form

an atomic beam, a slit is placed some distance away from the opening so that only atoms moving in a particular direction produce the beam. The atomic beam falls on a detector that can be moved and that can measure the number of atoms that impinge on it each second. The entire system is in vacuum so that there are no intermolecular collisions. If the beam is vertical, the Cs atoms fall in a region that is essentially an image of the collimating slit, as in the case of the light beam. However, if the beam is horizontal, the image is no longer sharp and it is asymmetric. Explain the asymmetry and point out whether atoms are found above or below the positions that would be expected if the image formed were that of the slit.

14. One mole of CH_4, a nonlinear molecule, at 20°C, is allowed to mix with one mole of CO_2, a linear molecule, at 25°C. The system is perfectly insulated so that no heat is lost. What is the temperature of the mixture when equilibrium is reached?

15. Diffusion is a random process of movement of molecules. That is, if a vessel is filled with a gas and a second gas is introduced at a single point in the container, molecules of gas #2 gradually diffuse through gas #1 until the two gases are uniformly mixed. Would you expect diffusion to increase or decrease with increasing temperature and with increasing pressure?

16. The viscosity of a fluid is a result of its resistance to flow. For a liquid, the viscosity is primarily a result of the attractive forces between the molecules, and since these forces are overcome when the individual molecules have a high kinetic energy, the viscosity decreases with increasing temperature. On the contrary, the viscosity of a gas (in which the attractive forces are nearly zero) increases with increasing temperature. Explain.

17. A particle is undergoing Brownian motion in the cytoplasm of an onion root-tip cell at 20°C. The radius of the particle is 1 μm or 10^{-4} cm. The positions of the particle at 10-second intervals are shown below. What is the viscosity in poise, which is the correct unit, of the cytoplasm?

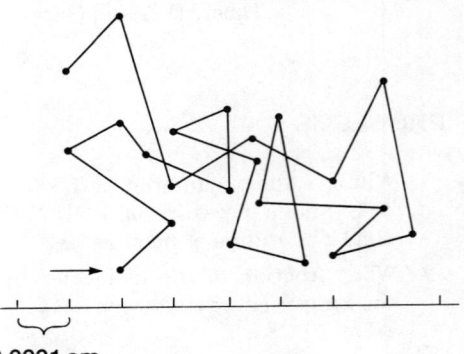

0.0001 cm

CHAPTER 3

INTERMOLECULAR FORCES

In Chapter 1 we analyzed an ideal gas—that is, a gas in which the molecules have no volume and in which there are no interactions between the molecules. In Chapter 2 we allowed the molecules to have volume and to collide with one another, and we examined some of the phenomena that result from the collisions. In this chapter we will see that not only is there a repulsive interaction that produces collisions but also an attractive interaction. This interaction forces us to modify the ideal gas law in order to account for the properties of real, rather than ideal, gases. This will enable us to understand such important processes as condensation and evaporation, solidification and melting, and the forces that determine the shape of macromolecules and complex aggregates of macromolecules.

3-1 FAILURE OF THE IDEAL GAS LAW: EVIDENCE FOR AN ATTRACTIVE FORCE

A gas is considered to be ideal if it obeys the relation $PV = nRT$ for all values of P and T. For most gases the ideal gas law is obeyed near atmospheric pressure and at room temperature. However, when P, V, and T are measured precisely over a wide range of P and T, it is found that no real gas obeys this law. One obvious failure of the law is that it predicts a continuous change in V as P is increased and as T is decreased, and hence does not take into account liquefaction—that is, at sufficiently low temperatures there is always a pressure above which the gas liquefies. This behavior can be seen in Figure 3-1, which shows a typical set of isotherms for a real gas. Observe that at the lower temperature there is a region (the horizontal part of the curve) in which the volume decreases at constant pressure; this decrease in volume is the change accompanying condensation or liquefaction of a gas. Let us consider the factors that must be involved in the condensation process. We recall that in the gaseous state the molecules always fill the containing vessel, no matter how large it is, and that in the liquid state, the volume of liquid is independent of the size of the container. Clearly, the molecules in the liquid state are closer to one another than the molecules in the gaseous state. Thus, the molecules of a liquid must be associated with one another in some way; in other words, there is an *attractive force* between the molecules of a liquid. Since the fundamental physical properties of each individual molecule should be the same in the liquid and the gaseous states, there should be attractive forces between the molecules of a gas, although the magnitude of the net attractive force in a gas may be weaker than in a liquid. This is not unreasonable, because all known attractive forces (for example, electrical and gravitational forces) between two particles decrease with increasing separation of the particles, and the average distance between molecules is much greater in the gas phase than in the liquid phase. If attractive forces exist between the molecules of all substances, then even if we only consider values of T far above the boiling point, the ideal gas law should not apply to any real gas and the discrepancy should be such that the volume of the gas at a particular pressure is less than that expected for an ideal gas. Furthermore, if the magnitude of the force depends upon the type of molecule, the extent of the discrepancy should not be the same for all molecules. This can be seen in Figure 3-2, which, in panel (a), compares the isotherm for N_2 with that of an ideal gas and, in panel (b), shows the ratio of the actual molar volumes (the volume of one mole of gas) of several gases to

FIGURE 3-1 Typical isotherms for a real gas. At temperatures T_1 and T_2 the isotherms resemble an ideal gas. At T_c (the critical temperature) an inflection point appears in the curve. The pressure and volume associated with this inflection point are the critical pressure P_c and critical volume V_c. At lower temperatures T_3 and T_4 there are regions at which the volume decreases without a change in pressure; in these regions condensation occurs. Note that once liquefaction is complete, the volume decreases only slightly with increasing pressure—that is, liquids are virtually incompressible. P_3 and P_4 are the vapor pressures at T_3 and T_4, respectively.

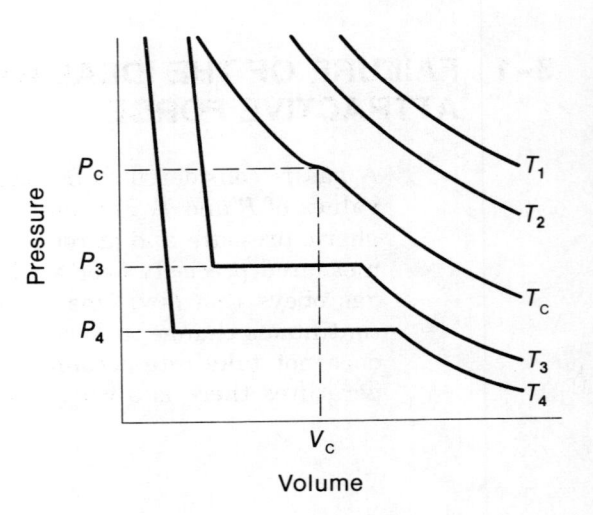

FIGURE 3-2 Comparison of real and ideal gases. (a) Isotherm for N_2 and an ideal gas. For pressures below the dashed line the real volume and attractive forces predominate; for greater pressures, repulsive forces predominate. (b) The ratio of molar volume of a real gas to that of an ideal gas for three real gases.

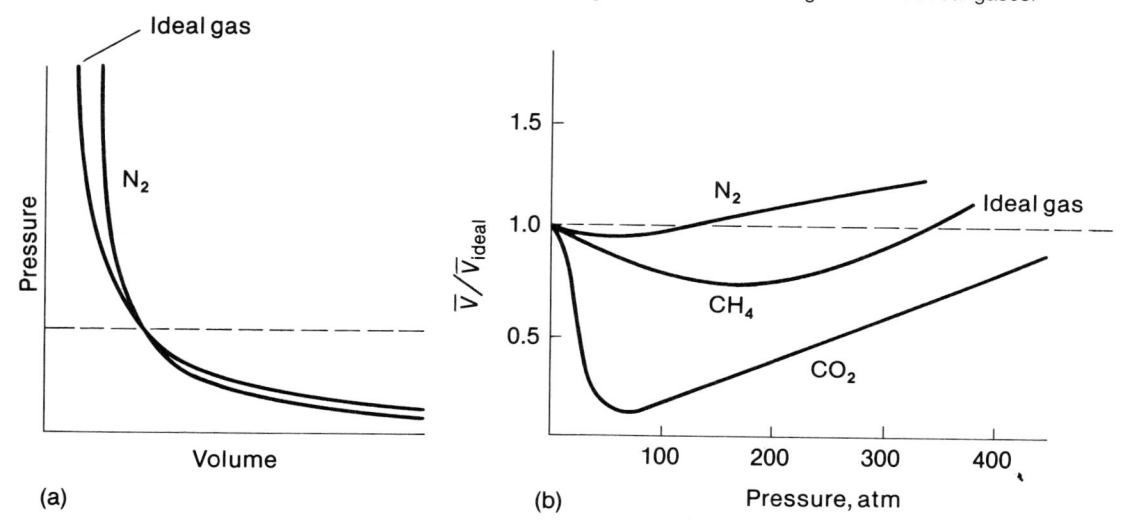

the molar volume of an ideal gas. Observe that the ratio is less than 1 for a wide range of pressures, as we have just predicted.

It should also be noted in Figure 3-2 that at high pressure the molar volume of N_2 and CH_4 is greater than that of an ideal gas—something seems to be keeping the molecules apart. A related phenomenon is present in some of the curves (T_3 and T_4) of Figure 3-1, in which we can see that once the liquid state has been reached at constant temperature, a very large pressure increase results in only a very small volume change; this is usually referred to as the incompressibility of liquids.* Both phenomena can be explained by the fact that the molecules have finite size, so that once two molecules are in contact, their centers cannot be pushed closer to one another. This size effect accounts for an obvious error in the ideal gas law, which states that at 0 K the volume of a gas is zero. In a real situation the volume occupied by a given number of molecules decreases substantially during liquefaction, but after liquefaction has occurred, the volume changes only slightly as T decreases. Clearly, in addition to intermolecular attractive forces a real gas law must incorporate the volume of the molecules and the associated repulsive force that must come into play when the molecular domains of two molecules overlap.

Two other phenomena also indicate that something is lacking in the ideal gas law. First, for a real gas there is usually a set of pressures and

---

*This is the principle used in a hydraulic pump such as that used in automobile service stations to raise a car off the ground. Clearly, a gas could not be used in such a pump.

temperature for which there is no volume change during liquefaction; this is called the *critical region*, and it will be discussed shortly. Second, the volume of some liquids (for example, water) increases during solidification. We will see later that this is a result of the partially ordered structure of water.

3-2 THE VAN DER WAALS GAS LAW AND ITS CONSEQUENCES

The first attempt to derive a new gas law was made by J.D. van der Waals in 1873. Many other modifications of the ideal gas law have also been derived. That of van der Waals, although it is not the one that most accurately describes the properties of real gases, is of particular interest because it describes in a simple way some of the important characteristics of the transition between the gaseous and the liquid states. It also illustrates quite clearly how one goes about developing a real gas law.

The derivation of the van der Waals equation begins by modifying the ideal gas law to include the molecular volume. We ignore the fact that liquefaction would occur before 0 K is reached and define the volume occupied by one mole of a particular gas as b.

The ideal gas law is then rewritten as

$$\bar{V} = b + \frac{RT}{P}, \tag{1}$$

in which $\bar{V}$ is the molar volume (the volume of one mole) at a temperature T. Since there is no molecular motion at 0 K, b might be close to the actual volume of one mole of molecules. Since molecules are not all the same size, b will not be a universal constant like R and every real gas will obey its own gas law. Thus Equation 1 is usually written as

$$P = \frac{RT}{\bar{V} - b}. \tag{2}$$

Since b must always be positive, the value of P corresponding to a given value of $\bar{V}$ and T should, for all gases, be greater than that predicted by the ideal gas law. This is not strictly the case, since we have not yet taken into account the attractive force; however, at high pressure and at low temperatures—that is, when the molar volume is small—it is generally true, as shown in Figure 3-3.

We now introduce the correction required by the existence of an attractive force. The pressure of a gas represents the force exerted by the gas on a unit area of the walls of the container. If there is an attractive force between the molecules, a molecule near the wall will be drawn to the bulk of the molecules in the container and therefore away from the wall. Thus, the effect of an attractive force should enter Equation 2 in a way that decreases P from the value of an ideal gas.

FIGURE 3-3 Isotherm illustrating excessive pressure of a real gas at small molar volumes.

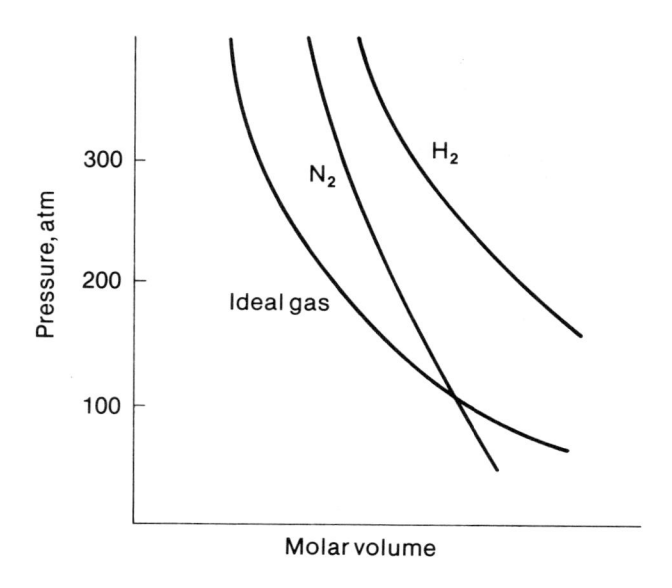

The mathematical form of the correction can be seen by the following reasoning. Consider two small-volume elements v_1 and v_2 within a vessel: each volume contains one molecule. Let the attractive force between v_1 and v_2 be f. If second, third, and nth molecules were added to v_2, the force between v_1 and v_2 would increase to $2f$, $3f$, and nf, respectively. If instead the number of molecules in v_2 remained one, and second, third, ..., nth molecules were added to v_1, again the force would increase to $2f$, $3f$, and nf, respectively. Thus the force between v_1 and v_2 is proportional to the concentrations of both of these two volumes. If we call these concentrations c_1 and c_2, respectively, the force between v_1 and v_2 is proportional to $c_1 c_2$. If a is the proportionality constant, the force is $a c_1 c_2$. Averaged over time, the concentration of molecules is the same throughout a gas; that is, $c_1 = c_2 = c$, in which $c =$ number of molecules per unit volume. Since concentration is proportional to $1/v$, the attractive force is proportional to $1/\bar{V}^2$ or equal to $a/\bar{V}^2$. As we have just stated, this force should enter Equation 2 to reduce the pressure by $a/\bar{V}^2$; that is, it should appear with a minus sign. Thus, we change Equation 2 to

$$P = \frac{RT}{\bar{V} - b} - \frac{a}{\bar{V}^2}. \tag{3}$$

This is called the van der Waals equation of state; in order to have the same form as the ideal gas law, it is usually written as

$$\left(P + \frac{a}{\bar{V}^2}\right)(\bar{V} - b) = RT. \tag{4}$$

Several isotherms corresponding to this equation are shown in Figure 3-4. Note that they resemble the isotherms in Figure 3-1 in that at some

FIGURE 3-4 Isotherms calculated from the van der Waals equation. At the temperature T_1 the term $a/\overline{V}^2$ is very small and the isotherm resembles that of an ideal gas. At the critical temperature T_c there is an inflection point; the prediction of this inflection point was one of the great triumphs of the van der Waals theory. At temperatures below T_c, for example at T_2, the calculated curve differs markedly from the actual curve shown in Figure 3-1. That is, instead of the curve having a constant P (the vapor pressure in the volume range $\overline{V}$ to $2\overline{V}$) the curve oscillates. Although such an isotherm is never obtained experimentally, it is interesting that the van der Waals theory predicts the existence, albeit metastable, of a superheated liquid (region 1–2 of curve) and a supercooled vapor (region 3–4 of curve); region 2–3 has not been observed.

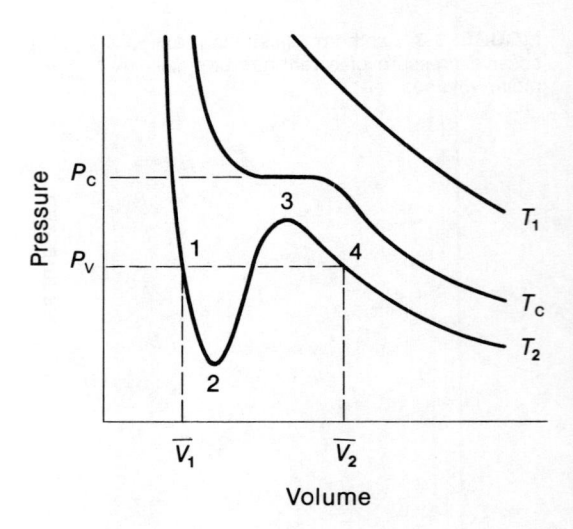

temperature T_c there is an inflection point. As explained in the legend to Figure 3-1, this temperature T_c is called the *critical temperature* and the corresponding values of P and V are called the *critical pressure* P_c and the *critical volume* V_c. The critical temperature is a temperature at which gas, liquid, and solid can coexist. Below T_c the theoretical curves depart significantly from experimental isotherms for real gases because Equation 3 is not accurate in this region. We will not analyze other more precise modifications of the gas law but will emphasize the important fact that just by introducing the idea that molecules occupy space and that they are mutually attractive, the existence of the critical state is predicted. We will see that measurement of T_c and P_c given some information about the values of a and b, although these values may not be as precise as we would want them to be.

At $T = T_c$ the PV curve is horizontal (that is, the slope is 0) so that the first derivative $(\partial P/\partial V)_T = 0$. Since the PV curve has an inflection point, the second derivative $\partial^2 P/\partial V^2 = 0$. Solving these equations and setting $T = T_c$, $P = P_c$, and $V = V_c$, one obtains

$$V_c = 3b, \qquad T_c = \frac{8a}{27Rb}, \qquad P_c = \frac{a}{27b^2}, \tag{5}$$

or

$$a = \frac{27(RT_c)^2}{64 P_c} \quad \text{and} \quad b = \frac{RT_c}{8P_c}. \tag{6}$$

If these values of a and b are substituted into Equation 4, there results the remarkable equality

$$\frac{P_c V_c}{T_c} = \frac{3}{8} = 0.375. \tag{7}$$

SEC. 3–2 THE VAN DER WAALS GAS LAW AND ITS CONSEQUENCES

For most real gases, the observed value is not 0.375 but 0.28. This could be taken to indicate a failure of the van der Waals formulation; however, the fact that nearly the same value is obtained for all gases suggests instead that the way the volume and the intramolecular forces are treated is approximately correct but that there remains at least another correction factor yet undiscovered.

From Equation 6 and measurement of P_c and T_c (which are not difficult measurements), one can calculate a and b. Since a and b are molar quantities and we are using the van der Waals formulation to discuss properties of individual molecules, we introduce the corresponding molecular parameters α and β, defined as $\alpha = a/\mathcal{N}^2$ and $\beta = b/\mathcal{N}$, in which $\mathcal{N}$ is Avogadro's number. With the help of statistical thermodynamics it can be shown that $\beta = 4\omega$, in which ω is the volume of the molecule if it were a rigid sphere. The radius $\mathcal{R}$ of this sphere is called the *van der Waals radius;* it is a useful parameter in discussing the size of molecules. Its value is related to T_c and P_c by

$$\mathcal{R} = \left(\frac{3kT_c}{128\pi P_c}\right)^{1/3}, \tag{8}$$

in which k is the Boltzmann constant.

To understand the meaning of α, it is necessary to know something about the attractive forces themselves. These forces, described in detail in a later section, are proportional to $1/r^6$, in which r is the distance between the centers of the spheres that each molecule occupies. If a molecule is assumed to be incompressible, then the shortest possible distance between the centers of two molecules, *the distance of closest approach,* is the diameter d of a molecule. Then, if W_0 is the potential energy between the two molecules when they are touching (that is, when $r = d$), the energy W of interaction between two molecules is

$$W = -W_0\left(\frac{d}{r}\right)^6. \tag{9}$$

From Equation 9 the potential energy of a single molecule due to all molecules of the same kind can be calculated, and it can then be shown that

$$a = W_0 b. \tag{10}$$

To examine the validity of a theory, it is convenient to write as many parameters as possible in terms of measurable quantities, so that from Equations 6 and 10,

$$W_0 = \frac{27RT_c}{8\mathcal{N}} = \frac{27}{8}kT_c. \tag{11}$$

It is instructive to calculate W_0 for carbon dioxide, helium, and water from the critical constants. For CO_2, $T_c = 304.2$ K at $P_c = 73.0$ atm; from

Equation 11, the value of W_0 is 14×10^{-21} J. A comparison of this value to thermal energy kT indicates whether W_0 is a large or small quantity. At room temperature (about 300 K), $kT \approx 4.1 \times 10^{-21}$ J and the translational energy of a gas molecule (according to the Law of Equipartition, Chapter 2), is $\frac{3}{2} kT$ or 6.2×10^{-21} J. Therefore, the potential energy between two CO_2 molecules at 300 K, even when they are in contact, is only about 2.2 times the average translational energy of one molecule. Thus, at 300 K, two CO_2 molecules do not stick together very well; this explains why CO_2 is a gas at room temperature. Below the boiling point, W_0 is about 3.3 times kT. For He, the value of $W_0 = 0.24 \times 10^{-21}$ J; this is about 1/15 the translational energy, which agrees with the fact that liquefaction of helium requires very low temperature. Indeed, the boiling temperature is 4.1 K, at which point W_0 is about 3.4 times the translational energy. Water, on the other hand, has a value of $T_c = 617$ K, at which point $W_0 = 38.5 \times 10^{-21}$ J, about 6.2 times the translational energy of a molecule at 300 K. This large factor is consistent with the fact that water is a liquid at room temperature. At the boiling point of water, W_0 is approximately 3.7 times the translational energy. Examination of a large number of substances shows that at the boiling point, W_0 is between 3.4 and 3.7 times the translational energy.

3–3 THE NATURE OF THE ATTRACTIVE FORCE

The attractive force between molecules is a result of interactions between electric dipoles and electric fields. In this section we examine this kind of interaction and the parameters needed to describe it.

The Dipole Moment

An electric dipole consists of a positive and a negative charge $+e$ and $-e$ separated by a distance r (Figure 3-5). The dipole moment μ is defined by

$$\mu = er \tag{12}$$

and is a vector quantity, the direction being by definition from the negative toward the positive charge. If a molecule possesses separate regions of positive and negative charge, its dipole moment is *permanent*. If the charged regions move with respect to one another (for example, if there are internal vibrations in the molecule), the molecule has a *fluctuating dipole moment*. If only the distance but not the direction varies, the dipole moment fluctuates in magnitude only. An example of a dipole that fluctuates in magnitude only is the heteronuclear molecule carbon monoxide. In this molecule the centers of positive and negative charge differ so that CO has

SEC. 3-3 THE NATURE OF THE ATTRACTIVE FORCE

FIGURE 3-5 An electric dipole.

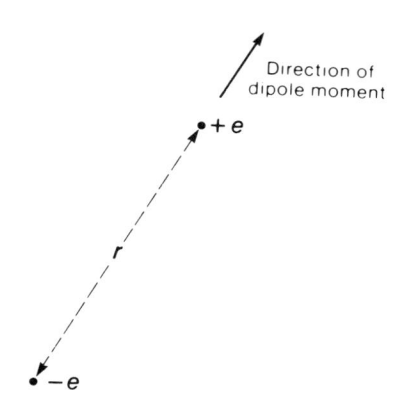

a permanent dipole moment. Furthermore the C—O bond vibrates (stretches) so that the dipole moment of CO fluctuates. In a nonlinear molecule consisting of three or more atoms, the movement of the charges can be such that both direction and magnitude vary.

All molecules do not have a permanent dipole moment. For example, in a homonuclear diatomic molecule such as H_2 or N_2, there is no possibility for permanent asymmetry of charge distribution, so there is no permanent dipole moment.

A symmetric, linear, triatomic molecule such as CO_2 also does not have a permanent dipole moment. Symmetric stretching—that is, with each O atom moving back and forth with respect to the C atom, both O atoms maintaining equal C—O distances at a given instant—also does not produce a dipole. However, any asymmetric stretching or asymmetric sideward displacement (for instance, the C moving out of line to produce a V-shaped molecule) can generate a fluctuating dipole moment.

A spherically symmetric atom such as He or Ne clearly cannot have a permanent dipole moment. However, the electrons in an atom are not stationary and move with respect to the positively charged nucleus, which is, therefore, not always precisely at the center of a sphere of negative charge. Thus, these atoms possess fluctuating dipole moments. An important difference must be noted between the fluctuating dipole moments of a CO molecule and an He atom. The dipole moment of a CO molecule fluctuates in magnitude only and does not change its sign. On the other hand, the dipole moment of He (and other noble gas atoms) fluctuates in magnitude *and* in direction. In fact, if the electrons are thought of as rotating around the nucleus, the dipole moment will also rotate. This idea of rotation is of course not exactly correct, but since the dipole moment can assume *all* possible directions within a very short period of time, the average dipole moment of an He atom is zero when averaged over a long period of time. Thus we say that such a symmetric molecule only has an *instantaneous dipole moment*. We will see, though, that two molecules having instantaneous dipole moments can attract one another.

Two dipoles attract if they are antiparallel and repel if they are parallel, as shown in Figure 3-6. This kind of interaction is in fact the source of the intermolecular attraction, as we will see shortly.

FIGURE 3-6 Attractive and repulsive configurations of two dipole moments. Repulsion occurs when the dipole moments are parallel; attraction occurs when they are antiparallel (that is, parallel but opposite in direction).

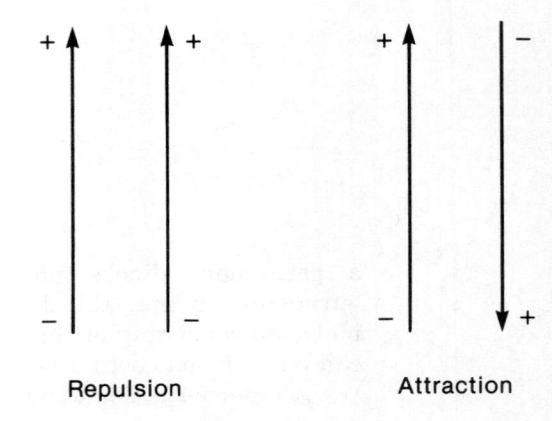

In the early formulation of the theory, the van der Waals attraction was thought to result from interactions only between permanent electrical dipoles. The reasoning went as follows: in a gas, the individual molecules move at random; however, if starting from a state of total disorder, such a collection of freely moving dipoles will, at any temperature above 0 K, gradually become partially oriented so that a small attractive force results. These interactions did not seem to be the sole source of the attraction, though, because as the temperature increases, the degree of orientation decreases; thus the molecular attraction should lessen with increasing temperature. Lessened attraction did not agree with the experimental facts, so that the view that the attractive force is a result of having a *rigid* electrical structure had to be discarded. Once it was realized that the electron cloud of a molecule can be deformed by another electric field, the concept of polarizability (the ability to be deformed), which will be discussed in the following section, was introduced; it was then easily shown that the combination of the electric field of one dipolar molecule and the polarization induced by this field in a second molecule also introduces an attractive force.

Polarizability

If a dipolar molecule is placed near an external positive charge, the molecule will orient itself so that its more negative portion is nearer to the

SEC. 3–3   THE NATURE OF THE ATTRACTIVE FORCE

external charge than is its more positive region. This is called *orientation polarization*. Furthermore, the positive region of the molecule is repelled by the external charge so that there is an increase in the separation of the positive and negative regions of the molecule. If, on the other hand, the molecule is unpolarized (not a dipole), it will become polarized in the electric field produced by the external charge because the positive and negative charges of the molecule will be distributed in the manner shown in Figure 3-7. This type of polarization is called *induced* or *distortion polarization* and will be seen to be an important factor in explaining the van der

FIGURE 3-7 Induced polarization. In panel (a) the dipole is infinitely far from a molecule whose 16 positive charges are located near the center of the electron cloud. In panel (b) the dipole is near a molecule whose electron cloud is distorted by the field of the dipole.

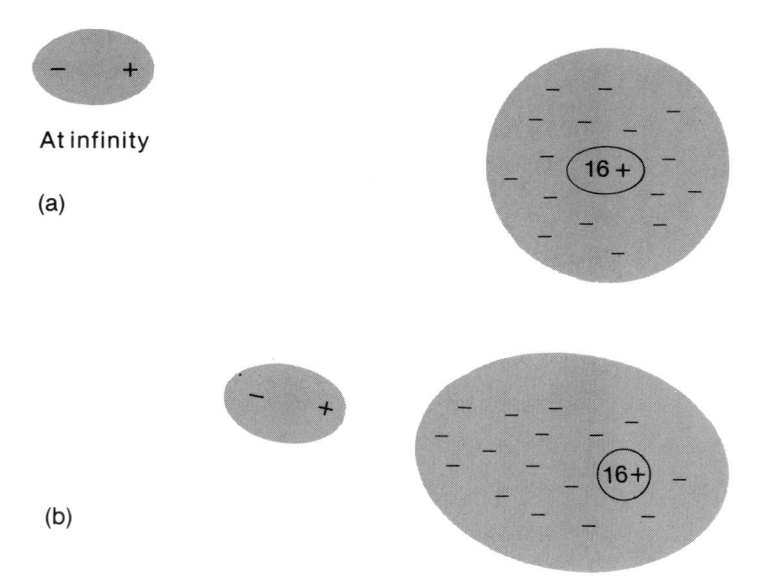

Waals attraction. The magnitude of the induced polarization of a molecule is given by a quantity called the *distortion polarizability*, α_0, defined by the relation

$$\mu = \alpha_0 E \qquad (13)$$

in which μ is the dipole moment induced by the applied electric field E. The distortion polarizability has the dimension of a volume and its numerical value is roughly that of the average volume of the molecule, so that larger molecules are generally more easily polarized than smaller molecules. The distortion polarizability is not a difficult parameter to measure, since for gases it is related to the refractive index χ by the simple equation

$$\alpha_0 = \frac{\chi^2 - 1}{4\pi n}, \qquad (14)$$

in which n is the number of molecules per cubic centimeter. This equation arises because the refractive index is essentially a measure of the interaction of the electrical field of the light wave with the electron cloud of the molecules being illuminated. For liquids a better value is obtained by using the Lorentz-Lorenz relation,

$$\alpha = \frac{3}{4\pi \mathcal{N}}\left(\frac{\chi^2 - 1}{\chi^2 + 2}\right)\frac{M}{\rho}, \tag{15}$$

in which $\mathcal{N}$ = Avogadro's number, M = the molecular weight, and ρ = the density of the liquid.

The Attraction Resulting from Dipole Interactions

Using the concepts of dipole moment and polarizability, we can derive a simple electrostatic theory of intermolecular forces that is valid for polar molecules. The reasoning is based upon only two facts: (1) if two polar molecules have the proper orientation, the positive end of one will attract the negative end of the other, and (2) the electric field of one dipole or polar molecule will induce a dipole moment in another molecule. A question that sometimes arises is this: since molecules are always in motion, it seems as if there should be as many pairs in the attracting parallel configuration as in the repulsive parallel configuration, so how can there be a net attractive force? This will be answered in the following argument, in which it will be seen that since the force enters as the second power of the field, there is no problem of field direction.*

Consider the configurations, shown in Figure 3-8, of two dipoles fixed in space, each having charges $+e$ and $-e$ that are separated by a distance a. The centers of the two dipoles are separated by a distance r, as shown in the figure. The force between them is determined in a standard way by assuming that one dipole is fixed and calculating the work required to bring the second dipole into position from infinity. This is most simply done by considering the charges separately; that is, we calculate the work required to bring each charge $+e$ and $-e$ from infinity (where the electric field of the stationary dipole is zero) to the positions r and $r + a$. From electrostatics we know that the force produced by a field E acting on a charge e is Ee and the work involved in moving a charge is the product of the force and the distance moved. Since the field strength depends upon the distance from the charge, the work must be calculated by summing the infinitesimal amounts of work required to move the charge through each infinitesimal distance at each point between the first dipole and infinity. This is expressed as a sum of two integrals,

---

*It is often found when dealing with populations that simple ways of thinking about averaging are incorrect. In this case it is the assumption that half are parallel and half are antiparallel that is incorrect.

SEC. 3-3 THE NATURE OF THE ATTRACTIVE FORCE

$$W = \int_\infty^r Ee\, dr + \int_\infty^{r+a} E(-e)\, dr, \quad (16)$$

in which the first and second terms refer to the charges $+e$ and $-e$, respectively. The terms can be combined as

$$W = -\int_r^{r+a} Ee\, dr. \quad (17)$$

The field strength E is a function of distance, but if a is very small (as it is in a molecule whose dimensions are in ångström units), E can be considered constant. Thus, the integral can be evaluated as

$$W = -Eea = -E\mu, \quad (18)$$

in which $\mu = ea$ is the dipole moment. Thus, for Figure 3-8, this is the potential energy of the dipole at the right having dipole moment μ in the field E that is produced by the dipole at the left.

FIGURE 3-8 Two dipoles: the figure shows the parameters necessary in discussing their interaction.

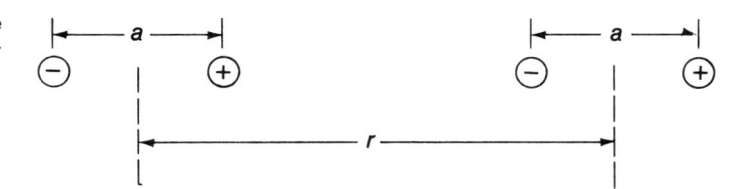

Since two molecules will not always have the orientation shown in Figure 3-8, the argument must be modified as follows. Consider molecule 1 with a permanent dipole moment; molecule 2 approaches molecule 1. The field of molecule 1 induces a dipole moment in molecule 2 either by orientation of a permanent dipole moment or by distortion of the electron cloud. The induced moment μ_2 will be in the direction of the field E_1 of molecule 1 and the strength of the induced dipole will be determined by α_2, the polarizability of molecule 2. Thus,

$$\mu_2 = \alpha_2 E_1. \quad (19)$$

To induce the dipole moment μ_2, a charge must be moved. The work done in this movement is evaluated as in Equations 16 through 18 and is equal to $\frac{1}{2}\alpha_2 E_1^2$. The total energy W_2 of molecule 2 in the field E_1 is the sum of both the work to induce the dipole moment and the potential energy due to its initial position, or

$$W_2 = -E_1\mu_2 + \tfrac{1}{2}\alpha_2 E_1^2. \quad (20)$$

Combining Equations 19 and 20 yields
$$W_2 = -\tfrac{1}{2}\alpha_2 E_1^2 .$$
The dipole induced in molecule 2 produces an electric field that acts on the dipole moment of molecule 1 so that molecule 1 possesses a potential energy W_1 resulting from the field of molecule 2:
$$W_1 = -\tfrac{1}{2}\alpha_1 E_2^2 . \tag{21}$$
The total interaction energy W of each molecule is a result of bringing all charges in from infinity, or
$$W = \tfrac{1}{2}(W_1 + W_2) , \tag{22}$$
in which the $\tfrac{1}{2}$ is introduced because we are considering the energy of just one molecule. If we consider a homogeneous gas or liquid, all molecules are identical, or
$$\alpha_1 = \alpha_2 = \alpha \quad \text{and} \quad E_1 = E_2 = E ,$$
so that
$$W = -\tfrac{1}{2}\alpha E^2 . \tag{23}$$
Since α is always positive and the electric field enters the equation as E^2 (which is necessarily positive), the interaction energy W is always negative. This means that the molecules have less energy when they are nearby than when one is at infinity (at which point $E = 0$). In other words, work must be done to move them apart; therefore, they tend to remain nearby; that is, they attract one another.

In order to evaluate W, the field E must be replaced by a measurable quantity. The most convenient quantity is the dipole moment itself. When this is done, Equation 23 becomes
$$W = -\frac{\alpha \mu^2}{r^6} , \tag{24}$$
in which r is the distance between the two molecules.* This is the potential energy of interaction; the force is the negative derivative with respect to r, or $F = -dW/dr$.

We would like to be sure that the interaction energy calculated in this way is actually the cohesive energy responsible for attraction between the molecules in a liquid. If this is the case, the calculated value of W should equal the energy that must be added to separate the molecules completely—in other words, to convert a liquid to a gas. Thus, W should be of the same order of magnitude as the heat of vaporization of most liquids. Typical values for μ and α are $\mu = 10^{-18}$ debye and $\alpha = 10^{-24}$, so that at a distance of 1 Å (that is, $r = 10^{-8}$ cm), $W = -10^{-12}$ erg per molecule or

---

*When a quantum mechanical calculation is performed, the more correct value of $W = -\tfrac{3}{2}(\alpha\mu^2/r^6)$ is obtained.

-60 kJ mol^{-1}, which is close to observed values for the heat of vaporization of many polar liquids. Thus, it seems reasonable that this interacting energy is an important factor in the cohesion of molecules in a polar liquid.

In the argument just presented, it was assumed that the molecules have a permanent as well as an induced dipole moment. Thus, the polarizability consists of two components, the distortion polarizability and the orientation polarizability. Since with increasing temperature the probability of orientation should decrease, we would expect to find the temperature in the expression for W. It can be shown in the theory of electrostatics that for any collection of dipoles of moment μ in a field E, the average dipole moment $\bar{\mu}$ per molecule in the direction of the field is

$$\bar{\mu} = \frac{\mu^2 E}{3kT}, \tag{25}$$

in which k is the Boltzmann constant and T is the absolute temperature. Since the orientation polarizability, α_μ, is defined by $\bar{\mu} = \alpha_\mu E$,

$$\alpha_\mu = \frac{\mu^2}{3kT}. \tag{26}$$

If α_0 is the distortion polarizability, the total polarizability, α, is $\alpha_0 + \alpha_\mu$, or

$$\alpha = \alpha_0 + \frac{\mu^2}{3kT}, \tag{27}$$

and Equation 24 becomes

$$W = -\frac{\mu^2}{r^6}\left(\alpha_0 + \frac{\mu^2}{3kT}\right). \tag{28}$$

As T increases, W becomes less negative and the attractive force decreases, as we would expect.

The London Dispersion Force

The interaction energies described in the preceding section are a satisfactory explanation when the molecules possess a permanent dipole moment. However, since our aim has been to explain departures from the ideal gas law such as liquefaction and critical temperatures, phenomena that also occur for nonpolar molecules, the theory presented so far is clearly inadequate. The idea of a fluctuating dipole, as in a symmetric linear molecule such as CO_2, enables us to apply the arguments just given with minor modifications, but for spherically symmetric molecules, such as CH_4 and the noble gases (He, A, Ne, and so on), a new principle is required.

This new principle was introduced in 1923 by Fritz London, who realized that even though the average field of a nonpolar molecule is zero, the molecule must have an "instantaneous field." The instantaneous field is easiest to envision in the hydrogen atom. A hydrogen atom consists of a single proton and one electron; at any instant, the electron and the positively charged proton constitute a dipole. This dipole continually changes direction as the electron whirls around the nucleus and, averaged over time, or at least over the period of one revolution, the net dipole moment is zero. However, if two hydrogen atoms are present, at any instant one atom will sense the electric field of the other. This argument can be extended even to an atom with a spherically symmetric distribution of electrons, such as argon. Although the electrons move relative to the nucleus in such a way that, averaged over time, the electron cloud is spherically symmetric, again at any given instant in the atom there is a separation of positive and negative charge and thus an instantaneous dipole moment. The orientation of the dipole moment changes continually so that the time-averaged dipole moment is zero. However, if two atoms having such fluctuating dipoles are brought near one another, the electronic motion of each will influence the motion of the other. That is, the fluctuating electric field of one of the dipoles will cause an orientation of the other dipole. The two fluctuating fields thus synchronize the electronic motion in the two atoms so that the dipoles remain in an attractive orientation.*

The form of the equation describing the interaction can be simply obtained from classical electrostatics. The electric field of a dipole with dipole moment μ has field intensity $E = \mu/r^3$ at a distance r from the dipole. Since the potential energy of polarization, W_p, is $-(\alpha/2)E^2$, then $W_p = -(\alpha/2)(\mu^2/r^6)$. The force $(-dW_p/dr)$ then will be proportional to the seventh power of the reciprocal distance. The argument that this generates an attractive force is again simply that the potential energy is negative.

A precise description of these forces, called *London dispersion forces,* requires a quantum mechanical derivation that is beyond the scope of this book. This involves considering each atom or molecule to be a harmonic oscillator (for example, a vibrating spring loaded at each end with weights) vibrating with a characteristic frequency ν_0 or a multiple thereof. The result of this calculation is that the potential energy W_d of interaction due to the dispersion forces is

$$W_d = -\frac{3}{4} h \nu_0 \left(\frac{\alpha_0^2}{r^6} \right), \tag{29}$$

in which h is the Planck constant. London also showed that for simple molecules, $h\nu_0$, expressed in units of energy, equals the ionization energy of the molecule. Thus, the complete expression for the interaction energy W_i is obtained by combining Equation 28, which expresses the sum of the

---

*Mathematically the attractive force results because the dipole moment appears in the equations as the average value of the square of the instantaneous dipole moment, which is always greater than zero. If the instantaneous dipole moment were averaged before squaring, it would of course be zero. Also, it is important to realize that the attraction does not require the two atoms or molecules to be identical. Even for different fluctuating dipoles, the instantaneous fields and dipole moments will interact to maintain the attractive orientation.

contributions due to orientation and deformation of permanent dipoles (if there is a permanent dipole), with Equation 29 for the dispersion energy W_d. This gives

$$W_i = -\frac{2\mu^2 \alpha_0}{r^6} - \frac{2\mu^4}{3kTr^6} - \frac{3}{4}\frac{\alpha_0^2 h\nu_0}{r^6}. \tag{30}$$

Surprisingly, the dispersion interaction almost always makes the major contribution to W_i even when there is a permanent dipole movement. Typical values of μ, α, and ν_0 are 10^{-18}, 10^{-24}, and 10^{-15}, respectively, so that the values of the three terms of Equation 30 are, from left to right, approximately 10^{-12}, 10^{-10}, and 10^{-5} for $r = 1$ Å. In fact, W_i depends primarily on the value of α_0. The value of α_0 is nearly proportional to the volume of the molecules (see below), which enters the expression as a square (α_0^2). Variation of $h\nu_0$ from one molecule to the next is first of all not very great and in any case has a smaller influence on the attractive force than does molecular volume. The relation between α_0 and molecular volume can be explained as follows. Larger molecules have more electrons than small molecules and thus have larger, more flexible electron clouds that are more easily deformed by an electric field. Polarizability is a measure of the ability to be deformed and therefore it increases with volume. Thus, since the value of α_0 contributes most significantly to the interaction energy, attraction between larger molecules is greater than between smaller ones. The magnitude of the attraction between the molecules of any substance determines whether that substance is a solid (strong attraction), liquid, or gas (least attraction). Thus, in a homologous series of molecules, the boiling point should increase with increasing molecular weight. This is evident with the simple alkanes (methane, ethane, propane, and so on), whose boiling points increase with increasing number of carbon atoms. Methane, the smallest molecule of the series, is a gas, but pentane, hexane, heptane, and other alkanes are liquids at room temperature and the very large hydrocarbons, such as polyethylene, are solids. Another example is the halogen series: at room temperature, the smallest molecule, fluorine, is a gas; bromine, which has a higher molecular weight, is a liquid; and iodine is a solid. Other examples will be given in a later section.

Evaluation of the van der Waals Constants *a* and *b*

The van der Waals equation differs from the ideal gas law by the constants a and b. Since it is the interaction energy and the size of the molecules that give nonzero values to a and b, it is informative to obtain expressions relating a and b to these parameters. This is done as follows.

For simplicity we factor out $1/r^6$ from Equation 30 to obtain

$$W_i = -\frac{A}{r^6}, \tag{31}$$

in which $A = 2\mu^2\alpha_0 + (2\mu^4/3kT) + (3\alpha_0^2 h\nu_0/4)$, and note that W_i is the interaction energy for *one pair* of molecules. The gas laws deal with huge numbers of molecules that are separated by a variety of distances. Thus we proceed by calculating an average value of the interaction energy for a pair of molecules and then sum over all pairs.

Consider a spherical vessel of radius **R** having volume $V = (\frac{4}{3})\pi \mathbf{R}^3$. There are N molecules in the container and hence N/V molecules per unit volume. We focus our attention on one molecule in the center of the vessel. The number of molecules at a distance between r and $r + dr$ from the central molecule is

$$dN = \left(\frac{N}{V}\right) 4\pi r^2 \, dr , \tag{32}$$

and the interaction energy of these molecules with the one at the center is $W_i \, dN$. The average interaction energy $\overline{W}_i$ with the one at the center is

$$\overline{W}_i = \int \frac{W_i}{N} dN . \tag{33}$$

To evaluate the integral we must remember that the molecules occupy space. If the radius of the molecule is $\sigma/2$, the distance of closest approach of the molecule centers is σ, or the molecular diameter; σ is then also the radius of a sphere whose volume is called the *excluded volume*, designated V_x—that is, the spherical volume of one rigid molecule that cannot be penetrated by the center of a second rigid molecule (Figure 3-9). This means that the integral must be evaluated from a lower limit of σ to **R**, the radius of the vessel. Therefore, using Equation 31, Equation 33 becomes

$$\overline{W}_i = -\frac{4\pi A}{V} \int_\sigma^R \frac{dr}{r^4} = \frac{4\pi A}{3V}\left(\frac{1}{\mathbf{R}^3} - \frac{1}{\sigma^3}\right) . \tag{34}$$

In a real situation, $\mathbf{R} \gg \sigma$, so that

$$\overline{W}_i = -\frac{4\pi A}{3V\sigma^3} . \tag{35}$$

FIGURE 3-9 The excluded volume. Two spherical molecules each having diameter σ are in contact. The center of molecule 2 cannot penetrate molecule 1 further than the sphere S that has diameter S. The volume V_x of the sphere S is the excluded volume.

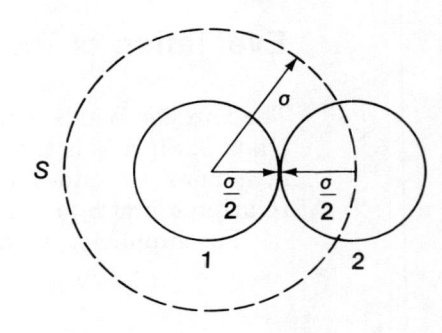

SEC. 3-3 THE NATURE OF THE ATTRACTIVE FORCE

Since this is the average interaction energy per pair of molecules, the total energy is obtained by multiplying by the number of pairs of molecules. This is $N(N - 1)$ or simply N^2, because N is very large. Since a pair of molecules XY is the same as the pair YX, the number of distinct pairs is $\frac{1}{2}N^2$. Thus, the total interaction energy W_{tot} is

$$W_{\text{tot}} = \frac{1}{2}N^2 \overline{W}_i = -\frac{2\pi N^2 A}{3\sigma^3 V} . \tag{36}$$

We convert this expression to energy per mole, $\overline{W}$, by noting that $\overline{W} = (\mathcal{N}/N)W_{\text{tot}}$, in which $\mathcal{N}$ is Avogadro's number and the molar volume $\overline{V} = (\mathcal{N}/N)V$, so that

$$\overline{W} = -\frac{2\pi \mathcal{N}^2 A}{3\sigma^3 \overline{V}} . \tag{37}$$

To evaluate the van der Waals constant a, we appeal to the First Law of Thermodynamics (Chapter 4) and note that the internal energy U of a real gas equals the internal energy of the gas, if it is ideal, plus the interaction energy $\overline{W}$. It can easily be shown that for a van der Waals gas $(\partial U/\partial V)_T = a/\overline{V}^2$ and that for an ideal gas $(\partial U/\partial V)_T = 0$, so that we differentiate Equation 37 to yield

$$\left(\frac{\partial \overline{W}}{\partial \overline{V}}\right)_T = \frac{2\pi \mathcal{N}^2 A}{3\sigma^3 \overline{V}^2} = \frac{a}{\overline{V}^2} . \tag{38}$$

Therefore,

$$a = \frac{2\pi \mathcal{N}^2 A}{\sigma^3} . \tag{39}$$

Thus, a depends upon the coefficient A of the interaction energy, as we intended to demonstrate. Referring to Equation 30 we note that if the molecule has a permanent dipole moment, a should depend upon temperature. Thus, we may assume that for such molecules the van der Waals equation would better describe the behavior of a gas if a were not considered to be a constant but a temperature-dependent parameter.

Equation 39 shows that a also depends upon σ^3 and hence the molecular volume. Since the parameter b enters the van der Waals equation via the volume, a and b must be related. To evaluate b, we note first (Figure 3-9) that the excluded volume $V_x = \frac{4}{3}\pi\sigma^3$. If i molecules are added one by one to a container of volume $\overline{V}$, the volume available to the first molecule is $\overline{V}$, to the second $\overline{V} - V_x$, and the volume V_i available to the ith is $\overline{V} - (i - 1)V_x$. The average available volume $(1/\mathcal{N}) \sum V_i = \overline{V} - b$. It can be shown that $(1/\mathcal{N}) \sum V_i = \overline{V} - (1/2) V_x$; thus

$$b = \tfrac{1}{2}\mathcal{N}V_x = \tfrac{2}{3}\mathcal{N}\pi\sigma^3 . \tag{40}$$

This can be combined with Equation 39 to yield

$$a = \frac{4\pi^2 \mathcal{N}^3 A}{3b} , \tag{41}$$

which relates a and b. The volume v_m of one molecule is

$$\frac{4\pi}{3}\left(\frac{\sigma}{2}\right)^3 = \frac{V_x}{8}$$

so that from Equation 40

$$b = \tfrac{1}{2}\mathcal{N}(8v_m) = 4\mathcal{N}v_m. \tag{42}$$

Inasmuch as $\mathcal{N} v_m$ is the volume of one mole of molecules, b is four times the volume of one mole of molecules.

If the pressure-volume dependence of a real gas is measured, the value of b can be determined experimentally. Once this is done, Equation 40 can be used to determine σ. This is one way to determine the radii of atoms and molecules. The values obtained are of particular interest in building molecular models, because they define the minimum separation of the centers of different atoms. The parameter $\sigma/2$ is usually referred to as the *van der Waals radius,* defined in Equation 8. Representative values for biologically significant atoms and groups are given in Table 3-1.

TABLE 3-1 Van der Waals radii for atoms and groups in biological molecules.

Atom or group	Radius or thickness, Å
H	1.2
N	1.5
O	1.4
P	1.9
S	1.85
C (aromatic)	1.7
C (carbonyl)	1.5
CH_3 group	2.0
NH_2 group	1.5

Strength of van der Waals Forces

The attraction produced by the interaction energy we have been discussing is frequently called a van der Waals bond. For commonly encountered molecules whose centers are separated by a sum of their van der Waals radii, the bonding energy is approximately 4 kJ mol^{-1}. The average thermal energy of molecules at room temperature is about 2.5 kJ mol^{-1}.

This places the van der Waals bond into the category of weak bonds. We will see in a later section that the strength of the van der Waals bond is such that a single bond is easily disrupted at temperatures slightly above those encountered in biological systems. However, clusters of van der Waals bonds—as occur, for example, when several portions of one molecule are bonded to several parts of another molecule—have ample stability. It is just this type of clustering that enables biological macromolecules to bind small molecules very tightly, as when, for example, an enzyme binds its substrate.

Modification of the van der Waals-London Theory

The theory we have described so far has involved two things: (1) a potential energy function of the form $W = -A/r^6$ in which A is a constant, and (2) the point of closest approach of two molecules, a distance that, in the case of spherical molecules, is equal to the diameter σ of the sphere. Thus, when $r = \sigma$, the spheres are in contact. This potential energy function is plotted in Figure 3-10(a). This curve shows a $1/r^6$-dependence for $r > \sigma$, but because of the simplifying assumption that the molecules are rigid spheres, W becomes infinitely positive at $r = \sigma$. This assumption does not properly describe real molecules because the electron cloud is deformable and considerable overlap of two electron clouds is possible (especially in the formation of covalent bonds). A more accurate description is that there is a repulsive force that should begin as the electron cloud of one molecule enters the domain of the other. Since the repulsive energy is only significant at very low values of r, a possible expression is one of the form $+B/r^n$ where $n > 6$ and B is a constant. It turns out that a value of $n = 12$ gives good agreement with experimental results. If we include the repulsive term, Equation 31 becomes

$$W = -\frac{A}{r^6} + \frac{B}{r^{12}}. \tag{43}$$

FIGURE 3-10 Interaction energy, W, as a function of the distance of separation, r: (a) van der Waals potential; (b) Lennard-Jones potential. In (b) the curve is asymptotic to $r = 0$.

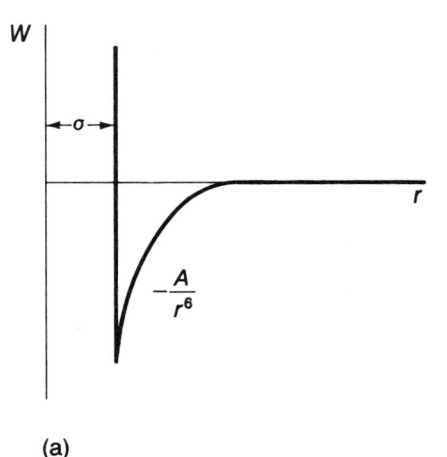

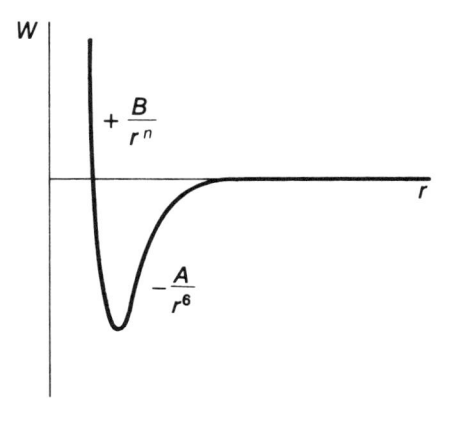

This is called the Lennard-Jones potential, or the 6–12 potential, and is shown in Figure 3-10(b). Note that the transition from attraction to repulsion is smoother with the Lennard-Jones potential than in Figure 3-10(a) and that infinite repulsion occurs at the center of the molecule rather than at a distance σ from the center, as it does with the van der Waals potential. Also, with the Lennard-Jones potential, the van der Waals radius is defined by the minimum of the potential energy function.

A more complete theory requires that we include terms for certain molecules that have higher moments, such as quadrupole moments; it turns out that this can be dealt with by including an additional attractive term such as C/r^8. Furthermore, for distances of several thousand ångström units, corrections must be made to account for the fact that when the instantaneous dipoles fluctuate, the accompanying fluctuation of the electric field is not propagated instantaneously but at the velocity of light. We shall not be concerned with these corrections in this book.

3–4 RELATION BETWEEN INTERMOLECULAR FORCES AND PHYSICAL PROPERTIES

The van der Waals Force and the Boiling Point

In this section we examine systematically the contributions of the various terms in Equation 30 for the interaction energy. We will do this by looking at the boiling point T_b for numerous compounds; T_b is an informative parameter because it indicates the strength of the cohesive force between the molecules of a liquid; that is, it is the temperature at which the translational kinetic energy of the molecules overcomes the negative potential energy of interaction. Hence the boiling point should increase as the interaction energy becomes more negative (that is, as α or μ increases).

We first consider the nonpolar, spherically symmetric, noble gases. As we have stated before, with increasing atomic number, the number of electrons increases, the electron cloud becomes looser and floppier, and α should increase. Therefore, T_b should also increase with atomic number. This trend can be seen by comparing the series shown in Table 3-2. T_b increases with atomic number, as expected.

TABLE 3-2 Boiling points of the noble gases: influence of atomic number.

Property	He	Ne	A	Kr	Xe
Number of electrons	2	10	18	36	54
$\alpha \times 10^{-24}$ cm^3	0.203	0.392	1.63	2.46	4.01
T_b, kelvin	4.216	27.3	87.3	119.9	165.1

SEC. 3-4 INTERMOLECULAR FORCES AND PHYSICAL PROPERTIES

A similar trend was observable with the nonpolar alkanes, as was described earlier.

If a series of molecules have permanent dipole moments but the number of electrons is nearly the same, the dispersion energy should be relatively constant and the magnitude of the dipole moment should be the major factor influencing T_b. That this is the case is shown in Table 3-3. In this case the presence of a dipole moment in isobutylene increases the boiling point over that of isobutane; the larger dipole moment of trimethylamine compared to isobutylene is also reflected in the higher value of T_b. Note that the effects on T_b are relatively small compared to those in Table 3-2 because the dipole moments are small. With larger dipole moments the changes are more dramatic, as shown in Table 3-4.

TABLE 3-3 Influence of small changes of the permanent dipole moment on the boiling point of unrelated compounds of similar weight.

Property	Isobutane	Isobutylene	Trimethylamine
Number of electrons	58	56	59
$\alpha \times 10^{-24}$ cm^3	8.36	8.36	8.08
μ, debye	0	0.49	0.67
T_b, kelvin	263	267	278

TABLE 3-4 Influence of the dipole moment on the boiling point of unrelated compounds of similar molecular weight.

Property	Propane	Dimethyl ether	Ethylene oxide
Number of electrons	44	46	44
$\alpha, \times 10^{-24}$ cm^3	6.4	6.0	5.2
μ, debye	0	1.30	1.90
T_b, kelvin	231	248	284

Comparisons of dimethyl ether with isobutylene and of ethylene oxide with trimethylamine (and of many other substances) show that the London forces are a more important factor than the effect of the permanent dipole moment. That is, T_b for the first of each pair is higher than T_b for the second, because in each case the first has more electrons and hence a greater polarizability than the second, even though in each case the second has a higher value of μ.

Evidence For Another Interaction—Hydrogen Bonding

The van der Waals forces are the most common forces responsible for intermolecular attraction. However, in biological molecules, another interaction, hydrogen bonding—also a weak interaction compared to covalent bond energy—plays a significant role in determining both intra- and intermolecular interactions and, in particular, the structure of macromolecules.

Molecules containing hydroxyl and amino groups have boiling points much higher than expected simply from the values of α and μ. For example, let us compare CH_3F ($\alpha = 3.84 \times 10^{-24}$ cm^3, $\mu = 1.81$ debye units) with CH_3OH ($\alpha = 3.0 \times 10^{-24}$ cm^3, $\mu = 1.70$ debye units). On the basis of the reasoning of the preceding section, one would expect the boiling point of CH_3OH to be slightly lower than CH_3F since the values of both α and μ for CH_3OH are lower than the values for CH_3F; however, CH_3OH boils at 338 K and CH_3F boils at 195 K. An equally striking comparison is water ($\alpha = 1.59 \times 10^{-24}$ cm^3, $\mu = 1.85$ debye units) and ethylene oxide ($\alpha = 5.2 \times 10^{-24}$ cm^3, $\mu = 1.90$ debye units). The value of α for ethylene oxide is so much higher than that of H_2O that one would expect the boiling point of ethylene oxide to be much greater than the boiling point of H_2O. However, the opposite is true—T_b for ethylene oxide is 284 K and that for H_2O is 373 K.

The existence of a value of T_b higher than expected from the values of α and μ implies that more energy must be put into the compound to maintain the gaseous state. This indicates that there must be an additional interaction between the molecules other than the van der Waals-London interaction. The attractive force we seek is that of the hydrogen bond (H bond). A hydrogen bond is formed by a proton interacting simultaneously with two atoms, each having a strong negative charge, and it is symbolized by a dotted line; for example, O—H $\cdots$ O or N—H $\cdots$ N. The H bonds of significance in biological systems are the following: O—H $\cdots$ O, N—H $\cdots$ N, O—H $\cdots$ N, and N—H $\cdots$ O. Strong H bonds also occur with F atoms in hydrogen fluoride; weak H bonds can occur with S and Cl atoms.

The theory of hydrogen bonding is poorly understood at the present time. A theory based upon a purely electrostatic interaction (that is, attraction between charges of opposite sign) has been extensively explored. This theory successfully explains some of the properties of H bonds yet is known to be inadequate, because it fails to explain the following.

Two charged groups whose charges have different signs are capable of attracting one another in aqueous solution. However, if a strong electrolyte such as NaCl is also present, the positively charged ions (Na$^+$) tend to cluster around the negatively charged group, and the negatively charged ions (Cl$^-$) cluster around the positive group (Chapter 8); the result is a shielding of the charges so that the charges no longer interact with one another. Thus, in aqueous solution, electrostatic bonds are weakened and ultimately eliminated as the concentration of the electrolyte is increased. However, the stability of a hydrogen bond is not significantly reduced in solutions of concentrated electrolytes. The H bond is clearly also not a covalent bond since hydrogen has only one orbital stable enough to form a bond yet the proton in an H bond interacts with two atoms. Furthermore, even though the proton interacts with two atoms, it remains covalently bound and is nearer to one of the atoms (the proton donor) than to the other (the proton acceptor). Typically if the covalent bond length with the donor is X ångström units, the distance from the proton to the other atom is $X + 0.8$ ångströms. An exception that has been widely studied is hydrogen fluoride, in which the proton is exactly midway between the F atoms, 1.13 Å from each F atom.

H bonds have a preferred direction. That is, the interaction is strongest when the two electronegative atoms and the hydrogen atom form a straight-line configuration. H bonds are usually not maintained in the gaseous state, in part because their strength depends upon the linearity of the array. Admittedly, the higher kinetic energy of molecules in the gaseous state than in the liquid state tends to disrupt all weak bonds but an H bond is broken more easily than van der Waals bonds because an initial collision can cause a bending oscillation that misaligns the bond, thereby making it extraordinarily susceptible to disruption in a second collision. Some H bonds are very stable. For example, the F atom can bind a proton so tightly that HF gas is H-bonded and exists as a zig-zag chain with the following structure:

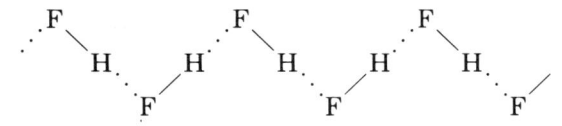

$$\begin{array}{cccccc} \cdot\cdot F & & \cdot\cdot F & & \cdot\cdot F & \\ \diagdown & & \diagdown & & \diagdown & \\ H\cdot\cdot & H\cdot & H\cdot\cdot & H\cdot & H\cdot\cdot & \\ & \diagup & & \diagup & & \diagup \\ & \cdot F & & \cdot F & & \cdot F \end{array}$$

Note that each sequence of F—H $\cdots$ F lies in a straight line since that is the most stable arrangement. Acetic acid, which can form *two* intermolecular H bonds, also remains H-bonded in the vapor phase, with the following dimeric structure (note again that the H bonds are linear):

$$\begin{array}{ccccc} H & & O\cdots H\!-\!O & & H \\ | & & \diagup\!\!\diagup & \diagdown & | \\ H\!-\!C\!-\!C & & & & C\!-\!C\!-\!H \\ | & \diagdown & \diagup\!\!\diagup & & | \\ H & & O\!-\!H\cdots O & & H \end{array}$$

We will see other examples later of several weak bonds acting together or *cooperatively* to form a strongly bound system.

Experimentally it is found that the energy of an H bond is in the range of 10 to 30 kJ mol^{-1}, which means that an H bond is stronger than a van der Waals bond. This is the reason that H bonds produce such marked effects on the boiling point, which usually represents a balance between the weaker van der Waals attractive forces and the disordering effect of translational kinetic energy.

The strength of H bonds enables us to understand a wide variety of apparent anomalies. For instance, in the liquid state, alcohols (R—O—H) can form structures of the form

$$\begin{array}{ccc} R & R & R \\ | & | & | \\ O\!-\!H\cdots O\!-\!H\cdots O\!-\!H \end{array}$$

and organic acids form structures such as that just shown for acetic acid. Sugars have numerous —OH groups also capable of H-bonding. These compounds all have unusually high boiling points. Many inorganic acids, for examples, HNO_3 and H_2SO_4, also have very high boiling points. Rewriting these formulas as NO_2OH and $SO_2(OH)_2$ clearly indicates the possibility of H-bonding.

Hydrogen bonds also have an effect on a thermodynamic quantity called the *entropy of vaporization* (see Chapter 5).* For most liquids the entropy of vaporization at the normal boiling point is about 85 $J\ K^{-1}\ mol^{-1}$. The entropy change in any process is a measure of the degree of disorder introduced by the process. Since gases are almost completely disordered (at least well above the critical point), the near constancy of the entropy of vaporization indicates that in a liquid the fraction of the molecules that are ordered must be nearly the same for most liquids. However, in a liquid in which many of the molecules are hydrogen-bonded, there is a higher degree of order (that is, the entropy is smaller) than in a liquid in which only van der Waals bonds are present. Hence a hydrogen-bonded liquid will suffer a greater entropy increase when it becomes a gas; in other words it will have a larger entropy of vaporization, as shown in Table 3-5; an asterisk denotes the hydrogen-bonded liquids.

TABLE 3-5 Entropy of vaporization, ΔS_{vap}, for various liquids in units of $J\ K^{-1}\ mol^{-1}$.

Compound	ΔS_{vap}	Compound	ΔS_{vap}
CCl_4	84.4	Water*	108.9
CS_2	83.8	Ethanol*	112.0
$CHCl_3$	82.0	Acetic acid*	62.1
Benzene	87.1	Hydrogen	
CH_2Cl_2	88.3	fluoride*	25.5

*Hydrogen-bonded liquids.

Hydrofluoric acid and acetic acid are also hydrogen-bonded liquids but are especially interesting because they have very small values of ΔS_{vap}, namely, 25.5 and 62.1 $J\ K^{-1}\ mol^{-1}$, respectively. This might lead one to conclude that in these cases the liquid is less ordered than a typical liquid. However, if the molecules of the gas were also ordered, there would be a small entropy of vaporization. This unusual situation is the case for both molecules since the hydrogen bonds are so strong that they persist in the gaseous state, as we stated earlier.

3–5 CHEMICAL BONDS

We have so far only considered weak bonds. The strong bonds or chemical bonds are more difficult to analyze because they require more detailed understanding of repulsive forces.

*Because entropy is a difficult concept, it is recommended that, to appreciate fully the discussion that follows, the student should reread this section after completing Chapter 5.

SEC. 3-5　CHEMICAL BONDS

As we have already seen, two atoms approaching one another experience an attractive force that increases as their separation decreases. At a distance equal to the sum of their van der Waals radii, the repulsive force becomes important. If the repulsive forces are overcome and the two atoms approach more closely than this critical distance, a chemical bond may result. There are two main types of chemical bonds, ionic bonds and covalent bonds, which differ as follows.

In an ionic bond electrons are transferred from one atom to another so that a positive and a negative ion are formed; these ions then attract one another electrostatically. In a covalent bond electrons are shared by two atoms. A property of the atoms called electronegativity determines whether an ionic or a covalent bond forms.

Electronegativity describes the ability of an atom to pull electrons toward itself; a more electronegative atom attracts electrons more strongly than an atom that is less electronegative. Electronegativity can in part be described theoretically by means of quantum mechanics and can be estimated from bond-dissociation energy, ionization potential, and electron affinities; however, a discussion of quantum mechanical theory is beyond the scope of this book.

The magnitude of the electronegativity of an element depends upon its position in the periodic table. For example, in any particular column of the periodic table, electronegativity decreases with increasing atomic number (Figure 3-11). This is because of the increased screening of the positive nuclear charge by the inner orbital electrons; that is, electrons of the outermost orbital are acted on by a smaller effective positive charge when

FIGURE 3-11 The periodic table. The arrows denote the directions of increasing electronegativity. The noble gases in the rightmost column are neither electronegative nor electropositive.

1 H																	2 He
3 Li	4 Be											5 B	6 C	7 N	8 O	9 F	10 Ne
11 Na	12 Mg											13 Al	14 Si	15 P	16 S	17 Cl	18 Ar
19 K	20 Ca	21 Sc	22 Ti	23 V	24 Cr	25 Mn	26 Fe	27 Co	28 Ni	29 Cu	30 Zn	31 Ga	32 Ge	33 As	34 Se	35 Br	36 Kr
37 Rb	38 Sr	39 Y	40 Zr	41 Nb	42 Mo	43 Tc	44 Ru	45 Rh	46 Pd	47 Ag	48 Cd	49 In	50 Sn	51 Sb	52 Te	53 I	54 Xe
55 Cs	56 Ba	57* La	72 Hf	73 Ta	74 W	75 Re	76 Os	77 Ir	78 Pt	79 Au	80 Hg	81 Tl	82 Pb	83 Bi	84 Po	85 At	86 Rn
87 Fr	88 Ra	89 Ar	†103 Lw														

*58-71
†90-102

there are more electrons between them and the nucleus. Thus, in a very weakly electronegative atom, the outermost electrons are far from the nucleus, and such an atom has a large atomic radius. Electronegativity also increases from left to right within each row of the periodic table with increasing charge of the nucleus. Of course, the nuclear charge and the number of electrons in the outermost orbit increase together, but because the number of electrons in the inner orbits remains constant, the screening is the same for elements within a row of the periodic table, and the electrons in the outer orbit are subjected to an increasing nuclear attraction. These two effects explain why fluorine (F) at the upper right of the periodic table is the most electronegative element and francium (Fr) at the lower left is the least electronegative. (The noble gases have been excluded from this discussion, as they rarely form chemical bonds.)

Two elements of very different electronegativity, for example, a strongly electronegative halogen and a weakly electronegative alkali metal, form an ionic bond because the single outer electron of the alkali metal is almost completely transferred to the incomplete outer shell of the halogen atom. Thus, KCl is an ionic compound. If two elements have nearly equal electronegativity, neither one attracts electrons more than the other and a covalent bond usually results. Thus, carbon, nitrogen, and hydrogen, which lie near the middle of the periodic table, form covalent bonds with one another.

Simple inorganic salts are generally held together by ionic bonds. The potential energy curves for such substances differ from the kind of curve observed for van der Waals forces in that the potential energy is zero at large separations for van der Waals interactions (Figure 3-10) but nonzero for ionic forces. For example, let us consider the case of NaBr. The ionization potential of sodium is 5.14 eV—that is, 496 kJ are required to change 1 mol of Na to Na$^+$. The energy obtained when an electron is added to bromine (the *electron affinity*) is 346 kJ mol^{-1}. Thus to form NaBr, the net energy required is 496 − 346 = 150 kJ mol^{-1}, which means that the separated ions have more energy (150 kJ mol^{-1}) than the atoms. Thus, at large distances, the potential energy is +150 kJ mol^{-1} rather than 0. As the ions approach one another, positive and negative charges attract, producing the usual Coulombic interaction energy, $-e^2/r$, in which r is the separation between the centers of the ions. As in the van der Waals interaction, when the ions are very close and the electron clouds overlap, a repulsive interaction (the repulsion of the negatively charged orbital electrons) takes over. The equation describing the potential energy is usually written in the form

$$W = -\frac{e^2}{r} + be^{-ar}, \qquad (44)$$

in which a and b are constants that are determined empirically and the term be^{-ar} represents the repulsive energy.

The theory of the covalent bond requires a difficult quantum mechanical analysis that is so complex that a precise mathematical solution has never been achieved. However, various approximations have allowed one to calculate accurate values for bond energies, which suggests that the reasoning utilized in setting up the formal mathematical description is probably correct. Once again there is an attractive and a repulsive force but this time the electron clouds significantly penetrate one another; that is, the distance between the nuclei is less than the sum of the atomic radii of the individual atoms. When the bond is formed, there is a buildup of electronic charge between the nuclei. The attraction then results from interaction between this highly negatively charged region and the two positively charged nuclei. The repulsive forces result from electrostatic repulsion between one nucleus and the other and between the inner orbital electrons of one atom and those of the other. The bond energies of covalent bonds are higher than those of ionic bonds, generally being in the range of 200 to 400 kJ mol^{-1}.

Remembering that the energies involved in the van der Waals interaction are about 4 kJ mol^{-1} and in hydrogen-bonding are from about 10 to 30 kJ mol^{-1}, we may distinguish between strong bonds (ionic and covalent bonds) and weak bonds. Such a distinction is more than academic because, whereas it is true that all molecules encountered in biological systems contain numerous strong bonds, biological properties are determined in a significant way by weak bonds. This is explored in the following section.

3–6 WEAK BONDS IN BIOLOGICAL SYSTEMS

The Role of H Bonds and van der Waals Forces

So far four types of bonds have been described: two strong ones—ionic and covalent bonds—and two weak ones—van der Waals bonds* and hydrogen bonds. Biological molecules, like all molecules in nature, consist of atoms primarily joined together by strong bonds. We have already mentioned that all molecules are subject to the van der Waals forces that determine the physical properties of most liquids and many solids. However, biological systems differ from other chemical systems in that an enormous role is played by the weak interactions in determining the properties of these molecules. These weak interactions are especially prevalent in biological macromolecules. For example, the shapes of biological macromolecules are determined almost entirely by H bonds and van der Waals bonds. Similarly, the binding of two macromolecules to one another and the binding of a small molecule to a macromolecule also only involve these weak bonds. This is discussed in greater detail in Chapter 15.

FIGURE 3-12 How van der Waals attraction occurs. (a) Atoms 1–5 form van der Waals bonds with atoms b, c, e, f, g, and i. (b) The addition of atom 6 allows a van der Waals interaction only with atoms b, f, and g. This substantially reduces the attractive force between molecules 1 and 2.

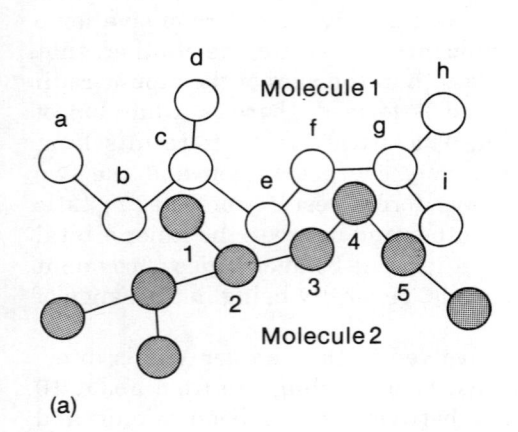

The energy of most hydrogen bonds ranges from 10 to 30 kJ mol^{-1}. This is 5 to 10 times greater than the average translational kinetic energy at physiological temperatures, so a single H bond is stable at these temperatures as long as it is possible for the three atoms engaged in the H-bonded unit to remain in the required linear array. The energy of a typical van der Waals bond is much lower, about 4 kJ mol^{-1} (only 50 percent greater than thermal energy at room temperature); thus, because of the broad range of translational energies in a population of molecules (see Chapter 2 for a discussion of the Boltzmann distribution), a single van der Waals bond is not very stable. Consequently, the van der Waals bonds are effective only when several atoms in one molecule are bound to several atoms in another molecule (or in the case of a macromolecule, with another part of the same molecule). It is this property of the van der Waals forces that confers enormous specificity in inter- and intramolecular interactions in biological systems. The van der Waals forces are not directional, as H bonds are, but are strongly dependent upon the separation of the atoms (remember the $1/r^6$-dependence); the force rapidly approaches zero as the separation exceeds the sum of the van der Waals radii. Thus, two clusters of atoms are usually able to attract one another if they have complementary shapes. Figure 3-12(a) shows two clusters that might interact strongly; in panel (b) a single extra atom increases the separation of the atoms so that the attraction is inadequate to overcome thermal disruption. It should be noted that if two clusters of atoms are arranged so that many

---

*Some authors distinguish between van der Waals forces and London forces, reserving the latter term only for the dispersion interaction. Collectively these are occasionally called van der Waals-London bonds—an awkward phrase. In this book the term van der Waals bonds refers to both.

van der Waals interactions occur, the total energy of binding can be as great as 40 kJ mol^{-1}, which is a relatively strong and stable interaction. It is this type of binding that occurs between an enzyme and a substrate. This is discussed in greater detail in Chapters 12 and 15.

The Role of Water: The Hydrophobic Interaction

The hydrogen bonds in water are important in many biological processes. Water is a strongly H-bonded liquid because it is highly polar and both the H and O atoms form H bonds. In each H_2O molecule the O atom can bind to two H atoms in two other H_2O molecules forming a tetrahedral structure. In ice the H bonds are so strong that the structure is very rigid. Liquid water has a looser structure because the bonds are continually being broken by thermal energy; however, at any given instant most water molecules participate in four H bonds.

H bonds are responsible for the solubility in water of many organic compounds such as those found in living systems. For example, a molecule such as glucose, having numerous OH groups, breaks water-water H bonds and forms glucose-water H bonds. These H bonds are stronger than the van der Waals bonds that would tend to make two glucose molecules attract one another. Thus, the glucose molecules are separated and are water soluble. For many organic molecules carrying single hydroxyl, keto, or amino groups, the number of H bonds per molecule is small and these are not as strong as the water-water bond. Such molecules—for example, the bases in DNA—have only limited solubility.

An important consequence of the existence of weak bonds is the *hydrophobic bond,* an interaction that is actually not a bond at all. A hydrophobic (literally, water-fearing) interaction refers to a clustering of nonpolar molecules in water. Consider the effect of adding a single benzene molecule to water. A benzene molecule cannot form H bonds with water molecules but, at the instant of being added to water, disrupts the ordered H-bonded water lattice. As the water molecules attempt to reform H bonds, an *ordered* structure around the benzene molecule results. Interestingly, the degree of order of the water molecules around the benzene molecule is *greater* than that present before the benzene molecule is added (see Chapter 5 for a more detailed discussion of this ordering phenomenon). As additional benzene molecules are added to the water, each will disrupt the water lattice and then cause excess ordering. Inasmuch as the creation of order is thermodynamically unfavorable (see Chapter 5), if two benzene molecules could be arranged in such a way as to minimize the increase in order, such an arrangement would be preferred. Since ordering happens at the surface of the benzene molecule and since clustering of the two benzenes reduces the surface-to-volume ratio, clustering will occur. As the number of molecules increases, the cluster will simply form a second phase; this accounts for the observed phenomena that benzene and water do not mix and that oil forms slicks on the surface of water.

72   CHAP. 3   INTERMOLECULAR FORCES

There is a slightly different interaction if the molecule has a polar and a nonpolar region. The polar region can H-bond to water to maintain solubility, yet the nonpolar region will be enveloped in an ordered array of water molecules to which it does not H-bond, as in the case of benzene. Examples of such molecules are fatty acids and the nucleotides in nucleic acids (Figure 3-13). The nonpolar regions will cluster to reduce the ordering of water. However, the polar region can maintain its H bonds to water so that the molecules will form an array in which the polar regions are external and the nonpolar portion is internal. For example, very long-chained fatty acids form surface films in which the polar carboxyl group of each molecule is in the water phase and the long, nonpolar chain protrudes

FIGURE 3-13 Formulae for a fatty acid and a nucleotide showing polar and nonpolar portions.

Fatty acid: palmitic acid

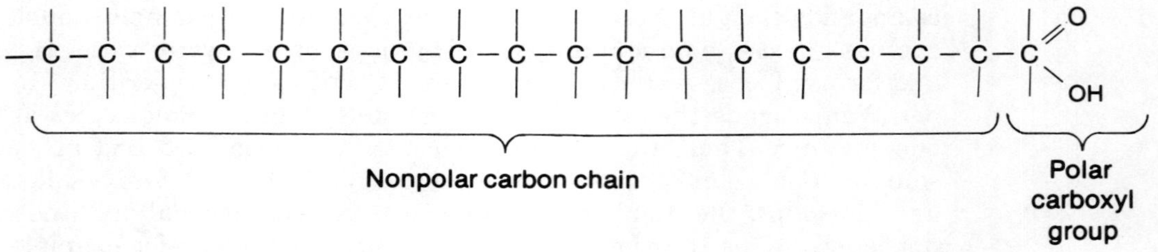

Nonpolar carbon chain | Polar carboxyl group

Nucleotide: adenosine -5'- phosphate

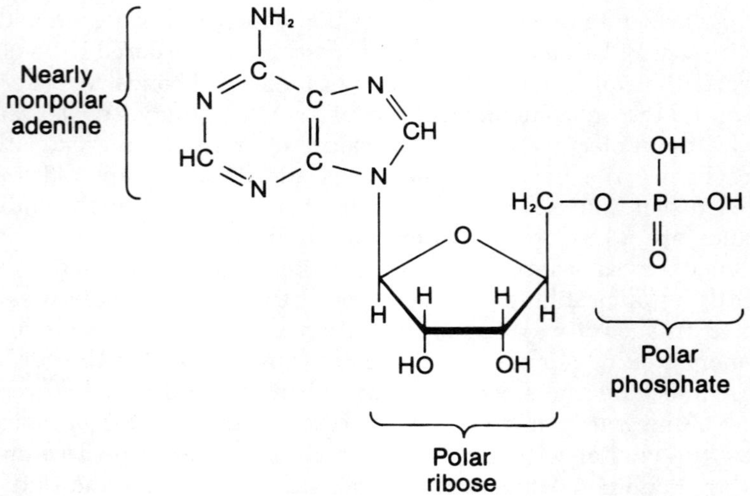

Nearly nonpolar adenine | Polar ribose | Polar phosphate

SEC. 3-6 WEAK BONDS IN BIOLOGICAL SYSTEMS

above the liquid surface into the air (Figure 3-14). Under certain circumstances the fatty acid molecules form spherical clusters *(micelles)* of molecules in which the carboxyl group is in contact with water and the nonpolar chain is inside the sphere. Both surface films and micelles are discussed in greater detail in Chapter 13.

FIGURE 3-14 A collection of fatty acids on a water surface. Each fatty acid has a polar and a nonpolar region. The polar region is soluble in water and therefore is in the aqueous phase. The nonpolar tail is insoluble and avoids water by entering the air phase. These tails tend to aggregate side by side because of attractive van der Waals forces.

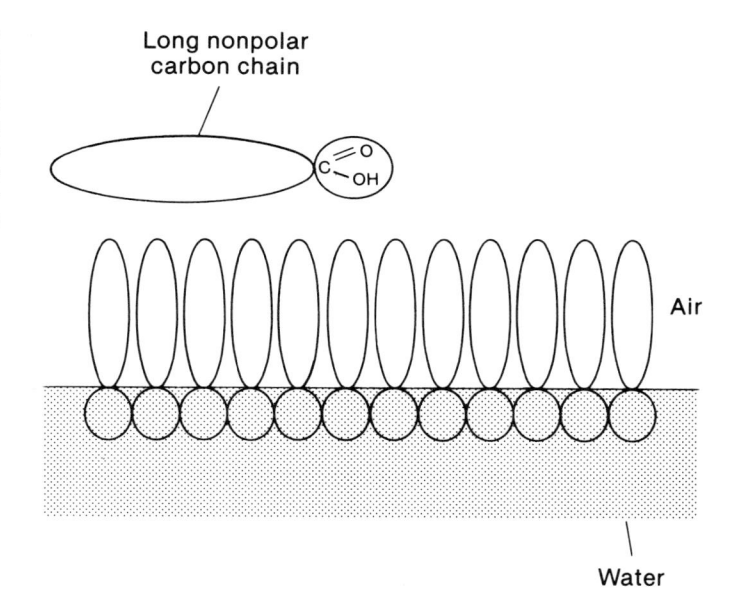

Nucleic acids are an especially interesting example of the role of hydrophobic interactions. Single-stranded ribonucleic acid molecules consist of a long chain of alternating sugar (ribose) and phosphate groups with a single organic base attached to each ribose (Figure 3-15(a)). The sugar-phosphate chain is highly charged and binds many H_2O molecules. The bases are organic heterocyclic ring systems to each of which a charged group is attached. The charge makes the base sparsely soluble in water but the rings disrupt the water structure significantly. These bases therefore tend to stack upon one another to minimize contact with water (Figure 3-15(b)). Thus, although there should be free rotation about the sugar-phosphate bonds, the base stacking creates a rather rigid structure. Naturally occurring deoxyribonucleic acid (DNA) molecules contain four different bases, adenine (A), cytosine (C), guanine (G), and thymine (T). The charged groups on these bases are such that H bonds can form between adenine and thymine and between cytosine and guanine. If two single-strand lengths of DNA have complementary base sequences (that is, if at a site of A or C on one strand there is a T or a G, respectively, on the other), the two strands can join to form a double-stranded structure in which the sugar-phosphate chains are exposed to the water, but the bases are

FIGURE 3-15 (a) Schematic diagram of an RNA molecule. The chain is an alternating copolymer of ribose and phosphate. (b) Two configurations of RNA. The ribose-phosphate chain is the continuous solid line; the short lines represent the bases. In the left drawing the bases are situated so that the planes of their rings are parallel; this is the stacked configuration. In the right drawing the bases have a random orientation. In RNA many of the bases are stacked.

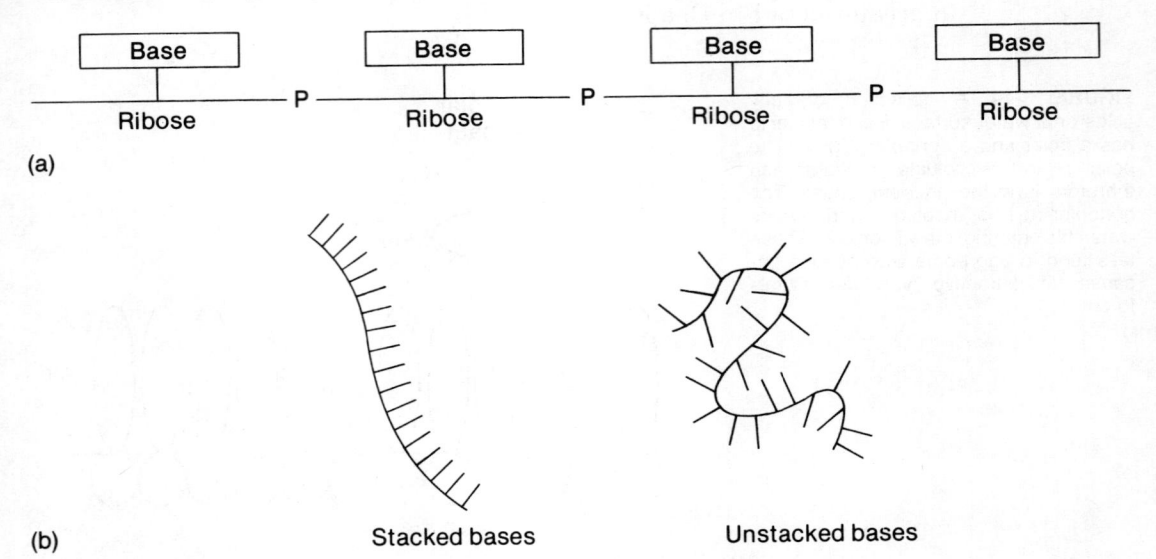

contained within the structure, stabilized by hydrophobic interactions. The role of hydrophobic interactions in nucleic acid structure is discussed also in Chapters 5 and 15.

The shapes of protein molecules are also determined in part by hydrophobic interactions (and also by H bonds). Protein molecules are long chains consisting of amino acids. Some of the amino acids contain strongly polar groups and others are highly nonpolar (Figure 3-16). Thus, there is a tendency for protein molecules to be folded in such a way that the nonpolar groups are clustered and buried within the molecule. This is discussed in greater detail in Chapter 15.

It is important to remember to distinguish between hydrophobic interactions and van der Waals bonds. Hydrophobic interactions are not bonds but merely arrangements of molecules based upon reducing contact with water. The van der Waals force is a true attraction that would occur whether or not water is present. A van der Waals force can of course stabilize or strengthen a hydrophobic interaction. More often, however, it adds specificity. That is, a protein molecule, for example, might contain numerous nonpolar regions that would tend to cluster. In the absence of van der Waals bonds the clustering might be random. However, if the van der Waals forces between certain groups or collections of groups are stronger than between others, the clusters with the stronger van der Waals bonds would be favored. This phenomenon has been referred to as "selective stickiness."

SEC. 3-6   WEAK BONDS IN BIOLOGICAL SYSTEMS   75

FIGURE 3-16 (a) Polar amino acids. (b) Nonpolar amino acids. (c) A folded protein. The open and solid circles represent polar and nonpolar amino acids, respectively. We have assumed that there are neither H bonds nor van der Waals forces so that the folding is determined only by hydrophobic interactions.

(a) Some polar amino acids ○

Serine

Asparagine

Glutamic acid

(b) Some nonpolar amino acids ●

Alanine

Leucine

Phenylalanine

(c) A hypothetical folded protein

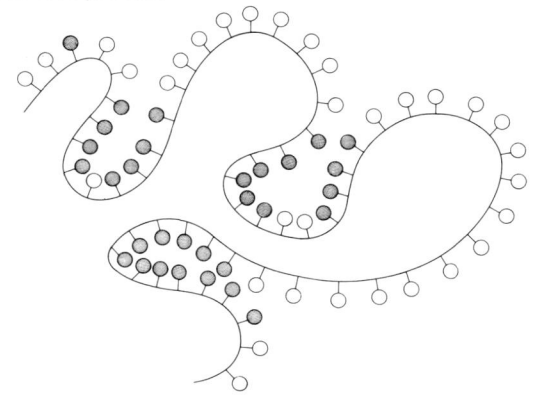

3–7 SUMMARY OF VARIOUS MOLECULAR INTERACTIONS

Table 3-6 lists the molecular interactions that have been discussed, their dependence on distance, and the range of energies observed for such interactions.

TABLE 3-6 Molecular Interactions and Their Properties

Type	Dependence on distance	Magnitude (kJ mol^{-1})
Covalent bond	No simple expression	200–800
Ionic bond	r^{-1}	40–400
Hydrophobic	No simple expression	10–30
Hydrogen bond	No simple expression	10–30
Total van der Waals	r^{-6} (mainly)	3–10
Induced dipole-induced dipole (dispersion)	r^{-6}	3–10
Ion-dipole	r^{-2}	3–10
Dipole-dipole	r^{-6}	0.5–3
Dipole-induced dipole	r^{-6}	0.4–3
Ion-induced dipole	r^{-4}	0.4–3
Thermal energy kT at 25°C	—	2.5

Note: Only kT and the dipole-dipole interaction are dependent on temperature. The dipole-dipole interaction varies as $1/T$.

REFERENCES

Atkins, P.W. 1978. *Physical Chemistry*. W.H. Freeman and Co. San Francisco.
Castellan, G.W. 1971. *Physical Chemistry*. Addison-Wesley. Reading, Mass.
Chu, B. 1967. *Molecular Forces*. Wiley-Interscience. New York.
Davis, K.S., and J.A. Day. 1961. *Water, the Mirror of Science*. Doubleday. Garden City, New Jersey.
Edsall, J.T., and J. Wyman. 1958. *Biophysical Chemistry*. Academic Press. New York.
Eisenberg, D., and W. Kauzmann. 1969. *The Structure and Properties of Water*. Oxford University Press. Advanced.
Hirschfelder, J.D., C.F. Curtiss, and R.B. Bird. 1974. *The Molecular Theory of Gases and Liquids*. Wiley. New York.
Pauling, L. 1960. *The Nature of the Chemical Bond*. Cornell University Press. Ithaca, New York.
Pimentel, G.C., and A.L. McClellan. 1960. *The Hydrogen Bond*. W.H. Freeman and Co. San Francisco.
Smith, J.W. 1955. *Electric Dipole Moments*. Butterworths. London.
Smyth, C.P. 1972. "Determination of Dipole Moments." In *Techniques of Chemistry*, edited by A. Weissberger and B.W. Rossiter, Vol. IV. Wiley-Interscience. New York.
Tanford, C. 1973. *The Hydrophobic Effect*. Wiley. New York.
Vinogradov, S.N., and R.H. Linnel. 1971. *Hydrogen Bonding*. Van Nostrand-Reinhold. New York. Advanced.
Webb, V.J. 1968. "Hydrogen Bond: Special Agent." *Chemistry*, 41, 17.
Yoder, C.H. 1977. "Teaching Ion-ion, Ion-dipole, and Dipole-dipole Interactions." *J. Chem. Educ.* 54, 402.

PROBLEMS

1. Which of the following molecules have significant dipole moments?

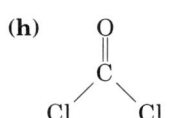

   (a) H₂C=CH₂ (ethylene)

   (b) H—C≡C—H

   (c) BF₃

   (d) H—N≡C—Ö:

   (e) H—Ö—Ö—H (hydrogen peroxide)

   (f) Cl—Sn—Cl

   (g) Cl₂Al(μ-Cl)₂AlCl₂

   (h) Cl—C(=O)—Cl

   (i) N≡C—C≡N

2. N₂O has a dipole moment. Can it be a linear molecule?

3. The boiling point of water is very high because of hydrogen-bonding. There is little or no hydrogen-bonding in H₂S, H₂Se, and H₂Te, whose boiling points are −68.7, −41.5, and −2°C, respectively. If water could not form hydrogen bonds, what would its boiling point be?

4. The dipole moment of chlorobenzene is 1.69 debye.

   (a) Estimate μ for o-, m-, and p-dichlorobenzene. Assume that the benzene ring retains perfect hexagonal symmetry and remember that μ is a vector quantity.

   (b) Is a calculation necessary to know μ for p-dichlorobenzene?

5. The dipole moment of nitrobenzene is 4.22 debye. Using the data of Problem 4, what is the dipole moment of p-chloronitrobenzene?

6. In another universe, the equation of state of a gas is $P = RT/(\bar{V} - b) - a/\bar{V}$. Do gases in this universe have critical points?

7. Acetone is a very volatile liquid. However, urea, H₂N—CO—NH₂, is a nonvolatile solid. What is the explanation for this?

8. The maximum solubility of benzene in water is 0.02 M. However phenol can make a 0.7-M solution. Why is phenol more soluble in water than benzene is?

9. Most organic compounds fail to absorb ultraviolet light in the wavelength range of 240–270 nm unless they contain aromatic rings. However, aqueous solutions of formaldehyde (HCHO) show weak ultraviolet absorption at room temperature. If the solution is boiled, the ability to absorb the ultraviolet light is eliminated, although after several hours at room temperature this ability is restored. Propose an explanation for this.

10. Two organic compounds A and B are capable of very weak hydrogen-bond formation although these bonds are totally disrupted at 30°C. A linear polymer of B (poly B) is prepared in which the functional groups that hydrogen-bond are not engaged in any of the bonds in the polymer chain. A single molecule A still binds poorly to poly B. However the dimer A—A binds more strongly and this bonding is not disrupted until 40°C. An octamer of A binds stably until 55°C; this is also the case for the binding of poly A to poly B. Explain the varying thermal sensitivity with the number of monomers in an A polymer.

11. What are the factors that cause ethane to have a higher boiling point than methane? Why is the boiling point of ethanol higher than that of ethane?

12. The vapor of formamide, HCONH₂, is not an ideal gas. Explain.

13. Glycerol, CH₂OH—CHOH—CH₂OH, is a liquid at room temperature and is completely miscible with water. What would you expect its solubility to be in ethanol, ether, and chloroform?

14. If anhydrous MgSO₄ is added at a low concentration to distilled water, the volume of the solution is less than the initial volume of the water—that is, the liquid shrinks. This is not an uncommon phenomenon with divalent salts, is less common with monovalent salts, and is rare with nonpolar compounds. Suggest an explanation.

15. The van der Waals radius of an oxygen atom is 1.40 Å. Would you expect the

PROBLEMS

separation of the oxygen atoms in the O_2 molecule to be greater than, less than, or equal to 2.80 Å?

16. The length of a typical bond between an O atom and an H atom is 0.95 Å. The van der Waals radius of an H atom is 1.2 Å; that of an O atom is 1.4 Å. Someone measures the distance between the two O atoms in a hydrogen-bonded OH group and an O atom, that is, O—H $\cdots$ O, and tells you it is 2.76 Å. Does this seem like a reasonable value?

CHAPTER 4

THERMODYNAMICS—THE FIRST LAW

Matter is composed of molecules in motion. The motion of these molecules, whether it be translational, rotational, or vibrational (see Chapter 2), contributes to the energy of the molecule. This energy of motion is called *kinetic energy*. A molecule may also possess *potential energy*—that is, the possibility of acquiring kinetic energy. Thus a rock thrown upward gains potential energy as it rises because it gains the ability to fall back toward the Earth. An electrically charged particle in an electric field has potential energy because it will move along the direction of the field. If there are several charged molecules, the electric field of one molecule can influence the movement of a second molecule; in such a case the two molecules are said to interact and to possess a potential energy of interaction or an *interaction energy*. In general, a collection of a large number of molecules possesses an amount of energy equal to the sum of the kinetic and potential energies of each molecule. The amount of kinetic energy of a collection of molecules can be detected by its temperature, as we saw in Chapter 2—the energy is greater as temperature is greater. The energy of these molecules can also be observed in the ability of the molecules, collectively as a system, to do work. For instance, the incessant bombardment of a movable object by the individual molecules of a substance in contact with the object will cause the object to move (that is, pressure can produce motion).

The bulk properties of matter are determined by the motion of molecules and the types of potential energy possessed by the individual molecules, and in theory it is possible to describe all properties of a quantity of matter by considering all of the energies and interactions of the separate molecules. However, it is often more convenient, quite simple, and elegant to describe the bulk properties of matter in terms of the relation between three easily measured quantities—namely, the temperature, pressure, and volume of the whole system—and the ability of the system to do work. This formulation of the properties of matter, which is based upon several empirical observations, is called thermodynamics; its extreme mathematical simplicity and its predictive power justify an effort to learn to use its (perhaps foreign) mode of reasoning.

A great deal of thermodynamics deals with properties of matter that seem to have little relation to biological systems. Thus, we shall not spend too much time in this book on the traditional examination of gases but concentrate more on their applications to the properties of solutions. Nonetheless, it is necessary to examine carefully the properties of very simple, somewhat abstract systems in order to develop the extraordinarily useful concepts of work, entropy, free energy, and chemical potential. The student is urged to learn these concepts thoroughly.

4–1 THERMODYNAMIC LAWS

Thermodynamics can be developed in several ways. One can begin with the theories of quantum mechanics and statistical mechanics and derive the basic thermodynamic functions by summing over the properties of individual molecules. This approach is satisfying but mathematically complex.* A second method is to start out with empirical "laws" of nature. This is the historical approach and has the virtue that the bulk properties of matter can be derived very easily without even referring to the fact that matter consists of molecules. This is a powerful method but we will modify such an approach slightly by continually reminding ourselves that we are in fact dealing with molecules; this is a useful procedure because it makes the laws of thermodynamics seem less abstract (their abstractness is the most common reason many students fail to understand thermodynamics). It is for this reason that the kinetic theory of matter was presented in an earlier chapter.

We can begin our approach to thermodynamics by stating two main laws briefly. The First Law of Thermodynamics states that

energy can neither be created nor destroyed—the energy of the universe is constant.

The Second Law has many forms. Two of the most important are that

the ability of an isolated system to perform work continually decreases, and it is never possible to utilize all of the energy of a system to do work,

---

*A student of mine once said that the use of statistical thermodynamics to understand molecular behavior is like using a cannon to kill a mosquito.

and,

> *in all real processes disorder increases.*

In this and the next chapter we will examine these statements in more detail.*

4-2 THE FIRST LAW OF THERMODYNAMICS

To understand the First Law of Thermodynamics one must begin with an examination of the concepts of *work, energy,* and *heat.*

Work

The simplest kind of work is done when a force acts on an object over some distance. Thus, when we lift an object or push a brick along a floor, we are doing work. These two processes, lifting and pushing, are not the same, though. When an object is lifted, it retains the capacity to do work in that if it is released, it will fall and can be made to operate some mechanical device during its fall. In the second case, an amount of work can be done when the object is moved, but when we stop pushing, the object remains in place. Thus, the difference is that the lifted object has potential energy—the energy we have expended in lifting the object is now possessed by the object and can be used to do more work. The pushed object has not gained potential energy.

The example of lifting illustrates an important property of work and energy—work can be converted to potential energy, which retains the capacity to do further work, so one type of work (lifting) can be converted to another type of work (gravitational). In the second example, the energy that we have put into the brick by moving it does not have the capacity to do more work. Thus, energy can also be converted to work irreversibly.

Another phenomenon can be observed if the temperature of the brick or the floor is measured after the brick has been pushed across the floor; the temperature of both will have risen. Thus, the work done has resulted in an increase in temperature or, looking at it in another way, the energy that we have put into the brick to move it has been converted to heat. The molecular basis of this conversion is the following. When the brick is pushed, we impart translational kinetic energy to the entire brick. If it were moving on a frictionless surface, we could stop pushing and the brick

---

*A simple statement of the First Law and the Second Law was widespread among physics students at one time: the First Law—you can't win (you can't somehow "beat the system" and end up with more than you began with); the Second Law—you can't even break even (you won't even get back everything you've put into the system).

SEC. 4-2 THE FIRST LAW OF THERMODYNAMICS

would continue to move. That is, it would then retain a capacity to do work because as long as it moved it could come in contact with some object and push it. However, on a surface having friction the exterior molecules of the brick collide with the molecules on the surface of the floor and then rebound. In these collisions the translational energy imparted to the brick is gradually converted to random motion (kinetic energy) of molecules in both the brick and the floor; as discussed in Chapter 2, this increase in the kinetic energy of the randomly moving molecules is observed by a corresponding increase in temperature.

It should now be clear that work is of two sorts—that which is converted to potential energy and that which is degraded to heat.* An intermediate case is of course also possible—namely, the conversion of some of the work to potential energy and of some to heat. For example, in electrical work, the movement of charges through a perfectly conducting wire can generate a magnetic field capable of lifting a block of iron, but movement of charges through resistance wire makes the wire red hot. In real situations the intermediate case is common. For instance, in an electric motor, the electric work is converted to potential energy and then to mechanical work but, because no wire is perfectly conducting and because the motor is not frictionless, heat is also generated.

Any system can be uniquely described by stating the values of particular parameters. For example, if values are given for the volume V and temperature T of n moles of a gas, we know its pressure P. Similarly, knowledge of P, V, and n tells us T. We say that we know the state of a system when a sufficient number of parameters are known to describe the system uniquely. A relation between variables that define the state of a system is called an *equation of state*. The ideal gas law $PV = nRT$ is an example of an equation of state. The state of a system can change if the value of any of these parameters is changed. We then say that the system has changed from an *initial state* to a *final state*. A process by which the state of a system changes is called a *path*. There are certain properties possessed by a system that are a function only of its state. One example is its energy. We will see other examples later. Such properties are called *state functions*. It should be noted that if the state of the system is uniquely defined, then the value of a state function—for example, the energy—does not depend on how the system got to that state. Thus, we can say that *the value of a state function after any change of state is independent of path*. We will now expand on these points.

When work is done by or on any system, the state of the system changes—for example, its volume or its temperature may increase or decrease. The amount of work done cannot be determined simply by knowing the initial and final state of the system; the work done depends on *how* the final state is reached. The thermodynamic statement of this fact is that

---

*We use the word degraded because, although the heat can be used to do work—for example, to expand a gas—the amount of work it can do is less than the amount of work initially done that caused the temperature rise.

the amount of work done depends upon the path. For example, suppose we wish to raise our brick to a table that is some distance away in order to increase the brick's potential energy. One way to do this would be to slide it up a frictionless inclined plane. If, instead of being frictionless, the plane were very rough, more work would be required, because both gravity and friction would have to be overcome. If the rough plane also undulated, even more work would be required. Will all of the additional work contribute to an increased potential energy? Because the ultimate potential energy of the system is determined by the final state of the system and is thus independent of the path, if extra work is done, this will produce heat. That is, the extra work done in following the undulating path cannot contribute to the ultimate potential energy reached because that is determined only by the height of the brick; the brick will fall with the same speed and produce the same force when hitting the floor, whether it got to the table by being lifted straight up to the table or by being slid up the plane. Thus, the work done in excess beyond that required to reach the height of the table appears as heat. (The student is urged to be sure to understand this example before reading further.)

We can make a simple mathematical formulation showing the dependence of work on path by examining the change in volume of an ideal gas that results from varying the external pressure. This kind of work is called pressure-volume or *P-V* work. Consider a cylinder (Figure 4-1) of uniform cross-sectional area A containing a piston at a distance x from the closed end of the cylinder. For simplicity we assume that there is no friction between the piston and the wall of the cylinder. The applied force f required to maintain the position of the piston is $f = PA$ where P is the pressure of the gas in the cylinder. If the piston is moved an infinitesimal distance dx, the amount of work dW done is

$$dW = f\,dx = PA\,dx = P\,dV. \qquad (1)$$

If work W is done to change the volume of V_1 to V_2, the total work is

$$W = \int_{V_1}^{V_2} P\,dV. \qquad (2)$$

FIGURE 4-1 A cylinder having cross-sectional area A fitted with a movable piston and containing a gas. If a force f is applied to the piston, the gas will exert a pressure P on the piston and on the walls on the cylinder in which $P = f/A$.

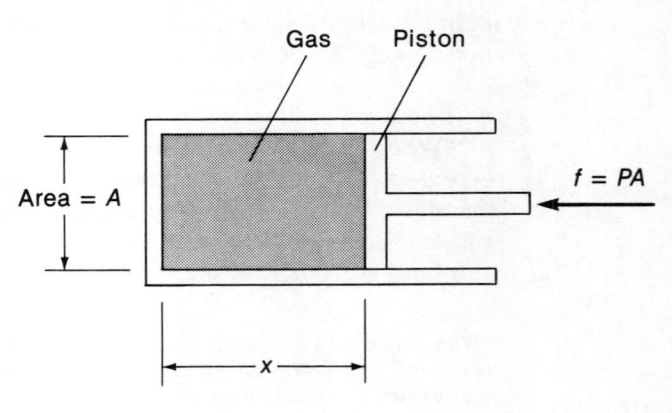

SEC. 4-2   THE FIRST LAW OF THERMODYNAMICS

FIGURE 4-2 Four different hypothetical paths for a gas whose volume changes from V_1 to V_2. The changes in volume can be accomplished by moving a piston as in Figure 4-1. For path 1 the work equals the area under AB and bounded by the V_1 and V_2 broken lines. For path 2 the work is the area under the curve AC. For path 3 the work is the area under the curve DB. For path 4, the work is the area under AFB *plus* the area bounded by A, E, and D; the area AED is the area EA minus the area under ED, the minus sign being a result of a *decrease* in volume.

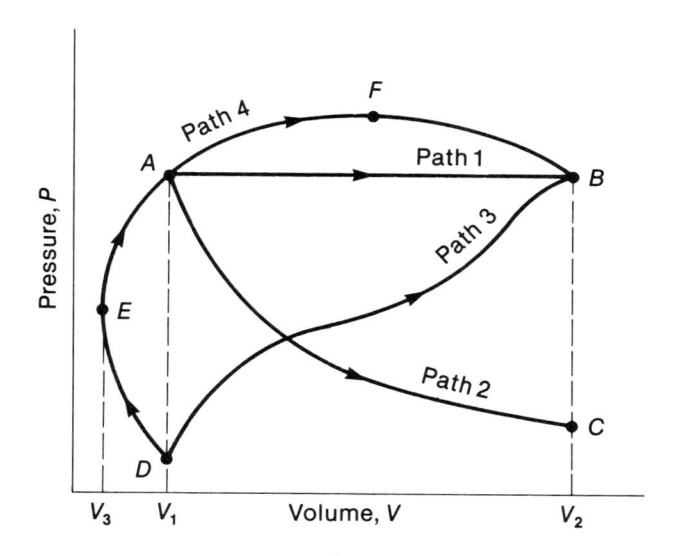

In order to evaluate this integral, one must know how P depends upon V. From the ideal gas law, $P = nRT/V$, it might seem that one need only substitute this expression for P and integrate. However, there is a subtlety here since, in doing so, we would have to assume that T is constant and we have seen that often when work is done, there is a temperature change. Furthermore, the observed change in temperature depends upon the path experienced by the system. The dependence of work on path can be seen in Figure 4-2. In each case shown in the figure we will calculate the work done when the volume increases from V_1 to V_2. When the pressure is constant (Path 1), the work is just the rectangular area under the curve AB. For Path 2 an isotherm is followed, and again the work is the area under a curve (AC) but this area is less than that for Path 1. For Paths 3 and 4 both pressure and temperature change. Note, however, that for Path 3 the work is simply the area under the curve DB. On the other hand, for Path 4, in which the volume decreases to V_3 before increasing, the work done is the area under the curve bounded by the curve AFB between V_1 and V_2 plus the area AED (because the contribution of EA to the work is positive and that of DE is negative). For a few simple cases W is easily calculated. For example, at constant pressure (Path 1) for a gas obeying any gas law,

$$W = \int_{V_1}^{V_2} P \, dV = P(V_2 - V_1) = P\Delta V . \tag{3}$$

For an ideal gas Equation 3 may also be written

$$W = nR\Delta T , \tag{4}$$

which indicates that if the pressure of an ideal gas is held constant, a volume increase is accompanied by a temperature increase.

For a reversible isothermal volume change of an ideal gas (Path 2),

$$W = \int P \, dV = nRT \int_{V_1}^{V_2} \frac{dV}{V} = nRT \ln \frac{V_2}{V_1}. \tag{5}$$

In Equation 2, as long as a relation between P, V, and T is known—that is, as long as there is an equation of state—the work can be calculated. Hence, for an isothermal change of a gas obeying the van der Waals equation of state (see Chapter 3, Equation 3),

$$W = RT \ln\left(\frac{V_2 - b}{V_1 - b}\right) + a\left(\frac{1}{V_2} - \frac{1}{V_1}\right), \tag{6}$$

in which a and b are the constants of the van der Waals equation.

There are many kinds of work other than pressure-volume work for which simple expressions can be derived. For our purposes, the most important are electrical work and chemical work, because these are encountered most frequently in biological systems. Some of these expressions for several kinds of work are the following.

A. Elastic Work

If an elastic wire or a spring of length x is stretched to length $x + dx$ by application of a force F, the work dW done on the system is

$$dW = F \, dx. \tag{7}$$

For a single small extension, F is constant and the integration is simple.

B. Electrical Work

If a charge q is moved in an electric field from a point at which the electrical potential is ϕ to another point of potential $\phi + d\phi$, the work dW done in moving the charge is

$$dW = q \, d\phi. \tag{8}$$

In a single movement from ϕ_1 to ϕ_2 the work W is*

$$q(\phi_2 - \phi_1) = q\Delta\phi.$$

This is a useful expression since it is the work done in moving one charge with respect to another charge whose location is fixed. If there are many charges moving—that is, if there is a flow of charge or current i between a fixed potential difference $\Delta\phi$ (which is equivalent to the voltage)—the work done per unit time is $i\Delta\phi$; this is the familiar equation in electrical

---

*Equation 8 actually defines the electrical potential. That is, if there is a charge at some point in space, there is an electrical field associated with that charge. At infinity the field is zero. The electrical potential of the second charge at some point in space is simply the work required to bring the charge from infinity to that point.

circuit theory stating that the watt is a measure of (work) = (current) × (voltage).

C. Gravitational Work

The movement of a mass through a difference in gravitational potential $\Delta\phi$ is $m\Delta\phi$. This is more conveniently expressed as

$$dW = mg\,dh\,, \tag{9}$$

in which g designates the gravitational constant and dh designates a change in height.

D. Work in Forming a Surface

A solid can be cut up into little pieces and liquid can be dispersed into droplets. Both processes clearly require that work be done on the substance being subdivided. The work is conveniently expressed in terms of the increase in surface area dA. The work done is

$$dW = \gamma\,dA\,, \tag{10}$$

where the proportionality factor γ is called the *surface tension*—the work required to produce unit area of surface. This will be discussed in detail in Chapter 13.

It is important to note that in any process many types of work might be applied to a system or done by the system. Work is an additive process, so that if all of the kinds of work just described are performed, the total amount of work done is simply the sum of the contributions made by each process.

A final point of great importance is the sign convention for work.

Work done on *(that is, applied* to*) a system is given a negative sign. Work is positive when it is done by a system to the system's surroundings.*

For example, if an applied pressure increase causes a decrease in the volume of gas, the work is negative because it is done on the gas. If a gas (the system) expands and pushes a piston (the surroundings), the work is positive. Note that the sign depends upon what is defined to be the *system:* in any process one must be consistent in definition.

Heat

We have seen in the preceding section that heat is often evolved when work is done and that this is observed as an increase in temperature. Heat may also be added to a system in order to enable the system to perform work. For example, in the piston arrangement of Figure 4-1, if the gas is heated, it will expand and do work by pushing the piston. We again need a sign convention to describe heat changes. We denote the heat transferred by Q and give it a positive sign when heat is added to the system;* similarly, if heat is transferred from the system to the environment, it is assigned a negative sign. Note that this is exactly the reverse of the sign convention for work, which we recall is positive when done on the system's environment. Also, it is important to observe that, as with work, the sign depends upon how the system and the surroundings are defined. These conventions are summarized in Table 4-1.

TABLE 4-1 Sign conventions for work and heat transfer.

Work:	
Done *on* the system	Negative
Done *by* the system	Positive
Heat:	
Transferred *to* the system from the environment	Positive
Transferred *from* the system to the environment	Negative

The heat transfer Q associated with a change from an initial to a final state also depends upon the path. This can easily be seen in the following example. If an object is lifted from the floor and placed on a table, work is done on the object, which now has more potential energy, but no heat is evolved (assuming no air friction). If, instead, the object is dragged up to the same table on an inclined plane with a rough surface, the change in potential energy is the same but heat is produced by friction between the object and the plane. Thus in one process no heat is produced but, in the other, heat is evolved. In the following section we will examine the amount of heat produced in several processes and see how it is related to the amount of work done.

Relation between Work and Heat: The First Law

As we have already stated, when a system moves from an initial state to a final state, the amount of work done and the heat change depend upon the path. We can look at this closely in the following example. Consider a gas under pressure in a cylinder fitted with a weighted piston. The external pressure on the cylinder is assumed to be very low. We define the gas as the system and all else as the surroundings. If the piston is released (for example, by decreasing the weight of the piston), the gas expands, pushes the piston, and performs work. When the process is complete, it is observed that the gas has cooled. The reason for the cooling of the gas is the following. The movement of the piston (as a response to the gas pressure) is a result of collisions between the gas molecules and the face of the piston.

---

*Note that Q and not ΔQ is used to denote heat transfer. This is often a source of confusion.

Thus, the kinetic energy possessed by the moving piston comes from the energy transferred from the molecules to the piston by elastic collisions. In giving up energy to the piston the gas molecules lose kinetic energy; since the kinetic energy is proportional to temperature, the temperature of the gas decreases. If the walls of the cylinder are thermally conductive and are in contact with a huge heat reservoir, heat will be taken up by the gas from the surroundings and the temperature of the gas will remain constant. Hence, in this process Q is positive because it is taken into the system from the surroundings.

Now we consider a similar process in which a volume of gas is allowed to expand slowly *into a vacuum;* conditions are chosen, however, so that the total volume change is the same as in the example just described for the piston. In this case no mass is moved during the expansion; thus, no work is done ($W = 0$); hence, no kinetic energy is lost and, as expected, it is observed that there is no temperature change. Thus, heat is neither added nor taken away—that is, $Q = 0$.

A very large number of systems and processes have been examined experimentally and it is always found that for a particular change in the state of the system, there are many possible pairs of values of W_i and Q_i, in which W_i and Q_i refer to the observed values for a particular path i. However, in comparing various paths, despite the fact that in general $W_1 \neq W_2 \neq W_3$, and so on, and $Q_1 \neq Q_2 \neq Q_3$, and so on, it has repeatedly been observed that the difference $Q_i - W_i$ is always the same for a given change in state. For example, suppose a gas is compressed by doing an amount of work W; at the same time the temperature changes from T_0 to T_n and the pressure changes from P_0 to P_n. The amount of work done and the heat change are measured and the difference is found to be $Q - W$. The same amount of an identical gas also at T_0 and P_0 is then carried through a process in which the gas passes through the states (T_1,P_1), (T_2,P_2), (T_3,P_3), ..., (T_{n-1}, P_{n-1}); finally it reaches the state (T_n,P_n). The work done and the heat produced is measured in each stage of the process and found to be (W_1,Q_1), (W_2,Q_2), ..., (W_n,Q_n). If one adds up all of the values of W_i and all of the values of Q_i (and keeps track of the signs) the difference between $\sum W_i$ and $\sum Q_i$ will always be observed to be $Q - W$. That is,

$$Q - W = \sum Q_i - \sum W_i = \sum (Q_i - W_i) . \tag{11}$$

Note, however, that in general

$$W \neq \sum W_i \quad \text{and} \quad Q \neq \sum Q_i . \tag{12}$$

It is only the *difference* $Q - W$ that is the same for all processes, no matter how complex, as long as the initial and final states are the same. That is, for each stage in the process,

$$Q_1 - W_1 = Q_2 - W_2 = \cdots Q_n - W_n . \tag{13}$$

This is the meaning of the statement that *the difference* $Q - W$ *is independent of path.*

This remarkable property of the difference $Q - W$ is true for all kinds of work, not only P-V work. It is given a special name, *the change in internal energy,* or ΔU, and is expressed as

$$\Delta U = Q - W . \tag{14}$$

This statement is the *First Law of Thermodynamics:* it is an empirical law and is not derived from fundamental physical principles. Three important points must be made about Equation 14:

1. The First Law provides the thermodynamic definition of the internal energy of a system.
2. The value of ΔU is independent of path, so that the internal energy of a system depends only upon the state of a system and never on how that state was achieved. A simple consequence of the independence of path is that for any cyclic process, the initial and final states of the system being the same, $\Delta U = 0$. In other words, even though a system might undergo many transformations in which work is done and heat is exchanged, nonetheless, when it is returned to its initial state, its internal energy will be exactly the same as it was before the transformation occurred.
3. In a cyclic process, initial and final states are the same, so that $\Delta U = 0$; thus, $Q = W$. This means that in a cyclic process heat and work are completely interconvertible. For example, if a gas contained in an insulated vessel is *slowly* compressed by a piston, the work of compression results in an increase in the temperature of the gas; if the piston is then slowly released, expansion occurs and the gas cools again. In the first stage of this cyclic process, the work done is reversibly converted to heat and in the second stage the heat is used to perform the work of moving the piston back to the starting position. The value of ΔU can be 0 for a noncyclic process also. For instance, when a gas expands freely into vacuum, both Q and W are 0 (see page 88), so that by Equation 14, $\Delta U = 0$.

The First Law of Thermodynamics is frequently stated in nonmathematical form in one of the following ways:

1. Energy can neither be created nor destroyed.
2. If a system is subjected to a cyclic process, the work produced in the surroundings equals the heat taken from the surroundings.

The first statement emphasizes the constancy of the energy of the universe although, as will be seen in the following chapter, it does not say whether the energy is all available to perform work. The second statement emphasizes the relation between energy and work.

It is important to note that the First Law does not make any statement about the *amount* of internal energy possessed by a system; rather it only defines ΔU, the *change* in internal energy that accompanies some process.

Indeed, by thermodynamic methods alone it is not possible to evaluate U; this requires the results of the theory of quantum statistics, which will not be discussed in this book. We will find this to be true of several other thermodynamic functions as well but that this does not affect the usefulness of thermodynamics.

Adiabatic and Isothermal Processes—Definition

In order to discuss the laws of thermodynamics, we will frequently analyze processes that occur either without heat exchange or with heat exchange but at constant temperatures.

In an *adiabatic* process, heat is neither added nor removed from the system. Thus $Q = 0$ and $\Delta U = -W$.

An adiabatic process can be carried out in two ways: the system might be thermally insulated so that heat exchange is impossible or the process might occur so rapidly that it is completed before heat exchange occurs. A very rapid compression is an example of the second type.

In an *isothermal* process, there is no temperature change (which does *not* mean that $Q = 0$). For a process to occur isothermally the system must be in good thermal contact with a heat reservoir that is so large that it can absorb or exchange heat without a detectable change in temperature (typically a thermostatted water bath in the laboratory). In this way, heat flow can occur either into or out of the system so that constant temperature results. Of course, no process can ever be truly isothermal, because heat exchange is not instantaneous.

Chemical reactions commonly occur under isothermal conditions in biological systems, since changes of temperature within an organism, when they occur, tend to be very small. (The higher animals, such as mammals, have evolved means of regulating their body temperatures.)

4-3 ENTHALPY OR HEAT CONTENT

Definition of Enthalpy

If we consider only chemical processes involving gases or uncharged systems (so that there can be no electrical work done), all work is P-V work. If the volume is constant, the work $W = P\Delta V = 0$, and, by Equation 14, the increase in internal energy is just the heat absorbed at constant volume, Q_V—that is,

$$\Delta U = Q_V. \tag{15}$$

It is fairly common in a chemical reaction for the volume to change, although typically the pressure remains constant at atmospheric pressure.

Under these conditions ΔU is often not an appropriate quantity to characterize chemical changes and another function that is more useful—namely, the *enthalpy, H*—is used.

If pressure is constant and the volume changes from V_1 to V_2 while the internal energy changes from U_1 to U_2, from Equation 14 there results

$$\Delta U = U_2 - U_1 = Q_P - P(V_2 - V_1) , \qquad (16)$$

in which Q_P denotes the heat exchange at constant pressure. This can be rearranged to

$$(U_2 + PV_2) - (U_1 + PV_1) = Q_P . \qquad (17)$$

The form of the expressions in the parentheses leads us to define the enthalpy, H, as

$$H = U + PV , \qquad (18)$$

so that Equation 17 becomes

$$H_2 - H_1 = \Delta H = Q_P . \qquad (19)$$

Therefore, when the only work is *P-V* work, *the increase in enthalpy is equal to the heat absorbed at constant pressure*. Like ΔU, ΔH is independent of path; thus it is a function of the initial and final state only, and must be zero for a cyclic process. We will see the utility of the enthalpy concept shortly.

Heat Capacity

As a system absorbs or gives off heat, the temperature of the system may change. The relation between the quantity of heat added and the number of degrees of temperature change is expressed by the heat capacity, C. This is defined as the quotient

$$C = \frac{Q}{\Delta T} . \qquad (20)$$

For an ideal gas, the heat capacity is a constant; however, for a real substance, C depends upon the temperature interval chosen, so that it must be defined in differential form as

$$C = \frac{dQ}{dT} . \qquad (21)$$

Since we commonly deal with processes that occur either at constant pressure or at constant volume, it is convenient to distinguish between the heat capacity C_V at constant volume and C_P, the value at constant pressure, and write

$$C_V = \frac{dQ_V}{dT} \qquad \text{and} \qquad C_P = \frac{dQ_P}{dT} . \qquad (22)$$

SEC. 4–3   ENTHALPY OR HEAT CONTENT

From Equations 15 and 18 these expressions become

$$C_V = \left(\frac{\partial U}{\partial T}\right)_V \quad \text{and} \quad C_P = \left(\frac{\partial H}{\partial T}\right)_P, \tag{23}$$

for which the partial derivative notation $(\partial X/\partial T)_Y$ means that X is a function of both T and Y and that in taking the derivative we are assuming that T is a variable and Y is a constant. (See Appendix III at the end of this book.) The value of C_P is always larger than C_V because, at constant pressure, part of the heat added is used in the work of expansion (that is, some of the heat is not used to raise the temperature) whereas at constant volume, all of the heat produces a temperature increase (because no work is done).

EXAMPLE 4–A   Increase in temperature of a gas owing to absorption of heat.

A sealed, insulated 1-l vessel at 0°C and 1 atm contains a minute amount of substances that react and produce 104.6 J of heat. We can calculate the temperature change of the air in the vessel from Equation 22. The volume of the vessel is constant, so we use the heat capacity at constant volume C_V, which is approximately 25.1 J K^{-1} mol^{-1} (or 6 cal K^{-1} mol^{-1}) for air. At atmospheric pressure, 1 liter of air is 1/22.4 = 0.0446 moles. Therefore, the heat capacity of the air in the vessel is (25.1)(0.0446) = 1.12 J K^{-1}. The temperature rise is then 104.6/1.12 = 93.3°C.

If the vessel were filled with water, the temperature change would be much less. One liter of water contains 55.55 moles. The value of C_V for water is about 71.13 J K^{-1} mol^{-1} (17 cal K^{-1} mol^{-1}). Thus, the heat capacity of the water in the vessel is (71.13)(55.55) = 3.95 kJ K^{-1} and the temperature rise is 104.6/3950 = 0.026°C. The high heat capacity of water is one of the reasons that the temperature of biological systems does not show a very large fluctuation as a result of numerous intracellular chemical reactions.

The concept of heat capacity allows one both to calculate the final temperature achieved when two different substances are placed in contact and to measure the heat transferred. How this is done is shown in the following examples.

EXAMPLE 4–B   The final temperature reached when two samples of the same substance at different temperatures are placed in contact.

A vessel containing 50 g of water at 10°C is placed in contact with a vessel containing 75 g of water at 25°C. At equilibrium both vessels will have the same temperature T. Clearly the 10°C sample will increase in temperature by $T - 10°C$ and the 25°C sample will decrease in temperature by $25 - T°C$. If no heat is lost to the surroundings, the heat Q_x gained by warming the 10°C sample must equal the heat Q_y lost by the 25°C sample. Thus $Q_x = (75/18)C_P(T - 10)$ and $Q_y = (50/18)C_P(25 - T)$. Setting $Q_x = Q_y$ yields $T = 16°C$. This is the same temperature that would be reached if the two samples had been mixed together.

Note that the value of C_P did not have to be known because it is the same for both samples. This is not the case in the following example.

EXAMPLE 4–C The final temperature reached when two different substances at different temperatures are placed in contact.

A 50-g block of a solid initially at 5°C is immersed in 75 g of water initially at 30°C. The molecular weight and value of C_P for the solid are 162 and 25 J K^{-1} mol, respectively. For water $C_P = 75.3$ J K^{-1} mol^{-1}. Using the notation of Example 4–B, $Q_x = (50/162)(25)(T - 5)$ and $Q_y = (75/28)(75.3)(30 - T)$, and with $Q_x = Q_y$, then $T = 29.4°C$.

EXAMPLE 4–D Calculation of the heat transferred to a solid placed in contact with a heat reservoir, and determination of C_P for the solid.

A 250-g amount of a solid of temperature 10°C and molecular weight 58 is placed in a 5-kg insulated water bath initially at 20.00°C. At equilibrium, the water bath is at a temperature of 19.93°C. The water bath has lost heat of the amount $(5000/18)(75.3)(20.00 - 19.93) = 1464$ J, and this amount of heat has been gained by the solid. The value of C_P for the solid is obtained by setting $(250/58)C_P(19.89 - 10)$ equal to 1464 J. Thus, $C_P = 34.35$ J K^{-1} mol.

Note that another way to measure the heat transferred is to pass an electrical current i through a wire having resistance r, through the water bath for a time t sufficient to restore the temperature of the water bath to 20.00°C. The electrical energy put into the water bath is i^2rt and this equals the heat transferred to the solid.

In Chapter 2 the properties of a molecule that affect the heat capacity were discussed. Since the temperature of a collection of molecules depends only on the total kinetic energy of the molecules, the heat capacity must depend on how much energy can be taken up by the molecules and retained in a form other than translational kinetic energy. The other storage modes are in intramolecular vibrations and in various types of rotation of the entire molecule. If the molecules interact with one another, as in the significant hydrogen-bonding between water molecules, the ability to take up energy with a minimal temperature rise is even greater.

Joules and Calories

Traditionally the joule was a unit of work or energy and the calorie was a unit of heat.

The calorie was originally defined in terms of the heating of water, a substance chosen as a standard for many systems of units. It had been observed that the energy that had to be added to one gram of a substance to raise the temperature 1°C depends upon the particular substance. This amount of energy differs from the heat capacity at constant pressure C_P only in that C_P refers to one mole rather than one gram of material. The heat capacity of one gram of material is called the *specific heat* of that material. Since water was almost always chosen as the reference substance (or heat absorber) in calorimetric experiments, the calorie was conveniently defined as the heat capacity of one gram of water. Alternatively, the specific heat of water was defined to be 1 cal K^{-1} g^{-1}. However, careful measurement showed that the heat capacity of water is not constant in the temperature range of 0–100°C (Figure 4-3), and the definition was thus changed to the following: the standard calorie is the heat required to raise the temperature of one gram of water from 15°C to 16°C or from 59°C to 60°C.*

The calorie is not an acceptable unit in the SI system of units. However, the calorie is still in such common use that it is important to know how to convert the calorie to the joule, the SI unit of energy. The equiva-

FIGURE 4-3 The variation of the specific heat of water with temperature at a pressure of one atmosphere.

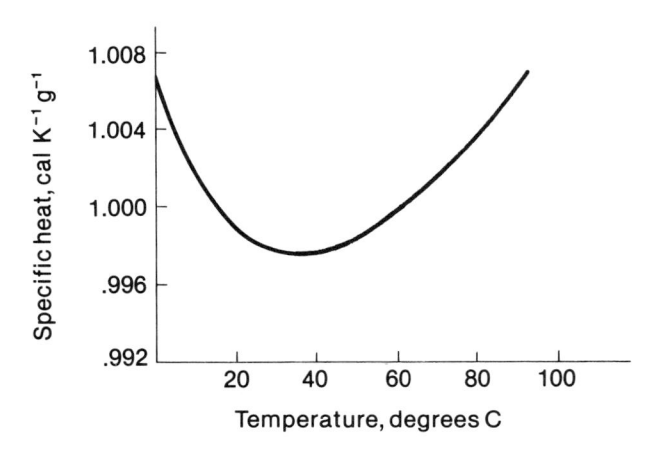

*In some applications the term *mean calorie* is used. This is 1/100 the energy to raise one gram of water from 0° to 100°C and equals 1.0014 standard calorie.

lence of work and heat was first established by an experiment in which a vessel of water was stirred by paddles under such conditions that the work done in agitating the water could be calculated. The rise in temperature of the water was correlated with the amount of work done. It was found that

one calorie is equivalent to 4.1840 joule,

a relation at one time called the *mechanical equivalent of heat*.

4–4 THERMOCHEMISTRY—THE APPLICATION OF THE ENTHALPY CONCEPT

The enthalpy change in a process in which no work is done is simply the heat added to or given off by a system. There are a great many physical changes (e.g., melting, boiling, dissolving) and chemical changes (e.g., formation of chemical bonds) in which no work is done.* Each process is characterized by a particular enthalpy change. Traditionally an enthalpy change is called "a heat," and the following kinds of notation are used to designate it: for an enthalpy change ΔH in a particular process (pro), one writes ΔH_{pro}. The process is given a three- or four-letter abbreviation and written as a subscript, though in some books a single letter abbreviation is used, and one may also find the process indicated in parentheses following the ΔH. Thus, we write ΔH_{vap}, ΔH_{sub}, and ΔH_{sol} for the enthalpies of vaporization (boiling), sublimation, and solution, respectively; for the melting process the symbol ΔH_{fus}, the enthalpy of fusion, is used; for chemical reactions such as combustion there is also ΔH_{comb}. In complex reactions in which several processes occur, a more descriptive notation is often used. Thus, in a process in which ice melts and the water then turns to vapor (vap) one might use $\Delta H_{\text{ice,H}_2\text{O}}$ and $\Delta H_{\text{H}_2\text{O,vap}}$; more generally, in a transition from solid (s) to liquid (l) to gas (g), one often sees $\Delta H_{\text{s,l}}$ and $\Delta H_{\text{l,g}}$. An older term still sometimes used for an enthalpy change is *latent heat,* deriving from the fact that in the 18th century the source of heat evolved in freezing, for instance, was not known and hence was latent or hidden.

For some processes the enthalpy change is positive and for some it is negative. In accord with the sign convention for heat changes (Table 4-1) and the fact that $\Delta H = Q_P$, a negative value of ΔH means that heat is given off to the environment and a positive value means that heat is taken up. The following terms are also used to describe enthalpy changes:

An *exothermic reaction* is one in which $\Delta H < 0$ (has a negative value) and heat is given off.

---

*This is true macroscopically but is not necessarily true on the atomic scale. However, thermodynamics deals only with the bulk properties of matter, so that in thermodynamic arguments one need not be concerned with discrepancies that might exist when a very small number of atoms or molecules is examined.

SEC. 4-4 THE APPLICATION OF THE ENTHALPY CONCEPT

An *endothermic reaction* is one in which $\Delta H > 0$ (has a positive value) and heat is absorbed.

At this point it is worth stating several reasons why evaluating enthalpies is of interest since often the subject appears to the student to be little other than an exercise in arithmetic. Clearly, it is valuable to be able to predict whether a reaction can occur or not. Our intuition might suggest that an exothermic reaction occurs spontaneously while an endothermic one does not. However, in Chapter 6 we will see that there is another important factor that must be considered and, in fact, the criterion for occurrence of a reaction is that the Gibbs free energy and not the enthalpy is negative. Nonetheless, the numerical value of the enthalpy is used in calculating the free energy, so it is necessary that one learns how to calculate enthalpies.

A second reason for examining enthalpies is that the optimal experimental conditions for a chemical reaction are often of great importance. For example, in Chapter 11 it will be seen that the rate of reaction increases with temperature. Thus, in a strongly exothermic process the components heat up and the process occurs rapidly; if gases are evolved, the reaction may occur with explosive violence. We will see shortly that the value of ΔH for most reactions can frequently be calculated without actually carrying out the reaction, so that such an effect can be predicted, and we can know in advance that cooling during the reaction is necessary. Similarly, in an endothermic reaction, if there is not a readily available heat source, the temperature may drop rapidly and the reaction time may be extraordinarily long. Again, this can be predicted and we can know to provide an adequate heat source.

To begin with, it is necessary to discuss what is called the *standard state* of a substance and to understand the units used to describe thermodynamic functions. This is done in the next two sections.

The Standard State

The enthalpy of a substance depends upon its temperature, its form (that is, gas, liquid, solid, crystalline state, and so forth), and the external pressure exerted on it. Actually, there is no way to determine an absolute enthalpy; only enthalpy changes are measurable. Thus, it is reasonable (1) to define for every substance a standard state and (2) to establish a reference point at which the enthalpy is considered to be zero. Since most chemical reactions are carried out at atmospheric pressure,

> *the standard state of a substance is defined as its most stable form at 1 atm pressure or, for a solute, a concentration of 1* M.

Temperature is not included in the definition of standard state, so tables of standard enthalpy changes (denoted $\Delta H°$) for particular processes must specify the temperature of the measurement. The zero enthalpy point is chosen to refer to the standard state of an element at 25°C—thus, *the enthalpy of every element in the standard state is defined to be zero.*

Since biochemical reactions are almost always studied at 37°C (and sometimes 30°C), it is necessary to convert tabulated values, which are almost always for 18°, 19°, 24°, or 25°C, to one or the other of these temperatures. How this is done will be described in a later section.

Enthalpies of Phase Transitions

When ice at 0°C is heated, it melts, but the temperature does not rise until all of the ice is converted to water. The heat that is added before the temperature rises is identical to the amount of heat given off when water at 0°C freezes. This same phenomenon occurs for all substances in a melting or freezing transition. The heat change for one mole of substance undergoing such a transition is the *enthalpy of fusion*, ΔH_{fus}. Likewise, when a liquid evaporates (or boils), heat is absorbed without temperature change and an identical amount of heat is produced when the vapor condenses. For one mole of a substance this heat change is called the enthalpy of vaporization, ΔH_{vap}. In general $\Delta H_{fus} \neq \Delta H_{vap}$. The particular value of each depends upon the substance examined and is a consequence of the cohesive force between the component molecules. Looked at in this way, the enthalpy of vaporization is roughly equivalent to the work done in overcoming the cohesive forces between the molecules of a liquid by moving them sufficiently far apart that they no longer constitute a net attractive force.

The boiling point T_b of a liquid is also related to intramolecular forces and, as might be expected, rises as the cohesive force increases (see Chapter 3, Tables 3-3 and 3-4). In fact, there is a rough proportionality between ΔH_{vap} and T_b. In 1884 Frederick Trouton observed that this was the case for a wide variety of substances and affirmed that

$$\Delta H_{vap}(\text{joule}) = 87.86\, T_b \qquad \text{or} \qquad \Delta H_{vap}(\text{calorie}) = 21\, T_b, \qquad (24)$$

a relation known as *Trouton's Rule*. The relation is actually an approximation but holds reasonably well for substances whose cohesive energy is solely a result of van der Waals and dispersion forces. For hydrogen-bonded liquids the proportionality constant is greater; for example, for methanol and water the values of the Trouton constant are 104.4 and 109.0 J K^{-1} mol^{-1}, respectively. Trouton's Rule is mostly of value in biochemistry for guessing whether a liquid is internally hydrogen-bonded. That is, if the constant is greater than 92, hydrogen-bonding is usually the cause.

There is no equivalent relation between ΔH_{fus} and the melting temperature T_m of compounds, because there are so many kinds of intermolecular forces present in solids.

The enthalpy change of a substance undergoing a transformation is independent of path; that is, it is a state function. This means that for a process consisting of several steps, the value of ΔH for the entire process is simply the arithmetical sum of the values of ΔH for each step. This additivity of enthalpies, when applied to physical and chemical reactions, is known as *Hess's Law;* this is a very useful principle.

EXAMPLE 4-E Calculation of the enthalpy change in converting one mole of ice at −10°C to steam at 120°C.

This process consists of several distinct steps. These are: (1) heating the ice to 0°C; (2) melting the ice to form water; (3) heating the water from 0° to 100°C; (4) boiling the water; and (5) heating the steam from 100° to 110°C. To calculate ΔH for the complete process, one need only evaluate ΔH for each step and add; for steps 1, 3, and 5, ΔH can be calculated from the relation $dH = C_P dT$, where C_P is the heat capacity of water. Values for ΔH_{fus} and ΔH_{vap} are available from standard tables. Thus the total enthalpy change ΔH_{tot} is

$$\int_{-10}^{0} C_P dT + \Delta H_{\text{fus}} + \int_{0}^{100} C_P dT + \Delta H_{\text{vap}} + \int_{100}^{120} C_P dT$$

or

$$\int_{-10}^{120} C_P dT + \Delta H_{\text{fus}} + \Delta H_{\text{vap}}.$$

The required values are $C_P = 75.3$ J K^{-1} mol^{-1} (or 18 cal K^{-1} mol^{-1}); $\Delta H_{\text{fus}} = 6.01$ kJ mol^{-1} (1.436 kcal mol^{-1}); and $H_{\text{vap}} = 40.67$ kJ mol^{-1} (9.72 kcal mol^{-1}). Thus $\Delta H_{\text{tot}} = 56.47$ kJ mol^{-1} (13.496 kcal mol^{-1}).

In this calculation we have assumed that C_P is independent of temperature and is the same for ice, water, and steam. This is often not the case, as will be discussed later.

Solids sometimes undergo transitions from one solid form to another solid form; the transition is usually a change in crystalline state that involves a molecular rearrangement. If the molecules are displaced with respect to one another, their cohesive force will be altered; thus heat is either evolved or taken up and the heat is equivalent to the work done in moving the molecules. An example of such a change is the transition of white tin (a tetragonal crystal) to gray tin (a cubic crystal), which occurs at 13.2°C. A heat change of this sort is called the *enthalpy of transition* and is usually denoted by indicating the two forms in a subscript—e.g., $\Delta H_{\text{tet,cub}}$. It is important in calculating ΔH for complex processes to include the enthalpy of transition. For example, in the heating of tin from 0° to 100° one would need the sum

$$\int_{0}^{13.2} C_P dT + \Delta H_{\text{tet,cub}} + \int_{13.2}^{100} C_P dT.$$

Since enthalpy changes are independent of path, $\Delta H = 0$ for any cyclic process; this means that for any phase transition, the enthalpy change in going from state 1 to state 2 is the negative of the change from state 2 to state 1. Thus, the reaction at the boiling point

$$H_2O(l) \rightarrow H_2O(g) ,$$

involves the heat change ΔH_{vap} and the reverse reaction $H_2O(g) \rightarrow H_2O(l)$ involves the change $-\Delta H_{vap}$. We could also use another notation: $\Delta H_{l,g} = -\Delta H_{g,l}$. The use of such reverse reactions will frequently be found to be useful in calculations. For example, a reverse-reaction notation would have been useful if in Example 4–E we had wanted to calculate ΔH for the process in reverse—that is, going from 120°C steam to −10°C ice. The value of ΔH for the total process would of course be the negative of the value calculated in Example 4–E.

Enthalpy Changes in Other Transitions

We have just seen that there is an enthalpy change associated with a phase transition. In biochemistry structural transitions of molecules and molecular arrays are of greater concern than phase changes; associated with these transitions is also a change in enthalpy. For example, when a muscle contracts or when two daughter chromosomes separate, there is a change in enthalpy. The biological transitions that are most amenable to thermodynamic analysis are those in which a molecule changes form. An important example of such a change is the folding of a newly synthesized linear protein chain to form a three-dimensional molecule that is biologically active.

The principle involved in the thermodynamic analysis of such transitions is that changes that are not visible can be made evident by examining a graph of either ΔH or C_P versus T. For instance, suppose ice at −20°C is placed in an opaque vessel and the vessel is placed in contact successively with heat reservoirs at various temperatures up to 130°C. Using the procedure described in Example 4–D to measure heat capacity, a graph can be made of ΔH or C_P versus the initial temperature of the reservoir. The graph will not be a smooth curve but will show discontinuities at 0°C and 100°C, as shown in Figure 4-4; this is of course because there are two transitions that occur between −20°C and +130°C. The important point of this trivial example is that the graph alone shows that there are two transitions in this temperature range even though neither the actual melting nor boiling can be seen because of the opacity of the vessel. Thus, a curve of ΔH or C_P versus T tells how many intermediate states there are in a particular change of form. Thus, if it is known that ice has been put into the vessel but that steam has been removed at 130°C, the curves of Figure 4-4 clearly indicate that there is an intermediate state in the process, and this of course is liquid water.

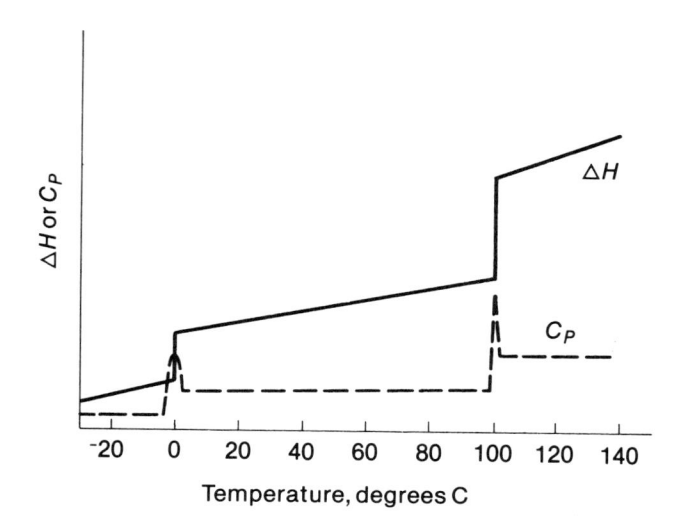

FIGURE 4-4 Graphs of ΔH and C_P versus temperature for water.

This type of analysis has been of great importance in the study of macromolecules and we will encounter it again in Chapter 15.

Heats of Solution and Dilution

If an ideal gas expands into vacuum, in an insulated vessel, there is no change in the internal energy because all types of energy—translational, vibrational, and rotational—are unaffected. This is not true of a real gas because the interaction energy due to the van der Waals forces decreases substantially as the distance between the molecules increases. The enthalpy or heat content also decreases. This decrease would also occur if two real gases A and B mixed at constant pressure and volume unless the cohesive force between individual molecules A and B is the same as that between individual molecules A and A and between individual molecules B and B. Thus, in any process in which mixing occurs there is an enthalpy of mixing whose magnitude and sign depend on the relative values of the cohesive forces. A liquid also necessarily possesses cohesive forces, so an enthalpy change occurs when two liquids mix. This is also the case when a solid dissolves in a liquid, and this change is often important when studying biological and biochemical systems.

When a solid dissolves in a solution, work is done to separate the molecules of the solid. If both the liquid form of the solid and the solution were ideal liquids (meaning that the interaction between the molecules of solid and liquid is the same as between the molecules of the liquid and between the molecules of the solid when in the liquid state), the formation of an ideal solution would be accompanied by an enthalpy change equal to the enthalpy of fusion (that is, conversion of the solute from the solid to the liquid state). This is sometimes the case for simple organic compounds dissolving in nonpolar organic solvents. However, usually when a solute dissolves, the solution is not ideal and there are significant changes in solvent and solute structure. Because of electrical effects the changes are particularly large when the liquid is polar and the solute is an electrolyte. This is discussed in detail in Chapters 7 and 8.

The enthalpy of solution may be negative (or exothermic—heat is given off) or positive (or endothermic—heat is taken up). For instance, dissolving anhydrous $CaCl_2$ in water is exothermic but dissolving CsCl in water is endothermic. The principal factors determining the sign of the enthalpy of solution are the relative interaction energies of solute and solute, solute and solvent, and solvent and solvent, and the *number* of solvent molecules interacting with each solute species. If the solute is a crystalline solid containing bound water, the loss of the water molecules affects the magnitude of the enthalpy change. For example, when anhydrous $CaCl_2$ dissolves in water, heat is given off, yet dissolving the hexahydrate $CaCl_2 \cdot 6H_2O$ produces an endothermic change.

The enthalpy of solution is denoted ΔH_{sol} and it is defined as the enthalpy change when one mole of solute dissolves in an infinite amount of solvent. A solute—for example, glucose—in an infinite amount of solvent—for example, water (aq)—is designated glucose (aq) and an equation describing the dissolving process is

$$\text{Glucose (s)} + H_2O \rightarrow \text{glucose (aq)}.$$

The term "infinite amount" requires some clarification. In a nonideal solution, the interactions between individual solute molecules clearly decrease as the solute concentration decreases and this is indicated by enthalpy changes accompanying dilution. Unfortunately, there are so many causes of nonideality of solutions that there is no simple equation relating the enthalpy of solution to the enthalpy of dilution (ΔH_{dil}). Furthermore, ΔH_{dil} does not depend in a simple way on the dilution factor; that is, the enthalpies of dilution for a particular solution diluted in one case from 2 M to 0.2 M and in another case from 0.2 M to 0.02 M (10-fold dilution) are not the same, nor are the values accompanying dilution from 2.1 M to 2.0 M and from 0.2 to 0.1 M (concentration decrease of 0.1 M) the same. Thus, when speaking of an enthalpy of dilution, we must designate the interval, for example, $\Delta H_{0.5\ M,\ 0.1\ M}$. Fortunately, when the solute concentration is fairly low (for example, 0.01 M or less), the enthalpy of dilution is usually very small compared to ΔH_{sol} and the solution can be thought of as infinitely dilute. Thus, whereas ΔH_{dil} is an important quantity in the analysis of many chemical reactions, it is usually of minor importance in biological reactions in which concentrations are sufficiently low that solutions are, for all practical purposes, infinitely dilute.

SEC. 4-4   THE APPLICATION OF THE ENTHALPY CONCEPT   103

Heats of Chemical Reactions

We have so far dealt with ΔH for physical processes. In biochemical systems we are mostly concerned with chemical transformations, which we now consider.

When chemicals react, the internal energy of the system changes as new bonds and new substances are formed. Unless all of the energy change is in the form of work done by or on the components (which is very unlikely), there will be a change in heat content. If a reaction occurs in a sealed vessel, there is no volume change and the heat change can be identified with the internal energy. However, *reactions are usually carried out at constant pressure so that the heat change is enthalpic*. Once again the change in enthalpy can be positive or negative—that is, the reaction may be exothermic, heat being given off (for example, in burning), or it may be endothermic.

In the discussion of enthalpy changes in phase transitions it was pointed out that *molar enthalpies are additive* and that this is a direct consequence of the fact that ΔH is independent of path. This principle (Hess's Law), which has already been stated for physical processes, is so useful in the calculation of reaction enthalpies that we restate it:

the enthalpy change of any reaction is the sum of the enthalpy changes of a series of reactions into which the overall reaction may theoretically be divided.

The most important point about Hess's Law is that the particular reactions into which the overall reaction is divided do *not* have to be on the usual reaction pathway and in fact do not have to be reactions that occur at all. For example, the synthesis of acetic acid could be written in either of the following ways:

(A) $C + 2H_2 \rightarrow CH_4$, ΔH_1;
     $C + O_2 \rightarrow CO_2$, ΔH_2;
     $CH_4 + CO_2 \rightarrow CH_3COOH$, ΔH_3.
(B) $2H_2 + O_2 \rightarrow 2H_2O$, ΔH_4;
     $2C + 2H_2O \rightarrow CH_3COOH$, ΔH_5.

In each case the enthalpy change in forming acetic acid from its elements would be the same; in case A it would be $\Delta H_A = \Delta H_1 + \Delta H_2 + \Delta H_3$; in case B it would be $\Delta H_B = \Delta H_4 + \Delta H_5$. *The important fact is that* $\Delta H_A = \Delta H_B$.

The principle behind calculating a heat of reaction is illustrated in the following example. Consider a reaction

$$A + 2B \rightarrow C + D,$$

whose heat of reaction we wish to calculate. If $x_1, x_2, \ldots, x_n$ are the elements contained in compound X, a hypothetical reaction

$$x_1 + x_2 + \cdots + x_n \rightarrow X$$

might be written; the enthalpy change ΔH_x is called the enthalpy or heat of formation of X from its elements; ΔH_{form} is also a general notation for this. We may write the following four hypothetical reactions whose heats of formation are indicated at the right of each equation for compounds A, B, C, and D:

(A) $\quad \sum a_i \rightarrow A, \quad\quad \Delta H_A$;

(B) $\quad 2 \sum b_i \rightarrow 2B, \quad\quad 2\Delta H_B$;

(C) $\quad \sum c_i \rightarrow C, \quad\quad \Delta H_C$;

(D) $\quad \sum d_i \rightarrow D, \quad\quad \Delta H_D$;

The first two reactions can be reversed and the sign of ΔH changed:

(A') $\quad A \rightarrow \sum a_i, \quad\quad -\Delta H_A$;

(B') $\quad 2B \rightarrow \sum b_i, \quad\quad -2\Delta H_B$;

Adding reactions (A'), (B'), (C), and (D) gives

$$A + 2B + \sum c_i + \sum d_i \rightarrow \sum a_i + 2 \sum b_i + C + D,$$

the heat of reaction of which is

$$\Delta H_C + \Delta H_D - \Delta H_A - 2\Delta H_B .$$

Since the equation is balanced, the sum of the elements on the left must equal the sum of the elements on the right—that is,

$$\sum c_i + \sum d_i = \sum a_i + 2 \sum b_i ,$$

so that the heat of reaction is $\Delta H_C + \Delta H_D - \Delta H_A - 2\Delta H_B$. We may generalize this by stating that

> to determine the heat of a reaction, one need only sum the heats of formation of the reactants, remembering to multiply each value of the heat of formation by the appropriate coefficient in the balanced equation.

This may also be written more formally as

$$\Delta H(\text{reaction}) = \sum p_i \Delta H_{\text{form}}(\text{products}) - \sum r_i \Delta H_{\text{form}}(\text{reactants}) , \quad (25)$$

in which p_i and r_i are the stoichiometric coefficients in the balanced equation.

It should be noted how we have added and subtracted chemical equations. Thermochemical equations obey all the laws of ordinary linear algebraic equations in several unknowns. Thus, if we know that in the oxidation of carbon at 25°C (298 K), or

$$C(\text{graphite}) + O_2(g) \rightarrow CO_2(g) ,$$

SEC. 4-4 THE APPLICATION OF THE ENTHALPY CONCEPT

$\Delta H_{298} = -393.5$ kJ mol^{-1}, then we may also state that for the decomposition of CO_2 to its elements, or

$$CO_2(g) \rightarrow C(\text{graphite}) + O_2(g) ,$$

$\Delta H_{298} = +393.5$ kJ mol^{-1}. If two moles were involved, as in

$$2CO_2(g) \rightarrow 2C(\text{graphite}) + 2O_2(g) ,$$

then $\Delta H_{298} = 2 \times 393.5 = 787$ kJ (total ΔH, not per mole).

Heats of formation are often difficult to measure directly because usually elements do not react directly to form the compound of interest or, if they do, the reaction occurs either extremely slowly or with explosive violence. For instance, the heat of formation of acetic acid could not be measured directly by trying to carry out the reaction

$$2C + 2H_2 + O_2 \rightarrow CH_3COOH ,$$

because the three elements C, H, and O would not combine in a calorimeter in a measurable time period and if they did, they would form thousands of different compounds.

It turns out that if the substance of interest burns easily, the most convenient way to evaluate ΔH_{form} is by studying the combustion reaction and calculating ΔH_{form} from the enthalpy of combustion ΔH_{comb}. How this is done is described in the following paragraphs.

In measuring a heat of combustion the substance is burned in an atmosphere of O_2 at high pressure so that the amount of O_2 never limits the reaction and so that the gases produced do not significantly alter the total pressure. The vessel is enclosed to prevent loss of the gases evolved and immersed in a large volume of water; the small temperature rise in the water measures the heat evolved. There is a subtlety here, though, because in a sealed vessel, a reaction occurs at constant volume and we have equated the heat of reaction with an enthalpy change that occurs at constant pressure. At constant volume no work is done and, by the First Law, the heat Q evolved corresponds to a change in internal energy ΔU. To calculate ΔH we use the relation $\Delta H = \Delta U + \Delta(PV)$ and assume, first, that the volume of the solid or liquid being burned is negligible compared to the volume of the gas phase and, second, that the gas behaves ideally—that is, $\Delta(PV) = RT\Delta n$, in which Δn is the change in the number of moles of gas during the reaction or the number of moles of gaseous products minus the number of moles of gaseous reactants. For simple organic compounds the gaseous products are H_2O, CO_2, and oxides of nitrogen, and the gaseous reactant is oxygen. Therefore, ΔH is calculated as $\Delta U + RT\Delta n = Q + RT\Delta n$.

As an example of this calculation consider the combustion of ethanol,

$$CH_3CH_2OH(l) + 3O_2(g) \rightarrow 2CO_2(g) + 3H_2O(l) .$$

The measured heat change at 18°C is -1370 kJ mol^{-1}. In this case there are three moles of O_2 initially and two moles of CO_2 at the end of the reaction. Thus Δn, which refers to gases only, is $2 - 3 = -1$. Thus

$$\Delta H_{291} = \Delta U + RT\Delta n = 1372 \text{ kJ mol}^{-1}.$$

Table 4-2 lists values for standard molar heats of combustion that will be used for later numerical examples and problems.

TABLE 4-2 Standard heats of combustion and heats of formation for various compounds and ions at 25°C in kJ mol^{-1}.

Substance	ΔH_{comb}	ΔH_{form}
Acetic acid (l)	−874.5	−484.1
Acetylene (g)	−1299.6	226.7
L-Alanine (s)	−1620.5	−560.5
Benzene (l)	−3267.5	49.0
Butyric acid (l)	−2163.5	−533.8
Carbon (s)	−393.5	0
Cl$^-$ (aq)	—	−167.1
Cl$_2$ (g)	—	0
Ethane (g)	−1559.8	−84.7
Ethanol (l)	−1366.8	−277.7
Ethylene (g)	−1411.0	52.3
α-D-Glucose (s)	−2803.0	−1273.0
L-Glutamic acid (s)	−2243.5	−1010.4
Glycerol (l)	−1655.4	−668.4
Glycine (s)	−973.5	−530.0
HCl (g)	—	−92.3
Hydrogen (g)	−285.8	0
H$^+$ (aq)	—	0
Methane (g)	−890.3	−74.9
Methanol (l)	−726.5	−238.7
OH$^-$ (aq)	—	−230.0
Oleic acid (l)	−11194.3	−748.5
Propane (g)	−2219.9	−103.9
Urea (s)	−631.7	−333.5
L-Valine (s)	−2919.6	−620.1
Water (l)	—	−285.8

Let us now examine how heats of combustion (which can be measured for any substance that reacts with oxygen) can be used to calculate a heat of formation. The principle will be the same as we have just seen in the addition of chemical equations. If one writes an equation for combustion of *each* reactant and of *each* product and adds these together to generate the overall reaction, the values of ΔH calculated in this way is, according to Hess's Law, the value of ΔH for the balanced equation. Thus,

the heat of any reaction equals the sum of the heats of combustion of the reactants minus the sum of the heats of combustion of the products (remember again to multiply by the appropriate stoichiometric coefficients).

SEC. 4-4   THE APPLICATION OF THE ENTHALPY CONCEPT   107

EXAMPLE 4-F   Calculation of the heat of formation of acetylene C_2H_2.

The reaction is

(A) $2C(\text{graphite}) + H_2(g) \rightarrow C_2H_2(g)$.

The combustion of each substance in the reaction and the values of ΔH_{comb} for each combustion reaction are

(B) $C(\text{graphite}) + O_2(g) \rightarrow CO_2(g)$, $\Delta H = -393.5$ kJ mol^{-1} ;
(C) $H_2(g) + \frac{1}{2}O_2(g) \rightarrow H_2O(l)$, $\Delta H = -285.8$ kJ mol^{-1} ;
(D) $C_2H_2(g) + \frac{5}{2}O_2 \rightarrow 2CO_2(g) + H_2O(l)$, $\Delta H = -1299.6$ kJ mol^{-1} .

In order to arrive at equation (A), we must multiply reaction (B) by 2 and write reaction (D) in the right-to-left direction. Thus,

$2C(\text{graphite}) + 2O_2(g) \rightarrow 2CO_2(g)$, $\Delta H = 2(-393.5)$ kJ ;
$H_2(g) + \frac{1}{2}O_2(g) \rightarrow H_2O(l)$, $\Delta H = -285.8$ kJ ;

and

$2CO_2(g) + H_2O(l) \rightarrow C_2H_2(g) + \frac{5}{2}O_2(g)$, $\Delta H = 1299.6$ kJ ;

which add to

$2C(\text{graphite}) + H_2(g) \rightarrow C_2H_2(g)$,

or

$\Delta H = \underbrace{2(-393.5)+(-285.8)}_{\text{Reactants}} - \underbrace{(-1299.6)}_{\text{Product}} = 226.8$ kJ mol^{-1} .

This process of calculating a heat of formation by decomposing the formation reaction into a suitable set of combustion reactions is sufficiently important that a second example is provided.

EXAMPLE 4-G   The synthesis of solid glycine from its elements.

The synthetic reaction is

(A) $\frac{1}{2}N_2(g) + 2C(s) + \frac{5}{2}H_2(g) + O_2(g) \rightarrow CH_2NH_2COOH(s)$, ΔH_A ,

which can be decomposed into three reactions:

(B) $2C(s) + 2O_2(g) \rightarrow 2CO_2(g)$, ΔH_B ;
(C) $\frac{5}{2}H_2(g) + \frac{5}{4}O_2(g) \rightarrow \frac{5}{2}H_2O(l)$, ΔH_C ;
(D) $\frac{1}{2}N_2(g) + \frac{5}{2}H_2O(l) + 2CO_2(g) \rightarrow CH_2NH_2COOH(s) + \frac{9}{4}O_2(g)$, ΔH_D .

Note that ΔH_B is twice the heat of combustion of carbon, and ΔH_C is the heat of combustion of H_2. Also, reaction D is the reverse of the combustion of glycine so that $\Delta H_D = -\Delta H_{\text{comb}}(\text{glycine})$. Therefore,

$\Delta H_A = \Delta H_{\text{form}}(\text{glycine}) = \Delta H_{\text{comb}}(\text{glycine}) + \Delta H_{\text{comb}}(CO_2) + \frac{3}{2}\Delta H_{\text{comb}}(H_2O)$.

Heats of Reaction in Dilute Solution

If two very dilute solutions of KCl and NaNO$_3$ are mixed, the heat of mixing is zero, because there are no interactions between any of the components. However, if AgNO$_3$ is used instead of NaNO$_3$, there is a heat change. The amount of heat evolved in the AgNO$_3$–KCl mixing is the same as if AgNO$_3$ and NaCl are mixed. The effect is entirely due to the precipitation of AgCl. If dilute NaOH is mixed with dilute HCl, 55.84 kJ mol^{-1} (13.345 kcal mol^{-1}) is evolved, a quantity of heat that is independent of both the cation of the base and the anion of the acid as long as the cation and anion do not interact. In each case, the heat change may be thought of as a result of removing ions from solution. Thus, there must also be an enthalpy change when a substance ionizes—that is, the heat of formation of an ion will be part of the heat change measured as a heat of solution.

It is not possible to measure directly the heat of formation of a single ion since one can only work with electrically neutral solutions. However, as we did in the definition of standard states, in which the enthalpy of an element is defined to be zero, the heat of formation of a convenient ion (H$^+$) in infinitely dilute solution is assigned zero value. We can then calculate the heat of formation of OH$^-$ from the reaction

$$H^+ + OH^- \rightarrow H_2O(l), \qquad \Delta H = -55.84 \text{ kJ mol}^{-1}.$$

The known value of $\Delta H_{form}(H_2O, l) = -285.8$ kJ mol^{-1} and by definition $\Delta H_{form}(H^+) = 0$, so that we may use Equation 25 to evaluate $\Delta H_{form}(OH^-)$ as follows:

$$\Delta H_{reaction} = -55.84 = \Delta H_{form}(H_2O,l) - \Delta H_{form}(H^+) - \Delta H_{form}(OH^-)$$
$$= -285.8 - 0 - \Delta H_{form}(OH^-)$$

so that

$$\Delta H_{form}(OH^-) = -229.96 \text{ kJ mol}^{-1}.$$

The heat of formation of any ion can be calculated in a similar way. Thus, to determine the heat of formation of the Cl$^-$ ion one begins with the reaction

$$HCl(g) \rightarrow H^+ + Cl^-,$$

whose heat of reaction is simply the heat of solution or 74.85 kJ mol^{-1}. Then, we use Equation 25 to obtain

$$\Delta H_{form}(H^+) + \Delta H_{form}(Cl^-) - \Delta H_{form}(HCl) = -74.85 \text{ kJ mol}^{-1};$$

substituting $\Delta H_{form}(H^+) = 0$ and $\Delta H_{form}(HCl) = -92.30$ kJ mol^{-1} (see Table 4-2), one obtains $\Delta H_{form}(Cl^-) = -167.15$ kJ mol^{-1}. In exactly the same way $\Delta H_{form}(Na^+)$ can be calculated from the heat formation of NaOH, using the value of $\Delta H_{form}(OH^-)$ calculated above.

In the calculations just presented for the heats of formation of ions it has been assumed that there is 100 percent ionization. This of course does

SEC. 4-4   THE APPLICATION OF THE ENTHALPY CONCEPT

not always occur and, if there is less ionization, it must be reflected in the heat evolved. For instance, we have just stated that 55.84 kJ mol^{-1} is evolved when equal amounts of a strong base and a strong acid are mixed. If a weak acid such as HOCl is used instead, the heat evolved is less, namely 41.84 kJ mol^{-1} in this case. Thus we may write the reaction

$$HOCl + OH^- \rightarrow H_2O + OCl^-, \qquad \Delta H = -41.84 \text{ kJ mol}^{-1}.$$

By subtracting the equation

$$H^+ + OH^- \rightarrow H_2O, \qquad \Delta H = -55.84 \text{ kJ mol}^{-1},$$

we obtain

$$HOCl \rightarrow OCl^- + H^+, \qquad \Delta H = 14.0 \text{ kJ mol}^{-1},$$

for which the heat of reaction is $-41.84 - (-55.84)$ or 14.0 kJ mol^{-1}. Since this reaction is nothing other than the ionization of HOCl, the heat of reaction is called the *heat of ionization*.

Chemical Bond Enthalpies

The formation of a chemical bond involves an enthalpy change. Since any compound consists of a collection of chemical bonds, the heat of formation of the compound should be related to the heat of formation of the bonds. Thus, the synthesis of ethane, C_2H_6, from its elements involves the formation of one C—C bond and six C—H bonds and we might write

$$2C + 6H \rightarrow C-C + 6(C-H),$$

for which the heat of reaction should be the heat of formation of ethane. Using Equation 25 to calculate the heat of reaction, we need only sum the heats of bond formation (called *bond enthalpies*) on the right and subtract the heats of formation of the elements on the left. One must be careful though, because there is a tendency to think that the values of ΔH_{form} for the elements on the left side of the reaction are all zero. However, we must remember that single atoms are rarely used to designate the standard state; for example, the standard state of gases other than the inert gases is a diatomic molecule and, of a solid element, it is a particular crystalline form. We will not go into the spectroscopic (and other) methods that are used to calculate the heats of formation of various atoms: suffice it to say that they are listed in tables such as Table 4-3. Values for bond enthalpies are also available, as in Table 4-4. Since ΔH for C—C, C—H, C, and H are -343, -415.9, -716.7 and -218 kJ mol^{-1}, respectively, the calculated heat of formation of ethane is

$$6(-415.9) + (-343) - 2(-716.7) - 6(-218) = -97.0 \text{ kJ mol}^{-1}.$$

The actual experimentally determined value is -84.7 kJ mol^{-1}. The agreement is not exact for several reasons. The main cause is that bond

TABLE 4-3 Heats of formation of various atoms and molecules at 25°C in kJ mol^{-1}.

Substance	ΔH_{form}	Substance	ΔH_{form}
Al(s)	0	I(g)	106.82
Br(g)	111.88	I$_2$(s)	0
Br$_2$(l)	0	I$_2$(g)	62.43
Br$_2$(g)	30.92	N(g)	472.71
C(graphite)	0	N$_2$(g)	0
C(diamond)	1.88	O(g)	249.15
Cl(g)	121.61	O$_2$(g)	0
Cl$_2$(g)	0	P(white)	0
F(g)	78.99	P(red)	−17.57
F$_2$(g)	0	P(g)	314.64
H(g)	1536.4	S(rhombic)	0
H$_2$(g)	0		

TABLE 4-4 Enthalpies of various chemical bonds at 25°C.

Bond	kJ mol^{-1}	Bond	kJ mol^{-1}
C—C	343	C—Cl	326
C=C	615	C—Br	272
C≡C	812	N—H	391
C—H	416	O—H	464
C—N	293	H—H	436
C=N	615	O—O	144
C≡N	819	N—N	159
C—O	351	P—O	368
C=O	724	P—H	321

enthalpies are not always the same for a given pair of atoms joined in different molecules or even in different parts of the same molecule. A second factor is that different molecules have different rotational and vibrational energies. The values of the bond enthalpies given in Table 4-4 are *average* values obtained from known heats of formation and are calculated by the process just used.

In calculations of the heats of formation of a compound, care must be taken to account for resonance or hybrid-bond character. For instance, the calculated value of ΔH_{form} for a keto form of a compound differs from that of an enol form by about −58 kJ mol^{-1}. The measured value of ΔH_{form} should reflect the ratio of the keto and enol forms. Thus, the departure of the measured from the calculated value is an indication of the equilibrium situation under standard conditions. For example, if ΔH_{form} for a keto form of a compound is −286 kJ mol^{-1} and for the enol form is −228 kJ mol^{-1}, then a mixture consisting of 75 percent keto and 25 percent enol would have a value of ΔH_{form} of $(.75)(-286) + (.25)(-228) = 271.5$ kJ mol^{-1}, which is -43.5 kJ mol^{-1} less than that of the enol form. In general, for any mixture of keto and enol forms, the fraction, f of ΔH_form that is in the keto form can be calculated from the measured value, ΔH_form (measured), by solving the equation

$$f\Delta H_\text{form}(\text{keto}) + (1 - f)\,\Delta H_\text{form}(\text{enol}) = \Delta H_\text{form}(\text{measured}) \,.$$

Temperature Dependence of Heats of Reaction

Tables of heats of various processes usually list values for 18°, 19°, 24°, or 25°C. It is frequently necessary to have values at other temperatures; for example, 30° and 37°C are the usual temperatures encountered in biochemistry.

The temperature dependence of ΔH is obtained as follows: consider a system initially having enthalpy H_1 at a temperature T_1; at a final temperature T_2, the enthalpy is H_2. The enthalpy change is $\Delta H = H_2 - H_1$. Differentiating this equation with respect to temperature but at constant pressure yields

$$\left(\frac{\partial(\Delta H)}{\partial T}\right)_P = \left(\frac{\partial H_2}{\partial T}\right)_P - \left(\frac{\partial H_1}{\partial T}\right)_P = C_{P,2} - C_{P,1} = \Delta C_P \,, \tag{26}$$

in which $C_{P,1}$ and $C_{P,2}$ are the heat capacities of the system at the initial and final state. At constant pressure $d(\Delta H) = \Delta C_P\, dT$, or

$$\Delta H(T_2) - \Delta H(T_1) = \int_{T_1}^{T_2} \Delta C_P\, dT \,. \tag{27}$$

For small differences in temperature, ΔC_P can be considered to be constant. Otherwise an empirical equation of the form

$$C_P = a + bT + cT^2 + \cdots \tag{28}$$

is used in which $a, b, c, \ldots,$ are experimentally determined constants for a particular substance. The values of $a, b, c, \ldots,$ are found in standard reference tables.

Calorimetry

Throughout this chapter we have referred to values of heats of various processes. These values are obtained experimentally by carrying out the process in a sealed chamber in which the heat change in the process is monitored, usually by measuring a temperature change. This technique is called *calorimetry*. The precise measurement of heat changes is quite painstaking and requires careful attention to detail.

The most common type of calorimetry uses an adiabatic system. This arrangement has the theoretical advantage that the total heat change can be readily measured because no heat is exchanged with the environment; in practice, of course, no system is perfectly adiabatic, so the experimenter must design the apparatus to minimize heat exchange. Figure 4-5 shows a

FIGURE 4-5 Schematic diagram of a bomb calorimeter. The entire apparatus is immersed in a water bath whose temperature is continually adjusted to equal that of the inner water volume shown.

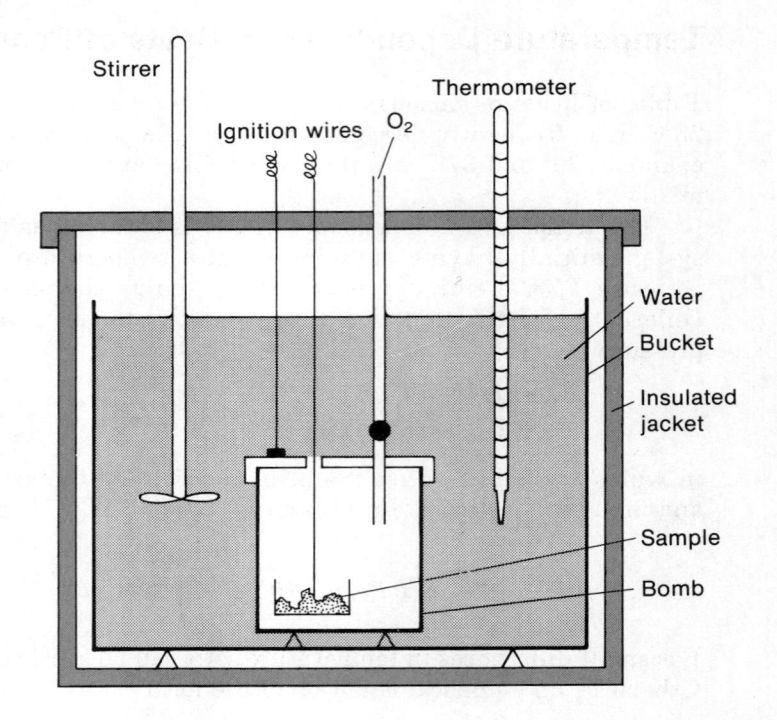

simple experimental set-up—a bomb calorimeter—used to determine the heat of combustion. The inner chamber or "bomb" contains a known mass of the sample to be burned. Pure oxygen is allowed to fill the chamber and the sample is ignited by a very brief electrical discharge. As the substance burns, heat is released. The bomb is immersed in a known volume of water, which is, in turn, in a container that is sufficiently insulated to simulate adiabatic conditions. The released heat causes the temperature of the water to rise. The amount of heat released equals the value of C_V of water multiplied by the temperature increase. Applying the First Law to the contents of the bomb, $\Delta U = Q_V + W = Q_V - P\Delta V$, which equals Q_V because V is constant. Thus, a simple measurement of the temperature change in the water gives the value of ΔU. Since the system is adiabatic, all of the heat evolved is, in theory, seen as a temperature increase of the water. Usually the temperature changes that are detected are so small—1–2°—that the small amount of leakage of heat through the insulating jacket that inevitably occurs causes a problem. For precise work, this leakage can be minimized by a simple trick. The entire jacketed system is immersed in a second

SEC. 4-4   THE APPLICATION OF THE ENTHALPY CONCEPT

volume of water, whose temperature is initially (before combustion) set to equal the temperature of the water inside the jacket. A heater is placed in the external water that is regulated by a sensing device that measures the difference in temperature between the two volumes of water. As heat is released during the combustion and the temperature of the internal water increases, the external water bath is heated to maintain a temperature difference of zero. With the temperatures equivalent, no heat can leak out of the inner volume of water.

The technique just described yields ΔU_{comb}. However, more commonly one wants the value of ΔH_{comb}. This is calculated in the standard way from Equation 18—$\Delta H = \Delta U + \Delta(PV)$. In any reaction in which all reactants and products are either solids or liquids, $\Delta(PV)$ is negligible compared to ΔH or ΔU and the term can be ignored—that is, ΔH and ΔU can be considered to be equal. However, in a combustion reaction, gases are involved and the term must be considered. Since the measurement is usually done near a pressure of 1 atm and the temperature changes are quite small, it is not unreasonable to assume that the gases are ideal; thus $\Delta(PV) = \Delta(nRT) = RT\Delta n$, in which Δn is the change in the number of moles of gas in the reaction. Thus, ΔH is calculated as $\Delta H = \Delta U + RT\Delta n$, using the measured value of ΔU.

EXAMPLE 4–H   **Determination of ΔU_{comb} and ΔH_{comb} for phenol.**

The reaction is

$$C_6H_5OH\ (s) + 7O_2\ (l) \rightarrow 6CO_2\ (g) + 3H_2O\ (l)\ .$$

A sample of 0.3214 g (= 0.00342 moles) of phenol (C_6H_5OH) was burned in a constant-volume bomb calorimeter. The temperature of the inner volume of water rose from 23.27°C to 25.25°C, so the temperature increase was 1.98°. The effective heat capacity of the calorimeter plus the water is 5268.2 J K^{-1}, so the heat evolved was $(1.98)(-5268.2) = -10.43$ kJ, or $\Delta U = -10.43/.00342 \doteq -3050$ kJ mol^{-1}. To calculate ΔH, note that the number of moles of gas changes from 7 (for O_2) to 6 (for CO_2). Thus, $\Delta n = -1$ and $\Delta H = \Delta U + (-1)RT$. The temperature needed is the initial temperature since products and reactants are to be compared in the same conditions. Thus, $\Delta H = -3050 + (-1)R(296.4) = -3052$ kJ mol^{-1}.

Other thermodynamic quantities may be measured using calorimeters of similar, but not identical, design. For example, to measure a heat of solution, the solute is contained in the calorimeter and the solvent, which is initially at the same temperature as the inner volume of water, is injected into the calorimeter. As the solute dissolves, temperature changes occur that are reflected in the temperature of the water jacket. With such an arrangement, one must consider the heat capacity of the solvent, when relating the amount of heat released or absorbed to the temperature change.

A variation of these techniques is needed to measure quantities such as the heat capacity, in which the parameter that is experimentally varied is the heat input. For example, to obtain a value for C_V one needs to add a known amount of heat (using conditions in which no work is done) and measure the temperature change that results. In such experiments the heat that is added is usually in the form of a known electrical current i flowing through a resistor with resistance R for a particular period of time t. The total amount of heat added is i^2Rt. A sample whose heat capacity is to be determined is placed in an insulated container whose temperature can be monitored. A known amount of (electrical) heat Q is passed through the sample and the value of ΔT is measured. The heat capacity is simply $Q/\Delta T$.

Special techniques are needed for studying biochemical processes because usually only very small quantities of material are available or a biochemical reaction (such as an enzymatic reaction) can only be carried out at low concentration; with the conditions used the temperature change is usually quite small. Such measurements are usually performed by *differential thermal analysis*. Two small calorimeters, both mounted in a single heat sink such as a block of metal, are used: one contains the sample in solution and the other, which serves as a reference, contains a solution that is complete except that the molecule of interest is absent. The properties of the two solutions are compared. The heat capacity can be measured with such an arrangement in a straightforward way: a particular amount of heat Q is supplied to both containers and the temperature of each solution is measured. The final temperature of the reference is subtracted from the final temperature of the sample to yield the temperature change ΔT caused by the molecules of interest. Again, $C_V = Q/\Delta T$. A modification of this technique—*differential scanning calorimetry*—is used to determine enthalpies of transitions of biological molecules. Again, two calorimeters are used, one containing a reference solution and the other containing the sample of interest. Heat is added continuously to both calorimeters and the temperatures are measured. A temperature regulator operates to maintain the temperatures of both calorimeters the same (to within 0.01°C) and the heat (or electrical power, as is usually the case) required to accomplish this equivalence is measured. The amount of power transferred to the reference is electronically subtracted from the amount supplied to the experimental sample, and the difference is plotted as a function of temperature. Not only can transition enthalpies be measured, but often transitions that were hitherto unknown are detected by discontinuities in the plot of ΔH versus T. Observation of such transitions gives important information about the presence of intermediate stages and intermediate compounds in biochemical processes. See Section 15-2 for applications of this technique to macromolecules.

REFERENCES

Abbott, M.M., and H.C. Van Ness. 1972. Schaum's Outline Series. *Theory and Problems of Thermodynamics*. McGraw-Hill.

Denbigh, K.G. 1971. *The Principles of Chemical Equilibrium*. Cambridge University Press. Cambridge, England.

Klotz, I.M. 1967. *Energy Changes in Biochemical Reactions*. Academic Press.
Klotz, I.M., and R.M. Rosenberg. 1972. *Chemical Thermodynamics*. Wiley.
Lehninger, A.L. 1971. *Bioenergetics*. W.A. Benjamin.
Schmidt-Nielsen, K. 1972. *How Animals Work*. Cambridge University Press. Cambridge, England.
Wall, F.T. 1974. *Chemical Thermodynamics*. W.H. Freeman.

PROBLEMS

1. In a particular process 3.5×10^9 erg are released. How many joules and how many calories are released?

2. What is the temperature when 100 g of water at 80°C is mixed with 50 g of water at 25°C, if there is no heat loss?

3. (a) What is the final temperature, if 50 g of ice at 0°C is mixed with 150 g of water at 0°C?

   (b) What is the final temperature, if 10 g of ice at 0°C is mixed with 50 g of water at 40°C?

   (c) Repeat part (b) with 100 g of ice.

4. How far do you have to drop a 1-kg weight so that it would raise the temperature of 1 liter of water into which it falls by 1°C?

5. What is the sign of Q in each of the following processes?

   (a) Evaporation of acetone.

   (b) Freezing of water.

   (c) Rise in temperature of water when concentrated (15 M) sulfuric acid is added.

   (d) Decrease in temperature of the solution when CsCl is dissolved in water.

6. One mole of a gas at 25°C expands reversibly at constant temperature from 10 atm to 1 atm.

   (a) How much work is done by the gas?

   (b) How much heat is evolved in the process?

7. One mole of a solid at a pressure of 1 atm melts and gives off 100 J. The densities of the solid and liquid are 2.09 and 2.00 g cm^{-3}, respectively. Calculate Q, W, ΔU, and ΔH. The molecular weight of the substance is 150.

8. A gas has the equation of state $PV = nRT + aPT - bP$, in which a and b are constants. One mole of such a gas expands reversibly and isothermally from V_1 to V_2. Derive an expression for the work done.

9. Ten grams of liquid water are vaporized at 100°C against a constant external pressure of 1 atm.

   (a) Calculate the work done, in calories.

   (b) Calculate the percent of the heat that vaporized the water and that was used to produce work, in calories. Use the value $\Delta H_{vap} = 40.58$ kJ mol^{-1}. The density of liquid water is 1 g cm^{-3}.

10. One mole of a monatomic ideal gas is compressed adiabatically by a 50-kg weight placed on a piston. The piston moves down 3.5 cm.

    (a) What is the temperature change of the gas?

    (b) What would the temperature change be if it were an ideal diatomic gas?

11. One mole of helium at 25°C and 1 atm pressure expands adiabatically until the pressure is 0.1 atm; the temperature is then 119 K. $C_V = 12.6$ J K^{-1} mol^{-1}.

    (a) How much heat is evolved in the expansion?

    (b) Evaluate the amount of work performed in the process.

12. Give a molecular explanation for the increase in temperature accompanying an adiabatic compression.

13. The definition of heat capacity at constant pressure, C_P, is $C_P = (\partial H/\partial T)_P$. Under certain circumstances one can write this

as $dH = C_P dT$. What are these circumstances?

14. Give a molecular explanation for the difference in the values of C_V and C_P. That is, why should constant pressure or constant volume affect the value of ΔT when a given amount of heat is absorbed?

15. The heat capacity of water is greater than that of steam and of ice. Explain.

16. What will be the final volume of one mole of an ideal gas initially at 0°C and 1 atm if 2000 J is given off to a heat reservoir during an isothermal expansion against a constant external pressure of 1 atm?

17. Three moles of an ideal gas for which $C_V = 12.53$ J K^{-1} mol^{-1} is heated from 25°C to 35°C in a sealed 50-liter container.

   (a) Calculate ΔU and ΔH.

   (b) Explain what ΔU is in molecular terms.

18. What is ΔH_{form} for gold at 25°C?

19. How much heat is required to convert 100 of ice at 0°C to steam at 100°C? Use the following values: $\Delta H_{fus} = 6.025$ kJ mol^{-1}; $C_V = 75.3$ J K^{-1} mol^{-1}; $\Delta H_{vap} = 40.58$ kJ mol^{-1}.

20. When 0.4 g of S is burned in a calorimeter in contact with 50 g of water, the temperature of the water increases by 17.8°C. What is the heat of combustion of S?

21. What is ΔH_{react} for each of the following reactions?

   (a) Burning of diamond to CO_2.

   (b) Hydrogenation of ethylene to yield ethane.

   (c) Oxidation of ethane to ethanol.

22. Acetone can be oxidized to acetic acid according to the reaction

   $$CH_3COCH_3 + 2O_2 \rightarrow CH_3COOH + CO_2 + H_2O$$

   At 25°C ΔH_{comb}(acetone) = -1787 kJ mol^{-1}, ΔH_{comb}(acetic acid) = -874.5 kJ mol^{-1}, and $\Delta H_{vap}(H_2O) = 44.01$ kJ mol^{-1}. What is the standard heat of reaction for the reaction?

23. At 20°C, ΔU for the combustion of one mole of acetylene is -1303 kJ and for one mole of benzene it is -3274 kJ. What is ΔH of formation for the synthesis of benzene from acetylene?

24. 10 g of malonic acid, $CH_2(COOH)_2$, is burned in O_2 in a calorimeter. At 25°C 83.8 kJ are evolved. What is ΔH_{comb} at constant pressure and at 25°C?

25. ΔU for the reaction $2A_2(s) + 5B_2(g) \rightarrow 2A_2B_5(g)$ is 63 kJ mol^{-1} at 25°C. What is ΔH at the same temperature?

26. Following are some data obtained for several liquids.

Substance	ΔH_{vap}, kJ mol^{-1}	Boiling temperature, T_b, in K
1	14.35	286
2	27.15	302
3	33.47	323
4	34.52	387
5	36.02	420
6	44.35	440

What can you say about intermolecular forces in these liquids?

27. When exercising, the human body generates heat. During strenuous exercise a typical value is 1000 joules per second (= 1000 watts). The body regulates its temperature by perspiring and vaporizing the perspiration. After such exercise one is usually thirsty, because the lost water must be replenished. How many grams of water must be drunk after engaging in such exercise for five minutes?

28. As a general rule, a moderately active male adult needs a number of nutritional Calories per day equal to 18 times his weight in pounds. The heat of combustion of a typical fatty acid is about 7 kcal g^{-1}. If a 150-pound man eats 2000 Calories per day, how long will it take him to lose 10

pounds? Note that the nutritional Calorie equals one ordinary kilocalorie. Assume that all weight loss is due to fat loss.

29. When the prevailing west wind in the northern hemisphere encounters a mountain, it is forced to rise. This occurs so rapidly that the change is nearly adiabatic.

(a) Does the air temperature rise or fall?

(b) The temperature change is smaller for moist air than for dry air. Explain.

(c) Once the air reaches the peaks of the mountains, it rapidly pours down the other side, causing a wind known in the Rocky Mountains as a *chinook* and in the Alps as a *foehn*. Does this wind on the lee side of the mountain make you want to put on a heavy coat or look for a swimming pool?

(d) If the mountain were perfectly symmetric, would the temperature be the same at the bottom of both sides of the mountain?

CHAPTER 5

THE SECOND LAW OF THERMODYNAMICS

The First Law of Thermodynamics states that the energy content of a system that is totally isolated from its surroundings is constant. Since the system may be the universe itself, the First Law implies that the energy of the universe is constant. The universe is not a continuous uniform substance, however, and thus may be thought of as a collection of lesser systems, each of which may have a different energy content. When systems are in contact, energy may flow from one to another either by means of the flow of heat or by one system performing work on the other. Unless two interacting systems are at equilibrium, energy flow is spontaneous; furthermore, it is always the case that only one of the two systems can be the energy donor, the other being the energy recipient. The Second Law of Thermodynamics is a generalization based upon numerous observations of the direction of flow; it is more abstract and more difficult to comprehend than the First Law but of enormous value because it provides the ultimate criterion for predicting the direction of all physical and chemical processes.

5–1 THE SECOND LAW

Basis of the Second Law in Experience

In our experience are many examples of changes in which the energy content of a system is either reduced or flows from one system having a higher internal energy to another having lower internal energy. For example, an object that is denser than air falls toward the Earth and comes to rest. In so doing, the object loses its potential energy, which is first converted to kinetic energy and ultimately to heat that is lost to the environment (that is, to the Earth and the atmosphere). It is never observed that an object resting on the ground suddenly becomes hotter than the Earth directly through the Earth's cooling and rises into the air.

Another example is that an object hotter than its surroundings spontaneously loses heat to the environment until both the object and its surroundings have the same temperature. That is, thermal equilibrium is reached by a flow of heat from a hot object to a cold object. The reverse, spontaneous transfer of heat from a cool object to a warm object, does not occur.

A third example is the movement of a piece of unmagnetized iron toward a magnet; it never moves away. The potential energy of interaction of the iron and the magnetic field is converted to kinetic energy and ultimately to heat if the iron collides inelastically with the magnet.

The two examples in which there is heat flow might lead to a conclusion that the internal energy of a system is reduced in any process that occurs spontaneously. However, this conclusion is false; a spontaneous process need not suffer a significant loss of internal energy. For example, a stretched rubber band (that is not overstretched and near the breaking point) snaps back spontaneously when released, yet no heat is transferred and no work is done, so that $\Delta U = 0$. Furthermore, the stretched rubber band has for all practical purposes the same probability (that is, a probability of nearly one) of contracting to the unstretched state as an object does of falling to the ground from a great height, in which case ΔU is large. This and many other examples clearly indicate that the sign of ΔU is not a criterion of spontaneity. This is not a new conclusion since we have already seen many examples in Chapter 4 of spontaneous chemical reactions in which ΔU or ΔH is positive, as well as examples in which the values are negative. Thus, we are forced to conclude that neither the evolution nor consumption of heat (nor an increase or decrease of temperature) determines whether a particular change in state occurs spontaneously; another criterion must be sought.

Spontaneity, Probability, and Order

In the foregoing discussion, the words *spontaneity* and *probability* were used. Everyone has an intuitive understanding of these words, but what is really meant by them? When a process is said to be spontaneous, we mean that it occurs with *very high probability* but not that it *always* occurs. For example, consider a closed vessel containing a gas. The molecules of the gas move in random directions and continually collide with one another. The direction in which one molecule moves after a collision with a second molecule is determined by chance; that is, the direction depends upon the angle of approach between the two molecules and their relative speeds and these are determined by chance because the molecules are moving in random directions. For any population of molecules, there is a certain probability that as a result of all of the intermolecular collisions and the collisions with the walls of the vessel, *all* of the molecules will at a particular time be moving in the *same* direction and all of the gas molecules will suddenly *all* be at the *same* end of the container. We do not need sophisticated mathematics to be sure that this is highly improbable; one would never say that such an event occurs spontaneously.

Consider now the probability if the vessel contained only two gas molecules instead of a large population; in this case it would not be so improbable to find both of the molecules frequently in one small region. In this example we see an important property of spontaneity—namely, that it is not only related to probability; whether or not a process occurs spontaneously may depend also upon the number of molecules in a system. This means that a rule describing the behavior of the bulk properties of matter may not apply if the volume of a system is so small that it contains only a small number of molecules. (When considering such a tiny volume it is customary to say that we are dealing with events on a molecular or atomic scale.)

Let us now examine a second example—the flow of heat from a hot object to a cooler one. Consider a container divided into two parts by a very thin, thermally conducting partition. On one side of the partition is a hot gas and on the other side is a cold gas. Experience tells us that the hot gas will spontaneously cool, giving heat to the cooler gas until the gases on both sides of the partition are at the same temperature. Remember that the temperature of a gas is a reflection of the average kinetic energy of the molecules in the gas, so that the molecules in the hotter gas have a higher average kinetic energy than those of the cooler gas. However, if we assume that the hot and cold gases are each at equilibrium internally before being placed in contact, then the energy distribution of each gas is a Boltzmann distribution (Chapter 2). This means that the kinetic energy has a large range of values; there are some molecules in the hot gas that have the same kinetic energy as some in the cold gas, although on the average the hot gas has a greater fraction of higher-energy particles than the cold gas. Heat exchange occurs between the two gases by collision of the molecules of each gas with the partition.

The molecular events that occur in this process of heat transfer are the following. A molecule (#1) having a high kinetic energy collides with a molecule (#2) in the partition, thus increasing the kinetic energy of this molecule significantly. The partition molecule (#2) then may have sufficient kinetic energy that when a molecule (#3) of the cold gas collides with the partition, the kinetic energy of the cold gas molecule will increase. It is possible of course that an especially energetic molecule in the cold gas can transfer its energy (via a molecule in the partition) to a less energetic molecule in the hot gas; however, the temperature of the hot gas will, *on the average,* fall because, according to the Boltzmann distribution, there are so many more highly energetic molecules in the hot gas than in the cold. Notice that a rise in the temperature of the hot gas and a cooling of the cold gas would be highly improbable: if the vessel contained only a few molecules, the Boltzmann distribution would not be meaningful; in such a sparse population collisions might occur such that one cold gas molecule having more energy than any of the individual molecules in the hot gas could transfer energy from the cooler to the warmer gas. We should conclude from this example that the spontaneous loss of heat from a hot body to a cold body is an overwhelmingly probable event when the number of molecules is large, but it might not occur if the number of molecules is very small. *High probability is reserved for a large population of molecules.*

The mixing of substances illustrates one other property of spontaneous processes. Consider an evacuated vessel at the opposite ends of which are two entrance ports. One of those ports delivers gas A and the other delivers gas B. Both gases are at the same temperature. As the gases enter the vessel, each will expand to fill the volume of the vessel and complete mixing will occur. It seems inconceivable that gas A should remain at one end of the vessel and gas B at the other because of the random motion of the individual molecules. What we mean by "inconceivable" is that the gases mix because mixing is the more probable event. Once mixed, it is also highly improbable that at any time the gases would again separate. This tendency to mix and remain mixed is also evident when, for example, sucrose dissolves in water. Assuming that the time is sufficient for complete dissolving and for diffusion to occur, the sucrose molecules will, even if there is no agitation of the liquid, ultimately be randomly distributed throughout the solution—at no time would it be very probable that all of the sucrose molecules would move to a single small region of the solution. Thus, a property of spontaneous processes that involve a large number of molecules is that *there is a tendency toward randomization or a loss of order.* The positions of the sucrose molecules are randomly distributed throughout the solution and the ordered arrangement of solid sucrose plus sucrose-free liquid water has been lost. Likewise, through thermal equilibration of the hot and cold gas, order is also lost; that is, prior to mixing, the energy of each gas is described by a particular Boltzmann distribution and these distributions are not the same because the two gases are initially at different temperatures. However, when temperature equilibrium is reached, this distinction is lost and randomization of kinetic energies occurs, resulting in a single Boltzmann distribution.

The tendency toward randomization of the energies and positions of the individual particles in a large population is the essence of the Second Law of Thermodynamics. We may then generalize from a large number of observations and state the Second Law in a rather uncommon form:

an isolated system tends to become more disordered in time, and the production or maintenance of order requires that work be done on the system.

This statement of the Second Law is a valuable one to have when considering biochemical and biological processes. This is because biological systems are highly ordered and thus could not arise spontaneously—both the creation and maintenance of this order is accomplished by an enormous amount of chemical work.

This statement of the Second Law does not provide a quantitative basis for discussing spontaneity. There are two approaches to obtaining this criterion. In one, the statistical mechanical approach, a procedure is developed for calculating the probability of a state. Except for a few very simple cases that will be discussed later, this is beyond the scope of this book. In the thermodynamic approach a new state function, called *entropy,* is defined, and it is argued that entropy must increase in any spontaneous process. Entropy is a complex concept, especially because it can be defined both as a logarithmic function of probability and as a ratio of transferred heat to temperature. The justification for these two definitions, which are discussed in detail later, is that they make accurate predictions. However, the definition of entropy and the criterion for spontaneity often seem very arbitrary.

In the following sections we examine both probability and the interchange of work and heat in an effort to make the definition of entropy seem reasonable. We begin with a historical and somewhat abstract discussion of a particular type of heat engine, which can be proved to be the engine having maximal efficiency. Calculating the efficiency of this engine, we discover that 100 percent efficiency cannot in theory be achieved. This phenomenon enables us to define entropy as a convenient thermodynamic state function; furthermore, it is useful to examine heat engines in an effort to understand the problems of developing more efficient ways to derive work from heat, problems that are of current interest in our energy-conscious society.

The Interchange of Heat and Work

In this section we examine the interconversion of heat and work. We know from experience that the energy of heat can be used to do work and that work can result in the production of heat. The First Law tells us something about this conversion—namely, that if the internal energy of a system is unchanged in some process, the heat change is numerically equivalent to the work done.

A simple machine can carry out this conversion. For example, a cylinder filled with hot gas and fitted with a piston can do work if the external pressure is less than the pressure in the cylinder. However, it can only do work once—that is, the piston will spontaneously be pushed out by the gas when the piston is not restrained, but work must be done to push the piston in again before the piston can do work a second time. Thus, to do work repeatedly, whether it is mechanical work or chemical work, the process must be cyclic—that is, the state of readiness must be restored at the end of each cycle. In this section we will explore the conditions required to do net work—that is, we will ask how the amount of work that is produced is related to the amount of work required to restore the state of readiness. In examining a cyclic process in which heat and work are interchanged, several new features of the universe, and a new state function, will become evident.

In a general process (and especially so in a cyclic process), the heat change Q has two components—the heat added to the system, and the heat given off by the system. Similarly the work is a combination of that done on the system and that done by the system. If we consider an engine as an example of a cyclic process, the heat added is the energy supply (the fuel) for the process. In a perfect engine all of the heat added would be converted to work with none of the heat wasted; such a perfect engine would be 100 percent efficient. An imperfect (or real) engine will fail to convert all of the added heat to useful work, so that its efficiency, which we define as the ratio of the work done by the system to the heat added to the system, will be less than 100 percent. The question to be answered is whether in a cyclic process 100 percent efficiency can ever be achieved. This question is most easily investigated by considering a cycle of expansion and compression of an ideal gas. We begin by examining a noncyclic process that is 100 percent efficient.

The work done in a *reversible* isothermal expansion of a gas is the maximal work that can be done by the gas. The idea of reversibility requires careful definition. By a reversible expansion is meant one that proceeds so slowly (actually, infinitely slowly) that there is no concern either about turbulence (and hence internal friction) or about inhomogeneity of temperature and pressure in the gas. In such an expansion, heat is slowly taken up by the system from its environment, and the expanding gas can push a piston and therefore perform work.

Because the process proceeds infinitely slowly and there is no temperature change, the molecules of the gas possess the same kinetic energy at all times, and $\Delta U = 0$ for the system. Therefore, by the First Law the heat absorbed equals the work done. Clearly this is the maximal work that can be done for that amount of absorbed heat and the efficiency of conversion of heat to work is 100 percent in a single, reversible isothermal expansion.* This is only true of an ideal gas.

---

*We can also prove that the work done in such an expansion is maximal even without resorting directly to the First Law. That is, in order that an expansion be reversible, it must occur against a force that is opposite in direction and only infinitesimally different in magnitude from the driving force. The opposing force must have only infinitesimally different magnitude because if the force were any greater than this, it would compress the first force instead of allowing it to expand and if it were less, the expansion would be too fast to be reversible.

In Chapter 4, Equation 5, it was shown that the work done in a reversible isothermal expansion of an ideal gas is $RT \ln(V_2/V_1)$, in which T is the temperature and V_1 and V_2 are the initial and final volumes. Less work will be done in a real gas than in an ideal gas because the cohesive forces between the molecules of a real gas will lessen the expansion and hence reduce the effect of a particular input of energy. Thus, in a reversible isothermal expansion, maximal work is done by an ideal gas.

Cyclic Processes

A gas that has just expanded cannot do work a second time by an isothermal expansion, unless it is again compressed. To avoid waste, the compression must also be done isothermally, because by the argument just given this requires minimal work. However, in this case the work that will be done on the gas is exactly equal to the work originally done by the gas; therefore, no net work is done in a single cycle of an isothermal expansion and compression, and all that is accomplished in such a cycle is to move heat first into and then out of the heat reservoir used to maintain constant temperature. This cyclic process is clearly not doing any useful work at all.

In order for a cyclic process to perform net work, it is essential that the work done on the system to restore the initial state be *less* than that done by the system in the work-producing step. Work done by the gas during isothermal expansion at the temperature $T_{\exp}$ is $W_{\exp} = RT_{\exp} \ln(V_2/V_1)$. Similarly the work done on the gas to compress it isothermally at the temperature T_{comp} is $W_{\text{comp}} = -RT_{\text{comp}} \ln(V_1/V_2) = RT_{\text{comp}} \ln(V_2/V_1)$. Clearly the only way that W_{comp} can be less than $W_{\exp}$ is if $T_{\text{comp}} < T_{\exp}$. That is, the temperature must be lowered during the compression.* In order to lower the temperature something must be done to remove heat.

At this point our development of the Second Law requires that we invoke our experience. That is, heat flows spontaneously only from high to low temperature and never in the reverse direction. Thus, an isothermal *expansion* can occur *only* if the heat reservoir[†] is at a *higher* temperature than the gas. Similarly, in an isothermal *compression* the heat reservoir must be at a *lower* temperature than the gas. Clearly we could lower the temperature *after* the compression by putting the gas in contact with a low-temperature reservoir but the heat being transferred would be wasted because it is not being used to do work.

A more efficient procedure would be to lower the temperature in such a

---

*It is reasonable to inquire why we are proposing an isothermal compression rather than some other means of compression. Suffice it to say that the cycle we are describing has been proved to be the most efficient cycle and that it uses an isothermal expansion and an isothermal compression step.

†By a heat reservoir one means a large mass of matter that is capable of giving up or taking up heat without a significant change in temperature. In practice heat reservoirs are usually large water baths because water has both a large heat capacity and a high thermal conductivity.

FIGURE 5-1 A *P-V* diagram for the Carnot cycle, consisting of four steps: 1, isothermal expansion; 2, adiabatic expansion; 3, isothermal compression; 4, adiabatic compression. One can imagine that the piston assembly is placed in contact with a high temperature reservoir for step 1, a low temperature reservoir for step 3, and is thermally insulated for steps 2 and 4.

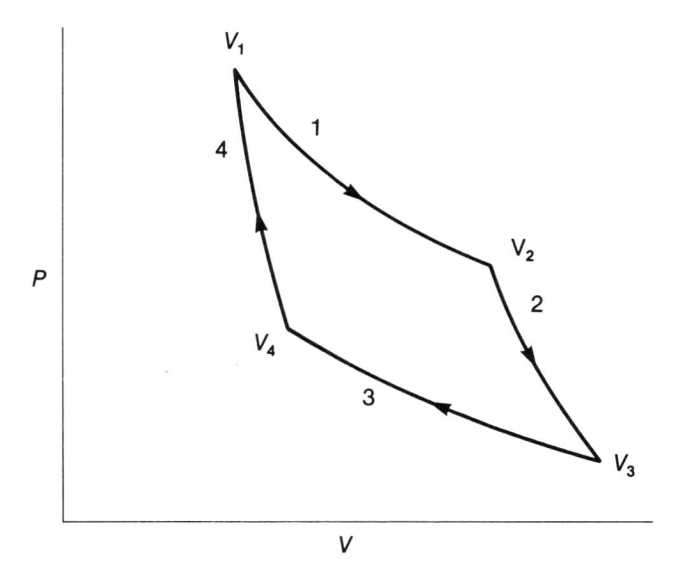

way that all of the heat energy is retained. The way to do this is to allow the gas that has just undergone the isothermal expansion to undergo an adiabatic expansion; that is, an expansion in which $Q = 0$. In such a process the work done by an ideal gas is $C_V(T_2 - T_1)$, in which T_2 is the lower temperature reached during the adiabatic expansion and C_V is the heat capacity at constant volume. Having done this, one can carry out the required isothermal compression. However, now we need a second heat reservoir at a temperature lower than that of the reservoir used during the expansion and even lower than T_2 because, during the isothermal compression, the gas must give off heat and heat can be transferred only to a cooler reservoir. The work done in this isothermal compression is now $RT_2 \ln(V_4/V_3)$. Now to complete the cycle we must raise the temperature back to T_1. Again, this is best done by an adiabatic compression in which the work done on the gas is $C_V(T_1 - T_2)$.

This process of alternating isothermal and adiabatic steps is called the *Carnot cycle* (Figure 5-1)*; it can be proved that the Carnot cycle is the most efficient way for a gas to perform work. The relevant values of ΔU, W, and Q are summarized below for the four steps of the cycle:

Step 1. Isothermal expansion:

$$\Delta U_1 = 0; \qquad Q_1 = W_1 = RT_1 \ln\left(\frac{V_2}{V_1}\right). \qquad (1)$$

Step 2. Adiabatic expansion:

$$\Delta U_2 = W_2 = C_V(T_1 - T_2); \qquad Q_2 = 0. \qquad (2)$$

---

*This is named after a French engineer, Nicolas Léonard Sadi Carnot, who first investigated its properties.

Step 3. Isothermal compression:

$$\Delta U_3 = 0; \qquad W_3 = Q_3 = RT_2 \ln\left(\frac{V_4}{V_3}\right). \qquad (3)$$

Step 4. Adiabatic compression:

$$\Delta U_4 = -W_4 = C_V(T_1 - T_2); \qquad Q_4 = 0. \qquad (4)$$

The total work done is $W_{\text{tot}} = W_1 + W_2 + W_3 + W_4$, and since $W_2 = -W_4$,

$$W_{\text{tot}} = +RT_1 \ln\left(\frac{V_2}{V_1}\right) + RT_2 \ln\left(\frac{V_4}{V_3}\right). \qquad (5)$$

To complete the analysis we need a relation between V_1, V_2, V_3, and V_4. This can be found by examining the two adiabatic changes V_2 to V_3 and V_4 to V_1. In an adiabatic ($Q = 0$) volume-change caused by an external pressure P, the First Law states that $dU = -dW = -P\,dV$. Then from the definition of C_V and the ideal gas law* we have

$$C_V dT = -RT \frac{dV}{V}. \qquad (6)$$

If at temperatures T_a and T_b, the volumes are V_a and V_b, respectively, Equation 6 can be written in integral form as

$$\int_{T_a}^{T_b} \frac{dT}{T} = -\frac{R}{C_V} \int_{V_a}^{V_b} \frac{dV}{V}. \qquad (7)$$

Evaluation of the integrals yields

$$\ln\left(\frac{T_b}{T_a}\right) = -\frac{R}{C_V} \ln\left(\frac{V_b}{V_a}\right), \qquad (8)$$

which provides the necessary relation between the temperatures and volumes in the adiabatic steps. In the two adiabatic steps of the Carnot cycle the change from T_1 to T_2 was accompanied by the change V_2 to V_3 and that from T_2 to T_1 by V_4 to V_1, so that Equation 8 can be written as

$$\ln\left(\frac{T_2}{T_1}\right) = -\frac{R}{C_V}\ln\left(\frac{V_3}{V_2}\right) \quad \text{and} \quad \ln\left(\frac{T_1}{T_2}\right) = -\frac{R}{C_V}\ln\left(\frac{V_1}{V_4}\right). \qquad (9)$$

Therefore

$$\frac{V_3}{V_2} = \frac{V_4}{V_1} \quad \text{or} \quad \ln\left(\frac{V_2}{V_1}\right) = -\ln\left(\frac{V_4}{V_3}\right), \qquad (10)$$

and Equation 5 becomes

$$W_{\text{tot}} = R(T_1 - T_2)\ln\left(\frac{V_2}{V_1}\right). \qquad (11)$$

---

*We are justified in using the ideal gas law because, as was said earlier, an ideal gas is the most efficient gas for performing work, and we are attempting to design a cycle that has maximal efficiency.

SEC. 5–1 THE SECOND LAW

The important conclusion of Equation 11 is that *the work done by the system depends upon the temperature difference between the reservoirs* and, as we saw before, *no work is done with only a single heat reservoir* (that is, if $T_1 = T_2$, then $W_{\text{tot}} = 0$).

The efficiency of the process is defined as the ratio of the work done by the system to the heat absorbed by the system:

$$\text{Eff} = \frac{\text{Work done by the system}}{\text{Heat absorbed by the system}}.$$

Let us formulate this in mathematical terms. Note that in the Carnot cycle, only in steps 1 and 3 is there a heat exchange. In step 1 the heat added to the system is $RT \ln(V_2/V_1)$. In step 3 there is an amount of heat that *must* be transferred to the low-temperature reservoir in order to make the cycle complete; however, this amount of heat is not considered when calculating the efficiency, because it is not *added* heat. Thus Eff, the efficiency of the cycle, is the ratio of the total work done, $R(T_1 - T_2)\ln(V_2/V_1)$, to the heat added, $RT_1 \ln(V_2/V_1)$, or

$$\text{Eff} = \frac{R(T_1 - T_2)\ln(V_2/V_1)}{RT_1 \ln(V_2/V_1)} = \frac{T_1 - T_2}{T_1} = 1 - \frac{T_2}{T_1}. \tag{12}$$

Equation 12 tells the whole story of conversion of heat to work in a cyclic process:

(*i*)   In order to do work two different thermal reservoirs are needed.

(*ii*) The efficiency of the cycle depends only upon the isothermal processes.

(*iii*) The efficiency increases as the temperature difference between the reservoirs increases.

(*iv*) Eff = 1 only if $T_2 = 0$ (when the cold reservoir is at absolute zero) or at $T_1 = \infty$.

These four facts simply state that as a result of the nature of heat exchange—namely, that heat flows spontaneously from high to low temperature (Carnot step 3)—no real cyclic process can ever be 100 percent efficient: some heat *must* always be wasted.*

A Carnot cycle can also be run in reverse as a *heat pump,* and one can calculate in similar fashion the efficiency of converting work to heat. The same arguments just used apply, with the signs of the heat and work reversed, and the efficiency is given by a similar expression.

The simple (though abstract) argument that follows, which again invokes our experience, shows that the heat-work cycle just described is the most efficient cycle possible—*no cycle is more efficient than the Carnot*

---

*Note how this powerful conclusion is based upon our empirical view of the universe. We have admittedly done a little bit of mathematics but the facts underlying the argument are the First Law, which we have already said is an empirical law, and the observation (and firm belief in its universality) that heat only flows spontaneously from hot to cold.

cycle. Let us imagine that there is such a more efficient engine working between the same two temperatures T_1 and T_2. If the Carnot cycle delivers work W, this engine would by definition deliver a greater value, W', from the same amount of heat, Q, taken from the hot reservoir. It could be more efficient only if it gives up less heat, q, than the amount Q_c, given up to the low-temperature reservoir by the Carnot cycle. This would be expressed as $= Q - q$. Let us couple the cycle that is engaged in by this more efficient engine to a Carnot cycle in which the work produced, which is $-W$, is equal to $(-Q + Q_c)$. This means that the two cycles are operated so that the work produced in the expansion of one cycle is used to produce compression in the other cycle, and vice versa, and the heat given off in one cycle is utilized in the other cycle. The total work W_{tot} obtained in the combined cycle is the sum of the work derived from each cycle, or,

$$W_{tot} = W' - W = Q_c - q . \tag{13}$$

This hypothetical cycle, more efficient than a Carnot cycle, was defined so that $W' > W$ and $Q_c > q$, so that work is done (that is, $W' - W > 0$) by extracting an amount of heat $Q_c - q$ from a thermal reservoir. Since $Q_c - q > 0$, the low-temperature reservoir would have to give up heat (cool) and simultaneously produce work. This is contrary to experience, in that no system has ever been designed that can operate in this way; thus, if there is a cycle more efficient than the Carnot cycle, it cannot function by cooling the low-temperature reservoir.

Suppose, instead, that the hypothetically more efficient engine is operated by withdrawing less heat, Z, from the high-temperature cycle than the amount, Q, drawn by the Carnot cycle, while producing the same amount of work, W, as the Carnot cycle. With such an engine, $W = Z - q$. If this engine is coupled to a Carnot cycle running in reverse as a heat pump (that is, one for which $-W = -Q + Q_c$), then, if the total work done is zero, $Z - q - Q + Q_c = 0$, and

$$Q - Z = Q_c - q . \tag{14}$$

The difference $Q - Z$ is heat transferred from a cold reservoir to a hot reservoir, and Equation 14 states that this is occurring without any work being done, which is equivalent to saying that there is spontaneous flow of heat from a cold object to a hot object. This again is contrary to experience, and we must conclude that a cycle cannot be made more efficient than a Carnot cycle by drawing less heat from the high-temperature reservoir than does a Carnot cycle. Thus, the two alternative hypothetical engines are not possible and we conclude that no cyclic process is more efficient than a Carnot cycle.

These arguments, both based upon experience, can be restated as another form of the Second Law:

1. *It is not possible by a cyclic process to take heat from a reservoir and use the heat to perform work without simultaneously transferring heat from a hot to a cold reservoir.*

2. *It is not possible to transfer heat from a cold to a hot reservoir unless in the same process work is converted to heat.*

3. *No cyclic process can convert heat to work with 100 percent efficiency* (unless one reservoir is at absolute zero, which is not possible). This third form of the Second Law is that referred to in Chapter 4 as "You can't break even."

It must be remembered that each of these is a statement of experience and not a consequence of deductive reasoning.

Biological systems, like all systems, must obey the Second Law. For instance, statement 2 says that if a cell or organism is at a higher temperature than its environment, as is the case with all warm-blooded animals, the temperature can be maintained only as long as work is done. In this case, it is chemical work resulting from breaking down various chemicals. Mechanical work also can cause a small temperature rise, as in shivering, or rubbing the hands together; the source of energy for this mechanical work is chemical, though. The Second Law also points out that cyclic biochemical reactions such as the tricarboxylic acid (Krebs) cycle cannot supply energy or be a source of work because they operate at constant temperature; a source of energy outside of the cycle must be provided. In fact, this is exactly what happens in the tricarboxylic acid cycle—molecules such as pyruvate and water are fed into the cycle, which only provides a mechanism for breaking these substances down and making the chemical bond energy available to do work.

5-2 THE ENTROPY CONCEPT

Irreversibility and Real Systems

The Carnot cycle consists of four steps that are reversible because each is carried out infinitely slowly. However, no real process consisting of compressions and expansions can be truly reversible since each step is carried out in a finite period of time. Furthermore, a real process utilizes real pistons, cylinders, walls, and so forth, so that there are irreversible losses owing to friction. Without examining the molecular basis of irreversibility we can argue that if a cycle contains an irreversible step, the efficiency of the cycle is less than that of a Carnot cycle. First, if there is one irreversible step in a cycle, the step cannot make the cycle more efficient than a Carnot cycle, because, according to the Second Law and the arguments just given, no cycle can be more efficient than a Carnot cycle. Second, a cycle containing an irreversible step cannot have the same efficiency as a Carnot cycle because, if it were that efficient, it could be driven in reverse with the same efficiency—that is, it would be reversible. Thus, a cycle having an irreversible step is less efficient than a Carnot cycle so that, for the general case, Equation 12 becomes

$$\text{Eff} \leq \frac{T_1 - T_2}{T_1}, \tag{15}$$

in which the $<$ sign applies to any irreversible case and equality applies only to the reversible case. The $<$ sign leads us to the most important consequence of the Second Law—the concept of *entropy*. This is explained in the following section.

Entropy

The work done in a Carnot cycle is the sum of the heat taken from the high-temperature reservoir, Q_h, and the heat given to the low-temperature reservoir, Q_l. Thus, the efficiency Eff in Equation 12 could have been written

$$\text{Eff} = \frac{W}{Q_h} = \frac{Q_h + Q_l}{Q_h}, \tag{16}$$

so that by Equation 15,

$$\frac{Q_h + Q_l}{Q_h} = \frac{T_h - T_l}{T_h}. \tag{17}$$

This may be rearranged to yield the equation

$$\frac{Q_h}{T_h} + \frac{Q_l}{T_l} = 0. \tag{18}$$

It can easily be shown that any reversible cycle in which work is done can be represented as a sum of some particular number of Carnot cycles. For any such cycle an equation such as Equation 18 can be written: if there are i reversible exchanges of heat Q_i at temperatures T_i, then we may write

$$\sum \frac{Q_i}{T_i} = 0. \tag{19}$$

In our discussion of the internal energy U and the enthalpy H we observed that in a cyclic process the sum of the values of ΔU and of ΔH is zero. This is because both ΔU and ΔH are state functions (that is, independent of path), so that whenever the initial state is restored, U and H achieve their initial values. The function Q is almost always path-dependent, so it is very rare that the equation $\sum Q_i = 0$ would be valid. The fact that $\sum Q_i/T_i = 0$ means that

the quotient Q/T *is a state function.*

That is, in a reversible process, simply by dividing Q by T, a function has been created that is independent of path. This function is called the *entropy*

SEC. 5–2   THE ENTROPY CONCEPT

and it is designated S. It is important to note that the entropy function has evolved from a discussion of the Carnot cycle, a process in which all steps are reversible. In defining entropy we must remind ourselves that Q is a heat change in a *reversible* step by writing Q_{rev}, so that the entropy change ΔS in a reversible process is correctly defined as

$$\Delta S = \frac{Q_{\text{rev}}}{T}. \tag{20}$$

Referring back to Equation 19, note that for the Carnot cycle, $\sum \Delta S_i = 0$, in which ΔS_i refers to the entropy change for one step of the cycle.

Any cyclic process that consists only of reversible steps can be written as a sum of an infinite number of infinitesimal processes. Thus, Equation 20 can be written in differential form as

$$dS = \frac{dQ_{\text{rev}}}{T}, \tag{21}$$

in which dQ_{rev} is an infinitesimal reversible heat change at temperature T. The change in the entropy of a system in going from one state to another is then obtained by integrating Equation 21, namely

$$S_2 - S_1 = \Delta S = \int_1^2 \frac{dQ_{\text{rev}}}{T}, \tag{22}$$

in which S_1 and S_2 are the entropy values for each state and the limits of integration are the initial and final values of whatever parameter (for example, pressure, or volume) is used to distinguish the two states of the system. It is important to realize that because Q_{rev} is dependent on path, the integral $\int dQ_{\text{rev}}$ is not necessarily zero for a cyclic process; however, because division of dQ_{rev} by T creates a state function,

$$\int \frac{dQ_{\text{rev}}}{T} = \int dS = 0$$

for any cyclic process.

The requirement for using a reversible path to express dQ is usually confusing. Since entropy is path-independent, one may argue that it should not matter how the system gets from state 1 to state 2, so that one need not worry about whether dQ is used for a reversible or an irreversible path. One can indeed evaluate ΔS for an irreversible path as long as the value of $(S_{\text{final}} - S_{\text{initial}})$ is known. But if one wishes to make use of the mathematical simplicity of Equation 22, then one must use dQ_{rev}, because only for a reversible path is the integral in Equation 22 equal to ΔS. This is because Equations 19 and 22 were derived from the treatment of the Carnot cycle, which is a cycle consisting of reversible heat transfers. This does not mean, though, that we cannot use Equation 22 to calculate ΔS for an irreversible process. Remember that S is a state function, so that ΔS is independent of path. This means that the value of $\Delta S = S_2 - S_1$ is the same for a reversible or an irreversible process in which a system whose entropy is S_1 is altered to a state in which the entropy is S_2. Thus, *in order to calculate ΔS for an irreversible process, one need only think of a hypothetical reversible process, expressible in simple mathematical terms, whose final state is that reached by the irreversible process.* This point demands careful thought by the reader.

The evaluation of entropy is best understood through examples. We present two examples at this point because the idea of irreversibility appears in them. Several more examples will be given in a later section.

EXAMPLE 5–A The entropy change of an ideal gas in a reversible expansion in a cylinder fitted with a piston.

From the First Law, $dQ_{rev} = dU + P\,dV = C_V\,dT + (RT/V)dV$. Since $dS = dQ_{rev}/T$, $dS = (C_V\,dT/T) + (R\,dV/V)$. If the initial state is defined by T_1 and V_1 and the final state by T_2 and V_2, the integration is

$$\Delta S = S_2 - S_1 = \int_{T_1}^{T_2} C_V \frac{dT}{T} + \int_{V_1}^{V_2} \frac{R}{V} dV$$
$$= C_V \ln\left(\frac{T_2}{T_1}\right) + R \ln\left(\frac{V_2}{V_1}\right), \tag{23}$$

if C_V is independent of temperature. If the expansion had been isothermal, T_1 would equal T_2 and the entropy change would be

$$\Delta S = R \, \ln(V_2/V_1). \tag{24}$$

In the following example we compare the process described in the previous example with one involving an irreversible change.

EXAMPLE 5–B Isothermal expansion into a vacuum.

In the previous example we noted that $dQ_{rev} = C_V\,dT + (RT\,dV/V)$. For an isothermal process $dT = 0$, so that $dQ_{rev} = (RT\,dV/V)$. According to Equation 22 we must use this value of dQ_{rev} even though the process of free expansion is irreversible. Thus, the entropy change is again $\Delta S = R \ln(V_2/V_1)$, as in the previous example. Note the important point that the fact that the process is irreversible is irrelevant because ΔS is independent of path.

These two examples can be used to illustrate another point that will be explored more fully later—that is, there is something essentially different about reversible and irreversible processes. We have just calculated the entropy change for the system and paid no attention to the surroundings with which the system is in contact. In a reversible expansion the heat taken up by the system is extracted from a reservoir whose temperature is only infinitesimally above that of the system. The expansion occurs so slowly that the heat change in the reservoir is equal in magnitude but different in sign from that of the system. Therefore the integral also is equal in magnitude but different in sign and we may state that ΔS_{syst}(system) = $-\Delta S_{surr}$(surroundings). Since the universe equals the system plus the surroundings, ΔS_{univ}(universe) = ΔS_{syst} + ΔS_{surr} = 0. In an irreversible expansion into vacuum the surroundings need not be affected in that the system can be isolated. Therefore, ΔS_{surr} = 0 and ΔS_{univ} = ΔS_{syst}. The essential difference between an irreversible and a reversible change in a system is the following.

The values of ΔS_{syst} for two processes, one reversible and one irreversible, are the same as long as the system moves from the same initial state to the same final state. However, the values of ΔS_{univ} are *not* the same.

In the following section we shall explore more fully the effect of irreversible processes on the universe.

Entropy Changes in the Carnot Cycle

The value of ΔS depends only on the initial and final states, so that in any cyclic process, ΔS = 0. This is of course true of the Carnot cycle, which is a cyclic process. In any cycle there may be temperature changes in the heat reservoirs for which there are associated values of ΔS. If ΔS_{reserv}(reservoir) $\neq$ 0, then ΔS_{univ} is also $\neq$ 0. The Carnot cycle illustrates this point. The Carnot cycle consists of two reversible isothermal steps and two reversible adiabatic steps. In a reversible adiabatic process Q = 0; there is no heat exchange between the system and the environment. Thus, from the definition of entropy, ΔS = 0 for the adiabatic steps. Referring to Equations 1, 2, and 10, and to the definition of entropy, ΔS_{univ} for the isothermal expansion must be equal but opposite in sign for ΔS_{univ} for the isothermal compression. Therefore ΔS_{syst} and ΔS_{surr} are both zero. We conclude then that in a Carnot cycle ΔS_{univ} = 0. In the following section we will see that this is not the case for all processes.

Entropy and Irreversibility

In our discussion of the First Law we saw examples of irreversible processes in which ΔU and ΔH are greater than zero and also those for which the values are less than zero. This is also true for ΔS with the single difference that will be shown immediately—that is, *for any irreversible process, the entropy of the universe is always greater than zero*. This is probably the most important statement of the Second Law.

Since any cyclic process can be broken down into a series of Carnot cycles, we can examine the general consequences of irreversibility by introducing a single irreversible step into the Carnot cycle. The work done in a Carnot cycle is equivalent to $Q_h - Q_l$, in which Q_h and Q_l are the heat transferred from the high-temperature reservoir and to the low-temperature reservoir, respectively. With an irreversible step the efficiency is reduced, so that from Equations 16 and 17 we may write

$$\frac{Q_h - Q_l}{Q_h} < \frac{T_h - T_l}{T_h} \tag{25}$$

and

$$\frac{Q_h}{T_h} + \frac{Q_l}{T_l} < 0 . \tag{26}$$

Let us consider the case in which irreversibility is introduced in the isothermal step of a Carnot cycle at temperature T_h. In the two adiabatic steps $Q_{rev} = 0$, so that in these steps there is no entropy change. In the two isothermal steps there are the entropy changes ΔS_h and ΔS_l, at T_h and T_l, respectively. Since entropy is independent of path (whether the path is a reversible one or an irreversible one) and because the initial state of the system is restored at the end of the Carnot cycle, it must be true that

$$\Delta S_h + \Delta S_l = 0 . \tag{27}$$

Subtracting Equation 26 from Equation 27 yields

$$\Delta S_h + \Delta S_l - \left(\frac{Q_h}{T_h} + \frac{Q_l}{T_l}\right) > 0 , \tag{28}$$

which can be rearranged as

$$\left(\Delta S_h - \frac{Q_h}{T_h}\right) + \left(\Delta S_l - \frac{Q_l}{T_l}\right) > 0 . \tag{29}$$

By definition the isothermal process at T_l is reversible, so that $\Delta S_l = Q_l/T_l$; therefore,

$$\Delta S_h > \frac{Q_h}{T_h} \tag{30}$$

in the irreversible step. This relation shows that there is a nonzero entropy change for any irreversible step even if there is no heat transferred.* A remarkable conclusion can be drawn from this statement by noting that if a system is thermally isolated from its surroundings, there can be no heat exchange. Therefore, for an irreversible process *carried out in isolation,*

---

*At first glance it may seem to be contradictory to state that ΔS, which depends on the value of the heat transferred, might be nonzero when the heat transfer is zero. However, it must be remembered that ΔS is defined in terms of the heat transferred in a *reversible* process between equivalent initial and final states. In this case, we are discussing an irreversible process in which $Q = 0$. Since $\Delta S > 0$ for the irreversible process, it would necessarily always be found that $Q_{rev} > 0$ for an equivalent reversible process.

SEC. 5-2 THE ENTROPY CONCEPT

$$\Delta S > 0 . \tag{31}$$

Given the fact that no change in an isolated real system is ever truly reversible, we may conclude that *any* change in a *real isolated* system is accompanied by an increase in entropy. We can then generalize this statement to the universe itself by noting that the entire universe is an isolated system (and in fact the only one); thus we must conclude that

> *accompanying any real change in the state of any system is an increase in the entropy of the universe.*

This statement is another expression of the Second Law and is the most important version for the life scientist.

We will now examine this conclusion in terms of our earlier discussion of spontaneous changes and randomness. Consider an isolated system that consists of a vessel containing two gases that are mixing. Clearly the state of the system changes until total mixing occurs. At this point equilibrium is reached and no further mixing can occur. All real processes are like this in the sense that they ultimately reach equilibrium; hence we conclude that the entropy of a system does not increase indefinitely but instead reaches a maximum at equilibrium. This can be said in another way:

> *the universe is gradually tending toward a state of maximum entropy.*

At equilibrium the molecules in the system of mixing gases are randomly arranged so that the state of maximum entropy is a state of randomness or maximal disorder. In this state there is no way to utilize any of the energy of the molecules; *the entropy function can then be thought of as a measure of the loss of our ability to make use of the energy.* In other words the entropy of a system *increases* as the system runs down.*

Entropy and Probability

We spoke earlier of a relation between probability and spontaneity and noted that the way that changes occur is determined by the laws of chance. We observed also that a state at which no further change spontaneously occurs—that is, the state of equilibrium—is a state of maximal probability. This can be stated in another way: *a system with a particular energy distribution changes until at equilibrium the most probable energy distribution is reached.* Since the system moves toward a state of maximal entropy, this state must also be the most probable distribution. We expect therefore that the entropy of any system in a particular state is a measure of the

---

*It might seem curious that an *increase* in a thermodynamic function is a measurement of the running down of the universe—that is, of the *decreasing* ability of energy to perform work. However, this is only a consequence of the sign convention for heat changes. If the convention had been reversed, entropy would decrease, which might be more satisfying.

probability of the system having the energy distribution corresponding to that state.

Clearly for further discussion we need a relation between entropy and probability. One way that this can be obtained is by examining once again the free expansion of an ideal gas.

In a free expansion of an ideal gas we have already seen that

$$\Delta S = R \ln\left(\frac{V_2}{V_1}\right).$$

Since $V_2 > V_1$, ΔS is greater than zero, as expected for an irreversible process. Now consider two vessels having volume V and V', connected by a tube (Figure 5-2) and containing $\mathcal{N}$ molecules (Avogadro's number). Intuition tells us that the concentration (molecules per unit volume) will be the same in both vessels, no matter what the relative sizes of the vessels are. What we mean by this statement is that this equality of concentration is the most probable distribution. The probability that a particular molecule is in the vessel having volume V is $V/(V + V')$—that is, the fraction of the total volume contained in that vessel. The probability, $\mathcal{P}$, of *all* $\mathcal{N}$ molecules being in that vessel and *no* molecules in the other vessel is

$$\mathcal{P} = \left(\frac{V}{V + V'}\right). \tag{32}$$

FIGURE 5-2 An experimental arrangement for free expansion into a vacuum. Initially the stopcock is closed. When it is opened, gas flows into the evacuated vessel and the number of particles per unit volume becomes approximately the same in the two vessels.

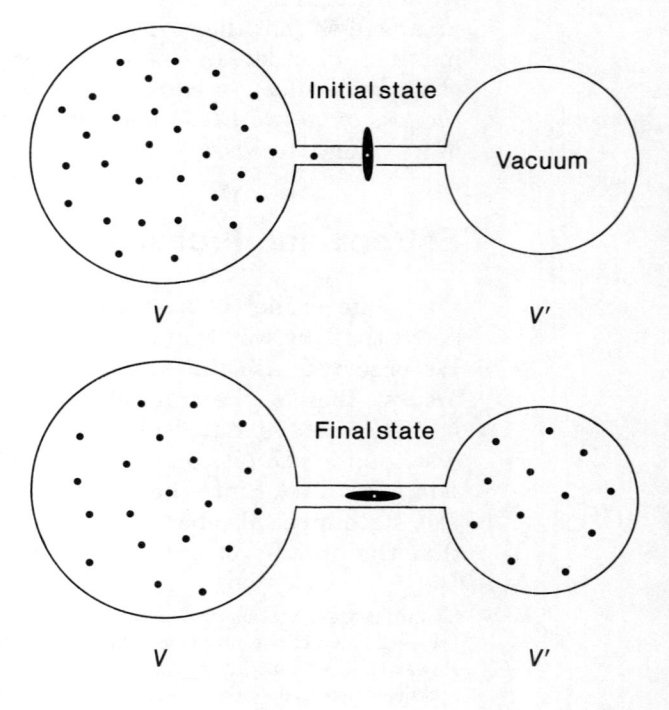

SEC. 5-2 THE ENTROPY CONCEPT

If the N molecules were placed in the first vessel and second vessel was empty, Equation 32 would also be the probability that the molecules would remain in the first vessel. The probability that a single molecule can be in *either* vessel is 1 and the probability that all N molecules are distributed in either vessel is $1^N = 1$. Thus, the ratio of the probabilities of the final distribution and the initial distribution is

$$\frac{1}{\mathcal{P}} = 1 \Big/ \left(\frac{V}{V+V'}\right)^N = \left(\frac{V+V'}{V}\right)^N. \tag{33}$$

Equations in which products and exponents appear are often simplified by converting to a logarithmic form. If this is done and the expression is multiplied by the Boltzmann constant k, one obtains

$$k \ln\left(\frac{1}{\mathcal{P}}\right) = k \ln\left(\frac{V+V'}{V}\right)^N = Nk \ln\left(\frac{V+V'}{V}\right) = R \ln\left(\frac{V+V'}{V}\right), \tag{34}$$

in which $Nk = R =$ the gas constant. Note that the final term is just the entropy change accompanying a free expansion from a volume V to a volume $(V + V')$ that is

$$\Delta S = k \ln\left(\frac{1}{\mathcal{P}}\right). \tag{35}$$

Thus, the entropy change accompanying free expansion of an ideal gas is simply the Boltzmann constant multiplied by the ratio of probabilities of the final (=1) and initial (=$\mathcal{P}$) distribution. Without resorting to the properties of ideal gases it can be shown that in general for any change of a real system

$$\Delta S = k \ln\left(\frac{\mathcal{P}_{\text{final}}}{\mathcal{P}_{\text{initial}}}\right). \tag{36}$$

This leads us to our final statement of the Second Law:

In an irreversible process the entropy of the universe increases. In a reversible process the entropy of the universe remains constant. At no time does the entropy of the universe decrease. If the entropy of a system decreases, the entropy of the surroundings always increases sufficiently that the total entropy change (that is, of the system plus the surroundings) either increases or remains unchanged.

The relation between entropy and probability can be seen in another way. (See Appendix IV at the end of the book for a review of probability.) The thermodynamic state of a system is a result of the positions and velocities of the constituent molecules. Consider a vessel containing N molecules. The vessel is arbitrarily divided up into N regions of equal size, each of which is large enough to hold all N molecules. Consider two possible distributions, one in which all N molecules are in a particular single volume and the other in which there is one molecule per region. The first arrangement can be done in only one way—namely, with all molecules in one place. The second arrangement can occur in an enormous number of ways. The probability of a particular arrangement then is proportional to the number of ways in which it can be achieved. Clearly, the disordered arrangement is much more probable than the highly ordered one and has a higher entropy.

The probability of various arrangements in part determines the absolute entropy of chemical compounds. For example, consider the formation, from the elements carbon and hydrogen, of n-butane and isobutane, both of which contain four carbon atoms and ten hydrogen atoms:

$$\begin{array}{cc} -\overset{|}{\underset{|}{C}}-\overset{|}{\underset{|}{C}}-\overset{|}{\underset{|}{C}}-\overset{|}{\underset{|}{C}}- & -\overset{|}{\underset{|}{C}}-\overset{|}{\underset{|}{C}}-\overset{|}{\underset{|}{C}}- \\ & -\overset{|}{\underset{|}{C}}- \\ n\text{-Butane} & \text{Isobutane} \end{array}$$

Let us imagine that these molecules are synthesized by first assembling the carbon atoms and then adding the hydrogen atoms. There is only one possible arrangement of the first three carbon atoms—namely, in a linear chain. The fourth carbon atom, however, can form a bond with either of the terminal atoms or with the central atom. There are six possible ways (three for each terminal carbon) to form butane but only two ways to form isobutane (Figure 5-3). Thus, starting from carbon and hydrogen, it is more probable that n-butane would form. Since entropy changes are independent of path and the entropy at the beginning of the reaction is the same for the butane path and the isobutane path, ΔS for the formation of butane must be greater than for the formation of isobutane. This is indeed the case, the difference being 15.5 J K^{-1} mol^{-1}, which does not differ significantly from $R \ln(6/2) = R \ln 3 = 9.2$ J K^{-1} mol^{-1}. A similar pattern is seen in the formation of n-pentane, isopentane, and neopentane. One would also expect the entropy change to increase as a homologous series is ascended, since a molecule containing more atoms can have more possible arrangements. This is indeed the case: for example, among the alkanes

$$\Delta S_{\text{form}}(\text{ethane}) - \Delta S_{\text{form}}(\text{methane}) = 43.1 \text{ J K}^{-1} \text{ mol}^{-1}$$

and

$$\Delta S_{\text{form}}(n\text{-pentane}) - \Delta S_{\text{form}}(n\text{-butane}) = 35.1 \text{ J K}^{-1} \text{ mol}^{-1},$$

and in the alcohol series

$$\Delta S_{\text{form}}(\text{ethanol}) - \Delta S_{\text{form}}(\text{methanol}) = 33.9 \text{ J K}^{-1} \text{ mol}^{-1}.$$

In each pair the numerical value noted is basically the entropy change resulting from addition of the final carbon atom.

FIGURE 5-3 The stepwise production of butane and isobutane by assembling carbon atoms.

Take one carbon atom. Then add a second to give any of the four positions of the first to give

$$-\overset{|}{\underset{|}{C}}-\overset{|}{\underset{|}{C}}-$$

Add a third carbon to any of the six positions above to give

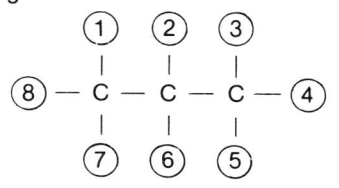

$$\overset{①②③}{\underset{⑦⑥⑤}{⑧-\overset{|}{\underset{|}{C}}-\overset{|}{\underset{|}{C}}-\overset{|}{\underset{|}{C}}-④}}$$

Then, to make butane, a fourth carbon atom can be added at any of the *six* positions 1, 3, 4, 5, 7, or 8. To make isobutane, only *two* positions, 2 and 6, can be utilized.

The Third Law of Thermodynamics

So far we have discussed only the distribution of the positions of molecules. The velocities of molecules are also distributed according to the Maxwell-Boltzmann distribution (see Chapter 2). We may now think about order and disorder with respect to velocities, also. For example, at 0 K the molecules are stationary, so that the velocity distribution is an orderly one (that is, all are at the same speed). As temperature increases, the range of velocities increases and greater disorder results. Thus, we expect entropy to increase with temperature. This is indeed the case and we shall calculate the dependence of S on temperature shortly. This consideration should enable one to understand what is frequently called the Third Law of Thermodynamics. *A perfect crystal (one in which all molecules are at their correct positions in the crystal lattice) is considered to be perfectly ordered with respect to position.*

At any temperature above 0 K, the molecules have kinetic energy (and also vibrational, rotational energy) so that there remains a degree of disorder owing to the many kinds of relative motion. At 0 K all motion stops and the molecules possess no energy (other than the zero point energy demonstrated by quantum mechanics and explained in Chapter 15). They can be thought of as being perfectly ordered with respect to velocity and energy. This state, a perfect crystal at 0 K, can arise in only one way. The significance of this can be seen by rewriting Equation 35 in a form calculable from statistical thermodynamics—namely,

$$S = k \ln \Omega , \tag{37}$$

in which Ω, the thermodynamic probability, is the number of probable arrangements. Thus, if there is only one way that a perfect crystal can arise, then $\Omega = 1$, $\ln \Omega = 0$, and $S = 0$. Thus, the Third Law is stated as *the entropy of a perfect crystal of each element or compound at 0 K is zero.*

We shall see shortly that the Third Law provides a method to calculate absolute entropies. Thus, for the calculations of entropy it is not necessary to define a standard state arbitrarily, as is done for internal energy and enthalpy.

5-3 EVALUATION OF THE ENTROPY OF A SYSTEM

Expressions for Entropy Changes

In this section we derive some expressions for entropy changes in both reversible and irreversible processes. The two most important points to remember are that the calculation by integration (Equation 22) requires that we use the heat change for a reversible path and that, since the entropy change depends only upon the initial and final states of a system, the value of ΔS will not depend upon the path, albeit hypothetical, that is chosen for the calculation.

A. Reversible Isothermal Expansion

We have already seen that in a reversible isothermal expansion of an ideal gas, $\Delta S = R \ln(V_2/V_1)$. This equation states that entropy increases with increasing volume at constant temperature. The expression just given applies only to an ideal gas, although the increase in entropy with volume is true for all systems in nature. This is because with increasing volume more space is available for the molecule so that there are more possible arrangements of position and hence greater disorder. The value of ΔS just stated is used for the gas alone, not for the gas plus the reservoir that is required to maintain constant temperature. The entropy change of the surroundings ΔS_{surr} is

$$\Delta S_{surr} = -\frac{Q_{rev}}{T} = -R \ln\left(\frac{V_2}{V_1}\right),$$

which is less than zero. At first glance one might worry that a negative value of ΔS violates the Second Law. However, the Second Law only says that $\Delta S > 0$ *for the universe*. Thus, *the entropy of any part of the universe may decrease as long as there is a compensating increase somewhere else.* Whenever order is created at some point in space a greater amount of disorder occurs elsewhere. For example, in the growth of a living cell or organism the order gained is compensated for by the breakdown of all of the nutrients used to run the metabolic processes.

B. Irreversible Isothermal Expansion of an Ideal Gas

It is not necessary to calculate ΔS for an irreversible isothermal expansion of an ideal gas because we know that the change in entropy depends only on the initial and final state. Therefore, ΔS is the same as for the reversible expansion because T is constant and the volume still changes from V_1 to V_2.

C. Reversible and Irreversible Heating or Cooling of a System

In a reversible temperature change the heat flow must, by definition, occur in infinitesimal increments so that $dS = dQ_{\text{rev}}/T$. From the definition $C_P = dQ/dT$, we may write $dQ = C_P\, dT$ and $dS = C_P\, dT/T$, so that

$$\Delta S = \int_{T_1}^{T_2} \frac{C_P}{T} dT. \tag{38}$$

If C_P is constant (which is only the case when $T_2 - T_1$ is small)

$$\Delta S = C_P \ln\left(\frac{T_2}{T_1}\right). \tag{39}$$

It has been found that the temperature-dependence of C_P can usually be represented by an empirical equation, $C_P = a + bT + cT^2 + \cdots$, in which case the entropy change is

$$\Delta S = a \ln\left(\frac{T_2}{T_1}\right) + b(T_2 - T_1) + \frac{c}{2}(T_2^2 - T_1^2) + \cdots, \tag{40}$$

in which $a, b, c, \ldots$, are empirical constants that can be found in standard tables. Similar expressions occur for constant volume processes, in which C_V replaces C_P.

We have continually emphasized the necessity of using Q_{rev} in evaluating ΔS by integration. However, once an equation such as Equation 38 is obtained, which does not include the heat change, it will apply to both reversible and irreversible processes. This is again because ΔS depends only on the initial and final states of the system and these states are characterized by volumes and temperatures. As an example of the use of Equation 38, we calculate the entropy change for the irreversible change of state resulting from putting a hot body at temperature T_h in contact with a cold body at T_c. We will assume that both bodies are made of the same material, so that C_P is the same for both and that each contains one mole of material. Clearly the heat lost by the hot body, $C_P(T_h - T)$, equals the heat gained by the cold body, $C_P(T - T_c)$, so that in terms of the final temperature T that is reached,

$$C_P(T_h - T) = C_P(T - T_c) \quad \text{and} \quad T = \frac{T_h + T_c}{2}. \tag{41}$$

The entropy change in cooling the hot body, ΔS_h, is $\Delta S_h = C_P \ln(T/T_h)$, and for the cold body, $\Delta S_c = C_P \ln(T/T_c)$. The total entropy change is

$$\Delta S_{tot} = \Delta S_h + \Delta S_c = C_P\left(\ln\frac{T}{T_h} + \ln\frac{T}{T_c}\right) = C_P \ln\frac{T^2}{T_h T_c}. \qquad (42)$$

It can easily be shown that $\Delta S_{tot} > 0$ for this process. By definition, $T_h > T_c$, so that $T_h - T_c > 0$ and $(T_h - T_c)^2 > 0$. This may be rewritten as $T_h^2 - 2T_h T_c + T_c^2 > 0$; by adding $4T_h T_c$ to both sides of the inequality, we obtain $(T_h + T_c)^2 > 4T_h T_c$. If we use the expression $T = (T_h + T_c)/2$, this inequality becomes $T^2 = [(T_h + T_c)/2]^2 > T_h T_c$. Therefore, $\ln(T^2) > \ln(T_h T_c)$, so that $\ln(T^2/T_h T_c) > 0$. From Equation 42 we once again find that $\Delta S > 0$.

D. Entropy Change in a Reversible Phase Transition

If a substance undergoes a phase transition (e.g., melting) at the transition temperature and at constant pressure (the most common condition), the reversible heat change Q_{rev} is simply the enthalpy of the transition ΔH_{trans} (e.g., ΔH_{fus} for the case of melting). Therefore

$$\Delta S_{trans} = \frac{\Delta H_{trans}}{T_{trans}}. \qquad (43)$$

If there is a temperature change in a process in which a phase change occurs, one must separately calculate ΔS for the temperature change and ΔS for the phase change.

EXAMPLE 5–C Calculation of ΔS for the reaction in which a phase transition is accompanied by a temperature change.

The reaction is

$$H_2O(s)_{0°C,\ 1\ atm} \rightarrow H_2O(g)_{100°C,\ 1\ atm}.$$

This can be broken down into the following reversible reactions:

(1) $H_2O(s)_{0°C} \rightleftarrows H_2O(l)_{0°C}$.
(2) $H_2O(l)_{0°C} \rightleftarrows H_2O(l)_{100°C}$.
(3) $H_2O(l)_{100°C} \rightleftarrows H_2O(g)_{100°C}$.

The first and third reactions are reversible because the phase changes at the transition temperature. The second is reversible because the temperature of liquid water can be freely raised or lowered. From Equation 43 the values of ΔS for the first and third reactions are $\Delta H_{fus}/T_f$ and $\Delta H_{vap}/T_b$, respectively, in which T_f is the freezing point (0°C) and T_b is the boiling point (100°C). The value of ΔS for the second reaction is given by Equation 39. Thus

$$\Delta S = \Delta H_{fus}/T_f + C_P \ln\frac{T_b}{T_f} + \Delta H_{vap}/T_b.$$

The values necessary for the calculation are $\Delta H_{fus} = 6008$ J mol^{-1} at 0°C, $\Delta H_{vap} = 40{,}670$ J mol^{-1} at 100°C, and $C_P = 75.3$ J K^{-1} mol^{-1} for liquid water. (Note that whereas we usually express ΔH in kJ mol^{-1}, here we use the smaller measure J. This is the convention in entropy determinations, because values of ΔS are generally small). Thus,

$$\Delta S = \frac{6008}{273} + 75.3 \ln \frac{373}{273} + \frac{40{,}670}{373} = 154.5 \text{ J K}^{-1} \text{ mol}^{-1}.$$

It is important to realize that the calculation just carried out is simple because, when the phase changes at the transition temperature, the process is reversible and we are justified in using ΔH_{trans}. However, if a phase transition occurs at another temperature, the transition is irreversible. As we have stated repeatedly, it is essential to set up an equivalent reversible process before the calculation can be performed. How this is done is shown in the following example.

EXAMPLE 5–D Calculation of the value of ΔS for the freezing of supercooled water at -10°C.

The reaction is

$$H_2O(l)_{-10°C} \rightarrow H_2O(s)_{-10°C}.$$

This is clearly an irreversible process because once the water has frozen at -10°C, it cannot be reversibly melted at that temperature. In order to calculate ΔS we need only set up an arbitrary set of reversible steps and add up the values of ΔS for each. One set of steps is

(1) $H_2O(l)_{-10°C} \rightleftarrows H_2O(l)_{0°C}$.

(2) $H_2O(l)_{0°C} \rightleftarrows H_2O(s)_{0°C}$.

(3) $H_2O(s)_{0°C} \rightleftarrows H_2O(s)_{-10°C}$.

Steps (1) and (3) are reversible because only a temperature change occurs; step (2) is reversible because it is a phase change at constant temperature.

The entropy change for this set of reactions is

$$\Delta S = C_P(l) \ln \frac{273}{263} - \frac{\Delta H_{fus}}{273} + C_P(s) \ln \frac{263}{273}, \qquad (44)$$

in which $C_P(l)$ and $C_P(s)$ are the heat capacities of $H_2O(l)$ and $H_2O(s)$, and have the values 75.3 and 2.05 J K^{-1} mol^{-1}, respectively. Thus $\Delta S = 75.3 \ln(273/263) - (6008/273) + 2.05 \ln(263/273) = -19.28$ J K^{-1} mol^{-1}. Note that $\Delta S < 0$ because in freezing the water becomes more ordered.

An important attribute of thermodynamic arguments can be seen by writing down a different set of reactions from that just shown. For instance, the same change of state would occur if the liquid is evaporated and expanded isothermally until the pressure of the gas (P_1) equals the vapor pressure of the solid (P_s), and then the vapor is condensed to a solid. This hypothetical reaction scheme is the following:

(1) $H_2O(l) \rightleftarrows H_2O(g)_{P_1}$.

(2) $H_2O(g) \rightleftarrows H_2O(g)_{P_s}$.

(3) $H_2O(g)_{P_s} \rightleftarrows H_2O(s)$.

These steps are reversible because each involves only a pressure change that is reversible, if it is made sufficiently slowly. Note also that step (3) is a sublimation step. If we assume that vapor is an ideal gas, the second step (the isothermal expansion) from V_1 to V_2 has an entropy change expressed as

$$\Delta S = R \ln\left(\frac{V_2}{V_1}\right) = R \ln \frac{P_1}{P_s}. \tag{45}$$

Therefore, the total entropy change is

$$\Delta S_{tot} = \frac{\Delta H_{vap,263}}{263} + R \ln \frac{P_1}{P_s} - \frac{\Delta H_{sub,263}}{263} \tag{46}$$

in which we note that the values of ΔH are for 263°C. Since

$$\Delta H_{sub,263} - \Delta H_{vap,263} = \Delta H_{fus,263}, \tag{47}$$

then

$$\Delta S_{tot,263} = -\frac{\Delta H_{fus}}{263} + R \ln\left(\frac{P_1}{P_s}\right). \tag{48}$$

Absolute Entropy

The Third Law states that for a system at absolute zero, perfect order exists, and that the entropy, which we denote S_0, is 0. This enables us to calculate the absolute entropy for any substance. (This differs from other state functions such as U and H for which no reference values exist.) Since

$$\Delta S = \int_{T_1}^{T_2} (dQ_{rev}/T) = \int_{T_1}^{T_2} C_P(dT/T), \quad \text{and} \quad S_0 = 0 \text{ at } T_1 = 0;$$

then S at any temperature T, that is, S_T, is

$$S_T = \int_0^T \frac{C_P dT}{T} = \int_0^T C_P d(\ln T). \tag{49}$$

In general, C_P is some function of T that must be determined empirically. Fortunately it is not necessary to know this function in mathematical form because the integral in Equation 49 is merely the area under a plot of

SEC. 5–3 EVALUATION OF THE ENTROPY OF A SYSTEM

either C_P/T versus T or C_P versus $\ln T$, with the limits of 0 and T. In fact the most common method of evaluating the absolute entropy of a substance is to measure C_P as a function of T and make the graphs just described. The single complication is that one must remember to take into account phase transitions (such as melting). Thus if a substance has a melting temperature $T_m < T$,

$$S_T = \int_0^{T_m} \frac{C_P(s)}{T} dT + \frac{\Delta H_{\text{fus}}}{T_m} + \int_{T_m}^{T} \frac{C_P(l)}{T} dT, \qquad (50)$$

in which $C_P(s)$ and $C_P(l)$ are the heat capacities of the solid and liquid respectively and ΔH_{fus} is the value at T_m. Other phase transitions such as changes in crystalline form (for example, from diamond to graphite) and vaporization must of course also be included.

Table 5-1 lists values of absolute entropies of elements and compounds in their standard states. It is interesting to note how molecular structure and entropy are related. This is expected, since entropy is a measure of

TABLE 5-1. Absolute entropies (J K^{-1} mol^{-1}) at 298 K and a pressure of 1 atm.

Solid elements		Monatomic gases		Diatomic gases	
Ag	42.55	He	126.06	H$_2$	130.58
Au	47.40	Ne	146.23	F$_2$	203.34
B	7.11	Ar	154.72	Cl$_2$	223.01
Ba	62.74	H	114.60	Br$_2$	245.18
Be	9.50	F	158.66	N$_2$	191.21
C (graphite)	5.73	Cl	165.10	O$_2$	205.01
C (diamond)	2.38	Br	174.89	HF	173.63
Ca	41.63	I	180.67	HCl	186.61
Na	51.30	N	153.18	HBr	198.32
Fe	27.28	C	157.99	HI	207.94
Si	18.83	O	160.96		
Zn	41.63				

	Liquids		Polyatomic gases	
	Br$_2$	152.30	H$_2$O	188.69
	H$_2$O	69.87	CO$_2$	213.8
	D$_2$O	75.94	H$_2$	205.43
	H$_2$O$_2$	109.62	NO$_2$	240.58
	Hg	76.02	NH$_3$	192.46

		Solid compounds			
AgCl	96.23	CaO	39.75	SiO$_2$ (quartz)	41.84
BaO	70.29	Ca(OH)$_2$	83.39	TiO$_2$ (rutile)	50.33
BaCO$_3$	112.13	CaCO$_3$	92.88	ZnO	43.51
BaSO$_4$	132.21	FeO$_3$	89.96	ZnCl$_2$	111.46

molecular disorder. For example, it is expected that for any particular substance the entropy of one mole of atoms of a solid is less then the entropy of a liquid and that this value should be in turn less than the entropy of a gas. This is indeed the case, as can be seen in the table. Note that the entropy per atom of a solid (that is, S_0 divided by the number of atoms in the molecule) tends to be less than 50 J K^{-1} mol^{-1}, whereas the value is usually higher for a liquid. With the exception of H_2 and gases containing hydrogen, the entropies of diatomic gases fall into the same range. It is noteworthy also that crystalline solids, such as quartz, silicon, graphite, and boron have a lower entropy than those whose degree of crystallinity is poor, such as calcium, iodine, $BaSO_4$, and $AgCl$. Furthermore, hard substances tend to have a lower entropy than soft ones; for example, the entropy of diamond ($S = 2.38$ J K^{-1} mol^{-1}) is less than that of graphite ($S = 5.73$ J K^{-1} mol^{-1}).

Residual Entropy

By using the methodology of statistical mechanics, absolute entropies can in theory be calculated for any substance. This methodology (which will not be discussed in this book) involves calculating the thermodynamic probability (Equation 37) from the possible energy states of the molecule. That is, one calculates, using either Boltzmann statistics or quantum statistics, the most probable arrangement in terms of the possible energy states (translational, vibrational, and rotational) of the molecule; this is equated to Ω, the thermodynamic probability. The data needed to evaluate the parameters are obtained from spectroscopic measurements. Usually the values determined from the spectroscopic data agree quite closely with the values obtained from heat capacity measurements, although sometimes the calculated values are higher than the observed values by a few J K^{-1} mol^{-1}. One possible explanation that must always be examined when this occurs is that in the range of temperatures used in the heat capacity measurement there is an unnoticed phase change. Another possibility, which is quite interesting, is that there is an apparent violation of the Third Law in that the substance may be somewhat disordered at 0 K. This is often the case; such a substance is said to possess *residual entropy*.

The cause of residual entropy can be seen in the following example. Consider a crystal containing one mole (i.e., $\mathcal{N}$ units) of a biomolecular ionic substance XA in which the atoms X and A have nearly the same size. In the crystal there may be so little difference, along a particular crystal axis, in the internal energies of the arrangements $XAXAXA\ldots$ and $XAAXXAAXX\ldots$, that both orientations occur in a solid at nearly the same frequency. When a liquid is cooled below the melting temperature, solidification does not occur infinitely slowly and both arrays may be frozen in and remain at 0 K. Using the equation $S = k \ln \Omega$, one can then calculate the residual entropy. In this case in which there are two possible arrays, each having the same probability, there are 2^N ways to achieve the same energy level; then

$$S = k \ln \Omega = k \ln 2^N = Nk \ln 2 = R \ln 2 = 5.56 \text{ J K}^{-1} \text{ mol}^{-1}. \quad (51)$$

Similarly, if there were three possible orientations, according to the Third Law, the crystal would not be perfect and the residual entropy would be $R \ln 3$.

Trouton's Rule

It is interesting that the entropy of vaporization, $\Delta S_{vap} = \Delta H_{vap}/T_b$, in which T_b is the boiling point, is roughly 88 J K^{-1} mol^{-1} for most liquids. This is known as Trouton's Rule. Variation occurs when the molecules of the liquid are hydrogen-bonded, and such variation is a reasonably good criterion for the presence of hydrogen bonds. A liquid that is hydrogen-bonded is more ordered than a liquid lacking hydrogen bonds, and therefore it has a lower absolute entropy. Hence the value of ΔS_{vap} for hydrogen-bonded liquids tends to be greater than 88 J K^{-1} (see Chapter 3 for a more detailed discussion of ΔS_{vap}).

Standard Entropy Change and Calculation of Reaction Entropy

It is important to remember that *both* elements and compounds have an absolute entropy greater than 0 at 298 K, the temperature frequently chosen as the standard state. Furthermore, the absolute entropy of a compound is not equal to ΔS_{form}, the entropy of formation of the compound from its elements. The entropy change ΔS in any reaction equals the sum of the entropies of the products minus the sum of the entropies of the reactants; that is,

$$\Delta S = \sum S(\text{products}) - \sum S(\text{reactants}), \quad (52)$$

so that the entropy of formation is

$$\Delta S_{form} = S(\text{compound}) - \sum S(\text{elements}). \quad (53)$$

It is common to refer to the *standard entropy change*, $\Delta S°$, of a reaction; this means that the reaction is to be written with all reactants in their standard states. Tables of absolute entropies usually list standard entropies, so that if a reaction is to be carried out under conditions other than those of the standard state, the values in the table must be converted to the appropriate values for the other conditions by using equations such as Equations 24 and 39. This means that to calculate a reaction entropy at a temperature other than 298 K one must know the values of C_P for each reactant and product.

However, if the reactants are in the standard state and the reaction evolves heat, so that the product is formed at a higher temperature than the initial temperature, one need only know C_P for the product, because the reaction can be written as a sequence of two reactions. The first reaction is the formation of the product with both the reactants and products in the standard state. The second reaction is one in which there is a rise in temperature of the product. These reactions are equivalent to the original reaction since the entropy change is a function of only the initial and final states.

EXAMPLE 5–E Calculation of the standard entropy change and the entropy change at 70°C for the decomposition of calcium carbonate.

The reaction is

$$CaCO_3(s) \rightarrow CaO(s) + CO_2(g).$$

At standard conditions (see Table 5-1)

$$\Delta S° = S°(CaO) + S°(CO_2) - S°(CaCO_3)$$
$$= 39.75 + 213.80 - 92.88$$
$$= 160.67 \text{ J K}^{-1} \text{ mol}^{-1}.$$

At 70°C = 343 K, we use Equation 38 and let

$$\Delta S = C_P \ln \frac{343}{298} = C_P \ln 1.15,$$

as long as we can assume that the C_P values are essentially constant from 298 K to 343 K. The values of C_P are 42.80, 37.11, and 81.88, for CaO, CO_2, and $CaCO_3$, respectively, and thus the respective increases in the values of the entropy are 5.98, 5.19, and 11.44; these must be added to the values of $S°$. Thus

$$\Delta S_{313} = \Delta S_{313}(CaO) + \Delta S_{313}(CO_2) - \Delta S_{313}(CaCO_3)$$
$$= 45.73 + 218.99 - 104.32$$
$$= 160.4 \text{ J K}^{-1} \text{ mol}^{-1}.$$

Note that the difference is not very great between the two temperatures; if there had been a phase change, the difference would have been much greater.

5-4 THE ENTROPY OF A SOLUTION IN WATER: AN EFFECT IN BIOLOGICAL SYSTEMS

When two substances are mixed together, order is lost (the substances are no longer separate) and entropy increases. Dissolving a solid in a solvent is essentially a mixing process and one might expect an entropy increase to accompany the formation of a solution. This is usually true when the solvent is an organic one. However, when water is the solvent and the solute is a somewhat soluble, organic, nonpolar molecule, more often than not, there is an entropy *decrease*. To explain this, we must examine the structure of water.

There is a tendency to think of a liquid as a structureless, random collection of molecules. However, the very fact that the constituent molecules interact sufficiently to maintain a nongaseous state suggests that a liquid might possess some internal order. In the case of water, the liquid phase is certainly structured because the molecules of water form hydrogen bonds with one another. That is, for a large fraction of the water molecules, the oxygen atom is hydrogen-bonded to two hydrogen atoms, each in a separate water molecule. This intermolecular hydrogen-bonding creates large clusters of water molecules that are highly ordered because the hydrogen bonds form at fixed angles with respect to the axes of the molecule. The cluster is a dynamic system in the sense that new water molecules continually join the cluster and molecules within the cluster leave.

The structure of water can be altered when a substance is dissolved in it. When a small ion having a large charge dissolves in water, five to six water molecules form a spherical shell around the ion and these water molecules are held in place rigidly. This causes a slight disruption of preexisting clusters of water molecules, but the ordering tendency of the ion overwhelms the disordering effect. The entropy of the water decreases when an ionic substance dissolves. There is also an entropy increase when the ions themselves dissociate and, depending on the ion, this may be less than or more than the entropy decrease of the water. Thus the entropy of the system may either increase or decrease in the formation of a solution. To a great extent the size of the ions determines the sign of ΔS.* For example, the more massive ions such as Cs^+, Br^-, and I^- are sufficiently large that they cause a significant disruption of the clusters without producing much order in their immediate vicinity, so that these ions cause a net entropy increase when dissolved.

A particularly interesting and biologically important phenomenon occurs when an organic molecule dissolves in water. That is, there is usually

---

*This does not violate the Second Law because, when the entropy decreases as a solution is formed, there will be a heat transfer that results in an entropy increase in the surroundings that more than compensates for the decrease. Thus, ΔS_{univ} will be positive.

a substantial decrease in entropy, indicating that order increases in the process. Organic molecules, especially if uncharged, interact very poorly with water molecules, so it is very unlikely that the ordering should be direct, as in the case of the ions just described. It is equally unlikely that the ordering is a result of mutual attraction between the organic molecules, because an entropy increase is observed when molecules such as methane or hexane, which mutually interact very poorly, are mixed with water. A variety of experiments indicate that the water molecules themselves become highly ordered. There is as yet no satisfactory explanation for the cause of the ordering, but we will mention several explanations that have been proposed.

The first idea is that the organic molecule keeps the cluster of water molecules intact for a longer time by interfering with the formation of new water-water hydrogen bonds at the periphery of a cluster.

The explanation which follows is considered to present a clearer picture. When a simple organic molecule enters the water lattice (or more particularly, one of the clusters), a hole in the network is created because hydrogen bonds are broken. The water molecules then reorient themselves in an attempt to re-form hydrogen bonds. The reformation of the array of water molecules does not produce a random, quasi-ordered structure but actually a more ordered array than that which existed before the organic molecule was added. This is because the hydrogen bonds must then obey another rule—namely, that the water molecules may not contact the organic molecule but must instead follow its contours.

Neither explanation may in fact accurately describe the state of affairs. Nonetheless it is clear that an organic molecule is surrounded by a highly ordered network of water molecules. This ordering of course implies that there is an entropy decrease when an organic molecule enters an aqueous phase and, in fact, values of ΔS of -120 to -40 J K^{-1} mol^{-1} are typically observed.

As the concentration of the organic molecule is increased, another interesting phenomenon occurs—namely, the clustering of the organic molecules. This is, as we will see, also the result of an attempt on the part of the solute molecule to minimize contact with water molecules. Let us first examine the state of two organic molecules, e.g., of methane, in an aqueous phase. Each molecule will, according to the second explanation, separately order the water lattice and cause an entropy decrease. If, however, the two methane molecules themselves cluster, the water structure is ordered to a lesser extent. This is because the surface of the molecule is the part that is avoiding water, so that clustering of the methane molecules reduces their total surface area. Thus when the molecules are clustered, the combined entropy decrease will be less than if the two methane molecules are separated. Since, according to the Second Law, an increase in entropy is thermodynamically favored, an effect such as clustering, which reduces the magnitude of an entropy decrease, will also be thermodynamically favored.

The tendency for organic molecules to form clusters when in aqueous solution is widespread in biological systems. It is called a *hydrophobic (fear of water) interaction*. It is sometimes called a *hydrophobic force* or *hydrophobic bond* but both of these terms are incorrect because there is neither a force nor a bond. A hydrophobic interaction merely describes the combined effect of the usual intermolecular forces and the hydrogen-bonding interaction that occurs in certain processes in aqueous solution. This interaction is also discussed in Chapter 3 (page 71).

The clustering of organic compounds in aqueous solution is especially prevalent with compounds containing rings, in which the intermolecular interactions themselves are somewhat stronger than between noncyclic organic molecules. It is often detected with optical absorbance techniques because it is usually the case that the absorbance of a ring compound is decreased by coming in close contact with a similar molecule. This was first observed in the dye industry in Germany in the nineteenth century when it was noted that the optical absorbance per mole of many organic dyes decreases as the concentration of the dye increases. In the case of ring compounds the clustering is usually called *stacking*. It is in fact precisely the stacking of the organic bases in molecules of DNA and RNA that maintains the helical structure of these molecules. Indeed, the addition of organic solvents to nucleic acids reduces the magnitude of the stacking interaction by an interaction of the organic solute with the organic solvent and causes a loosening and ultimate collapse of the structure of nucleic acid molecules. This is discussed in much greater detail in Chapter 15.

Hydrophobic interactions are also important in the structure of protein molecules. In a typical protein the polypeptide backbone is tortuously folded as a result of both hydrogen-bonding between different amino acids and strong hydrophobic interactions between the nonpolar side chains of the amino acids. The result is that these hydrocarbon side chains tend to be clustered deep within the molecule and thereby isolated from the water molecules. Another example of clustering is the formation of bilayer membranes of fatty acids and lipoproteins. This is discussed in detail in Chapter 14.

Appendix. Comparison of Isothermal, Adiabatic, Reversible and Irreversible Expansions

Expansion and compression of gases can be carried out isothermally or adiabatically (or, of course, neither) and reversibly or irreversibly. Since it is often difficult for a student to remember the essential features of each type of process, a summary is provided in this appendix.

The following points should be remembered about expansion and compression of gases.

1. If P_{ex} is the external pressure acting on a system, the work done *on* the

system is $-P_{ex}dV$; if in an expansion the volume changes from V_1 to V_2, $W = -\int_{V_1}^{V_2} P_{ex}dV$.

2. A change is reversible if either the external pressure P_{ex} or internal pressure P is adjusted so that $P = P_{ex}$. This means that for every infinitesimal change in one pressure, there is a corresponding infinitesimal change in the other. For this to occur, the expansion or compression must occur infinitely slowly. Thus, in practice, no change is truly reversible.

3. In an irreversible expansion P_{ex} is always less than the internal pressure. The expansion occurs at a finite rate and the work done is less than that done in a reversible expansion. In an irreversible compression $P_{ex} > P$.

4. In an isothermal expansion the temperature of the gas does not change. This constancy is maintained by transfer of heat into the gas from a heat reservoir, which may be thought of as the outside world. (In practice, the heat reservoir is usually a surrounding body of a liquid, such as water, that has a high heat capacity and whose temperature therefore changes minimally as heat is transferred to the gas. The reservoir is kept at constant temperature by a thermostating device of some kind.) For an ideal gas at constant temperature, the internal energy is independent of volume, so, by the First Law, as much heat leaks into the system as work is done as is drawn out; that is, $Q = \Delta W$. These changes are not equal for a real gas.

5. In an adiabatic expansion no heat enters the gas—$Q = 0$—though work is extracted. In practice, an adiabatic process is carried out in a thermally insulated chamber. Very rapid processes are frequently adiabatic if the process is completed before heat can leak in or out of the gas. In any adiabatic process in which work is done, by the First Law, $\Delta U = \Delta W$. In contrast with isothermal expansion, the temperature and the internal energy of the gas decrease.

These features are summarized in Table 5-2.

Work is done in both isothermal and adiabatic expansions *as long as there is an external pressure*. The maximum work is done when the process is reversible, and an isothermal expansion produces more work than an adiabatic expansion. This difference can be understood from the fact that in an isothermal process heat flows into the gas as the gas expands; thus, the internal pressure does not decrease as much as in an adiabatic expansion. This addition of heat in essence provides more fuel for the gas to do further work. The difference in the way that pressure decreases is illustrated in Figure 5-4, which shows the amount of decrease in pressure accompanying a change in volume from V_1 to V_2. The work done is, in each case, the area under the curve and this area is clearly greater for the isotherm than for the adiabat.

TABLE 5-2 Characteristics of work done on an ideal gas.

Type of work	W	Q	ΔU	ΔT
Expansion against constant pressure P_{ex}				
Isothermal	$-P_{ex}\Delta V$	$+P_{ex}\Delta V$*	0*	0
Adiabatic	$-P_{ex}\Delta V$	0	$-P_{ex}\Delta V$	$-P_{ex}\Delta V/C_V$*
Expansion against vacuum				
Isothermal	0	0*	0*	0
Adiabatic	0	0	0	0*
Reversible expansion or compression				
Isothermal	$-RT \ln(V_2/V_1)$	$+RT \ln(V_2/V_1)$	0*	0
Adiabatic	$C_V \Delta T$*	0	$C_V \Delta T$*	$T_1[(V_2/V_1)^{C_V/R} - 1]$*

*Entries marked with an asterisk apply only to an ideal gas; unmarked entries also apply to a real gas.

FIGURE 5-4 A *PV* diagram showing the difference between an isothermal and an adiabatic expansion.

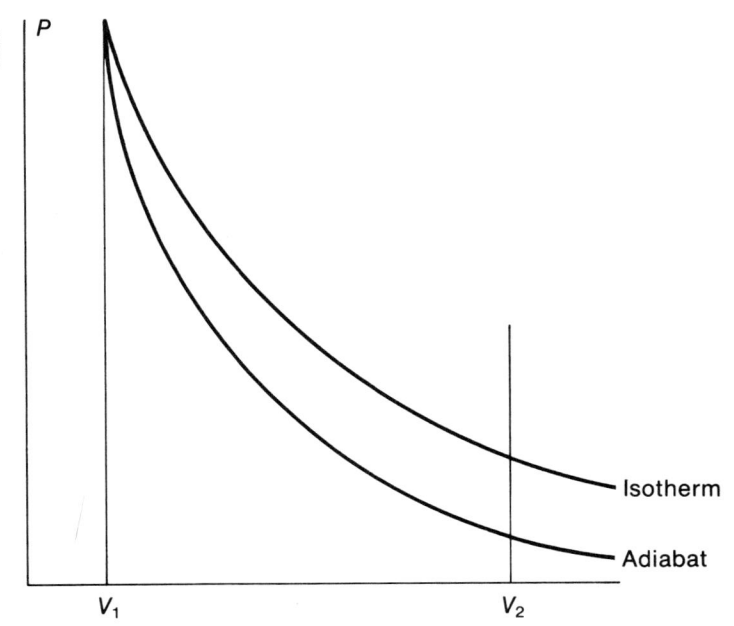

REFERENCES

Bent, H.A. 1965. *The Second Law*. Oxford University Press. Oxford.
Fast, J.D. 1963. *Entropy*. McGraw-Hill. New York.

Klotz, I. 1967. *Energy Changes in Biochemical Reactions.* Academic Press. New York.
Klotz, I.M., and R.M. Rosenberg. 1972. *Chemical Thermodynamics.* Wiley. New York.
Morowitz, H.J. 1970. *Entropy for Biologists.* Academic Press. New York.
Tanford, C. 1973. *The Hydrophobic Effect.* Wiley. New York.
Wall, F.T. 1974. *Chemical Thermodynamics.* W.H. Freeman and Co. San Francisco.
Wilkie, D. 1970. "Thermodynamics and Biology." *Chem. Brit.* 6, 477.

PROBLEMS

1. Which of the following processes are reversible and which are irreversible?

    (a) Compression of a gas in a cylinder by placing a heavy weight on a piston.

    (b) Compression of a gas in a cylinder by placing successively many very light weights on the piston.

    (c) Melting of ice at 5°C.

    (d) Precipitation of $BaSO_4$ by adding H_2SO_4 to $BaCl_2$.

    (e) Freezing of H_2O at 0°C.

2. Two metal balls each at a different temperature and in vacuum are connected by a conducting wire so that heat flows infinitely slowly from the hotter ball to the colder ball. Does the fact that the rate of heat flow is infinitesimal mean that the process is reversible?

3. Must the total entropy of the Earth continually increase? Explain.

4. For an adiabatic process, $Q = 0$. Thus $Q/T = 0$. Does this mean that $\Delta S = 0$ for an adiabatic expansion into a vacuum?

5. For which of the following processes is $\Delta S = \Delta H/T$?

    (a) An isothermal, reversible phase transition.

    (b) An adiabatic compression.

    (c) A process at constant temperature.

    (d) A process at constant volume.

6. Which of the following hands in five-card poker has the lowest "entropy"? A pair of aces, four aces, a flush, a royal flush, a straight flush.

7. In the growth of a plant cell, order is created in the sense that the molecules become arranged in a particular way determined by the structure of the cell. There is no doubt that the entropy of the cell decreases.

    (a) If the absolute value of ΔS_{cell} is written $|\Delta S_{cell}|$, what can be said about the value of $|\Delta S_{surr}|$?

    (b) Give several examples of positive entropy changes that occur in the surroundings.

8. Which compound of each pair has the larger value for absolute entropy?

    (a) Pentane, propane.

    (b) Acetic acid, formic acid.

    (c) Acetone, methylethyl ketone.

    (d) Glycine, acetic acid.

9. There is a finite probability, albeit small, that the molecules of an ideal gas at 1 atm and 20°C in a box with a volume of 1 m³ will suddenly move to one side of the box leaving the box half in vacuum.

    (a) What is the probability of this happening?

    (b) What would ΔS be for the event?

10. What is the value of ΔH and of ΔS for changing 1 g of ice at −10°C to water at 25°C. For ice and water, C_P = 37.2 J K^{-1} mol^{-1} and 75.3 J K^{-1} mol^{-1}.

11. One half mole of an ideal gas is in a vessel at a pressure of 4 atm. The gas is allowed to expand into an evacuated container. The final pressure is 1.5 atm. The initial and final temperatures are the same. What is ΔS for the gas?

PROBLEMS

12. What is ΔS if 100 g of water at 10°C are mixed with 200 g of water at 50°C? The specific heat of water is 4.184 J K^{-1} g^{-1}.

13. The photosynthetic reaction is

$$6CO_2(g) + 6H_2O(l) \rightarrow C_6H_{12}O_6(s) + 6O_2(g).$$

    Show that this reaction cannot occur spontaneously at 1 atm pressure and 25°C. The values of ΔH_{form} at 25°C for CO_2, H_2O, glucose, and O_2 are -393.5, -285.9, -1274.4, and 0 kJ mol^{-1}, respectively. The values of S are 214, 70, 212, and 209 J K^{-1} mol^{-1}, respectively.

14. This problem deals with the question of perpetual motion.

    (a) Is the orbital motion of the Earth around the Sun an example of perpetual motion?

    (b) Is the orbital motion of an electron around the nucleus an example of perpetual motion and, if so, does it violate the Second Law?

15. If one calculates the value of $\Delta S°$ for the highly exothermic reaction $2H_2 + O_2 \rightarrow 2H_2O$ at 25°C, $\Delta S°$ is found to be negative. How is this value of $\Delta S°$ reconciled with the fact that the reaction proceeds efficiently?

16. What is the entropy change at 25°C for the chemical reaction

$$\text{Leucine(s)} + \text{glycine(s)} \rightarrow \text{leucylglycine(s)} + H_2O(l).$$

    The standard entropies at 25°C are: leucine, 109.2; glycine, 103.5; leucylglycine, 281.2; H_2O, 69.91 J K^{-1} mol^{-1}.

17. A liquid whose heat capacity at constant pressure is 26 J K^{-1} mol^{-1} from 0° to 150°C is heated from 50° to 80°C. What is ΔS?

18. The heat of vaporization of a gas that boils at 53°C is 26.1 kJ mol^{-1}. What is the entropy of vaporization at 53°C at a pressure of 0.2 atm?

19. A cooling system cools a refrigerator by transferring heat to the air in the kitchen. The system is a 100-watt electrical unit (1 watt = 1 joule per second). The insulation of the refrigerator walls is of course not perfect and heat leaks into the refrigerator at a rate of 418 joules per second. What is the lowest temperature that the refrigerator can reach, if room temperature is 25°C.

20. A patent is filed for a cyclic engine that operates between two heat reservoirs at 25°C and 300°C and produces 2.5 J of work per 4.18 kJ of heat that is extracted from the high temperature reservoir. The claim is immediately rejected. Why?

21. A nuclear power plant generates 10^6 kW and operates at 50 percent of maximal efficiency. The reaction temperature is 300°C. A river whose temperature is 20°C and which flows at a rate of 200 cubic meters per second is used as a low temperature reservoir. What is the temperature rise of the river as a result of operation of the reactor?

22. When a polymer such as a protein or DNA is heated to a temperature in the range of 70° to 80°C, the polymer spontaneously undergoes significant changes in its properties. This process is called *denaturation* and is explained in Chapter 15; the transition may be written in the form of the following reaction: normal state → denatured state. It is always found that $\Delta S > 0$ for this process. What kind of molecular change probably occurs?

23. Water can be separated into the following six components by mass spectrometry: $^1H_2^{16}O$, $^1H_2^{17}O$, $^1H_2^{18}O$, $^2H_2^{16}O$, $^2H_2^{17}O$, and $^2H_2^{18}O$. What is the entropy change in such a separation of 100 grams of water? The natural isotopic abundances are the following: ^1H, 99.9844 percent; ^2H, 0.0156 percent; ^{16}O, 99.76 percent; ^{17}O, 0.04 percent; ^{18}O, 0.20 percent.

CHAPTER 6

FREE ENERGY

In the analysis of chemical and biological processes, changes of state and chemical composition of a system and the conditions of equilibrium are studied. The First Law introduces the idea of an energy balance sheet and tells us that in any process $\Delta U = Q - W$, in which ΔU is the change in internal energy, Q is the heat transferred, and W is the work done. It says nothing, though, about the direction of a reaction, or of the conditions that lead to equilibrium. The Second Law adds the constraint that the entropy of the universe must increase in any real process, and provides a means for calculating entropy changes. However, we still have not developed a criterion for determining the direction of a reaction. It is not the sign of ΔH, for, as was seen in Chapter 4, both exothermic and endothermic reactions that are spontaneous are common. The sign of ΔS_{system} is also not a useful criterion because although ΔS_{univ} must be greater than zero, ΔS_{system} can be positive or negative in a spontaneous reaction.

One approach to this problem is to seek another law of thermodynamics. However, this is not needed, because the criterion for spontaneity can be developed by insisting that a process satisfies both the First and Second Laws simultaneously. A particularly simple criterion for spontaneity is obtained by defining a new thermodynamic function that incorporates the requirements of both the First and Second Laws. This function is called the *Gibbs free energy;* it is also a state function and, hence, independent of path; it is denoted by G. This is also called the Gibbs function or, frequently in biochemistry, free energy*.

---

*In the older literature the Gibbs free energy was given the symbol F for *f*ree energy and some thermodynamic tables still use this symbol. The Gibbs free energy should not be confused with the Helmholtz free energy, $A = U - TS$.

6–1 THE GIBBS FREE ENERGY, G

Definition of G

The First Law requires that $-\Delta(U + PV)_{\text{system}} = Q_{\text{surr}}$, which may be written in terms of entropy, S, as

$$\frac{-\Delta(U + PV)_{\text{system}}}{T} = \frac{Q_{\text{surr}}}{T} = \Delta S_{\text{surr}} \tag{1}$$

at constant temperature and pressure. Therefore,

$$-\frac{\Delta(U + PV)_{\text{system}}}{T} + \Delta S_{\text{system}} = \Delta S_{\text{surr}} + S_{\text{system}} \geq 0 . \tag{2}$$

The inequality is of course due to the Second Law. Equation 2 can be rearranged to give

$$\Delta U_{\text{system}} + P\Delta V_{\text{system}} - T\Delta S_{\text{system}} \leq 0 . \tag{3}$$

The Gibbs free energy, G, is then conveniently defined as

$$G = U + PV - TS , \tag{4}$$

or, more conveniently, in terms of enthalpy $H = U + PV$, as

$$G = H - TS , \tag{5}$$

so that the statement

$$\Delta G \leq 0 \tag{6}$$

incorporates the requirements of both the First and Second Laws. Thus a decrease in the Gibbs free energy becomes the criterion for spontaneous change. If we write

$$\Delta G = \Delta H - T\Delta S \leq 0 , \tag{7}$$

it becomes clear that heat loss to the environment—that is, $\Delta H < 0$—also is not alone sufficient to determine the direction in which a process moves. Clearly two factors are involved, the value of ΔH and the value of ΔS; if $T\Delta S \geq \Delta H$, a process is said to be *entropy-controlled*. Note that ΔS is the entropy change *for the system* and that ΔS can be less than zero in a spontaneous process as long as ΔH is sufficiently large and negative. Of course, if ΔS_{system} is negative, ΔS_{surr} must be very large, and positive, so that always $\Delta S_{\text{univ}} > 0$. It is also worth noting that in biological reactions occurring in solution there is rarely a change in either pressure or volume, so that $\Delta H \approx \Delta U$. Thus, in trying to understand the molecular basis of various reactions, one can think in terms of changes in internal energy (that is, kinetic, rotational, and vibrational energy).

Spontaneity and ΔG

The Gibbs free energy (which we shall hereafter simply call free energy) is the most useful thermodynamic function for predicting whether a particular chemical or physical reaction will occur spontaneously. Furthermore, it is ideal for describing equilibria, because it simultaneously embodies the requirements of the First and Second Laws. It is valuable, therefore, to have a feeling for its relation to the amount of useful work a system can perform and to have an understanding of its meaning in molecular terms.

From Equation 4 for a change in a system from state 1 to state 2,

$$G_2 - G_1 = (U_2 - U_1) + (P_2V_2 - P_1V_1) - (T_2S_2 - T_1S_1) . \qquad (8)$$

For an isolated system, $U_2 - U_1 = Q + W$, so

$$G_2 - G_1 = Q + W + (P_2V_2 - P_1V_1) - (T_2S_2 - T_1S_1) . \qquad (9)$$

The most common situation encountered in chemical and biological systems is that of constant temperature and pressure, since in nature almost all reactions occur in equilibrium with the temperature and pressure of the atmosphere. Thus, in this book it will always be assumed that $P_1 = P_2 = P$, with $T = $ constant, unless otherwise noted. Considering these constraints and remembering that for the heat Q taken in by the system,

$$Q \leq T(S_2 - S_1) , \qquad (10)$$

we may write

$$-W - P(V_2 - V_1) \leq -(G_2 - G_1) . \qquad (11)$$

The work done by the system in displacing the environment is $P(V_2 - V_1)$. In a general system there may be other types of work, W' (for example, electrical work in a battery), and we define W' by

$$-W = -W' + P(V_2 - V_1) . \qquad (12)$$

Therefore

$$-W' \leq -(G_2 - G_1) . \qquad (13)$$

The inequality sign applies to an irreversible process and the equality sign, to a reversible one. Since the maximal amount of work done by a system at constant temperature and pressure occurs when a process is carried out reversibly, then

$$-W'_{max} = -(G_2 - G_1) . \qquad (14)$$

Thus, *the change in G is a measure of the maximal attainable work, not including volume changes.* Herein lies the value of the concept of the free energy. Since most processes encountered in nature do not involve PV changes but instead either interactions with forces (mechanical or electromagnetic, for example) or with changes in chemical state, the free energy measures the maximal work available from these processes. Note that this

SEC. 6-1   THE GIBBS FREE ENERGY, G

line of reasoning also provides a criterion for spontaneity. An isolated system at constant temperature cannot extract work from its surroundings, but it can spontaneously do work. Doing work utilizes free energy, so that G must decrease for a spontaneous reaction. In the absence of constraints, the systems will continue to change until G reaches a minimal value. At this state the system is at equilibrium. Thus:

$\Delta G < 0$ is the rule that determines the direction of change,

and

$\Delta G = 0$ is the condition for equilibrium.

Since P and T are constant in the laboratory and in most systems studied, these statements about ΔG are extremely valuable and important to remember.

The argument just given shows that the value of ΔG is a measure of the ability of a system to do work. More specifically, it is the amount of energy possessed by a system that the Second Law will *allow* to be converted to work. A molecular explanation of ΔG can be obtained by considering the probabilistic nature of entropy changes. In these terms ΔG can be redefined as ΔG = (change in energy) − (change in the degree of order). In any spontaneous process there is a tendency for internal energy to lessen and for order to decrease. However, a decrease in internal energy may sometimes only occur if a molecular rearrangement occurs that has greater order, so that the direction of change in a system must be determined by finding a balance between these two effects. A simple example demonstrates this competition. Consider a box containing equal numbers of white and black balls of equal size and mass. In the initial state, suppose the white balls are all on the bottom. If the box is shaken vigorously, the balls will redistribute themselves until ultimately there will be roughly the same number of black balls as white balls in the bottom layer. This random arrangement is the most probable distribution. Next, suppose the white balls are much denser than the black balls. In this case the initial state—the white balls on the bottom—is the state of minimal gravitational potential energy. There will be resistance to changing this state, even as the box is shaken, and the resistance will rise as the ratio of the densities of white and black balls increases. However, if the box is continually shaken, the balls will become displaced from their initial state: a random distribution will not result, though, because there always remains the *tendency* for more white balls than black balls to be at the bottom of the box. In this case the requirement for minimizing the gravitational potential energy opposes the randomizing tendency of continuous shaking. Thus, in a continuously shaking collection of more dense, white balls and less dense, black balls, most, but not all, of the white balls will be found in the bottom layer with most, but not all, of the black ones in the top layer.

When one thinks of ΔG as consisting of an energy-term minus an order- or probability-term, it is clear that if any process that occurs spontaneously is accompanied by a large *increase* in internal energy, there must also be a substantial increase in entropy. Furthermore the entropy increase must be greater at lower temperatures. If entropy increases in such a process, the system must become less ordered. If we examine such a change in molecular terms, we should expect to find that the initial state is highly ordered. Actually, in such an analysis we will find that instead of thinking of "order" it is more convenient to think of an entropy change as what results when a system undergoes a change from a state of somewhat restricted molecular motion to a state in which there is much more freedom of motion.

A simple example of this way of thinking is the melting of a solid at a temperature slightly above the melting temperature. We saw in our discussion of ΔH_{fus} (Chapter 4) that melting requires a large input of energy into the system—that is, the internal energy increases substantially. In the case of the ice–water transition, the increase in internal energy is approximately 5941 J mol^{-1} at 0°C (273 K). The entropies of ice and water at 0°C are 41.0 and 62.8 J K^{-1} mol^{-1}, respectively, so that melting involves $\Delta S = 21.8$ J K^{-1} mol^{-1}. For melting to occur spontaneously, the $T\Delta S$ term must exceed 5941 (which occurs at 273 K) in order that $\Delta G < 0$. This large increase in ΔS is of course a result of the rigid structure of ice and the great freedom of motion of water molecules in the liquid state.

In some cases the increase of freedom of motion is not so obvious as it is in melting. For instance, if one mixes solutions of Cu(NO$_3$)$_2$ and a typical protein, a complex spontaneously forms between the protein molecules and the Cu^{2+} ions and there is an increase of internal energy of 12,552 J mol^{-1}. At room temperature, ΔS must be about 42 J K^{-1} mol^{-1}. Since the Cu^{2+} ions and the protein molecules have become joined, they have lost freedom of motion, and one would expect ΔS to be less than zero. The compensating increase in entropy comes from disruption of the highly ordered water molecules that are bound to the Cu^{2+} ions and the protein molecules. Since the melting of ice is accompanied by $\Delta S \approx 21$ J K^{-1} mol^{-1}, and the reaction requires that ΔS be significantly greater than 42 (because of the decrease in ΔS accompanying binding), we can assume that there are considerably more than two moles of water molecules bound in a highly ordered structure per mole of Cu^{2+} and per mole of protein. This argument shows the necessity of clearly defining a system when a particular process is to be studied. That is, the reaction

$$\text{Protein} + \text{Cu}^{2+} \rightarrow \text{protein-Cu}^{2+}$$

does not describe what is occurring; either of the following two reactions,

$$\text{Protein-H}_2\text{O} + \text{Cu}^{2+} - \text{H}_2\text{O} \rightarrow \text{protein-Cu}^{2+} + \text{unbound H}_2\text{O}$$

or

$$\text{Protein-}m\text{H}_2\text{O} + \text{Cu}^{2+} - n\text{H}_2\text{O} \rightarrow \text{protein-Cu}^{2+} - a\text{H}_2\text{O} + b\text{H}_2\text{O},$$

would describe the process better. This particular example has been chosen to indicate that *whenever a large value of ΔS is observed and there is no obvious increase in freedom of motion, an additional component of the system should be suspected.*

In fact, since water is present in almost all biological reactions and because water molecules form highly structured systems, whenever a process that proceeds spontaneously has a value of $\Delta H > 0$, the role of H_2O molecules in the process should be investigated.

A large increase in internal energy and a high degree of order usually have the same molecular basis—namely, the presence of a strong cohesive force. For example, in ice the water molecules are tightly bonded by both van der Waals interactions and hydrogen bonds and a large amount of kinetic energy must be added to an ordered array of water molecules to overcome these forces. In the protein-Cu^{2+} example, there is a great deal of bound water because of the attraction between the negatively charged oxygen atoms in the protein molecule and the Cu^{2+} ion, and because of hydrogen-bonding of water to numerous groups of the protein molecule.

In a later section we will apply this explanation when we examine the molecular changes accompanying alterations in the shape of macromolecules such as proteins and nucleic acids.

Standard Free Energy Changes, ΔG°, and the Variation of ΔG with Local Conditions

The direction of a chemical reaction frequently depends on the concentration of the reactants. For example, if 10^{-4} mol of CaF_2 is added to a liter of 0.1 M NaCl, the CaF_2 will dissolve and Ca^{2+} ions and F^- ions will be produced that have concentrations of nearly $10^{-4}\ M$ and $2 \times 10^{-4}\ M$, respectively. Thus at these concentrations the reaction $CaF_2 \rightleftarrows Ca^{2+} + 2F^-$ proceeds rightward toward dissociation. However, if solutions of 0.01 M $CaCl_2$ and 0.02 M NaF are mixed, CaF_2 will precipitate; the reaction moves to the left at this concentration. Temperature also affects the direction of reactions; for example, ice melts at 10°C, but water freezes at -1°C. Therefore, since the value of ΔG must vary with concentration and temperature (and other experimental conditions as well), it is again convenient to define *standard conditions,* as described for ΔH. The standard free energy change, denoted $\Delta G°$, is defined as the value of ΔG at a concentration of 1 M of each reactant and at a pressure of 1 atm. (Note that the temperature is not specified.) Furthermore:

ΔG° is defined to be zero when an element is in its standard state (that is, at a pressure of one atmosphere).

When an element has two forms (for example, carbon may have the form of graphite or of diamond), one of these forms only (in this case, graphite) is chosen as the standard state.

Note that the definition of $\Delta G°$ in terms of a 1 M solution is entirely arbitrary. Alternately, we might have defined standard conditions as that of a pure substance (mole fraction = 1) at 1 atm. Indeed, this definition is used and is preferable in some cases. For example, for a reaction in which two gases combine to form a liquid, the product is a pure liquid, not a solution. Oxidation of a metal is another example in which there would be no reason to define the standard state in terms of a solution. The existence of two standard conditions leads to confusion in that tabulated values of $\Delta G°$ may refer either to one set of conditions or to the other and, therefore, one must be careful to note these conditions. It is proper—but not always done—to specify standard conditions by writing either $\Delta G°(1\ M)$ or, for a solid or liquid, $\Delta G°(s)$ or $\Delta G°(l)$, respectively. In the latter case, the symbols $\Delta G°_s$ and $\Delta G°_l$ are also seen. Note that for a substance that is soluble $\Delta G°(1\ M)$ and $\Delta G°(s)$ differ by the free energy change associated with forming a 1 M solution. The actual numerical difference between $\Delta G°(1\ M)$ and $\Delta G°(\text{pure})$ is usually fairly small (for glycerol, the values are -488.5 and -479.5 J mol^{-1}, respectively), but in many cases the difference may be significant. In this book, we refer almost exclusively to biochemical processes, which occur in solution, so when $\Delta G°$ is written, $\Delta G°(1\ M)$ is meant. Occasionally, when calculating the free energy of formation, the value of $\Delta G°$ for a pure substance will be used and designated accordingly.

The idea of a standard free energy change is often the source of confusion. Why the concept is needed is the following. The free energy is the sum of an entropy term and an enthalpy term. It is possible to calculate the absolute entropy of a substance at a particular temperature from the variation of heat capacity with temperature (Chapter 5) but the absolute value of the enthalpy cannot be evaluated. We can, however, measure a change in H, or ΔH, and we do so under standard conditions. Likewise, we can never determine G itself but only ΔG. As with ΔH, a reference point is needed so that ΔG values can be compared and that reference is defined as the standard state.

For a chemical reaction in which products are formed from reactants, ΔG is the sum of the values of the ΔG of formation (ΔG_{form}) for each product, $\Delta G_{\text{form}}(\text{product})$, minus the sum of $\Delta G_{\text{form}}(\text{reactants})$. This leads to the very important equation,

$$\Delta G° = \sum \Delta G°_{\text{form}}(\text{product}) - \sum \Delta G°_{\text{form}}(\text{reactant}). \tag{15}$$

Thus, to calculate $\Delta G°$ for a reaction we must evaluate $\Delta G°_{\text{form}}$ for each substance in the reaction. It is essential to realize that the choice of the standard state is arbitrary and totally unrelated to any molecular structure. Furthermore, it in no way affects the value of ΔG for any process. This is because, in the calculation of ΔG for a reaction, we subtract values of ΔG and not of G. The fact that the actual value of G need never be known is easily seen by the following argument.

In a reaction A + B → C, the free energy of formation ΔG_{form} of each substance is just the free energy of that substance minus the free energy of its constituent elements: that is,

$$G_{form}(A) = G(A) - \sum G_{el}(A),$$
$$G_{form}(B) = G(B) - \sum G_{el}(B), \quad (16)$$
$$G_{form}(C) = G(C) - \sum G_{el}(C).$$

in which $G(A)$ is the free energy of substance A and $\sum G_{el}(A)$ is the free energy of the elements in A. Furthermore,

$$\Delta G_{react} = G(C) - [G(A) + G(B)]. \quad (17)$$

The values of $G(A)$, $G(B)$, and $G(C)$ are, of course, unknown. Equation 17 can be rewritten in terms of Equation 16 to yield

$$\Delta G_{react} = \Delta G_{form}(C) + \sum G_{el}(C)$$
$$- \left[\Delta G_{form}(A) + \sum G_{el}(A) + \Delta G_{form}(B) + \sum G_{el}(B) \right]$$
$$= \Delta G_{form}(C) - [\Delta G_{form}(A) + \Delta G_{form}(B)]$$
$$+ \left[\sum G_{el}(C) - \left(\sum G_{el}(A) + \sum G_{el}(B) \right) \right]. \quad (18)$$

However, because the elements in A and B are combined to form C, then

$$\sum G_{el}(C) = \sum G_{el}(A) + \sum G_{el}(B). \quad (19)$$

Therefore Equation 18 becomes

$$\Delta G_{react} = \Delta G_{form}(C) - [\Delta G_{form}(A) + \Delta G_{form}(B)], \quad (20)$$

which is identical to Equation 19. Note that in this derivation no values of G need be known because they cancel out before Equation 20 is reached.

The main confusion that occurs when values of $\Delta G°$ are examined is that frequently $\Delta G° > 0$ for a reaction that you know is spontaneous. However, a value of $\Delta G° > 0$ just means that the reaction does not occur spontaneously under standard conditions. *The values of $\Delta G°$ are just reference numbers to be used when calculating ΔG for the conditions of interest.* A simple example shows exactly what is meant by this statement. Consider the synthesis of alanylglycine (Ala-Gly) from alanine (Ala) and glycine (Gly). The hypothetical reaction in the standard state is

$$\underbrace{Ala(aq)}_{1\ M} + \underbrace{Gly(aq)}_{1\ M} \rightarrow \underbrace{Ala\text{-}Gly(aq)}_{1\ M} + H_2O(l).$$

The value of $\Delta G°$ for this reaction is 17.28 kJ mol^{-1}, so this reaction is not spontaneous under standard conditions. However, one should *not* conclude that alanylglycine cannot spontaneously form when solutions of alanine and glycine are mixed. One can conclude, though, that the concentration of alanylglycine will not be 1 M in a solution containing 1 M alanine and 1 M

glycine: if ΔG is calculated* for a mixture that is 1 M alanine, 1 M glycine, and 0.1 M alanylglycine, the value will be 11.29 kJ mol^{-1}. This is still positive, yet smaller, suggesting that there may be a concentration of alanylglycine for which $\Delta G < 0$. If, for example, we mix 0.1 M alanine, 0.1 M glycine, and no alanylglycine, under these conditions, $\Delta G < 0$, and the formation of some low concentration of alanylglycine can occur spontaneously. The resulting concentration of alanylglycine can be calculated by setting $\Delta G = 0$, because when $\Delta G = 0$, a reaction will proceed neither to the right nor to the left, and all constituents of the reaction will stay at their equilibrium concentrations. When $\Delta G = 0$, one obtains $1.25 \times 10^{-5} M$ alanylglycine; this means that in a solution initially containing 0.1 M alanine and 0.1 M glycine, alanylglycine will form spontaneously until (at equilibrium) the concentration will be $1.25 \times 10^{-5} M$. A similar calculation could of course be made for lower (and unequal) concentrations of alanine and glycine. Note that the thermodynamic calculation only states what the final concentration will be but *does not tell how long it will take for these conditions to be reached.*

In order to perform the calculations of the preceding examples, it is necessary to be able to calculate values of ΔG at concentrations other than 1 M from the value of $\Delta G°$. How this is done is described in the following section.

Dependence of ΔG on Pressure and Concentration

The free energy is defined as $G = U + PV - TS$, from which, by a law of calculus, we may write

$$dG = dU + P\,dV + V\,dP - T\,dS - S\,dT. \quad (21)$$

From the First and Second Laws, we may substitute

$$dU = dQ - P\,dV = T\,dS - P\,dV, \quad (22)$$

so that

$$dG = V\,dP - S\,dT. \quad (23)$$

Since G depends only on the initial and final state, it is possible to consider a reaction in which the concentration change and temperature change consist of a sequence of two reactions, one for the concentration change and one for the temperature change. Thus we calculate the change in G at constant temperature ($dT = 0$), and later derive an expression for the temperature dependence. If $dT = 0$, Equation 23 becomes

$$dG = V\,dP. \quad (24)$$

If we first consider the case of an ideal gas for which $PV = nRT$, we obtain

$$dG = \frac{nRT}{P}\,dP. \quad (25)$$

---

*How to perform the calculations referred to in this paragraph is explained in the next section of this chapter.

SEC. 6-1 THE GIBBS FREE ENERGY, G

This can be integrated from the standard state ($P = P° = 1$ atm and $G = G°$) to an arbitrary value of P:

$$\int_{G°}^{G} dG = \int_{P°}^{P} \frac{nRT}{P} dP ,\qquad(26)$$

which is equivalent to

$$G - G° = nRT \ln \frac{P}{P°} = nRT \ln \frac{P}{1} = nRT \ln P ,\qquad(27)$$

if P is expressed in atmospheres. Actually, P can be expressed in any units as long as $P°$ is also expressed in the same units.

For an ideal or a dilute solution the partial pressure is proportional to the molarity c of the solute, so we might expect (and indeed it can be shown) that

$$G = G° + nRT \ln c ,$$

in which c is the molarity.*

As pointed out earlier, only ΔG, and not G, is measurable and $G - G° = \Delta G - \Delta G°$. Thus, this equation can be rewritten

$$\Delta G = \Delta G° + RT \ln c .\qquad(28)$$

An alternate form of Equation 28 utilizes the mole fraction X of the substance instead of the molarity:

$$\Delta G = \Delta G°_X + nRT \ln X .$$

As mentioned in the preceding section, we must be careful when using this equation, because $\Delta G°_X$ (or $\Delta G°(s)$) is not the same as $\Delta G°$ ($\Delta G°(1\ M)$).

Equation 28 says that the free energy of a substance depends not only on how many molecules are present, but also on their concentration—in particular, the free energy decreases with concentration. The following well-known phenomenon shows that this is reasonable. If a solution of a substance at high concentration is separated by a permeable membrane from a solution of the same substance at a lower concentration, solute molecules will flow from the more concentrated solution to the more dilute solution. This flow occurs spontaneously, so the free energy of the concentrated solution must be greater than that of the dilute solution. The magnitude of the free energy difference is shown by a more general version of Equation 28. If the free energies of n molecules of a solution at two concentrations c_1 and c_2 ($c_1 > c_2$) are G_1 and G_2 respectively, the equations $G_1 - G° = nRT \ln c_1$ and $G_2 - G° = nRT \ln c_2$ can be subtracted to yield

$$\Delta G_{\text{dil}} = G_1 - G_2 = nRT \ln \left(\frac{c_1}{c_2}\right)\qquad(29)$$

---

*For a solution that is not dilute, the concentration need only be replaced by the activity, a. (See Chapter 7 for the relation between activity and concentration.) It is commonplace in biochemistry, in which concentrations are usually quite low, to ignore departures from ideality and to refer to concentration instead of activity.

in which ΔG_{dil} is called the *free energy of dilution*. Equation 29 is the relation between free energy changes and concentration that we will use most often.

Equation 28 must not be used too casually, for in a system with several components the difference between the values of $\Delta G_{reaction}$ and of $\Delta G°$ produced by changes in concentration depends on the particular reaction being considered. For example, in a reaction at 298°C

$$A\ (s) \rightarrow B\ (0.1\ M)$$

with $\Delta G°_{298} = 10$ kJ mol^{-1}, indeed

$$\Delta G = \Delta G° + RT \ln 0.1 = 4.3 \text{ kJ mol}^{-1}.$$

Also, in a dimerization reaction, for example,

$$A\ (1\ M) \rightarrow A_2\ (0.5\ M),$$

Equation 28 would be correct. However, the equation does not apply in cases in which the reactants are not in their standard states—for example,

$$A\ (0.1\ M) \rightarrow B\ (0.1\ M)$$

with $\Delta G° = 10$ kJ mol^{-1}. Here,

$$\Delta G = \Delta G° + RT \ln (c_B/c_A) = 10 + 0 = 10 \text{ kJ mol}^{-1}.$$

The equation would also be inapplicable when the reactants are in their standard states but more than one product forms—for example,

$$A\ (s) \rightarrow B\ (0.1\ M) + C\ (0.1\ M),$$

for, then,

$$\Delta G = \Delta G° + RT \ln c_B c_C = 10 + RT \ln (0.1)(0.1)$$
$$= 10 - 11.4 = -1.4 \text{ kJ mol}^{-1}.$$

The reader should be sure to appreciate the differences between these reactions.

Now that we are going to calculate values of ΔG for a particular temperature and concentration, we need a notation that provides this information. The following is commonly used: $\Delta G_{T,c}$ is the value of ΔG at a temperature T and molar concentration c. Thus $\Delta G_{300,\ 0.1\ M}$ is the value of ΔG at 300 K and a concentration of 0.1 M. If we need to specify the process to which the values of ΔG correspond, we add a subscript to designate the process—for example, $\Delta G_{form,\ T,\ c}$ is the free energy of formation at temperature T and molarity c.

Change of Free Energy with Temperature

Since both ΔH and ΔS vary with temperature, we should expect that ΔG is also dependent on temperature. The precise relation between ΔG and temperature at constant pressure is easily derived. We start with the relation

SEC. 6-1 THE GIBBS FREE ENERGY, G

$$dG = V\,dP - S\,dT, \tag{30}$$

from which a law of calculus (see Appendix III at the end of the book) allows us to write

$$dG = \left(\frac{\partial G}{\partial T}\right)_P dT + \left(\frac{\partial G}{\partial P}\right)_T dP. \tag{31}$$

At constant pressure, $dP = 0$, so that Equations 30 and 31 combine to give

$$\left(\frac{\partial G}{\partial T}\right)_P = -S, \tag{32}$$

which can also be written

$$\left(\frac{\partial \Delta G}{\partial T}\right)_P = -\Delta S. \tag{33}$$

(This replacement of a state function X, Y, or Z by ΔX, ΔY, or ΔZ is commonly done and is always valid.)

If the entropy change is very small in a particular temperature range, one may simply write $\Delta G = -S\Delta T$ and use for S the value of the absolute entropy in that temperature range. That is, if ΔG_{T_1} is known and S is nearly the same at T_1 and T_2, then $\Delta G_{T_2} = \Delta G_{T_1} - S\Delta(T_2 - T_1)$. This is often not a good approximation, so we use the definition $\Delta G = \Delta H - T\Delta S$ and write

$$\left(\frac{\partial \Delta G}{\partial T}\right)_P = \frac{\Delta G - \Delta H}{T}. \tag{34}$$

This may be rearranged as

$$\frac{1}{T}\left(\frac{\partial \Delta G}{\partial T}\right)_P - \frac{\Delta G}{T^2} = -\frac{\Delta H}{T^2}. \tag{35}$$

The left side of the equation is simply the derivative of $\Delta G/T$, so that

$$\frac{\partial}{\partial T}\left(\frac{\Delta G}{T}\right) = -\frac{\Delta H}{T^2}. \tag{36}$$

If this is integrated between the temperatures T_1 and T_2, at which the free energy changes are ΔG_1 and ΔG_2, there results

$$\frac{\Delta G_2}{T_2} = \frac{\Delta G_1}{T_1} - \int_{T_1}^{T_2} \frac{\Delta H}{T^2}\,dT. \tag{37}$$

To evaluate the integral at the right, one needs to know how ΔH depends on T. (The simple case of ΔH independent of T will be discussed in a moment.) Often an empirical equation of the form

$$\Delta H = \Delta H_0 + \alpha T + \beta T^2 + \cdots \tag{38}$$

in which ΔH_0 is a value at some reference temperature, is known for the substance of interest, in which case Equation 37 becomes

$$\Delta G_2 = T_2 \left[\frac{\Delta G_1}{T_1} + \Delta H_0 \left(\frac{1}{T_2} - \frac{1}{T_1} \right) - \alpha \ln \frac{T_2}{T_1} - \beta(T_2 - T_1) \cdots \right], \quad (39)$$

from which ΔG_2 can be calculated.

A general expression for ΔG at any temperature can also be written. From Equation 35 we can write

$$\frac{\Delta G}{T} = - \int \frac{\Delta H}{T^2} + I(P), \quad (40)$$

in which $I(P)$ is some function of pressure that must be included since the partial derivatives were taken at constant pressure. Using Equation 38, one obtains

$$\Delta G = \Delta H_0 + I(P)T - \alpha T \ln T - \beta T^2 + \cdots. \quad (41)$$

The function $I(P)$ can be evaluated if one knows ΔG (and of course ΔH_0, α, β,...) at a particular temperature. Alternatively, since $\Delta G = \Delta H - T\Delta S$,

$$\Delta S = -I(P) + \alpha \ln T + \beta T + \cdots. \quad (42)$$

Then $I(P)$ can be evaluated if ΔS is known at a particular temperature.

It is interesting to observe that the values $I(P)$, ΔH_0, α, β, ... (which may be found in standard thermodynamic tables), depend upon the particular substance so that if ΔG is known at one temperature, the value that must be added or subtracted to evaluate ΔG at a second temperature is different for each substance. This is not the case for the effect of a concentration change since the increment is a term only in R, T, and the concentrations.

For relatively small temperature changes, for example, in an interval of 10°C, the preceding can be somewhat simplified. This simplification is especially useful in biochemical calculations, because the biochemical reactions that are studied occur in a range from 30°C to 37°C, whereas the thermodynamic constants listed in published tables are usually for a temperature of either 20°C or 25°C. Over a temperature range of 10°C, ΔH varies very little, so that we may assume that ΔH is constant and write, approximately,

$$\left[\frac{\partial (\Delta G/T)}{\partial T} \right]_P \approx \frac{\Delta G_2/T_2 - \Delta G_1/T_1}{T_2 - T_1} = -\frac{\Delta H}{T_{av}^2}, \quad (43)$$

in which $T_{av} = \frac{1}{2}(T_1 + T_2)$. Therefore

$$\Delta G_2 = \frac{T_2}{T_1} \Delta G_1 - \left(\frac{T_2 - T_1}{T_{av}^2} \right) \Delta H. \quad (44)$$

The interpretation of this equation is very important. If $\Delta H > 0$, a rise in

SEC. 6-1   THE GIBBS FREE ENERGY, G

temperature ($T_2 > T_1$) makes ΔG_2 more negative than ΔG_1, so that *increasing the temperature favors any reaction for which ΔH is greater than 0*. This simply means that if the reaction takes up heat (that is, it is endothermic), raising the temperature provides the needed heat and favors the reaction. If $\Delta H < 0$, the effect of heating is less favorable for the reaction, because heat is being put into a system that is attempting to release heat.

EXAMPLE 6-A   How is the combustion of glucose affected by an increase in temperature from 25°C to 37°C?

In the combustion of glucose (a basic biochemical process) at 25°C, $\Delta H_{298} = -2803$ kJ mol^{-1} and $\Delta G_{298} = -2862$ kJ mol^{-1}. At 37°C, $\Delta G_{310} = (310/298)(-2862) - [(310 - 298)/(304)^2](-2803) = -2975$ kJ mol^{-1}. ΔG_{310} is more negative than ΔG_{298} so that the reaction is more favorable at the higher temperature. (By "more favorable," one means that more of the product of the reaction is formed.)

Note that in Equation 40 the $-\Delta H$ term is merely a correction term that decreases the absolute value of the change in the first term. If ΔH is much less negative than ΔG, ΔH has a smaller effect. This is reasonable since $\Delta G << \Delta H$ when the positive entropy change is very great. Thus, as entropy plays a greater role in a reaction, the more a temperature increase will favor the reaction. This can also be seen from the basic equation $[\partial(\Delta G)/\partial T]_P = -\Delta S$, from which these arguments are derived. This equation simply states that if ΔS is large and positive, a positive ΔT must be accompanied by ΔG becoming more negative.

Several familiar phenomena confirm this argument. For example, when sucrose crystals go into solution, $\Delta S > 0$ because of the increased disorder as the molecules in the crystal are dispersed. As T increases, the solubility increases. A second example is the expansion of a gas within a cylinder caused by withdrawal of the piston. The gas expands in response to the reduced pressure. This process has a value of $\Delta S > 0$, because the gas is more disordered when it occupies a greater volume. Increasing the temperature aids in the expansion (that is, ΔG is more negative) and reduces the pressure drop required to cause a particular volume increase.

Pressure Dependence of ΔG

Although it is uncommon in living systems, there are many processes (for example, phase changes) that involve pressure changes. From Equation 23 we may write that, at constant temperature,

$$\left(\frac{\partial G}{\partial P}\right)_T = V \qquad \text{and} \qquad \left(\frac{\partial \Delta G}{\partial P}\right)_T = \Delta V, \qquad (45)$$

in which ΔV is the volume change occurring in a reaction. The effect of pressure on the boiling point of a liquid can be explained in terms of this equation. For vaporization, $\Delta V > 0$, because the gas phase always occupies more space than the liquid phase. Thus $[\partial(\Delta G)/\partial P]_T > 0$. At the boiling point, $\Delta G = 0$ for the vaporization process, because the gas and the liquid are in equilibrium. If P increases, $\Delta G > 0$, which means that the reverse process, condensation, rather than boiling, will have a value of $\Delta G < 0$ and will occur spontaneously. For boiling, $\Delta H > 0$, so that according to the previous section, ΔG decreases as T increases. Therefore, to decrease the positive value of ΔG resulting from a pressure increase to a value of $\Delta G = 0$, T must increase. This is indeed always the case—*as pressure increases, the boiling temperature increases.*

For both solids and liquids there is no volume change accompanying a pressure change (unless the pressure is very great, as may be found deep in the Earth or at the bottom of the sea). Thus, since $\Delta V = 0$, if $P > 0$, ΔG must equal 0. For solids and liquids, one need not be concerned with corrections in tabulated values of ΔG if the pressure is not 1 atm.

For reactions with gases, ΔG must depend upon pressure. For an ideal gas,

$$G_2 = G_1 + \int_{P_1}^{P_2} \frac{nRT}{P} \, dP = G_1 + nRT \ln\left(\frac{P_2}{P_1}\right) \qquad (46)$$

Since we only know values of ΔG, this must be written as

$$\Delta G_2 = \Delta G_1 + RT \ln\left(\frac{P_2}{P_1}\right). \qquad (47)$$

6–2 FREE ENERGY AND THE EQUILIBRIUM CONSTANT

We have seen how the free energy can be evaluated for a substance at a particular temperature and concentration and how the values obtained determine the direction of a chemical reaction. We have not yet shown how the values are related to the extent to which a reaction goes to completion. In the numerical calculation for the formation of alanylglycine, described earlier, there was an indication that the larger the absolute value of a negative ΔG, the greater would be the ratio of the concentrations of the products to the reactants. This was also the case in Example 6–A, in which an increase in temperature decreased ΔG, and in which we said that the reaction was favored. The precise relation between the value of ΔG and the extent of reaction can be seen by deriving the connection between ΔG and the equilibrium constant K.

At equilibrium the free energy change is zero. According to the view that the free energy change is a measure of the spontaneity of a reaction, this means that the forward reaction has the same probability of occurring

SEC. 6–2 FREE ENERGY AND THE EQUILIBRIUM CONSTANT

as does the back reaction. Another way of thinking about the fact that $\Delta G = 0$ at equilibrium is that there is no net reaction (which is what we mean by equilibrium) when the free energy of the reactants equals the free energy of the products.*

The equilibrium constant K of a reaction is defined in terms of ratios and products of the concentration of the constituents of the reaction. Free energy changes, on the other hand, are evaluated by addition and subtraction of values for each reactant and the product. Thus if there is to be a relation between free energy changes and K, we should expect that the free energy will be proportional to the logarithm of K. We will see that this is indeed the case.

The simple expressions that will now be derived are important for two reasons: (1) The equations show that a simple equilibrium measurement gives great insight about thermodynamic properties of a system; and (2) the equations provide one of the principal means of measuring ΔG for a chemical reaction.

As is often the case, thermodynamic expressions are easily derived for reactions between ideal gases at constant temperature. From the basic equation $dG = V\,dP - S\,dT$, there results at constant temperature

$$\left(\frac{\partial G}{\partial P}\right)_T = V. \tag{48}$$

By integrating from the standard state, at which the pressure and the free energy are $P°$ ($= 1$ atm) and $G°$, respectively, to P and G, and using the ideal gas law, one obtains

$$G - G° = RT \ln P. \tag{49}$$

In a gaseous mixture the pressure P is the sum of the partial pressures P_i of each of the i components. If n_i is the number of moles of the ith component,

$$P = \sum P_i = \frac{RT}{V} \sum n_i, \tag{50}$$

and for each component,

$$G_i - G_i° = RT \ln P_i. \tag{51}$$

Since for a chemical reaction between the i components, $\Delta G° = \sum n_i \Delta G_i°$, in which $\Delta G_i°$ is the free energy of formation of the ith component under standard conditions, we may write

---

*This statement sometimes causes confusion because it is not usual for the sum of the free energies of formation ΔG_{form} of the reactants to equal the sum of ΔG_{form} of the products. This apparent conflict is resolved merely by realizing that free energy is expressed in kilojoule *per mole* and the initial molar concentrations of the reactants are not the same as the molar concentration of reactants at equilibrium. By this we mean that if a reaction occurs, it is *always* possible to find concentrations of reactants and products for which $\Delta G = 0$.

$$\Delta G - \Delta G° = RT \sum n_i \ln P_i, \qquad (52)$$

in which ΔG is the free energy change for the reaction. When the pressures are the equilibrium values $P_{i,eq}$, then $\Delta G = 0$, and

$$-\Delta G° = RT \sum n_i \ln P_{i,eq}. \qquad (53)$$

For a typical reaction such as $aA + bB \rightleftarrows cC + dD$,

$$\sum n_i \ln P_{i,eq} = \ln \frac{(P_{C,eq})^c (P_{D,eq})^d}{(P_{A,eq})^a (P_{B,eq})^b}, \qquad (54)$$

which equals the equilibrium constant K of the reaction in terms of the partial pressures. Thus

$$-\Delta G° = RT \ln K. \qquad (55)$$

For an ideal solution, the vapor pressure of each component is proportional to its concentration (this fact is discussed in detail in Chapter 7); thus we should expect to be able to derive an identical expression in terms of concentrations, as will be seen in the following.

The desired expression relating ΔG and K for a reaction in solution will first be derived for a simple system of two components in equilibrium as, for example, in $A \rightleftarrows B$. An example of such a reaction is the isomerization of a sugar. The starting point is the equation

$$\Delta G = \Delta G° + RT \ln c, \qquad (56)$$

in which ΔG is the free energy of a component at concentration c having a standard free energy of formation $\Delta G°$. Denoting the concentrations of A and B by [A] and [B], respectively,

$$\Delta G_A = \Delta G_A° + RT \ln[A], \qquad (57)$$

and

$$\Delta G_B = \Delta G_B° + RT \ln[B]. \qquad (58)$$

At equilibrium there is no free energy change; that is, $\Delta G_A = \Delta G_B$, so that

$$\Delta G_B° - \Delta G_A° = RT \ln \frac{[B]}{[A]}. \qquad (59)$$

Since the standard free energy change for the reaction is $\Delta G°$ and by definition $K = [B]/[A]$, we have

$$\Delta G° = -RT \ln K,$$

which is identical to Equation 55. Note that $-RT \ln K \neq 0$, so $\Delta G° \neq 0$; this does not conflict with the equilibrium condition that $\Delta G = 0$, because $\Delta G°$ is the standard free energy change when all components are at a

SEC. 6-2 FREE ENERGY AND THE EQUILIBRIUM CONSTANT

concentration of 1 M, which is rarely the condition of equilibrium. The value of K defines the ratio of the concentrations of each component of the reaction at equilibrium and this ratio is related to $\Delta G°$ by Equation 59. *At equilibrium it is ΔG and not $\Delta G°$ that is 0.* For example, suppose the reaction were the isomerization of glucose (Glu) to galactose (Gal). The standard free energies of formation, $\Delta G°_{form}$(Glu) and $\Delta G°_{form}$(Gal), are respectively -917.22 and -923.53 kJ mol^{-1} at 30°C, so that $\Delta G° = -6.31$ kJ mol^{-1}. By Equation 59, $-6310 = -(8.314)(303) \ln ([B]/[A])$ and $[B]/[A] = 12.3$; that is, at this concentration-ratio, $\Delta G = 0$. For example, if $[A] = 10^{-3}$ M and $[B] = 1.23 \times 10^{-2}$ M, then $[B]/[A] = 12.3$; and then

$$\Delta G_A = -917.22 + (8.314 \times 10^{-3})(303) \ln 10^{-3} = -934.62 \text{ kJ mol}^{-1} ;$$

and

$$\Delta G_B = -923.53 + (8.314 \times 10^{-3})(303) \ln (1.23 \times 10^{-2})$$
$$= -934.62 \text{ kJ mol}^{-1} ;$$

both of these energies have the same values.

This reasoning can be extended to a more complex reaction, such as $A + B \rightleftarrows C + D$, to yield

$$\Delta G = \Delta G° + RT \ln \frac{[C][D]}{[A][B]}, \tag{60}$$

in which ΔG is the free energy difference between the given concentrations and the equilibrium concentrations: at equilibrium, $\Delta G = 0$, and $[A]$, $[B]$, $[C]$, and $[D]$ are the equilibrium concentrations, so that by using the definition $K = [C][D]/[A][B]$, we have once again

$$\Delta G° = -RT \ln K .$$

Note that for a reaction $aA + bB \rightleftarrows cC + dD$ we would have

$$\Delta G = \Delta G° + RT \ln \frac{[C]^c[D]^d}{[A]^a[B]^b}, \tag{61}$$

but since in this case $K = [C]^c[D]^d/[A]^a[B]^b$, the equation $\Delta G° = -RT \ln K$ remains valid.

We have so far derived expressions only for the conditions of constant temperature. The value of $\Delta G°$ changes with T in a quite simple way, as does K. The expression we will obtain will allow us to calculate K for a temperature other than that used in a measurement, or conversely, to calculate $\Delta G°$ at a temperature other than that used to determine K.

Starting with $-\Delta G° = RT \ln K$ rearranged as $\ln K = -\Delta G°/RT$, we may take the derivative of both sides with respect to T and obtain

$$\frac{d \ln K}{dT} = -\frac{1}{R} \frac{d(\Delta G°/T)}{dT}$$

$$= -\frac{1}{R} \frac{d}{dT}\left(\frac{\Delta H°}{T} - \Delta S°\right) . \tag{62}$$

If $\Delta H°$ and $\Delta S°$ are *both* independent of temperature*, then

$$\frac{d \ln K}{dT} = +\frac{\Delta H°}{RT^2}. \tag{63}$$

This equation is called the *van't Hoff equation*. It can be integrated to yield

$$\ln \frac{K_2}{K_1} = -\frac{\Delta H°}{R}\left(\frac{1}{T_2} - \frac{1}{T_1}\right), \tag{64}$$

in which K_1 and K_2 are the equilibrium constants at T_1 and T_2, respectively. Note that a plot of $\ln K$ versus $1/T$ gives a straight line of slope $-\Delta H°/R$.

The van't Hoff equation and its integrated forms are valuable for two reasons: (1) if K is known at one temperature and $\Delta H°$ is available, K can be calculated at another temperature; and (2) if K is measured at two temperatures, $\Delta H°$ can be calculated. This is in fact one of the more important ways to determine $\Delta H°$ experimentally.

The following point should also be noted—if $\Delta H° > 0$, K increases with temperature. Since K is the product of the concentrations of the reaction products divided by the product of the concentrations of the reactants, this means that the extent of reaction increases with increasing temperature if $\Delta H > 0$. This is easy to understand because, if a reaction needs to absorb heat, the extent of reaction will increase if the heat is provided through an increase in temperature.

It should also be noted that entropy changes play little or no role in the temperature dependence of K. This does not mean that entropy is not involved in the properties of the reaction because, of course, it is. Since

$$\ln K = -\frac{\Delta G°}{RT} = \frac{\Delta S°}{R} - \frac{\Delta H°}{RT},$$

the values of $\Delta S°$ determine a portion of the value of K that is temperature-*in*dependent and is based upon the actual molecular structures of the reactants and products. To this basic value is added the component dependent on both $\Delta H°$ and temperature.

EXAMPLE 6-B Calculation of $\Delta G°$, $\Delta H°$, and $\Delta S°$ at 25°C for the binding of a small molecule, X, to a protein, P.

Consider a reaction in which one mole of X binds to one mole of P to form one mole of the complex PX. The reaction is

$$P + X \rightleftarrows PX,$$

and $K = [PX]/[P][X]$. To determine K, various amounts of X are added to a

---

*This is generally a good assumption in biochemical analysis because the temperature of interest rarely differs by more than 15°C from the temperatures used in thermodynamic tables. If larger temperature changes are required, it is necessary to use Equations 38 and 42 for $\Delta H°$ and $\Delta S°$ prior to integration.

SEC. 6-2   FREE ENERGY AND THE EQUILIBRIUM CONSTANT   175

fixed amount of P and [PX] is measured by some means. The concentration of X is calculated as [added X] − [PX]. Rearranging the definition of K, we have [PX]/[P] = K[X], so that a plot of [PX]/[P] versus [X] yields a straight line having slope K. This process is carried out at several temperatures and the following data are obtained:

T, °C	$K \times 10^{-7}$
16	7.25
21	5.25
25	4.17
32	2.66
37	2.00

Evaluating $RT \ln K$ for each pair of numbers and averaging the values obtained yields, from the relation $-\Delta G° = RT \ln K$, a value of $\Delta G° = -43.5$ kJ mol^{-1}. The next step is to make a plot of $\ln K$ versus $1/T$. The slope of the straight line that results is $-\Delta H°/R$, or 5.435, so that $\Delta H° = -45.2$ kJ mol^{-1}. Since $\Delta S° = (\Delta H° - \Delta G°)/T$, then $\Delta S° = (-45.2 + 43.5)/298 = -5.7$ J K^{-1} mol^{-1} at 25°C.

When discussing chemical reactions, one often must make some statement about the extent of the reaction. Usually this is accomplished by giving the value of the equilibrium constant. This is quite informative because it is usually clear that if $K \ll 1$, the reaction is not very efficient, and if $K \gg 1$, the reaction will proceed strongly to the right. Many people express K in terms of a pK value (p$K = -\log_{10} K$); this has the advantage that the pK value is a simpler number to write than K is (for instance, if $K = 5.74 \times 10^{-6}$, the pK value is 5.24), but has the disadvantage that it is harder to calculate the extent of reaction in one's head because of the logarithmic nature of pK. Even more inconvenient is the practice of quoting values of $\Delta G°$. It is useful to have a table that relates the extent of reaction, K, pK, and $\Delta G°$. Table 6-1 gives these values for a reaction A $\rightleftarrows$ B.

6-3   SUMMARY OF METHODS FOR EVALUATING THERMODYNAMIC FUNCTIONS

Standard Free Energy Change

We have seen that this can be calculated from the following:

1. The values of ΔG_{form}.
2. The values of $\Delta H°$ and $\Delta S°$.
3. The value of K, by means of the relation $-\Delta G° = RT \ln K$.

TABLE 6-1 Relation between the extent of reaction and the values of K and $\Delta G°$ for a reaction $A \rightleftarrows B$.

Extent of reaction at equilibrium (fraction of reactant converted to product)	K (25°C)	pK	$\Delta G°$ (25°C) kJ mol^{-1}
0.0001	0.0001	+4.00	+22.80
0.001	0.001	+3.00	+17.11
0.01	0.01	+2.00	+11.42
0.1	0.11	+0.95	+5.48
0.3	0.43	+0.36	+2.09
0.5	1	0	0
0.8	4	−0.60	−3.43
0.95	19	−1.28	−7.28
0.99	99	−2.00	−11.38
0.999	999	−3.00	−17.11
0.9999	9999	−4.00	−22.80
0.99999	99,999	−5.00	−28.45
1.0	∞	$-\infty$	$-\infty$

In some cases the values needed for (1) and (2) are not available in standard reference tables and the equilibrium lies so far to one side that K cannot be measured, as required in (3). In this case a fourth procedure can sometimes be used that is based on the fact that G is a state function. This method is the following.

4. Calculation of $\Delta G°$ from equivalent reactions.

Let us consider the evaluation of $\Delta G°$ for a reaction $A \rightleftarrows B$ in which the amount of B is too small to measure. If one knows $\Delta G_1°$ for a reaction $A \rightleftarrows C$ and $\Delta G_2°$ for a reaction $B \rightleftarrows C$, one may write the reaction $A \rightleftarrows B$ as the sum of the other two reactions, for which the values of $\Delta G°$ are known; namely,

$$A \rightleftarrows C, \quad \text{with } \Delta G_1°,$$
$$\underline{C \rightleftarrows B, \quad \text{with } -\Delta G_2°,}$$
$$A \rightleftarrows B, \quad \text{with } \Delta G° = \Delta G_1° - \Delta G_2°.$$

A fifth method, which is indirect, is of great value for reactions that are strongly temperature-dependent:

5. Calculation of $\Delta G°$ for a reaction $X \rightleftarrows Y$ at a particular temperature, T, and a particular concentration, [X], when [Y] is immeasurable at the particular concentration of X and temperature of the reaction.

The method is to choose first a concentration and temperature at which both [X] and [Y] are measurable, determine K, correct K to the required temperature by using Equation 64, and finally calculate $\Delta G°$. This method is shown in the following example.

SEC. 6-3 EVALUATING THERMODYNAMIC FUNCTIONS

EXAMPLE 6-C Calculation of $\Delta G°$ for the reaction $X \rightleftarrows Y$ at 25°C, for which $[X] = 0.1\ M$ and $[Y]$ is $8.7 \times 10^{-6}\ M$, at 45°C, $10^{-4}\ M$ at 55°C, and $10^{-3}\ M$ at 65°C, respectively.

We prepare the following table.

[X]	[Y]	$K = [Y]/[X]$	$\ln K$	T, K	$1/T$
0.1	?	?	?	298	0.003356
0.1	8.7×10^{-6}	8.7×10^{-5}	-9.35	318	0.003145
0.1	1×10^{-4}	1×10^{-3}	-6.908	328	0.003049
0.1	1×10^{-3}	1×10^{-2}	-4.608	338	0.002958

Applying Equation 64, we plot $\ln K$ against $1/T$. Extrapolation to $1/T = 0.003356$ gives $\ln K = -14.73$. From $\Delta G° = -RT \ln K$, one obtains $\Delta G°_{298} = 36.48$ kJ mol^{-1}. Note also that if $\ln K = -14.73$, $K = 4 \times 10^{-7}$, and $[Y] = 4 \times 10^{-8}\ M$ at 25°C.

A sixth method in which $\Delta G°$ is evaluated from an electrochemical cell containing the reaction mixture is explained in Chapter 10.

Standard Enthalpy Change

Values of $\Delta H°$ can be calculated in three ways:

1. From the temperature dependence of K.
2. From values of $\Delta H°_{form}$.
3. From combustion in a bomb calorimeter.

Method 3 is becoming more widely used because of the development of very accurate microcalorimeters.

Standard Entropy Change

A value of $\Delta S°$ is usually calculated from $\Delta G°$ and $\Delta H°$ by using $\Delta S° = (\Delta H° - \Delta G°)/T$ or, if they are known, from absolute standard entropies.

6-4 APPLICATION OF FREE ENERGY AND BIOLOGICAL PROCESSES

Coupled Reactions

Many biochemical reactions have a very large positive value of ΔG at the conditions (for example, concentration and temperature) of the reaction in living organisms. These reactions do in fact occur but only when they are tightly coupled to a second reaction having a large negative value of ΔG, so that $\Delta G < 0$ for the pair of reactions. For example, consider the reactions

$$A \rightleftarrows B, \quad \text{with } \Delta G° = 23.01 \text{ kJ mol}^{-1},$$
$$B \rightleftarrows C, \quad \text{with } \Delta G° = -30.12 \text{ kJ mol}^{-1},$$

which couple through the common intermediate B. The net reaction $A \rightleftarrows C$ has a value for $\Delta G°$ of $23.01 + (-30.12) = -7.11$ kJ mol^{-1}. This of course is formally correct because we know that ΔG is a function only of the initial and final states (that is, of the state of A and of C). One must not make this statement too casually, because obviously the initial and final states must be connected. That is, the pair of reactions must be coupled. That is, the pair of reactions $A \rightleftarrows B$ and $C \rightleftarrows D$ cannot lead to the reaction $A \rightleftarrows D$ unless there is some way to get from B to C. Of course if there is a reaction $B \rightleftarrows C$, the coupled system consists of three reactions and the net ΔG must be the sum of the ΔG values of each. Clearly it is meaningful to speak of coupled reactions only when there is a continuous reaction pathway—that is, *when there are intermediates common to each successive pair of reactions.*

In judging whether a coupled reaction can occur, one must be careful to distinguish $\Delta G°$ from ΔG. For example, consider the reaction $A + B \rightleftarrows C + D$. We are interested in knowing whether we can achieve a minimal concentration of C and D, of 10^{-4} M for each at 25°C, if A and B are each at 10^{-2}-M concentration. Values of $\Delta G°$ are available for the following pair of reactions:

$$A \rightleftarrows B, \quad \text{with } \Delta G° = 23.01 \text{ kJ mol}^{-1},$$
$$B \rightleftarrows C, \quad \text{with } \Delta G° = -30.12 \text{ kJ mol}^{-1},$$

and

$$A + E \rightleftarrows C + F, \quad \text{with } \Delta G° = -1.00 \text{ kJ mol}^{-1},$$
$$D + E \rightleftarrows B + F, \quad \text{with } \Delta G° = -4.81 \text{ kJ mol}^{-1}.$$

The reaction

$$A + B \rightleftarrows C + D$$

is obtained by reversing the second reaction and summing; that is,

$$A + E \rightleftarrows C + F, \quad \text{with } \Delta G° = -1.00 \text{ kJ mol}^{-1}$$
$$\underline{B + F \rightleftarrows D + E, \quad \text{with } \Delta G° = +4.81 \text{ kJ mol}^{-1}}$$
$$A + B + E + F \rightleftarrows C + D + E + F, \quad \text{with } \Delta G° = +3.81 \text{ kJ mol}^{-1}$$

or simply

$$A + B \rightleftarrows C + D, \quad \text{with } \Delta G° = 3.81 \text{ kJ mol}^{-1}.$$

Thus the value of $\Delta G°$ for the coupled reaction is $-1.00 + (+4.81) = +3.81$ kJ mol^{-1}. This positive value of $\Delta G°$ means that the reaction does

SEC. 6-4 FREE ENERGY AND BIOLOGICAL PROCESSES

not occur spontaneously at a concentration of 1 M for each substance. To determine ΔG at the required conditions (in other words, not in the standard state) we use Equation 60 and substitute 10^{-2} M for [A] and [B] and 10^{-4} M for [C] and [D]:

$$\Delta G = \Delta G° + RT \ln \frac{[C][D]}{[A][B]} = -1900 \text{ kJ mol}^{-1}.$$

Thus the concentrations of C and D can be achieved and in fact they will be greater than required.

We can also calculate the concentration of C and D at equilibrium by using Equation 55. From the values of $\Delta G° = 3.81$ kJ mol^{-1} and $T = 298$ K, we obtain $K = 0.215$. Since $K = [C][D]/[A][B]$, if we substitute $x = [C] = [D]$ and $(10^{-2} - x) = [A] = [B]$, then [C] and [D] at equilibrium are each roughly 0.0032 M.

The Biochemical Standard State

The standard state of a substance in solution has been defined as a solution of the substance at a concentration of 1 M. Thus if the H$^+$ ion participates in a reaction, its concentration must also be 1 M for the purposes of calculating $\Delta G°$. Biochemical reactions typically occur near pH 7, however— that is, [H$^+$] = 10^{-7} M—and they occur in water. Thus many biochemists have adopted the convention that in the determination of standard free energy changes, the concentration of water should be ignored and, if the H$^+$ ion plays a role in the reaction, the standard state should be that at which all reactants have a concentration of 1 M except for [H$^+$], which will have a concentration of 10^{-7} M. Many biochemical tables reflect this convention. The biochemical standard free energy change is denoted $\Delta G°'$. It is often necessary to use a variety of pH values; the biochemical standard free energy change at a pH of X is denoted $\Delta G°'_{pH=X}$. When the subscript is omitted, a pH of 7 is implied. Some books use the more complete notation $\Delta G°'_{pH=7}$ when the pH value is 7; in this book we will not do that but will, instead, use no subscript when the pH value is 7.

The expression for ΔG in terms of $\Delta G°'$ differs from the equation utilizing $\Delta G°$. The new expression can be easily derived by considering the reaction H$^+$ + B $\rightleftarrows$ C + D. From Equation 60, we write the standard state of 1 M for each substance as

$$\Delta G = \Delta G° + RT \ln \frac{(1)(1)}{(1)(1)}.$$

At any other concentration, we would write

$$\Delta G = \Delta G° + RT \ln \frac{[C][D]}{[A][B]} - RT \ln \frac{(1)(1)}{(1)(1)}.$$

If B is the H$^+$ ion, the expression in terms of $\Delta G°'$ will be

$$\Delta G = \Delta G^{\circ\prime} + RT \ln \frac{[C][D]}{[A][B]} - RT \ln \frac{(1)(1)}{(10^{-7})(1)}$$

$$= \Delta G^{\circ\prime} + RT \ln \frac{10^{-7}[C][D]}{[A][H^+]}$$

$$= \Delta G^{\circ\prime} - 16.118 RT + RT \ln \frac{[C][D]}{[A][H^+]}, \qquad (65)$$

so that

$$\Delta G^{\circ\prime} = \Delta G^{\circ} + 16.118\, RT \,. \qquad (66)$$

An example of the use of $\Delta G^{\circ\prime}$ should remove any confusion.

EXAMPLE 6–D Properties of the reaction NADH + pyruvate + H$^+$ $\rightleftharpoons$ NAD$^+$ + lactate for which at 37°C the equilibrium constant K is 1.68 × 10^{11}.

We can calculate $\Delta G^{\circ\prime}$ from Equation 65 as

$$\Delta G^{\circ\prime} = -(8.314 \times 10^{-3})(310) \ln \frac{[NAD^+][lactate]10^{-7}}{[NADH][pyruvate][H^+]}$$

$$= 2.577 \ln(10^{-7} K) = -25.07 \text{ kJ mol}^{-1}\,.$$

If we were interested in the reaction under the conditions of [lactate] = 10^{-3} M, [pyruvate] = 5×10^{-5} M, [NADH]/[NAD$^+$] = 10^{-1}, and pH = 5, the value of ΔG would be

$$\Delta G = \Delta G^{\circ\prime} + (2.577) \ln \frac{[NAD^+][lactate]10^{-7}}{[NADH][pyruvate][H^+]}$$

$$= -25.07 + 2.577 \ln 2 = -23.28 \text{ kJ mol}^{-1}\,.$$

Free Energy of Solution

The free energy of solution for a particular substance is the value of ΔG required to dissolve the solid form of the substance and form a solution having a concentration c. This process is conveniently thought of as a sequence of two steps: (1) creating a saturated solution having the concentration c_{sat} and (2) diluting the saturated solution to the desired concentration. This two-step process is useful because *in a saturated solution, a solid is in equilibrium with the solution so that $\Delta G = 0$ for the formation of the saturated solution.* Thus the free energy of solution ΔG_{sol} is simply the free energy of dilution from c_{sat} to c, or,

$$\Delta G_{sol} = RT \ln(c/c_{sat})\,. \qquad (67)$$

This useful expression enables us to calculate the free energy of formation

ΔG_{form} at a concentration c; that is $\Delta G_{form,c}$. If ΔG_{form} is the free energy of formation of the solid itself, then from Equation 56 we have

$$\Delta G_{form,c} = \Delta G_{form} + RT \ln(c/c_{sat}) \ . \tag{68}$$

Note that if $c = 1\ M$, $\Delta G_{form,1M}$ is the same as $\Delta G^\circ_{form} = \Delta G_{form} - RT \ln c_{sat}$—that is, Equation 56.

If the substance ionizes in solution, the free energy of ionization must be included in Equation 67. This is done as follows. Consider a divalent acid that dissociates stepwise (see Chapter 9):

(1) $H_2A \rightleftarrows AH^- + H^+$;
(2) $AH^- \rightleftarrows A^{2-} + H^+$.

The total dissociation equation is

$$H_2A \rightleftarrows A^{2-} + 2H^+ \ ,$$

which has the equilibrium constant

$$K = \frac{[A^{2-}][H^+][H^+]}{[H_2A]} \ . \tag{69}$$

The equilibrium constants for steps 1 and 2 are

$$K_1 = \frac{[AH^-][H^+]}{[H_2A]} \quad \text{and} \quad K_2 = \frac{[A^{2-}][H^+]}{[AH^-]} \ , \tag{70}$$

so that

$$K = K_1 K_2 \ . \tag{71}$$

In general the free energy change in formation of the nth ion is $\Delta G^\circ_n = -RT \ln K_1 K_2 \cdots K_n$, so that

$$\Delta G^\circ_{ion} = \Delta G^\circ_{neutral} - RT \ln (K_1 K_2 \cdots K_n) \ .$$

Note that this equation is written with the standard free energy change at the standard concentration of $1\ M$. The value can be converted to any other concentration by subtracting the free energy of dilution.

In biochemical reactions, the H^+ concentration is usually determined by factors external to the reaction. Furthermore, it is often the case that the total concentration $[A_{tot}]$ of a dissociating substance is known rather than the concentration of the ions. Thus, it is convenient to calculate the free energy of dilution in terms of these parameters. Using the definitions of K_1 and K_2, one writes

$$[A^{2-}] = [A_{tot}]\frac{K_1 K_2}{[H^+]^2 + K_1[H^+] + K_1 K_2} \ . \tag{72}$$

Thus, the free energy of dilution of this ion from $1\ M$ to $[A^{2-}]$ is

$$\Delta G = RT \ln \frac{[A^{2-}]}{1} = RT \ln [A_{tot}]\frac{K_1 K_2}{[H^+]^2 + K_1[H^+] + K_1 K_2} \ , \tag{73}$$

and the free energy of an ion A^{2-} at any value of $[H^+]$ and of $[A_{tot}]$ is

$$\Delta G_A = \Delta G^\circ_{\text{neutral},1M} - RT \ln K_1 K_2$$
$$- RT \ln|A_{\text{tot}}| \frac{K_1 K_2}{|H^+|^2 + K_1|H^+| + K_1 K_2}. \quad (74)$$

Note that this equation applies only for a divalent cation such as CO_3^{2-}. If one is interested in the concentration of the first ionization species such as HCO_3^-, one must derive an equation such as that above for $[AH^-]$ instead.

Similar calculations can be performed for reactions in which a gas is in solution. However, in this case, we must be sure to remember that the standard state is at a pressure of 1 atm whereas the partial pressure of the gas may be less in the reaction. This can be seen in the following example.

EXAMPLE 6–E Under atmospheric conditions at 19°C, what is the free energy of the oxygenation of hemoglobin (Hb)?

The reaction is written

$$Hb(aq) + O_2(g) \rightleftarrows HbO_2(aq).$$

This reaction has been studied by using pure O_2 at atmospheric pressure and has a value for K of 85.5 atm^{-1} at 19°C, which by Equation 56 means that $\Delta G^\circ_{292} = -10.837$ kJ mol^{-1}. This is the value of ΔG for the reaction in which O_2 is dissolved, at a pressure of 1 atm (that is, the standard state in solution), and we would like to have the value of ΔG for normal atmospheric conditions. We must make use of the fact that at 19°C the concentration of O_2 in water is 0.00023 M when the water is exposed to air and that in air the partial pressure of O_2 is 0.2 atm. We then write the following set of reactions:

(1) $O_2(g, 1 \text{ atm}) \rightleftarrows O_2(g, 0.2 \text{ atm})$;
(2) $O_2(g, 0.2 \text{ atm}) \rightleftarrows O_2(aq, sat)$;
(3) $O_2(aq, sat) \rightleftarrows O_2(aq, 1 M)$.

For the first reaction, $\Delta G_{\text{form}} = 0$ at 1 atm (by definition, this is true for elements in the standard state); and the value of ΔG for the reaction is a result of the pressure change—that is, $\Delta G = RT \ln(0.2/1) = -3.93$ kJ mol^{-1}. In the second reaction, $\Delta G = 0$ because of the equilibrium between a substance and its saturated solution. For the third reaction, we consider the concentration change from the normal solubility to 1 M, so that $\Delta G = RT \ln(1/0.00023) = 20.38$ kJ mol^{-1}. These three reactions sum to

$$O_2(g) \rightleftarrows O_2(aq),$$

for which $\Delta G = 20.38 - 3.93 = 16.45$ kJ mol^{-1}. This is subtracted from $\Delta G^\circ = -10.837$ kJ mol^{-1} to yield $\Delta G = -27.287$ kJ mol^{-1} for the reaction of interest.

Free Energy of the Hydrophobic Interaction

The structure of many biological macromolecules, such as nucleic acids and proteins, is determined at least in part by hydrophobic interactions. This conclusion has in part been derived from knowledge of the strength of the interaction.

The free energy of a hydrophobic interaction can be calculated by using the methods of the preceding section.

If a linear polymer consists of nonpolar and polar monomers, then in a polar solvent the polymer spontaneously folds to minimize the contact of the nonpolar units with the solvent (see Chapters 5 and 17). This can be thought of as a tendency for the relatively insoluble groups to cluster and reduce the area of contact with the solvent. One says that the nonpolar components are held together by a hydrophobic interaction. The stability of this interaction should be understandable in terms of free energy changes. A nonpolar cluster is equivalent to a solution of a nonpolar group in a nonpolar solvent. The formation of a nonpolar cluster in water is like transferring a nonpolar group from water to a nonpolar solvent. Hence the free energy of forming the cluster is the same as the free energy of transfer of a nonpolar group from a polar solvent to a nonpolar solvent and it should be less than zero. An approximate calculation for the hydrophobic side chains of a protein shows that this is the case. (Actually we will calculate the free energy of transfer from a *non*polar to a polar solvent and show that it is positive.) Since the side chain cannot be studied in isolation, we assume that the transfer can be thought of as having two parts—(1) that of a nonpolar group R and (2) that of a polar portion of an amino acid. Since the latter portion can be approximated with glycine (which is the polar portion plus a single H atom), we use the thermodynamic parameters of this amino acid. Glycine and most other amino acids are sufficiently insoluble in nonpolar solvents that the concentrations cannot be evaluated. Thus we choose the partially nonpolar solvent ethanol (eth), realizing that if a trend is seen with this solvent, the trend will be greater with a more nonpolar solvent. We break down the transfer reaction into the following set of reactions:

$$\text{Gly(s)} \rightleftarrows \text{Gly(eth,sat)}, \quad \Delta G = 0 ;$$
$$\text{Gly(eth,1 } M) \rightleftarrows \text{Gly(eth,sat)}, \quad \Delta G = RT \ln c_{\text{sat,eth}} ;$$
$$\text{Gly(s)} \rightleftarrows \text{Gly(aq,sat)}, \quad \Delta G = 0 ;$$
$$\text{Gly(aq,sat)} \rightleftarrows \text{Gly(aq,1 } M), \quad \Delta G = RT \ln(1/c_{\text{sat,aq}}) .$$

The value of ΔG for the total transfer process is

$$\Delta G = RT \ln(c_{\text{sat,eth}}/c_{\text{sat,aq}}) .$$

Values of $c_{\text{sat,eth}}$ and $c_{\text{sat,aq}}$ are available for each amino acid. To calculate ΔG for the R group, we invoke the approximation that if R_1 is the side chain in amino acid #1, ΔG_1 is the free energy of transfer for the same amino acid, and

$$\Delta G_{R_1} = \Delta G_1 - \Delta G_{Gly} .$$

The result of such a calculation is shown in the following chart:

Amino acid	ΔG, kJ mol^{-1}	ΔG_R, kJ mol^{-1}
Glycine	-19.37	0
Alanine	-16.32	$+ 3.05$
Leucine	$- 9.25$	$+10.13$
Phenylalanine	$- 8.28$	$+11.09$
Isoleucine	$- 7.07$	$+12.43$

Thus we see that as the length of the hydrophobic side chain increases, the free energy required to transfer an amino acid from ethanol to water becomes more positive so that given a choice an amino acid will tend to dissolve in ethanol rather than water. Presumably if the solvent were more nonpolar, the ΔG_R values would be greater, as would be the tendency for an amino acid to dissolve in the nonpolar solvent. This means that if an amino acid solution in water were placed in contact with an immiscible nonpolar solvent, a significant fraction of the amino acid molecules would move to the nonpolar solvent; this fraction is greater for amino acids having longer hydrophobic side chains. This effect is entropy-driven because, as discussed in Chapter 5, the introduction of a nonpolar group into water results in the formation of a highly ordered array of water molecules surrounding the incoming molecule and therefore an entropy decrease. This ordering is a consequence of the strong tendency of water molecules to hydrogen-bond with one another. In ethanol, on the contrary, there is much less hydrogen-bonding between ethanol molecules, so there is little tendency for the solvent molecules to form an ordered shell around a nonpolar solute molecule. Therefore, dissolving an amino acid in ethanol is either accompanied by a very small, negative value of ΔS or, more commonly, by a positive ΔS (as usual, since the dissolved molecules are dispersed). Therefore the transfer of a nonpolar solute molecule from water to ethanol has a positive ΔS and is the favored process.

Activation Energy

It has been stated repeatedly that a reaction proceeds spontaneously if $\Delta G < 0$. However, we have said nothing about reaction rates. That is, a reaction might be completed in a microsecond or take thousands of years (as might be the case in geological processes such as the formation of rock). A further point that is not unrelated to the question of reaction rates is the following. It is well known that the oxidation of most compounds (for example, sucrose) does not proceed spontaneously at a measurable rate even though $\Delta G < 0$. In general, oxidation (or combustion) must be initiated by an input of energy, after which the reaction occurs rapidly. That is, wood must first be heated before it burns. The energy that must be added is called *activation energy* and denoted by E_a even though it is probably better thought of as an enthalpy. A common way to describe E_a is to say that there is an energy *barrier* between the reactants and the products.

There are two ways to think about the activation energy of a reaction. In the first, we consider that one of the reactants undergoes a conversion to a new substance when the activation energy is added; we say that the new substance is an activated form of the reactant or that the reactant is in an *activated state*. In these terms the reaction can be decomposed into two reactions—a first reaction having a positive ΔG, in which the substance is activated, and a second reaction, having a negative ΔG, in which the activated compound reacts with other reactants to form the products. A more traditional way (discussed more fully in Chapter 11) is to think about the activation energy in terms of the reactant velocities and the distribution of energies of the individual reactant molecules. The argument is the following. One assumes that the activated state must be reached in order to initiate the reaction and that the reaction rate (for example, the rate of formation of any product) is proportional to the number of activated molecules. In any population of molecules the energy per molecule obeys the Boltzmann distribution and the fraction of the population having energy equal to or greater than that of the activated state can be increased by adding energy (that is, heat) to the system—in other words, by raising the temperature. Svante Arrhenius drew an analogy to the equation that relates the equilibrium constant to the temperature (that is, $d \ln K/dT = \Delta H°/RT^2$) and proposed that the reaction rate constant k might depend on temperature in a way described by the equation

$$\frac{d \ln k}{dT} = \frac{E_a}{RT^2}. \tag{75}$$

He further assumed that for the most part E_a is independent of temperature, so that Equation 63 is equivalent to*

$$k = Ae^{-E_a/RT}, \tag{76}$$

in which A is constant. If $\ln k$ is plotted against $1/T$, there will be a straight line whose slope is E_a/R. Note also that the reverse reaction for which $\Delta G < 0$ also has a much greater E_a than the forward reaction.

It is important to realize that even though the spontaneous direction of a reaction is defined by the sign of ΔG, one cannot state that the reverse reaction does not occur. Thermodynamics does not deal with molecular changes but only the average properties of a system. The average direction

---

*This k is *not* the Boltzmann constant. Unfortunately the lower case k is also conventionally used to designate rate constants.

of a reaction can certainly be the result of both a forward and a backward reaction as long as the forward reaction predominates. In fact, the forward reaction predominates because it occurs more rapidly than the reverse reaction. Since the equilibrium constant K is not only defined in terms of the concentrations of products and reactants but also in terms of rates—that is,

$$K = k_1/k_{-1}, \tag{77}$$

in which k_1 and k_{-1} are the rate constants for the forward and back reactions—one can see that if $\Delta G°$ can be calculated from values in tables or measured independently and if either k_1 or k_{-1} is known, the other rate constant can be predicted from the relation $\Delta G° = -RT \ln K$.

Group Transfer Potential

There are numerous biochemical reactions for which $\Delta G > 0$ at the concentrations of the reactants that are present in living cells. Only at very high concentrations of the reactants is $\Delta G < 0$, yet these reactions do occur. Earlier in this chapter we discussed coupled reactions in which two reactions, one for which $\Delta G_1 > 0$ and a second for which $\Delta G_2 < 0$, are combined to produce a reaction for which $\Delta G_1 + \Delta G_2 < 0$. This coupling enables innumerable biochemical reactions to occur. The most common "helper" reaction is hydrolysis of so-called *high energy compounds,* of which the best studied is *adenosine-5'-triphosphate,* or, ATP (Figure 6-1). ATP is said to contain *high energy bonds.* This is a misnomer because, correctly speaking, a high bond energy means that the bond is very stable and not easily broken, whereas the relevant bonds in ATP are very easily broken. A correct statement is that the hydrolysis of ATP to form adenosine diphosphate (ADP) has a large and negative free energy of hydrolysis. That is, the equilibrium

$$ATP^{4-} + H_2O \rightleftarrows ADP^{3-} + H_2PO_4^-$$

Figure 6-1 Chemical structure of adenosine-5'-triphosphate (ATP). The "high energy phosphate bonds"—that is, those for which $\Delta G°'$ is large and negative—are indicated by the arrows.

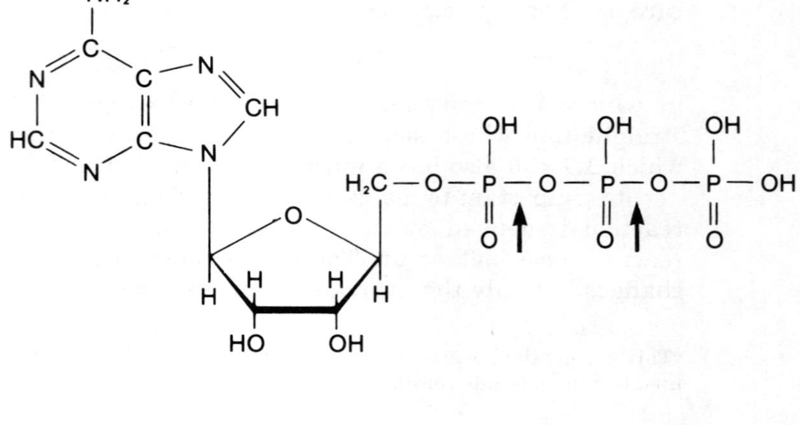

lies far to the right and has a value at 25°C of $\Delta G^{\circ\prime} = -32.22$ kJ mol^{-1}. Most of the biochemicals that have this large and negative free energy of hydrolysis are phosphate compounds. The hydrolysis always involves a transfer of phosphate to an acceptor, so the term *phosphate group transfer potential,* which is the negative of the free energy of hydrolysis at pH 7 in kJ mol^{-1}, is commonly used to characterize these reactions.* Table 6-2 lists these potentials for several commonly occurring compounds. Note that in using the free energy of hydrolysis, we have made the definition in terms of a standard acceptor, namely water.

TABLE 6-2 Phosphate transfer potentials.

Compound	Potential, kcal mol^{-1}	Potential, kJ mol^{-1}
Phosphoenolpyruvate	12.8	53.55
Creatine phosphate	10.3	43.10
Acetyl phosphate	10.1	42.26
ATP	7.7	32.22
Pyrophosphate	7.6	31.80
Arginine phosphate	7.0	29.29
Glucose-1-phosphate	5.0	20.92
Fructose-6-phosphate	3.8	15.90
Glucose-6-phosphate	3.3	13.90
Glycerol-1-phosphate	2.2	9.20

The term high energy phosphate (another reference to the misnomer we have just mentioned) is reserved for those compounds for which the potential is greater than 30 kJ mol^{-1} (7 kcal mol^{-1}).

The phosphate group transfer potentials enable one to predict whether a particular reaction can occur. For example, can glucose-6-phosphate transfer a phosphate to arginine under standard conditions? The reactions and their $\Delta G^{\circ\prime}$ values are

Glucose-6-phosphate + H$_2$O $\rightleftarrows$ glucose + H$_2$PO$_4^-$, with $\Delta G^{\circ\prime} = -13.81$,

and

Arginine + H$_2$PO$_4^-$ $\rightleftarrows$ arginine phosphate + H$_2$O, with $\Delta G^{\circ\prime} = +29.29$,

which yield $\Delta G^{\circ\prime} = -13.81 + 29.29 = 15.48$ kJ mol^{-1} for the phosphate transfer reaction. Thus under standard conditions (1 M for each reactant), arginine cannot be phosphorylated by glucose-6-phosphate because $\Delta G^{\circ\prime} > 0$. A reaction in which arginine phosphate phosphorylates glucose to yield glucose-6-phosphate has $\Delta G^{\circ\prime} = -15.48$ kJ mol^{-1}, so this reaction can

---

*These potentials are most frequently expressed in units of kcal mol^{-1}. Thus, they are given this way and in kJ in Table 6-2.

occur spontaneously. ATP has a higher group transfer potential than arginine phosphate, so ATP can phosphorylate arginine in a reaction having $\Delta G^{o\prime} = -32.22 + 29.29 = -2.93$ kJ mol^{-1}. It should be noted, though, that the fact that $\Delta G^{o\prime} < 0$ means only that the desired products will form when the particular reactants are mixed—no statement is made about the biochemical pathway (that is, about the intermediates in the reaction) or about the reaction rate, because the value of ΔG depends only on the initial and final states.

Of greater interest than these phosphate exchange reactions is the coupling of ATP hydrolysis to the conversion of one compound to another. We will see that this *can* occur by means of a phosphorylated intermediate and the transfer potential can indicate a *possible* pathway. Consider the enzymatic synthesis of sucrose from glucose and fructose. This reaction is

$$\text{Glucose + fructose} \rightleftarrows \text{sucrose} + H_2O, \quad \text{with } \Delta G^{o\prime} = +23.01 \text{ kJ mol}^{-1}.$$

Since $\Delta G^{o\prime} > 0$, the reaction will not proceed under standard conditions. If coupled to ATP hydrolysis for which $\Delta G^{o\prime} = -32.22$ kJ mol^{-1}, sucrose should be synthesized, because $\Delta G^{o\prime}$ would be $-32.22 + 23.0 = -8.78$ kJ mol^{-1}. One way in which this might occur is by means of the following intermediate reaction:

$$ATP^{4-} + \text{glucose} \rightleftarrows \text{glucose-1-phosphate} + ADP^{3-} + H^+,$$

for which $\Delta G^{o\prime}$ is $-32.22 + (+20.92) = -11.30$ kJ mol^{-1}. Note that we have used a positive value, $+20.92$ kJ mol^{-1}, because the glucose is being phosphorylated rather than hydrolyzed. This is then coupled to the reaction

$$\text{Glucose-1-phosphate + fructose} + H^+ \rightleftarrows \text{sucrose} + H_2PO_4^-,$$

which has $\Delta G^{o\prime} = +2.09$ kJ mol^{-1}, but the net free energy change is $-10.88 + 2.09 = -8.79$ kJ mol^{-1}.

We must emphasize again that the thermodynamics does not *prove* that glucose-1-phosphate is an intermediate, but only *suggests* a possible pathway. In fact, strictly from group transfer potentials, glucose-6-phosphate or fructose-6-phosphate might be intermediates; biochemical analysis is needed to determine the correct pathway. The point to be made is only that the formation of the phosphate-containing compounds can provide the free energy for a large number of biochemical reactions. A detailed discussion of this can be found in several of the references at the end of the chapter.

Utility of Free Energy Considerations in Studying Some Biological Processes

Thinking about the value of ΔG has led to some important discoveries in biochemistry, such as the existence of the high energy phosphate bond. These are described in the following examples.

SEC. 6-4 FREE ENERGY AND BIOLOGICAL PROCESSES

A. Synthesis of Urea

At one time it was thought that urea was synthesized in blood plasma from ammonia and CO_2. Two mechanisms for the synthesis of urea $(NH_2)_2CO$ were proposed:

(1) $2NH_3(aq) + H_2CO_3(aq) \rightleftarrows (NH_2)_2CO(aq) + 2H_2O(l)$;

(2) $2NH_4^+(aq) + 2HCO_3^-(aq) \rightleftarrows (NH_2)_2CO(aq) + 2H_2O(l) + H_2CO_3(aq)$.

By calculating the value of ΔG for each reaction, we can show that neither reaction can alone be responsible for urea production. That is, we will find that $\Delta G > 0$.

In order to carry out the analysis one needs to know the values for the free energy of formation of each substance in the reaction at 37°C (body temperature) and the concentrations of each substance in plasma. Available data for free energy of formation of urea is usually for the *solid* at 25°C, so it is necessary to calculate the values for a solution at 37°C. This is done by using Equation 69.

The free energy of the HCO_3^- ion can be calculated from the value for H_2CO_3 by using Equation 74. To obtain values in plasma one needs only to take into account the free energy of dilution. The values of ΔG of NH_3 and NH_4^+ in plasma are similarly calculated from tabulated values of ΔH_{form} and $\Delta G_{form,298}$ and the known concentrations in plasma.

The values in kJ mol^{-1} of ΔG_{form} in plasma at 37°C for each substance involved in reactions 1 and 2 are approximately: $NH_3(aq)$, -52.72; $NH_4^+(aq)$, -96.65; $H_2CO_3(aq)$, -636.80; $HCO_3^-(aq)$, -592.87; $H_2O(l)$, -235.14; urea(aq), -213.80. Therefore (by Equation 15) the value of ΔG for reaction 1 is $-213.80 + 2(-235.14) - 2(-52.72) - (-636.80) = +58.16$ and for reaction 2, $-213.8 + 2(-235.14) + (-636.80) - 2(-96.65) - 2(-592.87) = +58.16$. Both values are of course identical since the ions and undissociated molecules are in equilibrium.

Since the value of ΔG is positive, neither reaction can proceed spontaneously. Since urea is in fact made in cells, either a totally different reaction is used or the free energy gained by urea must be provided by another reaction that is coupled to one of these reactions. The latter conclusion is the correct one.

B. Synthesis of the Amino Acid Alanine

In the rat liver, alanine is synthesized from pyruvate and ammonia. The following reaction might be proposed:

$$\text{Pyruvate}^- + NH_4^+ \rightleftarrows \text{alanine} + \tfrac{1}{2}O_2.$$

At 1 M concentration for each of pyruvate$^-$, NH_4^-, and alanine, and with O_2 at a pressure of 1 atm, one can calculate that $\Delta G° = +165.14$ kJ mol^{-1}. This value is very high and indicates that the reverse reaction (that is, the oxidative deamination of alanine), proceeds spontaneously at these high concentrations. When converted to reasonable concentrations, ΔG is still positive. The extraordinary extent to which the forward reaction will fail to allow alanine synthesis can also be seen by converting the value of $\Delta G°$ to the equilibrium constant K by the equation $\Delta G° = -RT \ln K$. The value of K obtained is $10^{-28.9}$. Thus, if [pyruvate$^-$] and [NH$_4^+$] were 10^{-2} M, then [alanine] would be approximately 10^{-30} M.

Since alanine synthesis does occur, the proposed reaction above must occur in association with some reaction having a very large and negative ΔG. The synthesis of alanine was one of the many biochemical processes studied that led to the ideas of coupled reactions and the discovery of high energy bonds. Another example of this type follows.

C. Synthesis of the Peptide Bond

In liver extracts it has been observed that there are two reactions in which a peptide bond can form rapidly even at relatively low concentrations of reactants:

$$\text{Benzoate}^- + \text{glycine}^- \rightleftharpoons \text{hippurate}^- + H_2O\ ;$$
$$\text{Leucine} + \text{glycine} \rightleftharpoons \text{leucylglycine} + H_2O\ .$$

For both reactions, $\Delta G°$ is approximately $+10.46$ kJ mol^{-1}, indicating that $K \approx 10^{-2}$. Thus hippurate$^-$ at an initial concentration of 10^{-2} M would be hydrolyzed and reach an equilibrium concentration that is less than 10^{-4} M with [benzoate$^-$] and [glycine] approaching 10^{-2} M. However, experimental data show that the equilibrium occurs toward the right. One must again conclude, as with the synthesis of alanine, that the driving force for these reactions must be provided by some other reaction in which free energy is liberated. Reactions such as these led to the search for and discovery of the high energy phosphate bond.

REFERENCES

Atkins, P.W. 1978. *Physical Chemistry.* W.H. Freeman and Co. San Francisco.

Denbigh, K.G. 1971. *The Principles of Chemical Equilibrium.* Cambridge University Press. Cambridge.

Eisenberg, D., and D. Crothers. 1979. *Physical Chemistry with Applications to the Life Sciences.* W.A. Benjamin. Menlo Park, Calif.

Klotz, I.M. 1967. *Energy Changes in Biochemical Reactions.* Academic Press. New York. (This contains an intuitive approach that should be read by everyone. The concept of free energy is applied to familiar things.)

Klotz, I.M., and R.M. Rosenberg. 1972. *Chemical Thermodynamics.* John Wiley. New York.

Lehninger, A.L. 1965. *Bioenergetics.* W.A. Benjamin. Menlo Park, Calif.

Schmidt-Nielsen, K. 1972. *How Animals Work.* Cambridge University Press. Cambridge.

Spencer, J.N., O. Gordon, and H.D. Schreiber. 1974. "Entropy and Chemical Reactions." *Chemistry,* 47, 12.

PROBLEMS

1. What is the value of ΔG when 1 mol of ice is melted at 0°C and 1 atm?

2. What is the value of ΔG when 100 g of mercury is subjected to a pressure change from 1 atm to 100 atm at 25°C? The density of mercury is 13.534 g cm^{-3}. (*Hint:* liquids may always be considered to be incompressible unless pressure changes are enormous.)

3. Show that ΔS for a pressure change can be calculated from the thermal expansion factor $[(1/V)(\partial V/\partial T)]_P$.

4. Many components in blood are considerably concentrated in the urine. What is the free energy change per mole for a substance whose concentration in urine is 75 times that in blood? The temperature is 37°C.

5. If $\Delta G°_{form}$ for a particular substance in solution is -22.2 kJ mol^{-1} at 25°C, what is the value in a 0.01 M solution?

6. Without doing any calculations, state which of each of the following pairs has the higher free energy per mole, and tell how you know.

    (a) Ice at 5°C, or water at 5°C.

    (b) 1 M NaCl, or 0.01 M NaCl.

    (c) O_2 at a pressure of 10 atm, or O_2 at a pressure of 1 atm.

7. For the reaction $A(g) + 2B(g) \rightleftarrows 2C(g)$, at equilibrium the partial pressures of A, B, and C are 0.25, 0.65, and 0.54 atm, respectively, at 25°C. What is $\Delta G°$?

8. The following values of K are measured at the indicated temperatures: 60, at 400°C; 55.3, at 425°C; 49.1, at 465°C; 44.7, at 500°C. What is $\Delta H°$?

9. The reaction $AgCl(s) \rightarrow Ag^+(aq) + Cl^-(aq)$ has $\Delta H° = 65.52$ kJ mol^{-1} at 25°C. The solubility product at 25°C is 1.7×10^{-10}. What is the solubility of AgCl at 75°C?

10. A company is considering using the reaction $3C_2H_2(g) \rightarrow C_6H_6(g)$ as a means of synthesizing benzene at 25°C. The values of $\Delta G°_{form}$ for each substance are: C_2H_2, 209.2 kJ mol^{-1}; and C_6H_6, 124.5 kJ mol^{-1}. Assuming that there is no problem with either activation energy or a reaction rate, is it a good industrial method?

11. A reaction is characterized by the values $\Delta G° = 194.6$ kJ mol^{-1} and $\Delta H° = 232.2$ kJ mol^{-1}. Calculate K at 25°C and at 75°C and state the assumption about $\Delta H°$ that is used to obtain $K_{125°C} = 1.3 \times 10^{-24}$.

12. The values of $\Delta G°$ for $H_2O(g)$ and $H_2O(l)$ are -228.6 and -237.2 kJ mol^{-1} respectively. What is the vapor pressure of water at 25°C?

13. What are the values of $\Delta G°$ and K for $CO(g) + H_2O(g) \rightarrow CO_2(g) + H_2(g)$, at 25°C?

14. Given the fact that the pH of water at 25°C is 7, what can you say about the sign of $\Delta G°$ for the reaction $H_2O(l) \rightarrow H^+ + OH^-$?

15. In dilute solution the dissociation constant of acetic acid is $K = 1.8 \times 10^{-5}$. What is $\Delta G°$ for the dissociation reaction at 25°C?

16. The equilibrium constant for a reaction is 1.3×10^{-3} at 25°C. What is the value of $\Delta G°'$?

17. What are the values of ΔG, ΔS, and ΔH when one mole of O_2 and one mole of N_2 are mixed, when both gases are at 25°C and one atmosphere pressure?

18. The conversion of A to B in aqueous solution has a value of $\Delta G° = -28.45$ kJ mol^{-1} at 25°C. You prepare a 1 M solution of A and immediately measure [B]. A value of $[B] = 10^{-6}$ M is observed. How do you explain this value?

19. (a) Calculate $\Delta G°$ at 25°C for the reaction

    $$C_2H_4(g) \rightarrow 2C(graphite) + 2H_2(g).$$

    (b) Is the decomposition of C_2H_4 spontaneous at 25°C?

    (c) Explain why in fact C_2H_4 does not decompose at 25°C.

20. Glucose and O_2 are at concentrations of 0.005 M and 4.2×10^{-5} M, respectively, in interstitial fluid. Within living cells the concentrations are 10^{-4} M glucose, 1.4×10^{-5} M O_2, and 10^{-4} M CO_2. In the cells, glucose is oxidized by the reaction

Glucose + $6O_2 \rightarrow 6CO_2 + 6H_2O$.

The CO_2 is then released to blood in the veins at a concentration of 1.5×10^{-4} M. Calculate ΔG for the conversion of interstitial glucose to venous CO_2. The values of ΔG_{form} in kJ mol^{-1} that are needed are: 1 M glucose, -917.2; 1 M CO_2, -386.2; and H_2O, -237.2. The temperature is 37°C.

21. Consider the system shown in the figure below, which can do work by using the free energy of mixing of two ideal gases A and B. The left piston is permeable to A only and the right piston to B only. Note that if the pressure in each chamber is initially P, the final pressure is still P but the gases are at the partial pressures P_A and P_B. Show that

$$\Delta G = n_A RT \ln X_A - n_B RT \ln X_B,$$

in which X is the mole fraction.

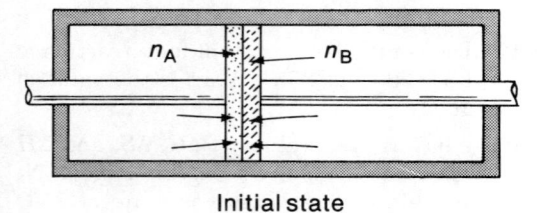

22. An enzymatic reaction catalyzed by enzyme 1, A $\xrightarrow{\text{Enz 1}}$ B, has a value of $\Delta G = 12.5$ kJ mol^{-1}. Another reaction, B $\xrightarrow{\text{Enz 2}}$ D, has a value of $\Delta G = -25$ kJ mol^{-1}. A third reaction, C $\xrightarrow{\text{Enz 3}}$ D, has a value of $\Delta G = -37.5$ kJ mol^{-1}.

(a) If a reaction mixture is prepared that contains A and enzymes 1 and 2, will D be found in an appreciable concentration?

(b) Will D be found if the mixture is A, enzyme 1, and enzyme 3?

23. The reaction

$$NH_4^+ + \text{glutamate} \rightarrow \text{glutamine} + H_2O$$

is an important process in living systems. At pH 7 and at 25°C the equilibrium constant K is

K = [glutamine]/([glutamate][NH_4^+])
   = 0.003.

(a) Calculate the equilibrium constant for the enzymatic reaction (catalyzed by the enzyme glutamine synthetase)

NH_4^+ + glutamate + ATP $\rightarrow$
    glutamine + ADP + PO_4^{3-}.

Use Table 6-2 (of phosphate transfer potentials).

(b) Do you think the enzyme has affected the value of K?

24. The compounds we have discussed that contain the so-called "high energy phosphate bonds" have very large negative free energies of hydrolysis. That is, in aqueous solution equilibrium lies very far in the direction of almost total breakdown. One might guess that aqueous solutions of these substances would be very unstable. In the laboratory, such solutions are somewhat unstable in the sense that if a solution is left sitting on a lab bench overnight, the compound becomes significantly hydrolyzed. However if left in the refrigerator, such solutions are very stable. Explain.

25. Most proteins have the property that their highly ordered structure is disrupted by raising the temperature above 60–65°C. A particular protein has the odd property that it is a relatively structureless linear polymer at 25°C but at 60°C it undergoes a transition to a highly ordered helix. At

45°C the ratio of [helix] to [linear] is 2 whereas at 60°C the ratio is 9.

(a) Calculate $\Delta H°$.

(b) Calculate $\Delta S(60°C)$.

(c) Give a possible molecular explanation to account for the sign of $\Delta H°$ and ΔS. (*Hint:* think about solvent-binding.)

26. When a polymer such as protein or DNA is heated to a temperature in the range of 70°C to 80°C, the polymer spontaneously undergoes significant changes in its properties. This process is called *denaturation* and is explained in Chapter 15; the transition may be written as this reaction:

    Normal state → denatured state.

    What are the signs of ΔH, ΔG, and ΔS for the process?

27. The Haber process is a way to optimize the exothermic reaction $N_2 + 3H_2 \rightarrow 2NH_3$. Would you expect this process to use high or low temperature and high or low pressure?

28. Consider the Katchalsky machine shown below. The pulleys of the machine are linked with a continuous belt made of a highly negatively charged polymer that has the following property. When the belt enters a solution having a high salt concentration, the high concentration of the ions in the interstices of the polymer reduces the repulsive electrostatic forces between the negatively charged groups so that the belt *contracts*. Since the two solutions shown in the figure have different concentrations, the pulleys will rotate as shown: the machine works because the section of belt dipping into solution A contracts, causing equal forces (arrows) to pull on pulley *C*. The *forces* are equal but the *torques* are not, because *C* is a double pulley with unequal radii; so the entire system rotates and this pulls a new section of belt into solution A, causing further contraction, force, and rotation. The machine goes round and round and the rotation can be used to do work such as lifting a weight or driving an electric generator. It is of course not a perpetual motion machine and eventually stops. What is the "fuel" that runs the machine and when will the machine run down and stop rotating?

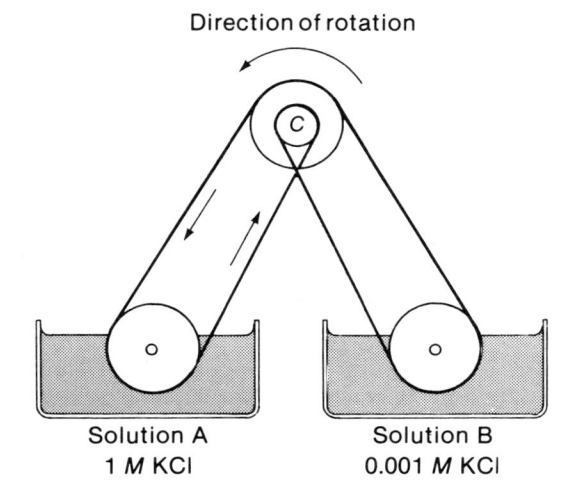

CHAPTER 7

SOLUTIONS OF UNCHARGED MOLECULES

In chemistry and biochemistry, the reactions and interactions of interest usually occur in solution. For this reason alone, it is important to understand the physicochemical properties of solutions. However, there is more to be gained. That is, measurement of some of the physical properties of solutions yields valuable information, such as the molecular weights of the components. In addition, examination of the forces and interactions that occur between the components provides information that is required to understand the molecular details of various reactions and processes that are of interest.

A *solution* is defined as a homogeneous single phase that consists of two or more substances mixed so uniformly that all parts of the solution have the same chemical and physical properties. This is a good definition as long as one does not look too closely at the surfaces and boundaries of the solution (which we consider in Chapter 12), or examine microscopic volumes in which local variations of the properties of adjacent microvolumes may occur. A solution is normally described by naming the *solute* (the component that is dissolved) and the *solvent* into which the solute disperses, and stating the solute concentration.* A solution may consist of a gas dispersed in a gas, a gas in a liquid, a liquid in a liquid, a solid in a liquid, or a solid in a solid; in this book we shall consider only gas-liquid, liquid-liquid, and solid-liquid solutions, because only these are biologically important.

---

*In some cases, for example when two liquids mix, the distinction between solvent and solute is ambiguous and by convention one defines the *major* component to be the solvent. This is, of course, also the convention when a solid dissolves, because then the liquid is the major component.

7-1 METHODS OF DESCRIBING CONCENTRATION

Several methods are commonly used to state the solute concentration. These are the *molarity,* the *molality,* the *mole fraction,* and the *weight percent.* They are defined as follows:

The molarity and molality of a component A are defined as

$$\left. \begin{array}{l} \text{Molarity of A} = \dfrac{n_A}{\text{liter of solution}}; \\[6pt] \text{Molality of A} = \dfrac{n_A}{\text{kilogram of solvent}}; \end{array} \right\} \quad (1)$$

in which n_A is the number of moles of A. If the solvent is water (or any solvent for which the density is 1 kg/l) and the concentration is low, molality and molarity are nearly equivalent.

The *mole fraction,* X_A, of A is defined as

$$X_A = \frac{n_A}{\sum n_i}, \quad (2)$$

in which n_i is the number of moles of the *i*th component. That is, if there are three components A, B, and C, then $X_A = n_A/(n_A + n_B + n_C)$.

The volume of a solution is both pressure- and temperature-dependent, whereas the weight is constant. Therefore, molality and mole fraction, which are both independent of P and T, are preferable to molarity in expressing concentration.

The *weight percent and weight per unit volume,* while not useful in a theoretical discussion, often arise in laboratory practice. The weight percent (wt. %) of A is

wt. % = (weight of A/total weight of solution) × 100.

The weight per unit volume (wt./V) is the weight of A per unit volume of *solution,* although on occasion one sees in scientific literature that the volume of the solvent is used.

In this book we will use the following convention for denoting the concentration. In theoretical equations where it is not necessary to distinguish between molarity and molality, concentration will be designated by a lower case c and, when it makes a difference, we will state whether the units are molar or molal: in stating a molar concentration, an italicized capital M will be used, and a bold-face capital **M** will be used for molal. Thus 0.1 M is 0.1 molar and 0.1 **M** is 0.1 molal. The symbol M will never be used alone to denote concentration—such a unitless M will be reserved for molecular weight. A lower case m will be used for mass (in grams or kilograms as necessary). The bracketed notation [A] means "the concentration of A" and may denote molar or molal units; it will never be used for *activity,* for which a lower case a will be employed. Other notation will be defined as needed.

7-2 THE CHEMICAL POTENTIAL

The thermodynamic properties of a substance in a mixture often differ from the properties of the same substance when it is pure. This is certainly true of volume—that is, the volume of a mixture of two liquids is rarely just the sum of the volumes of each component. For example, 50 cm^3 of water added to 50 cm^3 of ethanol yields 97 cm^3 of the mixture. In addition, there is the sugar syrup phenomenon: 1 cup of sugar + 1 cup of water gives only slightly more than 1 cup of sugar syrup. The free energy of a mixture, G, is also not the sum of the free energies of each of the components in the pure state; it is obvious that some free energy must be lost in the mixing process, because unmixing cannot occur spontaneously. There should be no doubt, though, that the free energy of the mixture does depend in some way on the *amount* of each component, because each component possesses its own free energy and makes some contribution to the total. Thus, G, which is already a function of T and P, must also be a function of n_i, the number of moles of the ith component. If we defined $\overline{G}_A$ as the free energy per mole of A *when all other components are present*, then

$$G = n_A \overline{G}_A + n_B \overline{G}_B + \cdots + n_i \overline{G}_i . \tag{3}$$

The term $\overline{G}_i$ is called the *partial molar free energy* of component i. We can obtain a simple expression for $\overline{G}_i$ as follows. If we have a system at temperature T and pressure P consisting of n_i moles of each of several components, the change dG in the free energy is

$$dG = \left(\frac{\partial G}{\partial T}\right)_{P,n} dT + \left(\frac{\partial G}{\partial P}\right)_{T,n} dP + \sum \left(\frac{\partial G}{\partial n_i}\right)_{T,P} dn_i . \tag{4}$$

For any system of constant composition (that is, when each $dn_i = 0$), $dG = -S\,dT + V\,dP$ (see Chapter 6), so that

$$dG = -S\,dT + V\,dP + \sum \left(\frac{\partial G}{\partial n_i}\right) dn_i , \tag{5}$$

which, at constant temperature ($dT = 0$) and constant pressure ($dP = 0$), becomes

$$dG = \sum \left(\frac{\partial G}{\partial n_i}\right)_{T,P} dn_i . \tag{6}$$

Comparison of Equations 3 and 6 indicates that the partial molar free energy is equal to $(\partial G/\partial n_i)_{T,P}$. This expression occurs so frequently in thermodynamics that it has been given a special name, the *chemical potential*, and it is denoted by μ_i. Thus

$$\mu_i = \left(\frac{\partial G}{\partial n_i}\right)_{T,P} , \tag{7}$$

SEC. 7-2 THE CHEMICAL POTENTIAL

and the condition for equilibrium in a mixture (that is, $dG = 0$), is, in this formulation,

$$\sum \mu_i dn_i = 0 . \tag{8}$$

This equation states that, at equilibrium, the chemical potential of one component of a solution cannot be varied without changing the chemical potential of at least one other component.

The chemical potential μ_A of any component A is a function of the mole fraction X_A of that component. The equation describing this function is one of the most important equations in physical chemistry, namely,*

$$\mu_A = \mu_A^\circ + RT \ln X_A , \tag{9}$$

in which μ_A° is the chemical potential of the component in the standard state. The standard state is defined as the pure substance at $P = 1$ atm; thus, $X_A = 1$ and $RT \ln X_A = 0$, and $\mu_A = \mu_A^\circ$. It is important to remember that μ_A is *not* proportional to X_A but increases with $\ln X_A$, a fact that is not intuitively obvious. In Chapter 6, we saw that two systems are in equilibrium when they possess the same free energy. The corresponding statement for systems that consist of several components is that when two systems are at equilibrium, the chemical potential of component A in one system is the same as the chemical potential of A in the second system. That is,

$$\left. \begin{array}{l} \mu_{A,I} = \mu_{A,II} ; \\ \mu_{B,I} = \mu_{B,II} ; \end{array} \right\} \tag{10}$$

and so forth, in which I and II refer to the two systems.

In order to write the expression for dG in Equation 4, we have only made use of the fact that G is a thermodynamic state function—that is, a function whose initial and final states are independent of path. Since this is also true of volume, we can obtain an expression for the partial molar volume $\bar{v}$ in the same way. Thus if we have a solution containing n_A moles of A and n_B moles of B, then dV, the increase in volume produced by adding dn_A moles of A and dn_B moles of B, is

$$dV = \left(\frac{\partial V}{\partial n_A}\right)_{n_B} dn_A + \left(\frac{\partial V}{\partial n_B}\right)_{n_A} dn_B .$$

The element $(\partial V/\partial n_A)_{n_B}$ is just the volume change induced by adding a small amount of A; this is defined as the *partial molar volume*, $\bar{v}_A$; that is

$$\bar{v}_A = \left(\frac{\partial V}{\partial n_A}\right)_{T,P,n_B} . \tag{11}$$

The value of $\bar{v}_A$ is not constant but depends both on the concentration of A and the identity of the solvent B. Furthermore it can be either positive or

---

*The reader is asked to accept Equation 9 without proof because of the great complexity of the proof. Equation 9 can also be taken as the definition of an ideal solution.

negative. For example, when small amounts of MgSO$_4$ are dissolved in water, the total volume *decreases,* owing to strong attractions between the solute ions and the solvent molecules. As the concentration of MgSO$_4$ increases, the solution volume increases. This is shown in Figure 7-1. Note that $\bar{v}_A < 0$ when the molarity is less than 0.07 and positive above that value. This is a common situation for solutions of ionic solids—that is, initially there is a contraction because of solute-solvent attraction, but ultimately the volume must increase because the solute molecules do take up space. Shortly after $\bar{v}_A$ becomes positive, it remains constant (that is, the slope of the curve in Figure 7-1 is constant) as the volume of the solute predominates in determining the solution volume. For very large molecules, such as proteins and other macromolecules, this linear behavior is approached at very low concentrations (for example, 10^{-4} *M*), because the volume of these huge molecules quickly becomes the major factor. Thus, for macromolecules, the partial molar volume is usually considered to be a concentration-independent constant.

FIGURE 7-1 Volume of a solution consisting of 1000 g of water at 20°C and MgSO$_4$ at the indicated molality. The partial molal volume of MgSO$_4$ at any particular concentration is just the slope of the curve at that concentration.

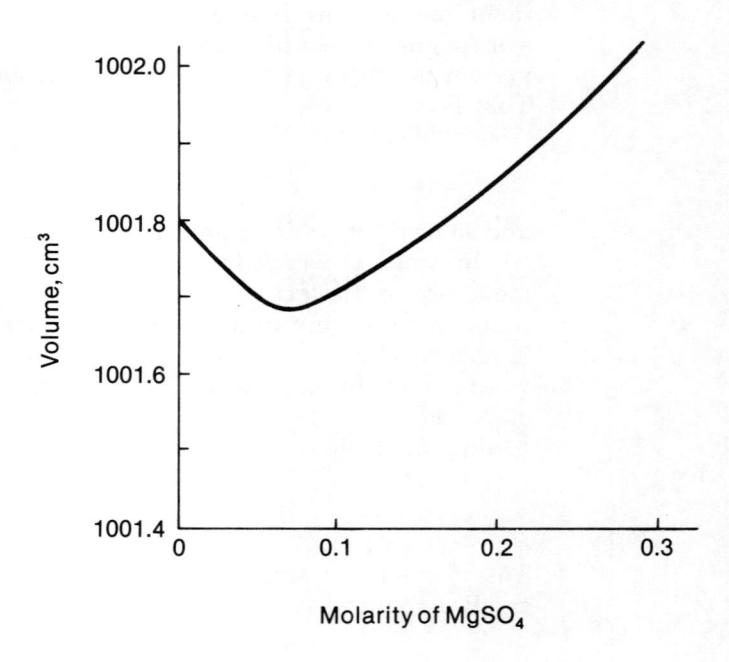

A curve such as that in Figure 7-1 is usually obtained by measuring the density of the solution (the weight in a fixed volume) rather than the volume. This is done because

$$\frac{\text{Weight of solvent} + \text{weight of added solute}}{\text{Density of solution}} = \text{volume of solution.}$$

7-3 IDEAL SOLUTIONS

Definition of Ideal Solutions

A solution is a homogenous mixture of two or more components between which there are cohesive forces, as we described in Chapter 2. In order for a solution to form—that is, for the solute and solvent to mix intimately—it is necessary that the cohesive forces between solute and solute and between solvent and solvent molecules be overcome. If this were not to occur, each component would remain separate and mixing would not occur. This reduction in cohesive forces can occur in several ways:

1. A solvent and a solute molecule might cohere more strongly than two solute molecules. In this case the solvent effectively pulls solute molecules into solution.
2. A solvent and a solute molecule cohere with the same strength that two solute molecules have when they cohere. In this case the net cohesive force on a particular solute molecule that is simultaneously interacting with one solute molecule and one solvent molecule is zero, so that a solute molecule would diffuse away from the main body of the solute. As this process continues, the spatial arrangement of solute and solvent molecules becomes random and a solution results.
3. A solute and a solvent molecule might cohere less than two solute molecules do, but the net force is reduced to the point that it is easily overcome by thermal motion. This is one of the reasons for the substantial increase in the solubility of most substances with rising temperature.

Let us now consider a solution consisting of a single type of solvent A and single solute B. In such a solution there are cohesive forces between A and A, B and B, and between A and B.* If these forces are all equal in magnitude, the spatial distribution of the individual molecules would be random and homogeneity of properties would persist (until the volumes of the solution approached molecular dimension, in which case the molecular fluctuations would cause local heterogeneity). Such a totally uniform solution, if it were capable of existence, is called an *ideal solution*.†

An important property of an ideal solution is the proportionality between vapor pressure and solute concentration, which we describe in the following section. This property may seem like a rather uninteresting one since, when thinking about chemical reactions in solution or about the

---

*We cannot assume, as in the case of an ideal gas, that there are no cohesive forces because in a condensed phase such as a liquid there must be cohesive forces, at least between solvent molecules.
†In textbooks published in Great Britain, an ideal solution is defined as one that satisfies Equation 9.

more interesting characteristics of solutions, the vapor pressure is certainly not one that would immediately come to mind. However, it will be seen that a close examination of the vapor pressure provides a particularly simple and convenient way to understand a variety of solution phenomena.

Raoult's Law

For any pure solid or liquid there is a tendency for molecules to escape into the gas phase. Some of the escaped molecules will, by random motion, collide with the surface of the substance and, owing to the cohesive forces, will rejoin the mass of the material. The pressure above the substance at equilibrium (when the rate of escape equals the rate of rejoining) is called the *vapor pressure*. In a solution, as well as in a pure substance, there remains this tendency for the components to escape to the vapor phase. If the cohesive forces between all components are the same (that is, if the solution is ideal), the tendency for a molecule to escape and to be captured is independent of its position—that is, of whether it is next to an A or a B. Thus the vapor pressure of both A and B will simply be proportional to the number of molecules of each. This can be expressed by the equations

$$P_A = X_A P_{A,\text{pure}} \quad \text{and} \quad P_B = X_B P_{B,\text{pure}}, \tag{12}$$

in which P_A is the vapor pressure of A above a solution, X_A is the mole fraction, and $P_{A,\text{pure}}$ is the vapor pressure of pure A. Equation 12 is known as *Raoult's Law*. An example of a system that obeys Raoult's Law is shown in Figure 7-2. The argument has been given for a two-component system but it is valid also for a system containing i components—that is, $P_i = X_i P_{i,\text{pure}}$. Whereas many studies have been carried out with liquid-liquid solutions (that is, one in which both solute and solvent are liquids), the kind of solution encountered most commonly in biological systems (and, in fact, in chemical systems also) is one in which the solute is a nonvolatile solid. In this case the vapor phase is for all practical purposes pure solvent vapor. That is, $P_{B,\text{pure}}$ is so small that it can be approximated as zero and the system is described by the single equation $P_A = X_A P_{A,\text{pure}}$. This is the most conventional use of Raoult's Law—namely, to describe the vapor pressure of the solvent*—and it will be used frequently in this way in this book.

In Appendix I at the end of this book a statistical derivation of Raoult's Law is presented.

Very few solutions actually obey Raoult's Law over the complete range of solubility of the solute. This is because absolute uniformity of cohesive forces is extremely unusual. If the cohesive force between unlike molecules A and B is smaller than that between A and A and that between B and B, the escape tendency of both solvent and solute molecules is higher than for

---

*We cannot emphasize strongly enough that Raoult's Law describes the effect of the *solute* on the vapor pressure of the *solvent*.

FIGURE 7-2 An example of a two-component solution that obeys Raoult's Law—that is, the curves labeled P_A and P_B are linear.

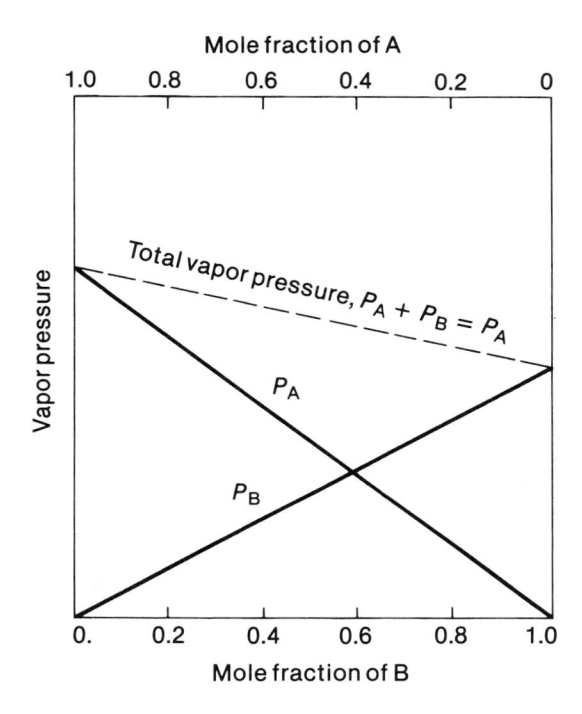

the pure substances, and it is said that there is a *positive deviation* from Raoult's Law, as is shown in Figure 7-3(a). The cohesive force between molecules A and B may, on the other hand, be greater than that between identical molecules, thus reducing the escape tendency of both; then it is said that there is a *negative deviation* (Figure 7-3(b)). Positive deviation often occurs in aqueous solutions of noncrystalline solutes, because water has a definite structure when it is pure and this structure is disrupted by the solute. For instance, pure water contains many ordered aggregates of water dipoles. Many solutes in large concentration interact with water in such a way that these ordered structures are disrupted; this decreases the cohesive force between the water molecules, allows their escapes and hence causes the vapor pressure of the water to increase. The reader should note that a simple measurement of the dependence of vapor pressure on the solute concentration gives some information about the nature of the intermolecular forces in a solution.

If a solution is sufficiently dilute, Raoult's Law will invariably be obeyed for the solvent. This is because the solute concentration is so low that a solvent molecule rarely encounters a solute molecule and thus the escape tendency is determined primarily by the cohesive force between solvent molecules. The range of application of Raoult's Law is for solute concentrations less than about 0.1 M and for solutes that are not charged and are not macromolecules. The effects of charge are discussed in Chapter 8 and the properties of solutions of macromolecules are described later in this chapter.

FIGURE 7-3 Positive (panel a) and negative (panel b) deviation from Raoult's Law. The dashed lines indicate Raoult's Law. Note that when the mole fraction of a component is very small, the vapor pressure of that component obeys Raoult's Law (see arrows).

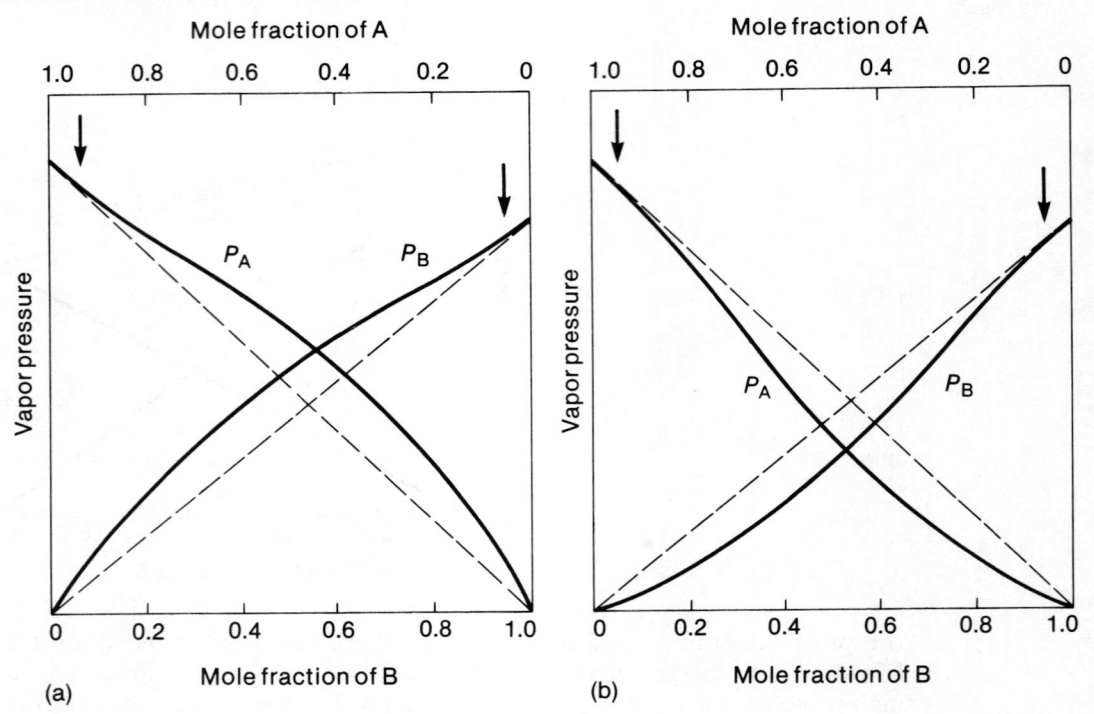

Solutions Containing Volatile Components—Henry's Law

In an ideal solution consisting of two volatile components, Raoult's Law is obeyed for both components. We have just pointed out that in a nonideal (real) *dilute* solution, Raoult's Law is still obeyed for the solvent, because the interactions are primarily between two solvent molecules. This is not usually true for the solute, because even in dilute solution, it is not the solute-solute interaction that predominates, but rather the solute-solvent interaction. However, for many volatile solutes, especially for gases, the vapor pressure of the solute depends on solute concentration in a straightforward way that resembles, but is not identical to, Raoult's Law. In a *very* dilute solution, a solute molecule B is on the average surrounded only by molecules of the solvent A. In this case, the escape tendency and thus the vapor pressure of B should be *proportional* to X_B, because a B rarely sees another B but only sees an A. However, the vapor pressure P_B of B is now determined by the cohesive force between A and B rather than between B and B, so that the proportionality constant is not P_B, as was the case in Equation 11, but a constant that we simply term K_B. Thus

$$P_B = K_B X_B .\qquad(13)$$

This is called *Henry's Law,* and it applies to the solute at a very low solute concentration. The value of K_B can be greater than or less than P_B, depending on whether the cohesive force between A and B is less than or greater than that between B and B. Note that the Henry's Law constant K_B is not a molecular constant that is characteristic of B alone. This is because it is the A–B interaction that determines the escape tendency. Thus the value of K_B depends on the solvent and any other component in the mixture that is at a high enough concentration to interact with B.

Equation 13 often gives rise to some confusion; that is, one expects that when $X_B = 1$, then P_B must be $P_{B,pure}$, which is not what the equation says. The explanation for this discrepancy is of course quite simple—*the equation is valid only when* X_B *is small*—that is, only when a typical B is surrounded only by A's.

We will say more about Henry's Law at a later time.* It is an important way to describe the exchange of gases in biological fluids such as blood.

7–4 COLLIGATIVE PROPERTIES

In this section we discuss four properties of solutions known collectively as *colligative properties*. These are the lowering of the vapor pressure of the solvent, the lowering of the freezing point of the solvent, the increase in the boiling point of the solvent, and the appearance of an osmotic pressure. In each case, these are a result of the presence of the solute. These phenomena are important to understand for several reasons. (1) From the magnitude of these properties, molecular weights can be calculated. (2) The values for a particular solute concentration enables one to measure the departure of the solution from ideality and hence to estimate the strength of intermolecular forces. (3) They originally provided one of the most important bits of evidence for the existence of ions, as will be shown in Chapter 8. (4) There are several biological phenomena in which osmotic pressure is important. In the following derivations, the reader should especially notice that these rather profound conclusions are based upon extremely simple consequences of the laws of thermodynamics.

The Lowering of Vapor Pressure

Raoult's Law states that *if two volatile substances are mixed, the vapor pressure of each is lowered by the presence of the other*. That is, $P_A < P_{A,pure}$

---

*It is worth remembering that Henry's Law deals with the vapor pressure of the solute, so there is no point to consider Henry's Law for a nonvolatile solute.

and $P_B < P_{B,pure}$, because X_A and X_B are always less than one. Thus for A, the lowering of the vapor pressure, or $\Delta P_A = P_{A,pure} - P_A$, is proportional to X_B; that is,

$$\Delta P_A = P_{A,pure} - P_A = P_{A,pure} - X_A P_{A,pure}$$
$$= P_{A,pure}(1 - X_A) = P_{A,pure} X_B . \tag{14}$$

Similarly, $\Delta P_B = P_{B,pure} X_A$.

It should be noticed that the *total* vapor pressure $P_A + P_B$ is *not* the sum of $P_{A,pure} + P_{B,pure}$, since $P_{A,pure} > (P_A + P_B) > P_{B,pure}$, as is shown in Figure 7-2. The fact that the vapor pressure depression is a function of the mole fraction of the solute indicates that it depends on the *number* of solute particles in the solution and not on their identity, as we have already mentioned. In the following sections we will see that this is true for each of the colligative properties.

Boiling Point Elevation and Freezing Point Depression

The boiling point of a liquid, T_b, is the temperature at which the vapor pressure equals the external pressure (usually atmospheric pressure). Thus the decrease in vapor pressure of a solvent produced by a nonvolatile solute must increase the boiling point because the solvent particles must have greater kinetic energy (that is, a higher temperature) in order to escape. The magnitude of the increase in the boiling point, ΔT_b, beyond the boiling point of the pure solvent, T_0, is calculated as follows.

At all temperatures the solvent vapor is in equilibrium with the solvent in solution. Thus, the partial molal free energy of the solvent vapor μ_{vap} must equal the chemical potential μ_A of the solvent in the solution. The chemical potential is related to the free energy of the *pure* solvent liquid in the standard state μ_A° by $\mu_A = \mu_A^\circ + RT \ln X_A$ (Equation 8), in which X_A is the mole fraction of the solvent. At the boiling point, $\mu_{vap} = \mu_A^\circ + RT \ln X_A$, or

$$\frac{\Delta G}{T} = \frac{\mu_{vap} - \mu_A^\circ}{T} = R \ln X_A , \tag{15}$$

in which ΔG is the free energy change when one mole of solvent boils at temperature T. By differentiating Equation 15 and by using the Gibbs-Helmholtz relation $(\partial(G/T)/\partial T)_P = -H/T^2$, we have at constant external pressure

$$\frac{d(\Delta G/T)}{dT} = -\frac{\Delta H_{vap}}{T^2} = R\frac{d}{dT}(\ln X_A) , \tag{16}$$

in which ΔH_{vap} is the enthalpy of vaporization. The normal boiling point T_0 occurs when $X_A = 1$ or $\ln X_A = 0$, so Equation 16 can be integrated between the limits of the normal boiling point and the boiling temperature

SEC. 7–4 COLLIGATIVE PROPERTIES

T that occurs when the mole fraction is X_A:

$$\int_{\ln X_A = 0}^{\ln X_A} d \ln X_A = \int_{T_0}^{T} \left(-\frac{\Delta H_{vap}}{RT^2} \right) dT .$$

Evaluating these integrals and rearranging terms yields

$$\ln X_A = -\frac{\Delta H_{vap}}{R}\left(\frac{T - T_0}{T_0 T}\right) = -\frac{\Delta H_{vap}}{R}\frac{\Delta T_b}{T_0 T} , \quad (17)$$

in which the boiling point elevation $\Delta T_b = T - T_0$. It is more convenient to write Equation 17 in terms of the mole fraction of the solute X_B than that of the solvent. This form of the equation is

$$\ln(1 - X_B) = -\frac{\Delta H_{vap}}{R}\frac{\Delta T_b}{T_0 T} . \quad (18)$$

At this point, we must remember that we have assumed that the solution is ideal, which occurs only when the solution is very dilute. By restricting Equation 18 to dilute solutions, several simplifications occur. First, the temperature change is very small, so that $T_0 T \approx T_0^2$, and second, $\ln(1 - X_B) \approx -X_B$, so that Equation 18 simplifies to*

$$\Delta T_b = \frac{RT_0^2}{\Delta H_{vap}} X_B . \quad (19)$$

It is conventional to write Equation 19 in terms of the molal concentration of B, namely $\mathbf{M}_B$. We note that $X_B = n_B/(n_A + n_B)$, which for dilute solutions is approximately n_B/n_A, which is equal to $M_A \mathbf{M}_B/1000$, in which M_A is the molecular weight of the solvent. Thus

$$\Delta T_b = \frac{RT_0^2 M_A}{1000 \Delta H_{vap}} \mathbf{M}_B = K_b \mathbf{M}_B , \quad (20)$$

in which we have substituted

$$K_b = \frac{RT_0^2 M_A}{1000 \Delta H_{vap}} . \quad (21)$$

The term K_b is called the *boiling point elevation constant*. Note that if ΔT_b is measured as a function of the molal concentration and if ΔH_{vap} is known, the molecular weight M_A of any solute can be measured. Since ΔH_{vap} is accurately known for water, M_B can be measured for any water-soluble compound that is sufficiently soluble that ΔT_b is large enough to be measured accurately. A discussion of the requirement for measuring M_B in this way will be given later in this chapter.

The freezing point depression (that is, the difference in degrees between the freezing point of the pure solvent and the lower value of the

---

*The approximation $\ln(1 - x) = -x$, when x is small, is often met with disbelief. A few sample values, which we list as $[x, -\ln(1 - x)]$, should be convincing: [0.3, 0.356], [0.2, 0.22], [0.1, 0.105], [0.05, 0.051], [0.03, 0.0305], [0.01, 0.01005].

freezing point of the solution) is calculated in a similar way. If we use the notations T_{frz} = freezing point of the pure solvent, ΔT_{frz} = the freezing point depression, X_B = mole fraction of the solute, and ΔH_{fus} = the enthalpy of fusion of the pure solvent, then we obtain the resulting equations, which are equivalent, respectively, to Equations 18, 19, and 20:

$$\ln(1 - X_B) = -\frac{\Delta H_{fus} \Delta T_{frz}}{RT_0 T} ; \qquad (22)$$

$$\Delta T_{frz} = \frac{RT_{frz}^2}{\Delta H_{fus}} X_B ; \qquad (23)$$

and

$$\Delta T_{frz} = \frac{RT_{frz}^2 M_A}{1000 \Delta H_{fus}} \mathbf{M}_B = K_{frz} \mathbf{M}_B ; \qquad (24)$$

in which $K_{frz} = RT_0^2 M_A/(1000\ \Delta H_{fus})$ is the freezing point depression constant. A sample calculation is given in the following example.

EXAMPLE 7–A Calculation of ΔT_b and ΔT_{frz} for a glucose solution.

Five g of glucose is dissolved in 1 kg (1 l) of water. The molecular weight of glucose is 180, so that the molality is 5/180 = 0.0278 **M**. The necessary constants for water are: boiling point, 100°C = 373 K; freezing point, 0°C = 273 K; ΔH_{vap} = 4.06 × 10^4 J mol^{-1}; ΔH_{fus} = 5.99 × 10^3 J mol^{-1}; molecular weight of water, 18; R = 8.314 J K^{-1} mol^{-1}. Thus

$$\Delta T_b = \frac{(8.314)(373)^2(18)}{1000(4.06 \times 10^4)}(0.0278) = 0.014 K ;$$

and

$$\Delta T_{frz} = \frac{(8.314)(273)^2(18)}{1000(5.99 \times 10^3)}(0.0278) = 0.052 K .$$

Three points should be noticed. First, we have used the absolute temperature. Second, the molecular weight is the weight in grams per mole. Third, if instead of glucose, we had a substance of unknown molecular weight, the value could be calculated from the experimentally observed values of either ΔT_b or ΔT_{frz}. The latter is preferable for aqueous solutions, because $\Delta T_{frz} > \Delta T_b$ and because the onset of melting is easier to identify than the beginning of boiling.

The Osmotic Pressure of Solutions

Consider a solvent separated from a solution by a semipermeable membrane that allows passage of solvent molecules but not solute molecules, as shown in Figure 7-4. By random motion, solvent molecules will move in

SEC. 7-4 COLLIGATIVE PROPERTIES

FIGURE 7-4 Osmotic equilibrium between a solvent and a solution separated by a membrane permeable to solvent (○) but not to solute (●) molecules. Solvent molecules move freely in both directions. Before equilibrium is achieved, there is a net flow of solvent into the solution. At equilibrium the leftward and rightward flow of solvent molecules is equal.

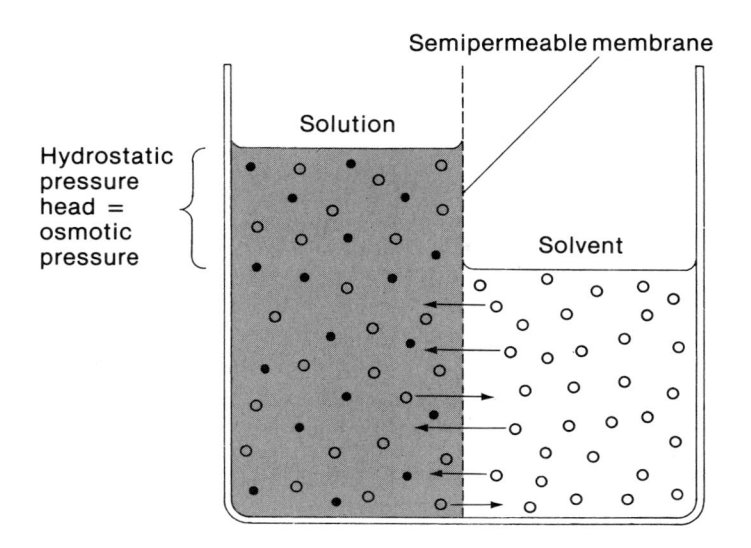

both directions through the membrane, but because the solvent concentration of the pure solvent is greater than the solvent concentration in the solution, in order to achieve equilibrium, the leftward flow is greater than the rightward rate. That is, the solvent will flow in such a way that the solution becomes more dilute because the tendency toward equilibrium is just an attempt to equalize the solvent concentration (in fact, the chemical potential) on the two sides of the membrane.

Accompanying this flow is an increase in the height of the liquid in the solution chamber, and a height will ultimately be reached at which the hydrostatic pressure resulting from the weight of the liquid (that is, resulting from the difference in height of the two liquids) will prevent further flow of solvent. The pressure reached at this point is called the *osmotic pressure* of the solution. Another way of stating this is that the *osmotic pressure is the external pressure that, if applied to the solution, prevents further flow of the solvent*. Two important points should be noted: (1) if the solution on the left side of the membrane were in an elastic sealed container, the pressure on the walls of the container would increase and the walls would stretch—that is, the osmotic pressure is a real pressure; (2) if a pressure is applied to the more concentrated solution and the pressure is greater than the osmotic pressure, then the solvent would flow from the more concentrated solution into the pure solvent.

The osmotic pressure is a result of the ability of only one component of a mixture to move, driven by an attempt to achieve equilibrium. The distinction between the mobile component and the immobile component is a result of the permeability properties of the membrane. However, flow does not require that the membrane be a *solid* semipermeable barrier—it is only the semipermeability that counts. Thus if a solution containing a nonvolatile solute is placed near a container of pure solvent, both in a closed system, the air that is in contact with both solute and solvent is like a semipermeable membrane in that solvent vapor molecules can evaporate from the surface of one liquid, move through the air, and condense on the other, whereas the solute, being nonvolatile, is immobile. That a movement of solvent akin to osmotic flow actually occurs in such a situation can be seen in the experimental arrangement that follows.

Consider a large container in which two jars are placed, one containing pure water and the other containing a solution (Figure 7-5). Both jars are

FIGURE 7-5 Transfer of solvent (flasks at right) to solution (left) through the vapor phase. Molecules evaporate from the solvent and condense on and thereby dilute the solution. The process continues until the solvent container is empty.

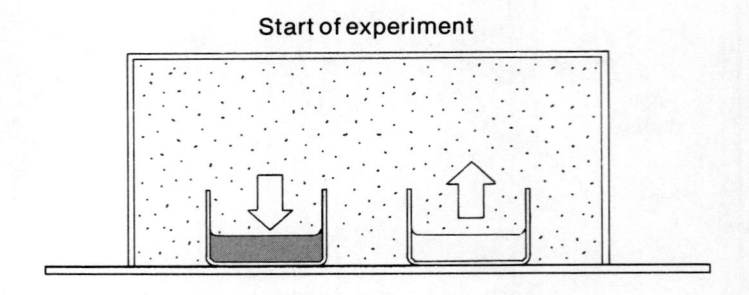

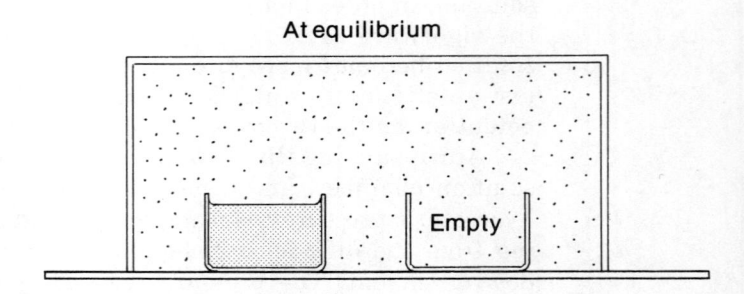

sealed and the container is evacuated. If the jar containing the water is then opened, the water will evaporate until the vapor pressure in the container reaches the equilibrium vapor pressure of water at the temperature of the system. Then the jar containing the solution is opened. Because of the depression of the vapor pressure of the solvent by the solute, the solution will not be in equilibrium with the water vapor; that is, water vapor will condense on the solution at a greater rate than the liquid evaporates from the solution. This adds volume to the solution and thereby lowers the solute concentration. However, when condensation occurs, water vapor is removed from the atmosphere of the container, so that the pure water will no longer be in equilibrium with the vapor and, thus, will evaporate in order to maintain the appropriate vapor pressure. The result is that the molecules leave the pure water and enter the solution and the process continues until the pure water is gone. Note the similarity between this phenomenon and the osmotic pressure—that is, the space above the two jars is equivalent to a semipermeable membrane in that solvent molecules can pass through this volume but the nonvolatile molecules cannot. Of course this system differs from that depicted in Figure 7-4 in that there is no pressure buildup that can stop the process. Nevertheless, this demonstrates how the vapor pressure depression can drive the flow of solvent. (The driving force is of course really the differences in chemical potentials.)

Suppose now that the solution is contained in an elastic semipermeable bag instead of a jar. In this case, as the solvent enters the solution by the process just described, the bag will expand and exert pressure on its contents. When this pressure reaches some critical value, the volume of the solution can no longer increase. This pressure is once again the osmotic pressure of the solution. We now have a curious result, because the vapor pressure is still that which is determined by the pure solvent but, because there is no longer any solvent flow, the solution must be in equilibrium with vapor at the higher pressure. We must conclude then that *exerting pressure on a solution increases the solvent vapor pressure.* A simple way to think about this phenomenon is that solvent molecules are squeezed out of the solution by the increased pressure. (This surprising phenomenon is encountered again in Chapter 12.) Thus *the osmotic pressure can be thought of as the pressure that, if exerted on a solution, makes the vapor pressure of the solution equal to that of the pure solvent.*

We would now like to prove these statements. At equilibrium, the chemical potentials of the vapor and the solvent must be equal, so that a change in one must be accompanied by a change in the other. If the partial molar volume of the solvent is $\bar{v}_l$, a pressure change, dP_l, causes a change in the chemical potential of $\bar{v}_l dP_l$. The corresponding change for the vapor is $\bar{v}_{vap} dP_{vap}$. If the vapor is an ideal gas, then $\bar{v}_{vap} = RT/P_{vap}$; thus, equating the chemical potentials, we obtain

$$\frac{dP_{vap}}{P_{vap}} = \frac{\bar{v}_l dP_l}{RT}. \qquad (25)$$

If P_{vap} is the vapor pressure when a pressure P is applied, and P'_{vap} is the value at a pressure P', we can integrate Equation 25 to obtain

$$P'_{vap} = P_{vap} e^{\bar{v}_l (P' - P)/RT}, \qquad (26)$$

showing that $P'_{vap} > P_{vap}$ when $P' > P$, or that the vapor pressure is increased by a greater external pressure.

The osmotic pressure is a consequence of there being a solute that is constrained to remain on one side of the semipermeable membrane. Since there is no osmotic pressure if the mole fraction of the solute is zero, one would expect the osmotic pressure to rise as the solute concentration increases. This is indeed the case, as shown in the following derivation.

Figure 7-4 indicates that solvent flow continues until the hydrostatic pressure equals the osmotic pressure Π. Thus Π is the pressure when equilibrium is reached. As always, the condition for equilibrium is that the chemical potential of the solvent A is the same on both sides of the membrane—that is, $\mu_{A,I} = \mu_{A,II}$. It is easily shown by studying the chemical potential of an ideal gas that the reduction in chemical potential of the solvent vapor from $P_{A,\text{pure}}$ for a pure solvent to P_A for the solution is $\Delta\mu = RT \ln(P_A/P_{A,\text{pure}})$. (A derivation of the relation between μ and P, namely, $\mu = \mu° + RT \ln P$ can be found in the thermodynamics texts listed at the end of this chapter.) At equilibrium this reduction is counterbalanced by the increase in μ_A caused by Π. Furthermore, we have just seen that the relation between a change $d\mu$ in chemical potential caused by a pressure change dP on the solvent is $d\mu = \bar{v}_A dP$, so that

$$\Delta\mu = \int_0^\Pi \bar{v}_A \, dP \ . \tag{27}$$

At equilibrium, μ_A in solution must equal $\mu_A°$ in the pure solvent, so that

$$\int_0^\Pi \bar{v}_A \, dP = -RT \ln \frac{P_A}{P_{A,\text{pure}}} \ . \tag{28}$$

Solutions are virtually incompressible, so that $\bar{v}_A$ is independent of pressure and

$$\int_0^\Pi \bar{v}_A dP = \bar{v}_A \Pi \ ;$$

hence

$$\bar{v}_A \Pi = RT \ln(P_{A,\text{pure}}/P_A) \ , \tag{29}$$

indicating, as we saw above, that the osmotic pressure is the external pressure that will raise the vapor pressure of the solution to that of the pure solvent. In practice these equations are rarely valid for real solutions, so we restrict ourselves to dilute solutions, in which case $\bar{v}_A$ is about the same as the molar volume $\bar{V}_A$, so that $\bar{V}_A = RT \ln(P_{A,\text{pure}}/P_A)$. This restriction allows us to apply Raoult's Law to obtain

$$\bar{V}_A \Pi = -RT \ln X_A \ . \tag{30}$$

Note that this is expressed in terms of the mole fraction of the solvent. It is usually more convenient to employ X_B; the approximation that $\ln(1 - y) = -y$ for small values of y enables us to write Equation 30 (which we remember is for dilute solutions or small values of X_B) as

$$\bar{V}_A \Pi = RT X_B \ . \tag{31}$$

This can also be written in terms of the concentration c by using the dilute solution approximation, $X_B \approx n_B/n_A \approx c$, as

$$\text{Osmotic pressure} = \Pi = RTc \ . \tag{32}$$

This equation is known as the *van't Hoff equation* for osmotic pressure; we shall see it again in another form in Chapter 8. Note that if c is measured in molarity and the osmotic pressure is to be expressed in atm, R must be given in units of l atm K^{-1} mol^{-1}.

EXAMPLE 7–B The osmotic pressure at 20°C of a solution containing 18 grams of glucose per liter of aqueous solution.

The molarity of the solution is $18/180 = 0.1$ M. We assume that this concentration is low enough that the solution is ideal. To express the osmotic pressure in atmospheres, we use $R = 0.082$ l atm K^{-1} mol^{-1}. Thus $\Pi = (0.082)(293)(0.1) = 2.40$ atm or 1824 mm Hg. Since a pressure difference of 1 mm Hg is easily measured, this value gives one an idea of the ease with which the osmotic effect is detected compared to the other colligative properties. For instance, the decrease in the freezing point of this solution is only 0.18°C. If the glucose had been 0.001 M, the osmotic pressure would have been an easily measurable 18.24 mm Hg whereas the decrease in the freezing point would have been 0.0018°C, a value whose measurement would certainly include a very large experimental error.

It should be realized that the osmotic effect is the pressure *above* the ambient pressure, which is usually atmospheric pressure.

Determination of Molecular Weight from the Colligative Properties

Each of the equations describing the colligative properties includes the concentration of the solute. In preparing a solution, one always knows the weight of the added solute, so that from the measured values of any of the colligative properties, molecular weights can be calculated. The most effective measurements are those using either the freezing point depression or determination of the osmotic pressure. How this is done and the limitations of the various techniques are shown in the following examples.

EXAMPLE 7–C Determination of the molecular weight of a sugar by using the freezing point depression of water.

The value of K_{frz} for water is 1.862. A sugar solution of 5 g/kg of water has a freezing point depression of 0.05°C. Therefore from Equation 24, $\Delta T_{frz} = 0.05°C = 1.862 c'$. The molality c' is $0.05/1.862 = 0.0268$ **M**, so that the molecular weight is $5/0.0268 = 172$. The limitations of the method are imposed by (1) the smallest value of ΔT_{frz} that can be measured accurately; (2) the maximal concentration for which the solution is an ideal solution (because Equation 24 only applies to an ideal solution); and (3) the solubility of the solute. The limits for the first two are roughly 0.01°C (with a 10 percent experimental error) and 0.05 **M**. The method fails for macromolecules above a molecular weight of about 10,000 because of the solubility limit. The maximal solubility of the *most* soluble proteins is about 100 g/kg, which, for a molecular weight of 10,000, is a concentration of 0.01 **M** and produces $\Delta T_{frz} \approx 0.01°C$. At this high concentration other sources of nonideality become important (mainly the large volume of the macromolecule), and it is necessary to extrapolate the measured value of ΔT_{frz} to zero concentration. Since the measured values obtained at low concentrations are very inaccurate (because they are so small), the extrapolated value is not very precise and the error in the molecular weight may be as great as 50 percent. Such an extrapolation is discussed in greater detail in the following section.

EXAMPLE 7–D Measurement of the molecular weight of a protein by the osmotic pressure method.

A solution consisting of 650 mg of a protein in 10 cm^3 (10 g) of water has a measured osmotic pressure of 0.12 atm at 25°C (298 K). The molarity is $\Pi/RT = 0.12/(0.0821)(298) = 0.0049$ M. This equals a concentration of 65 g/l, so that the molecular weight is 65/0.0049 = 13265. Owing to the lower limit of Π that can be measured accurately, and the maximal solubility, the maximal molecular weight of a protein accurately measurable in this way is about 30,000. For organic polymers, such as polystyrene, in solutions such as toluene, the maximal solubility is much greater and it is possible to use the method for molecular weights up to several hundred thousand.

There is also a practical lower limit to the molecular weight measurable by the osmotic pressure method. In the theory we have developed, we have assumed that the membrane is permeable to the solvent and perfectly impermeable to the solute. Various factors determine the impermeability of a membrane to a particular substance—one important factor is the relative cross-sectional area of the solute molecule and the area of the pores of the membrane. For molecules in the size range of sucrose (molecular weight, 342) it is virtually impossible to find or make a membrane that passes water but is a total barrier to the solute. There are membranes that have a permeability for molecules in this size range that is much reduced compared to that of the solvent, but "much reduced" is generally insufficient. Practically speaking, the lower limit is a molecular weight of about 1000.

The Effects of Impurities and Inhomogeneity of Molecular Weights on Molecular Weights Measured from the Colligative Properties

In measuring the molecular weight M of a molecule in solution, it is important to know whether the value obtained is that of the pure substance or whether there are several components in the solution that contribute to the value. Clearly if the freezing point depression method (or any of the

SEC. 7-4   COLLIGATIVE PROPERTIES

colligative properties) is used, all solute molecules contribute to the depression and an average value of M is obtained. With osmometry, this is not the case, because the measurement can be done in such a way that only those molecules that cannot pass through the membrane in use contribute to the measured value. For instance, if one is using osmometry to measure the molecular weight of a protein that is in 0.3 M NaCl, one need only have 0.3 M NaCl on the other side of the membrane, and the observed osmotic pressure will be determined only by the protein. Actually, if small molecule impurities are present, this need not even be done if the small molecules can pass freely through the membrane, because at equilibrium (that is, when a constant value of the osmotic pressure is observed), the small molecules will be at the same concentration on both sides of the membrane; thus, also in this case, only the protein contributes to the osmotic pressure. However, the former method—that is, having the small molecules initially on both sides of the membrane—is preferable, because the rate of migration of the small molecule may be much less than that of water, so that the osmotic pressure would appear to be constant before equilibrium is actually achieved, thus yielding an osmotic pressure that is higher than would be observed at true equilibrium; and, of course, this would give a value of the molecular weight lower than the correct value.

The contribution of small molecules is especially important when studying macromolecules. This is because the values of all of the colligative properties are determined by the *number* and *not the weight* of the solute molecules. Thus, even if a very small percentage of the weight of the molecules is contributed by very small molecules, the *number* of these extra molecules may be extremely large compared to the number of the macromolecules. There is also a second problem of great significance. Samples of macromolecules isolated from living cells are frequently heterogeneous with respect to molecular weight. This is because a fraction of the molecules may be broken during isolation. A similar problem exists for polymers that are synthesized in the laboratory or by an industrial process; that is, the result is often a sample having a range of values of the molecular weight. In any such mixture, the molecular weight measured by any of the colligative properties is an average value. The following example gives a numerical demonstration of this averaging process (which is discussed in greater detail in Chapter 13).

EXAMPLE 7-E What is the observed value for the molecular weight of a protein whose true molecular weight is 50,000, in the presence of a 2-percent-by-weight impurity whose molecular weight is 250, if the measurement is made by osmometry?

In a mixture of two components the average molecular weight M_{av} equals $(n_1M_1 + n_2M_2)/(n_1 + n_2)$, M_1 and M_2 being the molecular weights of the two components, and n_1 and n_2 being the number of each molecules. This type of average is called a *number-average molecular weight*, M_n. For equal weights, for every molecule of type 1 there are 50,000/250 = 200 molecules of molecule 2. Therefore, if 2 percent of the weight is molecule 2, then for 98 molecules of type 1 there are 2 × 200 = 400 molecules of type 2. Thus the observed value is M_{av} = [98(50,000) + 400(250)]/(98 + 400) = 10,040, roughly five times too low.

In a later section we will discuss the osmotic pressure of nonideal solutions and, in particular, solutions of macromolecules.

Difference between the Tonicity and the Osmotic Pressure of a Solution

The concept of tonicity arises when considering osmotic effects in living cells. Naturally occurring biological membranes are highly selective in their ability to pass molecules, discriminating by size, charge, and often the identity of the molecule. For example, some membranes are permeable to the Na^+ ion but not to the K^+ ion; others allow some sugars but not others to pass through. Furthermore, in biological systems, the membranes do not separate a pure solvent from a solution containing a single solute, but usually separate two solutions each containing many different types of solutes. The osmotic pressure is therefore a result of the relative permeability and the relative concentrations of each component. The effective osmotic pressure of a solution with respect to a particular membrane is called *tonicity*. It is *not* a colligative property because it depends on the chemical and physical properties of the dissolved particles rather than solely on the number of particles.

The distinction between tonicity and osmotic pressure can be seen by referring to Figure 7-6, in which various solutions are separated by a membrane that is permeable to water and to a solute A but not to either of the solutes B or C. Notice that in each of the three panels of the figure the solutions on the two sides of the membrane are at the same concentration and therefore would produce the same osmotic pressure if separated from the pure solvent by a membrane that is impermeable to the solute; they are *isosmotic*. In panel (a) there is no flow of either of the solutes across the membrane, because it is impermeable to both of them. The effective osmotic pressure across the membrane is zero, so that the solutions have the same tonicity; they are *isotonic*. In panel (b), solute A on the right side of the membrane can move to the left. Accompanying this motion will be a leftward flow of water that results in dilution of C. Since the flow is from right to left, the effective osmotic pressure is lower on the right, so that the tonicity is lower; the solution at the right is *hypotonic* compared to the *hypertonic* solution at the left. In panel (c), the concentration of the nondiffusible solute molecules is lower on the right than on the left and there is a leftward flow of both water and A; hence the solution at the right is again hypotonic.

SEC. 7-4 COLLIGATIVE PROPERTIES

FIGURE 7-6 Flow of solvent and solute resulting from differences in tonicity. Two solutions are separated, as indicated, by a membrane permeable to water and to solute A but not to solutes B or C. In panel (a) neither B nor C can move through the membrane so there is no flow of solvent. In panel (b) solute A moves leftward, thus increasing the solute concentration and the osmotic pressure on the left; this results in a leftward flow of solvent to reduce the solute concentration on the left. In panel (c), A again moves to the left, making the solute concentration on the left greater than that on the right. Therefore solvent flows leftward to dilute the solution.

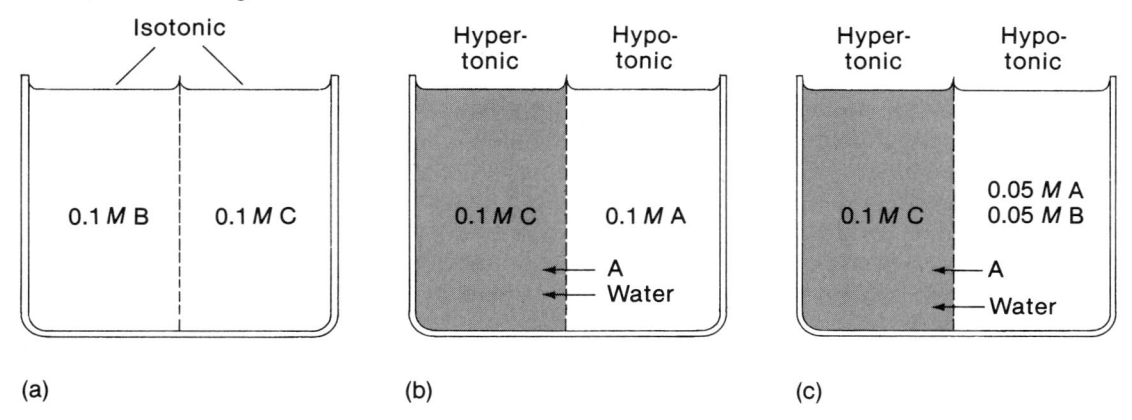

Note that the tonicity of a solution is mainly controlled by the concentration of the *impermeable* ion. Even if the contents of a cell are unknown, it is possible to prepare a solution that is isotonic to the cell contents if an impermeable solute is used. Isotonicity would be recognized by the absence of a net flow of liquid in either direction.

When two solutions of different tonicity are separated by a membrane, there is a flow of liquid. The rule governing the direction of flow is that

solvent flows from a solution of low tonicity to one of high tonicity.

This has been used to isolate intracellular molecules, as shown in the following examples.

EXAMPLE 7-F Effects of tonicity on living cells.

If a living cell is placed in distilled water or in a very dilute buffer (that is, in a hypotonic medium), the cell takes up water and swells. Red blood cells swell so much that the cell membrane becomes leaky and the hemoglobin inside the cells spills out. This is a useful method for measuring the hemoglobin content and, for other types of cells, for isolating intracellular proteins. For red blood cells, this process is called *hemolysis*. If cells are instead put into a concentrated salt solution (a hypertonic medium), water flows out of the cells and the cells shrink. This has been useful in studying cell wall structure in bacteria, yeasts, and plant cells, because the cell contents pull away from the somewhat rigid cell wall, making the cell wall easier to examine it by electron microscopy.

EXAMPLE 7-G Osmotic rupture of bacteriophages (bacterial viruses).

Most bacteriophages are freely permeable to most small molecules such as glycerol, sucrose, and NaCl. However, the passage of these molecules through the external protein layer to the internal DNA-containing regions is slow. For example, if *E. coli* phage T4 is placed in 2 M glycerol, it takes about one-half hour for the internal glycerol concentration to reach 2 M. If phages that have been equilibrated with 2 M glycerol are diluted rapidly into distilled water, the outward diffusion of the glycerol molecules is so slow that at the instant of dilution, the phages are in a highly hypotonic medium and they literally explode from the pressure within them. This is a valuable method for isolating DNA from phages and is called the *osmotic shock method*.

7-5 SOLUTION OF GASES IN LIQUIDS

Properties of Gaseous Solutions

Physically, there is little difference between a solution in which the solute is a gas and one in which the solute is a nonvolatile solid. In both cases, individual solute molecules are in a liquid environment and surrounded by solvent molecules. The principal difference, however, is that when the solute is a nonvolatile solid, the solute concentration is determined by the amount of solute added (up to the solubility limit), whereas for a gaseous solute the concentration is determined by the external pressure. If the solution is ideal, the concentration of gas above the solution (which is equal to the partial pressure P_i of the gas) is proportional to the concentration in solution—that is, Henry's Law is obeyed, or $P_i = K_i X_i$. This proportionality has led to the convention of describing the concentration of a gas in terms of its vapor pressure, which, for a gaseous solute, is always called the *tension*. Thus, when one says that the CO_2-tension of blood is 45 mm Hg, this means that the concentration of CO_2 in the blood is that which corresponds to a partial vapor pressure of 45 mm Hg.*

The dependence of the solubility of gases on temperature also differs from that of solid solutes. For a solid, the solubility generally increases as the temperature rises, since the cohesive force between solute molecules is weakened by their increased kinetic energy. When a gas dissolves, the gas molecules must be attracted to the solvent molecules because, otherwise, the gas molecules would prefer to be in the vapor phase. Thus, dissolving is

---

*Biologists continue to use mm Hg as the preferred unit for describing vapor pressure in biological systems.

SEC. 7-5 SOLUTION OF GASES IN LIQUIDS

in some ways akin to condensation; increasing the temperature allows thermal motion to overcome the cohesive force between gas and solvent molecules (just as the ability to condense decreases as the temperature rises), and the solubility of a gas decreases with increasing temperature—exactly the opposite situation from that with a solid solute. Of course, the vapor pressure increases with rising temperature, which means that the Henry's Law constant increases as the temperature rises.

Marked deviations from Henry's Law occur if the gas combines chemically with the solvent molecules, because the cohesive force between solute and solvent is generally much different from that between two solvent molecules. Three commonly encountered examples are ammonia, hydrogen chloride, and carbon dioxide in water. In the first case, the ammonia reacts with the H^+ ion to produce the stable NH_4^+ ion; in the second, HCl dissociates and the H_3O^+ ion forms; in the third, CO_2 reacts to form the bicarbonate ion, HCO_3^-. Actually one should not even think of these examples in terms of solubility but instead, as simple chemical reactions.

The solubility of a gas also depends upon the presence of other solutes, rising with some solutes and decreasing with others. For example, the presence of solutes that bind the solvent molecules decreases the solubility of most gases. This results from reducing the number of solvent molecules that are available for interacting with a molecule of the gas. The change in the solubility of a gas is rarely greater than a few percent when water is the solvent and the solute concentration is less than one molal because of the high molality (55.56 **M**) of water. The reduced solubility when high concentrations of strong electrolytes are present is called *salting out* and is explained in Chapter 8.

From a biological point of view, the most important effect on the solubility of a gas is the increase that occurs when a solute that reacts with or binds the gas is present. In principle, this is similar to the cases, just mentioned, of gases that react with solvent molecules. A striking example is the greatly increased solubility of O_2 in blood, compared to water, that results from the binding of O_2 by hemoglobin.

Each of the points we have just made is illustrated in the following example, in which the solubility of O_2, N_2, and CO_2 in blood is compared to the values expected from the partial pressures.

EXAMPLE 7-H Solubility of O_2, N_2, and CO_2 in blood.

The composition of air in the lungs is 15 percent O_2, 80 percent N_2, and 5 percent CO_2. The air is saturated with water vapor, which, at a total pressure of 760 mm Hg, has a partial pressure of 45 mm Hg. The partial pressures of O_2, N_2, and CO_2 are then $0.15 \times 715 = 107$, $0.80 \times 715 = 572$, and $0.05 \times 715 = 36$ mm Hg, respectively. From the Henry's Law constants, K_i, for each gas dissolved in water, the mole fraction of each, X_i, is calculated ($X_i = P_i K_i$). The solubility of each gas is reduced by 8 percent by the presence of the dissolved salts in the blood. Thus, each value X_i must be multiplied by $1.0 - 0.08 = 0.92$. The value X_i can be converted to cm^3 of gas per liter of water by multiplying by 55.56 (the molality of water) and 22,400 (the volume of 1 mol of gas in cm^3 at 1 atm and 25°C). If we express the results in the conventional way—namely, as cm^3 gas/100 ml solvent— the resulting values are: O_2, 0.31; N_2, 0.83; and CO_2, 2.4. However, the observed solubilities in arterial blood are: O_2, 19.5; N_2, 0.9; and CO_2, 49.7. The observed value for N_2, which does not interact significantly with any of the components of blood, is nearly the calculated value. However, the values for O_2 and CO_2 are about 60-fold and 20-fold greater, respectively. This is a result of the reaction of O_2 with hemoglobin and of both the binding of CO_2 to amino groups in many proteins and the conversion of CO_2 to the HCO_3^- ion.

Removal of a Gas or a Volatile Solid from a Solution

In an open container, a glass of water slowly evaporates as the liquid phase equilibrates with the atmosphere. This is because the volume of the atmosphere (or the laboratory) is so great that there are not enough molecules in the liquid to reach the value of the vapor pressure that is in equilibrium with the liquid at the particular temperature. A fan blowing on the surface of the water accelerates the evaporation by continually replacing the air above the liquid with air containing less water vapor. Consider now a solution of a small amount of ether in water. The ether is much more volatile than water, which means that pure ether has a higher vapor pressure than pure water; thus, the vapor above an ether-water mixture will be enriched for ether when compared to the relative amounts of each in the mixture. If the air is blown across the surface of an ether–water mixture, ether and water will evaporate to replace the molecules that have been removed. However, a greater fraction of the ether will enter the vapor phase than the fraction of water that evaporates. This means that the composition of the solution will change—specifically, the fraction that is ether will decrease. Thus, a volatile solvent can effectively be removed by blowing air across the liquid. If the air is saturated with water vapor, then no water will be removed and the solution can be freed of ether. The rate at which evaporation occurs depends upon the area of the liquid in contact with the gas phase. Thus, a volatile solute can be removed more rapidly by bubbling water-saturated air through the solution, because the bubbles vastly increase the surface area.

Note that a solution of a gas in a liquid can be depleted of a gas by bubbling a second gas through the liquid, because the gas in solution must always be in equilibrium with its partial pressure above the solution. Thus, bubbling pure O_2 through water can remove all of the N_2 from the water and, conversely, bubbling pure N_2 through the water removes all of the dissolved oxygen. This is easily demonstrated with blood. When O_2 is bubbled through blood, the color of the blood is bright red, which is the color of oxygenated hemoglobin. Bubbling N_2 through the blood changes the color to blue (as in venous blood), the color of deoxygenated hemoglobin.

7-6 FACTORS THAT DETERMINE THE SOLUBILITY OF A SOLID

When a sample of a solid substance dissolves, the molecules in the sample separate, which means that there must be a decrease in the net cohesive force between the individual molecules. This decrease can occur for three reasons: (1) the actual cohesive force is reduced; (2) the solute molecules are also attracted to solvent molecules; and (3) collisions with solvent molecules impart sufficient kinetic energy to the solute molecules that the cohesive force is insufficient to hold the solute molecules together. The first effect is important with solutions of charged molecules in a solvent having a high dielectric constant (see Chapter 8), but not for uncharged solutes and weakly polar organic solvents. The second effect is very common and is called *solvation;* a prime example of solvation is hydration by water molecules. Attraction to the solvent molecules is a requirement for an appreciable solubility, because if there is no cohesive force between solute and solvent molecules, no solute will go into solution. For an ideal solution, by definition, the solute-solute cohesive force equals the solute-solvent cohesive force, so that the different solubilities must have other origins. (In a real solution, of course, the relative strengths of the solvent-solute and the solute-solute interactions are extremely important.) The third item is an important factor in ideal solutions, as we shall see shortly.

As we saw in Chapter 3, the magnitude of the cohesive force in a solid is indicated in part by the temperature at which the solid melts—a low melting point implies that the cohesive force is small. A molecule that attracts a like molecule weakly should be quite easy to dissolve, because only a small cohesive force must be overcome. Thus, solutes with low T_{frz} values should, on this basis, have high solubility. Melting of a solid to a liquid is always accompanied by a heat change called the enthalpy of fusion, ΔH_{fus}. This may be thought of as the amount of kinetic energy that must be imparted to the molecules to overcome the cohesive force sufficiently that the molecules are free to move about. Thus, solutes having a lower value of ΔH_{fus} should be more soluble. That this reasoning is correct is evident by the fact that for an ideal solution we can obtain a simple expression for the mole fraction of the solute in a saturated solution as a function of T_{frz} and ΔH_{fus}. This expression is derived as follows.

A saturated solution is one in which, by definition, the solid solute is in equilibrium with the solution; that is, the chemical potential of the pure solid, μ_s, equals the chemical potential of the solute in the saturated solution. To obtain the latter quantity, we must for a moment think in a way that might seem odd, but in so doing we will see an essential facet of this process. Since the pure solid and the dissolved solute are in equilibrium, the vapor pressure of the solid must equal the vapor pressure of the solute. This might seem unreasonable, because the mole fraction of the solute in solution is not one, and hence, by Raoult's Law, the vapor pressure of the solute should be less than that of the pure solid. This apparent contradiction disappears if we think of the solution as consisting of solvent dissolved in the solute, which is possible because the distinction between solvent and solute is an arbitrary one. We now note that since the solution is a liquid, the solid must have undergone a solid-to-liquid phase change as it took up the other liquid. Thus, by Raoult's Law, the vapor pressure of the pure solid is the same as that of the pure *liquid* (at the same temperature) multiplied by the mole fraction of the solid component that we are viewing as a liquid. Note that at that temperature the solid would not be a liquid, so that we are in fact talking about a supercooled liquid. Nonetheless, the process of forming the solution may be thought of as consisting of two steps—melting of the solid and mixing of the supercooled liquid with the real liquid. This involved argument allows us to conclude that the chemical potential of the real solute (we now return to the original convention as to what is solvent and what is solute) is equal to the chemical potential of the liquid solute, μ_l, plus $RT \ln X_s$, in which X_s is the mole fraction of the solute. Therefore,

$$\ln X_s = -\frac{\mu_l - \mu_s}{RT} = -\frac{\Delta G}{RT}, \qquad (33)$$

in which $\mu_l - \mu_s = \Delta G$ is the hypothetical free energy change if the solute could melt at a particular temperature T. At the melting temperature T_{frz}, liquid and solid are in equilibrium (that is, $\Delta G = 0$), so that we can subtract $\Delta G_{frz}/RT_{frz}$ from the equation above (this is just an arithmetical trick) to yield

$$\ln X_s = -\left(\frac{\Delta G}{RT} - \frac{\Delta G_{frz}}{RT_{frz}}\right)$$

$$= -\left(\frac{\Delta H - T\Delta S}{RT} - \frac{\Delta H_{fus} - T_{frz}\Delta S}{RT_{frz}}\right). \qquad (34)$$

We now make the simplifying assumption that is often made in the analysis of phase transitions (see Chapter 4), namely, that H and S can be considered to be independent of T over a small range of T, so that when T is varied, ΔH and ΔS are zero except at T_{frz}.* Equation 34 then becomes

$$\ln X_s = -\frac{\Delta H_{fus}}{R}\left(\frac{1}{T} - \frac{1}{T_{frz}}\right) \qquad (35)$$

or

$$X_s = \exp\left(-\frac{\Delta H_{fus}(T_{frz} - T)}{RTT_{frz}}\right) = \exp(\Delta H_{fus}/RT_{frz})\exp(-\Delta H_{fus}/RT), \qquad (36)$$

*The validity of this assumption is based on the fact that ΔH and ΔS vary slowly with T except at a transition temperature and we are usually discussing events that occur in a small temperature interval.

which is the mole fraction of solute in a saturated ideal solution. Note that X_s increases as the melting temperature decreases or approaches T, and that it is greater as ΔH_{fus} is smaller, as we predicted. Furthermore, when $T < T_{\text{frz}}$, X_s increases as T becomes greater, which is in accord with the common observation that most substances increase in solubility as the temperature rises. Note that when $T = T_{\text{frz}}$, $X_s = 1$. This means that once the solid has melted, it becomes miscible with water and can increase in concentration indefinitely. This is what would be expected of an ideal solution, because, once liquefied, the solute no longer has cohesive forces to overcome.

The argument just given implies that the mole fraction at saturation depends on the properties only of the solute—not of the solvent. This is clearly incorrect; common experience dictates that the solubility varies widely with the nature of the solvent. The explanation for this discrepancy lies in the assumption of ideality—that is, that the cohesive forces between all components are identical. For example, in the case of sucrose as the solute in benzene (a solvent in which sucrose is very poorly soluble), the cohesive force between a sucrose molecule and a benzene molecule is nearly zero, so that an ideal solution cannot exist and the equations do not apply. Where, however, has the concept of ideality actually entered the derivation—inasmuch as we began the argument with chemical potentials? It is in the use of the mole fraction in the expression $\mu = \mu^\circ + RT \ln X$. In a later section, we will see that for a real solution a quantity called the *activity* should be used instead of the mole fraction, so that the role the solvent plays in determining the solubility is contained in the value of the solute activity.

A second problem that arises from assuming ideal conditions is that, in a real solution, especially at the high concentrations frequently approached at saturation, there is often a large contribution to the value of ΔS that has nothing to do with the melting transition. This value of ΔS is particularly important in high concentrations, in which order among solvent molecules might be created or disrupted. Thus, in an ideal solution, the value of X_s at saturation defines the solubility precisely, but in a real solution the relation between activity and solubility is more complex.

In conclusion, for concentrated solutions we need more complicated equations to account quantitatively for the solubility at saturation. However, if the saturated solution is so dilute that Raoult's Law is obeyed, one can predict the solubility. *Thus the equations we have given are valid for sparingly soluble substances.*

7-7 PARTITIONING—THE DISTRIBUTION OF A SOLUTE BETWEEN TWO IMMISCIBLE SOLVENTS

Consider two immiscible solvents that are in contact, each of which contains the same solute. The solute molecules are free to move across the interface from one solvent to the other, so that ultimately there will be an equilibrium in which the solute concentrations are somehow related to the solubility in each solvent. An expression for the distribution of solute molecules can be obtained by considering two points: (1) even if the solute is a solid, there will be some vapor of that solid above the solution; and (2) if the solute concentration is low, the value of the vapor pressure P_i is determined by Henry's Law, $P_i = K_i X_i$. If the two solutions are in contact, they will in general be in contact with the same vapor phase (although we could design an experimental system in which they are not), so that if K_1 and K_2 are the Henry's Law coefficients for solvents 1 and 2, and X_1 and X_2 are the mole fractions, then $K_1 X_1 = K_2 X_2$, and the ratio of the mole fractions is simply

$$\frac{X_1}{X_2} = \frac{K_2}{K_1} = K_D ; \tag{37}$$

K_D is called the Nernst distribution coefficient or the *partition coefficient*. In dilute solution, the mole fractions are approximately proportional to the concentration c, so that we may also write

$$\frac{c_1}{c_2} = K_D . \tag{38}$$

We demonstrate the use of this equation in the following example.

EXAMPLE 7–1 Extraction of a solute Q from 100 ml of a 0.15 *M* solution in water, using 100 ml benzene. For this solute, the value of K_D for benzene, compared to water, is 15.8.

One hundred milliliters of water (density = 1 g ml^{-1}, molecular weight = 18) contains $100/18 = 5.56$ mol of water and $0.1 \times 0.15 = 0.015$ mol of a solute Q. When 100 ml of benzene is added, the ratio of the mole fraction of Q in benzene to that in water must be 15.8. Let q be the number of moles of Q in the benzene at equilibrium; 100 ml of benzene is 88 g or 1.13 mol, so the mole fraction of Q is $q/(1.13 + q)$. The mole fraction of Q in water will be $(0.015 - q)/(5.56 + 0.015 - q)$. Thus $[q/(1.13 + q)]/[(0.015 - q)/(5.56 + 0.015 - q)] = 15.8$, so $q = 0.011$. Therefore, $0.015 - 0.011 = 0.004$ mol of Q remain in water, and the concentration has decreased by a factor of $0.015/0.004 = 3.75$. If the benzene is removed and replaced with 100 ml of fresh benzene, the concentration of Q will decrease by a second factor of 3.75 for a total reduction of $3.75^2 = 14.06$. If there are n extractions, the factor is $(3.75)^n$. Thus, if the final concentration of Q in water is to be less than $10^{-5} M$, this means that there must be fewer than 10^{-4} mol/100 ml of water, which requires a decrease of a factor of $0.015/0.0001 = 150$, which can be accomplished with four benzene extractions ($3.75^4 = 184$).

SEC. 7–7   PARTITIONING

The principle of partitioning can be seen in many biological phenomena, because a living cell or organism contains a large number of different liquids and solids. This is shown in the following example.

EXAMPLE 7–J   The anesthetic effect of ether.

Ether is a substance that is more soluble in organic solvents than in water. It exerts its biological effect by action on the cell membrane of nerve cells in the brain. If ether is inhaled, a small amount dissolves in the blood. As the blood enters brain tissue, which is extremely rich in lipids and fats, the ether partitions between the blood, which is aqueous, and membranes, which are lipid, and is concentrated in the brain, where the lipid content is very high. Note that the lipids remove ether from the blood. Because of the low solubility of ether in water, the blood is usually saturated when ether is inhaled. Thus, the removal of the ether by the lipids reduces the concentration to less than saturation and, in so doing, allows more ether to enter the blood. Thus, partitioning is, in this case, part of a mechanism of solute transport. This partitioning effect may account for the substantial decrease in the narcotizing effect of alcohol experienced when one eats fats (as, for example, in cheese). Ethanol is normally also concentrated in fatty tissue but, when one eats fats, the fat droplets in the blood may prevent the migration of the alcohol to nerve tissue.

In the laboratory, partitioning is of great use in removing unwanted substances from solutions, as can be seen in the following example.

EXAMPLE 7–K   Purification of DNA from bacteria.

When a bacterium is broken open, its contents spill out into the surrounding medium. If the resulting mixture is shaken with phenol, a liquid that is only slightly miscible with water, many components of the mixture distribute between the two phases. Lipids, some proteins, and, if the salt concentration is sufficiently high, a great deal of the RNA enter the phenol phase. (As an added bonus, most of the remaining proteins precipitate in the aqueous phase, and this precipitate is easily removed by centrifugation.) The phenol layer is removed and fresh phenol is added, which extracts more of the unwanted material from the aqueous phase. Because of the slight miscibility of phenol and water, the aqueous phase contains some phenol as well as the DNA. This must be removed because it interferes with most studies done with DNA. The phenol is also removed by partitioning, by adding successively several large volumes of ether with which water is only slightly miscible. Phenol partitions between ether and water such that a great excess is in the ether phase. Several ether treatments remove nearly all of the phenol. The small amount of ether that is dissolved in the water is then removed by bubbling air through the solution, as described earlier in this chapter.

The phenomenon of partitioning is utilized in three important techniques for separating one chemical from another—namely, partition chromatography, gas chromatography, and countercurrent distribution. These methods are described in detail in references given at the end of this chapter.

7–8 SOLUTIONS AT TEMPERATURES FOR WHICH THE COMPONENTS ARE NEAR THE MELTING POINTS—THE PHASE RULE

When a pure substance melts, the phase transition occurs over a very small temperature range and the transition is said to be sharp. In solution, the melting point of the solvent is depressed, as we have discussed, and the transition is still sharp (except for the case that we now discuss). Since the distinction between the solvent and the solute is arbitrary (that is, they are the major and minor components, respectively), there is really no physical difference between them, and hence, there must also be a depression of the melting point of the solute by the solvent. The melting point of a typical solid solute is so much higher than that of the solvent that this is usually not an especially interesting phenomenon. However, if both solute and solvent molecules have melting points that are near to one another, a phenomenon occurs that has some biological significance. In a solution, the melting points of each component are not only depressed by the presence of the other component, but also the melting transitions are often rather broad. To understand why this is sometimes so, we consider a relation known as the *Phase Rule,* which relates the number of phases, the number of components, and the number of physical variables of a system. The Phase Rule is a deceptively simple and abstract relation that has important predictive value. It is derived as follows.

Consider a solution consisting of many components. At equilibrium, the state of the system can be completely specified by giving the temperature, pressure, and the concentration of each component. If there are p phases and c components, the minimum number of variables that is sufficient to describe the system is pc (which tells us the concentration in each phase) + 2 (that is, the temperature and pressure, which are the same for all phases if they are at equilibrium). Actually $pc + 2$ of these variables are not needed, because the variables are not all independent. For instance, if X_i is the mole fraction of the ith component, then in any phase (for example, phase A) $X_1^A + X_2^A + \cdots + X_c^A = 1$, in which c is the number of

SEC. 7-8   THE PHASE RULE

components in that phase. Thus, if one knew all values of X_i but one, that one value could be calculated. This means that the number of variables needed to describe each phase cannot be greater than $c - 1$, and since this is true of every phase, the total number of variables needed to specify the entire system is $p(c - 1) + 2$. However, there are other constraints that limit the number of variables—namely, the fact that at equilibrium the chemical potentials of the ith component in one phase must equal the chemical potential of the ith component in each phase. Thus there are $p - 1$ equations for each of the c components, or $c(p - 1)$ more variables that need not be specified. We have now accounted for all of them and the number f of independently specifiable variables (also called the *degrees of freedom*) that remain is

$$f = [p(c - 1) + 2] - c(p - 1) = c - p + 2 \ . \tag{39}$$

This equation is called the Gibbs Phase Rule.

Let us now see what this abstract equation tells us. Consider a one-component ($c = 1$) system, pure water, and observe how many phases there are when P and T have various values. Figure 7-7 shows a *phase diagram*

FIGURE 7-7 Phase diagram for water, indicating the values of pressure and temperature at which various phases are present. If a particular pair of P-T values falls on any of the three curves, the phases on the two sides of that curve are in equilibrium. Thus, if $T = T_1$ and $P = P_1$, solid and vapor can coexist. Note that if P is less than that at the triple point, there is no temperature at which water can be liquid. Similarly the solid can never exist above the temperature of the triple point. There is no temperature at which vapor cannot exist if the pressure is low enough.

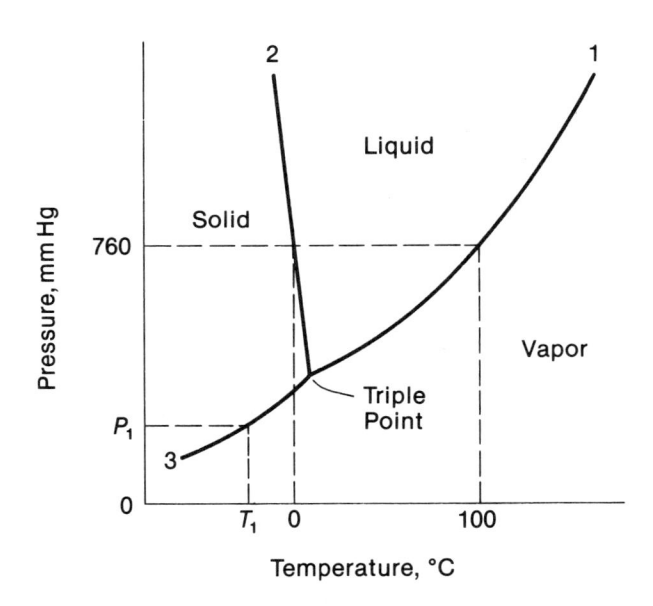

for water. Curve #1 is the vapor pressure curve, #2 is the melting curve, and #3 is the sublimation curve. The Phase Rule states that for this one-component system there are 2, 1, and 0 degrees of freedom when there are 1, 2, and 3 phases, respectively. For example, in the T-P region in which water is pure vapor ($p = 1$), T and P can be chosen independently; that is, $f = 2$. By "independently," we mean that if we choose a value of P, there can be many values of T. Along a line in which there are two phases that coexist ($p = 2$), if a particular value of P is chosen, then T is specified (we have no choice), and vice versa—that is, one variable uniquely describes the system ($f = 1$). At the single point on the curve at which solid, liquid, and gas coexist (the triple point), $p = 3$, and there is no degree of freedom because the triple point only occurs at a unique pair of values of P and T—we have no choice in the matter.

We now consider a two-component system consisting of a solute and a solvent. The Phase Rule tells us that at a unique combination of P and T (that is, when $f = 0$), $p = 4$, or four phases can coexist. For example, these phases might be solid solute, solid solvent, liquid solution, and vapor. However, we might choose to require that one variable ($f = 1$), for instance, the pressure, be different from the particular value to which we have just referred. In this case, three phases might be present. One possible set of three phases is solid solution, liquid solution, and vapor; another possible set is solid solute, liquid, solution and vapor. In both cases those sets can only occur at a unique temperature. Note that the Phase Rule does not say *which* phases are present—only the *number* of phases.

Let us now return to the question of the sharpness of melting points. The Phase Rule predicts that the melting point of a pure substance will be sharp. While melting occurs, both liquid and solid coexist ($p = 2$), so that $f = c - p + 2 = 1 - 2 + 2 = 1$. If we select the pressure (for example, as atmospheric pressure, which is the usual laboratory situation), then we have no further choice and melting occurs at a unique temperature; this is what is meant by a sharp transition. In a mixture of two components, c 2, and $f = 2 - 2 + 2 = 2$. Thus, if the value of the pressure is again determined, then the temperature of melting is no longer unique. That is, there are many temperatures at which solid and liquid phases can coexist—in other words, the melting point is not sharp. The phase diagram shown in Figure 7-7 should help make this argument clear.

So far, the phenomenon we have described is independent of how near each other the melting points of the solvent and solute are. Simply because a solution exists, both transitions are broadened. When the melting points differ by only a few degrees, the transitions overlap and a two-phase state exists throughout the temperature range that includes both transitions. If the two phases are finely dispersed and well mixed, the mixture will be fluidlike or semisolid. If a third component is added whose melting temperature is also within a few degrees of the other two components, the total transition increases in breadth. Thus, by having many components, the semisolid state can exist over a very wide range of temperatures.

The biological significance of this broad melting phenomenon can be seen by examining the structures and properties of biological membranes. These membranes, which consist primarily of proteins and lipids (see Chapter 13), are known to possess a high degree of fluidity. Their properties, of which the ability to transport sugars and to transport nerve impulses are important examples, are regulated to a great extent by variations in the degree of fluidity. Many of the lipid components have melting points in the range of from 5 to 40°C. Thus, if a membrane were to consist of a single component, in a lower organism that in its natural environment is exposed to a wide range of temperature, a liquid membrane might become solid or a solid membrane become liquefied at temperatures that the organism might encounter fairly often. However, by being a multicomponent system, a typical membrane instead has a very broad range of temperatures in which its fluidity does not change very much. Thus, having a multicomponent membrane has significant survival value for the organism.

Membranes also have selective permeability. One way that this might be achieved is for the membrane to be subdivided into several distinct regions. The Phase Rule indicates how a membrane consisting of two components might accomplish this. Most biological systems are subjected to varying temperatures, so that in our analysis, the temperature cannot be fixed—in other words, f must be at least one. With $f = 1$, if c is 2, the Phase Rule would allow the membrane to have three phases (and more, if $c > 2$). Let us consider what these phases might mean by examining two lipid molecules A and B that have the property that clusters of A, of B, and of intermixed A and B, each have distinct permeability properties. It might then be advantageous to a cell if these clusters could be maintained in a mixture. In fact, these clusters would be stabilized if, for example, the homogeneous clusters were in the solid phase and the heterogeneous cluster were a liquid solution. Another possibility is that each cluster could be part of a system consisting of three immiscible liquids. The Phase Rule tells us that either of these arrangements is thermodynamically possible, as long as there are two components. It does not, of course, state that the two components must separate into three phases, but only that if the physical properties of the lipids (for example, the melting temperatures and the solubilities) are such that phase separation occurs, there will be two phases.

7-9 NONIDEAL SOLUTIONS

Sources of Nonideality

We have so far discussed ideal solutions and the approach to ideality at low solute concentrations. We have implied, however, that most solutions are not ideal, so one might reasonably wonder why so much effort has been spent to understand something that occurs only rarely. The answer is simply this—as in the case of the ideal gas laws, the general principles and basic phenomena are most easily understood for the ideal solution; the lack of ideality only introduces quantitative rather than qualitative differences.

We explained that an ideal solution is one in which all intermolecular forces are identical. In a real solution containing dissimilar substances (for example, a liquid solvent and a solid or gaseous solute), these forces depend on the identity of each pair of molecules and this is the most basic cause of nonideality. Let us now review the interactions in a real solution. We consider a solution of two liquids, A, whose molecules attract one another strongly, and B, whose molecules attract one another weakly. In this solution, an A will sometimes be surrounded only by B's. If the force between an A and its surrounding B's is less than that between any two A's or any two B's, such an A will have a greater ability to escape to the vapor phase than will the A's of a pure-A solution, so that the vapor pressure of A in the mixture will be higher than that of pure A. Similarly, some B's are separated from one another by the A's. The two A's will attract one another more strongly than they would attract a B. Thus, the B's will have nothing with which to interact (not even another B) and can escape more easily from the mixture than from pure B. Hence the vapor pressures of both A and B increase in the mixture; this is an example of positive deviation from Raoult's Law, as we saw in Figure 7-3. On the other hand, if A and B interact more strongly than either A and A or B and B, the vapor pressures of both A and B are reduced and there is a negative deviation from Raoult's Law.

Nonuniformity of cohesive forces has two other effects—changes in volume and a nonzero heat of mixing. In the case of the positive deviation from Raoult's Law, the cohesive force between unlike components is less than that between like components. This results in a general loosening of the structure, so that the volume of the solution is greater than the sum of the volumes of the components. Furthermore, in order for mixing to occur, the like molecules must be separated and, to overcome the cohesive force, the kinetic energy of the molecules must be increased (because the solute-solvent cohesive force is not great enough to overcome the solute-solute interaction). This requires an input of energy so that on mixing there is an absorption of heat from the environment. When the deviation from Raoult's Law is negative, the cohesive force between unlike components is greater than between like components, and there is a tightening of the structure. By the same argument just given, on mixing there is a reduction in volume and an evolution of heat.

There are many other sources of deviation from ideality—for example, alteration of solvent structure by a solute, and concentration-dependent association of solute molecules. The first occurs often with water as a solvent, because water molecules are strongly associated with one another owing to the large dipole moment of water molecules and to hydrogen-bonding. We distinguish this association from the simple cohesive forces we have just been discussing, because it goes far beyond the mere attraction between two molecules. That is, water molecules associate to form large structures that contain many molecules. This strong association keeps the vapor pressure of water fairly low compared to what might be expected for a molecule of its size if there were no association. Depending on their properties, solute molecules of different types might increase or decrease the fraction of the water molecules that are associated. If the fraction were decreased, more water molecules would be free to escape to the vapor phase, the vapor pressure would be increased, and there would be a positive deviation from Raoult's Law.

Concentration-dependent association of solute molecules would cause an apparent deviation from ideality, because the concentration and the mole fraction would not be simply related. For example, with a nonassociating solute, doubling the amount of added solute results in a doubling of the number of independent solute units. However, if there is aggregation and the fraction of molecules that is aggregated increases with concentration, then the number of units would not double. Both Raoult's Law and Henry's Law, as well as each of the colligative properties, depend on the number of units, so that there would be a smaller effect on the solution properties than would be expected if there were no aggregation. If one knew the degree of aggregation at a particular solute concentration, the true mole fraction could be calculated and there would be no deviation from ideality, at least from this source. Since in general this is not known, there is an apparent deviation.

A significant cause of nonideality is the electrical interaction that occurs with charged solute molecules. Discussion of this phenomenon is deferred to Chapter 8.

All real solutes produce nonideal solutions, but at low concentrations the deviation from nonideality is not always apparent. In the case of most solutes that are small molecules, a solution behaves ideally at concentrations below about 0.05 M. However, for macromolecules such as proteins, nucleic acids, and polysaccharides in aqueous solution, or polymers such as polystyrene in organic solvents, nonideality may persist at concentrations as low as 10^{-4} M. This is because a single macromolecule is so large that it frequently affects an enormous number of solvent molecules. For instance, one DNA molecule might bind several million water molecules. Another complication resulting from the great size of polymers is the *excluded volume effect,* which is described in a later section.

Experimentally, how does one deal with nonideality? One way is to calculate the corrections, although this is not usually possible. A simpler procedure follows from the fact that each of the effects we have just described decreases as the concentration decreases. Thus, to obtain values of parameters (such as those of the colligative properties) that can be related to the properties of the individual molecules, one determines the values at various concentration and extrapolates to the value at a solute concentration of zero. An example of this is given in a later section.

Activity

So far we have derived for ideal solutions a small number of equations that relate various properties of solutions to the mole fraction or other measures of concentration. If we wish to have similar equations for nonideal solutions, we would immediately run into difficulty: to introduce each of the causes of nonideality into our simple mathematical expressions (if it could be done at all) would make the expressions very complex. In order to preserve the form of these equations, G.N. Lewis introduced a new concept, *activity*, a, which is related to the concentration c by the simple equation

$$a = \gamma c, \tag{40}$$

in which γ is the *activity coefficient*. With this relationship, the form of some of the equations can be preserved merely by using a instead of c, so that all of the complexity owing to nonideality is contained in the factor γ.*
Note that $\gamma = 1$ for an ideal solution. One must be especially careful in the use of activity coefficients, because the concentration c may be expressed as a mole fraction, molarity, or molality. Thus, when using tables of activities and activity coefficients, it is important to be sure that they are expressed in the appropriate units. The interconversion of units is described at the end of this chapter. Also, γ itself depends on concentration.

In considering real solutions, the two most important equations that should be remembered are the activity formulation of Equation 9—that is, the change in chemical potential μ with activity,

$$\mu = \mu° + RT \ln a = \mu° + RT \ln \gamma X, \tag{41}$$

and the activity formulation of Raoult's Law

$$P_A = a_A P_{A,\text{pure}}, \tag{42}$$

in which we must be careful that a_A is expressed as a mole fraction. Equation 41 is especially valuable, because it provides a means for measuring the activity of the *solvent* at a particular concentration. Equation 41 is often a source of confusion, because there is a tendency to think that it should give information about solute activity, since both solute-solvent interactions and solute-solute interactions may be responsible for the nonideality. Nonetheless, Equation 41 does *not* give direct information about the activity of the solute—*the measurement of the partial pressure yields the solvent activity only.*

In Appendix II at the end of the book, we show how the solute activity can be calculated from the solvent activity.

A sample calculation of the solvent activity is shown in the following example.

---

*From a strictly practical point of view γ can be regarded as a "fudge factor." However, theoretically it has significance. Remembering that μ is a free energy term, we can interpret the $RT \ln X$ term in the equation $\mu = \mu° + RT \ln X$ as signifying the work done in forming the mixture. If $\gamma \neq 1$, the term becomes $RT \ln \gamma X = RT \ln \gamma + RT \ln X$, and we can see that there is an extra term that represents the extra work that is done in forming the solution. This extra work is, of course, a result of the necessity of overcoming cohesive forces, and in forming specific arrangements of the solvent molecules. This concept of work is used in Chapter 8, in which γ is calculated for a solution of an electrolyte.

EXAMPLE 7-L Calculation of the activity and activity coefficient of water from a measurement of the vapor pressure.

A one-molal aqueous solution of a nonvolatile solute Q has a vapor pressure of 20.40 mm Hg at 25°C. The vapor pressure of pure water at 25° is known to be 23.75 mm Hg. Thus, according to Equation 41, the activity of water in this solution is $20.40/23.75 = 0.859$. To calculate the activity coefficient, we must remember that the activity when calculated from vapor pressure measurements is expressed in mole fraction units. The solution is 1 molal and therefore contains 1 mole of solute per 1000 g water or per $1000/18 = 55.56$ moles of water. Thus, the mole fraction of water is $55.56/(55.56 + 1.0) = 0.982$; hence $\gamma = 0.859/0.982 = 0.874$. Note that this activity coefficient is *not* a general property of *every* one-molal solution but applies only to a solution in which Q is present as the solute. In general, the value of the activity coefficient depends on the identity and concentration of the solute.

The activity of a solvent can be determined from any of the colligative properties, also. However, the freezing point and boiling point changes are of limited usefulness, because the measurement is restricted to temperatures near the freezing or boiling points. Osmotic pressure measurements can be carried out over a wide range of temperatures and therefore are of greater utility. The appropriate equation is

$$\ln a_A = -\frac{\Pi \overline{V}_A}{RT}, \tag{43}$$

in which $\overline{V}_A$ is the molar volume, and we must again remember to express the activity in units of mole fraction. Thus the activity of the solvent at a particular concentration of solute can be calculated directly from the measured osmotic pressure at that concentration, as is shown in the following example.

EXAMPLE 7-M Calculation by the osmotic pressure method of the activity and activity coefficient of water in a 0.05-*M* solution of a solute Q that cannot pass through a semipermeable membrane. The measured osmotic pressure of the solution is 3.1 atm at 25°C.

The molar volume of water is 0.018 l. The value of R in the appropriate units is 0.082 l atm K^{-1} mol^{-1}. Thus, by Equation 43 $\ln a_A = -(3.1)(0.018)/(0.082)(298) = -0.0023$, so that $a_A = 0.9977$. The mole fraction of water in a 0.05 *M* solution is $55.56/(55.56 + 0.05) = 0.9991$, so that the activity coefficient is 0.9986.

The Equilibrium Constant and Activity

In a nonideal system, an equilibrium constant is correctly written in terms of the activities rather than the concentrations of the components. The expression for the equilibrium constant K_a can be written for a reaction $pA + qB \rightleftarrows rC + sD$, in terms of activities, as

$$K_a = \frac{a_C^r a_D^s}{a_A^p a_B^q}. \tag{44}$$

This may also be expressed in terms of activity coefficients, γ, and concentrations, c, as

$$K_a = \frac{\gamma_C^r \gamma_D^s}{\gamma_A^p \gamma_B^q} \cdot \frac{c_C^r c_D^s}{c_A^p c_B^q}. \tag{45}$$

In biochemical reactions, the concentrations of all components are sufficiently low that it is nearly universal practice to ignore nonideality (even when it persists!) and to use molarity exclusively (not molality) to express concentration.

Experimental Correction for Nonideality

From what we have been saying, it is clear that the degree of nonideality is a concentration-dependent phenomenon. The associated problems can be avoided experimentally by obtaining data at various concentrations and extrapolating the data to zero concentration. Just as Raoult's Law is obeyed for all solutes at zero concentration, so are other properties of solutions.

Nonideality is an especially significant problem when using colligative properties to determine the molecular weight M of a solute. This is because experimental data are more accurate for the higher concentrations at which the equations, when written in terms of concentration, are incorrect. For example, when using these equations, it is almost always found that the observed molecular weight depends on concentration. This is particularly prevalent when studying macromolecules, for two reasons. (1) At high concentrations, the effects of the different cohesive forces, notably between solute and solvent, become larger. This reduces the effective concentration (that is, the activity is always less than the concentration) and thereby lowers the observed value of M. (2) The molecules may form aggregates (for example, dimers and trimers). The fraction of the number of molecules in aggregated form increases with concentration and the observed value of M, which depends on the *number* of solute units, decreases with increasing concentration. In theory, these problems could be dealt with by calculating activity coefficients, but this is extremely difficult. The simplest way to handle such problems is to extrapolate the observed value of M to zero concentration since in the limit of zero concentration, all solutions are ideal. This is shown in Figure 7-8.

FIGURE 7-8 Two types of experimental curves showing the dependence of the observed values of M as a function of the concentration. The dashed lines are the curves expected if the solution is ideal. The correct value of M for each solute is obtained by extrapolating the experimental points to $c = 0$.

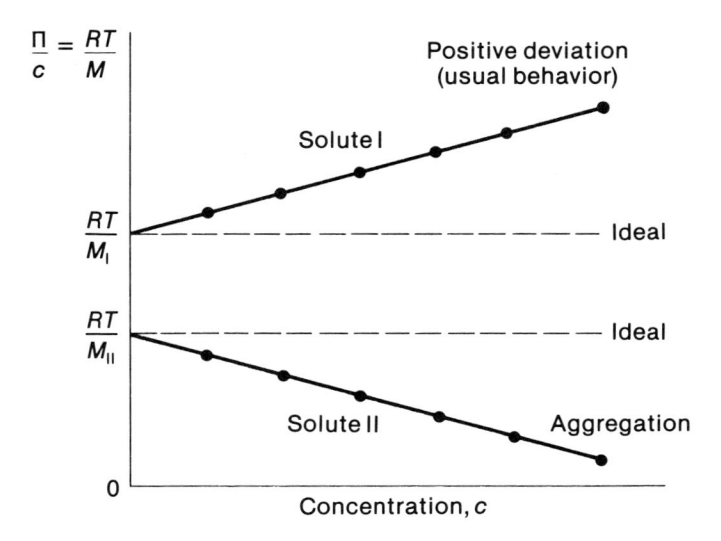

The Volume of Macromolecules as a Cause of Nonideality

Solutions of uncharged macromolecules frequently show strongly nonideal behavior. This may be caused by the aggregation-phenomenon discussed in the preceding section, but the principal reason is actually that the volume of the macromolecule is very large compared to the solvent molecules. This is called the *excluded volume effect,* and it is significant because it changes the volume of one mole of solvent molecules by decreasing the volume available to the solvent molecules. In our discussion of the van der Waals equation for gases in Chapter 2, we saw this excluded-volume effect and accounted for it by subtracting a constant b from the total volume of the gas. Similarly, the effect of attractive forces was included in a term a/V^2 added to the pressure. An analogous correction can be made for the osmotic pressure, which enables us to explain the lack of ideality of solutions of macromolecules. The data we seek to explain are shown in Figures 7-8 and 7-9.

We begin with Equation 27, $\Pi = cRT$, in which c is the molarity, which expresses the osmotic pressure of a dilute solution as a function of concentration. For exactly one mole of solute, this equation can be rewritten in the form

$$\Pi V_A = RTc'/M , \qquad (46)$$

in which V_A is the volume (not the molar volume) of the solvent, M is the molecular weight, and c' is the concentration in *grams* per liter. This equation is now in the form of the ideal gas law, so we introduce the constants a and b exactly as we did for a van der Waals gas, to obtain

$$\left(\Pi + \frac{a}{V_A^2}\right)(V_A - b) = \frac{RTc'}{M} . \qquad (47)$$

FIGURE 7-9 Deviation from ideality in osmotic pressure caused by the excluded volume effect.

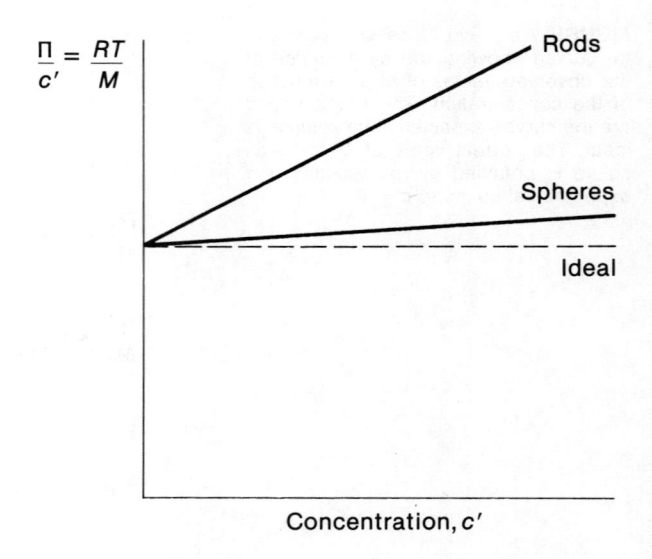

This equation can be rearranged to a form matching that used in Figure 7-9—namely,

$$\frac{\Pi}{c'} = RT\left[\frac{1}{M} + \frac{bc'}{M^2}\right] - \frac{ac'}{M^2}. \tag{48}$$

If for the moment we ignore b and consider a, we can understand the decrease in Π/c' with increasing c' for the case of association (Figure 7-8), because a is positive when there is attraction. If a is ignored, the increase of Π/c' with c' for molecules whose shape is either spherical or rodlike (Figure 7-9) is explainable by the positive value of b, which is a function of volume. Remembering that b is the excluded volume (see Chapter 2 for a discussion of the relation between b and the volume of a sphere), we can understand why, when Π/c' is plotted against c', the curve increases more rapidly for rods than for spheres. The excluded volume is not simply the volume of the molecule but the volume *per* molecule that cannot be penetrated as a pair of molecules in contact rotate about the point of contact. For two spheres, the radius of the sphere swept out by this motion is the sum of the radii of the spheres; for a rod, the radius of this sphere of exclusion is nearly equal to the sum of the lengths of the rod. Thus for two molecules of equal mass, a rod excludes more volume than a sphere.

We will discuss other properties of solutions of macromolecules in Chapter 15.

Interconversion of Activity Coefficients

The activity coefficient γ is defined as the ratio of activity to concentration. We have pointed out earlier that its numerical value depends on the units of concentration that are chosen. Furthermore, by definition, $\gamma = 1$ at infinite dilution. Since tables of activities refer to a particular concentration unit, it is valuable to be able to convert from one set of units to another. In the following, we use the subscripts X, **c** and c to refer to mole fraction, molality, and molarity, respectively. If the solution is dilute, the following equations may be used:

$$\frac{\gamma_X}{\gamma_c} = 1 + 0.001 c M_A \; ; \tag{49}$$

and

$$\frac{\gamma_\mathbf{c}}{\gamma_c} = \frac{\rho - 0.001 c M_B}{\rho_A} = \frac{\rho/\rho_A}{1 + 0.001 \mathbf{c} M_B} \; ; \tag{50}$$

in which ρ and ρ_A are the densities of the solution and the pure solvent, respectively, and M_A and M_B are the molecular weights of the solvent and solute, respectively, in grams per mole.

Fugacity: An Analog to the Activity

In an earlier section the activity was introduced as a way of handling the concentration of a real solution. Use of the activity simplifies the mathematics in most theoretical treatments in that equations for real solutions written with the activity have the same form as those for ideal solutions. A similar concept—the *fugacity*—exists for real gases. Although the fugacity is rarely needed in the life sciences, a brief description is presented here for the more thorough course.

The free energy G of an ideal gas is related to the pressure P by

$$G_{\text{ideal}} = G° + RT \ln P \; . \tag{51}$$

For a real gas the fugacity f is introduced merely be rewriting this equation as

$$G_{\text{real}} = G° + RT \ln f \; , \tag{52}$$

and the standard state is chosen at unit fugacity, just as for an ideal gas it is at unit pressure.

For an ideal gas the fugacity equals the pressure. Since all real gases become ideal at zero pressure, one may write

$$\lim_{P \to 0} f = P \; .$$

The analog of the activity coefficient is the fugacity coefficient γ, which is similarly defined as

$$\gamma = \frac{f}{P} \; .$$

As P approaches zero, γ approaches 1.

In general, an equation for an ideal gas that includes pressure (for example, the expression for an equilibrium constant) can be rewritten for a real gas simply by replacing P by f. It is important, however, to understand that whereas P is a directly measurable quantity, f usually has to be calculated from an equation of state. How this is done is shown in the following.

For any gas one may write—using molar quantities

$$\left(\frac{\partial \bar{G}}{\partial P}\right)_T = \bar{V}$$

or

$$d\bar{G} = \bar{V}\, dP$$

For an ideal gas, since $P\bar{V}_{ideal} = RT$,

$$d\bar{G}_{ideal} = \bar{V}_{ideal}\, dP = \left(\frac{RT}{P}\right) dP \; ; \tag{53}$$

for a real gas, we may only write

$$d\bar{G}_{real} = \bar{V}_{real}\, dP \; . \tag{54}$$

$\bar{V}_{real}$, the molar volume of a real gas, can either be measured directly or calculated from an empirical equation of state for a particular real gas.

Combining Equations 53 and 54 yields

$$d(\bar{G}_{real} - \bar{G}_{ideal}) = \left(\bar{V}_{real} - \frac{RT}{P}\right) dP \; . \tag{55}$$

At $P = 0$ (where $P = f$), $\bar{G}_{real} = \bar{G}_{ideal}$. Thus, integrating between $P = 0$ and an arbitrary pressure P yields

$$\bar{G}_{real} - \bar{G}_{ideal} = \int_0^P \left[\bar{V}_{real} - \left(\frac{RT}{P}\right)\right] dP \; . \tag{56}$$

We now assume that $\bar{G}^\circ$ is the same for real and ideal gases. This is not rigorously true, but since at a pressure of 1 atm most gases behave ideally, the assumption is reasonable. Then, by subtracting Equation 51 from Equation 52 and using this assumption to eliminate the two $\bar{G}^\circ$ terms, one obtains

$$\bar{G}_{real} - \bar{G}_{ideal} = RT \ln\left(\frac{f}{P}\right) = \int_0^P \left[\bar{V}_{real} - \left(\frac{RT}{P}\right)\right] dP \; , \tag{57}$$

which is the desired equation relating f and P. The following example shows how this equation is used.

EXAMPLE 7–N Calculation of the fugacity of a gas at 40.0 atm and 273°K, whose equation of state is $P(\bar{V} - b) = RT$ with $b = 0.0248$ liter mol^{-1}.

The equation of state is combined with Equation 57 to obtain

$$\ln\left(\frac{f}{P}\right) = \int_0^P \left(\frac{\overline{V}_{real}}{RT} - \frac{1}{P}\right)dP$$

$$= \frac{(0.0248)(40)}{(0.082)(273)} = 0.0443$$

Thus, $f = P\, e^{0.0443} = (40)(1.045) = 41.8$ atm.

REFERENCES

Castellan, G.W. 1971. *Physical Chemistry*. Addison-Wesley. Reading, Mass.

Cohn, E.J., and J.T. Edsall. 1965. *Proteins, Amino Acids and Peptides*. Hafner. New York. (This is the most complete treatment of the thermodynamics of solutions of biological substances that is available.)

Denbigh, K.G. 1971. *The Principles of Chemical Equilibrium*. Cambridge University Press. Cambridge.

Findlay, A., A. Campbell, and N.O. Smith. 1951. *The Phase Rule and its Applications*. Dover. New York.

Hildebrand, J.H. 1950. *Solubilities of Non-electrolytes*. Reinhold. New York.

Hildebrand, J.H., and R.L. Scott. 1962. *Regular Solutions*. Prentice-Hall. Englewood Cliffs, N.J.

Levine, I.N. 1978. *Physical Chemistry*. McGraw-Hill. New York.

McGlashlan, M.L. 1963. "Deviations from Raoult's Law." *J. Chem. Educ.* 40, 516.

Overton, J.R. 1971. "Determination of Osmotic Pressure." In *Techniques of Chemistry,* edited by A. Weissberger and B.W. Rossiter. Wiley-Interscience. New York.

Rioux, F. 1973. "Colligative Properties." *J. Chem. Educ.* 50, 490.

Tombs, M., and A.R. Peacock. 1975. *The Osmotic Pressure of Biological Macromolecules*. Oxford University Press. Oxford.

PROBLEMS

1. Sixty-five grams of a substance are dissolved in one liter of water to yield a solution that has a density of 1.020 g cm^{-3}. What are the values of the molarity, molality, and the mole fraction? The substance has a molecular weight of 250 and the temperature is 4°C, at which the density of water is 1 g cm^{-3}.

2. A 57.5 weight-percent solution of CsCl has a density of 1.700 g cm^{-3}. What are the values of the molarity, the molality, and the mole fraction?

3. Different amounts of a solute X are dissolved in 1000 g of water. It is found that up to a concentration of 0.5 molal, the volume V of the solution in cm^3 obeys the equation $V = (\mathbf{M} - 0.12)^2 + 1001.92$.

    (a) At what concentration is the partial molal volume zero?

    (b) What is the partial molal volume at a concentration of 0.2 $\mathbf{M}$?

4. A solution of 0.0013 M A in H$_2$O has a density of 0.9990 g cm^{-3}. The molecular weight M of A is 120. What is the partial molal volume of A at that concentration? The density of water is 0.9986 g cm^{-3} at the same temperature.

5. A particular two-component system formed by mixing 1.5 mol of A with 1.25

mol of B has a volume of 225 cm³. The molar volume of B is 120 cm³ mol⁻¹ in this solution. What is the molar volume of A in this solution?

6. A solvent A has a vapor pressure of 42.3 mm Hg at 20°C. A solution is prepared in which 8 g of B is dissolved in 62 g of A. The molecular weights of A and B are 84 and 234, respectively.

    (a) What is the vapor pressure of A in the solution, assuming that the solution is ideal?

    (b) Is the value in part (a) affected by whether B is volatile?

7. A solvent S having a molecular weight of 352 has a vapor pressure of 6.5 mm Hg. A solution is prepared in which 3.2 g of a solute B, whose molecular weight is 75, is dissolved in 200 g of S. What is the vapor pressure of S in the solution?

8. Twenty-five g of a liquid M, whose molecular weight is 150 and whose vapor pressure is 6 mm Hg, is mixed with 35 g of liquid N, whose molecular weight is 324 and whose vapor pressure is 4.3 mm Hg. The liquids are completely miscible. What are the vapor pressures of each solvent in the mixture?

9. At 100°C the vapor pressure of liquid A is 730 mm Hg and that of liquid B is 410 mm Hg. A mixture of the two liquids has a vapor pressure of 620 mm Hg. Assuming that the solution is ideal, what are the mole fractions of A and B?

10. Macromolecules such as proteins often produce extensive and highly ordered arrays of water molecules when in solution. Explain the effect that this might have on the Raoult's Law behavior of protein solutions at high concentration.

11. A solution of an unknown compound Q dissolved in water at a concentration of 0.060 g cm⁻³ produces a 1.2 percent decrease in the vapor pressure of water. What is the molecular weight of Q?

12. Complete the following sentence. At constant pressure, the vapor pressures of all liquids are equal at . . . .

13. The vapor pressure of a protein solution at 37°C has the following concentration dependence.

Concentration, g/100 cm³:	19.3	12.4	5.8
Pressure, atm:	0.044	0.025	0.011

What is the molecular weight of the protein?

14. The vapor pressure of a solution A is 105 mm Hg. One gram of a nonvolatile solute B is dissolved in 10 g of A; the vapor pressure of this solution is 93 mm Hg. What is the ratio of the molecular weights of A and B?

15. At 20°C, 0.328 mg of H_2 gas is dissolved in 200 g of water when the applied pressure is 1 atm.

    (a) What is the molarity of H_2 in water at an external pressure of 10 atm?

    (b) What is the Henry's Law constant for H_2 in water at 20°C?

16. When a diver goes deep under water, it is necessary to increase the air pressure in the diving suit in order to withstand the crushing pressure of the water. It is necessary that the diver surface very slowly to avoid the bends—that is, bubble formation in the blood. Explain why rapid surfacing causes the bends.

17. When drinking soda water, one invariably feels bloated and has the desire to belch. Explain this phenomenon.

18. A biological fluid in which gaseous N_2 is dissolved has a partial pressure of 120 mm Hg for N_2. What is the N_2 tension?

19. Fish fanciers know that chlorinated tap water should not be put into an aquarium. The standard technique is to dechlorinate the water by allowing it to stand overnight in an open pan. Why does this procedure remove the Cl_2?

20. Why must soda water be kept in a sealed container?

21. Explain in molecular terms why all solvents do not have the same freezing point depression constant, K_{frz}.

PROBLEMS

22. Consider the following P-T curve in which the solid line is for the pure solvent and the dashed line is for a solution.

(a) What phases are represented by the regions A, B, and C?

(b) What is represented by X?

(c) What is represented by Y?

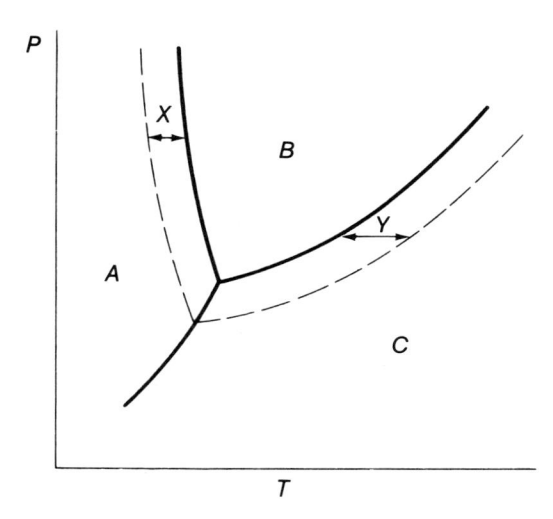

23. The freezing point depression constant for a particular solvent is 18.7°C. The molecular weight of the solvent is 118. Two g of a solid solute whose molecular weight is 174 are dissolved in 194 g of solvent. The freezing point depression is 0.55°C. What can you say about the state of the solute in this particular solvent?

24. A very small protein is dissolved in water (solution I) and produces a freezing point depression of 0.09°C. A solution of 0.02 M NaCl has a freezing point of -0.074°C. Surprisingly, the freezing point of a solution having the same protein concentration as in solution I, but also containing 0.02 M NaCl, has a freezing point of -0.12. Explain this result.

25. The molal freezing point depression of a particular solvent S is 14.1. A solution of 25.8 g of the solute A, whose molecular weight is 95, lowers the freezing point by 2.374°C.

(a) What discrepancy exists between these numbers?

(b) Assuming that A dimerizes somewhat in this solvent, what fraction of A is in the form of a dimer in this solution?

26. In the sense of the Phase Rule, state the number of independent components in each of the following systems.

(a) A mixture of N_2, H_2, and NH_3 at room temperature.

(b) The mixture in part (a) but at a temperature at which the equilibrium $N_2 + 3H_2 \rightleftarrows 2NH_3$ is established.

27. Two substances A and B, both at the same molar concentration, are on one side of a membrane that is impermeable to both A and B (side I). Side II contains pure water. At equilibrium, the level of the liquid on side I is 50 cm above the liquid level on side II.

(a) An enzyme is added that causes A and B to react to form AB. Very soon the reaction, which is irreversible, has gone to completion. The molar concentration of the enzyme is $< 10^{-7}$. At equilibrium, what will be the difference in the levels of the two liquids?

(b) What would be the difference in the liquid levels if AB were insoluble?

28. A vessel is divided into two regions by a membrane that is permeable to water but not to the water-soluble polymer, polyethylene glycol (PEG). Water is put in one compartment and solid powdered PEG is put in the other. The PEG is in contact with the membrane. Describe what happens as a function of time.

29. A vessel is divided into two regions by a membrane permeable to both water and sucrose. Water is put in one compartment and sucrose crystals in the other. The sucrose crystals are in contact with the membrane. Describe what happens as a function of time.

30. The osmotic pressure of blood is 7.6 atm at 37°C.

(a) What is its freezing point?

(b) What is the activity of water in blood?

31. Solutions A and B have osmotic pressures of 2.4 atm and 4.6 atm, respectively. What is the osmotic pressure of a solution prepared by mixing one volume of A and one volume of B? Assume that the solutions are ideal.

32. One gram of a protein dissolved in one liter of water has an osmotic pressure of 2.5×10^{-3} atmospheres at 20°C. Assuming that the solution is ideal, what is the molecular weight of the protein?

33. A particular protein is stable only if its suspending buffer is at a concentration of at least 10^{-3} M $MgSO_4$ and 0.5 M KCl. You have a sample of the protein in 10^{-2} M $MgSO_4$ and 1 M KCl and you wish to determine its molecular weight by osmometry.

(a) What solution should be placed in the compartment that does not contain the protein?

(b) Can the molecular weight be determined from a single measurement of osmotic pressure?

34. A solution containing 425 mg of a macromolecule in 5 cm³ of water has an osmotic pressure of 0.90 atm at 25°C. What is the molecular weight of the macromolecule assuming that there is no concentration dependence?

35. The osmotic pressure of a solution of a macromolecule is measured at 25°C when four different amounts of the macromolecule are dissolved in 1 cm³ of water. The values obtained are: 0.436 atm, 100 mg; 0.858 atm, 200 mg; 1.265 atm, 300 mg; 1.66 atm, 400 mg. What is the molecular weight of the macromolecule?

36. Consider a system for measuring osmotic pressure that consists of two aqueous chambers separated by a membrane permeable only to water. The left chamber is filled with water and the right chamber is filled with 0.15 M glucose. How much higher will the liquid level in the right chamber be than that in the left chamber when the system is at equilibrium? Assume ideality and a temperature of 20°C. The density of water at 20°C is 0.998 g cm^{-3} and the acceleration of gravity is 980 g cm² sec^{-2}.

37. A protein of unknown molecular weight is dissolved in one compartment of a system for measuring osmotic pressure. The concentration is 0.001 g cm^{-3} and the temperature is 25°C. The difference in the liquid levels in the two compartments is 6 mm. Assuming ideality, what is the molecular weight of the protein?

38. Consider the arrangement illustrated on page 232. Two compartments I and II both contain water and are separated by a membrane permeable to water but not to sugars or larger molecules. While peering down the vertical tube, which is five meters deep, you drop a wooden object that floats. You wish to get the object back. One way is to remove the semipermeable membrane and drain out the water but this will make a terrible mess. A bright physical chemistry student suggests dropping sugar down the tube leading to compartment II. Explain, using the concept of the chemical potential, how this will result in getting back the object.

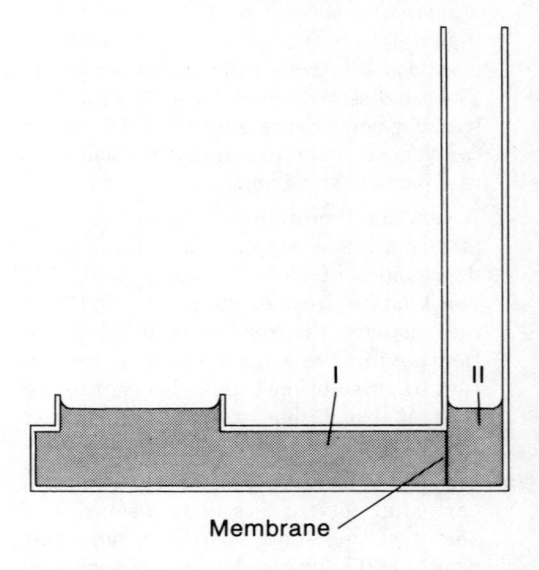

39. A membrane permeable to A, B, and C,

but not to D and E, separates solution I, which contains 0.05 M A, 0.8 M C, and 0.1 M E, from solution II, which contains 0.7 M B and 0.2 M F. What are the relative tonicities of the two solutions?

40. Many living cells contain small sacs called *vesicles*. These are droplets enclosed in a semipermeable membrane and frequently they contain a variety of enzymes. In an effort to study vesicles, it is often necessary to insert into a vesicle a particular chemical to which the vesicle is not permeable. The following technique frequently is successful. A substance X, which one hopes will enter the vesicles, is dissolved in distilled water. A suspension of vesicles is rapidly diluted into the solution of X in distilled water. Afterwards, X is often found in the vesicles. On the contrary, if X is dissolved in 0.1 M NaCl, and then the suspension of vesicles is added, no X appears in the vesicles. Explain why the technique works.

41. One-half liter of benzene is placed in contact with 10 cm³ of a 0.1 M aqueous solution of a substance P. The partition coefficient of P for benzene, compared to water, is 18.6. What is the concentration of P in benzene after equilibrium is achieved?

42. How many moles of Y are contained in 100 cm³ of benzene if 5.205 g of Y have been dissolved in a system consisting of 57.0 cm³ of water in contact with the benzene, if the partition coefficient of Y between benzene and water is 3.62? The molecular weight of Y is 243.

43. At 25°C, the distribution coefficient of solution X between $CHCl_3$ and water is 34. How much $CHCl_3$ must be added to 10 cm³ of a 0.1-M aqueous solution of X to reduce [X] to 0.0001 M?

44. What is the partition coefficient if 3.5 g of X in 125 cm³ of water is in equilibrium with 2.3 g of X in 50 cm³ of $CHCl_3$? The molecular weight of X is 143.

45. We have emphasized that chemical equilibrium must be described in terms of activities rather than concentrations. However it is almost universal practice in biochemistry to ignore the activity concept and work directly with concentration. This is because biochemical reaction mixtures are usually so complex that it is nearly impossible (at least, it would take a very large amount of work) to measure all of the activities. This has led to the tabulation of "concentration equilibrium constants." Suppose you wanted to predict the outcome of an uncharacterized biochemical reaction or to calculate various thermodynamic parameters; would you expect the discrepancy between activity and concentration to cause serious errors? To answer this question, you should consider the concentrations of reactants and products occurring in biochemical reactions; typical concentrations of enzymes and macromolecules are micromolar to millimolar and typical concentrations of reactants are millimolar. The total ionic concentration rarely exceeds 0.15 M.

46. The activity coefficient of water in a 2-M solution of a particular sugar is 0.72 at 20°C. What is the activity coefficient in molal units? The density of the solution is 1.16 g cm^{-3} and the density of water is 0.998 g cm^{-3}. The molecular weight of the sugar is 356.

47. A 0.2-molal solution of a substance X lowers the vapor pressure of a solvent from 86.2 to 81.5 mm Hg. The solvent has a molecular weight of 85.

    (a) What is the solvent activity in the 0.2-molal solution?

    (b) What is the activity coefficient?

48. One learns in elementary chemistry that the equilibrium constant K for a reaction equals the product of the concentrations of the products divided by the product of the concentrations of the reactants. However, in any real set of measurements, it is found that this expression is not constant at all. Why not?

49. Is the activity coefficient for a real substance always between 0 and 1?

CHAPTER 8

SOLUTIONS OF ELECTROLYTES

In Chapter 7 the properties of solutions of uncharged molecules were described. We were mainly concerned with ideal solutions because, for such solutions, thermodynamics provides simple equations that relate various experimental parameters to solute concentration. In reality, solutions are not ideal. This is mainly for two reasons: (1) the molecules of both solute and solvent take up space; and (2) there are van der Waals attractive forces between the solute molecules, the solvent molecules, and between the molecules of both solute and solvent. In dilute solutions, these factors are not especially important, so that as long as one does not look too closely, dilute solutions of nonelectrolytes behave as if they are ideal. In the life sciences, the solutions we encounter are usually fairly dilute. However, the solute molecules are almost always broken up into charged particles, and the presence of a charged solute molecule results in a dramatic loss of ideality that will be examined in this chapter. To begin our discussion, we will consider simple charged solutes—namely, *ions*—and describe the basic features of ions and electrolytes.

8-1 THE IONIC THEORY

Ions and Electrolytes

An atom or group of atoms that has lost or gained one or more electrons, so that it carries a charge, is called an *ion*. The number of charges on an ion

SEC. 8-1 THE IONIC THEORY

equals its *valence*—that is, the number of electrons that have been lost or gained. A substance that in solution is able to dissociate, at least in part, into ions is called an *electrolyte*. The main way that such a substance is recognized is that solutions of an electrolyte can conduct an electrical current when a voltage is applied. In an electrolyte, the ions exist both when the compound is in the solid and in the liquid state, but the ions are usually weakly paired. In solution, the pairs dissociate to freely moving positive and negative ions. The fraction of the ions that separate in solution is called the *degree of dissociation* and is denoted by α. Ionic compounds fall into two classes—the *strong electrolytes*, for which α is nearly one at all concentrations, and the *weak electrolytes*, for which α is usually less than one but increases with decreasing concentration.* Strong electrolytes always contain ionic bonds. Weak electrolytes, however, may be covalent compounds, as we will see below.

It is well known that for any particular compound α is not the same in all solvents, so that whether an electrolyte is classified as strong or weak depends on the solvent being used. However, the distinction usually refers to water as the solvent, and that convention is used in this book also. The property of a solvent that mainly determines the value of α is the dielectric constant D; that is, α is large with solvents such as water (a good dielectric) and nearly zero in a poor dielectric such as an organic solvent. This is because the electrical attraction between two charges decreases as D increases, and hence α increases, as we will see in the following section.

When an ionic compound dissociates in solution, the ions are usually bound to solvent molecules to some degree. This is called *solvation* or, in the case of water, *hydration*. This will be discussed more fully later. The strength of this bonding is not the same for all ions, but generally it is sufficiently weak that, by convention, an equation presented to describe the ionization of an electrolyte omits the solvent. Thus the dissociation of the ionic salts $NaCl$ and $Mg(NO_3)_2$ is described by the following scheme:

$$Na^+Cl^-(\text{solid}) \rightarrow Na^+ + Cl^-(\text{solution}) ;$$

and

$$Mg^{2+}(NO_3^-)_2(\text{solid}) \rightarrow Mg^{2+} + 2NO_3^-(\text{solution}) .$$

Covalently bonded neutral molecules also may ionize. This is of course not true of all covalent compounds; when covalent compounds do ionize, ionization almost always requires a reaction with a solvent molecule. In

---

*The increased dissociation of a weak electrolyte with decreasing concentration can be thought of in two ways. In the first it is simply a result of the constancy of the dissociation constant. This is not a molecular explanation, though. In the second we note that dissociation and association are in equilibrium. The probability of dissociation is independent of concentration, but for reassociation, the two ions must find one another. As the concentration decreases, reassociation becomes less probable because of the reduced probability of one ion colliding with another ion. Thus with dilution, reassociation becomes less probable than dissociation, and the fraction of dissociation decreases.

this case α is not only determined by the dielectric constant of the solvent but also by the electron-donating and electron-accepting properties of both the solvent and the solute molecules—that is, their acid–base properties (see Chapter 9). Because of the important role of the interaction with the solvent, the solvent must be designated in the chemical equation describing ionization. Thus, for the ionization of acetic acid and ammonia, we write

$$CH_3COOH + H_2O \rightarrow H_3O^+ + CH_3COO^-$$

and

$$NH_3 + H_2O \rightarrow NH_4^+ + OH^-.$$

Furthermore, α is almost always concentration-dependent, so that ionizable covalent compounds are weak electrolytes. Most of these dissociating covalent compounds are organic acids and bases. Salts of organic acids—for example, Na acetate—are interesting in that they dissociate completely in solution. However, when this occurs, the negative ion is not stable because the acid itself is a covalent compound that dissociates by solvation, as shown in the scheme above. Thus, because of an equilibrium between the acetate ion and the acetic acid, all of the Na^+ ions are free but only a fraction of the potential acetate ions are present in solution. This fraction is of course concentration-dependent. Furthermore, some salts of inorganic bases and organic acids—for example, lead acetate—ionize so poorly at commonly encountered concentrations that they are often considered to be nonionic.

The chemical and physical properties of solutions of electrolytes are determined almost entirely by the ions. For example, chemical reactions occur principally between ions, and the acidic and basic properties of most substances are mainly determined by the concentration of the H^+ (or, more correctly, H_3O^+) and OH^- ions. (We will discuss later some of the factors that affect the chemical activity of the ions.) The electrical conductivity of solutions is also determined by the concentration of the ions. When a voltage is applied to a solution through two electrodes, the electric current consists entirely of a movement of ions. A positively charged ion, or *cation,* moves toward the negative electrode or *cathode,* and a negative ion, or *anion,* moves toward the positive electrode or *anode.* This will be discussed in more detail later.

In Chapter 7 we noted that the colligative properties of a dilute solution are determined solely by the number of solute particles. Since the identity of the particles is unimportant, a particular number of un-ionized molecules or single ions should have the same effect. For a nondissociating molecule, the number of particles per mole of added solute is always known. This is not the case for a solution of an electrolyte, because the number of particles per mole may exceed Avogadro's number. Thus the contribution to any colligative property of an ideal 0.01-M solution of NaCl, which is 100 percent dissociated, should be twice as great as that of an ideal 0.1-M solution of glucose, which does not dissociate. We will have more to say about this later.

The Effect of the Solvent on the Degree of Dissociation

We have already stated that the degree of dissociation of a particular ionic compound is determined in part by the dielectric constant D of the solvent. To understand this, one must remember that the degree of dissociation is a result of an equilibrium between dissociation and reassociation. In the following we show that a high dielectric constant both increases the probability of dissociation and reduces the probability of reassociation.

The ions of an ionic compound have unlike charge and therefore attract one another. The force F between two charges (or ions) having charge q_1 and q_2 (in coulombs, C) that are separated by a distance r (in meters), in a medium of dielectric constant D, is given by Coulomb's Law,

$$F = q_1 q_2 / 4\pi\epsilon_0 D r^2,$$

in which $\epsilon_0 = 8.85 \times 10^{-12}$ J^{-1} C^{-1} m^{-1}. Thus, the attraction decreases as D increases. The energy dE required to change the distance between the charges from r to $r + dr$ is $dE = F\, dr = (q_1 q_2 / 4\pi\epsilon_0 D r^2) dr$, so that if the undissolved ions are separated by a distance a, the dissociation energy E_d (that is, the energy required to separate them totally) is

$$E_d = \int_a^\infty dE = q_1 q_2 / 4\pi\epsilon_0 D a \ .$$

The parameter a is the distance of closest approach of the centers of the ions and is called the *mean ionic diameter* of the two ions. The value of E_d decreases when D increases, meaning that less energy is required for dissociation in a solvent of high D. Water is a good dielectric ($D = 78$), so that dissociation of ionic compounds is a frequent event in aqueous solutions. In a solvent having a low value of D, the ions tend to form uncharged ion pairs. Larger values of D actually reflect the extent to which a *shielding* or *screening effect* is present. A solvent has a large value of D when the solvent molecules are highly polar; the presence of these dipoles between the ions reduces both the electrical interaction between the separated charged ions of the solvent and the probability of their reassociation.

In a weak electrolyte there is an equilibrium between undissociated molecules and ions; dissociation is mainly a result of collisions between molecules in the solute and the solvent. Since the thermal energy is kT, in which k is the Boltzmann constant and T is the absolute temperature, the degree of dissociation α is related to the ratio E_d/kT and approaches 1 when $E_d = kT$. It is not equal to 1 because the energy obeys the Boltzmann distribution; a precise calculation of the fraction of a population that is dissociated requires calculating the energy distribution for a particular temperature.

It should be noted that for a molecule having a mean ionic diameter a, there is theoretically a value of D of the solvent above which dissociation is complete. (There may, however, be no real substance having a sufficiently high value of D; similarly, for a solvent having a particular value of D, all molecules for which a is greater than a critical value will be dissociated at a particular temperature.) For water as solvent, the critical ion diameter is 0.35 nm (3.5 Å) at 25°C for monovalent ions.

For NaCl the mean ion diameter is only 0.276 nm (2.76 Å). However, below the limit of solubility, NaCl is 100 percent dissociated in water at 25°C, because there is a second important solvent effect—namely, *ion solvation*. For many solvents, of which water is an especially good example, the solvent molecules form relatively strong bonds with some ions. When this occurs, the ion is said to be *solvated*, or in the case of water, *hydrated*. For weakly ionic substances, solvation is often an integral part of the dissolving process. The way that hydration increases the degree of dissociation is as follows.

When an ion is hydrated, water molecules are oriented around the ion. This orientation causes a local increase in the polarity of the water and hence an increase in the dielectric constant. Thus, in the region of a hydrated ion, interionic attraction decreases and potentially paired ions will dissociate. Therefore, if some of the ions are hydrated, this facilitates dissociation of paired ions and thereby increases the degree of dissociation. This will be discussed more fully in a later section on hydration.

8–2 NEED FOR A THEORY OF SOLUTIONS OF ELECTROLYTES

Anomalous Values of the Colligative Properties

Studies of the colligative properties of solutions of electrolytes have indicated that their solutions have particular characteristics that distinguish them from solutions of nonelectrolytes. For instance the van't Hoff equation, $\Pi = cRT$ (Equation 32, Chapter 7), in which Π is the osmotic pressure and c is the molarity, and $\Delta T_{frz} = cK_{frz}$, in which ΔT_{frz} is the freezing point depression and K_{frz} is the freezing point depression constant, are not correct for electrolytes. They must be replaced by the equations $\Pi = icRT$ and $\Delta T_{frz} = icK_{frz}$, in which i is a concentration-dependent parameter *that has the same value in both equations*. (The factor i is also required to describe the boiling point elevation and the depression of the vapor pressure.) For an electrolyte, the value of i is always greater than one, for which one possible explanation is that the number of independent units per mole of dissolved solute is greater than Avogadro's number. This explanation is consistent with the hypothesis that the molecules of an electrolyte consist of subunits (namely, ions) that sometimes separate from one another when they are in solution. The value of i decreases as the concentration increases, so that under this hypothesis, the fraction of the subunits that dissociate must increase as the concentration decreases.* When solutions of a great variety of electrolytes are studied, three important observations are made.

First, i_0, the value obtained by extrapolating to zero concentration, is always greater than 1. Second, there are two classes of substances for which $i_0 > 1$: those for which the value of i_0 is always an integer, and those for which it is nonintegral. Third, in the case of an integral i_0, the value always equals the number of independent groups in the molecule; namely, $i_0 = 2$ for NaCl and KCl, $i_0 = 3$ for $CaCl_2$ and Na_2SO_4, and $i_0 = 4$ for $FeCl_3$ (Figure 8-1).

The explanation for these findings is of course now clear. The colligative properties are determined by the *number* of particles per unit volume

FIGURE 8-1 The i factor, calculated from the freezing point depression, as a function of molarity, c, for several substances. The value of i extrapolated to zero molarity is 1 for the nonelectrolytes sucrose and methanol and equals the number of ions for electrolytes.

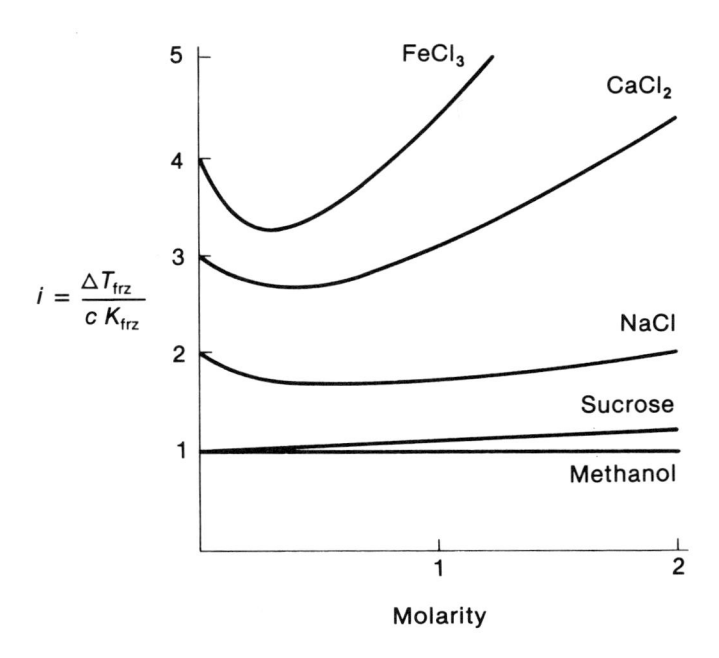

$$i = \frac{\Delta T_{frz}}{c \, K_{frz}}$$

*The degree of dissociation α can be related to i in the following way. If one molecule of a solute that can dissociate into n ions per molecule is dissolved and yields a solution of concentration c, the total number of particles per unit volume $= ci = c[(1 - \alpha) + n]$, so that the average number of particles per molecule is $i = (1 - \alpha) + n\alpha$. Thus $\alpha = (i - 1)/(n - 1)$. For a strong electrolyte, $i = n$ and $\alpha = 1$. For a weak electrolyte, $i < n$, so that $\alpha < 1$. The concentration dependence of α can be seen by considering a binary electrolyte AB whose dissociation is described by the equilibrium $AB \rightleftarrows A^+ + B^-$. If c_0 is the initial concentration of AB and K is the equilibrium constant, $K = (\alpha c_0)(\alpha c_0)/c_0(1 - \alpha)$, or $\alpha = \frac{1}{2}c_0 (\sqrt{K^2 + 4c_0 K'} - K)$.

of solution. Since in a solution of an electrolyte, the number of particles per mole of dissolved solute is greater than Avogadro's number $\mathcal{N}$, the use of the concentration c in the equations just given is inappropriate and c must in some way be replaced by a larger number; the parameter i, which is greater than 1, provides for this increase. For strong electrolytes, the number of particles per mole is simply $\mathcal{N}$ times the number of ions per molecule at all concentrations, so that i is integral for a strong electrolyte, which is 100 percent dissociated. In a weak electrolyte, the number of particles dissociating per molecule is an integer greater than 1 but, on the average, it is nonintegral because only *some* of the molecules are dissociated. It is clear also why the value of i increases with dilution, since the degree of dissociation increases as the concentration is lowered. The strong electrolytes present an anomaly, however, because, if it is true that dissociation of a strong electrolyte is complete at *all* concentrations, the value of i should be independent of concentration, and this is not so. We might of course simply conclude that total dissociation of a strong electrolyte does not occur, but we will see that another factor that has not yet been considered allows concentration dependence to be consistent with complete dissociation. However, before proceeding to the correct explanation, we will examine the electrical conductivity of solutions of electrolytes, since this phenomenon also presents peculiarities whose explanation is the same as that for the anomalous values of the colligative properties.

Anomalous Values of Electrical Conductivity

As we have mentioned before, the electrical conductivity of solutions of electrolytes is entirely a result of the movement of the ions, since only the ions can carry charge. Since conductivity is a measure of the number of charges transported through a standard volume under the influence of a standard electrical potential, it should be proportional to the concentration of the ions. Thus, for a weak electrolyte, dissociation increases as the concentration diminishes, so that the ratio of the conductivity to concentration should rise as concentration decreases; and this is in fact observed. For a strong electrolyte, since dissociation is presumably complete at all concentrations, the conductivity should be proportional to concentration or, in other words, the ratio of conductivity to concentration should be constant. However, once again, strong electrolytes give anomalous results and this ratio rises with decreasing concentration (Figure 8-2). The simplest explanation is that, for a strong electrolyte, dissociation varies with concentration, as is the case for a weak electrolyte. This is the same idea that was just suggested for the change in i with concentration. We now present another explanation that will in fact lead us to an understanding of the structure of ionic solutions.

FIGURE 8-2 The ratio of the conductance Λ of a weak electrolyte to the molarity c as a function of c for an aqueous solution of NaOH.

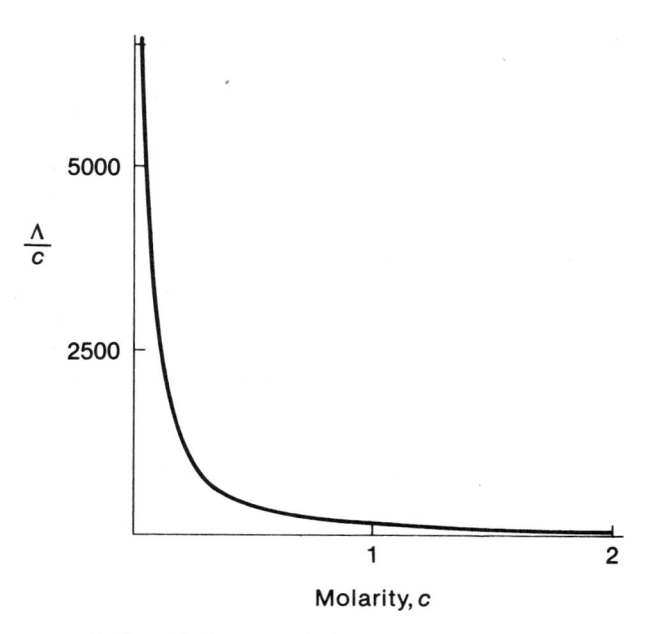

Kinetic Independence of Particles

In our discussions of colligative properties and of conductivity measurements, we have distinguished between undissociated molecules and ions. However, the real distinction should be between what might be called kinetically dependent and independent ions. In an undissociated molecule, the ions are constrained to move together at all times; that is, their motions are not independent. When the molecule is dissociated, we have assumed that the ions are free to move as they please and thus are kinetically independent. That is, with respect to the colligative properties, we consider each particle to move through the solution independently so that each ion contributes separately to the thermodynamic quantities that determine the numerical value of these properties; in the conductivity measurements, the positive and negative ions have been assumed to be free to move in opposite directions, each carrying charge. However, because the ions are charged, dissociated ions are not necessarily kinetically independent because the ions attract and repel one another. That is, the movement of one ion influences the movement of a second ion. Since the electrical interactions are greater with decreasing distance between the ions, these attractions and repulsions will be greater with rising concentration. Thus, the degree of independence is lower with increasing concentration and the solution behaves *as if* there are some undissociated molecules. What we are saying is that because the concentration does not include the notion of dependent motion, it is not the right parameter to use in the equations that describe the phenomena—in other words, a new parameter is needed that states the fraction of the molecules that are kinetically independent.

Solutions of nonelectrolytes also show a loss of ideality at high concentrations owing to attractive and repulsive forces. In order to keep the thermodynamic equations that describe solutions of nonelectrolytes in simple form—that is, independent of concentration—G.N. Lewis introduced the concept of *activity*. This was based upon the belief, which proved to be correct, that the activity calculated from one type of thermodynamic measurement—for example, colligative properties—would simplify equations dealing with other phenomena, such as electrical properties. Thus, a correct description of concentration-dependent phenomena requires an evaluation of activity as a function of concentration. We have mentioned several times that in biochemistry there is a tendency to equate activity and concentration, and a discussion of activity again at this point might seem superfluous. However, for electrolytes, activity has certain peculiarities that distinguish this from the activity of nonelectrolytes, and an understanding of these factors yields valuable information about molecular events that occur in ionic solutions. These peculiarities will be described in the following section.

8–3 THE CONCEPT OF ACTIVITY FOR ELECTROLYTES

The difference between the activity of a substance and the concentration of the substance is a measure of departure from ideality. As in the case of nonelectrolytes, the activity of an electrolyte differs from its concentration because of attractive and repulsive forces. However, for uncharged solutes, at concentrations below 0.01 M a solution behaves in a way that is not significantly different from an ideal solution. This is because, below this concentration, the molecules are well separated* and the van der Waals force decreases as the reciprocal of the sixth power of the distance between the molecules (Chapter 3). However, for electrolytes, nonideality persists at concentrations much lower than 0.01 M because of the $1/r^2$-dependence of electrical forces.

Definition of Activity of an Electrolyte

There are two complications that must be dealt with in defining the activity of an electrolyte. First, dissociation produces both positive and negative ions and both types are always present. Thus, we cannot talk about the effect of a positive ion without simultaneously considering the effect of the negative ion. Second, the difference between attraction and repulsion is not, as in the case of van der Waals forces, only dependent upon the distance between two particles, but depends also on the signs of the charges of two nearby ions.

---

*This statement does not apply to macromolecules because of their great size. Macromolecules will be discussed in a later section.

SEC. 8-3 THE CONCEPT OF ACTIVITY FOR ELECTROLYTES

The anomalies we have discussed arise primarily in the case of the strong electrolytes; therefore, in the discussion that follows, we will only consider a solution for which dissociation is nearly complete.

Several parameters are needed to evaluate the activity of an electrolyte. One would expect the number and charges of the ions to be important. Thus, we define the total number ν of ions per molecule as the sum of ν_+ positive ions and ν_- negative ions; that is,

$$\nu = \nu_+ + \nu_- . \tag{1}$$

In order to account for the requirement that both positive and negative ions are always present, it is convenient to define a *mean ionic activity*, which we designate as a. It can be shown that the following equations express the relation between a, the total activity of the electrolyte, and a_+ and a_-, the activities of the individual ions:

$$a = a_\pm^\nu = a_+^{\nu_+} a_-^{\nu_-} ; \tag{2}$$

and

$$a_\pm = (a_+^{\nu_+} a_-^{\nu_-})^{1/\nu} . \tag{3}$$

Equation 3 says that the mean ionic activity is the geometric mean of the individual ion activities. Of course we have no way of evaluating a_+ without evaluating a_-; therefore, this equation is not especially informative. However, it will be used later in one stage of the argument, and it does show the kind of averaging process that we are talking about.

For convenience, the theory we will present is formulated in terms of the activity coefficient, since this is more commonly measured than the activity. We defined the mean activity coefficient $\gamma_\pm$ and the mean ionic molarity $c_\pm$ by

$$a_\pm = \gamma_\pm c_\pm , \tag{4}$$

from which one can derive the relations

$$\gamma_\pm^\nu = \gamma_+^{\nu_+} \gamma_-^{\nu_-} \tag{5}$$

and

$$c_\pm^\nu = c_+^{\nu_+} c_-^{\nu_-}, \tag{6}$$

in which c_+ and c_- are the concentrations of the positive and negative ions, respectively. From the relations $\nu = \nu_+ + \nu_-$, $c_+ = \nu_+ c$, and $c_- = \nu_- c$, Equation 6 becomes

$$c_\pm = (\nu_+^{\nu_+} \nu_-^{\nu_-})^{1/\nu} c . \tag{7}$$

In calculating values of $c_\pm$ and $a_\pm$, it is important to remember to use the *numbers* of ions per molecule without regard to valence. Thus, for both KCl and MgSO$_4$, there is one each of the positive and negative ions per molecule, so that $\nu_+ = 1$, $\nu_- = 1$, and $\nu = 2$. Thus,

$$c_\pm = (1^1 \cdot 1^1)^{1/2} c = c .$$

However, for K$_2$SO$_4$, there are *two* K$^+$ ions and one SO$_4^{2-}$ ion per molecule, so that $\nu_+ = 2$, $\nu_- = 1$, and $\nu_3 = 3$; and thus

$$c_\pm = (2^2 \cdot 1^1)^{1/3} c = 1.59 c \ .$$

Since the properties of a chemical reaction are usually calculated from either the free energy or the chemical potential, it is convenient to introduce the expressions that have just been derived into the usual equation for the chemical potential, $\mu = \mu_0 + RT \ln a$. What we are in fact doing is replacing the activity in this equation by the appropriate value for an electrolyte. Equation 2 shows that this is accomplished merely by substituting the geometric mean ionic activity $a_\pm$ for a, so that the equation describing the chemical potential μ_e of an electrolyte is

$$\mu_e = \mu_0 + RT \ln a_\pm^\nu \ . \tag{8}$$

If we use Equations 4 and 6, Equation 8 becomes

$$\mu_e = \mu_0 + RT \ln(\nu_+^{\nu_+} \nu_-^{\nu_-}) + \nu RT \ln c + \nu RT \ln \gamma_\pm \ . \tag{9}$$

The significance of this equation is that for a solution having concentration c, each term is known except for the term including $\gamma_\pm$. That is, μ_0 is available from reference tables, ν_+ and ν_- are determined from the chemical formula of the compound in question, and c is defined by the experiment. Thus, the evaluation of ν at a concentration c merely requires determining $\gamma_\pm$. This can be done experimentally, as we will explain shortly, or can be calculated by the theory that we present shortly.

The Ionic Strength

Before proceeding, it is necessary to introduce a new parameter—the *ionic strength*, I. This parameter appears when considering solutions of charged particles for the following reason. Since we will be considering ions having a molal concentration c and a charge (or valence) z, we might expect that the equations will contain terms including the product cz as a means of describing the concentration of charge. Furthermore, because positive and negative ions always exist simultaneously, we should expect that there must be terms for both ions in the necessary equations. In the theory we shall present shortly, we will see that the appropriate form of the term is cz^2. Some time before this theory was developed, G.N. Lewis observed experimentally that numerous relations between particular properties of solutions of electrolytes and the electrolyte concentration gave linear graphs if the concentration was replaced by an expression containing cz^2. Lewis realized the significance of including all ions in a theory of electrolytes and defined the ionic strength I as

$$I = \tfrac{1}{2} \sum c_i z_i^2 \ , \tag{10}$$

in which the sum includes all ions. How one uses this equation is shown in the following example.

EXAMPLE 8–A Calculation of the ionic strength of 0.2 M NaCl, 0.5 M CaCl$_2$, and a solution containing both 0.2 M NaCl and 0.5 M CaCl$_2$.

For 0.2 M NaCl, $z_{Na^+} = 1$, $z_{Cl^-} = -1$, and $c_{Na^+} = c_{Cl^-} = 0.2\ M$. Thus,
$$I = \tfrac{1}{2}[(0.2)(1)^2 + (0.2)(-1)^2] = 0.2\ M\ .$$

For 0.5 M CaCl$_2$, $z_{Ca^{2+}} = 2$, $z_{Cl^-} = 1$, $c_{Ca^{2+}} = 0.5\ M$, and $c_{Cl^-} = 1\ M$. Thus,
$$I = \tfrac{1}{2}[0.5(2)^2 + 1(-1)^2] = 1.5\ M\ .$$

An important feature of the ionic strength is that, since it is a sum over all ions, *the ionic strength of a mixture is the sum of the ionic strength of each component*. Thus the ionic strength of the mixture, 0.2 M NaCl + 0.5 M CaCl$_2$, is $0.2 + 1.5 = 1.7$. This can also be seen by calculating the sum directly, using $c_{Cl^-} = 1.2$. That is, $I = \tfrac{1}{2}[(0.2)(1)^2 + (0.5)(2)^2 + (1.2)(-1)^2] = 1.7$.

For a weak electrolyte, the concentrations of each ion must be calculated from the equilibrium constant, as shown in the following example.

EXAMPLE 8–B What is the ionic strength of 0.2 M acetic acid (HAc)? The equilibrium constant is 1.8×10^{-5}.

To calculate the concentrations [H$^+$] and [Ac$^-$], we note that [H$^+$][Ac$^-$]/[HAc] = 1.8×10^{-5}. By the techniques described in Chapter 9,
$$[H^+] = [Ac^-] = 1.9 \times 10^{-3}\ M\ .$$

Since $z_{H^+} = 1$ and $z_{Ac^-} = -1$,
$$I = \tfrac{1}{2}[(1.9 \times 10^{-3})(1)^2 + (1.9 \times 10^{-3})(-1)^2] = 1.9 \times 10^{-3}\ M\ .$$

In the following section we will see the term for ionic strength appear as we attempt to evaluate $\gamma_\pm$.

Determination of the Activity Coefficient

There are numerous experimental techniques for determining $\gamma_\pm$. Several of these utilize measurements of colligative properties—in particular, the difference between the observed value and the value that would be expected if there were no concentration dependence. The values obtained from the freezing point depression are extremely precise, although they are confined to temperatures near the freezing point; this can be disadvantageous, because $\gamma_\pm$ varies significantly with temperature.* Similarly,

---

*There are ways to calculate the value of $\gamma_\pm$ at other temperatures, but the reliability of the calculations decreases as the temperature departs from the freezing and boiling points.

measurements of the boiling point elevation yield values of $\gamma_\pm$ only near the boiling point. Measurements of osmotic pressure and vapor pressure changes are useful over a broad range of temperatures but are not especially accurate. Two methods that do not make use of colligative properties are the most useful. One involves measurement of the solubility of an electrolyte. This gives very precise values but is possible only for weakly soluble salts. The other is a measurement of the electromotive force of electrical cells containing the electrolyte of interest. This is the best and most general method and will be described in Chapter 10. Details of the methods utilizing the colligative properties can be found in references given at the end of the chapter.

An understanding of the method of evaluating $\gamma_\pm$ from solubility measurements is informative because it shows why the ionic strength is a useful concept in the theory of electrolytes. Determination of $\gamma_\pm$ for a 1 : 1 electrolyte from solubility measurements proceeds as follows.* The solubility of a salt XY is defined by its solubility product K_{sp}, which is

$$K_{sp} = a_X a_Y = (\gamma_X c_X)(\gamma_Y c_Y) . \tag{11}$$

Using the definition $\gamma_\pm{}^2 = \gamma_+ \gamma_-$ and denoting the concentration at saturation as c_{sat}, or the solubility, we obtain

$$K_{sp} = \gamma_\pm^2 c_{sat}^2 \quad \text{or} \quad \frac{1}{2}\log K_{sp} = \log \gamma_\pm + \log c_{sat} . \tag{12}$$

Thus to evaluate $\gamma_\pm$, one need only measure K_{sp}. However, K_{sp} cannot be calculated from solubility data because, to do so, we need to know the activities and not the concentrations—that is, we need to know $\gamma_\pm$. At zero concentrations $\gamma_\pm = 1$, so that $K_{sp} = c_{sat}^2$. This is not really meaningful because, clearly, at zero concentration, there is nothing in solution. However, K_{sp} can be thought of as the value that would occur if the solution were ideal; and the evaluation at zero concentration then becomes just a trick to obtain the correct value. In order to perform this extrapolation we need to vary something other than the concentration of the solute. Thus we make use of the experimental observation that the solubility of a sparingly soluble salt XY can be varied by the addition of a strong, noninteracting electrolyte. It has been found that when the total ionic strength (that is, added solute plus the solute of interest) is small, then $-\log c_{sat}$ decreases linearly with $I^{1/2}$; that is,

$$-\log c_{sat} = A - BI^{1/2} , \tag{13}$$

in which A and B are constants. Combining this with Equation 12, we obtain $-\frac{1}{2} \log K_{sp} + \log \gamma_\pm = A - BI^{1/2}$; noting that as I approaches zero, $\gamma_\pm$ approaches 1 and $\log \gamma_\pm$ approaches 0, we see that then $A = -\frac{1}{2} \log K_{sp}$. Thus

$$-\log c_{sat} = \frac{1}{2}\log K_{sp} - BI^{1/2} , \tag{14}$$

---

*Calculations for the more general case of an electrolyte whose formula is $X_m Y_n$ can be found in references given at the end of the chapter.

SEC. 8-3 THE CONCEPT OF ACTIVITY FOR ELECTROLYTES

and a plot of $-\log c_{sat}$ versus $I^{1/2}$ yields $-\frac{1}{2}\log K_{sp}$ as the y intercept. Rearranging Equation 13 as $\log \gamma_\pm = -\log c_{sat} - (-\frac{1}{2}\log K_{sp})$ indicates that in this plot, $\log \gamma_\pm$ is the vertical distance between the y intercept and the ordinate corresponding to any value of I (see Figure 8-3).

FIGURE 8-3 Determination of $\log \gamma_\pm$ from solubility data. The solubility of a salt XY in NaCl solutions having various concentrations is measured. The data shown in the table (right) are obtained. The data are plotted as in the figure and $-\frac{1}{2}K_{sp}$ is obtained by drawing a straight line and extrapolating to $I^{1/2} = 0$; when no NaCl is present, $\gamma_\pm$ for XY can be calculated from $\gamma_\pm = \sqrt{K_{sp}}/c_{sat}$. To determine $\gamma_\pm$ for any value of I, for example, 0.56 M, we erect a perpendicular line from the x axis and draw two horizontal lines, as shown; the difference between the two horizontal lines is the value of $\log \gamma_\pm$. The value of $\gamma_\pm$ at [NaCl] = 0 could also be determined by erecting a line from the point labeled "No NaCl."

[NaCl]	c_{sat}	$I^{1/2}$	$-\log c_{sat}$
0	0.00101	0.032	6.897
0.5	0.0013	0.25	6.65
1.0	0.003	1.0	5.81
1.5	0.0043	1.41	5.45

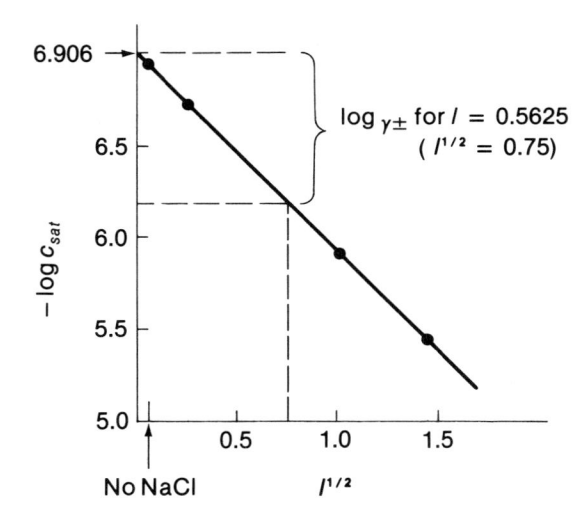

We emphasize again that I is the sum of the ionic strength of the added solute and the salt XY, so that to determine the value for $\log \gamma_\pm$ when there is no added solute, one does not extrapolate to $I = 0$, which would of course yield $\gamma_\pm = 1$, but to the value of I determined from the solubility. This calculation is carried out in the legend to Figure 8-3.

If one wishes to know the value of $\gamma_\pm$ in a particular solution, this can also be evaluated from the graph of Figure 8-3. However, this is not necessary because, from Equation 13,

$$K_{sp} = (\gamma_\pm)^2 c_{sat}^2 = (\gamma'_\pm)^2 c'^{2}_{sat},$$

so that

$$\gamma'_\pm = \frac{c_{sat}}{c'_{sat}} \gamma_\pm, \tag{15}$$

in which c'_{sat} is the concentration at saturation when the added salt is present and $\gamma'_\pm$ is the activity coefficient of the solute when the salt is present.

What is the physical significance of the increase in solubility? The principal conclusion of this analysis is that as ionic strength increases, $\gamma_\pm$ decreases, and solubility increases to maintain constancy of K_{sp}. The physical basis of the decrease in $\gamma_\pm$ with increasing I is explained in the next section. The reader should remember that the equations just given apply to binary salts whose ions have a single charge. For other salts the equations must be appropriately modified.

8-4 THE DEBYE-HÜCKEL THEORY

So far we have made several significant observations about solutions of strong electrolytes. First, with strong electrolytes there is an unexpected concentration-dependence both of the i factor in colligative property measurements and of electrical conductivity. Second, the role of the charge of the ions in determining the solubility of sparingly soluble electrolytes is by way of the ionic strength rather than simply a function of the charge itself. These features must be consequences of the charge of the ions because they do not occur in solutions of uncharged molecules. It is not surprising that ionic solutions have unusual properties, because electrical interactions are very strong. They are, in fact, so much stronger than the van der Waals interactions, which are the cause of nonideality of solutions of nonelectrolytes, that we might expect to see a significant departure from ideal solution-behavior at concentrations at which, for an uncharged solute, the solution would behave ideally. In order to explain these phenomena we need a theory that describes the state of ionic solutions in greater detail than we have seen up to this point. The generally accepted theory is that due to Peter Debye and Erich Hückel.

The mathematical derivation of the Debye-Hückel theory may seem formidable; in fact, it is not. In this book, long derivations have for the most part been avoided. In this case, the derivation will be presented, because in following the logic of it, the reader will become aware of its physical significance, its limitations, and the differences between this model and the theory of charged surfaces presented in Chapter 14.

The Debye-Hückel Limiting Law

The development of the Debye-Hückel theory begins by assuming that the term $\nu RT \ln \gamma_\pm$ in Equation 9 is a consequence solely of electrical effects—namely, the electrical attractions and repulsions between the charges. Two other important assumptions are also made: (1) that a strong electrolyte is

SEC. 8-4 THE DEBYE-HÜCKEL THEORY

totally dissociated; and (2) that the ions are tiny spheres. (Spherical symmetry significantly simplifies the mathematics.) The immediate goal of the theory is to develop a model for the structure of a solution of a strong electrolyte that allows us to calculate for any electrolyte a value of $\gamma_\pm$ that agrees with experimental measurements. The model proposed by Debye and Hückel is simply the following: *opposite charges attract one another so that positive and negative ions are not uniformly distributed—near a positive ion there are predominantly negative ions, and near a negative ion there are mainly positive ions.* We will amplify this picture later.

In order to solve the complex equations that arise in the theory, Debye and Hückel introduced the condition that the electrical potential energy is much less than the thermal energy (see Appendix VII). This condition is valid only when the concentration of ions is low and in fact *limits the validity of the theory to solutions whose concentrations are less than 0.01 M*. With this limitation, the following equation was derived:

$$\log_{10}\gamma_\pm = \left[\frac{(2\pi\mathcal{N})^{1/2}}{2.303}\left(\frac{e^2}{4\pi\epsilon_0 DkT}\right)^{3/2}\right]z_+z_-I^{1/2} = Az_+z_-I^{1/2} \qquad (16)$$

in which z_+ and z_- are the charges of the positive and negative ions respectively, D is the dielectric constant, I is the ionic strength, and A is the expression in the braces. This equation, known as the Debye-Hückel limiting law, is what is needed—a relation between $\gamma_\pm$ and measured concentration; remarkably, the concentration appears in the equation only as the ionic strength. In aqueous solutions and at 25°C (a common temperature in experiments) $A = 0.509$, so

$$\log_{10}\gamma_\pm = 0.509 z_+z_-I^{1/2} . \qquad (17)$$

Some of the factors in Equation 16 may be extracted and written

$$\kappa = \left(\frac{2\mathcal{N}e^2 I}{1000 DkT\epsilon_0}\right)^{1/2} . \qquad (18)$$

The significance of this expression will be discussed in the next section.

This equation enables one to calculate the activity of an electrolyte in solution as a function of I. It is important to realize that I is not necessarily the ionic strength of the electrolyte in question but rather the *total* ionic strength of the solution (as we will see in the calculation given below). This is because in electrical interactions only the sign and magnitude of the charge are important—not the chemical identity of the charged species. Thus the contribution of a particular electrolyte to some solution phenomenon (solubility or a colligative property) is affected not only by its own concentration but by all other charged particles that are present.

We would of course like to have experimental verification that the theory is correct. The principal evidence is based upon a demonstration that Equation 17 correctly describes the relation between $\gamma_\pm$ and I. Let us examine for a moment what Equation 17 says. If one has a dilute solution of an electrolyte, Equation 17 states that one needs only to calculate the ionic strength of the solution from the molar concentration in order to evaluate the activity coefficient. Thus, for 0.005 M MgSO$_4$, $I = \frac{1}{2}[(0.005)(2)^2 + (0.005)(2)^2] = 0.02$ and $\log_{10}\gamma_\pm = -(0.509)(2)^2(0.02)^{1/2} = -0.29$, so that $\gamma_\pm = 0.52$, and the activity of 0.005 M MgSO$_4$ is $(0.52)(0.005) = 0.0026\ M$. Furthermore, the equation says that if another electrolyte, for example, 0.06 M NaCl, is added, the value of I in Equation 17 is the *total* ionic strength of the solution or $0.02 + 0.06 = 0.08$ and for MgSO$_4$, $\log_{10}\gamma_\pm = -(0.509)(2)^2(0.08)^{1/2} = -0.576$ and $\gamma_\pm = 0.27$. Thus, in this case, the activity of 0.005 M MgSO$_4$ in the presence of 0.06 M NaCl is $(0.005)(0.27) = 0.00135\ M$.

Equation 17 has been shown to be correct by measuring $\gamma_\pm$ for dilute solutions of a variety of electrolytes at many concentrations. As we have mentioned, the product z_+z_- is always negative, so that a plot of $\log_{10}\gamma_\pm$ versus $I^{1/2}$ always has a negative slope, and indeed this has been observed in a great many measurements, as can be seen in Figure 8-4. The agreement between theory and experiment is not perfect though, in that there are significant departures at higher ionic strengths, as shown in the figure.

FIGURE 8-4 Measured dependence of log $\gamma_\pm$ of various electrolytes as a function of their ionic strength. Note that the theory predicts that the NaCl and KCl curves should be identical. The values are nearly the same when I is small but differ significantly for large values of I. This departure from theory is discussed at the bottom of this page.

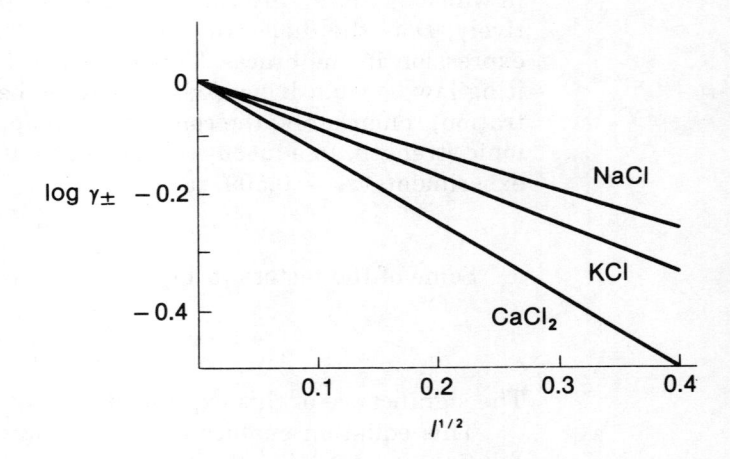

Such problems are not unexpected in view of several approximations that Debye and Hückel made in the derivation (see Appendix VII). The assumption causing the greatest difference between theory and experiment is the condition that the electrical interaction energy is small compared to thermal energy. Because electrical interactions act over considerable distances, this assumption is valid only if, on the average, the ions are well separated—in other words, in dilute solution. There are three other factors that have been neglected that also become increasingly important in more concentrated solutions. One is the so-called ionic size effect—that is, the existence of nonelectrical repulsive forces when the ions are near one another. Another factor is the assumption that the ions are all spherical, which was invoked in the geometric part of the argument. These two points account in part for the differences seen in Figure 8-4 for electrolytes as similar as NaCl and KCl. A third factor, which is probably the most significant, is the solvation interaction between the ions and the molecules of the solvent, which we have already said is often essential for allowing the ions to become separate in solution. With some solvent–solute systems, there are many bound solvent molecules per ion; the solvent molecules are highly oriented and often polarized, which effectively increases the dielectric constant and thereby decreases the interionic attraction. This means that the values of $\gamma_\pm$ would increase at higher concentrations, which is indeed the case, as can be seen in Figure 8-5.

FIGURE 8-5 Activity coefficients of various electrolytes as a function of concentration. By convention the concentration axis is $c^{1/2}$ rather than c. The data in Figure 8-4 would be contained in the early descending part of the curves in this figure.

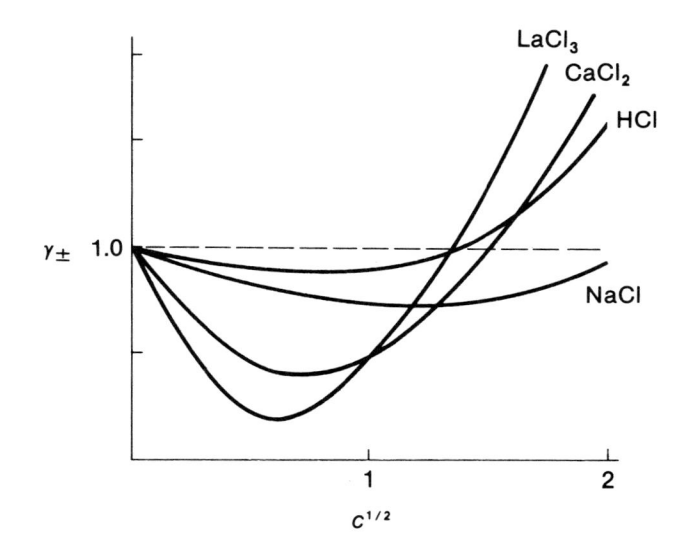

In view of the fact that agreement between theory and experiment is not perfect, one might reasonably ask what the value of the theory is.* There are two really important features of the Debye-Hückel theory. First, it provides a means of calculating the activity coefficient for a dilute solution and thereby explaining the anomalies in measurement of the colligative properties we have described. That is, for dilute solutions, at least, the appropriate value of the activity corresponding to a particular concentration can be calculated. Second, the Debye-Hückel theory provides an explanation for phenomena that depend on ionic strength. For instance,

---

*Recently, Kenneth Pitzer has discussed an improvement on the Debye-Hückel theory that allows the calculation of $\gamma_\pm$ over a broader range of ionic strength and gives good agreement with experiments for about 200 salts. The equation for ln $\gamma_\pm$ is very complex and lengthy, and includes several constants that have been determined experimentally. For biological systems the small improvement is not worth the additional complexity, but interested readers can consult the published report: K. Pitzer, 1977, "Electrolyte Theory—Improvements Since Debye and Hückel," *Accts. Chem. Res.* 10:371–377.

Equation 12 stated the relation between the solubility product K_{sp} and the solubility c_{sat}—namely, $\frac{1}{2} \log K_{sp} = \log \gamma_\pm + \log c_{sat}$. Combining this with Equation 17 yields $-\log c_{sat} = -0.509\, z^2 I^{1/2} - \frac{1}{2} \log K_{sp}$, which is equivalent to the empirical Equation 13.

One final point must be made concerning the validity of the Debye-Hückel theory. In Chapter 14 we discuss the structure of an ionic solution in the vicinity of a charged macromolecule and, more generally, in the vicinity of a surface carrying an electrical potential. The Debye-Hückel model does not lead to a correct description in this case because (1) there is not spherical symmetry and (2) the electrical potentials are too large to be small compared to kT. We will see that in this case there is a fixed layer of closely associated ions outside of which is a diffuse ionic cloud. The Stern model and the Gouy-Chapman theory, neither of which makes the second assumption, provide an adequate description and have good predictive value. These models are discussed in Chapter 14.

The Structure of an Ionic Solution According to the Debye-Hückel Theory

The principal significance of the success of the Debye-Hückel theory is that it helps us to form a picture of the structure of a dilute solution of an electrolyte. The theory begins with the assumption that strong electrolytes are totally dissociated into ions and that there are strong electrical interactions between these ions. The fact that the interaction energy is Boltzmann-distributed indicates that the ions are not uniformly distributed. Actually, the following simple argument leads to the same conclusion. Suppose the ions were distributed with perfect uniformity throughout the solution. Then, the probability of finding a positive ion near a particular ion would be the same as that of finding a negative ion in the same region. If this were the case, there would be no net electrical interaction energy, because every attractive configuration would be precisely balanced by a repulsive arrangement; however, the basic agreement between experiment and theory indicates that there is an electrical interaction. Such a perfectly random distribution is not physically reasonable anyway, since a negative ion has a greater probability of being near a positive ion than near a negative ion. Furthermore, if the ions were not in motion, the equilibrium configuration would be that of alternating positive and negative ions, as in a crystal. However, thermal motion disrupts this orderliness, so that the ionic distribution must be a balance between the tendency of the ions to be highly ordered and the tendency for thermal motion to disorder them. What is the final (though dynamic) distribution? Clearly, around each negative ion there are on the average more positive ions than negative ions and around a positive ion there will principally be negative ions. Thus, the physical picture of a solution of an electrolyte is that around each ion there is an "atmosphere" of oppositely charged ions. It can be

SEC. 8–4 THE DEBYE-HÜCKEL THEORY

shown that the mean radius of this atmosphere is $1/\kappa$, which is sometimes called the *Debye length* or the *Debye parameter*. In outline, one calculates the charge density in spherical shells of thickness dr at a distance r from the center of the sphere. This is the distribution function of the charge in the ion atmosphere. The derivative of this function is then set equal to zero and the resulting equation is solved for r, which is r_{max}. The result is

$$r_{max} = 1/\kappa . \tag{19}$$

That is, $1/\kappa$ *is the distance at which there is maximal probability of finding a charge in the atmosphere.* Since κ is proportional to ionic strength, at high ionic strength the ionic cloud is nearer to the central ion than at low ionic strength.

We can easily calculate the value of r_{max} for a 1 : 1 electrolyte (for example, NaCl) at different concentrations. Representative values are the following:

Molarity:	0.001	0.01	0.1	1.0
r, Ångström units:	96	30	9.6	3.0

These numbers represent the average separation between ions. They can be used to show that the theory ought to fail at concentrations above 0.01 M. Calculation of the electrical potential energy (see Appendix VII) for various values of r shows that when r exceeds 30 ångströms or, as indicated above, when the concentration exceeds 0.01 M, the electrical potential energy equals the thermal energy. It should be noticed also that because κ depends upon the dielectric constant D, the concentration at which the theory begins to fail is lower for solvents that have a value of D less than that of water.

One must be careful in thinking about the ionic distribution. Clearly, *each* negative ion cannot be surrounded by a cloud of positive ions, nor can each positive ion be immersed in a negative atmosphere. We must remember that the distribution is a dynamic one—that is, it is continually changing locally. Thus, what we mean to say is that at any given instant there are many ion clouds in solution and that, averaged over time, the number of positive clouds surrounding a negative ion equals the number of negative clouds surrounding positive charges. The entire solution must, of course, be electrically neutral, but locally there are charge inequalities. Furthermore, it must be realized that, surrounding a positive *cloud*, there will be negative ions, and around these negative ions there will be positive ions. This tendency decreases with distance. The main point is that the *ions are clustered* and the clusters account for several properties of ionic solutions, as we will see in a later section.

Phenomena Explained by the Debye-Hückel Theory

The basic findings of the Debye-Hückel theory are (1) that $\gamma_\pm$ decreases with increasing I, and (2) that the limiting slope of a plot of log $\gamma_\pm$ versus

$I^{1/2}$ is negative, proportional to the product $z_+ z_-$, and inversely proportional to $D^{3/2}$ and $T^{3/2}$.

The relation between $\gamma_\pm$ and I enables us to understand several other phenomena in which properties of solutions depend on ionic strength. We cannot always calculate $\gamma_\pm$ from I (for example, not if the concentration is too high), but we can with confidence use values of $\gamma_\pm$ determined experimentally at particular values of I. The important idea in the theory is that $\gamma_\pm$ is always less than one, which means that the interactions of one ion with other ions lowers the free energy of an ion in a solution of electrolytes. A lower free energy means that the ion is more stable in solution than it would be if it were uncharged. Thus any property that is thermodynamically related to the free energy change, which we know is dependent on the activity, should be understandable. Examples of such properties, which in fact show a dependence on I, are the effects of ionic strength on the pH of a weak acid or base and on the solubility of sparingly soluble salts; we will discuss these shortly. Other properties of solutions of electrolytes—properties which will not be discussed—are heat capacity, density, compressibility, and the coefficient of thermal expansion. Information concerning these can be found in references at the end of the chapter.

Recalling Equation 9 for the chemical potential in which the term $RT \ln \gamma_\pm$ is present, we can see from the negative value of $\ln \gamma_\pm$ that this term represents the extra stability of the ion in solution. What is the significance of this extra stability? Increasing the ionic strength decreases $\gamma_\pm$ and enhances the stability of an ion. Thus if there is an equilibrium between an undissociated molecule and its ions, the equilibrium will shift in the direction of increased ionization if the ionic strength is increased. In solutions of weak acids and of sparingly soluble salts, such an equilibrium does exist, so that increasing the ionic strength should increase the dissociation of an acid (that is, lower its pH) and should increase the solubility of a salt. These predictions can be stated quantitatively in the following way.*

In the dissociation of a weak acid, $HA \rightleftharpoons H^+ + A^-$, the equilibrium constant is

$$K = \frac{a_{H^+} a_{A^-}}{a_{HA}},$$

in which a is the activity. If we rewrite this in terms of activity coefficients and molarities, and substitute $\gamma_\pm^2$ for $\gamma_+ \gamma_-$, we obtain

$$K = \frac{\gamma_\pm^2 [H^+][A^-]}{\gamma_{HA}[HA]}. \tag{20}$$

This equation can be rewritten in terms of the initial concentration of the

---

*Note that we have said that the Debye-Hückel theory does not apply to weak electrolytes or to high concentrations, so one might ask how we are justified in discussing weak acids and 0.1 M KCl in the example to be given. The reason is that we are not calculating values of $\gamma_\pm$ for a particular concentration but only using experimentally determined values in our calculations.

SEC. 8–4 THE DEBYE-HÜCKEL THEORY

acid, c, and the degree of dissociation, α, after noting that $[H^+] = [A^+] = \alpha c$ and $[HA] = (1 - \alpha)c$, to yield

$$K = \frac{\gamma_\pm^2 \alpha^2 c}{\gamma_{HA}(1 - \alpha)} . \tag{21}$$

In a dilute solution, $a_{HA} = [HA]$, because HA is uncharged, so that $\gamma_{HA} = 1$. If we consider only poorly dissociated weak acids (that is, those having a small value of K), then α is small and $1 - \alpha$ can be approximated as 1. Thus, Equation 21 becomes

$$\alpha = (K/c)^{1/2} 1/\gamma_\pm . \tag{22}$$

If there are no ionic interactions, $\gamma_\pm$ would equal 1 for a dilute solution and the degree of ionization would simply be $(K/c)^{1/2}$. However, the Debye-Hückel limiting law tells us that $\gamma_\pm < 1$, so that $\alpha > (K/c)^{1/2}$, thus showing that α is *in*creased by the charge interaction. The limiting law also says that as I increases, $\gamma_\pm$ decreases and, hence, that α increases. We can look at this increase in α in two ways. First we can calculate from measured values of $\gamma_\pm$ the increase in α caused by the fact that the ions are charged, and compare it to the value it would have if there were no charges. For a weak acid such as acetic acid ($K = 1.8 \times 10^{-5}$), there is an increase of about 4 percent. A more interesting case is the effect of an added electrolyte that has no ion in common with the acid—as, for example, the effect of 0.1 M KCl on the ionization of dilute acetic acid. The limiting law tells us that we must be concerned about the significant contribution of 0.1 M KCl to the total ionic strength and that this will affect the activity of the ionic species present in acetic acid. We can calculate from the limiting law that $\gamma_\pm$ is 0.89; this number is not quite right because a concentration of 0.1 M is too high for the approximations made in the derivation to be valid—the correct value is 0.7. Thus, according to Equation 22,

$$\alpha = (K/c)^{1/2} 1/0.7 = 1.42(K/c)^{1/2} ,$$

which shows a 42 percent increase over what would be the value if the KCl were absent. This means a 42 percent increase in α_{H^+} or a *decrease* of about 0.15 pH units ($-\log 1.42 \approx -0.15$).

The analysis just given applies to any weakly dissociating ionic substance (for example, a buffer that ionizes according to the equation $HA \rightleftarrows H^+ + A^-$). The result of addition of KCl is strikingly different for a substance (a different type of buffer) whose ionization is described by the equation $BH^+ \rightleftarrows B + H^+$. In this case, $K = (a_B a_{H^+})/a_{BH^+}$, and a_B is, by the argument just given, equal to c_B. Substituting as above, one obtains $\alpha = 0.7 \, (K/c)^{1/2}$, so that there is an *increase* in pH of 0.15 pH units ($-\log 0.7 = 0.15$).

The influence of ionic strength on solubility occurs in a similar way. Consider a sparingly soluble salt XY having the equilibrium

$$XY \rightleftarrows X^+ + Y^- .$$

The solubility product constant K_{sp} is defined as

$$K_{sp} = a_X a_Y = (\gamma_+ c_+)(\gamma_- c_-) , \qquad (23)$$

in which c again represents a concentration. As we saw earlier, if we express the solubility c_{sat} in moles per liter, then $c_+ = c_- = c_{sat}$, and

$$K_{sp} = \gamma_+ \gamma_- c_{sat}^2 = \gamma_\pm^2 c_{sat}^2 . \qquad (24)$$

Since, by definition, $\gamma_\pm^2 < 1$, we know that the value of $c_{sat} = (1/\gamma_\pm)K_{sp}^{1/2}$ is always greater than $K_{sp}^{1/2}$. The limiting law tells us that the addition of any electrolyte decreases $\gamma_\pm$. That is, as I increases, $\gamma_\pm$ decreases, and c_{sat} increases. For 0.1 M KCl $\gamma_\pm = 0.7$, so that the solubility of a weakly soluble salt would be increased 42 percent by the addition of 0.1 M KCl. This phenomenon is called *salting in*.*

Let us consider for a moment the physical principle involved in the arguments just given. In both dissociation of a weak acid and solubilization of sparingly soluble salts, there is an equilibrium between charged ions and the uncharged, undissociated molecule. This equilibrium can be stated as

$$\frac{a_{un}}{a_1 a_2 \cdots a_n} = \frac{\gamma_{un} c_{un}}{\gamma_1 c_1 \gamma_2 c_2 \cdots \gamma_n c_n} = \text{constant} ,$$

in which the subscript "un" designates the undissociated species and the numbers refer to the ions. The Debye-Hückel limiting law tells us that as the ionic strength increases, the activity coefficient for a charged species decreases. The key point is that γ_{un} is unaffected by a change in I, because the undissociated molecule is not charged. Thus, in order for the ratio just stated to be constant while the ionic strength increases, the values of c_1, c_2, ..., must increase and c_{un} must decrease. In other words, *an increase in ionic strength increases the concentrations of the ions at the expense of the undissociated molecule.*

This phenomenon can be seen in a final example—the effect of ionic strength on the stability of complexes containing metal ions. Complex ions involving metal ions are relatively common in biochemistry. For example, Mg^{2+}, Zn^{2+}, Cu^{2+}, Fe^{2+}, and Fe^{3+} are frequently found bonded to nitrogen atoms or amino groups in protein molecules. This type of bonding ranges from simply being electrostatic to being weakly covalent. Very frequently, coordination compounds (see Chapter 9) are present, in which the metal ion is bound to several different groups—for example, in hemoglobin, iron

---

*It is important that one does not confuse salting *in* with two other phenomena, *salting out* (explained later in this chapter) and the *common ion effect*. In salting out, the solubility of uncharged or weakly charged substances, such as macromolecules, is decreased by the addition of a salt: the salt ions are so highly hydrated that they reduce the effective solvent concentration sufficiently that the hydration of the macromolecules, which is necessary for them to remain in solution, is in turn reduced below the point at which solubility is possible. In the common ion effect, there is a large decrease in the solubility of a salt with which the added electrolyte has a common ion; when the concentration of one of the ions is increased, this drives a dissociation reaction to the undissociated side.

SEC. 8–4 THE DEBYE-HÜCKEL THEORY

is bound to four nitrogen atoms in the porphyrin group of the heme and a fifth nitrogen atom in the protein globin. These metal-ion clusters dissociate to some extent and in the undissociated state are often uncharged. When this is the case, they are affected by ionic strength in the way just described—namely, as ionic strength increases, the compound dissociates and the concentration of the ions increases.

This decrease in the concentration of a complex with increasing ionic strength may also occur if the complex is charged, but whether this is the case depends on the charges of the components. Consider the case of a complex described by

$$A^{3+} + 4B^- \rightarrow AB_4^-.$$

If K is the equilibrium constant, the relevant equation is

$$K = \frac{[AB_4^-]}{[A^{3+}][B^-]^4} \cdot \frac{\gamma_{AB_4}}{\gamma_A \gamma_B^4}.$$

Taking the logarithm of both sides, and rearranging and substituting the Debye-Hückel limiting equation, we get

$$\frac{1}{K} \frac{[AB_4]}{[A][B]^4} = \log \frac{\gamma_A \gamma_B^4}{\gamma_{AB_4}}$$
$$= -0.51(3)^2 I^{1/2} - 4[0.51(1)^2 I^{1/2}] + 0.51(1)^2 I^{1/2}$$
$$= -6.12 I^{1/2}.$$

Thus, as I increases, the logarithm becomes more negative and $[AB_4^-]$ must decrease.

If, however, complex formation is described by

$$A^{3+} + 6B^{2-} \rightarrow AB_6^{9-},$$

the relevant equation becomes

$$\frac{1}{K} \frac{[AB_6]}{[A][B]^9} = -(.51)(3)^2 I^{1/2} - 6(.51)(2)^2 I^{1/2} + (.51)(9)^2 I^{1/2}$$
$$= +29.6 I^{1/2}$$

and an increase in ionic strength increases the stability of the complex.

This kind of reasoning leads to an understanding of the effect of ionic strength on reaction rates. That is, a multistep reaction frequently involves an intermediate whose concentration is determined by an equilibrium with the reactants—for example, $A + B \rightleftharpoons AB \rightarrow C + D$. In transition state theory (Chapter 11), the rate of an overall reaction is determined by the concentration of the complex AB. If any of the components A, B, or AB are charged, the concentration of AB will be affected by the ionic strength. The value of the charges and the stoichiometry of the equilibrium will, as we have just seen, determine whether [AB] and, hence, the overall reaction rate, increases or decreases with ionization. For a detailed discussion of this phenomenon, see Chapter 11.

8–5 OTHER CONSIDERATIONS ABOUT IONIC STRENGTH

Ignoring quantitative effects on activity coefficients and allied phenomena, the major effect of ionic strength is shielding. Consider two charged objects in water—for instance, macromolecules—at a fixed distance from one another. If the charges have a different sign, the objects will attract one another. However, if the water contains a high concentration of NaCl, many Na^+ ions will be attracted to the negatively charged object and reduce its effective charge, as far as the positively charged object is concerned. Similarly, the Cl^- ions will be attracted to the other particle. The net effect is to reduce the attraction between the two objects. If the macromolecules have the same charge, the repulsive force would also be substantially reduced. In short, *increasing the ionic strength of a solution decreases the electrostatic interaction between charged particles in the solution.* This effect plays a straightforward role in the structure and activity of biological molecules and is valuable in some separation techniques, as is described in the following examples.

EXAMPLE 8–C   The shape of a macromolecule.

Consider a hypothetical protein molecule containing one sequence of positive amino acids and one sequence of negative amino acids, in which there is no amino acid interaction other than electrostatic attraction. In a very dilute solution in which there is very little added electrolyte, these sets of amino acids will attract one another and the protein chain will have the shape shown in Figure 8-6. If the concentration of added electrolyte is significant so that ionic strength is high, there will be no attraction and a random arrangement will result.

EXAMPLE 8–D   The structure of DNA.

DNA consists of two strands of alternating sugars and phosphates held together in a double-stranded structure by hydrogen bonds between bases that are attracted to the sugars (Figure 8-7). The phosphate groups have a strong negative charge. In a very dilute DNA solution in which the concentration of added electrolyte is high and hence the ionic strength is also high, the repulsive force between the two strands is very weak. However, when DNA is dissolved in distilled water, the repulsive force exceeds the strength of the hydrogen bonds and the two strands become totally separated.

EXAMPLE 8–E   Reduction of the activity of many enzymes in solutions in which the concentration of added electrolyte is high.

The binding of many enzymes to their substrates (the molecule whose chemical reaction is catalyzed by the enzyme) includes both van der Waals

SEC. 8-5   OTHER CONSIDERATIONS ABOUT IONIC STRENGTH   267

FIGURE 8-6   Effect of ionic strength on the shape of a charged macromolecule.

Distribution of charges on extended protein chain

Structure in solution of very low ionic strength

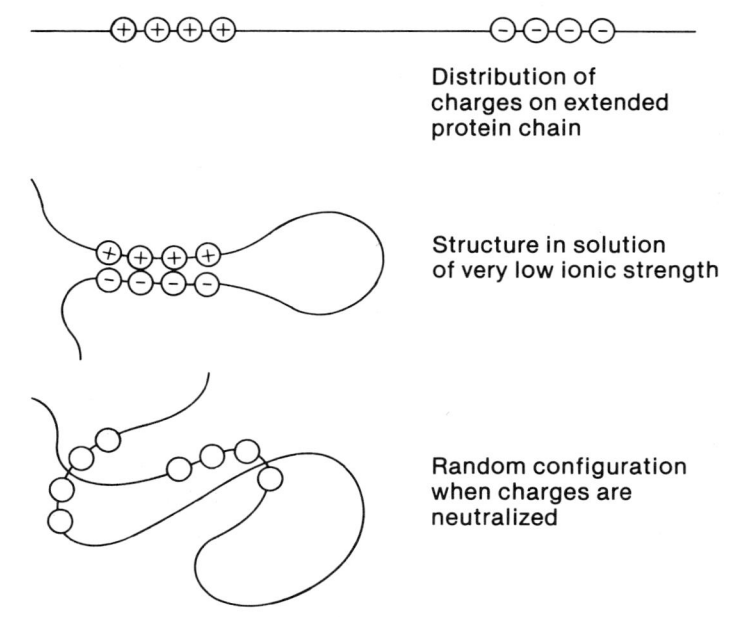

Random configuration when charges are neutralized

FIGURE 8-7   Structure of DNA in the presence and absence of ions. The solid lines represent molecules of the sugar deoxyribose. The squares are the bases, and the dotted lines are the hydrogen bonds. In panel (a) the Na$^+$ ions bind to and neutralize the negatively charged phosphates so that the hydrogen bonds can maintain the familiar, rigid, double-stranded structure. In panel (b), in the absence of the Na$^+$ ions, the strands repel one another and become flexible single strands. The charge repulsion keeps the phosphates well separated so that the molecule cannot fold back on itself.

(a)   In 0.1 M NaCl

(b)   In distilled water

and electrostatic attractive forces. When the concentration of added electrolyte is high, the electrostatic force is reduced, and either binding fails to occur or the substrate is not correctly oriented on the surface of the enzyme molecule. In many cases, there is no electrostatic attraction between the enzyme and the substrate and the high ionic strength resulting from the added electrolyte still reduces enzymatic activity; usually this is because, as shown in Example 8–C, the shape of the enzyme is changed at high ionic strength and the binding site on the enzyme no longer has the correct shape to maintain a strong van der Waals attraction.

EXAMPLE 8–F   Separation of two proteins with an ion exchanger.

An ion exchange resin consists of a solid support such as cellulose to which an ionizable substance is covalently attached. An example is diethylaminoethyl cellulose (DEAE-cellulose), which at pH 7 has one positive charge per monomer. DEAE-cellulose is an insoluble substance. If placed in water containing charged proteins, it will bind proteins having negative charges but not those having positive charges. Therefore, if a tube is packed with the powdered material, and a solution consisting of two proteins, one positively and one negatively charged, is allowed to flow through the material, the negatively charged protein will bind to the DEAE-cellulose whereas the other protein will pass through (Figure 8-8). In this way, the two proteins are separated. To recover the bound protein, one need only increase the ionic strength until the electrostatic attraction is too small for binding. Two different negatively charged proteins can also be separated if they have a different total charge or a different distribution of charged groups. The strength of binding depends upon the magnitude and distribution of the charge, so that it is often possible to select an ionic strength at which one protein can bind but the other cannot. This is the principle of *ion exchange chromatography*.

8–6   HYDRATION OF IONS

It has already been pointed out that the binding of water molecules by ions is important in maintaining separation of the ions in solution. In fact, this hydration is an important factor in determining solubility.

Two simple observations provide us with an idea of the extent of hydration of different ions: (1) in a particular group in the periodic table—for example, the alkali metal group—the rate of movement of the ions in an electric field increases with molecular weight, and (2) ions having low rates of movement in an electric field are very efficient at causing proteins to precipitate. Let us examine these two phenomena more carefully.

When a voltage is applied between two regions of a solution of an electrolyte, ions move in the electrical field, as described in Chapters 14

FIGURE 8-8 The principle of ion exchange chromatography. A tube is filled with the positively charged ion exchanger. In aqueous solution there are of course negative charges, for example OH⁻ ions, which maintain neutrality of the solution. In panel (a) a mixture of + and − charged proteins is applied to the column of material in the tube. If the ionic strength is low, the − proteins stick to the exchanger and the + proteins pass through as shown in panel (b). In panel (c), concentrated NaCl has been added. The Cl⁻ ions weaken the binding and ultimately displace the − proteins, which then flow out of the bottom of the tube.

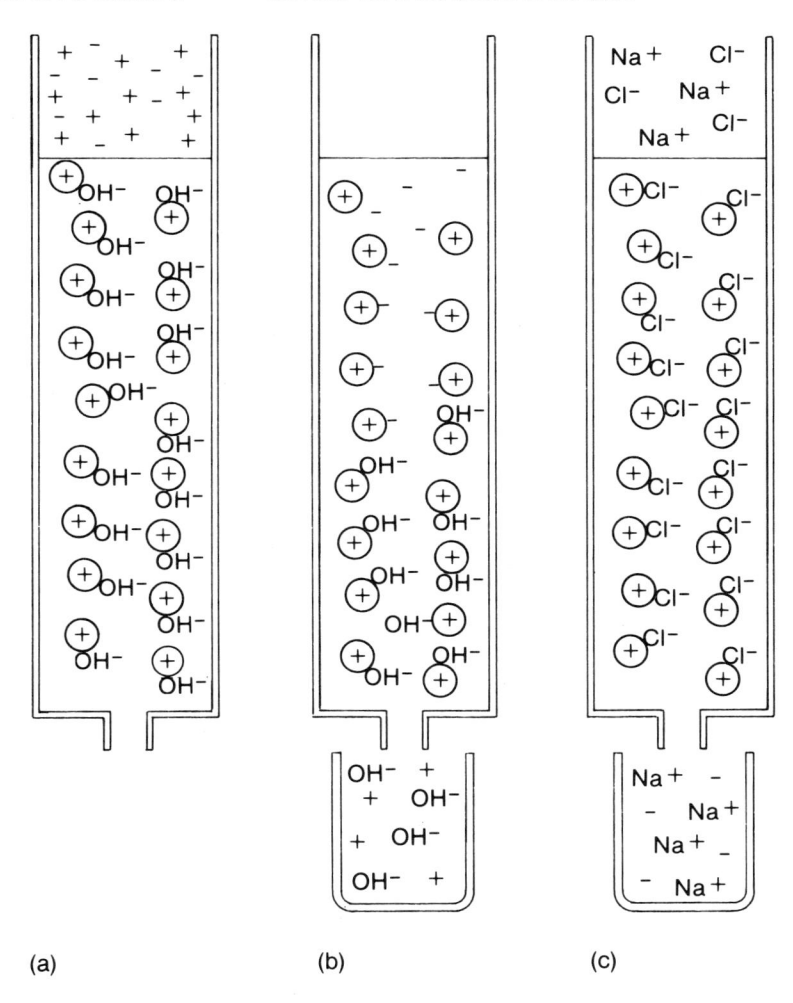

(a)   (b)   (c)

and 15. The positive ions move toward the negatively charged electrode and the negative ions move in the opposite direction. For a particular voltage, the two most important factors that determine the rate at which an ion moves are its charge and the friction it encounters when moving through the solvent. The rate of movement is proportional to the voltage, so one does not usually discuss the rate but instead the *mobility*—that is, the ratio of the velocity to the applied voltage (see Chapters 10, 11, and 15). The mobility of the divalent Mg^{2+} ion should be (and is) greater than that of the monovalent Na^+ ion because the Mg^{2+} ion has a larger charge.

The friction that a particle experiences in moving through a viscous liquid increases with the particle size and degree of asymmetry. For example, of two spheres having the same mass but different radii, the larger one will experience greater friction when moving through water; a rod having the same mass as a sphere will also encounter greater resistance to movement than does the sphere. Thus, it might be expected that of the univalent alkali cations (Li^+, Na^+, K^+, Rb^+, Cs^+), the Li^+ ion, which has the smallest ionic radius *when in an ionic crystal or in a gas,* would experience the least friction and therefore would have the highest mobility. Similarly, the Cs^+ ion, which has the largest radius of the ions in the group, should have the lowest mobility. However, this is not the case; the mobilities of the monovalent cations increase with atomic number: $Cs^+ > Rb^+ > K^+ > Na^+ > Li^+$. This is also true of the divalent alkaline earth cations—that is, $Ba^{2+} > Sr^{2+} > Ca^{2+} > Mg^{2+}$. This is because, when the ion is in solution, the ions having lower atomic numbers move as units having a large radius, since the smaller ions bind more water molecules than the larger ions. Li^+, the smallest alkali metal ion (in the crystal or gas phase) has the most intense local electric field in its vicinity because it is the smallest, and because the magnitude of the field of a charge varies *inversely* with the distance from the charge. Water molecules are dipoles, so that the negative region of a water molecule is attracted to a Li^+ ion. The number of water molecules bound to any ion depends on the electric field of the ion, and a smaller ion will be surrounded by more water molecules than a larger ion. These water molecules, which constitute what is called the *first hydration layer,* are oriented with their negative regions toward the ion so that they present their positive poles to the solution. Consequently, a second shell of water molecules is bound, although not as tightly as the first. This shell is also somewhat ordered (the ordering is of course antagonized by thermal motion), so that a third set of molecules joins the group. A large ion orients fewer molecules in the first layer than a small ion does. The radius of the sphere consisting of the smallest ion plus the first hydration shell is greater than that of the nonhydrated large ion, the net effect of which is that the radius of the complete hydration sphere of the Li^+ ion is greater than that of the Cs^+ ion. Therefore, a hydrated Li^+ ion encounters greater frictional resistance than a hydrated Cs^+ ion and a Cs^+ ion has a greater mobility than a Li^+ ion.

The solubility of many biological macromolecules—for example, proteins—depends on their ability to bind water molecules. If the ion concentration in a solution of protein molecules is increased, a concentration is often reached at which the protein precipitates. This phenomenon is called *salting out* and is a consequence of the binding of water molecules by the ions to such a large extent that there are not enough water molecules remaining to hydrate the protein molecule. Salting-out of proteins will be discussed in detail in the next section. For a particular concentration of electrolyte, the effectiveness of different ions in salting out increases with the number of water molecules in the hydration spheres of the different ions. It comes as no surprise then that for a particular valence, salting-out

SEC. 8-6   HYDRATION OF IONS

efficiency is greater for those ions having the lower mobility—that is, for those having the greater number of water molecules in their hydration spheres. Studies of salting-out efficiency have resulted in an ordering of a large number of ions in what is known as the *lyotropic series,* in which ions are traditionally ordered according to decreasing salting-out efficiency and decreasing extent of hydration. A portion of the lyotropic series for several ions of biological interest is the following:

Cations:
$$Li^+ > Na^+ > K^+ > NH_4^+ > Rb > Mg^{2+} > Ca^{2+} > Sr^{2+} > Ba^{2+}$$

Anions:
$$citrate^{3-} > SO_4^{2-} > acetate > Cl > Br > NO_3 > ClO_3 > I > CNS$$

The lyotropic series indicates the relative salting-out efficiency of various ions at a particular concentration. However, it does not tell which ion has the greatest ability to salt out because that is determined not only by the efficiency per unit concentration but also the maximum solubility. For example, Li citrate is more effective than K acetate at a concentration of 2 M, but K acetate, whose concentration is 25.8 M at saturation, is more effective overall than Li citrate, whose maximum solubility is 2.6 M.

The water-binding properties of some of these salts also has an interesting effect on the stability of the DNA molecule. A DNA molecule is a double-stranded molecule, each strand of which consists of a long chain of highly soluble, alternating phosphate groups and deoxyribose molecules to which are bonded relatively insoluble organic bases (refer again to Figure 8-7). The bases are not very soluble and an isolated base induces an ordering of the surrounding water molecules, as explained in Chapter 5. This ordering requires an entropy increase and hence is thermodynamically unfavorable. Thus the bases form clusters, which reduces the entropy increase. The most effective cluster, that is, one which minimizes contact with water, is a stacked array in which the planes of the base rings are in close contact. This stacking produces an orientation that allows a large number of hydrogen bonds to form between the bases on opposite strands and this results in the formation of the familiar double helix. If DNA is placed in a very concentrated solution (7 M) of $NaClO_4$, the helix dissociates into the two constituent polynucleotide strands. The explanation for this is the following. In 7 M $NaClO_4$ an enormous fraction of the water molecules is in the hydration spheres of the Na^+ and ClO_4^- ions. The deoxyribose and the phosphate groups in the DNA are so effective at binding water molecules that they remain in solution. However, there is no longer sufficient water to form an ordered shell around each base (although there is enough to keep the relatively insoluble bases in solution). Hence, any base which for an instant swings out from the helix will not have the entropic need to return to the stack. Thus, the stacking tendency is lost and thermal motion disrupts the orderly array; hydrogen-bonding between the strands is no longer facilitated and the two strands separate. (This effect is not seen with all salts—for example, not with NaCl—because it is not possible to reach the necessary high molarity with some salts.)

A particularly important example of hydration is the acidic H^+ ion, which exists tightly bound to one water molecule to form the hydronium ion H_3O^+ (see Chapter 9). There are many ways in which the existence of hydrated ions such as the H_3O^+ ion has been demonstrated; for example, in one interesting experiment it was shown that for every H^+ ion passing through the glass of the electrode of a pH meter, there is carried one H_2O molecule. The H_3O^+ ion is a particularly complex ion because each of the three protons is hydrogen-bonded to another H_2O molecule. Hydration is especially important with macromolecules such as proteins and nucleic acids in which case the so-called hydration sphere is a major factor in determining the three-dimensional structure of the macromolecule. This is described in some detail in Chapter 15.

We now return to the salting-out phenomenon and describe the great utility of salting out in the purification of proteins.

8–7 SALTING-OUT OF PROTEINS

As we have just mentioned, the solubility of proteins is significantly decreased when the ionic strength is very high. However, all proteins do not decrease in solubility to the same extent when a particular concentration of a salting-out electrolyte is reached. This is the case for two reasons. First, because of differences in structure, every protein does not require the same number of bound water molecules in order to remain in solution. Second, the strength of binding of water molecules is not the same for all proteins. The relative solubility of different proteins at a particular ionic strength can be used as a means of selective precipitation. How this occurs can be seen by combining the Debye-Hückel limiting law and Equation 14 to obtain

$$-\log \gamma_\pm = \log(S/S_0) = 0.509 z^2 I^{1/2} , \qquad (25)$$

in which the solubility $S = c_{\text{sat}}$ at ionic strength I and S_0 is the solubility when $I = 0$. This equation describes the salting-in phenomenon described earlier. Remembering that salting-out is a result of hydration, we must modify this equation to include the hydration term. We shall not go through the calculation but simply state the equation:

$$\log(S/S_0) = 0.509 z^2 I^{1/2} - K_s I \qquad (26)$$

in which K_s is an empirical constant called the *salting-out constant*. When the ionic strength is high (that is, when the conditions are appropriate for salting out), $K_s I \gg 0.509 z^2 I^{1/2}$, and Equation 26 may be written

$$\log(S/S_0) = -K_s I \qquad (27)$$

SEC. 8-7   SALTING-OUT OF PROTEINS

or
$$\log S = -K_s I + \log S_0 \ . \tag{28}$$

A plot of $\log S$ against I will have a slope of $-K_s$ and an intercept of $\log S_0$. It should be noticed that the meaning of S_0 has been changed slightly by the approximation just made, by ignoring the salting-in term—that is, S_0 is not a real solubility at $I = 0$ but merely the value of S extrapolated to $I = 0$. This is actually of no importance because the only part of the plot we are concerned with is the linear portion which we show in Figure 8-9. The main feature of these curves is that in a mixture of protein it is usually possible to find an ionic strength at which one protein can be precipitated and the others cannot. For instance, in Figure 8-9, protein A can be precipitated at an ionic strength of 3, then removed by centrifugation, and more salt can be added to reach $I = 4$, thereby precipitating protein B, which is removed by centrifugation, and so forth. Note that some proteins cannot be separated easily—for example, B and C—because their curves are too near to one another. Also, because K_s is not the same for all proteins, some may never be separable by this procedure—for example, proteins E and F.

FIGURE 8-9 Effect of ionic strength of ammonium sulfate on the solubility of different proteins.

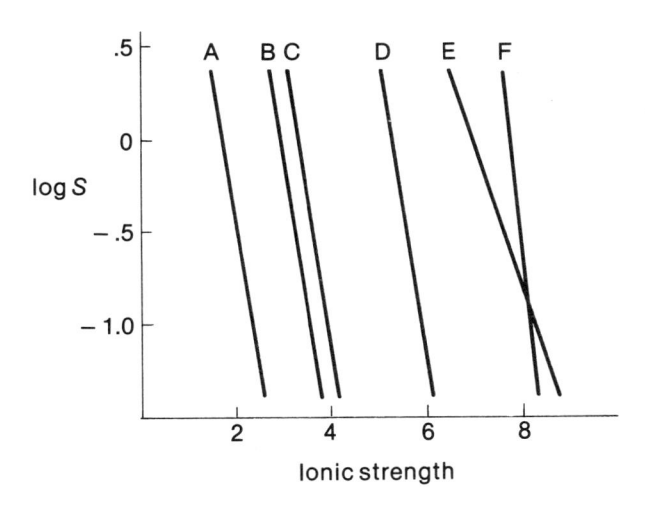

This procedure is invariably used as an early step in the purification of enzymes. Operationally, one adds the salt in small increments to a solution containing several types of protein molecules, and particular molecules precipitate when the concentration of the salt reaches different values.

The most commonly used compound for salting out protein molecules is $(NH_4)_2SO_4$; the method is known in biochemistry as *ammonium sulfate fractionation*. If one looks at the lyotropic series, one might reasonably ask whether Li citrate might not be better than $(NH_4)_2SO_4$. Indeed it might, but the choice of $(NH_4)_2SO_4$ is based upon four facts—its high solubility (5.34 M) at 0°C compared to Li citrate (2.6 M), its very low cost, the ease with which it can be removed, and an unknown stabilizing effect of the NH_4^+ ion on the biological activity of many protein molecules.

A typical protocol for purification of an enzyme from cellular material is the following. One adds ammonium sulfate to a cell extract to reach a concentration of 10 percent wt./V*; the precipitate is removed by centrifugation; more ammonium sulfate is added to reach 20 percent; the precipitate is removed; and so forth. After reaching a concentration of about 70 percent, each of the precipitates (called ammonium sulfate "cuts" in laboratory jargon) is redissolved and assayed for the enzyme of interest. Only the fraction containing enzymatic activity is retained. There is usually purification of about 10- to 20-fold as well as an increase in concentration of the protein by this procedure.

*In ammonium sulfate fractionation, the ionic strength is rarely stated. Conventionally the ammonium sulfate concentrations are expressed as weight percent per unit volume of the initial solution, since operationally the salt is added in that way in the laboratory.

REFERENCES

Bull, H.B. 1964. *An Introduction to Physical Biochemistry*. Davis. Philadelphia.
Castellan, G.W. 1971. *Physical Chemistry*. Addison-Wesley. Reading, Mass.
Cohn, E.J., and J.T. Edsall. 1965. *Proteins, Amino Acids, and Peptides*. Hafner. New York. (This is an unusually good and the most complete treatment of the thermodynamics of solutions of biological electrolytes.)
Driesbach, D. 1966. *Liquids and Solutions*. Houghton-Mifflin. Boston.
Harned, H.S., and B.B. Owen. 1958. *The Physical Chemistry of Electrolytic Solutions*. Reinhold. New York.
Levine, I.N. 1978. *Physical Chemistry*. McGraw-Hill. New York.
Robinson, R.A., and R.H. Stokes. 1959. *Electrolyte Solutions*. Academic Press. New York.

PROBLEMS

1. Calculate $c_\pm$ for Na acetate, $CuSO_4$, $Cu(NO_3)_2$, and $Al_2(SO_4)_3$.

2. Write Equation 9 in kJ mol^{-1} for 0.1 M $(NH_4)_2SO_4$ at 25°C.

3. What are the values of the ionic strength of the following salts: 0.08 M NaCl, 0.05 M $MgCl_2$, 0.02 M Na_2SO_4, and 0.05 M $Fe_2(SO_4)_3$?

4. (a) What is the ionic strength of a solution containing 0.01 M NaCl, 0.003 M Na_2SO_4, and 0.007 M $MgCl_2$?

   (b) What are the activity coefficients of any of the divalent ions in this solution?

5. Which solution would have a lower freezing point—0.5 M NaCl or 0.5 M $CaCl_2$?

6. The Debye length, κ, is a measure of the electrical potential of a charged object seen by an approaching charge—the so-called shielded Coulomb potential. The shielding decreases as κ increases.

   (a) What is the molecular explanation for this shielding?

   (b) Explain in molecular terms why κ increases with increasing temperature.

   (c) Why does κ increase as the dielectric constant of the solvent increases?

   (d) Why does κ decrease with increasing ion concentration?

PROBLEMS

(e) What are the relative values of κ for a solution of MCl_2, compared to MCl, if the molarities are equal?

7. In a very dilute solution of $CaCl_2$, some Ca^{2+} ions are surrounded by Cl^- ions and some Cl^- ions are surrounded by Ca^{2+} ions. According to the Debye-Hückel theory, what are the relative sizes of the two ionic atmospheres?

8. What is the expression for the limiting law at the biologically important temperatures of 30°C and 37°C?

9. The maximum solubility of a particular salt in water is 0.0023 M. At this concentration $\gamma_\pm = 0.98$. When 1 M NaCl is present, the maximum solubility is 0.0035 M. What is the value of $\gamma_\pm$ when 1 M NaCl is present?

10. What is the pH of acetic acid ($K = 1.8 \times 10^{-5}$) in the following solutions:

    (a) 0.05 M acetic acid.

    (b) 0.05 M acetic acid + 0.05 M KCl.

11. What is the activity coefficient of 0.01 M $CaCl_2$?

12. A sparingly soluble salt XY, whose maximum solubility is 3×10^{-4} M, is participating in a chemical reaction. In order to accelerate the reaction somewhat, you attempt to increase the solubility twofold by adding KCl. What concentration of KCl would be necessary?

13. A chemical reaction requires at one stage a collision between two positively charged molecules. Will the reaction rate increase or decrease as the concentration of added NaCl is increased?

14. The binding site of a particular protein molecule for an uncharged small molecule contains no charged amino acids. However, the addition of 1 M NaCl breaks down the binding. Propose an explanation.

15. We have seen that the solubility of a sparingly soluble salt is increased by the presence of an inert salt. In terms of the Debye-Hückel theory, this is a result of increasing the ionic strength and decreasing the activity coefficient, without affecting the solubility product constant. Now, explain in molecular terms why this is so, without resorting to mathematics.

16. How would you expect the number of water molecules bound to a protein to be affected by an increase in temperature?

17. A protein molecule in a very dilute aqueous solution of NaCl has several bound Na^+ ions. Not all potential binding sites have a Na^+ ion though. If the entire solution was prepared in 80 percent water–20 percent ethanol, and if the protein molecule itself was unaffected by the ethanol, would the average number of bound Na^+ ions be more or less than when the molecules are in pure water? Remember that the dielectric constant of water is higher than all other known solvents.

18. In considering the movement of solute molecules that is due to osmotic effects, one is usually concerned with the state at equilibrium. One could also make a statement about the rate of approach to equilibrium. It is a common observation that the movement of K^+ ions is greater than that of the Na^+ ion through most semipermeable membranes. Explain why this is so.

19. The enzyme RNA polymerase binds to DNA and is responsible for synthesizing RNA. In cells, the synthesis of RNA is initiated at unique sites on the DNA molecule. In the test tube it is found that in 0.2 M NaCl, four RNA polymerase molecules bind to a particular DNA molecule and synthesize the same RNA molecules that can be isolated from living cells. In 0.01 M NaCl about 50 RNA polymerase molecules bind to the DNA and a great many kinds of RNA molecules are synthesized. What can be said about the binding forces between DNA and RNA polymerase and the nature of the binding that occurs within living cells?

20. Chromatin, the substance of which chromosomes are made, consists of a DNA

molecule to which are bound several classes of protein molecules. In 0.01 M NaCl, chromatin is quite stable, yet in 1 M NaCl it dissociates almost completely to yield free DNA molecules and free protein molecules. What does this tell you about the nature of the forces that stabilize chromatin?

21. We have discussed precipitation of proteins by the addition of high concentrations of electrolytes—the salting-out phenomenon. From what you understand about this process, explain a similar phenomenon—precipitation of proteins by high concentrations of polyethylene glycol (-CHOH-CHOH-), a polymer that is not an electrolyte.

CHAPTER 9

ACID-BASE EQUILIBRIA

Of the chemical reactions occurring in living cells, the most widespread are those in which protons are added or taken away from donor and acceptor molecules (acid-base reactions) and those in which electrons are transferred (oxidation-reduction reactions). The former are the topic of this chapter; the latter are treated in Chapter 10. The importance of understanding acid-base reactions is threefold: (1) More biochemical reactions are proton-transfer reactions than are any other type. (2) A very large fraction of all enzymatic reactions utilize proton transfer as a means of catalyzing a reaction that at first glance does not seem to be an acid-base reaction. (3) The activity of almost all biological macromolecules is affected by pH or, more specifically, by the state of protonation of chemical groups in the macromolecule.

9–1 DISSOCIATING SYSTEMS

Acids and bases can be categorized in several ways. The most important distinction is whether an acid or base is *strong* or *weak*. An acid or base is called strong if it is 100 percent (or nearly 100 percent) dissociated at all concentrations: thus it is also a strong electrolyte. On the other hand, the degree of dissociation of a weak acid or base is concentration dependent, increasing as the concentration decreases. Furthermore, the range of the concentration dependence is enormous—for example, 10^{-3} M pyruvic acid is 89 percent dissociated, whereas carbonic acid is 0.024 percent dissociated. In this chapter we will be concerned almost entirely with the weak variety because these are the ones that are found in biological systems. Another distinction is based upon the number of protons that can be donated or accepted; if the number is greater than one, the compound is called *polyprotic*. The interesting thing about polyprotic compounds is that each proton dissociates at a different concentration. For instance, the triprotic acid H_3PO_4 dissociates in three steps as the concentration is decreased, releasing one proton at each stage.

Dissociation of Weak Acids and Bases in Water

The traditional view of acids and bases is that they are compounds that dissociate to release H^+ ions or OH^- ions, respectively. This view has several disadvantages. For example, in the dissociation of an acid $HA \rightleftarrows H^+ + A^-$, there is the implication that the proton is free in solution. This is rarely the case—the proton is usually bound to water as H_3O^+. Second, the traditional view fails to recognize the basic properties of a substance such as NH_3, which has no hydroxyl group. These problems are avoided in the Brønsted-Lowry theory, in which an acid and a base are defined as a proton donor and a proton acceptor, respectively. In this sense the dissociation of acetic acid HAc, $HAc \rightleftarrows Ac^- + H^+$, is instead written

$$HAc + H_2O \rightleftarrows Ac^- + H_3O^+,$$

in which the HAc is viewed as an acid, Ac^- is a base, and H_2O is a base. A pair such as HAc and Ac^- is called a *conjugate pair* in this theory—HAc is the conjugate acid and Ac^- is the conjugate base.

The accepting of a proton by the base NH_3 when it is dissolved in water,

$$NH_3 + H^+ \rightleftarrows NH_4^+,$$

is similarly described: NH_3 is the conjugate base and NH_4^+ is the conjugate acid.

The particular advantage of the Brønsted-Lowry theory is that a large number of chemical reactions, which in the traditional theory are quite different, can be analyzed by the same formalism as long as a molecule or group is classified by its proton-donating (acidic) and proton-accepting (basic) ability. Although we use this theory throughout this chapter, for simplicity of notation we will use $[H^+]$ to denote proton concentration rather than the more cumbersome notation $[H_3O^+]$.

Let us begin by reviewing a few of the calculations that are carried out for determining the concentrations of various species in aqueous solution. Consider an acid HA that dissociates to H^+ and Ac^-. The dissociation constant K_a' is

$$K_a' = \frac{[H^+][A^-]}{[HA]} \tag{1}$$

SEC. 9–1 DISSOCIATING SYSTEMS

in which the brackets denote concentrations. For a particular temperature, K_a' is a true constant—that is, a simple mathematical expression of the state of equilibrium that is always present between the components. Thus, if the concentration of any of the components is changed, the concentrations of the other components change to maintain a constant value of K_a'.

In Chapters 7 and 8 we pointed out that most expressions describing the properties of real solutions are valid only if the activity of each component is used, rather than the concentration. In biochemistry, concentration is used exclusively because the concentrations are generally so low that the activity and concentration are nearly the same.* Also, to avoid confusion the dissociation constant that is defined by Equation 1 is called the *apparent dissociation constant* and is written with a prime—that is, as K'—to distinguish it from K, which denotes the dissociation constant that is based on activity. In this chapter, we use K' exclusively. It must be borne in mind, then, that although we have just said that K' is a constant, this is true only in dilute solutions ($< 0.1\ M$). The reader must be cautious in reading the biochemical literature and obtaining values of dissociation constants from standard reference tables to note whether K or K' is being given.

Let us now see how to obtain information from K' about the concentrations of the various components in solution.

If the initial concentration of HA is c, then at equilibrium, when partial dissociation has occurred, the concentrations of each component are $[\mathrm{H}^+] = x$, $[\mathrm{A}^-] = x$, and $[\mathrm{HA}] = c - x$. Therefore

$$K_a' = \frac{x \cdot x}{c - x} = \frac{x^2}{c - x}, \qquad (2)$$

which, by the quadratic formula, has the general solution

$$x = \frac{-K_a' \pm \sqrt{K_a'^2 + 4cK_a'}}{2}. \qquad (3)$$

For acetic acid, $K_a' = 1.8 \times 10^{-5}$ moles per liter (mol l^{-1}) at 25°C, so that if the initial concentration is 0.05 M,

$[\mathrm{H}^+] = [\mathrm{Ac}^-] = 9.5 \times 10^{-4}\ M$ and $[\mathrm{HAc}] = 4.905 \times 10^{-2}\ M$.

When K_a' is very small, so that $x \ll c$, the approximation $c - x$ may be used, so that

$$x \approx \sqrt{cK_a'}. \qquad (4)$$

Using the same example, $[\mathrm{H}^+]$ and $[\mathrm{Ac}^-]$ would be

$$\sqrt{(0.05)(1.8 \times 10^{-5})} = 9.48 \times 10^{-4}\ M,$$

which is very near to $9.5 \times 10^{-4}\ M$.

---

*When I was a student, I was told that a biochemist is both an organic chemist who does not do an experiment if the necessary chemical cannot be bought and a physical chemist for whom all activity coefficients equal one.

To be precise in calculating the concentrations, one should take into account the [H⁺] that results from the dissociation of water. However, unless the concentration of acid is very low, this is a negligible effect, as can be seen in the following calculation. The dissociation of water is described by

$$H_2O \rightleftharpoons H^+ + OH^-,$$

for which the dissociation constant $K'_a = [H^+][OH^-]/[H_2O] = 1.8 \times 10^{-16}$ mol l⁻¹. However, since the concentration of water is huge—namely, 55.56 M—it is virtually unchanged in any dissociation reaction; thus, [H₂O] is combined with the dissociation constant to give what is called the ion product of water, K'_w, or

$$K'_w = [H^+][OH^-] = 1.8 \times 10^{-16} \times 55.56 = 1.01 \times 10^{-14} \text{ mol l}^{-1}, \quad (5)$$

which is universally quoted as $K'_w = 10^{-14}$. Thus, in pure water $[H^+] = [OH^-] = \sqrt{10^{-14}} = 10^{-7}$ M, which is so small compared to the [H⁺] that results from dissociation of the acid that it is negligible. The value of [OH⁻] is strongly depressed by the presence of the acid because it must always obey Equation 5. Thus in the example just given of 0.05 M HAc, in which $[H^+] = 9.5 \times 10^{-4}$ M, the value of [OH⁻] is $10^{-14}/[H^+] = 10^{-14}/9.5 \times 10^{-4} = 1.05 \times 10^{-11}$ M. In general, as the initial concentration of acid increases, [H⁺] increases and [OH⁻] decreases.

A similar but slightly different analysis is carried through for a base, as we see in the following example. Ammonia accepts a proton from water according to the scheme

$$NH_3 + H^+ \rightleftharpoons NH_4^+,$$

in which NH_3 is the conjugate base and NH_4^+ is the conjugate acid. Since water directly participates in the reaction, the equilibrium is written more completely as

$$NH_3 + H_2O \rightleftharpoons NH_4^+ + OH^-.$$

The equilibrium constant for this reaction is

$$K' = \frac{[NH_4^+][OH^-]}{[NH_3][H_2O]}.$$

However, because [H₂O] is constant, once again [H₂O] = 55.56 M is commonly included in the value, and one uses the symbol K'_b, which is defined as

$$K'_b = \frac{[NH_4^+][OH^-]}{[NH_3]}, \quad (6)$$

which has the value 1.79×10^{-5} mol l⁻¹. Since NH_4^+ is the conjugate acid, the value of K'_a for dissociation of NH_4^+ is often given, instead of K'_b. The constants K'_a are K'_b are related in a simple way. The dissociation reaction for the conjugate acid is

$$NH_4^+ \rightleftharpoons NH_3 + H^+,$$

so that

$$K'_a = \frac{[NH_3][H^+]}{[NH_4^+]},\qquad(7)$$

which can be combined with Equation 6 to yield

$$K'_a = \frac{[OH^-][H^+]}{K'_b}.$$

From Equation 5 we have

$$K'_a K'_b = K'_w.\qquad(8)$$

This relation is true for any conjugate pair. The reader is cautioned to remember that the reaction from which K'_a is calculated is not simply the reverse of the reaction from which K'_b is calculated.

Let us now return to the calculation of $[OH^-]$ when 0.1 mole of NH_3 is added to 1 liter of water. Using c for the initial concentration, and x for $[OH^-]$, we obtain

$$K'_b = \frac{x \cdot x}{c - x} = x^2,$$

so that x can be calculated from a modified form of Equation 3, in which K'_b is substituted for K'_a in the expression. If $x \ll c$, we can also use the approximation

$$x \approx \sqrt{cK'_b},\qquad(9)$$

which is valid in this case, so that

$$[OH^-] = [NH_4^+] = \sqrt{(0.1)(1.79) \times 10^{-5}} = 0.0013\ M.$$

In treating acid-base equilibria, a p notation is useful and very common in biochemistry. The p value is defined as the negative logarithm to the base 10 of a concentration or an equilibrium constant. Thus,

$$pH = -\log[H^+];$$
$$pOH = -\log[OH^-];$$

and

$$pK'_a = -\log K'_a.$$

For pure water, $[H^+] = [OH^-] = 10^{-7}\ M$, so that $pH = 7$, $pOH = 7$, and $pK = 14$. Since $[H^+][OH^-] = K'_w$, then $-\log[H^+] - \log[OH^-] = -\log K'_w$, and

$$pH + pOH = pK'_w = 14.\qquad(10)$$

The p notation is also used in the approximation in Equation 4, which becomes

$$-\log[H^+] \approx -\frac{1}{2}(\log K'_a - \log c),$$

or

$$\text{pH} \approx \frac{1}{2}(\text{p}K_a' - \log c) . \tag{11}$$

Thus, in the example just given of 0.05 M HAc, in which $K_a' = 1.8 \times 10^{-5}$, the p$K_a'$ is 4.756 and the pH is $\frac{1}{2}(4.756 - \log 0.05) = 3.02$, which is the same as $-\log(9.5 \times 10^{-4})$.

An equation similar to Equation 11 can be derived for a base—namely,

$$\text{pOH} = \frac{1}{2}(\text{p}K_b' - \log c) . \tag{12}$$

This is more conveniently written in terms of pH and the dissociation constant of the conjugate acid because it is more common to list K_a' values than K_b' values. In this form Equation 12 is

$$\text{pH} = \frac{1}{2}(\text{p}K_w' + \text{p}K_a' + \log c_b) = \frac{1}{2}(14 + \text{p}K_a' + \log c_b) , \tag{13}$$

in which it is important to note that pK_a' refers to dissociation of the acid and c_b is the concentration of the base. Thus, the pH of 0.05 M NH$_4$OH at 25°C is calculated as follows. With p$K_a' = 9.4$ and $c_b = 0.05$ M, pH is equal to $\frac{1}{2}(14 + 9.4 - 1.3) = 11.05$.

In using Equations 11, 12, and 13, it must be remembered that the equations may be used only when the conditions are met that make the approximation in Equation 4 valid—namely, when dissociation is weak. Operationally, this means that p$K_a' > 4$ and that the concentration of the conjugate acid is greater than 0.01 M.

Before proceeding, it is important that one reconsider the statement made earlier about the use of concentrations instead of activities. It is true, as we have said, that at the low concentrations of acids and bases that are usually encountered, the activity coefficient is nearly one. However, in the real systems encountered in biochemical studies, we are not working with a pure solution of an acid or a base but a mixture of many components. Some of these components might interact or even react with one of the species produced by dissociation, and this could affect the equilibrium significantly. This may or may not occur, though. However, in biological systems it is almost always the case that salts are present that raise the ionic strength to about 0.1. In our discussions of the Debye-Hückel theory, we saw that ionic strength has a significant effect on activity. Indeed, at an ionic strength of 0.1, activity coefficients are often considerably less than 1. In an effort to avoid the necessity of using activities, biochemists instead make use of *modified* dissociation constants—that is, values of K_a' and pK_a' that have been measured in the presence of sufficient NaCl (and sometimes other salts) to have an ionic strength of 0.1. Such values can be found in published tables and should be used when appropriate. In Table 9-1, we list many values of K_a' for weak acids and bases and, in some cases, for high ionic strength. Many of these values will be useful for the examples and problems in this chapter.

SEC. 9-1 DISSOCIATING SYSTEMS

TABLE 9-1 Dissociation constants for weak acids and bases at 25°C, and also at an ionic strength = 0.1 (marked *).

Acid or base	K_a'	pK_a'
Acetic acid	1.75×10^{-5}	4.75
	$2.19 \times 10^{-5}*$	4.66*
Ammonia	5.75×10^{-10}	9.24
	$3.98 \times 10^{-10}*$	9.4*
Carbonic acid I	4.32×10^{-7}	6.37
II	5.75×10^{-11}	10.24
	$9.55 \times 10^{-11}*$	10.02
Citric acid I	7.44×10^{-4}	3.13
II	1.74×10^{-5}	4.76
III	4×10^{-7}	6.40
Formic acid	1.78×10^{-4}	3.75
Fumaric acid I	8.04×10^{-4}	3.1
II	2.51×10^{-5}	4.60
Imidazole	1.0×10^{-7}	6.99
	$8.51 \times 10^{-8}*$	7.07*
Lactic acid	1.32×10^{-4}	3.88
Maleic acid I	1.23×10^{-2}	1.91
II	2.01×10^{-6}	5.70
Methylamine	2.29×10^{-11}	10.64
Phosphoric acid I	7.59×10^{-3}	2.12
II	6.31×10^{-8}	7.20
	$1.4 \times 10^{-7}*$	6.86
III	3.98×10^{-13}	12.40
tris-Hydroxymethyl-aminomethane (Tris)	8.54×10^{-9}	8.07

The values of the pK_a' or of K_a' tell immediately the relative strength of two acids, in which, by "strength," is meant the *degree of dissociation*. This can be seen as follows. If α is the fraction of molecules that dissociates, or

$$\alpha = \frac{\text{moles of HA dissociated}}{\text{total moles of species containing A}},$$

or

$$\alpha = \frac{[A^-]}{[HA] + [A^-]}, \quad (14)$$

then for a monoprotic acid HA whose initial concentration is c, the concentrations at equilibrium are:*

---
*For a polyprotic acid, H_nA, $[H^+] \neq [A^-]$. When dissociation is 100 percent, $n[H^+] = [A^-]$. However, Equation 14 remains a good definition for α since α is in terms of $[A^-]$.

$$[H^+] = [A^-] = c\alpha \quad \text{and} \quad [HA] = c - c\alpha = c(1 - \alpha).$$

Thus

$$K_a' = \frac{[H^+][A^-]}{[HA]} = \frac{c\alpha^2}{1 - \alpha}. \tag{15}$$

Clearly, for any value of c, if α approaches 1—that is, if the acid is a strong acid—K_a' is very large and if α is very small, K_a' must be small. This yields the following rule:

if an acid is strong, K_a' is large and pK_a' is very small.

EXAMPLE 9–A Which acid is stronger, acetic acid or formic acid? What is the degree of dissociation of each at a concentration of 10^{-3} M?

From Table 9-1, the pK_a' of acetic acid is 4.75 and that of formic acid is 3.75. Therefore formic acid is stronger than acetic acid. This means that if both acids were at the same concentration, a greater fraction of the formic acid would be dissociated.

To calculate α, we rearrange Equation 15 to

$$\alpha = \frac{-K_a' + \sqrt{K_a'^2 + 4K_a'c}}{2c}.$$

Thus, for formic acid, $K_a' = \text{antilog}(-3.75) = 1.8 \times 10^{-4}$, and $\alpha = 0.34$; for acetic acid, $K_a' = 1.8 \times 10^{-5}$, and $\alpha = 0.12$, so that at 10^{-3} M the dissociation of formic acid is about three times as great as that of acetic acid.

As a rule of thumb, an acid can be considered weak if the pK_a' value is above 2 and relatively strong if less than 2. Few organic acids have a pK_a' less than 1 other than halogenated acids, such as dichloracetic acid.

Dissociation of Salts of Weak Acids or of Weak Bases

All salts dissociate to some extent in water, producing solutions having a wide range of pH. For example, $Ca(NO_3)_2$, a salt of a strong acid and a weak base, is acidic in solution—its pH is less than 7. On the other hand, Na acetate, a salt of a strong base and a weak acid, has a pH that is greater than seven. The resulting pH is a result of the relative strength of the acid and of the base and it is easily calculated, as we show in the following analysis.

When NaAc is dissolved in water, it dissociates completely into Na^+ and Ac^- ions. However, the Ac^- ions will not all remain free because they must also satisfy the equilibrium with the H^+ ions of water defined by the dissociation equation for HAc. As the Ac^- ions combine with the H^+ ions,

SEC. 9–1   DISSOCIATING SYSTEMS

more water dissociates. The total reaction may be written

$$Ac^- + H_2O \rightleftharpoons HAc + OH^-.$$

The equilibrium constant, which in this case is called the *hydrolysis constant*, K_h', is

$$K_h' = \frac{[HAc][OH^-]}{[Ac^-]} = \frac{[HAc][OH^-][H^+]}{[Ac^-][H^+]} = \frac{K_w'}{K_a'}. \tag{16}$$

Note that according to Equation 8, $K_h' = K_b'$, so that we can easily calculate K_b', since $K_a' = 1.8 \times 10^{-5}$ and $K_w' = 10^{-14}$—that is, $K_h' = 10^{-14}/(1.8 \times 10^{-5}) = 5.5 \times 10^{-10}$. If c is the initial concentration of NaAc and $x = [HAc] = [OH^-]$, then $5.5 \times 10^{-10} = x^2/(c - x)$. If $c = 0.1\,M$, then $x = 7.4 \times 10^{-6} = [OH^-]$, and $[H^+] = 10^{-14}/(7.4 \times 10^{-6}) = 1.35 \times 10^{-9}$; the pH = 8.87, and the solution is basic.

It is easy to see that as the acid becomes weaker, the pH of the salt solution increases. That is, as K_a' becomes smaller, K_h' becomes larger and $x = [OH^-]$ increases; thus $[H^+]$ decreases and pH = $-\log[H^+]$ rises.

The pH of the salt solution can also be calculated from Equation 13, in which pH = $\frac{1}{2}(14 + pK_a' + \log c_b)$, and in which pK_a' refers to the acid and c_b is the initial concentration of the salt. This is possible because the Ac^- ion is the proton acceptor and thus the base in the Brønsted-Lowry sense. Thus, in the calculation just presented, in which $c_b = 0.1\,M$ and $K_a' = 1.8 \times 10^{-5}$ or $pK_a' = 4.25$, the pH = $\frac{1}{2}(14 + 4.75 + \log(0.1)) = 8.87$, as just calculated.

For completeness we will also go through the calculation for the salt of a weak base and a strong acid, for example, NH_4Cl. The NH_4^+ ion partly dissociates to NH_3 and H^+, and the dissociation constant is

$$K_a' = \frac{[NH_3][H^+]}{[NH_4^+]}.$$

Since $[NH_3] = [H^+]$ and the $[H^+]$ produced by water itself is small,

$$K_a' = \frac{[H^+]^2}{[NH_4^+]}.$$

If c is the initial concentration and $x = [H^+]$, we obtain again $K_a' = x^2/c - x$, as in the case of the salt of a strong base and weak acid. Also, if $x \ll c$, one obtains

$$K_a' = \frac{x^2}{c} = \frac{[H^+]^2}{c},$$

or

$$pH = \frac{1}{2}(pK_a' - \log c). \tag{17}$$

It is important to remember to use pK_a' *for the conjugate acid of the weak base and not* pK_b' *for the base.* For the NH_4^+ ion, pK_a' is 9.4, so that for a 0.1-M solution of NH_4Cl, the pH is $\frac{1}{2}(9.4 - \log 0.1) = 5.2$.

9-2 ACID-BASE MIXTURES

Titration Curves

If small aliquots of a strong base such as NaOH are added to a solution of weak acid, [H$^+$] decreases and hence the pH rises. This is because the OH$^-$ ions combine with the H$^+$ ions to form water. To maintain equilibrium the acid responds to this decrease of [H$^+$] by releasing more protons. One can think of the process as one in which OH$^-$ ions pull off protons from the undissociated acid; the concentration of OH$^-$ needed to pull off a particular number of protons depends upon the strength of binding of a proton in the molecule. Thus a weak acid, which dissociates poorly because it holds its protons more tightly than a stronger acid, requires a higher [OH$^-$] before it starts releasing protons. When proton-release occurs, the OH$^-$ ions are neutralized; thus the pH remains relatively constant. Ultimately, as [OH$^-$] continues to increase, the acid concentration becomes so low that it cannot contribute enough protons to neutralize the OH$^-$ ions and the pH begins to rise. A graph that describes this process—that is, a plot of pH versus molar equivalents of added OH$^-$ is called a *titration curve*. Several curves for different weak acids are shown in Figure 9-1. All of these titration curves have nearly the same shape but are displaced to higher pH values with increasing pK_a'. At the midpoint of each titration curve, which is both the inflection point of the curve and the point at which 0.5 equivalents of OH$^-$ have been added to 1.0 equivalents of acid, the pH is equal to the pK_a'. Why this is so is shown in the following section.

The Henderson-Hasselbalch Equation

So far, in our analysis of the dissociation of an acid HA, it has always been true that [H$^+$] = [A$^-$]. However, when OH$^-$ are added, this is not the case; in fact [A$^-$] > [H$^+$]. Furthermore, when 0.5 equivalents of OH$^-$ is added, half of the acid is neutralized and [HA] = [A$^-$]. If K_a' = [H$^+$][A$^-$]/[HA] is rearranged as

$$[H^+] = K_a' \times \frac{[HA]}{[A^-]},$$

by taking logarithms of both sides and using the p notation, we obtain

$$\boxed{pH = pK_a' + \log\frac{[A^-]}{[HA]}}, \qquad (18)$$

which is an important equation known as the Henderson-Hasselbalch equation. At the midpoint of the titration curve, [A$^-$]/[HA] = 1, so that pH = pK_a'.

The Henderson-Hasselbalch equation may also be written in the more general form

$$\text{pH} = \text{p}K_a' + \log\frac{[\text{proton acceptor}]}{[\text{proton donor}]}. \tag{19}$$

Equation 18 is important for three reasons:

1. It indicates that the $\text{p}K_a'$ of any acid can be measured from its titration curve. This is in fact the most common way to determine $\text{p}K_a'$.
2. Once the $\text{p}K_a'$ is known, the ratio $[\text{A}^-]/[\text{HA}]$ can be calculated for any pH.
3. If K_a' is known, the pH of a solution of an acid can be set to a chosen value merely by adjusting $[\text{A}^-]$, for example, by adding more $[\text{A}^-]$ in the form of a salt. This point will be examined more carefully later.

A similar analysis of a weak base titrated with acid leads to the equation

$$\text{pH} = \text{p}K_a' + \log\frac{[\text{B}]}{[\text{BH}^+]}, \tag{20}$$

in which $\text{p}K_a' = 14 - \text{p}K_b'$ and the $\text{p}K_a'$ refers to the conjugate acid of the base.

Buffers

We now turn our attention to the region of each curve in Figure 9-1 in which the pH does not vary much as OH^- is added. In this region the solution is said to be *buffered*. Buffers are extremely important to living systems and in laboratory science because they enable a system to withstand a change in the concentration of an acid or a base. This is essential because the rates of all enzymatic and many other reactions are dependent upon pH. Inspection of the curves in Figure 9-1 shows that buffering occurs principally when the pH is within one pH unit of the $\text{p}K_a'$.

Let us now see how a buffer can be prepared. The most obvious procedure is to add 0.5 equivalents of base to 1.0 equivalent of an acid. However, Equation 18 indicates that the same thing can be accomplished by adding equimolar amounts of a monoprotic acid and its salt. Thus, both of the mixtures of $0.1\,M$ HAc + $0.1\,M$ NaAc and $0.01\,M$ HAc + $0.01\,M$ NaAc have a pH of 4.75 (the value of $\text{p}K_a'$ for HAc). (In a moment we will discuss the difference between these two mixtures.) According to the Henderson-Hasselbach equation, a solution can be prepared at any pH by choosing an appropriate ratio of $[\text{A}^-]$ to $[\text{HA}]$. However, the solution will not always be a buffer—only if the pH is $\text{p}K_a' \pm 1$ will there be buffering, as we said before. Thus to prepare a buffer having a pH of 7, a NaAc-HAc mixture would be inappropriate because, as can be seen in Figure 9-1, addition of a small amount of base produces a large change in pH at this pH. To prepare a buffer whose pH is 7 one needs an acid whose $\text{p}K_a'$ is nearly 7.

FIGURE 9-1 Titration curves for three weak acids. The pK_a' values are the intercepts of the dashed lines.

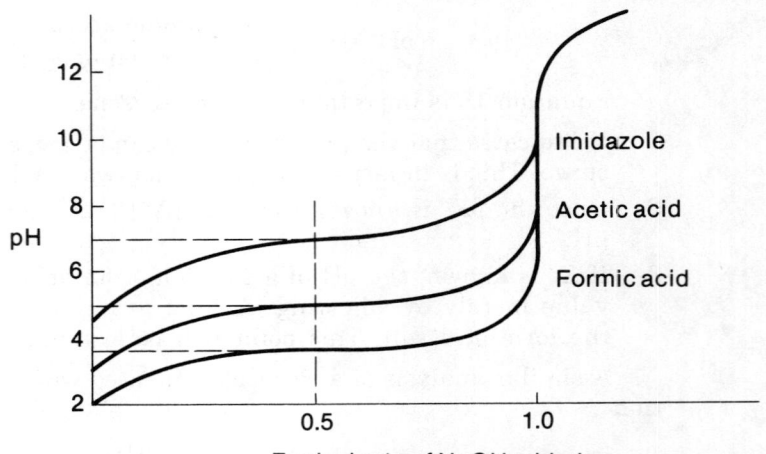

In selecting a buffer for a particular experimental arrangement, it is important not only to choose a combination that gives the right pH but also to choose the buffer concentration that would enable it to withstand the addition of a particular amount of acid or base. Clearly, if a buffer were at the pK_a' value but its concentration were 0.01 M, it could not maintain its pH if 0.05 M acid were added. This introduces the idea of the *buffering capacity* (also called the *buffer index*). This is defined as the molar amount of acid or base that must be added per liter of buffer to produce a change of one pH unit. For addition of base to acid, the buffer index $\mathscr{B}$ is

$$\mathscr{B} = \frac{d[\text{OH}^-]}{d(\text{pH})} ; \tag{21}$$

that is, the derivative of [OH$^-$] with respect to pH. An expression for $\mathscr{B}$ can be calculated as follows.

Let c_S and c_A be the initial concentrations of the salt and the acid, respectively. Addition of acid or base will change the values of [A$^-$] and [HA] but it will always be true (as long as the addition does not change the volume of the solution significantly) that

$$[\text{A}^-] + [\text{HA}] = c_S + c_A. \tag{22}$$

Combining this with $K_a' = [\text{H}^+][\text{A}^-]/[\text{HA}]$ yields

$$[\text{A}^-] = \frac{(c_A + c_S)K_a'}{[\text{H}^+] + K_a'} . \tag{23}$$

The buffering capacity, $\mathscr{B}$, is then

$$\mathscr{B} = \frac{d[\text{OH}^-]}{d(\text{pH})} = \left(\frac{d[\text{A}^-]}{d[\text{H}^+]}\right)\left(\frac{d[\text{H}^+]}{d(\text{pH})}\right), \tag{24}$$

SEC. 9–2   ACID-BASE MIXTURES

in which $d[A^-]/d[H^+]$ is the derivative of $[A^-]$ with respect to $[H^+]$. From Equation 23,

$$\frac{d[A^-]}{d[H^+]} = -\frac{(c_A + c_S)K'_a}{([H^+] + K'_a)^2}, \qquad (25)$$

and, because $d(\ln x) = (1/x)dx$ and $-2.303 d(-\log x) = (1/x)dx$, then

$$\frac{d[H^+]}{d(\text{pH})} = -2.303[H^+]. \qquad (26)$$

Combining Equations 23, 25, and 26 yields

$$\mathcal{B} = \frac{2.303(c_A + c_S)K'_a[H^+]}{([H^+] + K'_a)^2}. \qquad (27)$$

Thus, once having selected the desired pH, in other words, the value of $[H^+]$, $c_A + c_S$ can be calculated on the basis of how much of a pH-change can be tolerated if a particular amount of base is to be added.

EXAMPLE 9–B What concentration of an NaAc-HAc buffer will allow a maximal change of 0.1 pH unit from a pH of 4 if 0.01 M NaOH is to be added?

The desired buffering capacity is $0.01/0.1 = 0.1$. The pH of the buffer is 4, so that $[H^+] = 10^{-4}$ M. The pK'_a is 4.75, so that $K'_a = 1.78 \times 10^{-5}$ M. Therefore, from Equation 27,

$$0.1 = \frac{2.303(c_A + c_S)(1.78 \times 10^{-5})(10^{-4})}{(10^{-4} + 1.78 \times 10^{-5})^2},$$

so that $c_A + c_S = 0.34$ M. From Equation 18, $4.00 = 4.75 + \log(c_S/c_A)$, so that $c_S/c_A = 0.18$. Thus $c_S = 0.05$ M and $c_A = 0.29$ M.

The maximal buffering capacity $\mathcal{B}_{max}$ for a particular buffer system can also be calculated. This occurs when the pH equals the pK'_a, or when $[H^+] = K'_a$. Substituting into Equation 27 yields the maximal buffering capacity for a particular concentration $c = c_A + c_S$, as

$$\mathcal{B}_{max} = \frac{2.303 c K'_a [H^+]}{(K'_a + [H^+])^2} = 0.576c.$$

It is informative to see a plot of $\mathcal{B}$ as a function of pH and the relation between this curve and a titration curve. This is shown in Figure 9-2.

In some experiments the total ionic strength of a buffer is of greater importance than the buffering capacity, because many biochemical reactions are very dependent on ionic strength. The ionic strength can of course always be increased by addition of a salt that is not a component of the buffer system—for example, by adding KCl to a NaAc-HAc buffer. If the

FIGURE 9-2 Relation between a titration curve and the buffering capacity.

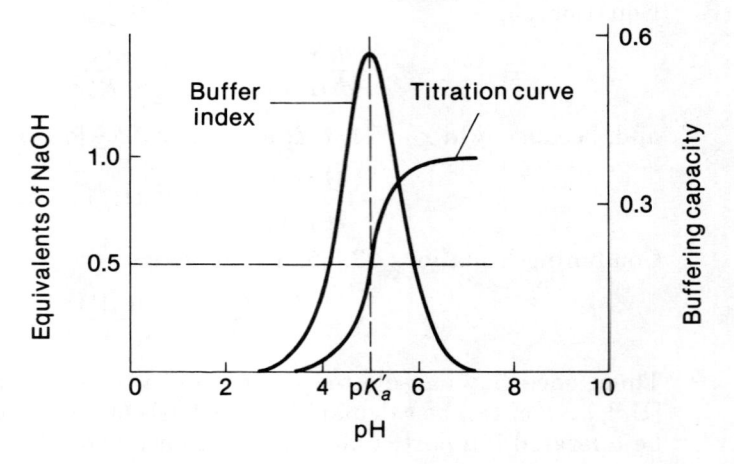

ionic strength must be kept low, however, it is desirable to be able to prepare a buffer having a particular ionic strength and pH. We now show how this is done—again with the NaAc-HAc system.

EXAMPLE 9–C Preparation of a NaAc-HAc buffer at pH 4.5 and with ionic strength of 0.05.

The expression for the ionic strength I is $\frac{1}{2}\sum z_i^2 c_i$ in which z_i and c_i are the charges and concentrations of each species. The species that are present are Na^+, Ac^-, OH^-, and H^+ ions. Since the acid dissociates only slightly, $[Ac^-]$ is just the concentration of acetate that is added, and $[Na^+] = [Ac^-]$. At an ionic strength of 0.05, the concentrations of the components will be near that value, so that $[H^+]$, which, for a pH of 4.5, is 3.2×10^{-5} M, and $[OH^-]$, which is 3.1×10^{-10} M, can be ignored. Thus,

$$I = \tfrac{1}{2}([Na^+] \cdot 1^2 + [Ac^-] \cdot 1^2) = \tfrac{1}{2} \cdot 2\, c_S = c_S.$$

Thus, since I is to be 0.05, $c_S = 0.05$. The value of c_A is calculated from Equation 18; that is, $4.5 = 4.8 + \log(c_S/c_A)$, so that $c_A = 0.1$ M. Note that the ionic strength is determined almost entirely by the salt concentration because a poorly dissociating acid does not contribute many ions to the solution. This is the case except when the desired ionic strength is very low (in which case the buffering capacity is usually so low that there is no buffering), and if the pK_a' is less than about 2.5, because a strong acid does provide the solution with a significant number of ions.

In the examples given so far, we have used Equation 18 to calculate the pH when a salt is added to the acid. Of equal interest is the result of

SEC. 9–2 ACID-BASE MIXTURES

adding base to an acid. We go through the appropriate calculation in the following example.

EXAMPLE 9–D What is the pH resulting from adding 0.05 mole of solid NaOH to 500 ml of 0.3 M HAc?

Equation 18 requires that we obtain the ratio [NaAc]/[HAc]. Initially there is no NaAc; however, it is formed by adding NaOH and the amount formed is easily calculated. The amount of acid present initially is 0.15 mole. Since the amount of added NaOH, 0.05 mole, is less than the amount of HAc, the amount of NaAc produced is 0.05 mole. The amount of HAc remaining is 0.15 − 0.05 = 0.10 mole.

Thus, converting to concentrations, [NaAc] = 0.10 M and [HAc] = 0.20 M, so that the

$$\mathrm{pH} = \mathrm{p}K'_a + \log\frac{[\mathrm{NaAc}]}{[\mathrm{HAc}]} = 4.75 + \log\frac{0.1}{0.2} = 4.45 \ .$$

Calculations such as that in the previous example can sometimes be facilitated by writing Equation 18 in terms of the degree of dissociation, α, which, from Equation 14, can be rearranged as

$$\frac{[\mathrm{A}^-]}{[\mathrm{HA}]} = \frac{\alpha}{1-\alpha}, \tag{28}$$

so that Equation 18 becomes

$$\mathrm{pH} = \mathrm{p}K'_a + \log\frac{\alpha}{1-\alpha}. \tag{29}$$

This equation can also be used in Example 9–D. The number of moles of HA dissociated is the number of moles of NaOH that is added, or 0.05. Neutralization does not affect the number of moles of the species containing A, so this is just the initial value of [HA], or 0.15. Thus, $\alpha = .05/.15 = \tfrac{1}{3}$ and $\alpha/(1-\alpha) = \tfrac{1}{2}$ and the pH = 4.75 + log $\tfrac{1}{2}$ = 4.45, in agreement with the previous calculation.

The calculation of the pH resulting from the addition of a strong base to a weak acid often causes confusion, because one must be careful to remember to use the appropriate values for the conjugate acid in Equation 18. How this is done can be seen in the following example.

EXAMPLE 9–E What is the pH resulting from the mixing of 100 ml of 0.2 M NH₄OH and 100 ml of 0.03 M HCl. The value of K'_b for the reaction NH$_4^+$ + OH$^-$ ⇌ NH₃ + H₂O is 1.8 × 10^{-5}.

To begin with, we use Equation 8 to calculate $K'_a = K'_w/K'_b = 10^{-14}/1.8 \times 10^{-5} = 5.5 \times 10^{-10}$; thus, p$K'_a$ = 9.26. The acid in this reaction is the NH$_4^+$ ion. It is poorly ionized in the NH₄OH but the NH₄Cl produced by neutralization is totally ionized. Thus [NH$_4^+$] can be taken to be equal to [HCl] or

$\{100/(100 + 100)\}(0.03) = 0.015\ M$. The salt is the proton acceptor, or NH_3, whose concentration is $\{100/(100 + 100)\}(0.2) - \{100/(100 + 100)\}(0.3) = 0.085\ M$. Thus $pH = 9.26 + \log(0.085/0.015) = 9.26 + 0.75 = 10.01$.

Polyprotic Acids

Many acids contain more than one H^+ ion and, almost always, each H^+ ion has a different pK_a'. Such acids are called *polyprotic acids*. The notation used in discussing a polyprotic acid is a Roman numeral subscript for each successive dissociation step. For example, in a diprotic acid, H_2A, the pK_a' values for dissociation of the first and second protons are designated pK_I' and pK_{II}', respectively. The example that we will discuss is phosphoric acid, H_3PO_4, which dissociates in three steps:

$$H_3PO_4 \rightarrow H_2PO_4^- + H^+, \qquad pK_I' = 2.12;$$
$$H_2PO_4^- \rightarrow HPO_4^{2-} + H^+, \qquad pK_{II}' = 7.2;$$

and

$$HPO_4^{2-} \rightarrow PO_4^{3-} + H^+, \qquad pK_{III}' = 12.4.$$

Phosphoric acid has a titration curve consisting of several steps (Figure 9-3). The plateau value for each step corresponds to the pK_a' for that particular dissociation reaction.

FIGURE 9-3 Titration curve for H_3PO_4, a triprotic acid. The intersection points of the dashed lines define the regions of the curve from which the pK_a' values are evaluated.

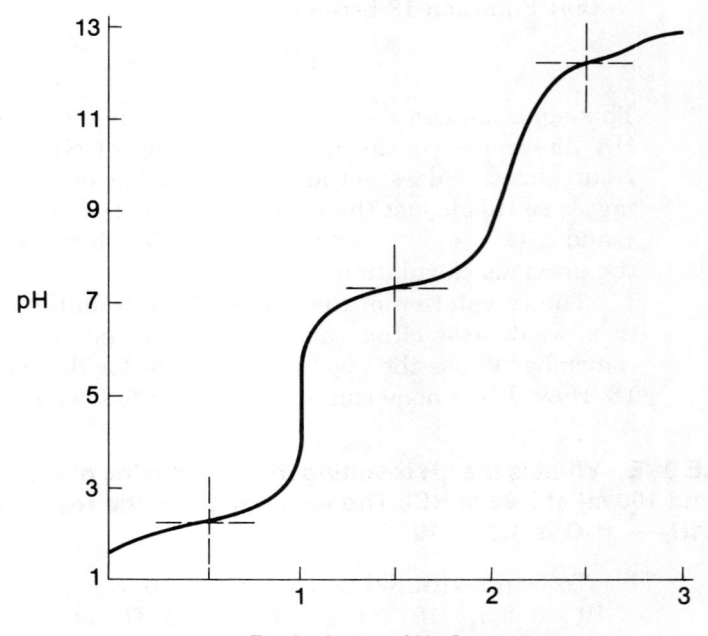

SEC. 9-2 ACID-BASE MIXTURES

At any pH value all of the dissociated species are present. However, at very low pH, $[PO_4^{3-}]$ is very low (and negligible) and $[H_2PO_4^-]$ is high; at high pH, $[PO_4^{3-}]$ predominates and $[H_2PO_4^-]$ is negligible. In general, at any particular pH, two of the three ions predominate and the third is present only in an insignificant concentration. The pK_a' values indicate those that predominate. For instance, from the Henderson-Hasselbalch equation, we have

$$\log \frac{[H_2PO_4^-]}{[H_3PO_4]} = pH - 2.12,$$

$$\log \frac{[HPO_4^{2-}]}{[H_2PO_4^-]} = pH - 7.2,$$

and

$$\log \frac{[PO_4^{3-}]}{[HPO_4^{2-}]} = pH - 12.4,$$

so for a pH at or near each pK_a' only the two ions in one of the equations are present at high concentration. Figure 9-4 shows the fraction of each species present at various pH values. This type of concentration distribution is typical for polyprotic acids. Because there are several pK_a' values, a polyprotic acid system can buffer in several different ranges of pH. For example, an equimolar mixture of NaH_2PO_4 and Na_2HPO_4 will yield a buffer whose pH is 7.2, whereas an equimolar mixture of Na_3PO_4 and Na_2HPO_4 will buffer at pH 12.3. (This high-pH buffer is a common alkaline buffer in nucleic acid research.)

Polyprotic acids may be subdivided into two classes—those in which successive ionization steps are dependent on one another and those in

FIGURE 9-4 The fraction of phosphoric acid that is present as each of the four species, as a function of pH.

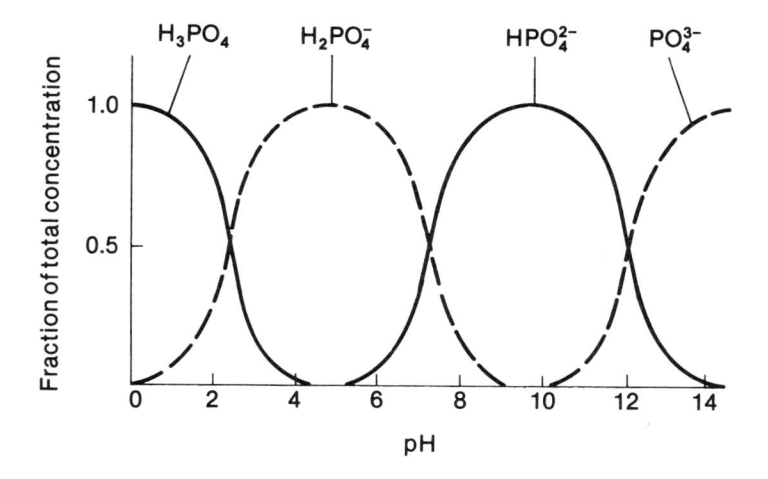

which they are independent. The distinction is based upon whether or not the protons in the undissociated state are attracted by the same charge or nearby charges. For instance, in the case of H_3PO_4, each proton is associated with a negatively charged oxygen, each of which is attached to a single phosphorous atom (Figure 9-5). Thus, if one proton leaves the molecule, a second proton, in order to escape, must overcome the attractive force of both the unassociated oxygen, which is no longer neutralized, and the oxygen to which it is associated. This means that the second proton has considerably greater difficulty leaving than the first one did or, in more quantitative terms, K'_{II} is much smaller than K'_{I}. A similar argument applies to K'_{III}. This is clearly the case for the phosphate system, in which K'_{I}, K'_{II}, and K'_{III} are 7.5×10^{-3}, 6.3×10^{-8}, and 5.01×10^{-13}, respectively, and for the carbonate systems, in which $K'_{I} = 4.5 \times 10^{-7}$ and K'_{II} 5.6×10^{-11}. An example of a system in which the ionizations are independent is a polycarboxylic acid such as citric acid (Figure 9-5), in which K'_{I}, K'_{II} and K'_{III} are 8×10^{-4}, 4×10^{-5}, and 3×10^{-6}. The most significant difference between these two classes of polyprotic acids is the number of charged species present at a particular concentration. In the case of a dependent system such as phosphate, there are never more than two anions present in appreciable quantities. This is because, in the sequence of anions that become present as the concentration decreases, such as $H_2PO_4^-$, HPO_4^{2-}, and PO_4^{3-}, and PO_4^{3-} would not begin to appear in any significant way until all of the $H_2PO_4^-$ had dissociated to form HPO_4^{2-}, as is seen in Figure 9-4. This is not true of the independent class of polyprotic acids, in which all ionized forms may be present at the same time. The case of independent ionizations is quite complex to treat mathematically and will not be discussed in this book; the appropriate analysis can be found in references given at the end of the chapter. Instead, we shall next examine some of the simpler calculations for the dependent case.

As we have just stated, if the dissociation constants are sufficiently different, only the components that constitute the equilibrium for a single

FIGURE 9-5 Structure of three acids. The asterisk denotes the ionizable proton.

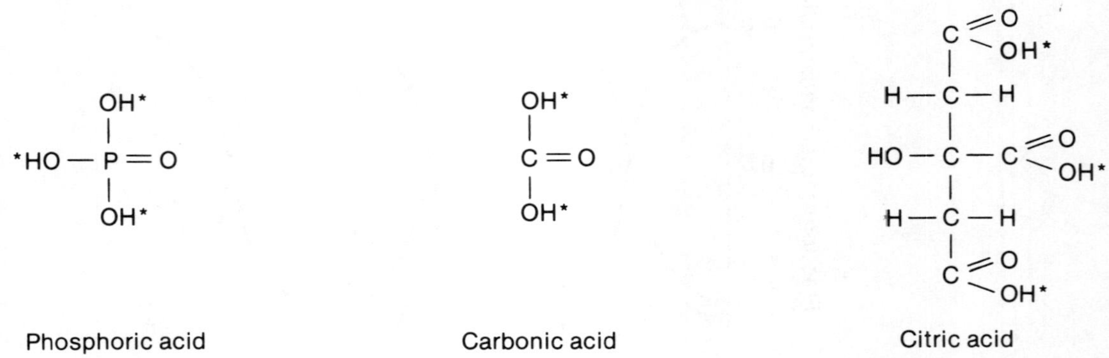

Phosphoric acid

Carbonic acid

Citric acid

SEC. 9-2 ACID-BASE MIXTURES

dissociation step are present. Therefore, if the undissociated acid, such as H_3PO_4, is put in water, the pH is determined entirely by the first dissociation step. Similarly, if Na_3PO_4 is put in water, the pH is determined only by the final step.

EXAMPLE 9-F What is the pH of 0.1 M H_3PO_4?

The first step in the dissociation is $H_3PO_4 \rightleftarrows H_2PO_4^{2-} + H^+$, for which K_I' is 7.5×10^{-3}. Let $x = [H_2PO_4^{2-}] = [H^+]$; then $7.5 \times 10^{-3} = x^2/(0.1 - x)$ or $x = 0.024$, and the pH of 0.1 M H_3PO_4 is 1.62.

The calculation of the pH of a solution of NaH_2PO_4 is slightly more involved, because the $H_2PO_4^-$ ion can both accept a proton and donate a proton to water. This is always true of the intermediates in the dissociation of a polyprotic acid (for example, the statement is also true for the HCO_3^- ion). Such a substance has both acid and basic properties and is called *amphoteric*. An amphoteric substance X must always satisfy two equilibria, namely,

$$XH \rightleftarrows X^- + H^+ \quad \text{and} \quad XH + H_2O \rightleftarrows XH_2^+ + OH^-.$$

The equilibrium constant for the first reaction is simply $K_{a,I}' = [X^-][H^+]/[XH]$, but for the second reaction it is the basic constant $K_{b,II}'$, which can be written in terms of K_w' as follows:

$$K_{b,II}' = \frac{[XH_2^+][OH^-]}{[XH]} = \frac{[XH_2^+][OH^-][H^+]}{[XH][H^+]} = \frac{[XH_2^+]K_w'}{[XH][H^+]}. \tag{30}$$

We now invoke the principle of electroneutrality—namely, $[H^+] + [XH_2^+] = [OH^-] + [X^-]$, and combine this with the above equation to yield

$$[H^+] = \left(\frac{K_w' + K_{a,I}'[XH]}{1 + K_{b,II}'[XH]/K_w'} \right)^{1/2} \tag{31}$$

The use of this equation is illustrated by the following example.

EXAMPLE 9-G What is the pH of 0.1 M KH_2PO_4?

The equivalent of XH is HPO_4^{2-}, for which $K_{a,I}' = 6.3 \times 10^{-8}$. The equivalent of XH_2^+ is $H_2PO_4^-$, whose acid-dissociation constant is 7.5×10^{-3}. The equation calls for $K_{b,II}'$. However, $K_{b,II}'/K_w'$ is the reciprocal of the acid-dissociation constant, $K_{a,II}'$. Thus

$$[H^+] = \left[\frac{10^{-14} + (6.3 \times 10^{-8})(0.1)}{1 + (0.1/7.5 \times 10^{-3})} \right]^{1/2} = 2.1 \times 10^{-5} \, M,$$

and the pH is 4.68.

For many acids and for the concentrations usually encountered, $K'_w \ll K'_{a,I}[XH]$ and $K'_w \ll K'_{b,II}[XH]$, so that Equation 31 simplifies to

$$[H^+] = \left(\frac{K'_{a,I}[XH]K'_w}{K'_{b,II}[XH]}\right)^{1/2} = \left(\frac{K'_{a,I}K'_w}{K'_{b,II}}\right)^{1/2}. \tag{32}$$

This equation can be used in Example 9–G, in which case $10^{-14} \ll 0.1 \times 6.3 \times 10^{-8}$, and $10^{-14} \ll 0.1 \times 7.5 \times 10^{-3}$, so that

$$[H^+] = [(6.3 \times 10^{-8})(7.5 \times 10^{-3})]^{1/2},$$

which again equals 2.1×10^{-5} M. It should be noted that when $K'_w \ll K'[XH]$, the pH is independent of the concentration of XH. This can be seen for 0.5 M KH_2PO_4 by performing the calculation without the approximation. This yields pH 4.67, which is nearly identical to the value (4.68) just calculated. However, as the concentration decreases, the approximation becomes less valid. For example, the precise calculation yields a pH of 4.78 for 0.01 M KH_2PO_4 (actually not much of a difference) and 5.12 for 0.001 M KH_2PO_4.

Equation 32 can also be written

$$pH = \frac{1}{2}(pK'_I + pK'_{II}), \tag{33}$$

in which K'_I and K'_{II} are the acid-dissociation constants for the two dissociation reactions. It is instructive to derive this relation in a slightly different way. We note that for a diprotic acid, the ion HA^- must satisfy the equilibria

$$H_2A \rightleftarrows H^+ + HA^- \quad \text{and} \quad HA^- \rightleftarrows H^+ + A_2^-,$$

and $[HA^-]$ must satisfy the two equations

$$[HA^-] = \frac{[H_2A]K'_I}{[H^+]} \quad \text{and} \quad [HA^-] = \frac{[H^+][A^{2-}]}{K'_{II}},$$

so that

$$[H^+]^2 = \frac{K'_I K'_{II}[H_2A]}{[A^{2-}]},$$

or

$$pH = \frac{1}{2}(pK'_I + pK'_{II}) + \frac{1}{2}\log\frac{[A^{2-}]}{[H_2A]}, \tag{34}$$

which is related to the Henderson-Hasselbach equation. An amphoteric substance in solution without any other added ions has equal probability of picking up or releasing a proton. Thus, if the pK'_a values are far enough away from that of water, and if the amount of added HA^- is high enough that it is large compared to the number of OH^- and H^+ ions produced by dissociation of water, the concentrations $[A^{2-}]$ and $[H_2A]$ will be approximately equal. Hence for an amphoteric substance, Equation 34 simplifies to Equation 33. Since $pK_I' = 2.12$ and $pK_{II}' = 7.20$ for the two steps in which $H_2PO_4^-$ is involved, the pH of the 0.1-M-KH_2PO_4 solution discussed in Example 9–G is $\frac{1}{2}(2.12 + 7.20) = 4.66$, which agrees with the values calculated from the more general equation.

Equation 33 also indicates that, just as for a monoprotic acid, a solution of a known pH can be prepared by adding the appropriate amount of $[A^{2-}]$ and $[H_2A]$. For example, for H_3PO_4, the two components could be Na_3PO_4 and Na_2HPO_4. However, if $[PO_4^{3-}]$ equals $[H_2PO_4^-]$—that is, if the chosen pH is midway between the two pK' values—there will be no buffering. This can be seen by referring to the titration curve in Figure 9-3: note that at this midvalue the pH is changing most rapidly with a small change in the amount of added acid or base. This is just opposite from the situation with a monoprotic acid in which the buffering capacity is maximal for equimolar concentrations of acid and salt.

Useful Buffer Systems

Buffers are important in research as well as in maintaining intracellular pH in living cells. The two most important buffers in living cells are the bicarbonate system (H_2CO_3, HCO_3^-), which buffers blood, and the phosphate system ($H_2PO_4^-$, HPO_4^{2-}), which buffers muscle and other cell types. We have already discussed the phosphate system and will describe the interesting properties of the bicarbonate buffer in the following section. In this section we concentrate on buffers that are useful in the laboratory.

In any laboratory experiment a buffer is necessary because all biological systems are sensitive to changes in pH. Enzymes are active only in a relatively narrow range of pH (usually about three pH units) and their activity can change 100-fold or more when the pH varies by only one pH unit from the pH associated with maximal activity. Also, the shape of most biological macromolecules changes slightly, but significantly, with small changes in pH; complete loss of structure frequently occurs if the pH drops below 2 or goes above 10.5.

The initial requirement that a useful buffer must meet is good buffering capacity. This of course means that the buffer should only be used near its pK_a'. Second, it should not require a very high ionic strength to achieve good buffering capacity; this once again means working very near the pK_a'. Finally, it should be nonreactive and noninhibitory to biochemical reactions. These are stringent requirements and impose many limitations. For instance, many enzymes are inhibited by phosphate ions. Furthermore, substances such as citrate and phosphate, which precipitate many divalent cations, cannot be used when studying the innumerable enzymatic reactions that require Mg^{2+}, Ca^{2+}, or other divalent cations. The most common buffer used in biochemical research today is *tris*-hydroxymethylaminomethane, $H_2NC(CH_2OH)_3$; this buffer is also called simply Tris buffer. This has a pK_a' of 8.1, although it is widely used at pH 7. Why it is so popular is unclear, because Tris buffer has four problems: (1) the pH varies more with temperatures than do other buffers; (2) the pH depends on the buffer concentration, so that if a buffer is prepared at a particular pH and concentration, the pH will decrease by about 0.1 pH unit per tenfold dilution; (3) Tris reacts with certain electrodes used in pH meters, resulting in an incorrect measurement of pH; and (4) many biological systems react with the hydroxyl and primary amine groups (such as chloroplasts, several of whose biochemical reactions fail to occur in Tris buffer). Nonetheless, for reasons that are not obvious, Tris remains in widespread use, a situation that should gradually change as laboratory workers become aware of the great variety of buffers that are available. Some of the other commonly used buffers and their useful range of pH are listed in Table 9-2.

TABLE 9-2 Useful buffers and their applicable range of pH.

Buffer	pH
Glycine-glycine HCl	1.5–3.5
Citric acid-NaOH	2.5–5.0
Acetate-acetic acid	3.8–5.8
NaH_2PO_4-Na_2HPO_4	6.0–8.0
Tris	7.0–8.5
Borax-boric acid	7.0–9.0
Borax-NaOH	9.0–11.0
Na_2HPO_4-Na_3PO_4	11.0–12.0

An interesting set of pH 6–8 buffers known as Good buffers (they *are* good, but actually they are named after N.E. Good, who developed them) are particularly nonreactive and noninhibitory to most enzymes. They are gradually coming into wider use. Their structures, names (several of which are acronyms), and pK_a' values are listed in Table 9-3.

Bicarbonate Buffer: A Biologically Important Polyprotic System

The major intracellular buffer is HPO_4^{2-}; however, in blood and in the interstitial fluid of all vertebrates, bicarbonate is the prevalent buffer. This buffer has several distinctive features, certainly the most remarkable being that the pK_a' of H_2CO_3 is 3.8, yet the (H_2CO_3, HCO_3^-)-system buffers effectively at pH 7.*

*In standard tables and earlier in this chapter, a pK_a' value of 6.4 for H_2CO_3 is quoted. Although this is the proper value to use in calculations, we will see that this is actually a composite figure, and that the true $pK_a' = 3.8$.

SEC. 9–2　ACID-BASE MIXTURES

TABLE 9-3　Some of the Good buffers.

Structure	Designation	pK_a'
O(CH$_2$CH$_2$)$_2$N$^+$HCH$_2$CH$_2$SO$_3^-$	MES	6.15
NaO$_3$SCH$_2$CH$_2$N(CH$_2$CH$_2$)$_2$N$^+$HCH$_2$CH$_2$SO$_3^-$	PIPES	6.8
O( )N$^+$HCH$_2$CH$_2$CH$_2$SO$_3^-$	MOPS	7.01
(HOCH$_2$)$_3$CN$^+$H$_2$CH$_2$CH$_2$SO$_3^-$	TES	7.5
HOCH$_2$CH$_2$N$^+$H(CH$_2$CH$_2$)$_2$NCH$_2$CH$_2$SO$_3^-$	HEPES	7.55
(HOCH$_2$CH$_2$)$_2$N$^+$HCH$_2$COO$^-$	Bicine	8.35

Let us see why this is so. CO_2 is present in the atmosphere and is slightly soluble in water; furthermore, H_2CO_3 is in equilibrium with the dissolved CO_2. Thus the total concentration of H_2CO_3 is determined by the partial pressure of the CO_2 in the gas phase. In order to maintain a stable pH of 7 with a pK_a' of 3.8, the ratio [HCO_3^-]/[H_2CO_3] must be high—namely, 1584, according to Equation 18. Since [HCO_3^-] is not especially high in biological fluids, [H_2CO_3] must be very low, which would result in a very small buffering capacity. Thus, it seems unlikely that in the course of evolution, the (H_2CO_3, HCO_3^-)-system should have been selected in view of the necessity in most organisms to maintain the limits of the pH of blood between ± 0.1 pH units. However, the reason that the buffering capacity would be small at very low concentrations of H_2CO_3 is that the H_2CO_3 could be depleted by small amounts of base. This actually does not occur, because there is a virtually unlimited reservoir of CO_2 in the gas phase, and this is capable of replacing any H_2CO_3 that is consumed. Thus, if NaOH were added to blood, H_2CO_3 would be converted to HCO_3^-, and the pH would rise. However, the H_2CO_3 that is neutralized by the NaOH is immediately replenished from the gas phase, and this greatly increases the buffering capacity.

Looking at this process in a more quantitative way, we note that we are actually dealing with two equilibria in this system—namely, the solubility of CO_2 and the dissociation reaction, as shown below:

$$CO_2 + H_2O \rightleftarrows H_2CO_3, \qquad K_g' = 2.6 \times 10^{-3},$$

and

$$H_2CO_3 \rightleftharpoons H^+ + HCO_3^-, \qquad K_a' = 1.73 \times 10^{-4}.$$

The equilibrium constant for the first reaction is K_g'. It is not possible to distinguish experimentally between dissolved CO_2 and H_2CO_3, so that one usually writes $[H_2CO_3]_{total} = [CO_2] + [H_2CO_3]$. The equilibrium constant for the first dissociation step K_I' for the complete system is

$$K_I' = \frac{[H^+][HCO_3^-]}{[H_2CO_3]_{total}} = \frac{[H^+][HCO_3^-]}{[H_2CO_3] + [CO_2]}$$

$$= \frac{[H^+][HCO_3^-]}{[H_2CO_3](1 + [CO_2]/[H_2CO_3])} = \frac{K_a'}{1 + (1/K_g')},$$

which, when we substitute the values for K_a' and K_g', yields $K_I' = 4.45 \times 10^{-7}$ or $pK_I' = 6.36$. This is the value for 25°C in water that appears in published reference tables. At 37°C in blood, the $pK_I' = 6.10$ (the dependence of pK on temperature will be discussed in the next section). Thus, the Henderson-Hasselbach equation is

$$pH = 6.10 + \log\frac{[HCO_3^-]}{[CO_2] + [H_2CO_3]},$$

which shows that this system is capable of buffering at a pH of 7.

Dependence of the pH of a Buffer on Temperature

We have so far used variables of K_a' only for a temperature of 25°C. Since biochemical reactions frequently occur at higher temperatures (30–37°C), it is important to be able to convert these values to 37°C. The necessary equation can be found in Chapter 6, and it provides the dependence of an equilibrium constant on temperature;

$$\ln\frac{K_2'}{K_1'} = -\frac{\Delta H^\circ}{R}\left(\frac{1}{T_2} - \frac{1}{T_1}\right), \tag{35}$$

in which K_1' and K_2' are the equilibrium constants at the temperatures T_1 and T_2, respectively, ΔH° is the enthalpy of ionization, and R is the gas constant. This equation is more conveniently expressed in terms of the pK'; namely,

$$pK_1' - pK_2' = \frac{-\Delta H^\circ}{2.303 R}\left(\frac{1}{T_2} - \frac{1}{T_1}\right). \tag{36}$$

From the Henderson-Hasselbach equation it is clear that the pH of a buffer will change by the same amount as the pK changes.

Example 9-H What is the pH at 37°C of a phosphate buffer whose pH = 7.0 at 25°C?

In this pH-range, the relevant pK'_a is pK'_{II} or 7.20; for this ionization step, $\Delta H° = 3.77$ J mol^{-1}. Thus, from Equation 36,

$$\Delta pK' = 7.20 - pK'_{37} = -\frac{(3770)}{(2.303)(8.314)}\left(\frac{1}{310} - \frac{1}{298}\right)$$

or $\Delta pK' = 0.026$. Substitution in Equation 18 shows that the pH also decreases by 0.026 and has the value of $7.0 - 0.026 = 6.97$ at 37°C.

9-3 AMINO ACIDS AND PROTEINS

The amino acids have the general structure

$$H_2N-\underset{R}{\underset{|}{\overset{H}{\overset{|}{C}}}}-C\overset{\displaystyle O}{\underset{\displaystyle OH}{}}$$

in which R represents one of twenty or more different chemical groups. The chemical structures for the naturally occurring amino acids are shown in Figure 15-1 of Chapter 15.* The amino acids differ from all of the substances that have been discussed up to this point in this chapter, in that they contain *both* an acidic and a basic group—namely the carboxyl OH and the amino NH_2 groups. A typical weak acid such as acetic acid has the property that at very low pH the carboxyl group is poorly ionized and the molecule has little or no net charge, whereas at high pH it is dissociated and carries a negative charge. The converse can be said for a weak base—it is positively charged at low pH and uncharged at high pH. An amino acid must be charged at both low and high pH. At low pH it must bear a positive charge, because the amino acid group is protonated, and at high pH it must be negatively charged, because the carboxyl group is ionized. At intermediate values of pH, there are two possibilities—either the amino group and carboxyl group are both uncharged or they are both charged. In both cases, there should be a pH at which there is no net charge. A great deal of evidence indicates that the second alternative is correct—that is, at pH values near the pH at which there is no *net* charge, the amino and carboxyl groups are both charged; that is, the molecule is a dipole. The word *zwitterion* (from the German *zwitter*, meaning "hybrid") is used to describe the doubly charged but neutral amino acid. The result of having both an acidic and a basic group is that a single amino acid such as glycine (NH_2CH_2COOH) has two pK'_a values—one for each group. This is shown in Figure 9-6, which shows a titration curve for glycine and the forms of

---
*The reader should consult this table before continuing.

FIGURE 9-6 Titration curve for glycine. The major species present in each region of the curve is shown. Note that in the midrange, the neutral molecule is not uncharged but is a dipole.

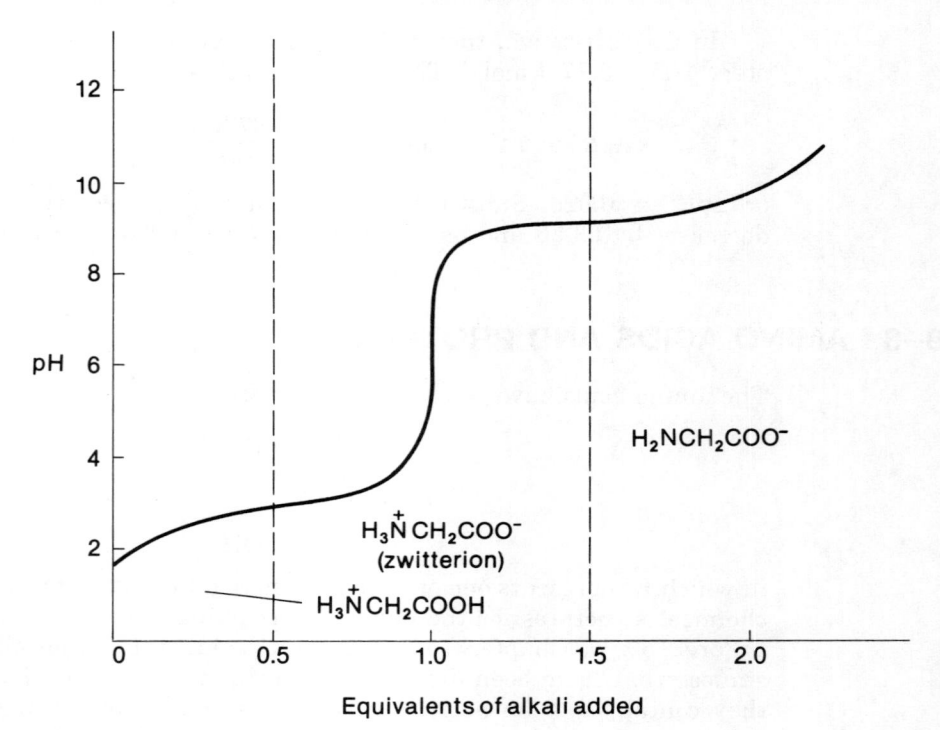

glycine present at various pH values. The pH midway between the two pK'_a values is called the *isoionic point*. In Chapter 12, we discuss the isoelectric point, or, the pH at which a particle will not move in an electric field. This is not always the same as the isoionic point, because often at the isoionic point, other ions, such as the Na$^+$ ion, bind to the amino acid. However, the isoionic point and the isoelectric point are very near one another and, in fact, the term isoelectric point is commonly used to refer to both the true isoelectric point and the isoionic point.

The titration of the terminal amino group and the terminal carboxyl group is actually of little interest in biochemistry; these groups do not exist in proteins except at the amino and carboxyl termini of the polypeptide chain. This is because usually both groups combine to create the peptide bond.

Amino acids contain other titratable groups, though—namely, the ionizable groups in the side chains. For example, aspartic acid and glutamic acid are dicarboxylic acids and each has a carboxyl group that is not engaged in a peptide bond. Other acidic groups are the phenolic hydroxyl group of tyrosine and the sulfhydryl group of cysteine. Lysine contains a second amino group that is always unbonded: other basic groups are the imidazole group of histidine and the guanidyl group of arginine. These

FIGURE 9-7 Titration curve of the acid form of glutamic acid with NaOH.

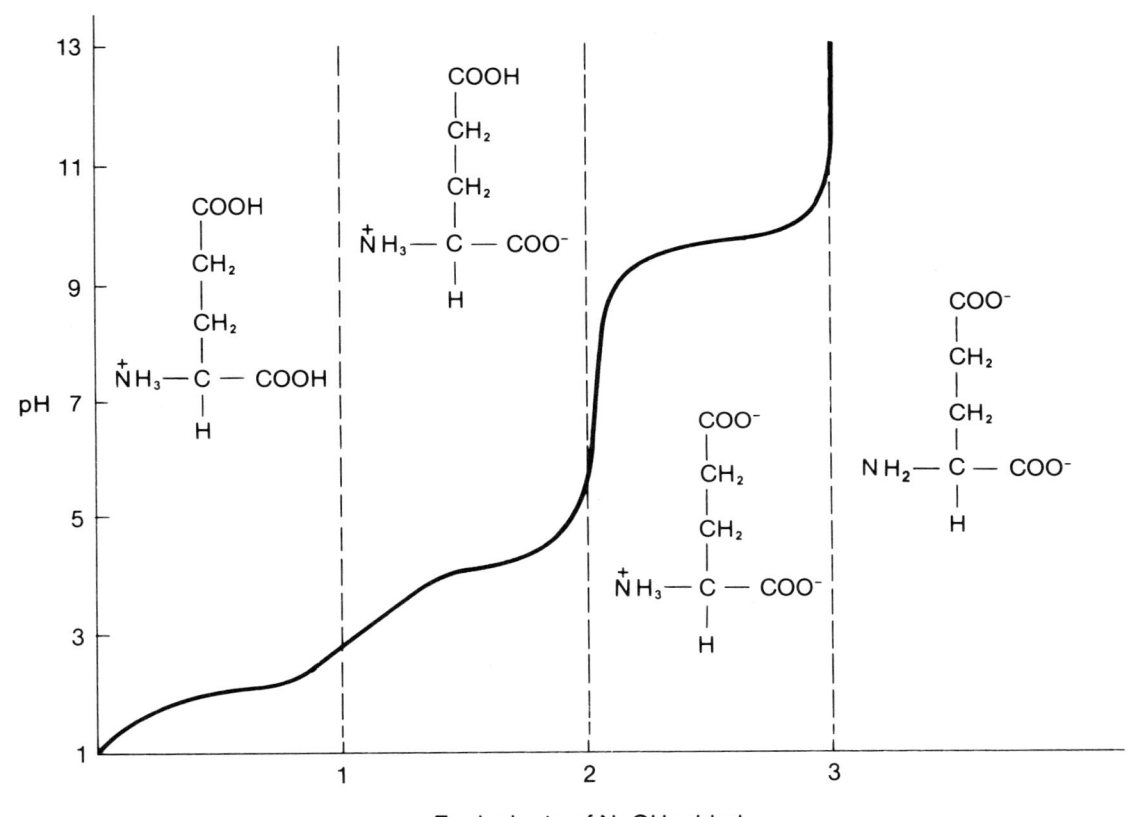

extra groups change the titration curves considerably, as is shown in Figure 9-7, and give rise to several different charged species. It should be noted that all of the different pK_a' values are not evident in the curve. This is because the ionizations are of the independent type discussed in Section 9-2; thus the pK_a' values, and hence the inflection points in the curve, are very near to one another.

These additional titratable groups are valuable in the study of protein structure because they are the only titratable groups present and can be titrated only when they are on or near the surface of the protein. Table 9-4 lists the pK_a' values for the various amino acid side chains. These values will be needed to interpret the titration curves discussed in this section.

Inspection of Table 9-4 indicates that the pK_a' of a chemical group is not an absolute characteristic of the group but depends upon its location in a molecule. For instance, the pK_a' of the carboxyl groups in aspartic acid and glutamic acid are not the same. Furthermore, the pK_a' of the α-carboxyl groups differ in each amino acid—for example, the value is 1.71 in cysteine, 2.20 in tyrosine, and 2.35 in glycine. This is because the charge

TABLE 9-4 The pK_a' values for titratable groups present in proteins.

Group	pK_a'
α-Carboxyl	1.7–2.4
β-Carboxyl (aspartic acid)	3.90
γ-Carboxyl (glutamic acid)	4.07
Imidazole (histidine)	6.04
α-Amino	9–10
Phenolic hydroxyl (tyrosine)	10.13
Sulfhydryl (cysteine)	10.78
ϵ-Amino (lysine)	10.79
Guanidyl (arginine)	12.48

distribution (electronic and nuclear) of the entire molecule determines the value of the electrical potential that attracts an ionizable proton. Thus, we should not be surprised to learn that when the amino acids are assembled in a protein, the pK_a' value of a particular functional group differs from the value of the same group in a free amino acid. Although the value of the pK_a' of a particular group depends mostly on the immediate environment of the group, there is a general trend. That is, at low pH the protein has a net positive charge, owing to the presence of amino groups, and this charge causes a slight repulsion of all protons. This facilitates ionization and, hence, the pK_a' values of the various groups that ionize at low pH are usually lower when the amino acid is in a protein than for the free amino acid. Similarly, at high pH, a protein has a net negative charge, and this increases the attraction of those protons that ionize at high pH, thereby raising their pK_a'. The result is that the entire titration curve of a protein covers a wider range of pH than does a mixture of free amino acids. There are also other effects on the pK_a' of a particular group when that group is part of a protein. For example, ionization of the imidazole group of histidine is affected by the electric fields of nearby amino acids, which may either be adjacent to the histidine in the polypeptide chain or nearby because of the folding of the chain. Thus the portion of the complete protein-titration curve that reflects imidazole titration is really the sum of many slightly different titration curves; this broadening of the regions of the titration curve corresponding to some amino acids makes it difficult to distinguish some of the different regions unambiguously. Nonetheless, a considerable amount of information about the structure of a protein can be gained from titration studies, as we will see shortly. The range of pK_a' values for each of the chemical groups in a protein is rather large, as is shown in Table 9-5. These should be compared to the values that are given in Table 9-4.

The fundamental rule of titration of proteins is that

a group can be titrated only if it is in contact with the solvent.

TABLE 9-5 The range of pK_a' values for the ionizable groups in proteins.

Group	Range of pK_a' values
Carboxyl	2.5–5.0
Imidazole	6.0–8.0
Terminal amino	6.0–8.0
Side chain amino	9.0–10.5
Phenolic OH	9.0–10.5
Sulfhydryl	9.0–10.5
Guanidyl	12.0–12.5

There are several causes for the inability of a group to come in contact with the solvent; the three most important ones are listed below. The first virtually removes the possibility of titration. The second and third make it very difficult but still possible; the difficulty is reflected in a very large change in the value of pK_a'.

1. A protein may be folded in such a way that a particular polar amino acid is separated from the solvent by a cluster of nonpolar amino acids. Such an amino acid is not titratable until there is a change in the shape of the protein that exposes the amino acid to the solvent.

2. The amino acid may be very near a large number of nonpolar amino acids (in a so-called *hydrophobic pocket* or *hydrophobic cleft*) but still sterically accessible to the solvent. The nonpolar amino acids create a region having a very low dielectric constant in which ionization occurs with great difficulty; this results in a substantial change in the pK_a'.

3. A titratable group may be participating in a hydrogen bond; this does not prevent titration but does alter the pK_a' significantly. Thus for a hydrogen-bonded acidic group, ionization is more difficult than when the group is in the unbonded state, so that the pK_a' is increased; conversely, a potential proton acceptor (for example, an amino group), when hydrogen-bonded, finds it more difficult to bind a proton, and the pK_a' of that group is depressed.

These points can be used to gain information about protein structure. For instance, if treatment of a solution of the protein by a physical agent (such as temperature) or a chemical reagent (one that breaks hydrogen bonds, such as urea, or breaks disulfide bonds, such as β-mercaptoethanol) later alters the titration curve of the protein, it is clear that the protein has undergone a change in shape. Furthermore, from the particular pK_a' values that change, one knows the amino acids that are inaccessible. For example, consider a hypothetical protein containing 50 alanines and 1 histidine. This protein has three titratable groups—a terminal amino group, a terminal carboxyl group, and an imidazole. If the titration curve indicates three pK_a' values—namely, 3, 6.5, and 8.5—one could conclude that the protein is folded in such a way that the histidine (whose pK_a' is 6.5) is external. However, if the only transitions observed were found at pH 3 and 8.5, one could readily conclude that the histidine is internal. In this example, information is gained from qualitative data—namely, whether a transition is or is not present. More commonly, the quantitative aspects of the titration curve are used to give information. That is, since known amounts of acid or base are added, from the number of equivalents added one knows how many groups are titrated. Suppose we are studying a protein consisting of 50 alanines and 2 histidines. If only one histidine is external, a pH-6.5 transition would be seen; however, if both are external, it would take twice as much added base (that is, two equivalents) to complete the transition. Hypothetical titration curves for the case of one and two external histidines are shown in Figure 9-8.

FIGURE 9-8 Hypothetical titration curves for a protein having two histidines as the only ionizable groups. The dashed curve is obtained if one histidine is inaccessible to the solvent.

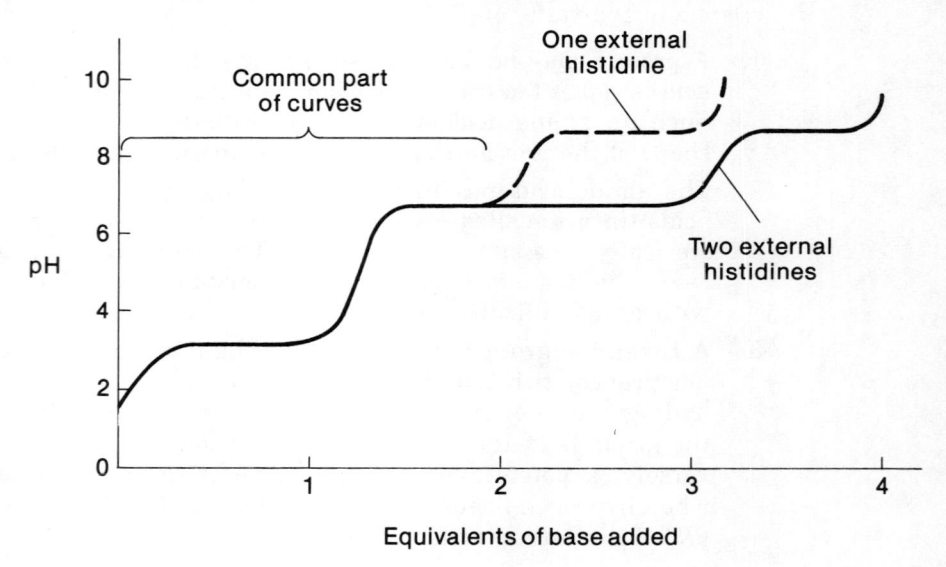

The titration of real proteins is complicated by denaturation (see Chapter 15). Denaturation refers to an unfolding or loss of the defined three-dimensional structure of a macromolecule, which results from exposure to various physical agents or chemical reagents. The way in which denaturation affects the titration curve is the following. The structure of a protein is determined by many factors (see Chapter 15), an important one of which is hydrogen-bonding. As a result of the three-dimensional structure, potentially titratable groups are inaccessible to the solvent, as we have just said. When the groups engaged in hydrogen bonds are titrated, the hydrogen bonds break, the structure of the protein changes markedly, and usually all groups become accessible to the solvent. Thus, in an alkaline titration, when the pH is reached at which the structure changes, all groups that had been sterically protected from titration at a lower pH are immediately titratable. For example, a protein initially at a pH of 5 might contain numerous histidines which are internal, so that they are not titrated when the pH passes through the range of pH 6 and 6.5. However, at a pH of 10.5 all of the hydrogen bonds involving the amino groups of lysine are broken. If there are three buried histidines and five lysines, the transition at pH 10.5 will appear to involve the removal of not five but eight protons (five from the lysines and three from the histidines). Usually a simple criterion can be used to detect denaturation. If there is no structural alteration, the titration curves should be reversible. That is, if alkali has been added, the addition of acid should cause the pH to drop in a way that follows the titration curve. For example, if the pH values of a protein solution are 6.3 and 9.2 when 7 and 18 equivalents of base have been added, respectively, then as long as denaturation has not occurred up to pH 9.2, the addition of $18 - 7 = 11$ equivalents of acid (that is, a *back titration*) should reduce the pH to 6.3. This would not occur if denaturation had occurred and buried groups titrating between pH 6.3 and 9.2 had been released, unless the denaturation happened to be reversible.

We now calculate the titration curves for several different proteins to indicate the kind of data that are obtained, and then examine the titration curve of a fourth protein and try to deduce some of its structural features.

1. *The titration curve of a tripeptide, alanyl glutaminyl lysine.* At very low pH, this molecule is in its totally protonated form and has the structure

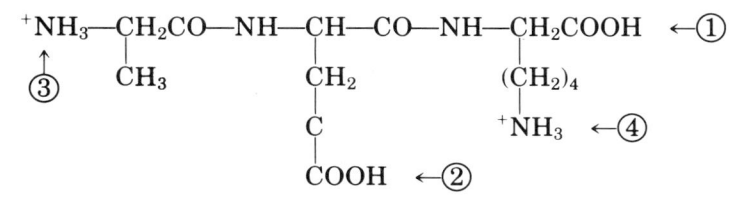

$$^+NH_3-CH_2CO-NH-CH-CO-NH-CH_2COOH \leftarrow ①$$

with ③ pointing to the terminal $^+NH_3$, side chains CH_3, $CH_2-CH_2-C-COOH$ (② pointing to COOH), and $(CH_2)_4-^+NH_3 \leftarrow ④$

Its four ionizable groups indicated by arrows and the approximate pK_a' values are

① Terminal COOH   3.3
② Glutamyl COOH   3.8
③ Terminal NH_3^+ 8.0
④ Lysyl NH_3^+ 10.0

Each group requires one equivalent of base for titration and the resulting pH equals the pK_a' when it is half titrated. Thus, the titration curve will have plateaus at each of the four pK_a' values; furthermore, the middle of each plateau will occur when $\frac{1}{2}$, $1\frac{1}{2}$, $2\frac{1}{2}$ and $3\frac{1}{2}$ equivalents of base are added. Thus the curve shown in Figure 9-9 will result.

FIGURE 9-9 The titration curve of the tripeptide alanine-glutamine-lysine.

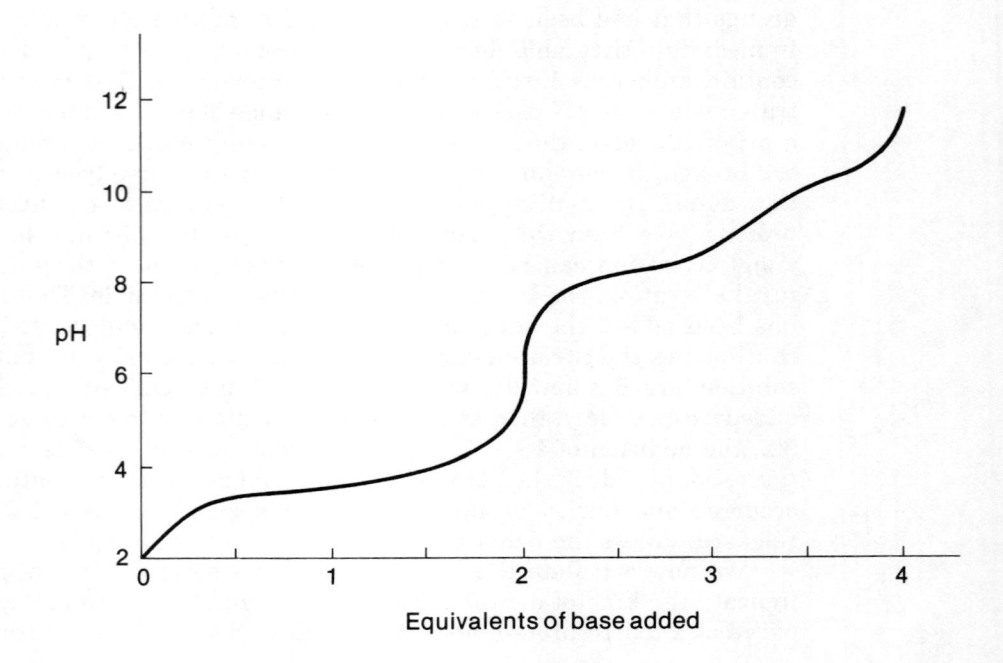

2. *The titration curve of a denatured protein containing 282 amino acids of which 8 are aspartic acid, 49 are glutamic acid, 18 are histidine, 11 are lysine, 15 are cysteine, 15 are tyrosine, and 10 are arginine.* The ionizable groups and their average pK_a' values are:

1 Terminal COOH	3.3
8 Aspartyl COOH	3.8
49 Glutamyl COOH	3.8
18 Histidine imidazole	6.3
1 Terminal NH_3^+	8.0
11 Lysyl NH_3^+	10.0
15 Cysteyl SH	10.1
15 Tyrosyl OH	10.1
10 Arginine guanidyl	12.5

The single terminal COOH having $pK_a' = 3.3$ will not be resolved from the 57 aspartyl and glutamyl carboxyls ($pK_a' = 3.8$) nor will the lysyl NH_3^+, cysteyl SH, and tyrosyl OH be resolved. Thus $1 + 8 + 49 = 58$ protons will have a $pK_a' \approx 3.8$; 18 will have $pK_a' = 6.3$; 1 (which will contribute too little to be seen) will have $pK_a' = 8.0$; $11 + 15 + 15 = 41$ will have a $pK_a' \approx 10.1$; and 10 will have $pK_a' = 12.5$. The midpoint of the plateau occurs when half of the groups corresponding to each pK_a' are titrated. Therefore, the midpoints are:

SEC. 9-3 AMINO ACIDS AND PROTEINS

$$\frac{1}{2}(1 + 8 + 49) = 29, \qquad \text{for } pK_a' = 3.8;$$

$$29 + 29 + \left(\frac{1}{2} \cdot 18\right) = 67, \qquad \text{for } pK_a' = 6.3;$$

$$67 + 9 + \frac{1}{2} = 76.5, \qquad \text{for } pK_a' = 8.0;$$

$$76.5 + \frac{1}{2} + \left(\frac{1}{2} \cdot 41\right) = 97.5, \qquad \text{for } pK_a' = 10.0\text{--}10.1;$$

$$97.5 + \left(\frac{1}{2} \cdot 41\right) + \left(\frac{1}{2} \cdot 10\right) = 123, \qquad \text{for } pK_a' = 12.5.$$

At $123 + 5 = 128$ equivalents of base, all groups are neutralized, and the curve will rise sharply. Figure 9-10(a) shows this titration curve.

3. *A protein having the same amino acid composition as in 2, but in which 10 histidine imidazole and 8 glutamyl groups are buried deep in the protein and unavailable to the solvent, and 14 of the 15 cysteines are linked via —S—S— bridges to form 7 cystines. The protein denatures at pH 11.*

The histidine imidazole groups and the 8 glutamyl COOH groups are not titrated until after denaturation occurs. The 14 cysteines are never titrated. Thus, the pK_a' values that are observed are the same as in the denatured protein, but they will occur at different points. The midpoint of the plateau and pK_a' values are:

$$\frac{1}{2}(1 + 8 + 49 - 8) = 25, \qquad \text{for } pK_a' = 3.8;$$

$$25 + 25 + \frac{1}{2}(18 - 10) = 54, \qquad \text{for } pK_a' = 6.3;$$

$$54 + 4 + \frac{1}{2}(1 + 11 + 1 + 15) = 72, \qquad \text{for } pK_a' = 10.0\text{--}10.1;$$

$$72 + 14 + 18 + \left(\frac{1}{2} \cdot 10\right) = 109, \qquad \text{for } pK_a' = 12.5.$$

When more equivalents are added (of the remaining $13\frac{1}{2}$ corresponding to $pK_a' = 10.1$) the pH reaches 11, denaturation occurs, and 18 more groups become available. These release protons, and the rate of increase of pH drops slightly and then remains nearly constant until all of these are titrated. Then the pH rises slowly and at 109 equivalents, the midpoint of the plateau at $pK_a' = 12.5$ is reached. At 114 equivalents the titration is complete and the pH rises abruptly. The titration curve is shown in Figure 9-10(b); it should be compared to that in Figure 9-10(a).

4. *A protein has 14 aspartic acids, 18 glutamic acids, 10 histidines, 16 lysines, 8 cysteines, and 6 arginines. Its titration curve is shown in Figure 9-11.*

FIGURE 9-10 (a) The titration curve for a protein molecule containing the following number of ionizable amino acids: aspartic acid, 8; glutamic acid, 49; histidine, 18; lysine, 11; cysteine, 15; tyrosine, 15; arginine, 10. (b) The same protein, but none of the cysteines are titratable and 10 histidines and 8 glutamic acids are unavailable until pH 11, when denaturation occurs. See text for details.

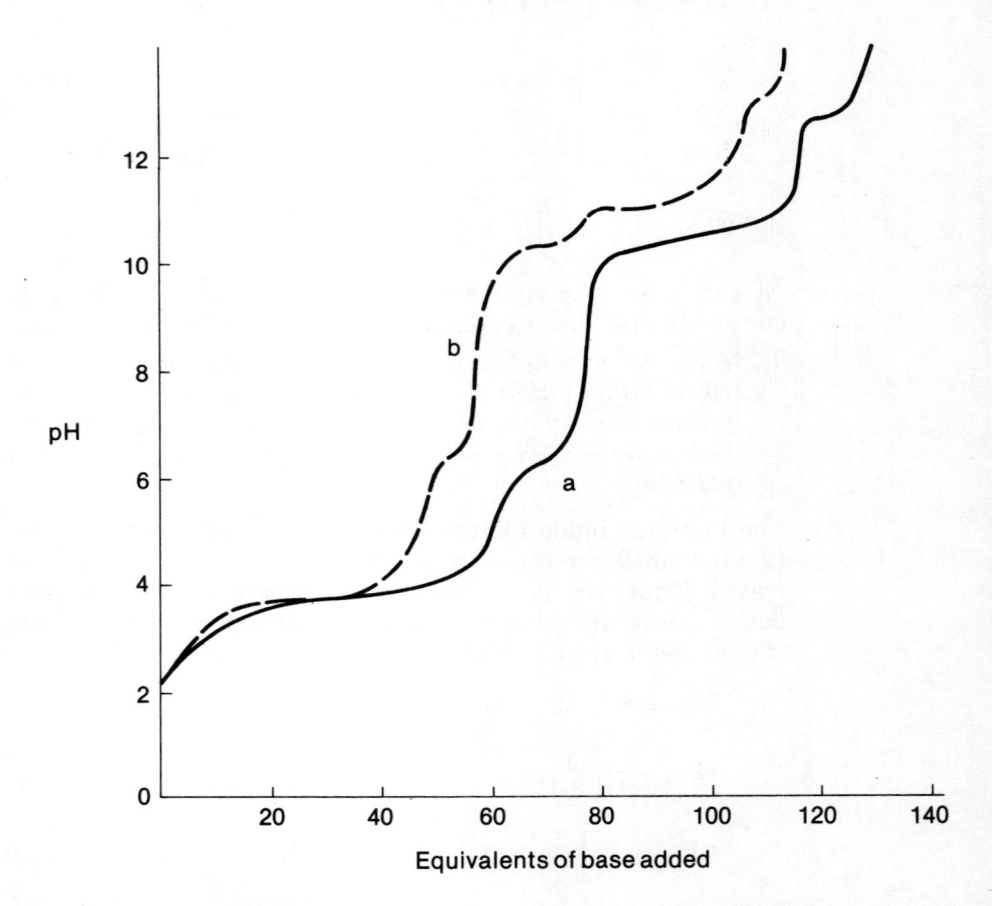

The first plateau at $pK_a' = 3.8$ involves only 26 protons. Thus $32 - 26 = 6$ amino acids are unavailable. Then 5 are involved at $pK_a' = 6.3$. Since there are 10 histidines, 5 must be buried. At pH 8.5, 7 amino acids are titrated. Only one of these can be the terminal amino group; therefore the others must have been made available as a result of the uncovering of buried groups—this indicates that the protein undergoes a shape change at pH 8.5. At pH 10, only 16 protons take part in this transition, instead of $16 + 8 = 24$, so that again there must be buried groups. These could be either or both of cysteine or lysine. An interesting variation of the titration enables one to determine the fraction of each that is buried. That is, another sample of the protein is titrated in the presence of the formaldehyde, a reagent that reacts with lysyl amino groups and prevents them from binding protons. Hence lysine does not take part in the titration. Let us assume

FIGURE 9-11 Titration curve for a hypothetical protein described in the text. The dashed curve shows the only change that occurs if formaldehyde is present throughout the titration.

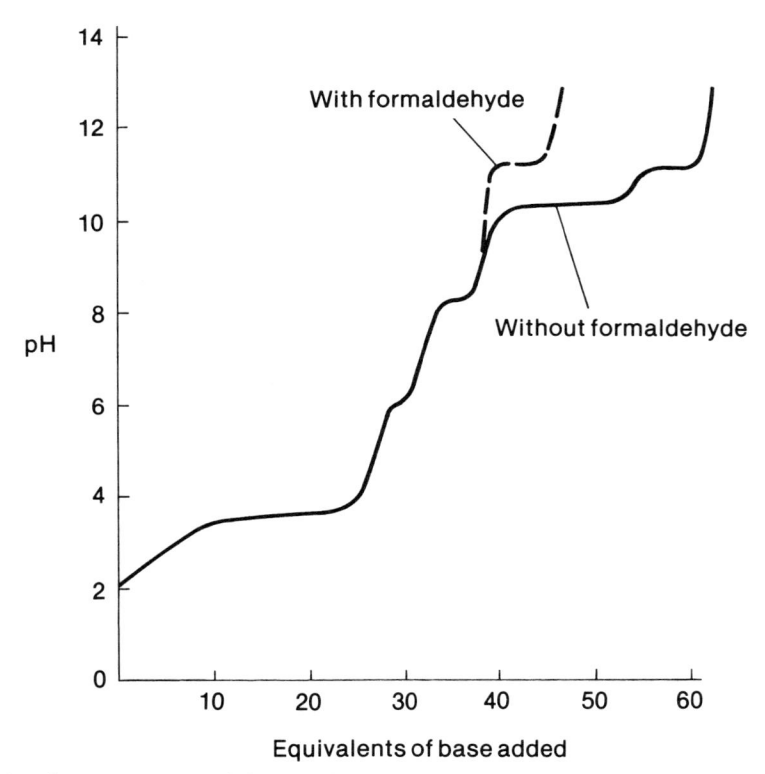

that in the presence of formaldehyde, no pH-10 transition is seen. If cysteine were exposed to the solvent, its SH groups should be seen. However, they are not, so that the 8 protons that are missing in the titration (without formaldehyde) must be cysteine. These might be buried, but it is equally possible that the 8 cysteines are linked to form four disulfide bonds. This could be checked by adding a reagent that cleaves disulfide bonds and repeating the titration. At pH 11, 5 amino acids are titrated; these are presumably the five histidines that did not appear at pH 6.3. Thus, from the titration data we have learned five things: (1) there may be four disulfide bonds; (2) a shape change occurs at pH 8.5; (3) denaturation probably occurs at pH 11; (4) there are five internal histidines; and (5) six amino acids (which may be glutamic acids or aspartic acids) are internal below pH 8.5 and external above that pH.

A great deal of information can be obtained when titrations are coupled with spectroscopic measurements, because the spectra of several of the amino acids show differences when buried and unburied. For further information, the reader should consult the references given at the end of the chapter.

9-4 METAL ION (COORDINATION) COMPLEXES

A large number of biochemicals and proteins form stable complexes with metal ions. For example, the proteins hemoglobin and the cytochromes both contain tightly bound Fe^{3+} ions; numerous hydrolytic enzymes and the dehydrogenases contain Zn^{2+} ions; Cu^{2+} is associated with many oxidative enzymes; and vitamin B12 contains Co^{3+} ions. Even more common is the requirement by many enzymes for divalent cations (usually Mg^{2+} or Ca^{2+}) in order to make the substrate accessible to enzymatic catalysis. For example, Mg^{2+} is required in nearly all enzymatic reactions in which phosphate participates. In inorganic chemistry, a variety of compounds also exist in which there are stable bonds between metal ions and nonionic substances, as in compounds such as $Na_3Fe(CN)_6$, which ionizes to yield Na^+ and $Fe(CN)_6^{3-}$ ions; and the $Cu(NH_3)_4^{2+}$ ion, which results from bubbling NH_3 through a solution of $CuSO_4$. The structure of these complex ions and metal-bound biochemicals cannot be explained by the ordinary type of chemical bond. However, the acid-base theory of G. N. Lewis enables us to understand these compounds. This theory also explains how certain acid-base color indicators can change color when exposed to compounds that neither accept nor donate protons.

The Brønsted-Lowry theory extends the classical definition of an acid and a base by defining an acid as a proton donor and a base as a proton acceptor. In the Lewis theory, *an acid is an acceptor of an electron pair* and *a base is a donor of an electron pair*. Thus, in a reaction such as the protonation of ammonia,

$$:NH_3 + H^+ \rightleftharpoons (NH_4)^+ ,$$

the Brønsted base and acid are the NH_3 molecule and the NH_4^+ ion, respectively, whereas in the Lewis theory the NH_3 is the base and the H^+ ion is the acid. Even though the two theories take slightly different viewpoints, both are equally able to explain the reaction. However, for the reaction

$$2 :NH_3 + Cu^+ \rightleftharpoons Cu(NH_3)_2^+$$

the Brønsted theory can make no comment because there is no proton transferred; the Lewis theory explains the reaction as an acid-base reaction between the electron-pair-accepting Cu^+ (the Lewis acid) and the electron-pair-donating NH_3 (the Lewis base). The bond that is formed is called a *coordinate bond* and the $Cu(NH_3)_2^+$ ion is called a *coordination compound*. The Lewis theory also explains the formation of compounds such as $B(NH_3)F_3$ as a reaction between BF_3, the acid, and NH_3, the base. *The coordinate bond is a bond in which an electron pair is shared by the Lewis acid and the Lewis base.*

The Lewis theory is of special value in biochemistry, in which hydration or the binding of water molecules is so prevalent. Water possesses an unshared electron pair and is capable of binding to numerous cations that can accept a pair. For example, the Cu^{2+} ion in water exists predominantly as the $(Cu^{2+}(H_2O)_4)$-hydration complex.

The value of this approach is that any molecule or ion possessing an unshared electron pair qualifies as a Lewis base (for example, CO, H_2O, I^-, SO_4^{2-}), and any potential acceptor (for example, the Na^+, Cu^+, Ag^+, Mg^{2+}, Zn^{2+}, Fe^{2+}, Co^{2+}, Al^{3+}, and Co^{3+} ions) qualifies as a Lewis acid.

A special term has been adopted for a donor that shares an electron pair with a cation—it is called a *ligand*. This word is also used in a more general sense, as will be described in Chapter 14.

Will any Lewis base react with any Lewis acid? In a sense, the answer is yes, but as in any reaction between a weak base and a weak acid, the extent of reaction is expressed by the equilibrium constant. This is of course true also of the formation of a coordination compound.

Thus the reaction

$$A + :B \rightleftarrows A:B$$

is described by

$$K = \frac{[A:B]}{[A][:B]}$$

In coordination chemistry K is not called an equilibrium constant but is the *stability constant,* or, more commonly, the *formation constant.* Frequently two identical ligands bind to an acceptor (as in the formation of the $Cu(NH_3)_2$ complex); that is,

$$A + 2 :B \rightleftarrows B:A:B ,$$

and the formation constant is

$$K = \frac{[B:A:B]}{[A][:B]^2}$$

Clearly the strength of binding is expressed by the magnitude of the formation constant, as in any other equilibrium.

TABLE 9-6 Coordination numbers for various metal ions.

2	4	6
Cu^+, Ag^+	Na^+, Cu^+, Cu^{2+} Zn^{2+}, Co^{2+}, Mg^{2+} Al^{3+}	Fe^{3+}, Co^{3+}, Al^{3+}

The number of ligands with which a particular metal ion can coordinate (the coordination number of the ion) depends upon many factors, and a discussion of this is beyond the scope of this book. In Table 9-6 are listed the coordination numbers of the metal ions frequently found in coordination compounds in biochemical systems. It should be noted that these are all even numbers; odd numbers have been observed but are not very common. In biochemicals and in macromolecules the most common atoms that are coordinated with metal ions are N, O, S, and P. Typically the N will be in an amino group or a heterocyclic ring; an O is usually a carboxyl O^- or a phosphate O^-; S coordinates only when in methionine.

It is not uncommon for a ligand molecule to have several binding sites. For instance, an amino acid can utilize both its NH_2 group and its carboxyl O^- group to coordinate with metal ions. For example, glycine can form a copper-glycine complex with the Cu^{2+} ion. Since the coordination number of Cu^{2+} is four and each glycine can satisfy only two of these, the Cu^{2+}-glycine complex consists of one Cu^{2+} and two glycines:

Similar structures occur with Ni^{2+} and Co^{2+} and with amino acids of related structures, such as alanine and leucine. A peculiar terminology is used to describe the number of binding sites on a ligand—each site is called a *tooth*. That is, a ligand is monodentate, bidentate, or tridentate, if it contains one, two, or three binding sites, respectively, and so forth.

When a protein binds a metal ion, it is common for many amino acids to participate. Thus, in hemoglobin, which is coordinated to the Fe^{3+} ion, four of the coordination bonds are to the heme moiety, one is to the imidazole of a single histidine, and the sixth binds the oxygen (or water, if O_2 is not present). In the enzyme carboxypeptidase, the Zn^{2+} is bound to two imidazole nitrogens, one glutamic acid carboxyl, and one water. It should be realized that binding of a metal ion can only occur when the protein molecule has the correct three-dimensional configuration, since otherwise the binding sites may be too far apart. In a few cases the metal ion contributes to the stability of the structure by maintaining certain groups near one another.

If a metal ion is presented with a solution containing two different ligands, as in any equilibrium reaction, it will form a coordination compound preferentially with that ligand for which the formation constant is larger. Furthermore, since a coordination compound is in equilibrium with its constituents, if a compound A:B having a formation constant of K_1 is presented with a ligand C for which A:C has a larger constant K_2, A:B will dissociate and A:C will predominate. This can be seen in the following example.

SEC. 9-4 METAL ION (COORDINATION) COMPLEXES

EXAMPLE 9-1 An aqueous solution of a Zn-protein complex having $K_1 = 2 \times 10^8$ is mixed with an equimolar amount of a Zn-complexing agent for which $K_2 = 3 \times 10^{10}$. What is the distribution of Zn between the two complexes?

The second complex predominates, so that if the molarity of the initial Zn-protein and the Zn-complexing agent is m, the approximate concentration of the second Zn complex is $\sqrt{m/K_2}$. Thus, the concentration m' of the Zn-protein complex must satisfy the equation $m'/(\sqrt{m/K_2})^2 = K_1$ or $m'/m = K_1/K_2 = 6.6 \times 10^{-3}$. Another way to think about it is that K simply describes the probability that Zn is in a particular complex.

If a molecule contains two ligands A and B and if a metal ion can bind to A with a constant K_1 and to B with a constant K_2, then, if it can bind simultaneously to both A and B, it will do so, with a constant that is normally much larger than the greater of the two numbers. This is because binding to one ligand facilitates binding to the second. For example, Zn^{2+} can form a complex with acetate having $K = 10.7$ and with ammonia having $K = 1.15 \times 10^9$, but when binding to ethylenediaminetetraacetate, it binds simultaneously to the NH_2 and acetate groups with $K = 3.2 \times 10^{16}$.

Figure 9-12 Structure of ethylenediaminetetraacetic acid (EDTA) coordinated to the Fe^{3+} ion.

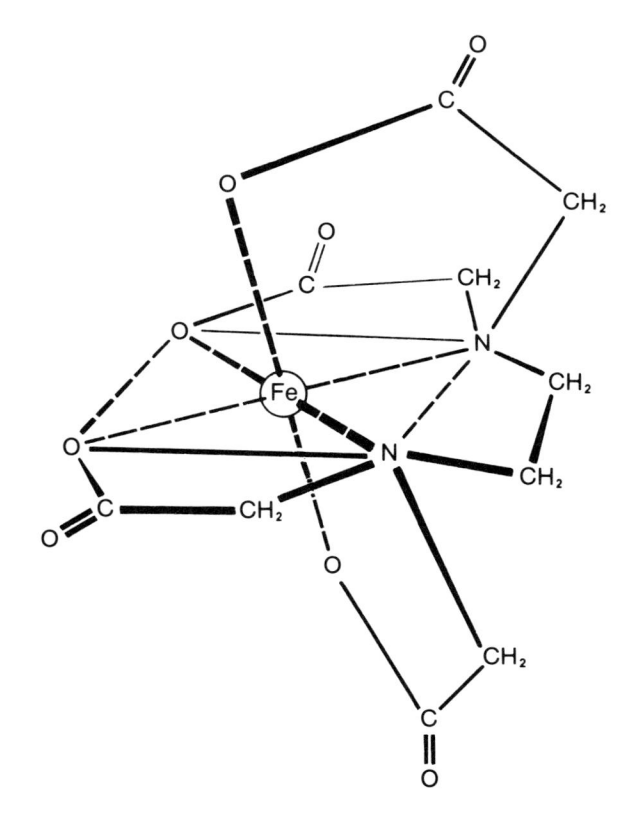

Multidentate ligands are called *chelating agents* (from the Greek *chele* or "claw," because a chelating agent holds ions very tightly). Two simple examples of such compounds are the citrate ion and ethylenediamine, the first being tridentate (having three carboxyl O⁻ groups) and the second being bidentate (having two NH_2 groups). The more ligands that are available (up to six, which is the maximal coordination number that has been encountered), the higher will be the formation constant, as long as the ligands are sufficiently near one another that a simple metal ion can bind to them.

The most common chelating agent used in biochemical research is the Na salt of ethylenediaminetetraacetic acid, or EDTA, as it is commonly called (the Na salt is used because the acid form is rather insoluble.)* This compound complexes through two nitrogen and four oxygen atoms, as shown in Figure 9-12. The steric arrangement of these groups is such that all six coordination bonds are easily made and, as a result, the formation constant for most divalent cations is extremely high ($\approx 10^{16}$). Thus, when EDTA is present, the concentration of most divalent cations is extremely low even at very low concentrations of EDTA. This is shown in the following example.

EXAMPLE 9–J What is the concentration of free Mg^{2+} ions in a solution prepared with 0.02 *M* NaEDTA and 0.01 *M* $MgCl_2$?

The relevant equilibrium constant is

$$10^{16} = \frac{[Mg_2EDTA]}{[Mg^{2+}]^2[EDTA^{4-}]}.$$

The formation constant is so high that there is very little free Mg^{2+}; thus, by the stoichiometry, the concentration of Mg_2EDTA is approximately 0.5[added Mg^{2+}] or 0.005 *M*. Thus, $[EDTA^{4-}]$ will be 0.02 − 0.005 = 0.015 *M*. Then $[Mg^{2+}] = [0.05/0.015 \times 10^{16}]^{1/2} = 5.5 \times 10^{-9}$ *M*.

EDTA is an extraordinarily valuable reagent for biochemical studies because of its powerful ability to bind divalent cations. For example, it is able to withdraw divalent cations from most metal-ion-requiring enzymes and thereby serves as a strong inhibitor of enzyme activity. It is commonly used in the isolation of DNA and RNA molecules because in its presence potentially destructive nucleases (enzymes that hydrolyze nucleic acids), most of which require the Mg^{2+} ion, are inactive. Furthermore, its formation constant with monovalent cations is so small that the concentration of these ions, which are often needed for the stability of macromolecules, is unaffected. EDTA is sometimes added in small quantities as a preservative to certain medications and foods, since it prevents the growth of most microorganisms by inhibiting many enzymes.

*In books published before 1970 the names *versene* and *sequestrene* were used for EDTA.

There are also chelating agents for which the formation constants for various divalent cations are significantly different. For example, there are agents that bind Ca^{2+} but not Mg^{2+}, and these agents are useful in inhibiting Ca^{2+}-requiring enzymes in a mixture of enzymes in which the activity of a Mg^{2+}-requiring enzyme is desired. Also some Mg^{2+}-requiring enzymes are inhibited by minute concentrations of Ca^{2+} ions, which can be removed with such agents.

REFERENCES

Albert, A., and E.P. Sergeant. 1962. *Ionization Constants of Acids and Bases.* Wiley. New York.
Barrow, G.M. 1974. *Physical Chemistry for the Life Sciences.* McGraw-Hill. New York.
Bull, H.B. 1964. *Introduction to Physical Biochemistry.* Davis. Philadelphia.
Cohn, E.J., and J.T. Edsall. 1965. *Proteins, Amino Acids, and Peptides.* Hafner. New York.
Dweyer, R.P., and D.P. Mellon. 1964. *Chelating Agents and Metal Chelates.* Academic Press. New York.
Edsall, J.T., and J. Wyman. 1958. *Biophysical Chemistry,* Volume 1. Academic Press. New York.
King, E.J. 1965. *Acid-Base Equilibria.* Pergamon. Elmsford, N.Y.
Steinhardt, J., and J.A. Reynolds. 1969. *Multiple Equilibria in Proteins.* Academic Press. New York.
Tanford, C. 1961. *Physical Chemistry of Macromolecules.* Wiley. New York.
Tanford, C. 1962. "Interpretation of Hydrogen Ion Titration Curves of Proteins." In *Advances in Protein Chemistry,* Volume 17. Academic Press. New York.

PROBLEMS

1. The pK_a' of lactic acid is 3.86. What is the value of the equilibrium constant?

2. An 0.02-M solution of an acid is 0.01 percent ionized. What is the pK_a''?

3. What molarity of acetic acid ($pK_a' = 4.73$) would have a pH of 3 at 25°C? (Let c be the desired molarity.)

4. What is the concentration of the OH^- ion in 0.01 M HCl?

5. The pK_a' for a weak base is 10.8. What is the value of K_b''?

6. What is the pH of 0.1 M acetic acid?

7. What is the pH of 0.1 M NH_4OH?

8. What is the pH of 0.1 M NH_4Cl?

9. What is the degree of dissociation, α, of 0.05 M lactic acid?

10. What is the pH of 0.01 M sodium acetate?

11. What is the pH that results from mixing 5 ml of 0.2 M NaOH and 100 ml of 0.05 M formic acid?

12. What is the pH of a solution containing 0.15 M methylamine and 0.05 M HCl? The pK_b' of methylamine is 3.36.

13. Which of the following buffers would be useful for an enzymatic reaction that is carried out between pH 6 and 8 and that requires that the Ca^{2+} ion be present at a minimum concentration of 0.001 M? The pK_a' values are in parentheses: acetate-acetic acid (4.8); citrate-citric acid (3.1, 4.7, 6.4); phosphate (2.2, 7.2, 12.4); and Tris (8.1).

14. What is the pH of a solution of glycine at the isoelectric point? The two ionization constants are 4.66×10^{-3} and 1.66×10^{-10}.

15. A normal adult eliminates about one liter of urine per day. The pH of urine is 6. How many equivalents of acid are excreted?

16. You have 100 cm³ of four different solutions—HCl, acetic acid, formic acid, and HNO_3—each at a concentration of 0.1 M. Which solution requires the greatest amount of base for complete titration?

17. A fumaric acid buffer has a pH of 4.95 at 25°C. What is the pH at 37°C?

18. Calculate the following for 0.1 M formic acid containing 0.05 M Na formate:
    (a) What is the pH?
    (b) What is the buffering capacity?
    (c) What would be the pH if 1 ml of 1 M HCl were added to 100 ml of this buffer?

19. What concentration of sodium formate–formic acid will allow a maximal change of 0.2 pH unit for a pH of 4 if 0.01 M HCl is to be added? The pK_a' is 3.77.

20. What is the maximal concentration of a pH-5.0 acetate buffer that could have an ionic strength of 0.02? What would be the buffer capacity?

21. What is the simplest way to prepare a pH-4.8 buffer that is 1 M with respect to K^+ ions, contains no Na^+ ions, and in which the concentration of the acetate ion is not to exceed 0.05 M?

22. H_2SO_4 is totally dissociated into H^+ ions and HSO_4^- in aqueous solution. What are the concentrations of H^+, HSO_4^-, and SO_4^{2-} ions in 0.2 M H_2SO_4? The value of K_a' for H_2SO_4 is 1.3×10^{-2}.

23. What is the pH of 0.01 M H_3PO_4?

24. Fifty ml of a solution consisting of 0.1 M NH_3 and 0.2 M methylamine is to be adjusted to pH 9.50. How many milliliters of 0.1 M HCl must be added to achieve this pH? The pK_b' values are 4.75 for ammonia and 3.26 for methylamine.

25. The density of solid glycine is 1.607 g cm⁻³; that is, its molar volume is 46.71 cm³. The density of an isomer, glycolamide, is 1.390 g cm⁻³ (molar volume = 54.01 cm³). Propose a molecular explanation for this density-difference.

26. The volume of a one-molal solution of glycine depends on pH. At what pH would you expect the volume to be minimal?

27. Draw a curve to show the titration of the hexapeptide alanine-glycine-lysine-lysine-tyrosine-methionine.

28. Draw the titration curve for a protein containing 3 alanines, 5 arginines, 8 aspartic acids, 5 cysteines, 10 histidines, 15 lysines, and 10 tyrosines.

29. A protein has 6 aspartic acids, 9 glutamines, 8 histidines, 6 lysines, 4 glycines, and 3 arginines. What information about the structure of the protein can be gained from the titration curve shown below?

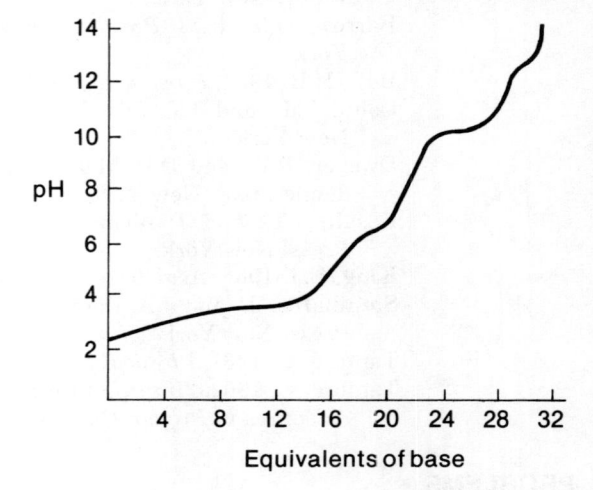

30. A protein containing 8 lysines has the following titration curves obtained in the presence and absence of formaldehyde. What can you say about the position of the lysines in the molecule?

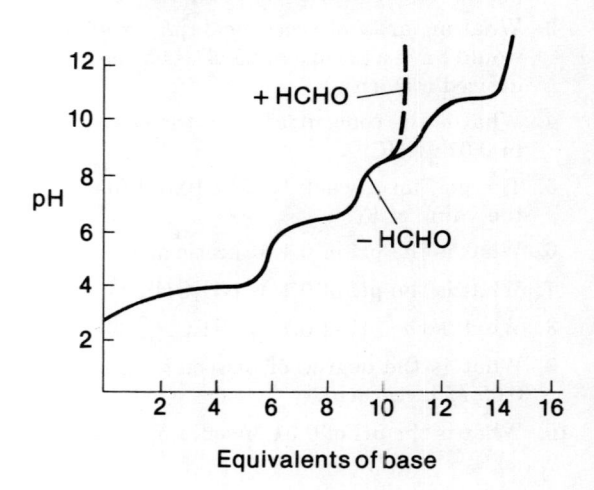

PROBLEMS

31. What is the concentration of free Ca^{2+} ions in a solution containing 0.01 M EDTA and 0.003 M $CaCl_2$?

32. In cases of lead poisoning, $CaNa_2EDTA$ is injected into muscle tissue. What is probably the principle underlying this technique?

33. Ethyl ether is slightly soluble in water. Could it be used as a chelating agent for substances in aqueous solution?

34. Would you expect oxalic acid to be a chelating agent?

35. Many bacterial viruses and some complex protein aggregates are almost totally dissociated if EDTA is added. What does this tell you about their structure?

36. When the density of DNA is being measured in concentrated CsCl, EDTA is always added to the solution. Without this addition, the measured value is variable from one experiment to the next. Explain this variability.

CHAPTER 10

ELECTROCHEMISTRY AND OXIDATION-REDUCTION REACTIONS

In previous chapters we have discussed the value of the free energy concept in predicting the direction of physical and chemical reactions. Several methods were described for determining the values of the standard free energy. In this chapter, the easiest and most precise method for measuring free energies of chemical reactions, as well as entropy changes and activity coefficients, will be described. If the reaction can be carried out in an electrochemical cell—for example, a battery—the value of the free energy of the reaction can be calculated from the voltage of the cell. We will see that this procedure is especially valuable in the study of both simple biochemical oxidations and reductions and complex, multistep processes such as the respiratory chain. Finally, the methodology described is the basis for such important laboratory techniques as pH measurement and the determination of the concentration of a particular ion in a complex mixture.

Electrochemical cells are used to study oxidation and reduction reactions. The phenomena that are observed are voltages and the flow of electrons from one substance to another. Thus, prior to beginning the analysis, some of the more important chemical and physical concepts are reviewed in the following section.

10-1 REVIEW OF BASIC ELECTROCHEMICAL CONCEPTS

There are seven basic facts that will be needed about the flow of electrons. These are the following.

1. Electrons move only if there is a difference in electrical potential.
2. Electrons flow from a negative electrical potential to a positive electrical potential.
3. Electrons move easily only through good electrical conductors such as metals, but do not flow through water or ionic solutions. In solution, the flow of charge is the motion of ions, not electrons.
4. At equilibrium, a solution cannot have a net charge nor can one *macro*scopic region of a solution have a positive charge and another part have a negative charge.* That is, the charged units in a solution of electrolytes will always, on the average, be distributed so that the solution and each macroscopic region is electrically neutral. Thus, if an electron, or any other charged particle, flows in or out of one part of a solution, there will be a compensating movement of charge elsewhere so that electroneutrality is achieved at equilibrium. This is called the principle of electroneutrality.
5. Oxidation refers to the acquisition of electrons by an oxidizing agent and loss of electrons by the oxidized material. When an oxidizing agent has taken up electrons, it is said to have become reduced.
6. Reduction refers to loss of electrons by a reducing agent and gain of electrons by the reduced material. When a substance has donated electrons, the substance is said to have become oxidized.
7. Oxidation and reduction reactions must occur together. That is, if one substance in a reaction is oxidized, another substance in the same reaction must be reduced.

Point 7 gives rise to a useful way to think about oxidation-reduction reactions, namely, the half-reaction concept. This is explained in the following section.

10-2 ELECTROCHEMICAL REACTIONS AND ELECTROCHEMICAL CELLS

Half-Reactions

If a Zn rod is placed in a solution of $CuSO_4$, metallic Cu deposits on the Zn rod and Zn^{2+} ions appear in solution. Thus the electrically neutral reaction

$$Zn + Cu^{2+} \rightarrow Zn^{2+} + Cu$$

---
*On a molecular scale, fluctuations can produce a local imbalance of charge. This is discussed in Chapter 14.

has occurred. In this reaction electrons are transferred from Zn to Cu^{2+}; this transfer can be expressed by writing the total reaction as the sum of the following two *half-reactions:*

$$Zn \rightarrow Zn^{2+} + 2e^-$$
$$\underline{Cu^{2+} + 2e^- \rightarrow Cu}$$
$$Cu^{2+} + Zn \rightarrow Cu \quad + Zn^{2+}$$

These half-reactions do not occur in isolation. That is, if the Zn rod is placed in a solution of $ZnSO_4$, there is no net reaction. Some Zn atoms may occasionally go into solution as Zn^{2+} ions but because the release of two electrons would violate the requirement for electroneutrality of the solution, the same or another Zn^{2+} ion would immediately pick up two electrons and be deposited on the rod as a Zn atom. From a thermodynamic point of view, the cause of the failure of this half-reaction is that the movement of two electrons away from a Zn nucleus constitutes electrical work and no work is being done on the system to induce this movement. Admittedly, the positively charged Zn^{2+} ion attracts the electrons, but the total free energy change in the movement will always be zero; this is because as soon as an electron starts to move, it will be attracted back to the ion it has just left. However, under certain experimental conditions, two half-reactions can be coupled and this movement of electrons can occur.

Consider now a metallic copper rod placed in a solution of $ZnSO_4$. In contrast with the previous arrangement, there is no reaction. That is, Cu does not go into solution as Cu^{2+} ions, nor is Zn deposited on the Cu rod. Thus, the reaction $Zn + Cu^{2+} \rightarrow Zn^{2+} + Cu$ proceeds only to the right, as has just been described. One might conclude from examining the half-reactions that Zn can only give up electrons and the Cu^{2+} ion can only accept electrons. However, this does not agree with experimental observations, because there are reactions in which Cu can go into solution and Zn^{2+} ions can be deposited as metallic Zn. What can be said is that the Cu^{2+} ion holds two electrons more tightly than does metallic Zn, so that if only Zn, Cu, Zn^{2+}, and Cu^{2+} are present, the reaction proceeds as written.

The directionality of this reaction can also be described by the free energy concept. The total reaction goes spontaneously to the right so that, as the reaction is written, $\Delta G < 0$. Each of the half-reactions also has an associated value of ΔG, at least one of which must be negative. We shall soon see how to determine the value of ΔG for a half-reaction, using the idea that when an electron is donated or accepted, the movement of charge generates an electrical potential that can be measured.

The analysis of oxidation-reduction reactions is made simpler if a convention is adopted for the direction of half-reactions. This convention is the following:

All half-reactions are written as reductions—that is, the electrons are accepted by the oxidizing agent.

SEC. 10-2 ELECTROCHEMICAL CELLS

Thus the particular half-reactions we have been discussing are written

$$Zn^{2+} + 2e^- \rightarrow Zn$$
$$Cu^{2+} + 2e^- \rightarrow Cu$$

and oxidation by molecular oxygen would be written

$$O_2 + 2e^- \rightarrow 2O^-.$$

This convention can be rememberd by the simple mnemonic of the letter "R":

Reduced form is at the Right.

We now proceed in the following section to the analysis of these reactions by electrical means.

The Galvanic Cell

An arrangement consisting of a Zn rod placed in a solution of $ZnSO_4$ is called a half-cell. The Zn rod is called an *electrode* and the hypothetical half-reaction occurring in this cell is $Zn \rightarrow Zn^{2+} + 2e^-$. Another half-cell in which a Cu electrode is placed in a solution of $CuSO_4$ could also be set up. In neither case would there be a chemical reaction, though. Consider now the following experimental arrangement (Figure 10-1) in which the $ZnSO_4$ and $CuSO_4$ solutions are connected by two conducting paths—namely (1) a wire between the two electrodes and (2) a porous plate to prevent mixing of the solutions. In such a system, which is called a *galvanic cell,* it is found

FIGURE 10-1 An electrochemical cell. The Cu electrode is at a higher potential than the Zn electrode and therefore attracts electrons, resulting in a flow of electrons through the wire in the direction indicated.

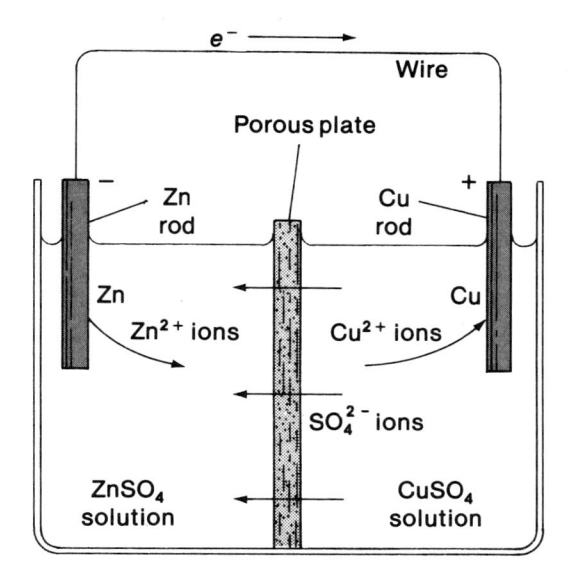

experimentally that Zn from the Zn rod goes into solution as Zn^{2+} ions and Cu^{2+} ions in solution are converted to metallic copper, which deposits on the Cu rod. This is different in physical arrangement from the deposition of Cu on a Zn rod in a $CuSO_4$ solution, but the chemical reactions are identical, namely, $Zn + Cu^{2+} \rightarrow Zn^{2+} + Cu$. An important feature of this arrangement is the following. If the connecting wires are replaced by a high-resistance voltmeter, a difference in electrical potential, a voltage called the *electromotive force* or emf, is observed between the Zn and Cu electrodes. Furthermore, as long as the electrical circuit is complete, electrons flow through the wire or the voltmeter. The movement of electrons is in a single direction—namely, from the Zn electrode to the Cu electrode, as shown in the figure. Both the voltage and the electron flow decrease in time as the concentrations of the Zn^{2+} ions increase and the Cu^{2+} ions decrease. These changes continue until the emf is zero and there is no flow of electrons. What is happening is as follows. Both the connecting wires and the electrodes are metals and thus contain many highly mobile electrons. (A metal has been described as a sea of electrons in which nuclei and ions move about.) Both Zn^{2+} and Cu^{2+} ions in solution are capable of attracting these electrons. However, the removal of an electron from the metal would violate the condition of electroneutrality and would require that another electron enter the metal. The Cu^{2+} ion attracts an electron more strongly than a Zn^{2+} ion does, so that, on the average, the Cu^{2+} ions will be more successful than Zn^{2+} ions at attracting electrons. The deficiency of electrons induced in the wire will immediately be replaced by a Zn atom giving up two electrons and entering the solution as a Zn^{2+} ion. Thus there is a net flow of electrons from the Zn electrode to the Cu electrode. However, if this were the only thing that occurred, the $ZnSO_4$ solution would accumulate Zn^{2+} ions and become positively charged and the $CuSO_4$ solution would become negatively charged by virtue of having an excess of SO_4^{2-} ions. Thus, to preserve electrical neutrality, SO_4^{2-} ions flow through the porous plate from the $CuSO_4$ to the $ZnSO_4$, one SO_4^{2-} ion for every two electrons passing through the wire. Thus the flow of charge consists of movement of electrons through the conductors and the movement of negatively charged ions through the solution.

The motion of the electrons through the wire constitutes electrical work, so that there must be an electrical potential between the two electrodes. Furthermore, the potential of the Cu must be positive compared to that of the Zn, because the negative electrons are flowing from the Zn to the Cu. This means that the potential for the half-reaction $Cu^{2+} + 2e^- \rightarrow Cu$ must be larger than that for the half-reaction $Zn^{2+} + 2e^- \rightarrow Zn$. Here we have seen the value of the half-reaction concept.

Two important points can be made about galvanic cells:

1. The reaction that occurs at the electrode with the higher potential is a reduction. In the cell just discussed, the Cu^{2+} accepts two electrons and thus is reduced.

2. The reaction that occurs at the electrode with the lower potential is an oxidation. In this example, Zn gives up two electrons and thus is oxidized.

These two points can be summarized by the statement that

the half-reaction occurring at the electrode with the higher potential takes electrons away from the half-reaction occurring at the electrode at the lower potential.

This statement is extremely important because it relates the sign of the emf between the electrodes to the direction of the chemical reaction. That is, if the Zn electrode were grounded (i.e., so that its potential is zero), the emf between the Zn and Cu electrodes would be positive for the reaction $Zn + Cu^{2+} \rightarrow Zn^{2+} + Cu$, but if the reaction were instead to go spontaneously from right to left, the emf would be negative. It should be clear, then, that if the emf of a galvanic cell could be predicted, the direction of a reaction would be known. This is analogous to the prediction of the direction of a reaction from the value of ΔG. In our study of free energy changes, it was shown that the value of ΔG for a complete reaction could be determined by addition of the values of the free energy of formation of the components. We shall see that the same type of thing can be done with galvanic cells if some means can be found to obtain the potential for each half-reaction. Just as in assigning values of ΔG, this will require the definition of standard states and the assignment of a potential of zero volts for a particular half-reaction. How this is done is described in the following section.

Standard Electrode Potentials

Let us consider a variation of the cell we have been discussing. In this variant the Cu rod is replaced by platinum (Pt). In this arrangement Zn will still go into solution as Zn^{2+} ions in one half-cell, and Cu^{2+} ions will still be converted to Cu in the other half-cell. The difference is only that Cu will deposit on the Pt rod, whose function is only as an inert electrical conductor that delivers the electrons to the Cu^{2+}-containing half-cell. The Pt itself does not undergo any permanent chemical reaction and is called an *inert* electrode. We will make extensive use of the Pt electrode in the analysis of galvanic cells.

Before proceeding to an analysis of other cells, it is necessary to introduce a standard notation for describing galvanic cells. The following notation is used to describe a half-cell:

Electrode material|species in solution;

when an inert electrode is used, the following notation is frequently useful:

Inert electrode|material associated with the electrode|species in solution .

The vertical line denotes a phase boundary in each case. Thus, if Cu is placed in $CuSO_4$, this is written $Cu|Cu^{2+}$ and the half-reaction is written $Cu^{2+} + 2e^- \rightarrow Cu$. The Pt-$CuSO_4$ half-cell mentioned in the preceding paragraph is written $Pt|Cu^{2+}$.*

To describe a galvanic cell, two half-cells must be combined. Thus the galvanic cell we constructed initially (Figure 10-1) is denoted

$$Zn|Zn^{2+}|Cu^{2+}|Cu .$$

(In some books, the phase boundary between two liquids is denoted by a vertical dashed line—that is, in the cell just described, between Zn^{2+} and Cu^{2+}—but we will not use this notation.) Often the concentrations of the electrolytes are also designated. Thus, if the solutions were $1\ M\ ZnSO_4$ and $0.1\ M\ CuSO_4$, the cell would be written $Zn|1\ M\ Zn^{2+}|0.1\ M\ Cu^{2+}|Cu$.

Let us now consider the cell shown in Figure 10-2. In this cell one half of the vessel contains a Cu electrode immersed in $CuSO_4$, and in the other half is a platinum disc over which H_2 gas is bubbled, immersed in H_2SO_4. The Pt is an inert substance. The half-cell is denoted $Pt|H_2(gas)|H^+$ and the half-reaction is $H_2 + 2e^- \rightarrow 2H^-$. The electrode in this arrangement is called a Pt-H_2 electrode; it functions because H_2 adsorbs strongly to the surface of Pt, as described in Chapter 14. With this cell there is also a flow of electrons through the wire from the Pt-H_2 electrode to the Cu electrode. H_2 dissociates to increase the H^+ concentration, Cu^{2+} ions leave the $CuSO_4$, depositing Cu on the Cu electrode, and SO_4^{2-} ions flow from the $CuSO_4$ solution to the H_2SO_4. Thus, by the reasoning used previously, the poten-

*For convenience, in this chapter a hyphen will be used in place of a vertical line if only some components of a half-cell or an electrode are referred to.

FIGURE 10-2 An electrochemical cell in one half of which is a Pt-H_2 electrode formed by passing bubbles of gaseous H_2 over a platinum disc. The Cu electrode in the other half is at a higher potential than the Pt-H_2 electrode, so that electrons flow through the connecting wire from the Pt-H_2 electrode to the Cu electrode.

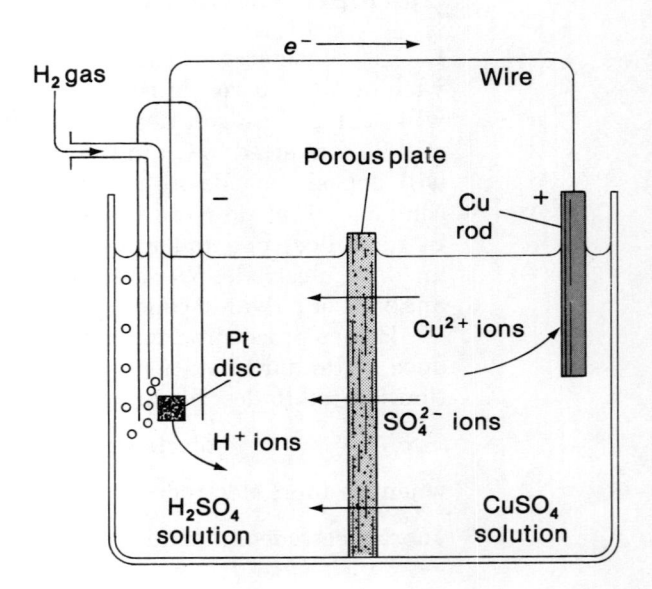

SEC. 10–2   ELECTROCHEMICAL CELLS   327

FIGURE 10-3 Two electrochemical cells in series. The two Pt-H$_2$ electrodes are attached to one another and the Zn and Cu electrodes are connected. Electrons and ions move as indicated. Note that in the cell at the left, H$^+$ ions pick up electrons to become H$_2$ gas while in the cell at the right, H$_2$ gas gives up electrons and dissociates to form H$^+$ ions.

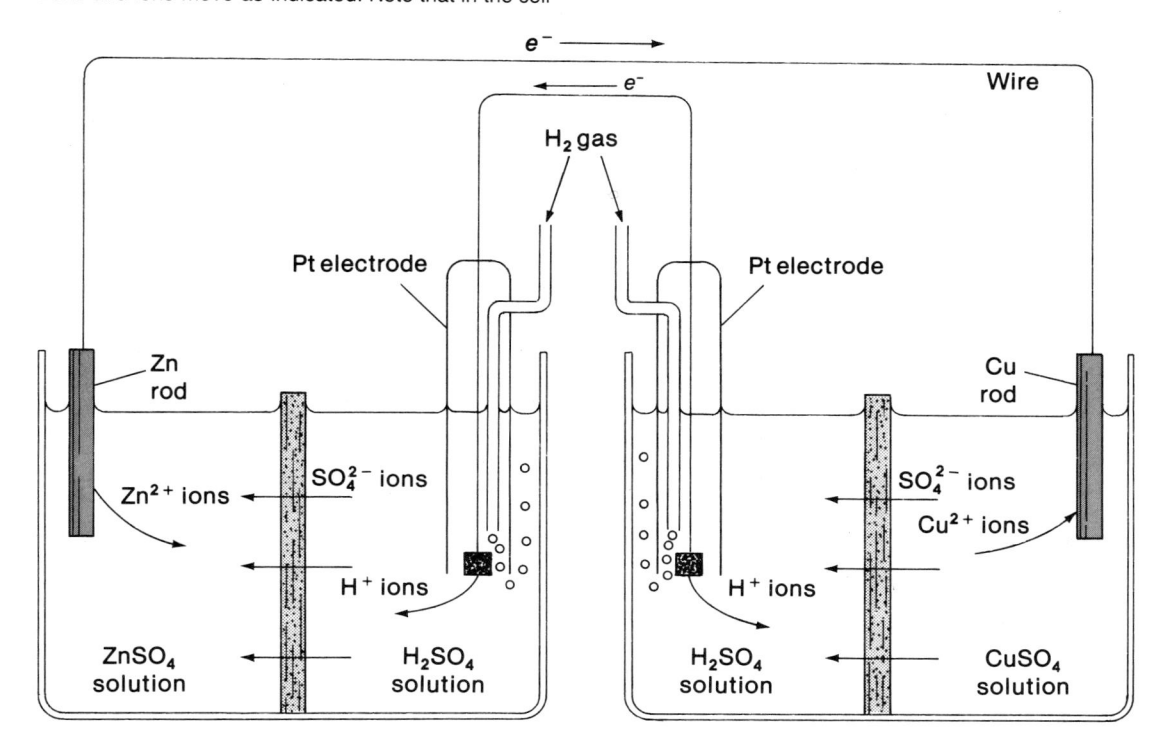

tial of the half-reaction $Cu^{2+} + 2e^- \rightarrow Cu$ must be greater than that for $2H^+ + 2e^- \rightarrow H_2$. However, if the half-cell containing the Cu and CuSO$_4$ were instead to contain a Zn electrode immersed in ZnSO$_4$, electrons would flow from the Zn electrode to the Pt-H$_2$ electrode, Zn would go into solution, and H$_2$ gas would be produced. Thus, the reverse half-reaction $2H^+ + 2e^- \rightarrow H_2$ would occur in this cell and it would have a higher potential than the half-reaction $Zn^{2+} + 2e^- \rightarrow Zn$. An important feature of these cells is that when the Cu|Cu^{2+} half-cell is replaced by the Zn|Zn^{2+} half-cell, the sign of the emf between the two electrodes in the cell is reversed. This is in accord with the direction of electron flow being determined by the relative values of the potentials of each half-cell.

Let us now examine the result of linking two cells in series (Figure 10-3) in such a way that the two Pt-H$_2$ electrodes are connected and the Zn and Cu electrodes are also linked. This system is

$$\{Zn|Zn^{2+}|H^+|H_2|Pt\} - \{Pt|H_2|H^+|Cu^{2+}|Cu\} \;.$$

Electrons will again flow from Zn to Cu and, furthermore, the measured emf will be the same as for the Zn|Zn^{2+}|Cu^{2+}|Cu cell first described. This illustrates the additivity of the emf—exactly the same phenomenon as the fact that two 1½-volt flashlight batteries connected in series combine to give a 3-volt unit.

We now consider another circuit in which the two cells just discussed are in series but in which the polarity of one is reversed—namely, the arrangement

$$\{Zn|Zn^{2+}|H^+|H_2|Pt\} - \{Cu|Cu^{2+}|H^+|H_2|Pt\}.$$

In this circuit electrons still flow in the same clockwise direction as in Figure 10-3; however, the flow is less and the combined emf of the two cells is smaller. Let us denote the emf of the $Zn|Zn^{2+}|H^+|H_2|Pt$ cell as $\mathcal{E}_1$ and that of the $Cu|Cu^{2+}|H^+|H_2|Pt$ cell as $\mathcal{E}_2$. Thus, when the two cells are in a series as in Figure 10-3, the total voltage is $\mathcal{E}_1 + \mathcal{E}_2$; when the Cu-H_2 cell is reversed, it contributes $-\mathcal{E}_2$ to the sum, which now yields a net positive voltage of $\mathcal{E}_1 - \mathcal{E}_2$. Thus, the fact that there is still a positive voltage with the inverted Cu-H_2 cells means that $\mathcal{E}_1 - \mathcal{E}_2 > 0$ or simply that $\mathcal{E}_1 > \mathcal{E}_2$.

At this point, we should observe that in both arrangements the total emf in the series is independent of the value of the potential of the Pt-H_2 electrode. That is, if the potentials, ϕ, of the Zn, Cu, and H_2 half-cells are ϕ_{Zn}, ϕ_{Cu}, and ϕ_{H_2}, the observed voltages (which, of course, are potential differences) are $\mathcal{E}_1 = \phi_{Zn} - \phi_{H_2}$ and $\mathcal{E}_2 = \phi_{H_2} - \phi_{Cu}$, so that $\mathcal{E}_1 - \mathcal{E}_2 = \phi_{Zn} + \phi_{Cu} - 2\phi_{H_2}$.

In this analysis we have implicitly been using the Pt-H_2 electrode as a reference electrode. Clearly it is desirable that the total voltage observed for any order of connecting the cells be independent of the potential of the reference electrode. This can be accomplished simply by an assignment of the value of zero for the potential. This is not unreasonable, because the absolute potential of a half-cell is not measurable—only potential differences can be measured. Thus,

the potential of any half-reaction is defined as the voltage measured when the appropriate half-cell is connected to a $Pt|H_2|H^+$ half-cell.

Two problems arise in the determination of these potentials:

1. The emf of a galvanic cell depends on the concentrations of the electrolytes. Thus, if standard values of the emf are to be obtained, standard conditions must be defined. Since we will later relate the potentials to free energy changes, it is convenient to choose the standard conditions to be the same as those for evaluating $\Delta G°$. Thus standard electrode potentials are the potentials of solutions at a concentration of 1 molar and at a pressure of 1 atmosphere. Furthermore, the measurements are always carried out at 25°C.

2. The emf of a galvanic cell is not constant in time. This is because these cells are not systems in equilibrium—the concentrations of the constituents continually change until the current and voltage become zero. Hence it is necessary to measure the voltage in such a way that

FIGURE 10-4 A potentiometric circuit for measuring the voltage of an electrochemical cell. A fraction of the voltage of a battery is selected to oppose the voltage of the cell. When the voltages cancel, the galvanometer indicates that no current is flowing. The value of the fraction of the battery voltage is known and equals the cell voltage.

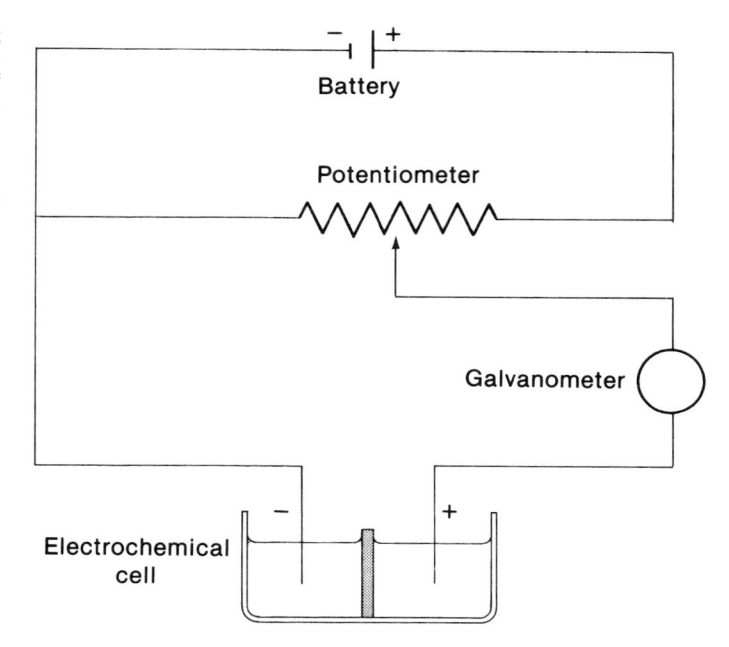

the conditions do not change. In order to avoid the change in concentrations, the measurement is performed by using a potentiometer (Figure 10-4). This is merely a circuit in which a very high resistance is placed between the electrodes in order that the electrical current is very small; then the voltage required to stop the flow of current is determined. If standard conditions and a temperature of 25°C are employed, this voltage is called the *standard electrode potential* and it is designated by $\mathscr{E}°$. Because of the convention that all half-reactions are written as reductions, the potentials are *standard reduction potentials*.

By setting up an appropriate galvanic cell, the standard electrode potential of any half-reaction can be measured with respect to a Pt-H_2 electrode. A list of standard reduction potentials for half-cells is given in Table 10-1. Note that the choice of the Pt-H_2 electrode as that having a value of 0.00 volt means that some potentials have a positive sign and some have a negative sign. It should also be realized that the sign is entirely arbitrary and depends on the way that the reaction is written. Thus, reaction $Zn^{2+} + 2e^- \rightarrow Zn$ has a standard potential of -0.763 volt, whereas, if written $Zn \rightarrow Zn^{2+} + 2e^-$, the potential would be $+0.763$ volt.

The Use of Standard Reduction Potentials

When all the reactants shown in Table 10-1 are in the standard state (1 M and 1 atm), the table allows us to predict the chemical reactions that occur

TABLE 10-1 Standard electrode reduction potentials at 25°C.

Half-cell*	Half-reaction	$\mathscr{E}°$, volt
Pt|Ce^{3+},Ce^{4+}	Ce^{4+} + e^- → Ce^{3+}	+1.61
Pt|Mn^{2+},MnO$_4^-$	MnO$_4^-$ + 8H$^+$ + 5e^- → Mn^{2+} + 4H$_2$O	+1.51
Pt|Cl$_2$|Cl$^-$	Cl$_2$ + 2e^- → 2Cl$^-$	+1.36
Pt|Cr^{3+},Cr$_2$O$_7^{2-}$	Cr$_2$O$_7^{2-}$ + 14H$^+$ + 6e^- → 2Cr^{3+} + 7H$_2$O	+1.33
Pt|O$_2$|H$^+$	O$_2$ + 2H$^+$ + 2e^- → H$_2$O	+1.229
Hg|Hg^{2+}	Hg^{2+} + 2e^- → Hg	+0.854
Ag|Ag$^+$	Ag$^+$ + e^- → Ag	+0.799
Pt|Fe^{2+},Fe^{3+}	Fe^{3+} + e^- → Fe^{2+}	+0.771
Pt|I$_2$|I$^-$	I$_2$ + 2e^- → 2I$^-$	+0.54
Cu|Cu$^+$	Cu$^+$ + e^- → Cu	+0.52
Pt|O$_2$|OH$^-$	O$_2$ + H$_2$O + 4e^- → 4OH$^-$	+0.40
Pt|Fe(CN)$_6^{4-}$,Fe(CN)$_6^{3-}$	Fe(CN)$_6^{3-}$ + e^- → Fe(CN)$_6^{4-}$	+0.36
Cu|Cu^{2+}	Cu^{2+} + 2e^- → Cu	+0.337
Hg|Hg$_2$Cl$_2$|Cl$^-$	Hg$_2$Cl$_2$ + 2e^- → 2Hg + 2Cl$^-$	+0.28
Ag|AgCl|Cl$^-$	AgCl + e^- → Ag + Cl$^-$	+0.22
Pt|Cu$^+$,Cu^{2+}	Cu^{2+} + e^- → Cu$^+$	+0.15
Pt|Sn^{2+},Sn^{4+}	Sn^{4+} + 2e^- → Sn^{2+}	+0.15
Pt|H$_2$|H$^+$	2H$^+$ + 2e^- → H$_2$	0.00
Pt|D$_2$|D$^+$	2D$^+$ + 2e^- → D$_2$	−0.0034
Fe|Fe^{3+}	Fe^{3+} + 3e^- → Fe	−0.036
Pb|Pb^{2+}	Pb^{2+} + 2e^- → Pb	−0.13
Sn|Sn^{2+}	Sn^{2+} + 2e^- → Sn	−0.14
Ni|Ni^{2+}	Ni^{2+} + 2e^- → Ni	−0.255
Cd|Cd^{2+}	Cd^{2+} + 2e^- → Cd	−0.40
Fe|Fe^{2+}	Fe^{2+} + 2e^- → Fe	−0.44
Cr|Cr^{3+}	Cr^{3+} + 3e^- → Cr	−0.744
Zn|Zn^{2+}	Zn^{2+} + 2e^- → Zn	−0.763
Mn|Mn^{2+}	Mn^{2+} + 2e^- → Mn	−1.18
Al|Al^{3+}	Al^{3+} + 3e^- → Al	−1.66
Mg|Mg^{2+}	Mg^{2+} + 2e^- → Mg	−2.37
Na|Na$^+$	Na$^+$ + e^- → Na	−2.71
Ca|Ca^{2+}	Ca^{2+} + 2e^- → Ca	−2.87
Cs|Cs$^+$	Cs$^+$ + e^- → Cs	−2.92
K|K$^+$	K$^+$ + e^- → K	−2.93
Li|Li$^+$	Li$^+$ + e^- → Li	−3.05

*When two substances in a half-cell are separated by a comma, they are contained in the same phase.

in a galvanic cell, the sign of the emf, and the direction of electron flow. In addition, we can predict the outcome of a chemical reaction between the constituents of the half-cells, if the constituents were to be mixed. The rules are these:

1. The oxidizing agent of a half-reaction that is higher in Table 10-1 will oxidize the reduced form in any half-reaction that is lower in the table.

SEC. 10–2   ELECTROCHEMICAL CELLS

2. To obtain the emf of the cell for the spontaneous reaction, reverse the sign of $\mathscr{E}°$ for a half-reaction that is lower in the table and add it to $\mathscr{E}°$ for a higher half-reaction. That is,

$$\mathscr{E}°_{total} = \mathscr{E}°_{higher} + (-\mathscr{E}°_{lower}) .$$

3. The higher half-reaction draws electrons away from any lower half-reaction. Thus electrons flow through the connecting wire away from the electrode corresponding to the lower half-reaction. A mnemonic for this uses the letter "L": *electrons Leave the Lower.*

In the Zn-Cu example we have been discussing, the half-reactions and their potentials are

$$Zn^{2+} + 2e^- \rightarrow Zn, \quad -0.763 \text{ volt}$$

and

$$Cu^{2+} + 2e^- \rightarrow Cu, \quad +0.337 \text{ volt} .$$

The larger potential for the Cu half-reaction says that Cu can draw electrons from Zn. Thus, Zn must be converted to Zn^{2+} and Cu^{2+} will be converted to Cu, and the total reaction must be $Zn + Cu^{2+} \rightarrow Zn^{2+} + Cu$. To obtain this reaction requires that the Zn-Zn^{2+} half-reaction be subtracted from the Cu-Cu^{2+} half-reaction, so that the emf of the cell is $0.337 - (-0.763) = 1.100$ volt (if both solutions are 1 molar). The Zn will be the negative electrode and the Cu will be the positive electrode, and the flow of electrons will be from Zn to Cu.

An important point can be made if we consider a cell in one half of which a Cu electrode is immersed in $Cu(NO_3)_2$ and in the other half of which a Ag electrode is immersed in $AgNO_3$. (We have chosen the nitrate instead of the sulfate so that we do not have to consider the fact that Ag_2SO_4 is not sufficiently soluble to achieve a concentration of 1 M.) The half-reactions and potentials are

$$Ag^+ + e^- \rightarrow Ag, \quad +0.799 \text{ volt}$$

and

$$Cu^{2+} + 2e^- \rightarrow Cu, \quad +0.337 \text{ volt}.$$

The first half-reaction has a higher potential than the second, so that Ag can draw electrons from Cu. Thus, Ag^+ is converted to Ag and Cu to Cu^{2+} or

$$2Ag^+ + Cu \rightarrow 2Ag + Cu^{2+} .$$

To obtain this reaction, one must subtract the Cu half-reaction from the Ag half-reaction, so that the voltage produced by the cell is $0.799 - 0.337 = 0.462$ volt. The Cu is the negative electrode, the Ag is the positive electrode, and the electrons flow from the Cu to the Ag. Note that in summing the half-reactions we have used the first half-reaction twice. That is,

$$2Ag^+ + Cu \rightarrow 2Ag + Cu^{2+}$$

is the sum of the reactions

$$2\{Ag^+ + e^- \rightarrow Ag\} \quad \text{and} \quad \{Cu + 2e^- \rightarrow Cu^{2+}\}.$$

The question that arises is this: should we have doubled the potential for the first half-reaction and written the emf of the cell as $2(0.799) - (0.337) = 1.261$ volt? Turning to experiment, we note that the measured voltage is in fact 0.462 volt, so that apparently it does not depend upon how the reaction is summed. This is because *the potential is a function of the nature of the reactants, not of their number.* For the same reason the voltage of a cell is independent of the volume of liquid and the dimensions of the container. A more complete argument concerning this point will be made later when we relate the electrode potential to $\Delta G°$.

Oxidation-Reduction Reactions That Do Not Include Metals or Other Elements

Oxidation-reduction reactions do not occur only between metals and cations. For example, Fe^{3+} and Fe^{2+} ions can be interconverted by reduction and oxidation, respectively. The potential for the half-reaction $Fe^{3+} + e^- \rightarrow Fe^{2+}$ can be easily measured in a galvanic cell consisting of the standard $Pt|H_2|H^+$ half-cell and a half-cell consisting of Pt foil immersed in a solution containing both $1\ M$ $FeCl_2$ and $1\ M$ $FeCl_3$. The inert Pt foil is used because if an Fe electrode were used, the interconversion of Fe to both Fe^{3+} and Fe^{2+} would be measured rather than the potential of interest. Table 10-1 indicates a potential of 0.771 volt for the half-reaction $Fe^{3+} + e^- \rightarrow Fe^{2+}$; thus, the Pt electrode (or, more accurately, because two species are present in the solution, the $Pt|Fe^{2+}, Fe^{3+}$ electrode), is at a lower potential than the $Pt-H_2$ electrode, electrons flow to the foil Pt electrode through the wire from the $Pt-H_2$ electrode, H^+ ions are produced, and the Fe^{3+} ion is reduced to the Fe^{2+} ion in the region around the foil Pt electrode. Thus, we know that bubbling H_2 gas through a solution of $FeCl_3$ results in the reaction

$$H_2 + 2Fe^{3+} \rightarrow 2H^+ + 2Fe^{2+}.$$

The use of two inert electrodes allows us to extend the range of reactions that can be studied beyond simple metal-cation and cation-cation reactions. For example, consider a cell in which one half-cell is the $Pt|H_2|H^+$ system but the other contains $1\ M$ $H_2Cr_2O_7$ into which is placed a foil Pt electrode. The $Cr_2O_7^{2-}$ ion can be converted to the Cr^{3+} ion, so that the half-reactions that could occur in each cell are

$$2H^+ + 2e^- \rightarrow H_2, \quad 0.00 \text{ volt},$$

and

$$Cr_2O_7^{2-} + 14H^+ + 6e^- \rightarrow 2Cr^{3+} + 7H_2O, \quad 1.33 \text{ volt}.$$

The total potential for the reaction is positive, so that it occurs as written. Examination of Table 10-1 further shows that the $Cr_2O_7^{2-}$ ion can convert Fe^{2+} to Fe^{3+}, because the $Cr_2O_7^{2-}$ reduction half-reaction has a higher potential than the $Fe^{3+} + e^- \rightarrow Fe^{2+}$ half-reaction. On the other hand, if the $Cr_2O_7^{2-}$ ion is mixed with the Ce^{4+} ion, neither is oxidized nor reduced; rather, $Ce^{4+} + Cr^{3+}$ result in the formation of Ce^{3+} and $Cr_2O_7^{2-}$.

In biochemistry, one is not usually interested in the simple inorganic reactions just described. However, the examples we have given in which the inert Pt electrode was used indicate how to measure the potentials for biochemical half-reactions. For example, consider the conversion of maleic acid to succinic acid:

$$\begin{array}{c} HC\!-\!COOH \\ \| \\ HC\!-\!COOH \end{array} + 2H^+ + 2e^- \rightarrow \begin{array}{c} CH_2COOH \\ | \\ CH_2COOH \end{array}$$

This could be studied by using a half-cell containing both maleic acid and succinic acid into which is placed a Pt electrode, and the voltage between this half-cell and the $Pt|H_2|H^+$ half-cell could be measured. This type of reaction will be discussed in greater detail later in the chapter.

It should be noted that this reaction includes the H^+ ion. Many chemical reactions include the H^+ or OH^- ions and H_2O, and thus are strongly pH-dependent. They are never carried out under standard conditions ($[H^+]$ = 1 M) because they are enzyme-catalyzed and the enzymes are inactive at such low pH. The convention is that the potentials are measured at a standard pH of 7.0, giving rise to biochemical standard values of $\mathscr{E}°$ (pH 7) analogous to the biochemical free energy change denoted $\Delta G°'$.

10–3 REFERENCE ELECTRODES

A $Pt\text{-}H_2$ electrode is somewhat dangerous: it is extremely sensitive to small impurities in the H_2 gas and to the gas pressure. This has led to the development of reference electrodes whose potentials with respect to the $Pt\text{-}H_2$ electrode are accurately known. To understand how these electrodes work, let us consider a silver electrode onto which is deposited AgCl, a very insoluble salt. This electrode, which is denoted by $Ag|AgCl|Cl^-$, is commonly called a Ag-AgCl electrode. It is a useful and frequently encountered electrode. This electrode can be put into 1 M HCl to make a standard half-cell. If this is coupled to a $Pt|H_2|H^+$ half-cell, a voltage of 0.222 volt appears between the electrodes, with the Ag-AgCl electrode at the higher potential. The half-reaction occurring at this electrode is $AgCl + e^- \rightarrow Ag + Cl^-$. Since this potential is higher than that of the $Pt\text{-}H_2$ electrode, the electrons flow from the $Pt\text{-}H_2$ electrode to the Ag-AgCl electrode. The source of electrons is the dissociation of H_2, which increases the H^+-ion concentration in the HCl; the electrons are then picked up by the

AgCl, which releases Cl⁻ to the HCl, maintaining electroneutrality, and adds Ag to the silver already present. Simple chemical considerations indicate that if the H⁺-ion and Cl⁻-ion concentrations are very high, the half-reactions $H_2 \rightarrow 2H^+ + 2e^-$ and $AgCl + e^- \rightarrow Ag + Cl^-$ will proceed to the right less effectively. Note, however, that the Pt-H₂ electrode is only affected by the H⁺-ion concentration and the Ag-AgCl electrode is sensitive only to the Cl⁻ concentration. Thus, if the material in the cell were Na₂SO₄ and NaCl, the observed voltage would depend only on the NaCl concentration. This means that if we were to choose the Ag-AgCl electrode as a reference, the observed potential would always depend upon the concentration of the Cl⁻ ion. One way to avoid this problem is to include a standard concentration of Cl⁻ in all reactions—obviously not an ideal arrangement. However, if the electrode were in contact with an inert solution containing Cl⁻, for instance KCl, and this solution were in turn in contact (but without mixing) with the solution containing the reactants of interest, the problem would be solved. Figure 10-5 shows an arrangement that accomplishes this. That is, a solution of KCl is contained within a glass sleeve and the KCl solution is in contact with the solution of interest by means of a porous plug. The Ag-AgCl electrode is immersed in the KCl solution within the sleeve. In order to avoid changes in potential that might result from leakage of KCl out of the unit and hence a change in KCl concentra-

FIGURE 10-5 A schematic diagram of a Ag-AgCl reference electrode. The Ag is coated with a layer of AgCl. This is immersed in saturated KCl. Contact with the solution to be studied is through a porous plug.

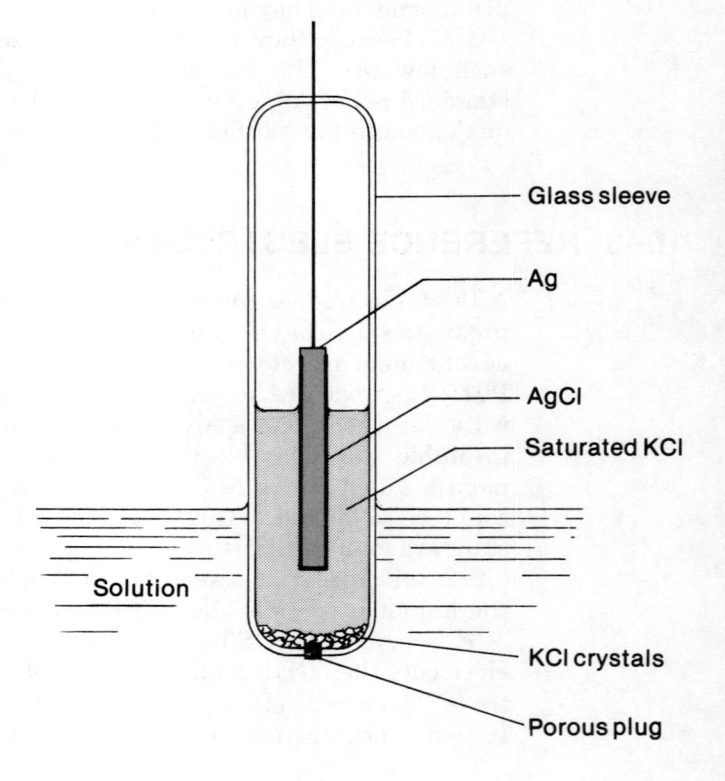

tion, the KCl in the sleeve is a saturated solution containing excess KCl crystals. Once the Ag-AgCl electrode has been calibrated against a Pt-H_2 electrode, it can serve as a reference in any reaction.

The Ag-AgCl reference electrode can also be used without a sleeve containing KCl, as long as the Cl^- concentration is known. (In a later section we will show how to correct the observed potential for the Cl^- concentration.) This electrode is very useful because it can be made extremely small—for example, it can be a nearly microscopic Ag wire coated with a very thin layer of AgCl. This has had widespread use in neurophysiology because the electrode can be made small enough that it can be inserted into a single nerve cell and used to measure membrane potentials.

Another important point to be noticed in the description just given is that an electrode may contain more than one nonmetallic component and these components can be insoluble. It is noteworthy also that the Ag-AgCl electrode could be replaced by a Pt rod coated first with Ag and then with a second layer, of AgCl, and the same voltage would be observed, because the Pt does not take part in any of the reactions. This principle will be seen in the following discussions of another reference electrode.

The most popular reference electrode is the *calomel* electrode shown in Figure 10-6. It is denoted $Hg|Hg_2Cl_2|Cl^-$; the half-reaction is $Hg_2Cl_2 + 2e^- \rightarrow 2Hg + 2Cl^-$. Since Hg is a liquid, it cannot be coated with the insoluble salt Hg_2Cl_2. Thus, the electrode contains a small pool of Hg into which is inserted a Pt wire. Above the Hg is a paste consisting of Hg_2Cl_2, KCl, and saturated KCl; this paste is called calomel. KCl crystals are present for the same reason as in the Ag-AgCl electrode. With saturated KCl, the potential is 0.245 volt compared to the Pt-H_2 electrode.

10-4 STANDARD ELECTRODE POTENTIALS AND THERMODYNAMIC QUANTITIES

Electrode potentials can be used to determine the direction of a particular oxidation-reduction reaction and the voltage of a galvanic cell, as long as the components are at an activity of one molar. This latter requirement arises because the standard electrode potentials are recorded at one molar and because potentials vary significantly with concentration. Invariably, information is needed about other concentrations, and this is easily obtained. In examining the concentration dependence we will also see how electrochemical measurements can be used to measure ΔG, activity coefficients, pH, and the concentration of particular ions in solution.

Evaluation of the Standard Free energy Change $\Delta G°$

In a galvanic cell electrons move because of the difference in potential between the two electrodes. This motion requires that work be done on the electrons. If the potential difference or emf is $\mathscr{E}$, the work done by the chemical reaction that results in the movement of N electrons is

$$\text{Electrical work} = Ne\mathscr{E} . \tag{1}$$

This is available work and is therefore equal to the change in free energy of the reaction. It is convenient to express Equation 1 on a molar basis. This is done by introducing the faraday, $\mathscr{F}$, which is the charge of one mole of electrons, and n, the number of moles, to obtain

$$\Delta G = -n\mathscr{F}\mathscr{E} . \tag{2}$$

The negative sign in this equation results from the two conventions that if a reaction occurs spontaneously, $\Delta G < 0$ and $\mathscr{E} > 0$. Note that the units and constants of Equation 2 are the following: $\mathscr{E}$ is given in units of V (volt) the charge of the electron is 1.6×10^{-19} C (coulomb), $\mathscr{F}$ is 96,500 C, and ΔG is expressed as J mol^{-1}.

In any chemical reaction for which the standard free energy change is

FIGURE 10-6 A schematic diagram of a calomel reference electrode. An inner sleeve holds a pool of Hg in which is the calomel paste. A Pt wire is in contact with the Hg and is isolated from both the calomel and the KCl solution by another glass sleeve. The calomel is in contact with saturated KCl, which, in turn contacts the solution of interest via a porous plug.

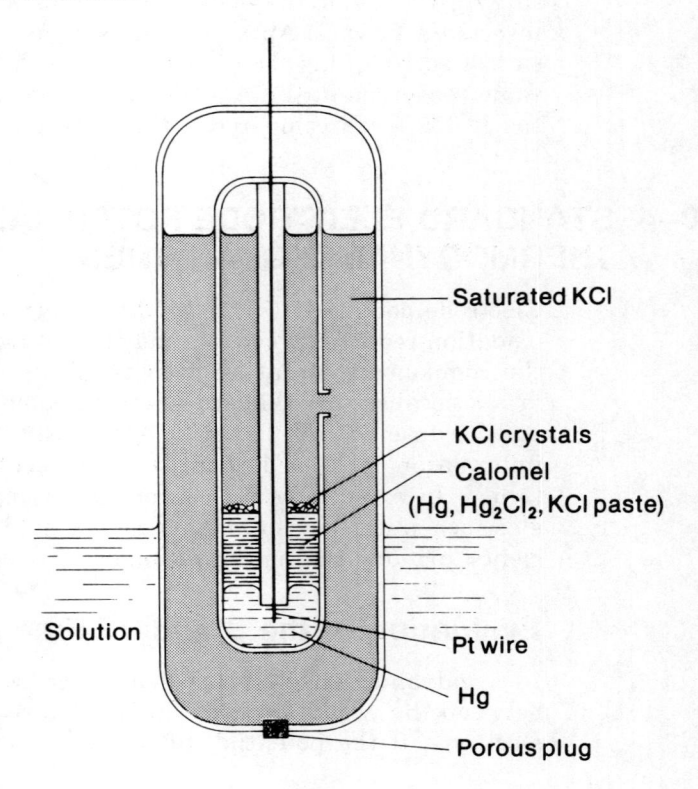

SEC. 10-4 STANDARD ELECTRODE POTENTIALS

$\Delta G°$, the free energy change ΔG in particular conditions is

$$\Delta G = \Delta G° + RT \ln Q ,\qquad(3)$$

in which Q represents the proper quotient of activities (for example, concentrations, or pressures). Combining Equations 2 and 3 leads to

$$-n\mathcal{F}\mathcal{E} = \Delta G° + RT \ln Q .\qquad(4)$$

The value of $\mathcal{E}°$ is defined as the electrode potential that occurs when all constituents are at unit activity, namely, when $Q = 1$ and $\ln Q = 0$. Therefore

$$\Delta G° = -n\mathcal{F}\mathcal{E}° .\qquad(5)$$

This equation, which is called the Nernst equation, indicates that $\Delta G°$ for a particular chemical reaction can be evaluated from a simple measurement of $\Delta \mathcal{E}°$ of a galvanic cell in which the reaction is occurring.

In Chapter 6, it was shown that

$$\Delta G° = -RT \ln K ,\qquad(6)$$

in which K is the equilibrium constant for the reaction. Combining Equations 5 and 6 yields

$$\mathcal{E}° = \frac{RT}{n\mathcal{F}} \ln K ,\qquad(7)$$

which shows that the equilibrium constant K is also easily measured in a galvanic cell. Combining Equations 4 and 5 yields the highly useful equation

$$\mathcal{E} = \mathcal{E}° - \frac{RT}{n\mathcal{F}} \ln Q .\qquad(8)$$

This equation allows the calculation of $\mathcal{E}$ and hence of ΔG for any reaction at any concentration of reactants. To use this equation correctly requires understanding the meaning of Q and of n. The value of Q is obtained by multiplying the activities of all of the reaction products and dividing by the activity of each reactant. Thus, for a reaction $A + B \rightarrow C + D$, $Q = a_C a_D / a_A a_B$. The value of n is equal to the number of electrons transferred in the reaction. Hence for a reaction $A + B^{2+} \rightarrow B + A^{2+}$, the half-reactions are $A \rightarrow A^{2+} + 2e^-$ and $B^{2+} + 2e^- \rightarrow B$, so that two electrons are transferred and $n = 2$. Similarly, for the reaction $C + D^{3+} \rightarrow D + C^{3+}$, $n = 3$. Finally, consider the oxidation of lactic acid ($CH_3CHOHCOOH$) by Fe^{3+} to form pyruvic acid ($CH_3COCOOH$). The half-reactions are

$$\begin{array}{c}CH_3\\|\\CHOH\\|\\COOH\end{array} \rightarrow \begin{array}{c}CH_3\\|\\C{=}O\\|\\COOH\end{array} + 2H^+ + 2e^-$$

and

$$Fe^{3+} + e^- \rightarrow Fe^{2+} .$$

The complete reaction is

$$\text{Lactic acid} + 2\text{Fe}^{3+} \rightarrow \text{pyruvic acid} + 2\text{H}^+ + 2\text{Fe}^{2+},$$

and $n = 2$.

The use of the preceding equation is shown in the following examples. In Example 10–A, Equation 8 is used to predict the voltage of a galvanic cell whose components are not in standard conditions. In Example 10–B, Equations 5 and 8 are used to calculate $\Delta G°$.

EXAMPLE 10–A What is the voltage of a cell containing a Zn|ZnSO$_4$ half-cell and a Cu|CuSO$_4$ half-cell in which the concentration of ZnSO$_4$ is 0.1 M and that of CuSO$_4$ is 0.2 M and the temperature is 25°C? Assume activity equals concentration.

From Table 10-1, the voltage at standard conditions (that is, 1 M activity) is $0.337 + 0.763 = 1.1$ volt. The chemical reaction in the cell is $\text{Zn} + \text{Cu}^{2+} \rightarrow \text{Cu} + \text{Zn}^{2+}$. This requires a transfer of two electrons, so that for every mole of Zn oxidized to Zn^{2+}, two faradays of charge move. Use of Equation 8 with $n = 2$ yields

$$\mathcal{E} = \mathcal{E}° - \frac{RT}{2\mathcal{F}} \ln \frac{[\text{Cu}][\text{ZnSO}_4]}{[\text{Zn}][\text{CuSO}_4]}.$$

By definition the activity of any element in its standard state is 1, so that

$$\mathcal{E} = 1.1 - \frac{(8.314)(298)}{2(96500)} \ln \frac{(1)(0.1)}{(1)(0.2)} = 1.109 \text{ volt}.$$

Of course, for a precise calculation of $\mathcal{E}$, activities instead of concentrations must be used.

EXAMPLE 10–B What is $\Delta G°$ at 25°C for the reaction $\text{A}^{3+} + \text{B}^+ \rightarrow \text{A}^{2+} + \text{B}^{2+}$. The concentrations are: 0.05 M; ACl$_2$, 0.2 M; BCl, 0.4 M; BCl$_2$, 0.2 M.

With a calomel electrode, a voltage of 0.842 volt is measured. Thus, compared to a Pt-H$_2$ electrode, $\mathcal{E} = 0.842 - 0.241 = 0.601$. The reaction involves transfer of one electron, so that $n = 1$. Rearranging Equation 8,

$$\mathcal{E} = 0.601 - \frac{RT}{\mathcal{F}} \ln \frac{(0.2)(0.2)}{(0.5)(0.4)} = 0.524 \text{ volt}.$$

Using Equation 5,

$$\Delta G° = -n\mathcal{E}°\mathcal{F} = (1)(0.524)(96500) = 5.06 \times 10^4 \text{ J mol}^{-1}.$$

Temperature Dependence of $\mathscr{E}$ and the Evaluation of Entropy and Entropy Changes

The emf of a galvanic cell can decrease with temperature, a fact that is well known to anyone who has attempted to start an automobile on a cold winter morning. Equation 8 includes the temperature in one term and appears to provide a direct means for calculating $d\mathscr{E}/dT$. However, Equation 8 is deceptive, because $\mathscr{E}°$ is also temperature dependent. The value of $d\mathscr{E}/dT$ can be calculated from ΔS for the reaction. However, since $d\mathscr{E}/dT$ is easily measured simply by measuring the emf at various temperatures, the equation that will be derived is primarily used as a means of evaluating ΔS.

The temperature dependence, $d\mathscr{E}/dT$, is an easily measurable quantity and can be used to evaluate ΔS and ΔH from electrical measurements. We have seen in Chapter 6 that at constant pressure

$$\Delta S = -\frac{d(\Delta G)}{dT}, \tag{9}$$

which, from Equation 2, can be written

$$\Delta S = n\mathscr{F}\frac{d\mathscr{E}}{dT}. \tag{10}$$

This equation allows a simple measurement of ΔS for any reaction. If $\Delta S°$ is required, this is simply $n\mathscr{F}d\mathscr{E}°/dT$. Thus, if a reaction in which two electrons are transferred is set up in a galvanic cell and $d\mathscr{E}/dT$ is found to be 0.0023 volt K^{-1}, then $\Delta S = 2(96500)(0.0023) = 444$ J K^{-1} mol^{-1}. In general, $d\mathscr{E}/dT$ is found to depend on T because ΔS is a function of T. However, for a small range of temperature, ΔS is constant.

Equation 10 provides an electrochemical means for evaluating ΔH. From the equation $\Delta H = \Delta G + T\Delta S$,

$$\Delta H = -n\mathscr{E}\mathscr{F} + n\mathscr{F}T\frac{d\mathscr{E}}{dT} = -n\mathscr{F}\left(\mathscr{E} + T\frac{d\mathscr{E}}{dT}\right). \tag{11}$$

To obtain $\Delta H°$, $\mathscr{E}$ is merely replaced by $\mathscr{E}°$ in the equation above. Thus, if a reaction in which two electrons are transferred has an emf of -1.26 volt at 25°C and a value of $d\mathscr{E}/dT = 0.0014$ vol K^{-1},

$$\Delta H = -2(96500)[-1.26 + 298(0.0014)] = -163 \text{ kJ mol}^{-1}.$$

Note that the value of $\mathscr{E}$ is required for the measurement of ΔH but not for ΔS.

In an earlier section, the question of adding electrode potentials was discussed. Recall that it was explained that if the stoichiometry of a complete reaction requires doubling all of the components of a half-reaction, the standard electrode potential of the half-reaction is not doubled.

This can easily be shown by converting the electrode potentials $\mathscr{E}_1$ and $\mathscr{E}_2$ for both half-reactions to ΔG_1 and ΔG_2 by using Equation 5, $\Delta G =$

$-n\mathscr{F}\mathscr{E}$. The value ΔG for the complete reaction is $\Delta G = \Delta G_{1r} + \Delta G_2 = -\mathscr{F}(n_1\mathscr{E}_1 + n_2\mathscr{E}_2)$. The question is: what are the values of n_1 and n_2? These do in fact represent the number of electrons transferred by each half-reaction. However, it must be remembered that the total equation is always balanced, so that $n_1 = n_2 = n$. Thus, for the total cell, n is not $n_1 + n_2$ but is the *same* for each half-cell; that is, the electrons gained at one electrode are lost at the other. This means that $\Delta G = -n\mathscr{F}(\mathscr{E}_1 + \mathscr{E}_2)$ or $\mathscr{E} = -\Delta G/n\mathscr{F} = \mathscr{E}_1 + \mathscr{E}_2$, and the total emf for the cell is just the sum of the electrode potentials.

Measurement of Activity Coefficients

Equation 8 includes a term containing activities. For example, for a cell containing a standard Pt-H_2 electrode and a Ag-AgCl electrode, both immersed in HCl,

$$\mathscr{E} = \mathscr{E}° - \frac{RT}{\mathscr{F}} \ln \frac{(a_{H^+})(a_{Cl^-})}{(a_{HCl})^{1/2}}. \tag{12}$$

The activity of H_2 at a pressure of 1 atm is 1. As pointed out in Chapter 8, neither a_{H^+} nor a_{Cl^-} can be measured separately, so the mean value must be used. Thus we rewrite Equation 12 in terms of the concentration c (M) and the mean activity coefficient $\gamma_\pm$, noting that $c_{H^+} = c_{Cl^-}$, to yield

$$\mathscr{E} = \mathscr{E}° - \frac{2RT}{\mathscr{F}} \ln c - \frac{2RT}{\mathscr{F}} \ln \gamma_\pm. \tag{13}$$

Thus, by measuring $\mathscr{E}$ at an HCl concentration of c, the value of $\gamma_\pm$ corresponding to that HCl concentration can be determined.

To determine the activity coefficient of $ZnCl_2$ at a particular concentration, the same scheme can be used, except that a Zn electrode would be used instead of a Pt-H_2 electrode and the solution in the cell would be $ZnCl_2$ instead of HCl.

The fact that activity and not concentration is required in Equation 8 brings up the question of how $\mathscr{E}°$ is measured. Since $\mathscr{E}°$ is the value when the activity is 1 molar and since we would not know how to prepare a solution of unit activity without knowing the activity coefficient, whose determination requires $\mathscr{E}°$, there seems to be a paradox. One solution is to determine the activity coefficient in another way. However, this is not necessary. We merely rearrange Equation 13 as

$$\mathscr{E} + \frac{2RT}{\mathscr{F}} \ln c = \mathscr{E}° - \frac{2RT}{\mathscr{F}} \ln \gamma_\pm$$

and remember that as c approaches 0, $\gamma_\pm$ approaches 1 and therefore that $(-2RT/\mathscr{F}) \ln \gamma_\pm$ approaches 0. Thus, $\mathscr{E}$ is measured at various values of c and a plot of $[\mathscr{E} + (2RT/\mathscr{F}) \ln c]$ versus c is made by extrapolation to $c = 0$, and $\mathscr{E}°$ is obtained. Details of how these plots are done can be found in references at the end of the chapter.

10-5 CONCENTRATION CELLS AND ION-SELECTIVE ELECTRODES

A Simple Concentration Cell

We have seen how measurements of the voltage of a cell can yield values for various thermodynamic parameters. Perhaps even more commonly, such measurements are used to determine concentrations, as in pH measurement. How this is done is described in this section.

Consider the two Pt|H$_2$|HCl|AgCl|Ag cells in Figure 10-7. The two cells contain different concentrations of HCl and, furthermore, they are arranged to oppose one another. If the concentrations were equal, there would be no net voltage between the two Pt-H$_2$ electrodes. However, the voltage of each cell depends on the HCl concentration, so that there is a voltage, whose value is described by Equation 8. Assuming that the H$_2$ is at a pressure of one atmosphere, that activity equals concentration, and using [ ] to denote concentration and the subscripts 1 and 2 to denote the left and right cells in the figure, we obtain

$$\mathscr{E} = \left(\mathscr{E}^\circ_{Ag,AgCl} - \frac{RT}{\mathscr{F}}\ln[Cl^-]_1 + \frac{RT}{\mathscr{F}}\ln\frac{1}{[H^+]_1}\right)$$
$$- \left(\mathscr{E}^\circ_{Ag,AgCl} - \frac{RT}{\mathscr{F}}\ln[Cl^-]_2 + \frac{RT}{\mathscr{F}}\ln\frac{1}{[H^+]_2}\right)$$
$$= \frac{RT}{\mathscr{F}}\ln\frac{[H^+]_2[Cl^-]_2}{[H^+]_1[Cl^-]_1} = \frac{2RT}{\mathscr{F}}\ln\frac{[HCl]_2}{[HCl]_1}.$$

The important point about this arrangement is that if the concentration in one cell is fixed, a simple measurement of $\mathscr{E}$ enables the concentration in the other cell to be measured. This principle could be used to measure pH. Similarly, two opposing cells each containing a Ag-AgCl electrode and some reference electrode could be used to measure [Cl$^-$] in any solution; also [SO$_4^{2-}$] could be measured with cells having a Hg-Hg$_2$SO$_4$ electrode. These are all examples of *ion-selective* electrodes. Such electrodes were at one time widely used for measuring ion concentrations but they have been supplanted by a great variety of commercially available electrodes containing membranes that are permeable only to a single ion. These are described in the following section.

Measurement of pH and Ion Concentration

Let us consider a variation of the concentration cell of Figure 10-7 in which, instead of having two cells opposed, there are two HCl solutions in contact via a porous membrane, and each contains a Pt-H$_2$ electrode. The observed voltage once again is dependent upon the HCl concentration in

FIGURE 10-7 A concentration cell. The cell on the right has a higher potential than the cell on the left and therefore electrons flow from left to right through the wire connecting the Pt-H$_2$ electrodes.

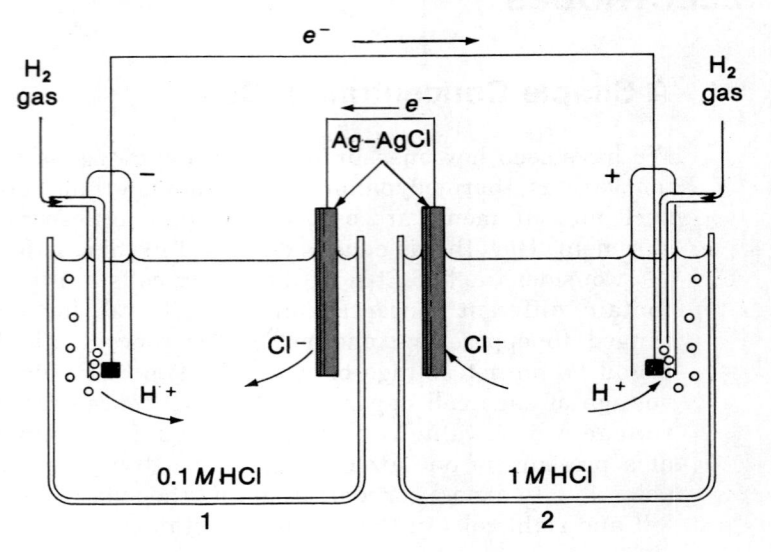

each region. The situation is more complicated though, because for every H$^+$ ion that moves from the region of higher concentration to that of lower concentration, there must be a similar movement of a Cl$^-$ ion; the significant point is that the Cl$^-$ ion encounters more friction in moving through water than does a H$^+$ ion. Since electroneutrality must be maintained, the movement of the H$^+$ ion is impeded, so that the potential is not a function of the concentration of the H$^+$ ion alone but of the nature of the anion. Clearly if one wishes to know H$^+$ ion concentrations, it is much more convenient to have an experimental arrangement in which the identity of the anion is irrelevant. An arrangement that would provide this is a porous plug with a membrane that is permeable to the H$^+$ ion and not to any anion. Furthermore, if the membrane was also impermeable to all other cations, the H$^+$ ion concentration could be measured in the presence of any other ion. Such a system is the *glass electrode,* a remarkable device whose detailed mechanism of action is not really well understood. A typical glass electrode and an arrangement for measuring the H$^+$-ion concentration is shown in Figure 10-8. A Ag-AgCl electrode is in a solution of 0.1 M HCl. Its potential is constant because, as we have seen before, it is determined only by the Cl$^-$-ion concentration. The HCl solution is contained in a glass tube that is terminated with a very thin-walled bulb of glass having a special composition. The glass electrode is immersed in a solution whose pH is to be determined and a reference electrode, which is almost always a calomel electrode, completes the cell.

To understand how the glass electrode is affected by the H$^+$-ion concentration, one must consider the chemical and physical structure of glass for

FIGURE 10-8 Glass and reference electrodes of a pH meter. [SOURCE: After *Physical Biochemistry*, First Edition, by David Freifelder. W.H. Freeman and Company. Copyright © 1976.]

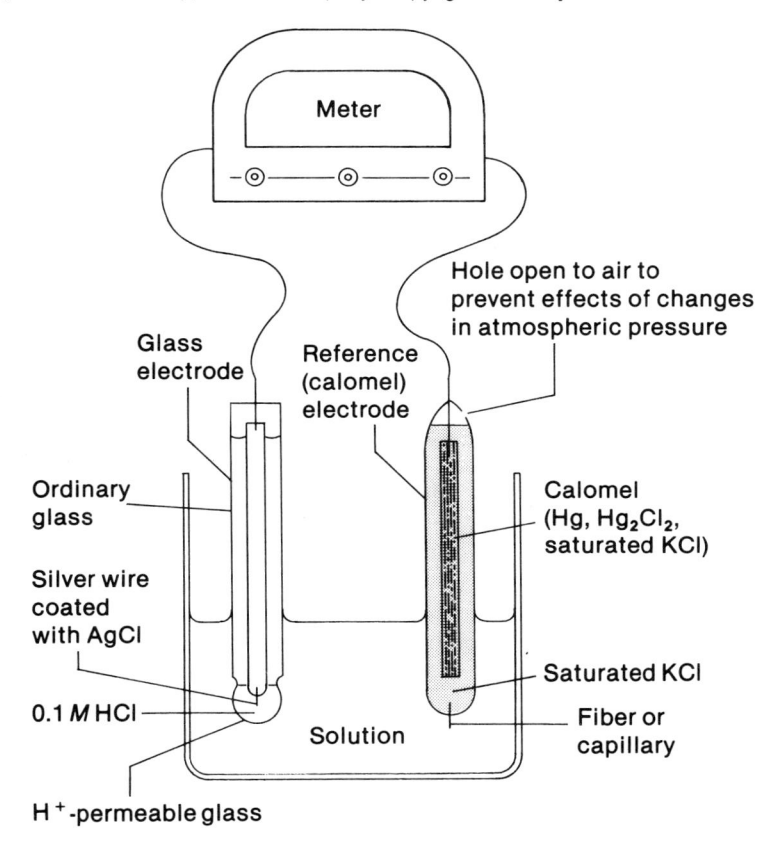

a moment. Glass is a complex negatively charged network of covalently bound silicon and oxygen atoms, in whose interstices are many positive ions (e.g., Na^+, K^+, Ca^{2+}, and so forth), so that an electrically neutral structure is created. The alkali metal ions are slightly mobile in the network, so that the glass has a small electrical conductivity. Prior to first use, a glass electrode is soaked for several hours in water or dilute acid. During this period, monovalent cations near the surface of the glass are replaced by H^+ ions. When a glass electrode is placed in a solution containing H^+ ions, an equilibrium is rapidly established between the bound and unbound H^+ ions; that is, there exists a number of bound ions that is determined by the concentration of the H^+ ions in solution. Thus, the number of bound ions on both sides of the glass membrane is determined by the H^+ concentrations on both sides. This produces an electrical potential across the glass, as in a concentration cell. There are numerous complications in deriving an equation relating the observed voltage and the H⁺-ion concentration because of several other potentials such as the liquid junction potential (see Chapter 14). If all of these effects are combined with the potential of the reference electrode, the observed voltage, V, at 25°C is found to be

$$V = V^* + 0.0592 \log_{10}[H^+] , \qquad (14)$$

in which V is measured in volts, $[H^+]$ in molarity, and V^* is a constant equal to the potential of the reference electrode plus the potentials resulting from the other effects. Note that the logarithm is to base 10. Since

$$pH = -\log_{10}[H^+] , \qquad (15)$$

this equation becomes

$$V = V^* - 0.0592\, pH, \qquad \text{at 25°C} . \qquad (16)$$

The value of V^* is in practice determined by measuring the voltage that appears when the electrodes are placed in a solution of known pH. Actually, in using commercial pH meters one is unaware of measuring the voltage, since the meter dials are labeled "pH" rather than "voltage."

The principle of the glass electrode is used in the fabrication of electrodes capable of measuring the concentration of other ions. We saw earlier that the concentration of any ion can in principle be determined by using the concentration cell shown in Figure 10-7. However, to determine the concentration of an alkali metal ion—Na⁺, for example—would require a Na|Na⁺ half-cell with a pure Na electrode. Such an electrode could not be prepared because it would react with water. The problem of reactivity can be alleviated somewhat by use of a Hg-Na amalgam. However, a better solution is to use ion-selective glass electrodes—that is, a glass that exchanges at the surface with only a single ion type. The H⁺-ion-sensitive glass of the pH meter is prepared by fusing SiO_2, Na_2O, and CaO in the appropriate ratios. For the detection of Na⁺ ions a glass prepared from SiO_2, Al_2O_3 and Li_2O can be used; it is sensitive only to the Na⁺ ion and an equation similar to that of Equation 14 can be derived for [Na⁺].

Glasses having other compositions allow the selective measurement of the univalent cations Li⁺, K⁺, Rb⁺, Cs⁺, Ag⁺, Cu⁺, Tl⁺, and NH_4^+.

The essential characteristics of an ion-selective glass electrode are that it consists of a water-insoluble substance that has some ion conductivity and is capable of adsorbing a particular ion. Thin crystals of insoluble salts sometimes have these properties and with such crystals an electrode of the type shown in Figure 10-9 can be made. For instance, an electrode containing a slice of a LaF_3 crystal can be used to measure [F⁻]. Crystals of Ag_2S and AgX (in which X is a halide) are used to measure [S²⁻] and [X⁻]. Liquid membranes can also be used; here an inorganic salt in an organic solvent is held in the pores of a porous glass or plastic. For example, a liquid membrane of $Ca[(C_{10}H_{21}O)_2PO_2]$, calcium decyl phosphate, is sensitive to [Ca²⁺]. At present, electrodes are available for measurement of the

FIGURE 10-9 An ion-selective membrane using a crystal membrane.

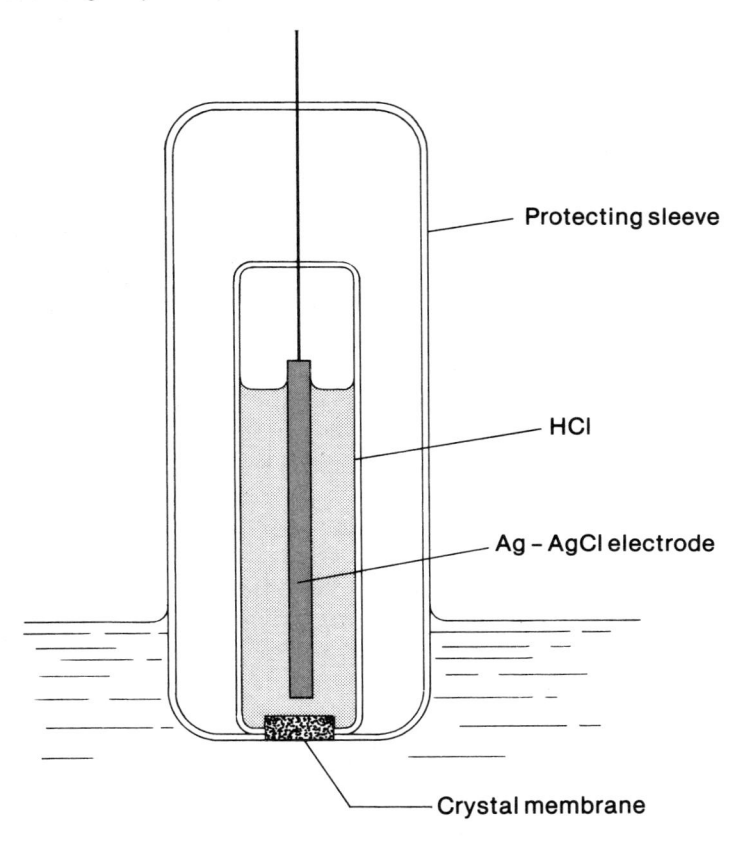

- Protecting sleeve
- HCl
- Ag – AgCl electrode
- Crystal membrane

concentrations of all of the commonly encountered univalent and divalent cations, and numerous anions.

In chemistry and biochemistry, the principal use of the ion-selective electrode is in titration and to determine concentration changes in reactions. An important use in biological research is the determination of ion concentrations in fluids such as blood and cell extracts, which contain a large number of different ions. Of special interest are glass microelectrodes that can be used to determine ion concentrations in living cells. That has been an important method for determining the changing Na^+ and K^+ concentration in various biological processes.

10–6 USE OF REDUCTION POTENTIALS TO CALCULATE CONCENTRATION AT EQUILIBRIUM

Biological systems contain a large number of oxidation-reduction systems. The values of $\mathscr{E}°$ for each can, of course, always be measured by using

Equation 8. One merely measures the potential of a solution containing a known concentration of oxidant and reductant; then $\mathcal{E}°$ is

$$\mathcal{E}° = \mathcal{E} + \frac{RT}{n\mathcal{F}} \ln \frac{[\text{reductant}]}{[\text{oxidant}]},$$

or, more generally,

$$\mathcal{E}° = \mathcal{E} + \frac{RT}{n\mathcal{F}} \ln \frac{[\text{electron donor}]}{[\text{electron acceptor}]}.$$

What is most often desired, though, is information about the concentrations of all components when two oxidation-reduction systems are present in the same system. For example, for a mixture of electron-transport proteins, cytochromes b and c, one might want to know what the concentrations of the reduced and oxidized forms of each will be at equilibrium. This information can be obtained as follows. We consider two general half-reactions 1 and 2,

$$O_1 + e^- \rightarrow R_1, \qquad \mathcal{E}_1°,$$
$$O_2 + e^- \rightarrow R_2, \qquad \mathcal{E}_2°,$$

having standard potentials $\mathcal{E}_1°$ and $\mathcal{E}_2°$ and in which O and R refer to the oxidant and reductant respectively. Let us assume that $\mathcal{E}_2° > \mathcal{E}_1°$ so that the net reaction is

$$R_1 + O_2 \rightarrow R_2 + O_1.$$

The equilibrium constant K is $K = [O_1][R_2]/[R_1][O_2]$, which is all that is needed to calculate the concentrations. From Equation 8 written for each component,

$$\mathcal{E} = \mathcal{E}_1° - \frac{RT}{\mathcal{F}} \ln \frac{[R_1]}{[O_1]},$$

and

$$\mathcal{E} = \mathcal{E}_2° - \frac{RT}{\mathcal{F}} \ln \frac{[R_2]}{[O_2]}.$$

When combined with the expression for K, there results

$$\ln K = \frac{\mathcal{F}(\mathcal{E}_2° - \mathcal{E}_1°)}{RT}, \qquad (17)$$

from which K is easily evaluated. The concentrations are then determined in the following way. If C_1 and C_2 are the concentrations of R_1 and R_2 when initially added to the mixture, then at any time

$$C_1 = [O_1] + [R_1]$$

and

$$C_2 = [O_2] + [R_2].$$

From the stoichiometry of the reaction $[O_1]$ always equals $[R_2]$. Thus, by setting $c = [O_1] = [R_2]$, then

$$K = \frac{c^2}{(C_1 - c)(C_2 - c)} \tag{18}$$

so that c can be calculated from the known values of C_1 and C_2.

10-7 POTENTIALS OF BIOCHEMICAL REACTIONS

Biochemical reactions are almost always strongly dependent on pH. Furthermore, the standard state of the Pt-H_2 electrode, for which the activity of the H^+ ion is 1 (pH = 0), is inappropriate for biological work since no biological reaction proceeds at such a low pH. The pH-dependence of biochemical reactions is of two types: (1) an effect on the reaction rate, because the enzymes needed to catalyze the reaction are active only in a small range of pH; and (2) an effect on the equilibrium, because the H^+ or OH^- ion actually participates in the reaction. The second point is the important one for our purposes, because it has led to the definition of the biochemical standard state, namely, 1 molar concentration of all components except for the H^+ ion, which is at 10^{-7} M or pH 7. (This definition was also used in the listing of standard free energy changes for biochemical reactions, as was shown in Chapter 6.) The standard electrode potentials for this state are denoted $\mathscr{E}°'$; several representative values are listed in Table 10-2. How these values are used is shown in the following: first,

TABLE 10-2 Standard reduction potentials for reactions of biochemical interest at pH 7 and 25°C.

Reaction	$\mathscr{E}°'$, volt
$\frac{1}{2}O_2 + 2H^+ + 2e^- \rightarrow H_2O$	+0.816
Cytochrome a $Fe^{3+} + e^- \rightarrow$ cytochrome a Fe^{2+}	+0.29
Cytochrome c $Fe^{3+} + e^- \rightarrow$ cytochrome c Fe^{2+}	+0.25
Methemoglobin $Fe^{3+} + e^- \rightarrow$ hemoglobin Fe^{2+}	+0.14
Ubiquinone + $2H^+ \rightarrow$ ubiquinone H_2	+0.10
Cytochrome b $Fe^{3+} + e^- \rightarrow$ cytochrome b Fe^{2+}	+0.08
Fumarate + $2H^+$ + $2e^- \rightarrow$ succinate	+0.031
FAD + $2H^+$ + $2e^- \rightarrow$ $FADH_2$	−0.06
Oxalacetate + $2H^+$ + $2e^- \rightarrow$ malate	−0.102
Acetaldehyde + $2H^+$ + $2e^- \rightarrow$ ethanol	−0.163
Pyruvate + $2H^+$ + $2e^- \rightarrow$ lactate	−0.19
NAD^+ + $2H^+$ + $2e^- \rightarrow$ NADH + H^+	−0.320
Uric acid + $2H^+$ + $2e^- \rightarrow$ xanthine	−0.36
Acetyl CoA + $2H^+$ + $2e^- \rightarrow$ acetaldehyde + CoA	−0.41
H^+ + $e^- \rightarrow \frac{1}{2}H_2$	−0.414
Gluconate + $2H^+$ + $2e^- \rightarrow$ glucose + H_2O	−0.45

we note that the value of $\mathcal{E}^{\circ\prime}$ for the Pt-H$_2$ electrode is no longer zero. This can be seen by writing Equation 8 for $\frac{1}{2}$ H$_2 \to e^- +$ H$^+$ at 25°C (the standard temperature):

$$\mathcal{E} = \mathcal{E}^\circ - \frac{RT}{\mathcal{F}} \ln \frac{[\text{H}^+]}{[\text{H}_2]^{1/2}} = 0 - \frac{2.303RT}{\mathcal{F}} \log_{10} \frac{1}{10^{-7}} = -0.414 \text{ volt.} \quad (19)$$

Thus $\mathcal{E}^{\circ\prime} = -0.414$ volt, as shown in Table 10-2. Note that from the definition of pH,

$$\mathcal{E}^\circ = -\frac{2.303RT}{\mathcal{F}}(\text{pH}) = -0.059(\text{pH}) \,. \quad (20)$$

Let us now see how to convert the potential of a pH-dependent reaction to pH 7. Consider the reaction

$$\text{A} + \text{H}^+ + e^- \to \text{AH} \,,$$

which has a standard potential of $\mathcal{E}^\circ$ when [H$^+$] = 1 M. Thus at pH 7 and 25°C the potential $\mathcal{E}$ is

$$\mathcal{E} = \mathcal{E}^\circ - \frac{RT}{\mathcal{F}} \ln \frac{1}{10^{-7}} = \mathcal{E}^\circ - \frac{2.303RT}{\mathcal{F}} \cdot 7 = \mathcal{E}^\circ - 0.414 \,.$$

Thus $\mathcal{E}^{\circ\prime} = \mathcal{E}^\circ - 0.414$. This may be generalized for any reaction in which the H$^+$ ion is a reactant and whose standard potential is $\mathcal{E}^\circ$:

$$\mathcal{E}^{\circ\prime} = \mathcal{E}^\circ - 0.059 \,(\text{pH}) \,. \quad (21)$$

If a value of $\mathcal{E}^{\circ\prime}$ is given, it can be converted easily to another pH. For example, suppose $\mathcal{E}^{\circ\prime}$ is 0.15 and we wish to know the value of the potential at pH 8: since $\mathcal{E}^{\circ\prime} = 0.15 = \mathcal{E}^\circ - 0.059(7)$, then $\mathcal{E}^\circ = 0.15 + 0.414 = 0.564$, and $\mathcal{E}^\circ(\text{pH } 8) = 0.564 - (0.059)8 = 0.092$. More generally, the value of $\mathcal{E}^\circ$ at pH X is

$$\mathcal{E}^\circ(\text{pH } X) = \mathcal{E}^{\circ\prime} + 0.059(7 - X) \,. \quad (22)$$

Let us now use the values in the table to determine the direction of the reaction at pH 7 in a solution initially containing 10^{-3} M acetaldehyde, 0.1 M ethanol, 10^{-3} M NAD$^+$ (nicotinamide adenine dinucleotide), and 10^{-3} M NADH$^+$. The reactions and values of $\mathcal{E}^{\circ\prime}$ are

(1) NAD$^+$ + H$^+$ + 2$e^- \to$ NADH, $\qquad \mathcal{E}^{\circ\prime} = -0.32;$
(2) Acetaldehyde + 2H$^+$ + 2$e^- \to$ ethanol, $\qquad \mathcal{E}^{\circ\prime} = -0.163.$

These values must be converted from standard conditions to the particular concentrations required. Thus,

$$(1) \quad \mathcal{E}_1 = -0.32 - \frac{RT}{2\mathcal{F}} \ln \frac{10^{-1}}{10^{-3}} = -0.38 \,;$$

$$(2) \quad \mathcal{E}_2 = -0.163 - \frac{RT}{2\mathcal{F}} \ln \frac{10^{-3}}{10^{-3}} = -0.163 \,.$$

The potential for reaction 2 is more positive, so that NAD$^+$ will be reduced

SEC. 10–7 POTENTIALS OF A BIOCHEMICAL REACTIONS

to NADH and the ethanol reduced to acetaldehyde. Thus the reaction is

$$NAD^+ + ethanol \rightarrow NADH + acetaldehyde + H^+.$$

The reaction continues to the right until the concentrations of each component are such that $\mathscr{E}_1 = \mathscr{E}_2$ (that is, equilibrium is reached). We now calculate the equilibrium constant. Two electrons are transferred, so that $n = 2$.

From the values of $\mathscr{E}^{\circ\prime}$ for each reaction, the standard potential at pH 7 for the total reaction is $-0.32 + (+0.163) = -0.157$ volt. We next note that $\mathscr{E}^{\circ\prime} = \mathscr{E}^\circ - (RT/n\mathscr{F})\ln 10^{-7} = \mathscr{E}^\circ - (RT/2\mathscr{F})\ln 10^{-7}$, and thus

$$\mathscr{E} = \mathscr{E}^\circ - \frac{RT}{2\mathscr{F}}\ln\frac{[NADH][acetaldehyde][H^+]}{[NAD^+][ethanol]}.$$

Combining these equations yields

$$\mathscr{E} = \mathscr{E}^{\circ\prime} + \frac{RT}{2\mathscr{F}}\ln 10^{-7} - \frac{RT}{2\mathscr{F}}\ln\frac{[NADH][acetaldehyde][H^+]}{[NAD^+][ethanol]}$$

$$= -0.157 - \frac{RT}{2\mathscr{F}}\ln\frac{[NADH][acetaldehyde][H^+]}{[NAD^+][ethanol](10^{-7})}.$$

Note that the only effect of using $\mathscr{E}^{\circ\prime}$ instead of $\mathscr{E}^\circ$ is to put 10^{-7} in the denominator of the logarithmic term. Thus, at equilibrium when $\mathscr{E} = 0$ and at pH 7 (which was the condition stated in our problem),

$$\ln\frac{[NADH][acetaldehyde]}{[NAD^+][ethanol]} = -12.2,$$

or

$$\frac{[NADH][acetaldehyde]}{[NAD^+][ethanol]} = 4.9 \times 10^{-6}.$$

This is the equilibrium constant, but it is only valid at pH 7. If we wanted the equilibrium constant at pH 7.5, we would substitute $10^{-7.5}$ for $[H^+]$ to obtain

$$+0.157 = -\frac{RT}{2\mathscr{F}}\ln\frac{[NADH][acetaldehyde]10^{-7.5}}{[NAD^+][ethanol](10^{-7})},$$

or

$$\ln\frac{[NADH][acetaldehyde]}{[NAD^+][ethanol]} = -11.05;$$

the equilibrium constant at pH 7.5 is then 1.58×10^{-6}.

In chapter 6, we discussed the biochemical standard free energy change denoted $\Delta G^{\circ\prime}$. This is simply related to $\mathscr{E}^{\circ\prime}$ since just as $\Delta G^\circ = n\mathscr{F}\mathscr{E}^\circ$, similarly $\Delta G^{\circ\prime} = -n\mathscr{F}\mathscr{E}^{\circ\prime}$.

10-8 THE TERMINAL OXIDATION CHAIN IN LIVING CELLS

An important energy-storage compound in living cells is adenosine triphosphate (ATP). The hydrolysis of this compound, which has a value of $\Delta G°$ of -30.5 kJ mol^{-1}, is a major source of free energy for numerous biochemical processes. Living cells devote an enormous fraction of their metabolic effort to the synthesis of ATP. In general, biochemical oxidations occur by combining two half-reactions with sufficiently different electrode potentials that the work done, $-n\mathcal{F}\mathcal{E}$, is larger than the free energy required to synthesize some essential compound.

Many important oxidation-reduction reactions use a powerful oxidizing agent, nicotinamide adenine dinucleotide (NAD$^+$), to oxidize some reduced metabolite. The basic reaction is the following:

$$\text{Metabolite} + \text{NAD}^+ \rightarrow \text{oxidized product} + \text{NADH} + \text{H}^+,$$

in which NADH is the reduced form of NAD$^+$. Following this, a series of reactions occurs called the *terminal electron-transport chain* (in which O$_2$ is the final electron acceptor), and the reduced NADH is oxidized, reforming NAD$^+$. This multistep process is the principal source of the free energy used to generate ATP. The entire NADH $\rightarrow$ NAD$^+$ oxidation can be written as the sum of the two half-reactions

$$\tfrac{1}{2}\text{O}_2 + 2\text{H}^+ + 2e^- \rightarrow \text{H}_2\text{O}, \qquad \mathcal{E}°'(\text{pH} = 7) = 0.815 \text{ volt}$$
$$-(\text{NAD}^+ + \text{H}^+ 2e^- \rightarrow \text{NADH}), \qquad \mathcal{E}°'(\text{pH} = 7) = -0.320 \text{ volt},$$

for which the total reaction is

$$\text{NADH} + \tfrac{1}{2}\text{O}_2 + \text{H}^+ \rightarrow \text{H}_2\text{O} + \text{NAD}^+, \qquad \mathcal{E}°'(\text{pH} = 7) = 1.135 \text{ volt}.$$

Two electrons are transported per mole of NADH, so that the free energy change in the reaction is

$$\Delta G°' = -2\mathcal{F}(1.135) = -2.19 \times 10^5 \text{ J mol}^{-1}$$

This is sufficient to allow synthesis of 7 moles of ATP, whose free energy of formation at pH 7 is $+30.5$ kJ mol^{-1}. Actually, only three moles of ATP are formed, because a certain amount of the free energy is utilized in reactions that are wasteful from the point of view of synthesis but efficient because they are fast. By this we mean the following: a reaction sequence A $\rightarrow$ B $\rightarrow$ C might produce sufficient free energy at each step to generate two molecules of ATP, but the step A $\rightarrow$ B might be very slow. An alternate sequence,

$$A \xrightarrow{1} X \xrightarrow{2} Y \xrightarrow{3} Z \xrightarrow{4} B$$

might be very fast, but the free energy evolved in each of the steps 1, 2, and 3 might not be sufficient to produce an ATP molecule. An examination of the actual sequence of ATP synthesis and the values of $\mathcal{E}°'(\text{pH } 7)$ for each step in it clearly indicate the only steps in which ATP could be

FIGURE 10-10 An electrode potential diagram for the components of the terminal electron transport chain.

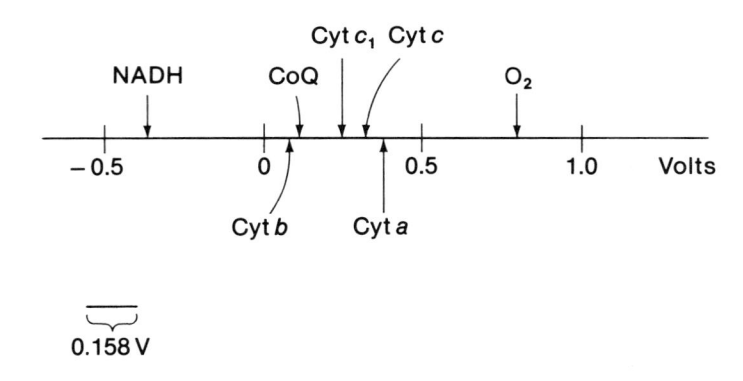

generated. The scheme below shows the electron carriers in the sequence:

$$\text{NADH} \xrightarrow{1} \text{CoQ} \xrightarrow{2} \text{Cyt } b \xrightarrow{3} \text{Cyt } c_1 \xrightarrow{4} \text{Cyt } c \xrightarrow{5} \text{Cyt } a \xrightarrow{6} O_2$$

in which CoQ means coenzyme Q and Cyt x means cytochrome x. The result of this scheme is the oxidation of NADH to NAD^+ and the reduction of O_2 to water. The scheme can also be shown as an electrode potential diagram, as in Figure 10-10. The minimal change in potential equivalent to 30.5 kJ mol^{-1} is 0.158 V, represented as the length of the line shown in the figure, so that it is clear that ATP could be generated only in reactions 1, 3, and 6. The advantage of an analysis such as this is that if the ATP-generating steps are not known, the analysis limits the possibility to particular steps which can then be examined in detail in the laboratory.

REFERENCES

Bates, R.G. 1964. *Determination of pH.* Wiley. New York.
Bockris, J.O'M., and A.K.W. Reddy. 1974. *Electrochemistry for Ecologists.* Plenum. New York.
Castellan, G.W. 1971. *Physical Chemistry.* Addison-Wesley. Reading, Mass.
Edsall, J.T., and J. Wyman. 1958. *Biophysical Chemistry,* Volume I. Academic Press. New York.
Eisenman, G. 1967. *Glass Electrodes for Hydrogen and Other Cations. Principles and Practice.* Dekker. New York.
Hobey, W.D. 1972. "Biogalvanic Cells." *J. Chem. Educ.* 49, 413.
Hush, N.S. 1971. *Reactions of Molecules at Electrodes.* Wiley-Interscience. New York.
Ives, D.J.G., and G.J. Janz. 1961. *Reference Electrodes.* Academic Press. New York.
Latimer, W.M. 1952. *The Oxidation States of the Elements and Their Potentials in Aqueous Solutions.* Prentice-Hall. Englewood Cliffs, N.J.
Lehninger, A.L. 1975. *Biochemistry.* Worth. New York. Chapter 18.
Levine, I.N. 1978. *Physical Chemistry.* McGraw-Hill. New York.

PROBLEMS

1. What is the standard potential of the following electrochemical cells at 25°C?
   (a) Mn|Mn(NO$_3$)$_2$|Pb(NO$_3$)$_2$|Pb.
   (b) Ag|AgNO$_3$|Cu(NO$_3$)$_2$|Cu.
   (c) Zn|ZnSO$_4$|MgSO$_4$|Mg.
   (d) Zn|Zn(NO$_3$)$_2$|AgNO$_3$|Ag.

2. Which of the following reactions proceed spontaneously in the standard state?
   (a) Mn + CrCl$_3$ → Mn^{2+} + Cr.
   (b) MnO$_2$ + I$^-$ + H$^+$ → Mn^{2+} + I$_2$ + H$_2$O.
   (c) Ni + Cr$_2$O$_7^{2-}$ + H$^+$ → Cr^{3+} + Ni^{2+} + H$_2$O.

3. A cell consists of two Pt electrodes, one placed in Cu$_2$SO$_4$ and the other in KMnO$_4$, with the solutions connected by a porous plug. What is the standard potential of the cell, the direction of flow of electrons, and the cell reaction?

4. What is the cell reaction for the cell
   Ag|0.2 M AgNO$_3$|0.001 mM CuSO$_4$|Cu.
   Is the reaction spontaneous at 25°C?

5. What is the voltage of a cell formed when standard half-cells of FeSO$_4$ and Cl$_2$ in water are connected? What is the direction of electron flow?

6. Consider the reaction
   $$2Cr + 3Cu^{2+} \rightarrow 2Cr^{3+} + 3Cu.$$
   a) What are the half-reactions?
   (b) Referring to Table 10-1, what is the standard cell potential?
   (c) Express the equilibrium constant in terms of the components of the reaction.
   (d) What is the numerical value of the equilibrium constant at 25°C?

7. An electrical cell consists of a 0.01-M solution of ZnSO$_4$ in contact through a porous plug with 0.5 M ZnSO$_4$. The electrodes in both solutions are Zn.
   (a) What is the voltage at 25°C?
   (b) What physical change occurs at each electrode when the external circuit is completed?
   (c) At equilibrium what is the voltage of the cell and the concentrations in the two solutions?

8. Consider the cell
   Cu|Ag|AgCl|CdCl$_2$|Cd|Cu,
   in which Cu is inert.
   (a) What are the half-cell reactions and what is the total reaction?
   (b) What is $\mathcal{E}°$ for this configuration?
   (c) How many electrons are involved in the cell reaction?
   (d) What is the voltage of the cell if the CdCl$_2$ is at a concentration of 0.01 M? Assume that activity equals concentration and that the temperature is 25°C.
   (e) What is the direction of the reaction that occurs in the cell?

9. Consider the cell Pt|H$_2$|HCl|AgCl|Ag|Pt.
   (a) What are the half-reactions in the cell?
   (b) What is the total chemical reaction in the cell and how many electrons participate in the reaction?
   (c) Would either $\mathcal{E}°$ or $\Delta G°$ be affected by replacing the Pt by Pd, another inert metal?

10. What is $\mathcal{E}°$ for the reaction Zn + Cl$_2$ → Zn^{2+} + 2Cl$^-$?

11. Many flashlight batteries contain ZnCl$_2$, NH$_4$Cl, and MnO$_2$. The reaction producing the current is 2MnO$_2$ + Zn + NH$_4^+$ → Zn^{2+} + Mn$_2$O$_3$ + NH$_3$(aq) + OH$^-$. If a battery contains 2 g MnO$_2$ and a light bulb draws a current of 5 milliampere, how long can the light burn?

12. Consider the cell discussed in the text, in which Zn and Cu electrodes are placed in a solution containing 1 M ZnSO$_4$ and 1 M CuSO$_4$, for which $\mathcal{E}° = 1.10$ V. If current is allowed to flow until no more can flow, what is the concentration of CuSO$_4$ remaining in the solution? Assume that there is enough Zn and Cu in the electrodes that neither is ever totally dissolved. The temperature is 25°C.

PROBLEMS

13. Fill the blanks in the following table in the correct order:

ΔG	$\mathscr{E}$	Reaction
> 0	_____	_____
$= 0$	_____	_____
< 0	_____	_____

In the column labeled "reaction," the terms to be inserted are spontaneous, not spontaneous, and equilibrium. In the $\mathscr{E}$ column, insert > 0, < 0, or $= 0$.

14. The temperature coefficient of a cell is -1.28×10^{-4} V K^{-1}. At 25°C, $\mathscr{E}°$ is 0.36 V. If the reaction in the cell involves two electrons, what are the values of $\Delta G°$, $\Delta H°$, and $\Delta S°$ for the reaction in the cell at 25°C?

15. What is $\Delta G°$ at 25°C for the reaction Zn + 2CuCl (1 M) → 2Cu + ZnCl$_2$ (1 M)?

16. The values of $\mathscr{E}°$ under standard conditions are listed for the following reactions. In each case, calculate $\mathscr{E}°'$ at pH 7.0.
    (a) Fe^{3+} + e^- → Fe^{2+}, $\mathscr{E}°$ = 0.771.
    (b) I$_2$ − 2e^- → 2I$^-$, $\mathscr{E}°$ = 0.536.
    (c) Oxaloacetate + 2H$^+$ + 2e^- → malate, $\mathscr{E}°$ = 0.318.
    (d) Pyruvate + 2H$^+$ + 2e^- → lactate, $\mathscr{E}°$ = 0.224.
    (e) Fumarate + 2H$^+$ + 2e^- → succinate, $\mathscr{E}°$ = 0.433.

17. What is the equilibrium constant at 25°C for the following reaction?

    4Na$_4$Fe(CN)$_6$ + 2H$_2$O + O$_2$ → 4Na$_3$Fe(CN)$_6$ + 4NaOH.

18. Calculate the solubility product for AgCl from the electrode potential at 25°C.

19. Determine $\gamma_\pm$ for PbCl$_2$ at 25°C in a cell Pb|0.05 M PbCl$_2$|AgCl|Ag, if $\mathscr{E}$ = 0.522 V.

20. What is the concentration of Cu^{2+} ions in a cell consisting of CuSO$_4$, a Cu electrode, and a calomel electrode, if the cell potential at 25°C is 0.020 V?

21. An early proposal about the state of the univalent mercury ion was that it existed as a dimer, Hg$_2^{2+}$, rather than as a single ion Hg$^+$. To test this idea, the electrochemical cell Hg|solution 1|solution 2|Hg was prepared, in which solution 1 contained 2.63 g HgNO$_3$ per liter in 0.1 M HNO$_3$ and solution 2 contained 0.263 g HgNO$_3$ per liter. The measured voltage of the cell at 18°C was 0.0289 V. What is the form of the ion in solution?

22. A cell having one calomel electrode and a Ca^{2+}-sensitive electrode,

    Hg|Hg$_2$C$_2$O$_4$| CaC$_2$O$_4$

    yields a voltage at 18°C of 0.3243 V when the electrodes are in 0.01 M NaNO$_3$ + 0.01 M Ca(NO$_3$)$_2$. When immersed in a second solution, again containing 0.01 M NaNO$_3$ and an unknown amount of the Ca^{2+} ion, the voltage at 18°C is 0.3111 V. What is the concentration of Ca^{2+} in the second solution?

23. A cell contains a solution of MCl$_3$. A total charge of 2400 coulomb of electricity passes through the cell and, as a result, 2.05 gram of the metal M deposit on the cathode. What is the atomic weight of M?

24. A particular electrical current yields 2.1 g of O$_2$ in 10 min. How long will it take for the same current to deposit one-half mole of iron from FeSO$_4$?

25. At high pH the measured pH of NaOH is less than the true value, whereas that of KOH corresponds to the H$^+$-ion concentration. At near-neutral pH the measured value of a 1-M Na phosphate buffer is lower than that for a K phosphate buffer. Propose an explanation for the sodium error.

26. The potential of a calomel half-cell is 0.2443 V. When this electrode and a Pt-H$_2$ electrode are placed in a particular solution at 25°C, the measured voltage is 0.305 V. What is the pH of the solution?

CHAPTER 11

CHEMICAL KINETICS

Chemical reactions involve the breakage and formation of covalent bonds. A question of great interest is how the reaction proceeds; that is, we wish to know which chemical groups interact to yield the products of the reaction, whether there are intermediate steps in the reaction, and what the intermediate compounds might be. One valuable approach to obtaining answers to these questions is to study, under a variety of conditions, the rates of consumption of the reactants and of appearance of the products. The basis of this approach is the fact that when a reaction mechanism is known, the rate of disappearance of any reactant and the rate of formation of each product can be calculated. The converse is not true—that is, the evaluation of these rates does not unambiguously point to the reaction mechanism. Nonetheless, experimentally observed rates do limit the number of possible mechanisms and tend to direct one's thinking along the right lines. It has been said that no kinetic result can ever prove a mechanism—rather, the result indicates whether a researcher is justified in proposing a particular mechanism.

The study of reaction rates is called chemical kinetics. The theory of chemical kinetics is a mathematical theory but, as we will see, the mathematics is very simple. Furthermore, the variables in the equations we shall derive are only the numbers and concentrations of reactants and products and no particular reference is made in the theory to the nature of the bonds that form. Thus the theory also applies to the numerous weak bonds that are present in biochemical systems.

11-1 THE CONCEPT OF REACTION MECHANISMS

A balanced chemical equation identifies the reactants that are consumed and the final chemical species (the products) that are produced, and states what the relative concentrations of reactants and products would be, were the reaction to go to completion. However no indication is given of the pathway by which the reaction occurs. For example, the reaction described by a chemical equation could occur in a single step, but it also might proceed by a sequence of steps not indicated in the balanced equation. In general, apparent simplicity of a reaction is not an indication of the number of steps; an example of a single-step reaction is $A + B \rightarrow AB$, in which one molecule of A and one of B collide and, without the presence of any intermediate compound, a covalent bond forms. Such a single-step reaction is called an *elementary reaction*. The reaction $A_2 + B_2 \rightarrow 2AB$ could also be an elementary reaction if, in a single collision between the diatomic molecules A_2 and B_2, the interatomic bonds are broken and the A—B bonds form. However, a reaction of this type would usually occur by means of a multistep pathway. For example, the apparently simple reaction

$$H_2 + Br_2 \rightarrow 2HBr$$

requires three steps, namely

$$Br_2 \rightarrow 2Br\cdot$$
$$H_2 + Br\cdot \rightarrow HBr + H\cdot$$
$$H\cdot + Br_2 \rightarrow HBr + Br\cdot$$

in which the dot indicates the single unpaired electron in the atom. This complex mechanism is required because the H_2 molecule is unable to overcome an energy barrier and initiate the reaction. Instead the reaction sequence makes use of the partial instability of the Br_2 molecule to yield a single Br atom that is capable of reacting with H_2.

In some reactions, it is immediately obvious that there must be a multistep pathway. For instance, in the reaction

$$3Fe^{2+} + HCrO_4^- + 7H^+ \rightarrow 3Fe^{3+} + Cr^{3+} + 4H_2O ,$$

the rightward reaction would require the simultaneous collision of eleven components (a highly unlikely event!) if it were a single-step reaction. Thus in a reaction of this type there must be intermediate steps and compounds. Often the intermediates are unstable substances (for example, Cr^{4+} and Cr^{5+} ions in this case, and H· and Br· atoms in the previous case) and have only a transitory existence.

Chemists and biochemists are interested in how reactions occur—that is, in the reaction mechanisms. In part this is because of the desire of scientists to solve puzzles, but there are other reasons. Understanding a reaction mechanism enables a chemist to select reaction conditions that can improve the yield of a desired product or avoid hazards such as explosions. For the biochemist, some insight is provided into the complexity of a living cell in which tens of thousands of reactions are occurring simultaneously, each neatly regulated so that required substances are made at the right time and in the right amount. In a more practical vein, an understanding of a mechanism can lead to the development of drugs that enhance or inhibit particular biochemical reactions; this has in recent years been an important approach used by pharmaceutical firms.

In the following section we describe some properties of simple reactions and indicate how one goes about determining the reaction kinetics. Later we will indicate how a mechanism is guessed from the kinetics.

11–2 RATE LAWS, ORDER, AND MOLECULARITY

The *rate* or *reaction velocity* of a chemical reaction is customarily expressed as the change in concentration per unit time of either a reactant or a product. *By convention the rate is always positive* even when it is stated as the decrease in concentration of a reactant. A reaction does not have an absolute rate even under specified conditions, because the numerical value of the rate depends on the substance measured. For example, in a reaction A + B → C, each of the quantities $\Delta[A]/\Delta t$, $\Delta[B]/\Delta t$, and $\Delta[C]/\Delta t$, in which t is time, has the same numerical value, but in the reaction 2A + B → C, this is not the case—that is, $\Delta[A]/\Delta t = 2(\Delta[C]/\Delta t)$ and the stated rate depends on the choice of the substance A, B, or C whose concentration-change is presented to describe the rate. The choice of A, B, or C is generally determined only by the experimental technique used in the measurement—that is, *the substance chosen is usually that which can be measured most easily*.

In determining a rate, one measures the concentration of some component as a function of time. However, this alone gives little information because, as will be seen, one needs to know how this change is affected by changing the concentration (1) of other components that appear in the balanced equation and (2) of substances that are not in the balanced equation but which one might guess to play a role in the reaction sequence (for example, H^+ ions). These effects are expressed formally in the following way.

Let [P] be the concentration of a product P that results from a reaction involving reactants A, B, ..., N. An equation relating the reaction rate $d[P]/dt$ and the concentrations [A], [B], ..., [N] is called a *rate law*. The substances A, B, ..., N may be any substance, including P itself, that affects the value of $d[P]/dt$.

A general expression for the rate is

$$\frac{d[P]}{dt} = k[A]^a[B]^b \ldots [N]^n \tag{1}$$

in which k is a positive number called the *rate constant*. A similar equation

SEC. 11-2 RATE LAWS, ORDER, AND MOLECULARITY

could also be written for a reactant, but since the reactant will decrease and k is defined to be positive, *a minus sign must appear in the equation.*

The notation used to designate rate constants is often a source of confusion, especially because the symbol k is also used for the Boltzmann constant. (To avoid this problem, in this chapter the Boltzmann constant will be written in bold face, as **k**, in the few places in which it appears.) If the overall reaction consists of a series of reaction 1, 2, ..., n, the rate constant for each reaction is denoted by $k_1, k_2, \ldots, k_n$. In a reversible reaction A ⇌ B, there are two rate constants, one for the *forward* (rightward) reaction and one for the *back* (leftward) reaction. A minus sign is added to the subscript for the back reaction. Thus, if the first step in a sequence is reversible, the rate constants are designated k_1 and k_{-1}, respectively; similarly, k_3 and k_{-3} would be used for a third step, if the step were reversible.

The *order* of a reaction whose rate law is Equation 1 is defined as the sum of the exponents; namely, $a + b + \cdots + n$. For example, the reaction $N_2O_5 \rightarrow 4NO_2 + O_2$ is described by the rate law $d[N_2O_5]/dt = -k[N_2O_5]$ so that this is a first-order reaction. (Note that there is a minus sign in the equation because we have expressed the rate in terms of the concentration of a reactant rather than the product.) The reaction $5Br^- + BrO_3^- + 6H^+ \rightarrow 3Br_2 + 3H_2O$ obeys the rate law $d[Br_2]/dt = k[Br^-][BrO_3^-][H^+]^2$, and thus it is a fourth-order reaction. It is often informative to state the order in terms of a particular reactant. Thus we might say that the reaction just given is first order in $[Br^-]$, first order in $[BrO_3^-]$, second order in $[H^+]$, and fourth order overall.

A reaction might also be of zeroth order, that is, the rate is independent of the concentration of any component. How this can occur will be described in a later section.

For an *elementary* irreversible reaction the order is related to the coefficients in the balanced equation by the following rule:

the coefficient of X in the balanced equation is the exponent of [X] in the rate equation.

This is illustrated for the following cases:

(1) A → P.

The rate equations are $d[P]/dt = k[A]$ and $d[A]/dt = -k[A]$.

(2) A + B → P.

The rate equations are $d[P]/dt = k[A][B]$ and $d[A]/dt = d[B]/dt = -k[A][B]$ and the reaction is second order. Thus for A + A → P, $d[P]/dt = k[A][A] = k[A]^2$. Note that since $d[P]/dt = -\frac{1}{2}d[A]/dt$, then $d[A]/dt$ also equals $\frac{1}{2}k[A]^2$.

(3) 2A + B + C → P.

This is not likely to be an elementary reaction because four molecules have to collide but, when it is elementary, the rate equations are $d[P]/dt = $

$k[A]^2[B][C]$ and $d[A]/dt = d[B]/dt = d[C]/dt = -k[A]^2[B][C]$ and the reaction is fourth order.

If a reaction is reversible, the rates of both the forward and the back reactions must be considered. Thus, for the simple reaction

$$2A \underset{k_{-1}}{\overset{k_1}{\rightleftarrows}} B + C,$$

the rate law is $d[B]/dt = k_1[A]^2 - k_{-1}[B][C]$. A reaction of this type is not considered to be an elementary reaction because there are two reactions in which the product participates.

It is generally true that when the exponents in an experimental rate law do *not* equal the coefficients of the balanced equation, the total reaction *must* occur in a sequence of steps. If they are equal, the reaction *may* occur in a single step but this is not necessarily the case.

For most reactions that are observed, the exponents in the rate equation do not equal the coefficients in the balanced equation, so that it becomes the problem of the experimenter to determine the correct sequence of reactions. This puzzle solving requires both guesswork and an understanding of the likelihood of certain reactions.

The number of molecules that come together defines the *molecularity* of the reaction. In an elementary reaction it is the molecularity that determines the exponents in the rate equation. In a multistep reaction sequence, each step has a particular molecularity. Thus in the reaction $2HI \rightarrow H_2 + I_2$, two HI molecules come together with sufficient energy to rearrange the bonds, and the reaction is bimolecular. In the reaction $2NO + O_2 \rightarrow 2NO_2$, two NO molecules and one O_2 molecule must collide, so this is a trimolecular reaction.

> *Stating the order of a reaction gives information about the experimental rate equation; the molecularity tells about the reaction mechanism.*

When a balanced equation is the sum of two consecutive and essentially *irreversible* reactions, the reaction rate usually depends on the rate of only one of the two steps—this step is called the *rate-determining* step. A reaction sequence does not necessarily have a unique rate-determining step because in a sequence, a reactant may disappear at a rate that differs from the rate of appearance of the product. Thus, in order to designate a step as rate-determining, it is essential to state whether the reaction is being described in terms of the loss of a reactant (the more usual situation) or in terms of the synthesis of a product. For instance, consider a reaction sequence in which the first step is slow and the second step is fast. The rule is this:

> *the rate-determining step for synthesis of the product is the slow reaction, whereas the step determining the loss of the reactant is always the step in the sequence in which the reactant first appears.*

As we stated above, this rule is valid only when each step in the sequence is irreversible. In a later section we discuss the effect of a reversible step.

The use of this rule in describing a reaction is given in the following example.

EXAMPLE 11–A **Catalysis of the decomposition of H_2O_2 by iodide.**

The overall reaction is

$$2H_2O_2 + I^- \rightarrow 2H_2O + O_2 + I^-.$$

This reaction occurs by means of a sequence of two steps; namely,

$$H_2O_2 + I^- \rightarrow H_2O + IO^- \quad \text{(slow)}$$
$$IO^- + H_2O_2 \rightarrow H_2O + O_2 + I^- \quad \text{(fast)}$$

The production of O_2 is determined by the rate of the first step because the first reaction is the slow step. The loss of H_2O_2 is also governed by the first reaction, which is the first step in which H_2O_2 is involved. Also, the reaction is bimolecular, because no more than two molecules ever collide. Each step is irreversible, so that only reactants will appear in the rate law. In addition, since all of the reactants appear in the first step, which is rate-determining, only the reactants in the balanced equation influence the rate, and the rate law is $d[H_2O_2]/dt = -k_2[H_2O_2][I^-]$, and the reaction is second order overall. Note that we might have chosen to express the reaction velocity as $d[O_2]/dt$. In that case, the rate law would be $d[O_2]/dt = \frac{1}{2}k_2[H_2O_2][I^-]$.

We now turn our attention to the mathematical expressions for reactions of different order.

11–3 RATE LAWS FOR REACTIONS OF DIFFERENT ORDER, AND THE DETERMINATION OF THEIR RATE CONSTANTS

The order of a reaction is determined experimentally by measuring the concentration of a reactant or product as a function of time, while the concentrations of the components of the reaction are varied. One then takes the data and attempts by trial and error to fit it to theoretical expressions for reactions of particular order and special characteristics. In this section we derive a few of these expressions and then list several other expressions that are important. The rates are arbitrarily written in terms of a reactant, but these are easily converted to product rates, as we have just shown.

Zeroth-Order Reaction

In a zeroth-order reaction, the rate is independent of the concentration of any component. The rate equation is

$$-\frac{dc}{dt} = k, \tag{2}$$

in which c is the concentration of the reactant and k is the zeroth-order rate constant. This can be integrated to yield

$$c_0 - c = kt, \tag{3}$$

in which c_0 is the initial reactant concentration and c is the concentration at time t. Thus k can be evaluated by making a plot of $(c_0 - c)$ versus t (Figure 11-1); if the reaction is zeroth order, a straight line will result, having a slope k.

FIGURE 11-1 A plot of c_0-c versus time for a zeroth-order reaction. The curve is independent of the value of c. For instance, if $c_0 = 100$ mM, then at $t = 1, 2, 3, 4,$ and 5 minutes, $c = 90, 80, 70, 60,$ and 50 mM, and $c_0-c = 10, 20, 30, 40,$ and 50 mM, respectively. If $c_0 = 67$ mM, then at $t = 1$ to 5 minutes, $c = 57, 47, 37, 27,$ and 17 mM, and c_0-c is again 10, 20, 30, 40, and 50 mM.

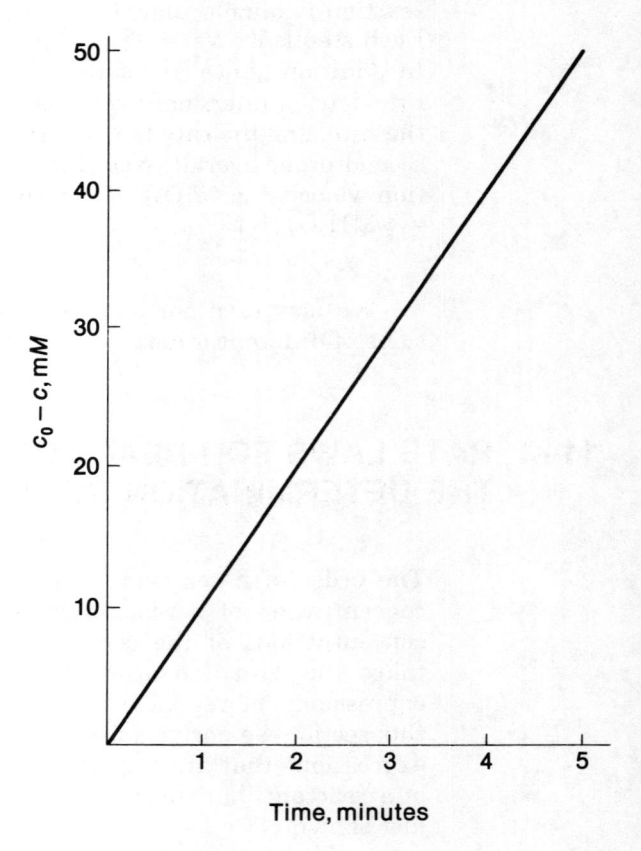

FIGURE 11-2 A semilogarithmic plot of concentration as a function of time for a first-order reaction. The half-life can be determined for any pair of concentrations that differ by a factor of two, as indicated.

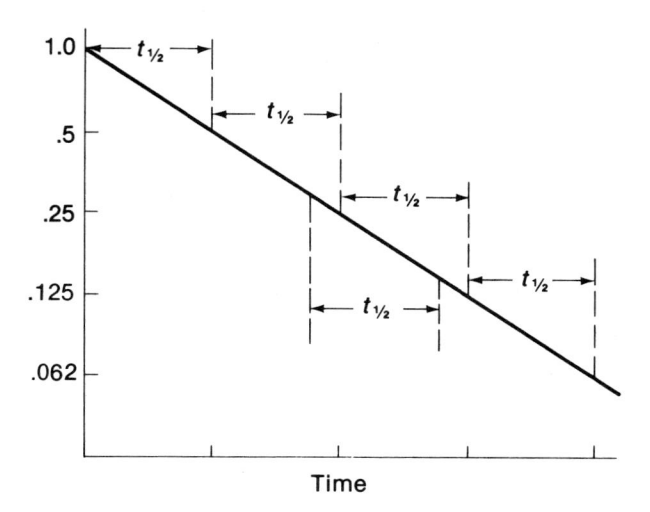

First-Order Reaction

A first-order reaction is one in which the rate of loss of a reactant A is proportional to the concentration [A]. That is,

$$\frac{dc}{dt} = -kc, \qquad (4)$$

in which k is the first-order rate constant. Integration from the value of c_0 at $t = 0$ to c at time t yields

$$\ln\left(\frac{c}{c_0}\right) = -kt, \qquad (5)$$

and a plot of $\ln(c/c_0)$ versus t gives a straight line having slope $-k$ (Figure 11-2). Another plot can also be used to demonstrate that a reaction is first order. For example, suppose that in a first-order reaction A → B + C it is difficult experimentally to measure [A] as a function of time, whereas the measurement of [B] might be simple. We need only note that $d[A]/dt = -d[B]/dt$, because every A is converted to a B. Thus, just as a plot of $\ln([A]/[A_0])$ against t gives a straight line whose slope is $-k$, a plot of $\ln[B]/[B_0]$ against t gives a straight line whose slope is $+k$.

An important property of a first-order reaction can be seen by examining the half-life, $t_{1/2}$, of a chemical reactant—that is, the time required for the concentration of a reactant to decrease to half its initial value. We can see by examining Equation 5 that the half-life is the time at which $c = \frac{1}{2}c_0$, so that

$$t_{1/2} = \frac{\ln 2}{k} . \qquad (6)$$

Note that $t_{1/2}$ is independent of concentration and can be determined for any pairs of concentrations c_1 and c_2 such that $c_2 = \frac{1}{2}c_1$ (Figure 11-2). This independence of concentration does not occur for higher order so that *if* $t_{1/2}$ *is independent of concentration, the reaction is first order.* (Shortly we will discuss the experimentally valuable pseudo-first-order kinetics for which $t_{1/2}$ is also concentration-independent.)

Second-Order Reaction

In a second-order reaction of the type A + B → C or A + B → C + D, one form of the rate equation is

$$\frac{d[A]}{dt} = -k[A][B] . \qquad (7)$$

Note that because of the stoichiometry of the reaction, the amount of A that is consumed in the reaction equals the amount of B that is consumed. Thus, the rate equation might also be written

$$\frac{d[B]}{dt} = -k[A][B] . \qquad (8)$$

In order to obtain the integral form of the rate equation we use, for convenience, a variable x that is the concentration of *either* A or B that has been consumed in the reaction. Using a and b to denote the initial concentrations, we obtain

$$\frac{dx}{dt} = k(a - x)(b - x) . \qquad (9)$$

Integration between the limits of 0 and x and between 0 and t yields

$$\frac{1}{a - b} \ln \left[\frac{b(a - x)}{a(b - x)} \right] = kt . \qquad (10)$$

In this case, the rate constant is evaluated from a plot of $\ln[(a - x)/(b - x)]$ against t.

Note that Equations 6 through 9 apply only to reactions in which one mole of A reacts with one mole of B. The effect of stoichiometry can be seen by comparing these equations with those obtained for a second-order reaction in which two moles of B are consumed per mole of A; that is, A + 2B → C + D. In this case, when x moles of A have reacted, $2x$ moles of B have reacted and the rate law in differential form is

$$\frac{dx}{dt} = k(a - x)(b - 2x) , \qquad (11)$$

and in integral form,

$$\frac{1}{2a - b} \ln\left[\frac{b(a - x)}{a(b - 2x)}\right] = kt .$$ (12)

If the reaction involves two different molecules at the same initial concentration or two identical molecules (as in 2A → B), Equation 10 cannot be used because, when $a = b$, $kt = 0/0$. The rate equation in that case may be written either as

$$\frac{d[A]}{dt} = -k[A]^2 \quad \text{or} \quad \frac{dx}{(a - x)^2} = k\, dt ,$$ (13)

which integrates to

$$\frac{x}{a(a - x)} = kt .$$ (14)

As in the case of a first-order reaction, some information can be gained from measurement of the half-life. For a second-order reaction of the form A → P, the integrated form of the rate law is also $kt = x/[a(a - x)]$. The value of $t_{1/2}$ is obtained when $x = \frac{1}{2}a$, so that

$$t_{1/2} = \frac{1}{ka} .$$ (15)

Thus for a second-order reaction, $t_{1/2}$ depends on the initial concentration. This can be used to test for second-order behavior as follows. At any time in the course of the reaction a concentration a_i is selected as an initial concentration and the time $t_{1/2,i}$ required to reach a concentration of $\frac{1}{2}a_i$ is measured. Each value of $t_{1/2,i}$ is plotted against $1/a_i$. If the reaction is second order, a straight line having a slope of k results (Figure 11-3).

Reactions of Higher Order

Reactions of higher order are not uncommon. Equations such as those already shown can be derived and will be found in the references given at the end of the chapter. Usually these equations are not used because various means that are somewhat simpler are used to determine the order. The most fruitful method is described in the next section, on pseudo-order reactions.

Sometimes the method of measuring the half-life is useful. A general equation for $t_{1/2}$ can be derived for a reaction of order n (n can be positive, negative, or even fractional) in which either one reactant or several reactants are mixed in equimolar proportions and react mole for mole. The general rate equation is

$$\frac{dc}{dt} = -kc^n ,$$ (16)

FIGURE 11-3 Determination of k for a second-order reaction by the method of half-lives. In the large graph the concentration of the reactant is plotted as a function of time. The half-times for successive periods are listed below the curve. In the insert, each of these values is plotted against the reciprocal of the starting concentration for each successive interval. Each point in the insert is derived from the point in the large graph having the same Roman numeral.

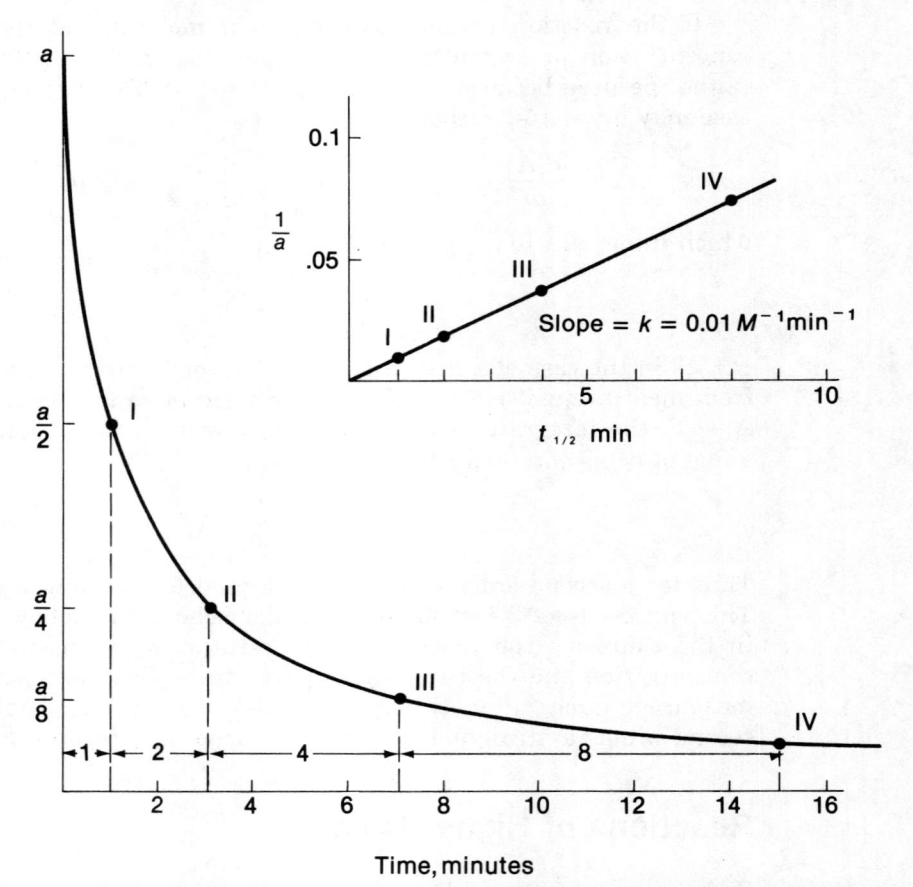

in which c is the concentration of any reactant at time t. The integrated form is

$$c^{1-n} - c_0^{1-n} = (n - 1)kt \qquad (17)$$

in which c_0 is the initial concentration. Setting $c = \tfrac{1}{2}c_0$ in order to evaluate $t_{1/2}$, there results

$$t_{1/2} = \frac{2^{n-1} - 1}{(n - 1)k} c_0^{1-n} . \qquad (18)$$

(Note that this equation cannot be used for $n = 1$ since $t_{1/2}$ reduces to $0/0$.) In logarithmic form, this is

$$\ln t_{1/2} = \frac{2^{n-1} - 1}{(n - 1)k} - (n - 1) \ln c_0 . \qquad (19)$$

Thus a plot of ln $t_{1/2}$ versus ln c_0 yields a straight line with slope $(1 - n)$; k can also be determined from the y intercept.

Pseudo-Order Reactions

Whenever a reactant is present in great excess, its concentration is virtually constant during the course of the reaction, because only a very small fraction of it is consumed. Thus in a second-order reaction $A + B \rightarrow C$ having a rate law $d[A]/dt = -k[A][B]$, if B is in excess, [B] is constant and the rate law is written in terms of an observed, concentration-specific rate constant k_{obs}; that is, $d[A]/dt = -k_{obs}[A]$ in which $k_{obs} = k[B]$. Under these conditions, the reaction is said to be pseudo-first order and k_{obs} is the pseudo-first-order rate constant. Note that k_{obs} is not a true constant because it depends on [B].

A third-order reaction having a rate law $d[B]/dt = k[A]^2[B]$ can also be made pseudo-first-order if [A] is in great excess. If the rate is written $d[A]/dt = k[A]^2[B]$, it will be pseudo-second-order if [B] is in excess.

The study of some complex reactions can be considerably simplified by this technique, since it enables one to observe separately the behavior of a small number of components. For instance, in the third-order reaction just mentioned, one can determine the exponent of [A] when B is in excess and the exponent of [B] when A is in excess. There is one experimental problem in using this method—since one component must be at a very high concentration, an impurity in that component may reach a sufficiently high concentration in the reaction mixture that it can influence the rate. Nonetheless, with precautions, this method is probably one of the most valuable procedures that is available to the experimenter for simplifying the determination of a rate constant.

Pseudo-order is a natural occurrence in biochemical reactions in which water is one of the reactants, because water is at a concentration of about 55 M; that is, it is always in excess. For example, in the conversion of sucrose to glucose and fructose by the enzyme invertase,

$$\text{Sucrose} + \text{H}_2\text{O} \xrightarrow{\text{invertase}} \text{glucose} + \text{fructose},$$

the reaction can be said to be second order* as characterized by the rate equation

$$\frac{d[\text{sucrose}]}{dt} = -k[\text{sucrose}][\text{H}_2\text{O}]. \tag{20}$$

Since the change in [H_2O] is undetectable in such a reaction, the rate equation will appear to be

$$\frac{d[\text{sucrose}]}{dt} = -k_{obs}[\text{sucrose}]. \tag{21}$$

---
*For the moment we are ignoring the role of the enzyme. Since its concentration is also important, the reaction is in fact third order.

Initial Rates

The measurement of initial rates is a means of facilitating the determination of a rate constant. The principle is that if one examines the reaction just when it is beginning, the concentrations of the reactants are essentially constant. For example, in a reaction such as $A + B \rightarrow P$, which has a rate law $d[P]/dt = k[A][B]$, at an early time

$$k = (1/ab)\, d[P]/dt = -(1/ab)\frac{d[A]}{dt} = -\left(\frac{1}{ab}\right)\frac{d[B]}{dt},$$

in which a and b are the concentrations of A and B at $t = 0$. Experimentally this is a difficult procedure because it requires an accurate measurement of either very small concentrations of P or very small changes in the initial concentration of A or B.

11–4 RATE EQUATIONS FOR MORE COMPLEX REACTIONS

So far we have been considering rate equations and methods applicable to a simple one-step reaction or to a series of irreversible reactions in which only one reaction is rate-determining. However, it is relatively uncommon in general and especially rare in biochemistry that reactions follow a simple course. The three most common causes of complexity are a result of there being (1) a sequence of reactions, one or more of which may be reversible; (2) reactions that compete with the formation of the product; and (3) parallel reactions. In the following we analyze some of these cases.

Reversible Reactions

Implicit in the formalism for each case that has been presented so far is that the reaction goes to completion; that is, equilibrium lies very far to the right. This means that given sufficient time the concentration of some reactant should reach zero. It is a common observation, though, that after a long period of time there are still significant concentrations of all reactants even when equilibrium has not been achieved (that is, all concentrations are still changing). Whenever this is observed, it should arouse suspicion that the reaction is reversible. We will now derive the necessary equations for two common situations: (1) both the forward and the back reactions are first order, and (2) both are second order. The complex case, in which one is first order and the other is second order can be found in references at the end of the chapter.

A. Both Reactions Are First Order

Consider a first-order reaction $A \rightarrow B$ for which the rate constants for the rightward and leftward reactions are k_1 and k_{-1}, respectively. If the initial concentrations of A and B are a and b, respectively, and after a time dt, x moles of A are converted to x moles of B, so that $[A] = a - x$ and $[B] = b + x$, then

$$\frac{dx}{dt} = k_1(a - x) - k_{-1}(b + x), \qquad (22)$$

which integrates to the formidable equation

$$\ln\left[\frac{k_1 a - k_{-1} b}{k_1 a - k_{-1} b - (k_1 + k_{-1})x}\right] = (k_1 + k_{-1})t. \qquad (23)$$

This single equation is of course insufficient to solve for the values of both k_1 and k_{-1}. However, this can be done in one of two ways.

1. The reaction can be allowed to go to equilibrium and the equilibrium constant K can be calculated. This will aid us in the following way. Equation 23 can be written in exponential form as

$$\frac{k_1 a - k_{-1} b - (k_1 + k_{-1})x}{k_1 a - k_{-1} b} = e^{-(k_1 + k_{-1})t} \qquad (24)$$

At equilibrium (that is, at $t = \infty$), the exponential term is zero, so that

$$k_1 a + k_{-1} b - (k_1 + k_{-1})x = 0, \qquad (25)$$

which can be rearranged to

$$\frac{b + x}{a - x} = \frac{k_1}{k_{-1}}. \qquad (26)$$

However, $a - x$ and $b + x$ are the equilibrium concentrations of A and B, respectively, and by definition $[B]_{eq}/[A]_{eq} = K$, the equilibrium constant. Thus

$$K = \frac{k_1}{k_{-1}}. \qquad (27)$$

Therefore, from either Equation 25 or Equations 23 and 26 taken together, we have two simultaneous equations that enable us to solve for k_1 and k_{-1}.

2. The initial value of dx/dt can be measured. That is, for x nearly equal to zero, Equation 22 becomes

$$\frac{dx}{dt} \approx k_1 a - k_{-1} b, \qquad (28)$$

which also provides a second equation relating k_1 and k_{-1}.

B. Both Reactions Are Second Order

The simplest reaction of this type is $2AB \rightarrow A_2 + B_2$. If $[AB] = c_0$ at $t = 0$ and, at time t, $[AB] = c_0 - c$, then at time t, $[A_2] = [B_2] = \frac{1}{2}c$ and the rate equation is

$$\frac{dc}{dt} = k_1(c_0 - c)^2 - k_{-1}\left(\frac{1}{2}c\right)^2. \tag{29}$$

This can be integrated to yield an equation that is very complex algebraically. However, there is an alternative method, which we describe here, that is often simpler. The reaction is allowed to reach equilibrium, and the equilibrium value of [AB], that is, c_{eq}, is measured. At equilibrium there are no changes in concentration, so that $dc/dt = 0$. Therefore

$$\frac{c_{eq}^2}{4(c_0 - c_{eq})^2} = \frac{k_1}{k_{-1}} = K. \tag{30}$$

This can be inserted into the rate equation to yield

$$\frac{dc}{dt} = k_1\left[c_0^2 - 2c_0c + \left(1 - \frac{1}{4}K\right)c^2\right]. \tag{31}$$

One can then do one of two things: (1) measure dc/dt at $c = 0$ (by extrapolation or by determining the initial slope of the graph dc/dt versus c), in which case $(1/c_0^2)(dx/dt) = k_1$; or (2) make a plot of c versus t, obtain dc/dt for various values of c from the graph, and plot dc/dt against the bracketed expression to calculate k_1.

Competing or Parallel Reactions

Substances sometimes can react to give more than one product; also, a single substance can decompose by several pathways to give either the same or different products. When either of these situations occurs, we say that there are parallel reactions. If the parallel reactions are not recognized, the interpretation of the kinetics will be incorrect.

In general, *when parallel reactions exist, the reaction with the greater rate is usually dominant.* For instance, with a pair of reactions A → B and A → C having rate constants k' and k'', respectively, if $k' > k''$, the principal product would be B. In biochemical systems, in which almost all reactions are enzyme-catalyzed, it is rare—although there are examples—that a single enzyme catalyzes both of the reactions A → B and A → C. However, it is not uncommon for a living cell to possess two different enzymes, each of which can catalyze one member of a pair of parallel reactions. This is a major problem in the study of certain reactions in crude, unfractionated cell extracts.

In the case of the parallel reactions, if each reaction is first order, the rate equations are

$$\frac{d[B]}{dt} = k'[A] \quad \text{and} \quad \frac{d[C]}{dt} = k''[A]. \tag{32}$$

Therefore,

$$-\frac{d[A]}{dt} = k'[A] + k''[A] = k_{obs}[A], \tag{33}$$

in which $k_{obs} = k_1' + k_1''$. If a is the concentration of A at $t = 0$, then at any time, $[A] = a\, e^{-k_{obs}t}$, and from this expression k_{obs} can be determined. Therefore,

$$\frac{d[B]}{dt} = k'[A] = k'a\, e^{-k_{obs}t}, \tag{34}$$

and

$$[B] = \frac{ak'}{k_{obs}}(1 - e^{-k_{obs}t}). \tag{35}$$

For product C,

$$\frac{d[C]}{dt} = k''[A] = k''a\, e^{-k_{obs}t}, \tag{36}$$

and

$$[C] = \frac{ak''}{k_{obs}}(1 - e^{-k_{obs}t}). \tag{37}$$

Similar but more complex equations apply to a system containing more than two parallel reactions. These equations can be found in references given at the end of the chapter.

An interesting type of parallel reaction is that in which two different pathways lead to the same product. This is shown in the following example.

EXAMPLE 11-B Two pathways leading to synthesis of one product.

The product is the ion I_3^-; the reaction is

$$H_2O_2 + 3I^- + 2H^+ \rightarrow I_3^- + 2H_2O.$$

When the pH is very low (for example, $[H^+] > 0.5\,M$), the rate is found to be proportional to the product $[H^+][H_2O_2][I^-]$, yet at pH ≈ 3, the rate depends upon $[H_2O_2][I^-]$. This clearly indicates that I_3^- is generated by two pathways, one in which the H^+ ion participates and the other in which there is a direct interaction between the H_2O_2 molecule and the I^- ion. Thus the rate equation must be a sum of two terms:

$$\frac{d[I_3^-]}{dt} = k'[H_2O_2][I^-] + k''[H^+][H_2O_2][I^-]. \tag{38}$$

It should be noted that if the reaction had only been studied at pH > 3, the rate equation would have been simple and the reaction occurring at low pH might not have been discovered.

The possibility of alternate pathways must always be kept in mind; a simple rate equation clearly cannot be taken as evidence that no other pathway exists, if the data are obtained for only one set of conditions.

Consecutive Reactions

It is especially common in biochemical systems that in the conversion of A to C there is an intermediate compound B. Such a reaction sequence,

$$A \xrightarrow{k_1} B \xrightarrow{k_2} C,$$

is characterized by two first-order rate constants k_1 and k_2. As long as the reaction A → B is not reversible, the concentration of A is governed only by the rate equation

$$\frac{d[A]}{dt} = -k_1[A]; \tag{39}$$

that is, the first reaction is the rate-determining step for consumption of A. However, the concentration of the intermediate B is determined by the rate of production of B *and* the rate of conversion of B to C—that is,

$$\frac{d[B]}{dt} = k_1[A] - k_2[B]. \tag{40}$$

The concentration of C is, as we have mentioned before, determined by the reaction that is the slow step. Equation 39 has a simple exponential solution. If only A were present at $t = 0$, at a concentration a, then $[A] = a e^{-k_1 t}$. This can be substituted into Equation 40, which, for the initial value of $[B] = 0$ (only A was present at $t = 0$) has the solution

$$[B] = a\left(\frac{k_1}{k_2 - k_1}\right)(e^{-k_1 t} - e^{-k_2 t}). \tag{41}$$

The rate equation for [C] is obtained by noting that at all times $[A] + [B] + [C] = a$, because the concentrations are expressed in molarity. Thus $[C] = a - [A] - [B]$, and, by recombining with Equations 39 and 41, yields

$$[C] = a\left[1 + \left(\frac{1}{k_1 - k_2}\right)(k_2 e^{-k_1 t} - k_1 e^{-k_2 t})\right]. \tag{42}$$

Graphs showing [A], [B], and [C] as a function of time are shown in Figure 11-4.

Let us now consider an example in which the A → B reaction is very much slower than B → C; that is, $k_2 \gg k_1$. Examining Equation 42, we see that $e^{-k_2 t} \ll e^{-k_1 t}$, so that $e^{-k_2 t} \approx 0$. Also $k_2/(k_1 - k_2) \approx -1$, so that Equation 42 becomes

$$[C] = a(1 - e^{-k_1 t}), \tag{43}$$

which, as we have said, shows that the rate is governed only by the smaller rate constant k_1.

The case of $k_2 > k_1$, in which k_2 is large but not enormous compared to k_1, is of special interest and is common in biochemistry. This is the case of a *reactive intermediate*. In this case, after a period of time there is a

FIGURE 11-4 The concentration of the reactant A, intermediate B, and product C for two consecutive first-order reactions when the first reaction is about ten times as fast as the second reaction. [A] drops rapidly and [B] rises rapidly at early times. [C] increases more slowly because its production is controlled by [B]. The dashed line shows the total concentration of all reactants; this is the initial concentration of A and the final concentration of C.

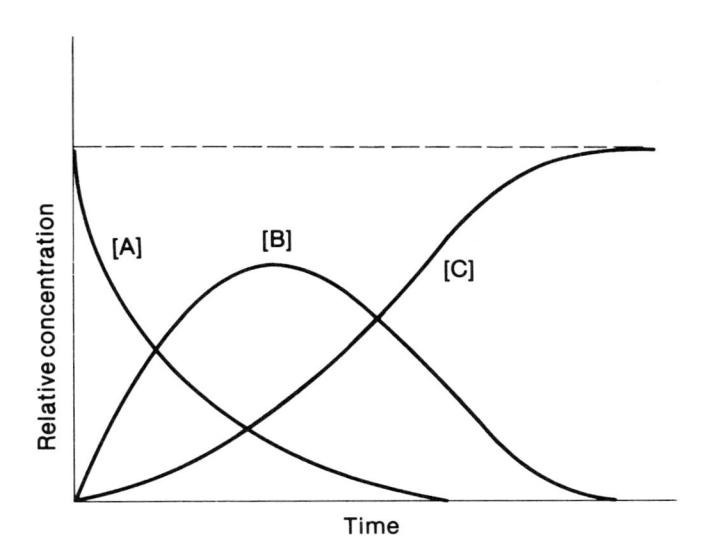

significant amount of B and Equation 41 can be approximated as

$$[B] \approx a\left(\frac{k_1}{k_2 - k_1}\right) e^{-k_1 t} . \qquad (44)$$

Equation 39 can be solved to yield $[A] = a\, e^{-k_1 t}$, so that

$$\frac{[B]}{[A]} \approx \frac{k_1}{k_2 - k_1} \approx \frac{k_1}{k_2} . \qquad (45)$$

Since $k_2 > k_1$, A reacts slowly compared to B, so that for short reaction times [A] is constant. Therefore [B] is nearly constant because k_1/k_2 is constant or $d[B]/dt = 0$. This situation is a very important one. That is,

> when a reaction achieves a state in which the reactive intermediate is roughly at a constant concentration, it is said to be in a steady state.

The mathematical analysis of this state is called the *steady-state approximation*. An example of this is shown in Figure 11-5.

Working under conditions of a steady state (that is, by determining experimentally when [B] is changing very slowly and then measuring [A] and [C] as functions of time) is a valuable method for obtaining approximate values for k_1 and k_2. This can be seen by combining Equations 43 through 45 to yield

$$\frac{d[C]}{dt} = k_2[B] \approx k_1[A] . \qquad (46)$$

If the data obtained obey Equation 46, it is likely that both reactions are first order and that k_1 and k_2 can be measured.

Seeking steady-state conditions also simplifies the analysis of a very

FIGURE 11-5 The concentration of the reactant A, the intermediate B, and the product [C] for two consecutive first-order reactions when the second reaction is about ten times as fast as the first reaction. In the broad region in which [B] is nearly constant, the steady-state approximation is valid. This graph should be compared to that in Figure 11-4.

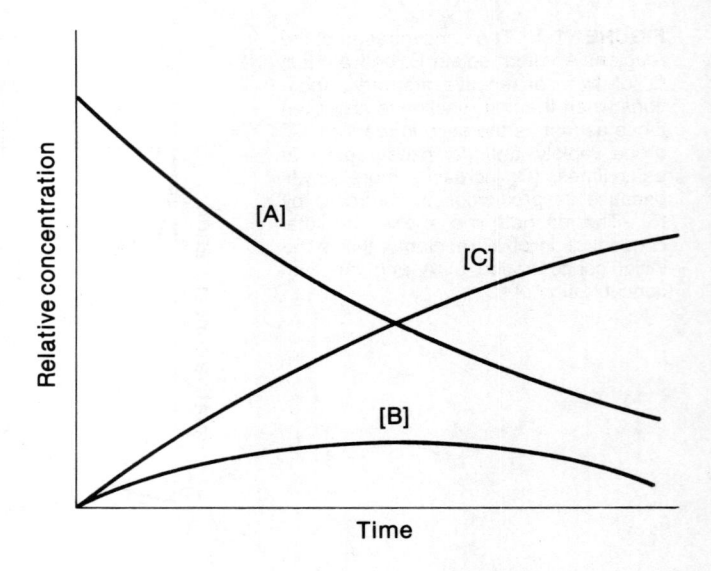

common type of reaction—the bimolecular, consecutive reaction in which the first step is reversible—namely,

$$A + B \underset{k_{-1}}{\overset{k_1}{\rightleftarrows}} AB \overset{k_2}{\longrightarrow} C.$$

We shall explore this when discussing enzyme kinetics.

Consecutive reactions should be suspected for all third-order reactions. For instance, in a reaction of the type $2A + B \rightarrow C$ that is experimentally shown to be third order (that is, the rate of production of C is proportional to both [B] and $[A]^2$), if the reaction were to occur in a single step, it would have to be a termolecular reaction, requiring the simultaneous interaction of three molecules. This is not very likely. Two examples of a more reasonable pathway are the following:

$$A + A \rightarrow A_2 \quad \text{and} \quad A_2 + B \rightarrow C;$$
$$A + B \rightarrow AB \quad \text{and} \quad AB + A \rightarrow C.$$

Consecutive reactions are extremely important in biochemistry, because all enzyme-catalyzed reactions, no matter how simple, consist of consecutive reactions. We will discuss enzymatic reactions in the following chapter.

11–5 FAST REACTIONS

Many reactions occur too rapidly to measure changes in concentration as a function of time. There are two principal techniques available to determine the rate constants for such reactions: the stirred-flow system, and relaxation. The latter is the most accurate and most prevalent but the stirred-flow system has the advantage that an intermediate having a very short lifetime often becomes detectable.

Flow Systems

Consider a reaction vessel having volume V and with an inlet and outlet tube as shown in Figure 11-6. The vessel is equipped with a stirrer so that all components will mix rapidly. A reactant is made to flow into the vessel at a particular flow rate u and concentration A_1. The fluid leaving the vessel will be at a concentration A_t that varies with time and may be greater than or less than A_1. The rate of change of the number of molecules of A in the volume is the sum of two terms: the one due to the chemical reaction we write as rV, in which r is the reaction rate per unit volume; the other is equal to the number of molecules of A entering the vessel (uA_1) minus the number of molecules leaving the vessel (uA_2). Therefore,

$$\frac{dn_A}{dt} = rV + uA_1 - uA_t. \tag{47}$$

FIGURE 11-6 A stir-flow reaction vessel.

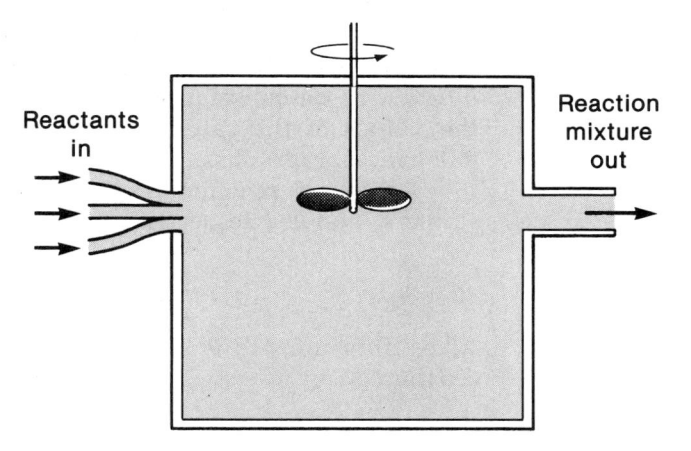

Rather than solving the equation exactly, it is common, experimentally, to allow the system to reach a steady state in which A_t has the unvarying value A_2. In this state $dn_A/dt = 0$, and

$$r = \frac{u}{V}(A_2 - A_1). \tag{48}$$

But r by definition is $-d[A]/dt$. Thus by suitable choice of u and A_1, the reaction rate $-d[A]/dt$ can be measured simply by measuring A_2. In this way the order of a particular component of a reaction can be determined by studying each reactant A_1 separately and noting how $d[A_i]/dt$ depends on $[A_i]$.

The stirred-flow system has another important use—namely, in the identification of a reaction intermediate whose average lifetime is so short that its concentration is too low to be detected. By appropriate choice of A_1 and u, the concentration of an intermediate can be increased. At the higher concentration, an intermediate might be studied spectroscopically if the reaction were carried out in a flow cell of a spectrophotometer or other spectral instrument.

Relaxation Methods

Consider a system that is in equilibrium. Appropriate measurements of concentration yield the equilibrium constant but give no information about the forward and reverse rate constants, whose quotient is the equilibrium constant. If any of the parameters that determine the equilibrium state—for example, temperature or pressure—is altered, the composition of the system will change until a new state of equilibrium is achieved. The process of moving toward a new equilibrium state is called *relaxation*. Measurements of this equilibrium state yield the new equilibrium constant. However, the kinetics by which the second equilibrium state is reached are determined by the rate constants; thus, measurement of these *relaxation kinetics*, in combination with the value of the equilibrium constant, yield the values of the rate constants. How this is done is explained in the following paragraphs.

Consider a reaction $A + B \rightleftarrows P$ having forward and back rate constants k_1 and k_{-1}, respectively. The expression for the reaction rate is

$$\frac{d[P]}{dt} = k_1[A][B] - k_{-1}[P] \ . \tag{49}$$

At equilibrium, [P] is constant; i.e., $d[P]/dt = 0$, so that the equilibrium constant K is

$$K = \frac{k_1}{k_{-1}} = \frac{[P_{eq}]}{[A_{eq}][B_{eq}]} \ .$$

If the temperature is suddenly increased by 10°C (for example, by discharge of a capacitor through the reaction mixture, a technique called the *temperature-jump method*), K will have a new value, because

$$\left(\frac{\partial \ln K}{\partial T}\right)_P = \frac{\Delta H°}{RT^2} , \tag{50}$$

as was shown in Chapter 5. This will necessitate changes in all concentrations—namely, $\Delta[A]$, $\Delta[B]$, and $\Delta[P]$—from the first equilibrium value. In this reaction $\Delta[P] = -\Delta[A] = -\Delta[B]$. Denoting by a, b, and p the concentrations in the new state of equilibrium, and assuming that the changes introduced by the temperature-jump are so small that $(\Delta[A])^2 \ll (\Delta[A])$,

SEC. 11-5   FAST REACTIONS

$(\Delta[B])^2 \ll \Delta[B]$ and $(\Delta[P])^2 \ll \Delta[P]$, we may write

$$[A] = a + \Delta[A] = a - \Delta[P],$$
$$[B] = b + \Delta[B] = b - \Delta[P],$$

and

$$[P] = p + \Delta[P],$$

so that

$$\frac{d[P]}{dt} = \frac{d(p + \Delta[P])}{dt} = \frac{d(\Delta[P])}{dt}. \tag{51}$$

In the new equilibrium state, $[p]$ = constant, so that $d[p]/dt = 0$. Thus

$$\frac{d(\Delta[P])}{dt} = \frac{d[P]}{dt}$$
$$= k_1 ab - k_{-1}p - k_1\{[A]\Delta[P] + [B]\Delta[P] - (\Delta[P])^2\} - k_{-1}\Delta[P].$$

Since $(\Delta[P])^2$ is very small and from Equation 49 the first two terms equal zero, we may write

$$\frac{-d(\Delta[P])}{dt} = \{k_1(a + b) + k_{-1}\}\Delta[P], \tag{52}$$

which is usually written

$$\frac{-d(\Delta[P])}{dt} = \frac{\Delta[P]}{\tau}, \tag{53}$$

in which τ is called the relaxation time and has the value

$$\tau = \frac{1}{k_{-1} + k_1(a + b)}. \tag{54}$$

Integration of this equation yields

$$\Delta[P] = \Delta[P_0]e^{-t/\tau},$$

in which $\Delta[P_0]$ is just the initial concentration of P minus p. Thus a measurement of concentration as a function of time yields the value of τ because τ is the time required for $\Delta[P]$ to reach $1/e$ times the initial value.

A single measurement of τ is clearly insufficient to measure k_1 and k_{-1}, since we have only a single equation with two variables. However a and b can be selected by the experimenter. Thus if Equation 5 is rearranged to $1/\tau = k_{-1} + k_1(a + b)$, and $1/t$ is measured for several values of $a + b$, a plot of $1/\tau$ versus $(a + b)$ yields k_1 as the slope and k_{-1} as the y intercept.

The analysis just presented has been carried out for several types of reactions. As long as the displacements from the initial equilibrium concentrations are small so that the square terms, such as $(\Delta[A])^2$, can be neglected, it is always found that the relaxation kinetics are first order—

TABLE 11-1 Expression for the relaxation time, τ, for several types of single-step reactions.

Reaction	τ
$A \rightleftarrows B$	$\tau = \dfrac{1}{k_1 + k_{-1}}$
$A + B \rightleftarrows P$	$\tau = \dfrac{1}{k_{-1} + k_1(a + b)}$
$2A \rightleftarrows A_2$	$\tau = \dfrac{1}{4k_1 a + k_{-1}}$
$A + B \rightleftarrows C + D$	$\tau = \dfrac{1}{k_1(a + b) + k_{-1}(c + d)}$
$A + B + C \rightleftarrows P$	$\tau = \dfrac{1}{k_{-1} + k_1(ab + bc + ac)}$

that is, they have the $e^{-t/\tau}$ term. Expressions for τ depend on the particular reaction, though; these reactions and expressions are listed in Table 11-1. An example of the use of the temperature-jump method follows.

EXAMPLE 11–C Determination of the rate constants for a reaction $A + B \rightleftarrows P$, in which P absorbs blue light and whose absorbance is proportional to the concentration of P.

Figure 11-7(a) shows the absorbance of P as a function of time after a temperature jump that occurs in 1 microsecond. The total concentration of A and B at equilibrium is 10^{-5} M. The initial value of the absorbance of P is 0.34; after a very long time, the new equilibrium value of 0.25 is reached. Thus $\Delta[P_0] = 0.34 - 0.25 = 0.09$. The data in Figure 11-7(a) are replotted on semilog paper to obtain what is shown in Figure 11-7(b). A straight line results, so that τ, the time required to reach $(1/e)(0.09) = 0.033$, is 75 milliseconds. In order to determine k_1 and k_{-1} the experiment was repeated, using mixtures for which $a + b$ took other values. The data obtained were: $a + b = 5 \times 10^{-5}$ M, $\tau = 37.1$ milliseconds; and $a + b = 10^{-4}$ M, $\tau = 22.7$ milliseconds. A plot of these data, for which the second entry of Table 11-1 was used, is shown in Figure 11-8. The slope and intercept are shown in the figure, so that $k_{-1} = 9.93$ M^{-1} sec^{-1} and $k_1 = 3.4 \times 10^5$ M^{-1} sec^{-1}.

Many reactions, of course, consist of multiple steps. This can be recognized when a plot such as is shown in Figure 11-7(b) is made, in that a straight line will not result. Multistep reactions having more than two rate constants can be analyzed by the relaxation procedure but the analysis, which

FIGURE 11-7 Data for Example 11–C.

(a)

(b)

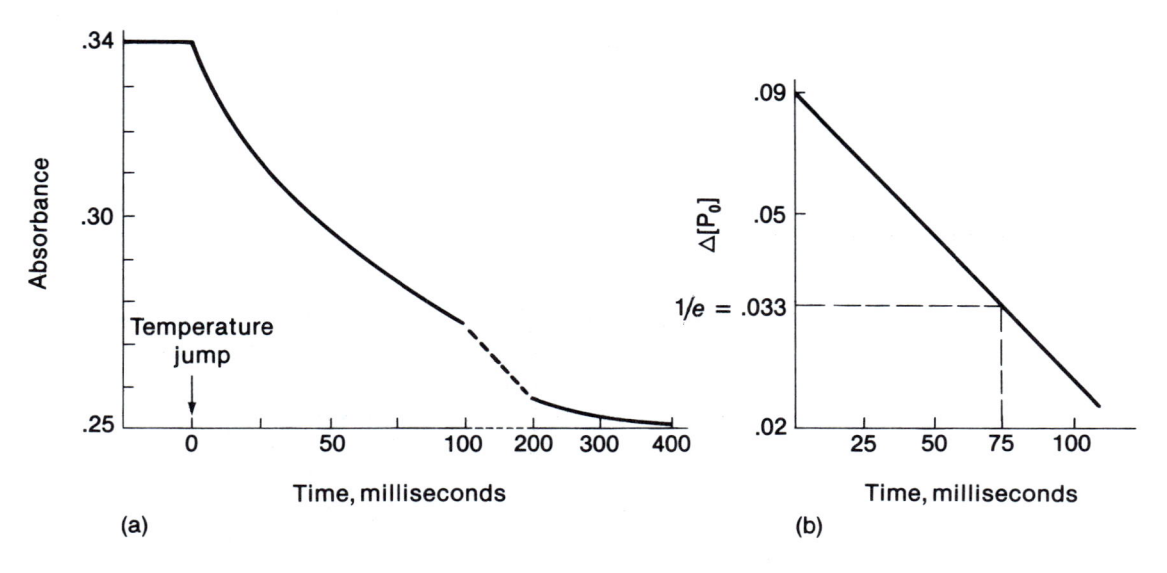

FIGURE 11-8 Data for Example 11–C.

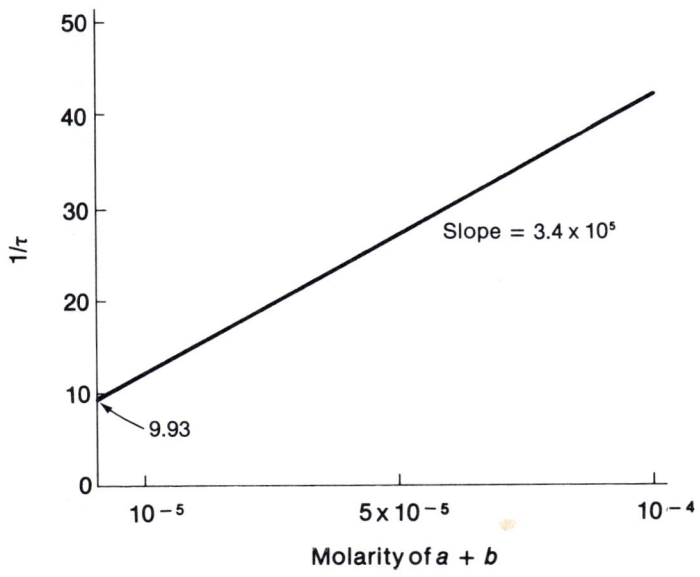

is considerably more complex than the analysis for single-step reactions, is beyond the scope of this book. The interested reader can consult the references at the end of this chapter.

11-6 CHAIN REACTIONS

Experimentally determined rate equations often include fractional exponents, for example, $\frac{1}{2}$ and $\frac{3}{2}$, which suggests that the mechanism is somewhat complex. When this occurs, it is also discovered in the rate equation that the reaction is inhibited by increasing the concentration of the product. This is called *product inhibition* or *end-product inhibition* and this almost always means that the reaction is a *chain reaction;* that is, a reaction sequence in which a component formed in a late step in the sequence is used to initiate an early step. A well-studied example of a chain reaction is the apparently simple reaction

$$H_2 + Br_2 \to 2HBr ,$$

which has the complex empirical rate law

$$\frac{d[HBr]}{dt} = \frac{c_1[H_2][Br_2]^{1/2}}{1 + c_2([HBr]/[Br_2])} , \tag{55}$$

in which c_1 and c_2 are constants. In the following, we show how such a strange equation can arise. The reaction has been shown to proceed via the following sequence:

$$(1) \qquad Br_2 \xrightarrow{k_1} 2Br\cdot$$

$$(2) \qquad Br\cdot + H_2 \xrightarrow{k_2} HBr + H\cdot$$

$$(3) \qquad H\cdot + Br_2 \xrightarrow{k_3} HBr + Br\cdot$$

$$(4) \qquad H\cdot + HBr \xrightarrow{k_4} H_2 + Br\cdot$$

$$(5) \qquad Br\cdot + Br\cdot \xrightarrow{k_5} Br_2$$

Reaction 1 is the *initiation step* resulting from thermal dissociation. Steps 2 and 3 are the *propagating steps* in which two HBr molecules are produced and Br· is formed to continue the cycle. Step 4 is an *inhibiting* or *competing reaction* in which the product of the overall reaction combines with the H· needed in the cycle. Note that the existence of step 4 is equivalent to step 2 (or reaction 2) being reversible. Step 5 is a *chain-breaking* or *termination step*, since the Br· needed to maintain propagation is removed; this is of course just the back-reaction of step 1.

The rate of formation of HBr is a result of its production in reactions 2 and 3 and its destruction in reaction 4; therefore,

$$\frac{d[\text{HBr}]}{dt} = k_2[\text{Br}][\text{H}_2] + k_3[\text{H}][\text{Br}_2] - k_4[\text{H}][\text{HBr}] \ . \tag{56}$$

In the case that the rate constants are such that [H] and [Br] are nearly constant, which applies to this set of reactions, once the chain is established, [H] and [Br] can be evaluated by the steady-state approximation—that is,

$$\frac{d[\text{H}]}{dt} = k_2[\text{Br}][\text{H}_2] - k_3[\text{H}][\text{Br}_2] - k_4[\text{H}][\text{HBr}] = 0 \ ; \tag{57}$$

and

$$\frac{d[\text{Br}]}{dt} = 2k_1[\text{Br}_2] - k_2[\text{Br}][\text{H}_2] + k_3[\text{H}][\text{Br}_2] + k_4[\text{H}][\text{HBr}] - 2k_5[\text{Br}]^2$$
$$= 0 \ . \tag{58}$$

These two equations are solved for [H] and [Br] and substituted into Equation 56 to yield

$$\frac{d[\text{HBr}]}{dt} = \frac{2k_2(k_1/k_4)^{1/2}[\text{H}_2][\text{Br}_2]^{1/2}}{1 + \{k_4[\text{HBr}]/k_3[\text{Br}_2]\}} \ , \tag{59}$$

which is equivalent to the rate law in Equation 55 if we substitute $c_1 = 2k_2(k_1/k_4)^{1/2}$ and $c_2 = k_4/k_3$.

It should be noted that both the Br and H atoms have an unpaired electron. Such substances are known as *free radicals* and are denoted by a dot, as with Br· and H·. Free radicals are common intermediates in organic reactions and in biochemical pathways, and should be sought when an empirical rate law has fractional order.

11–7 EFFECT OF TEMPERATURE ON THE REACTION RATE

Thermodynamic arguments given in the late nineteenth century led to a relation between the value of the equilibrium constant K of a reaction and the temperature. The essential equation was

$$\frac{d \ln K}{dT} = \frac{\Delta U}{RT^2} \quad \text{or} \quad \ln K = -\frac{\Delta U}{RT} + \text{constant} \ , \tag{60}$$

in which ΔU is the internal energy change. Since K is related to the reaction rates k_1 and k_{-1} of the forward and reverse reactions, respectively, it was proposed by Svante Arrhenius in 1889 that the rate constant shows a similar dependence on temperature, which, when written in exponential form, is

$$k = A\, e^{-E_a/RT}, \tag{61}$$

in which A is a constant and E_a is called the *activation energy*. This equation is frequently written as

$$\frac{d \ln k}{d(1/T)} = -\frac{E_a}{R}, \tag{62}$$

because a plot of $\ln k$ versus $1/T$ is linear and E_a can be calculated from the slope. Such a plot is called an *Arrhenius plot*.

This exponential relation means that the reaction velocity is very strongly dependent on temperature and at low temperature the reaction proceeds very slowly. This can be seen in the effect of temperature on the decomposition of N_2O_5, as shown in Table 11-2. In this table the rate constant and corresponding half-life of N_2O_5, $t_{1/2}$, are shown for various temperatures. The value of E_a calculated from these data is 101.3 kJ mol^{-1}. The enormous reduction in rate at low temperatures should be noted.

TABLE 11-2 The rate constant for the decomposition of N_2O_5 as a function of temperature.

Temperature, °C	k, sec^{-1}	$t_{1/2}$
−100	3.5×10^{-18}	6.3×10^9 year
−50	3.1×10^{-11}	700 year
0	8×10^{-7}	10 day
+50	8.9×10^{-4}	13 min
+100	0.15	4.6 sec
+150	7.6	0.091 sec

When Arrhenius introduced the concept of activation energy, he did not make any assumptions about its significance. The fact that E_a is constant indicates that it does represent some type of energy that is important for the reaction. One possible explanation is that it is the minimal amount of energy that must be possessed by a pair of molecules in order that they react. A simple view of this energy is that it represents the translational energy of the molecules (which would obey the Boltzmann distribution and hence would have an exponential dependence on temperature). By this explanation, E_a would be the kinetic energy of the molecules that, in an inelastic collision, would be used to form a new chemical bond. An alternate explanation is that E_a is a free-energy term, including both enthalpic and entropic contributions. The former idea has been developed in the collision theory of chemical reactions. This theory has been moderately useful in explaining how some chemical reactions occur in the gas phase. However, in its simplest form, the collision theory predicts reaction rates that are too large, and to save the theory, it is necessary to introduce arbitrary steric factors to include the idea that a collision may be successful only when the two reactants are in the correct relative orientation. This complication in the theory is called the *steric problem* and is illustrated in Figure 11-9. Since the steric factor cannot be calculated, the collision theory is severely limited in its usefulness. For a reaction in the liquid state, the theory is of no value because there is no good model for collisions in liquids. In 1935, Henry Eyring introduced a new idea that successfully explains how many chemical reactions occur; this is called the *transition-state theory* and it is explained briefly in the following section.

FIGURE 11-9 An example of the steric problem in a collision between molecules A and B. The dark areas must combine for the reaction to occur.

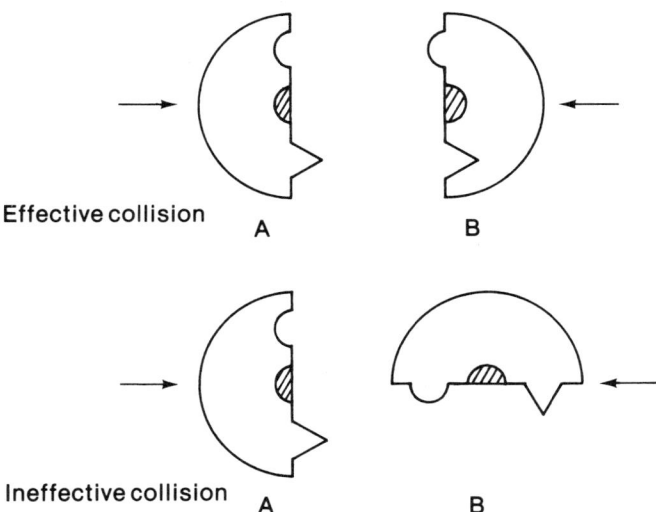

11-8 TRANSITION-STATE THEORY

The basic idea of the transition-state theory is that when two molecules approach, they form (with a certain probability) an activated complex called the *transition state*. Once having formed, the activated complex can do one of two things: (1) it can return to the original state of two distinct molecules, which will then move away from one another; (2) it can be internally rearranged to form new bonds and product molecules. How the second possibility might occur can be seen in the following simple example. Consider an atom A approaching a diatomic molecule BC held together by a single bond, the B—C bond. (More generally, B and C could be atoms bonded in a large molecule consisting of many atoms.) When A is so close to BC that the orbitals of A overlap the orbitals of BC, the B—C bond begins to have increased energy and stretch. A potential bond may even form between A and B. If A has enough energy, it can very closely approach B, stretch the B—C bond very far, and form a relatively strong A—B bond. There will be a point at which the activated composite structure, which we designate (ABC)*, has so much energy that an infinitesimal decrease in the length of the A—B bond will carry the activated structure through the transition state and the B—C bond will break. What is the source of this infinitesimal decrease in bond length? We know that all bonds possess vibrational energy and that the activated complex will have many modes of vibration. In transition-state theory it is assumed that the complex flies apart when the energy of one of its modes of vibration becomes a translational energy, so that what was once one of the bonds in the complex becomes the line of centers between the separating fragments.

The course of a reaction is often described pictorially as a potential energy diagram in which the variable is called the *reaction coordinate* (Figure 11-10), a quantity that is a function of the interatomic distances A—B and B—C. The highest point between A—B and B—C is the transition state; at this point the energy can be lowered by moving either rightward or leftward. In some cases, the intermediate in a reaction is more or less stable ("metastable") and a small amount of energy has to be provided to restore the original reactants or to convert the intermediate to products.

Figure 11-10 Potential energy curves for the dissociation of A—B and B—C (panels a and b), and for the reaction between A—B and C to yield B—C without complete dissociation of A—B (panel c). TS represents the transition state. The reaction coordinate represents the lowest energy pathway from A—B to B—C.

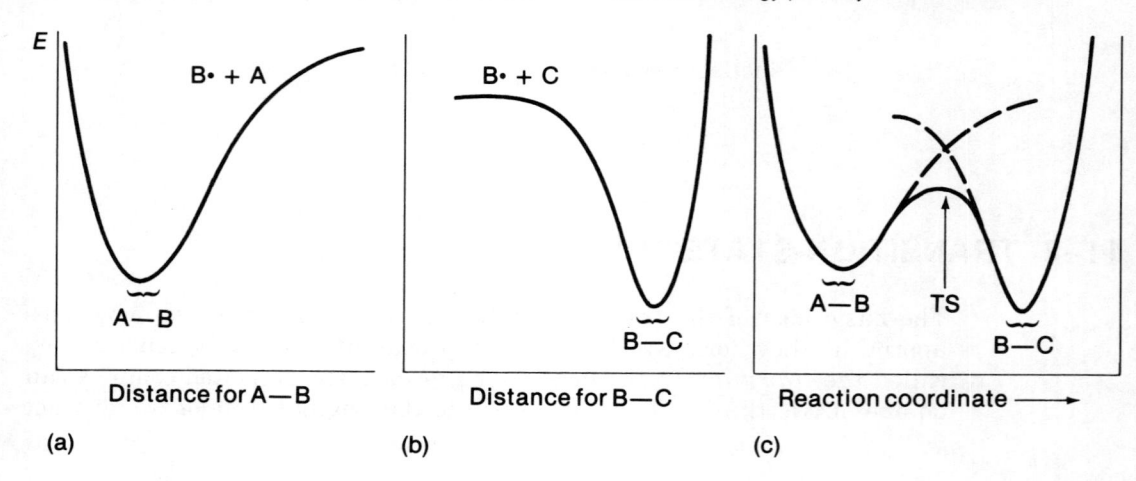

SEC. 11-8 TRANSITION-STATE THEORY

FIGURE 11-11 Energy diagram for a reaction between A—B and C to yield B—C in which the intermediate is either a free radical (B·) or an ion (B⁺). This differs from the diagram in Figure 11-10(c) in that here complete dissociation of A—B occurs before bond formation with C begins.

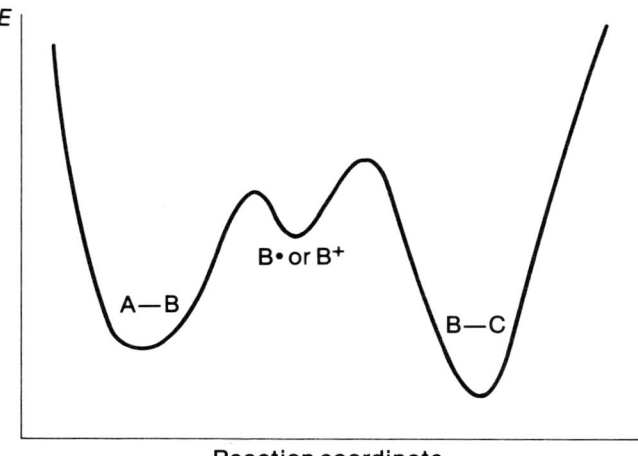

This process is shown in Figure 11-11, in which the intermediate state is the dip in the transition-state diagram. This occurs when the intermediate is a free radical or an ion formed by dissociation of a reactant.

The concentration of the transition state [(ABC)*] can be expressed in terms of an equilibrium constant:

$$K^* = \frac{[(ABC)^*]}{[A][BC]}.$$

Application of quantum mechanical principles yields the following equation for the rate constant k for the formation of the product:

$$k = \frac{K^* \mathbf{k} T}{h} \kappa, \qquad (63)$$

in which $\mathbf{k}$ is the Boltzmann constant, h is the Planck constant, and κ is a constant called the transmission coefficient (that can rarely be calculated and is usually taken to be 0.5 or 1) representing the probability that (ABC)* will fly apart. This equation is known as the *Eyring equation*. The quotient $\mathbf{k}T/h$ has the following significance. In the quantum theory the energy of a vibration having frequency ν is $h\nu$; for a reaction to occur this must be roughly equal to the thermal energy $\mathbf{k}T$ available to cause this vibration. Thus $\mathbf{k}T/h = \nu$. The rate constant k is proportional to ν because if an atom vibrates ν times per second, there are ν opportunities per second for the bond to be broken.

The expression $k = K^*(\mathbf{k}T/h)\kappa$ can also be given in thermodynamic terms, by using the relation between the free energy and the equilibrium constant, $\Delta G^* = -RT \ln K^*$. Thus

$$k = \kappa(\mathbf{k}T/h)e^{-\Delta G^*/RT}. \qquad (64)$$

Since $G = H - TS$, we may define ΔH^*, an enthalpy of activation and ΔS^*, an entropy of activation, by the relation $\Delta G^* = \Delta H^* - T\Delta S^*$. Hence

$$k = \kappa(\mathbf{k}T/h)e^{\Delta S^*/R}e^{-\Delta H^*/RT}. \tag{65}$$

It can be shown that $\Delta H^* = E_a - RT$. In most cases $E_a \gg RT$, so we may write

$$k = \kappa(\mathbf{k}T/h)e^{-E_a/RT}e^{\Delta S^*/R}. \tag{66}$$

Equation 66 does not have exactly the same temperature dependence as Equation 62. However, the variation of k with temperature is primarily determined by the rapidly varying factor $e^{-E_a/RT}$, so that experimentally the equations yield the same result. The advantage of the formulation in Equation 66 is that from a measurement of k and E_a, the entropy of activation ΔS^*, can be calculated. This is a useful number because entropy changes are related to configurational changes and the sign of ΔS^* indicates whether the transition state is more or less complex than the reactants and enables one to make an intelligent guess about the structure of the transition state. For example, when two molecules collide, they are very near one another so that there is a decrease in randomness and therefore a decrease in entropy. The determining factor in the sign of ΔS^* is how tightly bound the activated complex is—the more tightly bound, the lower the entropy. In some reactions, it is found that the value of ΔS^* is nearly the same as ΔS for the reaction (as measured from standard thermal analysis); in such a case one must conclude that the principal entropy change is in the formation of the complex and hence that the structure of the complex must be very similar to the structure of the products. This information is often valuable in trying to determine a reaction mechanism.

11-9 DIFFUSION-CONTROLLED REACTIONS

In order for two molecules to react they must come in contact with one another—that is, they must collide. In solution, owing to the high concentration of the solvent, each solute molecule collides mainly with solvent molecules. For two solute molecules to reach one another, they must move through the solvent; the movement is in almost all cases by diffusion.* This movement is much slower in a liquid than in a gas and the frequency of reactant-encounters is very small compared to the collision frequency in a gas. However, once two reactant molecules have diffused to the same position, they have the same difficulty moving away from one another that they had in approaching one another. Hence, a pair of solute molecules in an encounter remain near one another for a much longer time than do two molecules in a gas—this has been called the *cage effect*. This tendency to remain together means that if E_a and ΔS^* were zero, an encounter would almost always result in reaction. Of course, E_a and ΔS^* are rarely zero.

*The proton moves exceptionally fast because it can jump from one water molecule to another by a relay mechanism. This is called *facilitated diffusion* because the proton moves faster than by simple diffusion, which is a random process.

The source of the activation energy in a reaction in solution is slightly different from that in a gas; the collision or bouncing of solvent molecules against the two nearby reactant molecules often provides the necessary energy if the reactants themselves do not have enough kinetic energy.

In this section, we consider the special case in which E_a and ΔS^* are sufficiently small that the transition state almost always forms, and in which the conversion of the transition to products is very rapid. This is the case in which *every collision leads to reaction*. When this occurs the rate of reaction may be determined mostly by the rate at which reactants diffuse together. Such a reaction is said to be a *diffusion-controlled* reaction. Since molecules move extremely rapidly through a solvent (although not as fast as when gaseous), diffusion-controlled reactions are very fast and the rate constant ranges from approximately 10^9 to 10^{11} M^{-1} sec^{-1}. If the reactants are considered to be spheres of radius r_A and r_B moving through a solvent of viscosity η, it is possible to calculate the rate constant k. We shall not go through the derivation, but the result is

$$k = \frac{2RT}{3000\eta} \frac{(r_A + r_B)^2}{r_A r_B}. \tag{67}$$

The first point to be noticed is the dependence of k on the solvent viscosity—that is, as the solvent viscosity increases, k decreases. This is reasonable because, with greater viscosity, the reactant molecules experience greater resistance to movement and therefore move more slowly. Viscosity dependence is an important criterion for identifying a diffusion-controlled reaction. This can be tested by incorporating high concentrations of a viscous substance, such as glycerol, into the reaction mixture and observing whether k varies inversely with viscosity. (It is of course necessary to ascertain that the glycerol does not participate chemically in the reaction.) The second important point is that k depends on the ratio of the radii of the reactants and not on their masses. This is because the increased target size for collision of larger molecules exactly compensates for their slower diffusion. Notice also that if the reactants have the same radius, $r_A = r_B$, then $k = 8RT/3000\eta$ is the minimal value of k. If $r_A < r_B$, k is greater than if $r_A = r_B$, because of the combination of the high mobility of A and the large target size of B. Thus, in diffusion-controlled reactions in which one reactant is a small ion and the other is a large molecule, the rate constant is very high.

Another useful equation for analyzing a diffusion-controlled reaction is the following, which is an expression for the constant A in Equation 61. This is

$$A_{\text{diff}} = \frac{4\pi(r_A + r_B)(D_A + D_B)\mathcal{N}}{1000}, \tag{68}$$

in which D_A and D_B are diffusion coefficients (Chapter 15)—not the dielectric constant D used in earlier equations. For a typical pair of small molecules whose radii are 3 Å and for which $D \approx 1.5 \times 10^{-5}$ cm^2 sec^{-1},

Equation 68 gives a value of $A_{\text{diff}} = 10^{10}$ l mol^{-1} sec^{-1} = 10^{10} M^{-1} sec^{-1}. In the limiting case of $E_a = 0$, $A_{\text{diff}} = k$, so that this number would be the maximal value of a rate constant that would be observed if every encounter led to a reaction—that is, if the reaction were controlled only by diffusion. Such reactions are very unusual, as might be expected, because E_a is rarely zero and there is usually a negative entropy of activation. Some reactions such as those between spherically symmetric, small molecules, for which orientation effects are minimal, approach this limit—an example is $NH_4^+ + OH^- \rightarrow NH_4OH$, for which $k = 3 \times 10^{10}$.

We can estimate the expected value of k for an enzymatic reaction, if it is a diffusion-controlled reaction. We consider a reaction in which a typical sugar ($r \approx 4$ Å) is converted to a product by an enzyme ($r \approx 40$ Å). Values for D_{sugar} and D_{enzyme} are about 6×10^{-6} cm^2 sec^{-1} and 2×10^{-7} cm^2 sec^{-1}, respectively. In this case $A_{\text{diff}} = 2.1 \times 10^{10}$ M^{-1} sec^{-1}. An enzyme-substrate reaction of this sort would normally have a value of $k < 10^7$, indicating that such a reaction is rarely diffusion-limited. This is because there are usually large entropic effects in the formation of the enzyme-substrate intermediate, as we will discuss later.

EXAMPLE 11–D A biological reaction that seems to exceed the diffusion-controlled limit.

In the bacterium *E. coli,* the metabolism of the sugar lactose is regulated by a protein called the *Lac repressor* that binds to a particular region of the *E. coli* DNA molecule called the *Lac operator*. When the Lac repressor is bound to the Lac operator, the enzymes needed for metabolism of lactose are not made. When lactose is provided to the cells, the Lac repressor binds a compound related to lactose (called *allo*-lactose), then falls off of the operator, and a sequence of events is initiated that results in synthesis of β-galactosidase, the enzyme that is required for lactose metabolism. When the lactose is exhausted, the free Lac repressor molecules bind again to the Lac operator and thereby turn off synthesis of β-galactosidase. In aqueous solutions containing purified Lac repressor and purified DNA, it has been observed that the binding reaction is extremely fast and has a rate constant of 10^{10} M^{-1} sec^{-1}. From the molecular weight of the repressor molecule and the size of the operator, the radii in Equation 67 are about 5 Å, which yields a rate constant of about 10^8 M^{-1} sec^{-1}. Thus the rate of binding seems to exceed the limit set by the rate of diffusion. Clearly there is some factor that greatly accelerates the binding reaction. There are basically two possible explanations. (1) There is a long-range attractive force that brings the two molecules together, so that molecules do not collide as a result of random motion. An example of such a force would be an electrostatic attraction, if the repressor and the operator have charges of different sign. (2) One of the radii in Equation 67 is much greater than 5 Å. For example, the repressor could bind weakly to almost any region of the DNA molecule and slide (or diffuse) along the DNA molecule until it encounters the operator and binds tightly. This is the explanation that is currently accepted. Migration of small molecules along large molecules probably occurs in many biological processes and must be considered whenever exceedingly high reaction rates are encountered.

The temperature dependence of diffusion-controlled reactions is like that of other reactions. This should be expected since, if a solute molecule is to push its way through a network of cohering solvent molecules, it must break the van der Waals bonds holding the solvent molecules together. The ability to do this should depend upon its kinetic energy and this energy will be distributed according to the Boltzmann law. In fact, this is the case, and it can be shown that the diffusion coefficient has an $e^{-E/RT}$-dependence on temperature. Similar reasoning shows that the viscosity varies as $e^{+E/RT}$; that is, viscosity decreases with increasing temperature. Since η enters the expression for k in the denominator, k will vary as $e^{-E/RT}$. Thus the rate constant for a diffusion-controlled reaction has Arrhenius-type behavior.

11–10 REACTIONS BETWEEN IONS IN SOLUTION

If reactants possess an electric charge (as is almost always the case in biological systems), other factors must be considered in the theoretical calculation of a rate constant. Clearly, if two molecules have the same charge, the probability of an encounter will be lower than if they are uncharged, because of electrostatic repulsion. Similarly, if they have unlike charges, they will attract one another and the probability of an encounter will be greater than that between uncharged particles. We will see that this introduces several new factors into expressions for rate constants—namely, the charge of the reactants, the dielectric constant of the solvent, and the ionic strength of ions that do not even participate in the reaction.

In the case of a diffusion-controlled reaction, the charge of the ions and the dielectric constant of the solvent affect the rate merely by altering the probability that two reactants come into contact. The correction to the rate constant is the rather complicated multiplicative factor that follows:

$$P = \frac{z_A z_B e^2}{D(r_A + r_B)\mathbf{k}T} \frac{1}{\exp[z_A z_B e^2/D(r_A + r_B)\mathbf{k}T] - 1}, \tag{69}$$

in which z_A and z_B are the charges of the two reactants, e is the charge of an electron, and D is the dielectric constant relative to the vacuum. The reader should satisfy himself that if D is large (that is, if the charges are shielded from one another), P approaches 1, and that if D is small, a negative product of z_A and z_B leads to a value of $P > 1$ and a positive product leads to a value of $P < 1$; of course, if either z_A or z_B is 0, $P = 1$.

The effect of nonreactive ions will be by way of the screening effect that we encountered in the Debye-Hückel theory. For a simple reaction

$$A + B \rightleftarrows (AB)^* \xrightarrow{k_2} C,$$

in which A and B have charges z_A and z_B and in which $[(AB)^*]$ is determined by an equilibrium constant K^*, the equation for the observed overall rate constant k_{obs} is

$$\log_{10} k_{obs} = \log k_2 K^* + 1.018 z_A z_B I^{1/2} . \tag{70}$$

This equation indicates that a plot of $\log k_{obs}$ versus $I^{1/2}$ is a straight line with slope $1.018 z_A z_B$, which has been amply confirmed by experiments.

If both ions have a charge of the same sign, k increases with I, whereas if the signs differ, k decreases with I. This is called the *primary kinetic salt effect* and can be easily understood. If the signs are the same, the ions repel one another and the reaction is slow, because the molecules have difficulty approaching one another. The presence of many ions produces a cloud of charge throughout the solvent, effectively screening the charged reactants from one another and allowing the rate to increase and approach that for uncharged species. Similarly, if they have unlike charge, the rate is especially high and the shielding restores it to the uncharged situation.

This effect of ionic strength must always be borne in mind when reactions between ions are being studied. If no inert species (*counterions*) are present, the ionic strength decreases continually as the reaction proceeds. Thus, the value of k_{obs} changes with time and if this is not recognized, an incorrect value will be determined. To avoid this, it is conventional in the laboratory to add a molar excess of an inert salt (typically NaCl) so that I is effectively constant throughout the reaction.

The sign of the slope of a plot of $\log k_{obs}$ versus $I^{1/2}$ tells whether the reactants that form the transition state have the same or different charges. This is important information to have when proposing a mechanism because often components that do not appear in the balanced equation are those that form the transition state. For instance, the decomposition of hydrogen peroxide, $2H_2O_2 \rightarrow 2H_2O + O_2$, is greatly facilitated by HBr. The H_2O_2 reaction does not include ions, yet the rate of decomposition is affected by the addition of an inert electrolyte. The slope of $\log k_{obs}$ versus $I^{1/2}$ is negative, indicating that two components having unlike charge must interact during formation of the transition state. The H^+ and Br^- ions would be the most likely candidates, and this would be investigated further. It is clear, though, that an interaction *solely* between H_2O_2 and an H^+ ion could not be responsible for forming the transition of ionic strength, because this would be independent of ionic strength. The existence of the primary kinetic salt effect and the sign of the $\log k_{obs}$ versus $I^{1/2}$ plot in no case provides proof for a proposed reaction but rather (1) indicates whether ionic components are involved in the reaction; (2) whether components have like or unlike charges; and (3) is capable of disproving a proposed mechanism. Thus, like kinetic results, the ionic-strength studies tell the researcher whether he or she is correct in proposing a particular mechanism.

Another point must be made about ionic reactions. Often a stoichiometric equation shows an apparent reaction between ions of like charge, for example

$$2Fe^{3+} + Sn^{2+} \rightarrow 2Fe^{2+} + Sn^{4+}.$$

It would be expected that in the absence of counterions, such a reaction would be very slow, because the positively charged reactants would repel one another. However, in many cases the reaction is fast, suggesting that it uses a pathway in which ions of like charge do not react. To understand the results, we need only remember that in a solution of Fe^{3+} and Sn^{2+} ions there must also be negative ions, and that these may participate in the reaction. In fact, when the salts are chlorides, it is found that the reaction depends on the chloride concentration in a way that is not explained by the effect of ionic strength alone. Indeed this reaction proceeds by the mechanism

$$SnCl_2 + 2Cl^- \rightarrow SnCl_4^{2-}$$
$$SnCl_4^{2-} + Fe^{3+} \rightarrow Fe^{2+} + SnCl_4^-$$
$$SnCl_4^- + Fe^{3+} \rightarrow Fe^{2+} + SnCl_4.$$

Note that in this mechanism, $SnCl_2$ is assumed to be incompletely ionized and that there is no reaction that involves an interaction of like charges.

REFERENCES

Benson, S.W. 1960. *The Foundation of Chemical Kinetics*. McGraw-Hill. New York.
Campbell, J.A. 1965. *Why Do Chemical Reactions Occur?* Prentice-Hall. Englewood Cliffs, N.J.
Castellan, G.W. 1971. *Physical Chemistry*. Addison-Wesley. Reading, Mass.
Fersht, A. 1977. *Enzyme Structure and Mechanism*. W.H. Freeman and Co. San Francisco.
Gardner, W.C. 1969. *Rates and Mechanisms of Chemical Reactions*. W.A. Benjamin. Menlo Park, Calif.
Glasstone, S., K.J. Laidler, and H. Eyring. 1941. *The Theory of Rate Processes*. McGraw-Hill. New York.
Hammes, G.G. 1978. *Principles of Chemical Kinetics*. Academic Press. New York.
Jencks, W.P. 1972. *Catalysis in Chemistry and Enzymology*. McGraw-Hill. New York.
King, E.L. 1963. *How Chemical Reactions Occur*. W.A. Benjamin.
Levine, I.N. 1978. *Physical Chemistry*. McGraw-Hill. New York.
Moore, W. 1972. *Physical Chemistry*. Prentice-Hall. Englewood Cliffs, N.J.
Sheehan, W.F. 1970. "Along the Reaction Coordinate." *J. Chem. Educ.* 47, 853.

PROBLEMS

1. A reaction has the rate law $d[P]/dt = [A][B]^2[C]$. What is the order of this reaction with respect to each of A, B, C, and D?

2. Which of the following reactions are *not* likely to be elementary?
   (a) $A + B \rightarrow AB$.
   (b) $2A + B_2 \rightarrow 2AB + B$.
   (c) $A_2 + B + 2C \rightarrow AC + BC$.
   (d) $A + B + C + D \rightarrow ABCD$.

3. State the molecularity of the following reactions if each is an elementary reaction.
   (a) $A_2 + B \rightarrow A_2B$.
   (b) $A_2 + B \rightarrow AB + A$.
   (c) $A_2 + 2B \rightarrow A_2B_2$.
   (d) $A_2 + 2B \rightarrow 2AB$.
   (e) $A_2 + 2B \rightarrow AB_2 + A$.

4. For each of the following reactions write the rate equation in terms of appearance of the product assuming that each is an elementary reaction.
   (a) $A_2 + B \rightarrow A_2B$.
   (b) $A_2 + 2B \rightarrow A_2B_2$.
   (c) $A + B + C \rightarrow ABC$.
   (d) $A_2 + B_2 \rightarrow 2AB$.

5. What is the relation between $d[\text{reactant}]/dt$ and $d[\text{product}]/dt$ for each of the following elementary reactions. The reactant and product to be considered are underlined.
   (a) $\underline{A_2} + B_2 \rightarrow \underline{A_2B_2}$.
   (b) $\underline{A_2} + B_2 \rightarrow \underline{2AB}$.
   (c) $\underline{A} + B_2 \rightarrow \underline{AB} + B$.
   (d) $\underline{A_2} + 2B \rightarrow \underline{A_2B_2}$.

6. A rate equation is written for each of the following elementary reactions. State whether $k < 0$ or $k > 0$ for each reaction.
   (a) $A_2 + B \rightarrow A_2B$.
       $d[A_2B]/dt = k[A_2][B]$.
   (b) $A + B_2 \rightarrow AB + B$.
       $d[A]/dt = -k[A][B_2]$.
   (c) $A_2 + 2B \rightarrow A_2B_2$.
       $d[A_2B_2]/dt = k[A_2][B]^2$.

7. A reaction $A + B \rightarrow C$ is studied under two different experimental conditions. When [A] is $0.8\ M$ and $[B] < 0.01\ M$, the observed rate law is $d[C]/dt = k[B]^2$. When $[B] = 1\ M$ and $[A] < 0.01\ M$, the observed rate law is $d[C]/dt = k[A]$. Also, $d[C]/dt$ is independent of [C] over a wide range of [A], [B], and [C]. Other factors such as pH do not affect the rate. What is the probable rate law?

8. A reaction $A + B \rightarrow C$ is studied by measuring the initial velocity, v, at various initial concentrations a and b of A and B, respectively. The measured initial velocities are the following

a, in M	b, in M	v
0.01	0.05	.0031
0.05	0.05	.0155
0.05	0.1	.0155
0.07	0.02	.0217
0.1	0.01	.031
0.1	0.05	.031
0.07	0.1	.0217

What is the rate law?

9. A reaction $A + 2B \rightarrow C$ is studied. The hypothesis that the reaction is elementary and trimolecular is being tested by measuring the initial velocity, v, as a function of the initial concentrations a and b of A and B, respectively. The data are the following:

a, in M	b, in M	v
.001	.004	3.2×10^{-5}
.001	.008	6.4×10^{-5}
.002	.004	1.28×10^{-4}
.002	.008	2.56×10^{-4}
.003	.004	2.88×10^{-5}

Is the hypothesis correct?

PROBLEMS

10. The reaction $NH_2NO_2 \rightarrow N_2O(g) + H_2O$ is a first-order reaction. From 50 mg of NH_2NO_2, 6.19 cm^3 of NO_2 are evolved at a temperature of 15°C and a pressure of 760 mm Hg after 70 minutes. What is the half-life, $t_{1/2}$, for NH_2NO_2?

11. A reaction $A + B \rightarrow C$ is first order in both A and B and second order overall when measured as $d[A]/dt$. If A and B are initially at the same concentration, 90 percent of the initial amount of A remains after 50 seconds.

    (a) How long would it have taken to reach half the initial value of [A]?

    (b) How long would it take to reach half the initial value of [A] if the concentrations of both A and B were doubled?

12. In a reaction $A + B \rightarrow C$, the following data were obtained.

t, in min	[C]
0	0
20	0.0095
30	0.0140
40	0.0222
60	0.0261
80	0.0333

    Initially $[A] = [B] = 0.2\,M$. Assuming that this is an elementary reaction, evaluate the rate constant.

13. A gas, Q, is heated at 200°C and it decomposes. The decomposition is reflected as an increase in pressure because the number of moles increases. The pressure is measured as a function of time and it is found that the rate of increase of pressure changes as the concentration of Q changes. Determine the order of the reaction for the data below.

[Q], in M	Rate of pressure increase, M min^{-1}
0.0413	0.621
0.0028	0.052

14. The following reaction is studied: $A_2 + B \rightarrow AB + A + H^+$. The rate of production of AB as a function of the initial molar concentration of A_2, B, and H^+ are given below.

$[A_2]$	[B]	$[H^+]$	$d[AB]/dt$, M sec^{-1}
2×10^{-4}	0.1	0.01	3×10^{-5}
5.9×10^{-4}	0.1	0.01	3×10^{-5}
3.1×10^{-4}	0.3	0.01	8.9×10^{-5}
2×10^{-4}	0.3	0.03	2.7×10^{-4}

    (a) What is the order of the reaction with respect to each of A_2, B, and H^+?

    (b) What is the rate equation?

    (c) What is the rate constant?

15. The decomposition of A in the first-order reaction $A \rightarrow B$ is 54.8 percent complete in 20 minutes at 45°C. What is the rate constant for the reaction?

16. Two mechanisms are proposed for the reaction

    $$2AB + C_2 \rightarrow 2ABC.$$

    These are

    (a) $AB + C_2 \rightarrow ABC_2$,
    $ABC_2 + AB \rightarrow 2ABC$.

    (b) $2AB \rightarrow A_2B_2$,
    $A_2B_2 + C_2 \rightarrow 2ABC$.

    Can these mechanisms be distinguished by measuring the order of the reaction?

17. The reaction $A + B \rightarrow C + D$ is found to have the rate equation $d[C]/dt = k[A]^2$. Suggest a mechanism.

18. A reaction $A \rightarrow B$ proceeds by means of two mechanisms. The rate law is $d[B]/dt = k[A] + k'[A][H^+]$, in which $k = 3.7 \times 10^{-3}$ sec^{-1} and $k' = 6.7 \times 10^{-2}\,M^{-1}$ sec^{-1}. At which pH are the rates along each path the same?

19. The growth of bacteria is a kinetic process. Bacteria multiply by reaching a critical size and then dividing in half. In a growth experiment the following data are obtained:

Time, t, in minutes	Bacteria per ml
0	3.4×10^6
25	6.8×10^6
50	1.3×10^7
75	2.6×10^7
100	5.2×10^7
125	10^8

(a) What is the doubling time for a typical bacterium?

(b) What is the cell concentration at 200 min?

(c) How long will it take the culture to reach a concentration of 1.6×10^9 bacteria per ml?

(d) What is the rate constant for the process?

20. A reaction A + B → C, which is not an elementary reaction, is carried out and [A] is measured. Specifically, at various times $[A]_t$ is measured and the time to reach half that value is measured. The following data are obtained:

[A]	Time, sec	[A]	Time, sec
1.880	0	0.192	158
0.940	48	0.096	206
0.554	72	0.051	250
0.277	120	0.025	298

What is the reaction order and what is the value of k?

21. A reaction A + B → C has the rate law $d[A]/dt = -k[A][B]$. In a mixture that is initially 0.1 M A and 0.1 M B, the reaction is 10 percent complete in 90 sec. How long will it take to reach 50 percent completion?

22. When double-stranded (ds) DNA molecules are heated above a certain critical temperature, the hydrogen bonds that join the two strands of the double helix are broken and the two strands unwind, resulting in single-stranded (ss) DNA. What would you guess to be the order of the reaction dsDNA → ssDNA?

23. A reaction A + B → C proceeds by two paths, one of which is pH-dependent, and has a rate law $d[C]/dt = k[A] + k'[A][H^+]$. If at pH 2 the rates of each reaction are equal, what can you say about the values of k and k'?

24. A cell extract is prepared by breaking open bacteria. An enzymatic reaction is detected in which a sugar A is oxidized at a particular velocity, v. The enzymatic activity is partially purified from the cell extract. When purified, the reaction rate is 0.1 v.

(a) Give at least two explanations for this change in velocity.

(b) Suppose the reaction in the crude extract is second order and that in the more pure sample it is first order. What might be an explanation for this?

25. There are two pathways for the biosynthesis of X, namely A + B → X and A_2 + 2B → 2X. If each pathway is an elementary reaction, write an expression for $d[X]/dt$.

26. A particular reaction A + B → AB occurs only when ultraviolet (UV) light shines on gaseous A and B. AB is a solid that falls to the bottom of the reaction vessel. Initially the vessel contains an equal number of moles of the two gases, which are both ideal gases. The pressure is 500 atm. The vessel is illuminated with UV light and the pressure is measured as a function of time. The temperature is maintained constant throughout the reaction. The amount of AB found as a function of the time of irradiation is the following: $t = 0$, none; $t = \frac{1}{2}$ min, 16 mg; $t = 1$ min, 32 mg; $t = 2$ min, 64 mg; $t = 5$ min, 160 mg. Also, it is found that $d[A]/dt = -d[AB]/dt$.

(a) What is the order of the reaction?

(b) How can you explain the order in view of the apparent bimolecularity of the reaction?

27. The reaction A + B → E consists of two steps:

(1)   A + B → C + D   (slow)
(2)   C + D → E   (fast)

(a) Which step is rate-determining for the production of E?

(b) Which step determines $d[A]/dt$?

(c) What is the molecularity of the reaction?

(d) What is the rate law for $d[B]/dt$?

(e) What is the rate law for $d[E]/dt$?

28. Consider the reaction sequence
$$A \underset{k_{-1}}{\overset{k_1}{\rightleftarrows}} B \underset{k_{-2}}{\overset{k_2}{\rightleftarrows}} C \underset{k_{-3}}{\overset{k_3}{\rightleftarrows}} D .$$

If the $B \rightleftarrows C$ reaction is rate-limiting, what relations (<, >, or =) must exist between the various rate constants?

29. Which step is probably rate-determining for A and for AD in the following reaction sequence? Assume that each step is irreversible.

(i) $\quad C + C \rightarrow C_2 \quad$ (fast).

(ii) $\quad A + C_2 \rightarrow AC + C \quad$ (fast).

(iii) $\quad C + D_2 \rightarrow CD + D \quad$ (slow).

(iv) $\quad AC + CD \rightarrow AD + C_2 \quad$ (fast).

30. A reaction $2A_3 \rightarrow 3A_2$ is being studied. Two mechanisms are proposed.

(i) $\quad A_3 \underset{k_{-1}}{\overset{k_1}{\rightleftarrows}} A_2 + A \quad$ (fast).

$\quad A + A_3 \overset{k_2}{\rightarrow} 2A_2 \quad$ (slow).

(ii) $\quad 2A_3 \overset{k}{\rightarrow} 3A_2$.

(a) What are the rate laws for the two different mechanisms?

(b) How can these mechanisms be distinguished?

31. Consider the reaction
$$H^+ + HNO_2 + \phi NH_2 \overset{Br^-}{\longrightarrow} \phi N_2^+ + 2H_2O$$
which is catalyzed by the Br^- ion in water and in which ϕ is the phenyl group. The proposed mechanism is

$H^+ + HNO_2 \underset{k_{-1}}{\overset{k_1}{\rightleftarrows}} H_2NO_2^+ \quad$ (fast),

$H_2NO_2^+ + Br^- \overset{k_2}{\longrightarrow} ONBr + H_2O \quad$ (slow),

$ONBr + \phi NH_2 \overset{k_3}{\longrightarrow}$
$\quad \phi N_2^+ + H_2O + Br^- \quad$ (fast),

and the observed rate law is $d[\phi N_2^+]/dt = k[H^+][HNO_2][Br^-]$.

(a) Calculate $d[\phi N_2]/dt$, assuming that the second step is rate-limiting.

(b) Repeat (a), assuming that $H_2NO_2^+$ is an intermediate and that the steady-state approximation is valid.

(c) Under what conditions will the steady-state approximation yield the observed rate law?

32. A reaction $A + 3B + 2C \rightarrow D + E + F$ has the rate law $r_{foreward} = d[F]/dt = k_{foreward}[A][B][C]$. What is the rate r_{back} for the back reaction?

33. Radioactive decay is a first-order process. The decay rate r (decays per minute) is described by $\ln r = -kt$. This equation is the basis of the radiocarbon dating method. A biological sample in contact with the atmosphere possesses ^{14}C, a radioisotope, in an amount such that the decay rate is 15.3 decays per minute per gram of carbon. Once it is no longer equilibrated with atmospheric CO_2, the decay rate drops according to the equation just given. A log buried at an archeological site contains ^{14}C having an activity of 3.2 decays per minute per gram of carbon. How old is the log? The half-life (the time in which the activity drops to half of its initial activity) of ^{14}C is 5670 years.

34. Why does it take longer to boil an egg at high altitude than at sea level?

35. What is the activation energy of a first-order reaction having a rate constant of 2.3×10^{-3} sec^{-1} at 20°C and 5.8×10^{-2} sec^{-1} at 50°C?

36. The rate constant of a reaction is determined at several temperatures. The values observed are: 0°C, 20.8; 10°C, 23.9; 20°C, 27.1; 30°C, 30.6; 40°C, 34.2. What is the activation energy?

37. An example of the primary kinetic salt effect is the increase of the velocity of a reaction between two reactants with like charges. This phenomenon is widespread in simple chemical reactions. However, in

many instances, with enzymes, the reaction rate increases in accord with Equation 69 for concentrations of NaCl between 0.01 M and 0.1 M, yet at very high ionic strength, for example, 1 M, the reaction rate decreases substantially. Give a possible explanation for this decrease.

38. A bimolecular reaction is studied as a function of ionic strength, I. One of the reactants in the rate-determining step is an ion A^+. Two hypotheses have been given in which the other reactant is either an ion B^- or an ion C^{2-}. From the data given below, decide which ion is the second reactant.

k	I
1.600×10^2	0.032
1.876×10^2	0.021
2.317×10^2	0.010
2.66×10^2	0.005

39. The activation energy of a reaction is found to be 34 kJ mol^{-1}. Furthermore, the constant A in the Arrhenius equation (Equation 61) is 1.6×10^{11}. Is the transition state a very complex molecule?

40. In transition-state theory, reactions are sometimes described as a path on a potential energy diagram. Consider the reaction $A + BC \rightarrow AB + C$ and the diagram shown for this problem, in which each contour line moving away from the origin represents an increased potential energy.

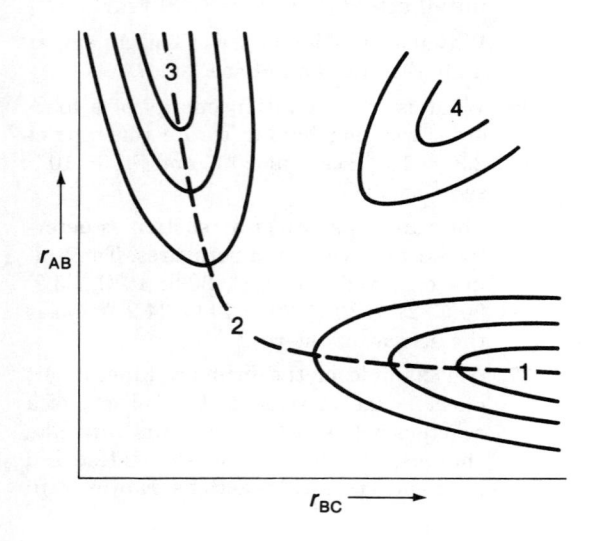

What molecules are present at positions 1, 2, 3, and 4?

41. Consider a reaction described by the reaction diagram shown below.

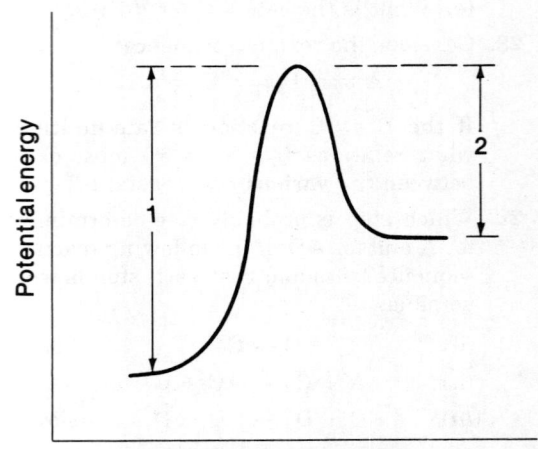

(a) What does 1 represent?
(b) What does 2 represent?
(c) Is the forward reaction exothermic or endothermic?

42. Identify the components A, B, C, D, and E in the following reaction diagram:

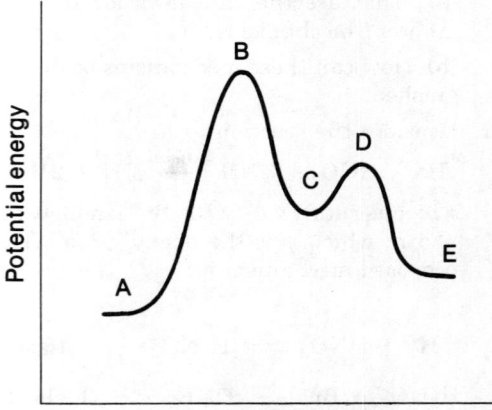

43. The mechanisms of many enzymatic reactions consist of a series of consecutive

reactions in which the enzyme E and the substrate S interact to form the ES complex, S is converted to the product P, so that ES becomes EP, and EP dissociates to form E and P. Locate E + S, ES, EP, and E + P in the following diagram.

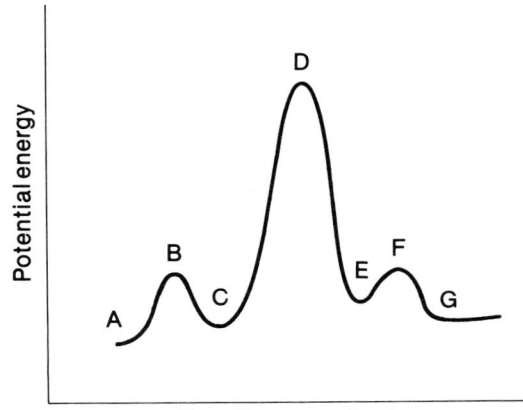

44. Consider the following two diffusion-controlled reactions: (1) The reaction between two iodine atoms to form I_2 when dissolved in hexane, and (2) the combination of two identical free radicals to form a single molecule in aqueous solution—that is, X· + X· → X_2. The viscosities of hexane and water at 25°C are 0.0033 and 0.01 poise, respectively. What is the ratio of the rate constants for the two reactions?

45. From Equation 67 it appears that the second-order rate constant in a diffusion-controlled reaction is proportional to the temperature. However this is not true because diffusion-controlled reactions show Arrhenius-type behavior. Is there any discrepancy between the equation and this fact?

46. If a reaction is diffusion-controlled, it must be the case that there is no activation energy for the reaction between two reactants. However, as mentioned in the previous problem, the second-order rate constant shows approximately an Arrhenius-type behavior and an activation energy can be calculated. For what is the activation energy required?

CHAPTER 12

CATALYSIS AND ENZYME KINETICS

Most chemical reactions proceed far too slowly to be of use in biological systems.* All reaction rates depend on the concentration of the reactants, and reaction rates can be increased by increasing the concentration, but since the maximal concentration of each of the thousands of intracellular components cannot be very high, in biological systems, a direct increase in concentration is not a feasible procedure. However, if the reactants can be adsorbed to a surface, the local concentration can be very high, and this is a possible means to increase reaction rates. There are other ways to accelerate the formation of a product from reactants—for example, by eliminating or reducing an activation energy or by providing an alternative reaction mechanism.

Substances that accelerate reactions without being consumed are called *catalysts*. Several types of catalysis will be discussed in this chapter. The most spectacular variety, *enzymes,* are able to increase reaction rates by up to 10^{15}-fold. Some information concerning how this is accomplished can be gleaned from kinetic analysis, as we will see.

12–1 CATALYSIS

The rate of a reaction can be increased by a substance that does not appear in the balanced equation. Such a substance is called a *catalyst*. Included in

---

*For example, bacteria divide every half hour; without catalysts, however, the oxidation of sufficient glucose to complete one bacterial life cycle would take hundreds of years.

the definition of a catalyst is the requirement that it not undergo a *permanent* chemical change; if it does change in the course of the reaction, it is regenerated without loss. The physical properties of the catalyst may be altered, though; for example, when MnO_2 catalyzes the decomposition of $KClO_4$, it is converted to a fine powder in the process.

It is important to realize that *a catalyst cannot affect the equilibrium of a reaction;* it only affects the rate of approach to equilibrium. A simple thermodynamic argument based on the First Law proves this point. Consider a gas reaction in which there is a volume increase. At equilibrium the gas occupies a certain volume. The gas is in a vessel fitted with a piston and the catalyst is contained within the vessel in a container that is opened and closed by the piston as it moves down and up, respectively. Suppose the catalyst were to shift the equilibrium so that the volume would increase. If the catalyst were exposed to the gas, the volume of the gas would increase, the piston would be pushed out, and the catalyst would then be covered and kept from catalyzing the reaction. Then the volume would decrease, the piston would move in, and the catalyst would once more be exposed. Thus, we would have a perpetual motion machine if the catalyst were to shift the equilibrium of a reaction; and such a machine is not possible, by the Second Law.

FIGURE 12-1 Energy diagram showing the reduction in the free energy of activation (ΔG^*) for a reaction when uncatalyzed and catalyzed.

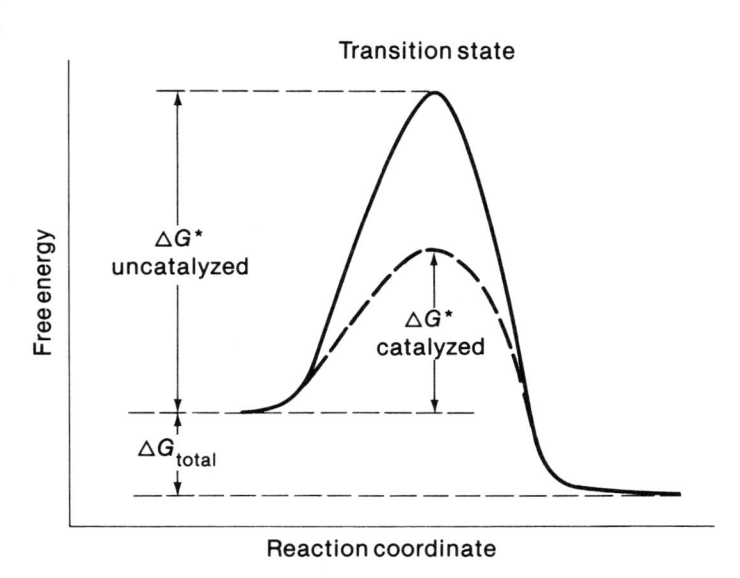

The mechanism by which a catalyst increases the reaction rate is by decreasing ΔG^*, the free energy of formation of the transition state, as depicted in Figure 12-1. This is usually done by providing an alternate pathway for the reaction. Invariably an activated complex consisting of the catalyst and a reactant is formed and this decomposes spontaneously or reacts with some other substances to yield the final products. How the rate changes is shown by Equation 64 of Chapter 11. Thus, at 20°C, for the decomposition of H_2O_2, the free energy of activation, ΔG^*, is 75 kJ mol^{-1} for the uncatalyzed reaction, 54 kJ mol^{-1} when the reaction is catalyzed by platinum, and 29 kJ mol^{-1} when it is catalyzed by the enzyme catalase. Thus the increase in rate by platinum is $e^{75/2.43}/e^{54/2.43}$ or 5.7×10^3, and the increase by catalase is $e^{75/2.43}/e^{29/2.43} = 1.7 \times 10^8$.

Catalysts can be classified as *heterogeneous,* in which the catalyst is introduced in a different phase (usually solid) from that of the reactants (usually liquid or gas), and *homogeneous,* in which the reactants and catalyst form a single phase.

Heterogeneous Catalysis

Heterogeneous catalysts act by adsorbing reactants onto their surfaces. The nature of the surface is quite important; for example, in the presence of aluminum oxide at 300°C, ethanol is converted to ethylene and water; yet at the same temperature and in the presence of copper, the products are acetaldehyde and hydrogen gas.

All heterogeneous catalytic reactions probably proceed through the following steps:

1. Diffusion of reactants to the surface of the catalyst.
2. Adsorption of reactants to the surface. This may be *physisorption,* resulting from van der Waals forces between the molecule in the surface and the reactant molecule, or *chemisorption,* in which strong (often covalent) bonds form. (See Chapter 13 for other characteristics of physi- and chemisorption.)
3. Migration of one reactant to the adsorption site of another reactant often, but not always, occurs. This of course does not occur in a unimolecular reaction, such as a decomposition or dehydrogenation.
4. Chemical reaction on the surface.
5. Desorption of adsorbed products.
6. Diffusion of products away from the surface.

Steps 1 and 6 are extremely rapid in gas reactions and are rarely rate determining. In liquids they are much slower but still sufficiently fast that they have little influence on the rate. The same is probably true of steps 2 and 5. If the reactants are physisorbed or weakly chemisorbed, the binding is weak and migration occurs freely; step 4 is usually rate determining in that case. If there is strong chemisorption, any one of steps 2–5 might be rate limiting.

SEC. 12-1   CATALYSIS

An instructive example of heterogeneous catalysis is the hydrogenation of a C—C bond of propene by a platinum catalyst:

$$-\overset{|}{\underset{|}{C}}-\overset{|}{C}=\overset{|}{C}- + H_2 \rightarrow -\overset{|}{\underset{|}{C}}-\overset{|}{\underset{|}{C}}-\overset{|}{\underset{|}{C}}-.$$

H_2 molecules adsorb to the platinum surface and are rapidly broken down to H atoms that remain on the surface but migrate rapidly. The propene adsorbs and forms two bonds with the metal—one bond from each carbon atom on the two sides of the double bond—so that the double bond is broken during adsorption. A migrating H atom collides with the bonded propene, breaks one of the bonds between the carbon atom and the metal, and forms a C—H bond. A second H atom comes along and repeats the process and the newly formed propane is released. Thus the reaction proceeds rapidly for two reasons: the H_2 molecule forms an active complex that spontaneously breaks down to H atoms, and the propene forms a new compound whose reaction with H atoms has a lower free energy change than the reaction between gaseous H_2 and propene. The rate increase by the platinum catalyst is extremely high because the components are chemisorbed, a process that involves chemical changes in the reactants. If both propene and H_2 had been physisorbed, there would have been an increased rate because of the greater effective concentration of reactants on the catalyst surface, but the rate increase would not have been nearly as great. As we will see later, similar effects occur on the surface of an enzyme.

Heterogeneous catalyzed reactions may have positive, negative, and fractional orders. In contrast with uncatalyzed reactions, the value of the order with catalyzed reactions is almost always a function of the parameters of the adsorption process. This can be seen by examining the simplest case—a unimolecular reaction in which the product does not bind to the catalyst. In this case the reaction order depends on the concentration of the reactant and the extent of its adsorption. We shall consider three cases: (1) A high concentration of a reactant that is weakly adsorbed to the catalysts; (2) a low concentration of a reactant that is weakly adsorbed; (3) an intermediate concentration of a reactant that is weakly adsorbed. The significant factor that affects the kinetics is the fraction of the surface area (or, more precisely, the fraction of potential adsorption sites) that is occupied.

Case 1. If the concentration of a reactant is sufficiently high, the adsorptive power is unimportant and we can assume that all sites are filled. This will also be true if adsorption is very strong and the concentration is high enough that the sites are saturated and there remains some excess reactant. In the case of strong adsorption, the rate is independent of concentration and the reaction is zeroth order. Thus the order of a reaction studied at high reactant concentration can give some information about the strength of the binding to the catalyst. For example, both platinum and gold catalyze the decomposition of HI but with platinum the reaction is first order and with gold it is zeroth order. Thus, HI binds more tightly to gold than to platinum.

Case 2. If the concentration [A] is low and adsorption is weak, only a fraction θ of the sites is filled and only a fraction of the reactants is bound. We assume that the Langmuir adsorption equation* describes the binding; that is,

$$\theta = \frac{b[A]}{1 + b[A]} \tag{1}$$

in which b is a constant. If [A] is small, $\theta \approx b[A]$, and increasing the concentration of reactants increases θ linearly. The reaction rate $-d[A]/dt$ is proportional to θ; that is, $-d[A]/dt = k\theta = bk[A] = k_1[A]$, so the reaction is first order.

Case 3. At intermediate concentrations of reactants in the presence of a catalyst that shows weak or intermediate adsorption, it has been observed that the fraction θ of sites filled always obeys an equation $\theta = b[A]^n$, in which n is an empirical number between 0 and 1. Thus the reaction order will be fractional, since $-d[A]/dt = k_n[A]^n$.

The kinetics of a biomolecular reaction $A + B \rightarrow C$ can be obtained by a simple extension of the preceding reasoning. We assume that the adsorption sites of A and B are the same. The Langmuir equation yields

$$\theta_A = \frac{b_A[A]}{1 + b_A[A] + b_B[B]} \quad \text{and} \quad \theta_B = \frac{b_B[B]}{1 + b_A[A] + b_B[B]}. \tag{2}$$

If the binding is weak, $\theta_A \approx b_A[A]$ and $\theta_B \approx b_B[B]$, and the rate equation is

$$\frac{d[C]}{dt} = k\theta_A\theta_B = kb_Ab_B[A][B] = k_2[A][B]. \tag{3}$$

If the binding of A is weak ($b_A \ll 1$) but that of B is strong ($b_B[B]$ is large), the approximations become

$$\theta_A = \frac{b_A[A]}{1 + b_B[B]} \quad \text{and} \quad \theta_B = \frac{b_B[B]}{1 + b_B[B]}, \tag{4}$$

and the rate equation is

$$\frac{d[C]}{dt} = k_2 \frac{[A][B]}{(1 + b_B[B])^2}, \tag{5}$$

which shows a complex order with respect to B.

The cases just described demonstrate that the order is dependent on reactant concentration. In these examples it was assumed that the product does not bind to the catalyst. However, it is often the case that the product

---

*See Chapter 13 for a detailed discussion of the Langmuir model for adsorption.

does bind to the catalyst and in so doing it competes with the reactants for adsorption sites. The effect of this in a unimolecular reaction, $A \to D$, can be analyzed by reasoning similar to that just used for the case of two reactants. The fraction of sites occupied becomes

$$\theta = \frac{b_A[A]}{1 + b_A[A] + b_D[D]}. \tag{6}$$

If D is very weakly adsorbed, $b_D[D] \approx 0$ and the situation reduces to that described in cases 1 to 3 above. However, if D adsorbs more strongly than A, the rate equation is

$$-\frac{d[A]}{dt} = \frac{k'[A]}{1 + b_D[D]}, \tag{7}$$

and it is clear that as D is produced, the reaction rate drops continuously—that is, the product is inhibitory to the reaction.

For the bimolecular reaction $A + B \to D$, the general rate equation is

$$\frac{d[D]}{dt} = \frac{k[A][B]}{(1 + b_A[A] + b_B[B] + b_D[D])^2}, \tag{8}$$

which, for the special case of very strong binding of D, is

$$\frac{d[D]}{dt} = \frac{k[A][B]}{(1 + b_D[D])^2}, \tag{9}$$

which again shows strong inhibition by the product.

This inhibitory effect is referred to as *poisoning* of the catalyst by the product. Poisoning can occur in many ways and it is of widespread occurrence in industrial processes. It is frequently a result of impurities in the catalyst or in the reactants, or products made in minor reactions that occur in the reaction mixture. In enzymatic reactions, inhibition by the product is common in synthetic reactions and is an important means of regulating enzyme activity. In enzymology, this phenomenon is usually called *feedback inhibition* or *end-product inhibition*.

There are three important ways in which this inhibition occurs. The two most prevalent are *competitive inhibition*, which is explained later in this chapter, and *allosteric inhibition*, which is explained in detail in Chapter 16. In allosteric inhibition, the product of the reaction binds to a particular site on the enzyme. As a result of this binding, the enzyme molecule undergoes a shape change in which the binding site for the reactant molecule is no longer capable of binding the molecule tightly and the catalytic activity is markedly reduced.

The third mechanism has never been named and we call it inhibition by cofactor binding for want of a better name. This occurs in some enzyme reactions in which the enzyme has a bound cation (for example, Mg^{2+}) when in the active catalytic form. When such a bound ion is essential for enzymatic activity, the ion is called a *cofactor*. In some cofactor-requiring reactions the product of the reaction binds the cofactor more tightly than does the enzyme, so that once the product is synthesized, the cofactor is removed from the enzyme and the number of active enzyme molecules decreases with time. This kind of inhibition is easily recognized because the reaction rate continually decreases with time but can be increased to the initial rate by adding more of the cofactor.

Homogeneous Catalysis

In homogeneous catalysis, a reaction is carried out in a single phase. The general mechanism involves the formation of an intermediate from a reactant and the catalyst. Catalysis is again the result of reducing the free energy of activation.

We can illustrate the importance of catalysis by first examining the mechanism of a simple reaction when uncatalyzed and then observing the effect of a catalyst.

Consider the hydrolysis of an ester, a reaction that occurs frequently in biological systems. The transition state is a structure in which a positive charge is formed on the incoming H_2O molecule and a negative charge is formed on the carbonyl oxygen:

This conversion requires a large input of energy in order to maintain the separation of the unlike charges. The acetate ion is a catalyst for this reaction, because it can pick up one of the water protons and eliminate the charge on the water oxygen:

The separation of charge in a reaction intermediate can also be stabilized by an electrostatic interaction in which the stabilizing agent does not even bind to one of the charged components; the stabilization can result from the electric field of the catalyst. This is called *electrostatic catalysis*. To understand how this works we must remember the basic equation of electrostatics. The interaction energy E between two point charges q_1 and q_2 (expressed in coulombs), separated by a distance r (in meters), in a medium of dielectric constant D, is $E = q_1 q_2 / 4\pi\epsilon_0 Dr$, in which ϵ_0 has the value 8.85×10^{-12}. Thus the energy of a positive and negative charge separated by 3 Å (a typical separation in the formation of an intermediate) is -46.0 kJ mol^{-1} in vacuum. If the charges are in water, $D = 81$ and $E = -5.7$ kJ mol^{-1}. Thus water itself might serve as an electrostatic catalyst in this sense. One usually thinks of the molecules of the dielectric as being between the interacting charges, while the dielectric is decreasing the electrostatic interaction. However, this need not be the case, because the molecules of the dielectric are polarizable. Thus the solvent molecules surrounding a pair of charges can be polarized—that is, dipoles can be induced and the electrostatic field from these dipoles can partially neutralize the electric field produced by the separation of charges. This is one of the ways that an enzyme can act as a catalyst. Enzymes are almost always highly folded protein molecules. Most of the polar amino acids in the protein chain are on the surface of the molecule, but some are deep in crevices in the surface. These crevices are sometimes called *polar pockets*. A reaction intermediate may diffuse into such a crevice and find itself in a region of high polarizability that reduces the interfering electric field.

There is another way in which an enzyme functions as an electrostatic catalyst. Suppose an active site in the crevice of an enzyme molecule contains a carboxylate ion (for example, on the amino acid, aspartic acid). The electric field generated by this negative ion is sufficient even at a distance of a few ångström units to neutralize the charge of a carbonium ion (which is sometimes found in intermediates). If the active site contains water molecules ($D = 81$), the water can, by virtue of the high dielectric constant, diminish the electric field of the carboxylate and render it ineffective as an electrostatic catalyst. (Water might of course also stabilize the charge separation within the intermediate, but if the intermediate is large enough, the water molecules may be too far away for this to occur.) However, many enzymes are designed so that the active site contains a large group of nonpolar amino acids (for example, leucine). These amino acids have a lower dielectric constant but, more important, the nonpolar cluster usually excludes polar molecules such as water from the crevice. By having the reactive intermediate in close contact with nonpolar amino acids but adjacent to an ion such as carboxylate, the neutralizing field of the ion is greater than if water were allowed.

Each of the types of catalysis just described involves stabilization of the transition state that normally occurs in the uncatalyzed reaction. Although there is a change in mechanism in certain cases, in that the transition state may be slightly different in this type of catalysis, it is generally considered that the same mechanism is maintained. However, a catalyst might also function by forming a covalent bond with a reactant, thereby providing an alternative pathway whose rate constant is very large, as we saw in the case of the surface catalysts. This kind of catalysis is called *covalent* catalysis.

Enzymes contain many reactive groups and the enzyme molecule itself often forms a covalent bond with the substrate. In many cases this enzyme-substrate compound is a very unstable intermediate that rapidly decomposes to give the product and the regenerated enzyme.

In the remainder of this chapter we consider reactions catalyzed by enzymes. Again the goal of the biochemist is to determine the reaction mechanism. A useful approach is to determine the kinetics of the reaction and seek substances that influence the reaction velocity. The study of enzyme kinetics and its peculiarities is thus an important branch of biochemistry. As we will see in the following section, enzyme kinetics are somewhat different from the kinetics of other types of reactions and are described by a specialized theory.

12–2 ENZYMATIC REACTIONS

Living cells contain a very large number of distinct proteins (about 5000 for a typical bacterium). At least half of these molecules are enzymes—that is, protein catalysts of biochemical reactions. Enzymes have two rather remarkable properties: (1) they enable chemical reactions to reach extraordinarily high rates, up to 10^{15} times greater than the uncatalyzed reaction, and (2) they are very selective, often catalyzing only a single intracellular reaction.

One of the aims of biochemistry is to understand how enzymes work—how, in general, they achieve their selectivity and their high reaction rate and, for a specific enzyme, what the detailed chemical mechanism for the reaction is. There are several approaches to accomplishing these goals. One of these is made possible by the fact that it is usually possible to synthesize many compounds that can act as a *substrate* (the substance acted on by the enzyme) for a particular enzyme, and a study of the efficiency with which the reaction is catalyzed for a particular substrate is often very informative. The reaction rate is one measurement of this efficiency.

Kinetics of Enzymatic Reactions

A significant property shared by all enzymes and many other catalysts is that they show a saturation of catalytic activity as the substrate concentration is increased. A typical curve relating the reaction velocity V (in enzyme kinetics, the derivative notation is usually replaced simply by V), and the substrate concentration, is shown in Figure 12-2. Usually V refers to the *initial* rate, because (1) many enzymes are inhibited by the

FIGURE 12-2 The reaction velocity V as a function of the substrate concentration [S] for an enzyme obeying Michaelis-Menten kinetics. V_m is the maximal velocity reached and K_m, the Michaelis constant, is the value of [S] yielding a velocity of $\frac{1}{2} V_m$. The inset shows how V is measured; V is the negative of the slope of the dashed line, which is an extension of the *linear* portion of the curve.

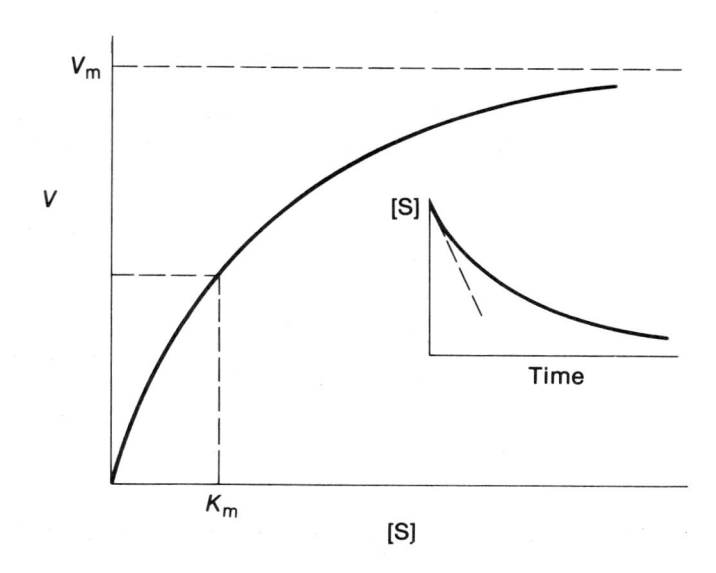

product of the reaction, so that the rate is measured while the product concentration is very low, and (2) the reaction is often reversible.

These saturation kinetics are observed because, as we have seen before in our discussion of catalysis, once all catalytic sites are occupied, no further increase in rate is possible. The curve in Figure 12-2 is characterized by two numbers; the maximal velocity achieved when the substrate has saturated the enzyme—that is, V_m (sometimes also written V_{max}), and the substrate concentration K_m that gives half-maximal activity. K_m is called the *Michaelis constant*. A satisfactory explanation for the curve in Figure 12-2 was given by Leonard Michaelis and Maud Menten in a model that stated that the conversion of a substrate S to a product P, catalyzed by an enzyme E, proceeds by formation of a complex ES between the enzyme and the substrate—that is,

$$E + S \underset{k_{-1}}{\overset{k_1}{\rightleftharpoons}} ES \overset{k_2}{\rightarrow} E + P \tag{10}$$

By this model, saturation means that all of E is converted to ES. Since the product P is formed only from ES, it is clear that when [ES] is maximal, the velocity is also maximal.

This model enables one to explain another characteristic of enzymatic reactions. When [S] is very high, V_m is proportional to the amount of added enzyme, $[E]_{tot}$. This is reasonable because V_m must be the rate at which ES breaks down to form the product, so that when $[ES] = [E]_{tot}$ (that is, at saturation),

$$V_m = k_2[ES] = k_2[E]_{tot} . \tag{11}$$

Thus if $[E]_{tot}$ is increased n-fold, still maintaining $[S] > [E]_{tot}$, then V_m will also increase n-fold. In other words, the rate is proportional to $[E]_{tot}$ and the rate is first order with respect to $[E]_{tot}$. The constant k_2 is also denoted k_{cat} and is called the *turnover number*. This is because k_{cat} tells us how many times the enzyme acts per second or how many times it passes through the ES complex and is regenerated ("turns over"), yielding the product. Thus, if a reaction mixture contains an enzyme concentration of x molar and the product is formed at a rate such that its concentration increases by y molar per second, when the enzyme is saturated, $k_{cat} = y/x$ per second. This means that y/x substrate molecules are converted to product molecules every second. Correspondingly, the time required for a single conversion is x/y seconds. Turnover numbers for most enzymes usually range from 1 (a poor enzyme) to 10^4. A few enzymes have turnover numbers above 10^5. The ability of a cell to produce a particular amount of substance by an enzymatic reaction in one life cycle is proportional to both the turnover number and the number of enzyme molecules per cell.

EXAMPLE 12–A Calculation of k_{cat} for an enzyme, E.

An enzyme, at a concentration of 10^{-6} M, converts a substrate to a product P with the following kinetics:

[P], M:	0	0.024	0.047	0.073	0.093
t, minutes:	0	$\frac{1}{2}$	1	$1\frac{1}{2}$	2

Plotting [P] as a function of time, the slope $\Delta[P]/\Delta t$ is 0.048 M min^{-1} or 8×10^{-4} M sec^{-1}. Since $[E] = 10^{-6}$ M, $k_{cat} = 8 \times 10^{-4}/10^{-6} = 800$ sec^{-1}.

K_m is the substrate concentration that gives a velocity equal to $\frac{1}{2}V_m$. Thus K_m is the concentration at which half of the enzyme is in the form of ES—that is, at $[S] = K_m$, $[ES] = \frac{1}{2}[E]_{tot}$.

When [S] is far below K_m, it is observed that the rate of product formation increases linearly with increasing [S]. That is, the reaction is first order with respect to [S]. The rate is also always proportional to $[E]_{tot}$, which is reasonable according to the reaction just stated, because doubling $[E]_{tot}$ should double [ES]. Thus, at low values of [S], the reaction velocity is

$$V = k'[E][S] . \tag{12}$$

It has been observed that

$$k' = k_{cat}/K_m , \tag{13}$$

a fact that must be included in any model of enzyme activity. In summary, the rate equation for an enzyme reaction must include a transition from a second-order reaction dependent on $[S][E]_{tot}$, when [S] is small, to a first-order reaction dependent only on $[E]_{tot}$, when [S] is large. Furthermore, the rate constants are k_{cat} for the first-order reaction and k_{cat}/K_m for the second-order reaction in which K_m is the value of [S] when $V = \tfrac{1}{2}V_m$. Before deriving the rate equation it is worth interpreting in molecular terms what these parameters of an enzymatic reaction signify. Since k_{cat} is the rate constant for conversion of ES to the product plus the enzyme, it is a measure of how active the enzyme is in carrying out the reaction *after the enzyme has bound the substrate*. The constant k_{cat}/K_m is instead a measure of the total activity of the enzyme, which includes the ability of the enzyme to bind a particular substrate. We will interpret these and other parameters in greater detail after we have derived the rate equation.

To derive the rate equation for enzyme reactions, we refer again to the mechanism proposed by Michaelis and Menten, namely

$$\mathrm{E + S} \underset{k_{-1}}{\overset{k_1}{\rightleftarrows}} \mathrm{ES} \overset{k_2}{\rightarrow} \mathrm{E + P} .$$

Let us consider a reaction mixture in which the substrate concentration [S] is greater than the enzyme concentration [E]. Note that we are not assuming that [S] is so large that all of the enzyme molecules are in the ES form, but only that [S] is sufficiently larger than [E] that [S] does not rapidly become so small that [S] < [E]. We choose an excess of substrate because this is the most common condition used to study enzymatic reactions (or any catalyzed reaction), as the nature of catalysts allows the catalyst to be present at a very low concentration. Under such conditions, the concentration of ES very rapidly becomes nearly constant. The near constancy of [ES] means that the rate of formation of ES equals its rate of breakdown or, in other words, that steady-state kinetics can be applied to the reaction. In general

$$\frac{d[\mathrm{ES}]}{dt} = k_1[\mathrm{E}][\mathrm{S}] - k_{-1}[\mathrm{ES}] - k_2[\mathrm{ES}] . \qquad (14)$$

In the steady-state approximation, $d[\mathrm{ES}]/dt \approx 0$, so that

$$[\mathrm{ES}] = \frac{[\mathrm{E}][\mathrm{S}]}{(k_{-1} + k_2)/k_1} . \qquad (15)$$

We now make the definition that

$$K_m = \frac{k_{-1} + k_2}{k_1} , \qquad (16)$$

so that Equation 15 becomes

$$[\mathrm{ES}] = \frac{[\mathrm{E}][\mathrm{S}]}{K_m} . \qquad (17)$$

Substituting $[\mathrm{E}]_{tot} = [\mathrm{E}] + [\mathrm{ES}]$, we obtain

$$[\mathrm{ES}] = [\mathrm{E}]_{tot} \frac{[\mathrm{S}]}{[\mathrm{S}] + K_m} . \qquad (18)$$

Since the product is only formed from ES, the rate of product formation is

$$V = k_2[\mathrm{ES}] . \qquad (19)$$

Combining this with Equation 18 yields

$$V = k_2[E]_{tot}\frac{[S]}{[S] + K_m}, \qquad (20)$$

which is one form of the desired rate equation. Usually the equation is rewritten to contain V_m. When [S] is very large, all of the enzyme is in the form of ES—that is, $[ES] = [E]_{tot}$ and $[E] \approx 0$. Under these conditions, $[S]/([S] + K_m) \approx 1$ and $V_m = k_2[E]_{tot}$, so that Equation 20 can be written

$$V = V_m\frac{[S]}{[S] + K_m}. \qquad (21)$$

This is called the *Michaelis-Menten equation*. It describes the reaction rate at all values of [S] as long as $[S] > [E]_{tot}$.

This equation can be seen to have the features of the experimental curve that we earlier discussed. These are the following:

1. At high substrate concentration, that is, when $[S] > K_m$, then $V = V_m$. In other words, the rate is independent of [S]. This is because all enzyme molecules are in the ES complex whose concentration determines the rate.
2. At low substrate concentration, that is, when $[S] < K_m$, $V = (V_m/K_m)[S]$. The rate is proportional to [S]. This is because [ES] is proportional to [S] and [ES] determines the rate.
3. When $V = \frac{1}{2}V_m$, $[S] = K_m$; thus K_m is the substrate concentration at which the reaction rate is half-maximal.

When an enzyme is being studied, the first goal is to evaluate K_m and V_m. There are two common ways to do this. By one method, Equation 21 is rearranged to

$$\frac{1}{V} = \frac{1}{V_m} + \frac{K_m}{V_m}\frac{1}{[S]}. \qquad (22)$$

By this equation, a plot of $1/V$ versus $1/[S]$ gives a straight line having a y intercept of $1/V_m$, an x intercept of $-1/K_m$, and a slope of K_m/V_m. This is called a Lineweaver-Burk plot (Figure 12-3). As discussed later in Chapter 16, this plot has the disadvantage that the accuracy of the plot is determined by the less precisely evaluated points at low values of [S]; the more accurate measurements at high [S] cluster and are less valuable in determining the linear graph.

A superior method (because points are usually more equally spaced) of plotting data is the Eadie-Hofstee plot. Equation 21 is written as $V = V_m - K_m V/[S]$ and V is plotted against $V/[S]$. The y intercept is V_m, the x intercept is V_m/K_m, and the slope is $-K_m$ (Figure 12-3).

The Michaelis-Menten equation is based on the two-step reaction we have described. In order to make use of this equation, one needs to interpret the equation in molecular terms. There are two important reaction

SEC. 12–2 ENZYMATIC REACTIONS

FIGURE 12-3 A Lineweaver-Burk plot and an Eadie-Hofstee plot for the same data.

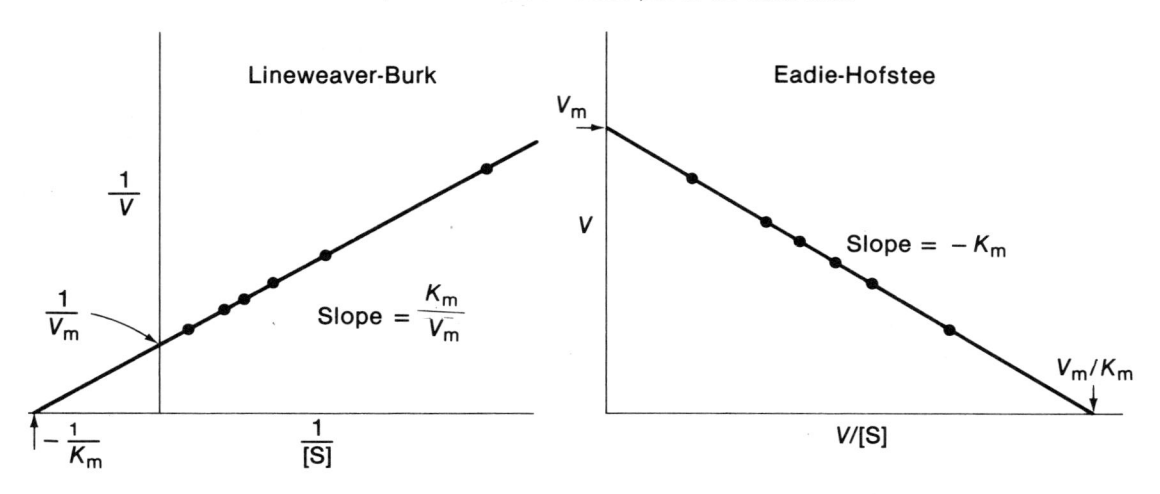

mechanisms for which the velocity obeys the equation—these differ by whether ES breaks down faster to give E and S ($k_{-1} > k_2$) or to give products and the free enzyme ($k_2 < k_{-1}$). We will now interpret the equation for these two cases.

Case I. Where $k_{-1} \gg k_2$. In this case we assume that ES is at equilibrium with E and S and that ES dissociates more often to yield E and S than to yield the product. The dissociation constant K_s for this equilibrium is

$$K_s = \frac{[E][S]}{[ES]} = \frac{k_{-1}}{k_1}. \qquad (23)$$

From the definition of K_m we can then write for the case $k_{-1} \gg k_2$ that

$$K_m = \frac{k_{-1} + k_2}{k_1} \approx \frac{k_{-1}}{k_1} = K_s.$$

As always, the observed rate of production of P is proportional to [ES], which is, in this case, an equilibrium concentration and is determined by [S] and $K_s = K_m$. In this case K_m is a measure of the strength of binding of the enzyme and the substrate—that is, if K_m is large, the binding of S to E is weak. This interpretation applies to a great many enzymatic reactions.

Case II. Where $k_2 \gg k_{-1}$. In this case the rate of dissociation of ES to E and S is small, so that products are usually formed. For all practical purposes, the reaction sequence can be thought of as consisting of two irreversible reactions

$$E + S \xrightarrow{k_1} ES \xrightarrow{k_2} P + E.$$

The total reaction rate is always determined by [ES]. At low values of [S], ES is formed by a second-order reaction whose rate is proportional to

[E][S]. At high values of [S], [ES] is constant and the rate is determined by the first-order degradation of ES to the products. Thus, as we have said earlier, as [S] is increased, there is a gradual change in the rate-determining step from the second-order to the first-order reaction. Note that the change in the rate-determining step is in this case entirely a result of the fact that the second-order reaction depends on [S] and the first-order one does not. This case differs from case I, in which the saturation behavior is a result of an equilibrium. In this case, since $k_2 > k_{-1}$,

$$K_m = \frac{k_{-1} + k_2}{k_1} \approx \frac{k_2}{k_1} \neq K_s.$$

This means that the second-order rate constant for the reaction when [S] is small is $k_1 = k_2/K_m = k_{cat}/K_m$ and K_m is the value of [S] at which the change in the rate-determining step from second order to the first-order reaction is half complete.

It is important to realize that cases I and II are two possible mechanisms for which the reaction velocity obeys the Michaelis-Menten equation. Simply by showing that the equation is obeyed gives no information about the mechanism that applies to a particular reaction. This information is gained from other types of experiments designed to measure the value of k_{-1} or k_1/k_{-1}. A discussion of this can be found in references given at the end of the chapter.

Enzyme Inhibition

A study of the inhibition of enzymes by various reagents often yields valuable information about the mechanisms of enzyme action and the chemical nature of the catalytic site. Inhibition is also an important means of intracellular regulation of enzyme activity and of the action of various drugs.

Inhibitors may be classified as reversible and irreversible. In the latter case, inhibition usually occurs by covalent modification of functional groups on the enzyme, and provides information about the chemical identity of groups in the active site. Irreversible inhibition is studied by chemical rather than kinetic procedures and is outside of the scope of the book. Reversible inhibition, in which the inhibitor combines rapidly and reversibly with the enzyme or the enzyme-substrate complex, can be analyzed within the framework of Michaelis-Menten kinetics. There are three types of reversible enzyme inhibition: competitive, uncompetitive, and noncompetitive. These have the properties that follow.

1. Competitive Inhibition

A competitive inhibitor competes with the normal substrate for binding at the active site of the enzyme. It forms an enzyme-inhibitor complex (EI) by the reaction

$$E + I \rightleftharpoons EI.$$

SEC. 12–2   ENZYMATIC REACTIONS

This reaction is not different from the formation of ES—rather, inhibition occurs because EI cannot proceed to the next step in the reaction. Thus, the inhibitor is not chemically changed.

In competitive inhibition, if [S] is made large enough, the enzyme will be saturated with S only. This is because S and I bind at the same site. A higher concentration of S is required to achieve half-maximal velocity— that is, K_m increases and depends on the value of [I]. The formation of EI is characterized by an inhibition constant K_I, which is comparable to K_s, the dissociation constant for ES. The inhibition constant is defined by

$$K_I = \frac{[E][I]}{[EI]}, \tag{24}$$

and it can be shown that for any value of [I], the new Michaelis constant is

$$K_{m,I} = K_m\left(1 + \frac{[I]}{K_I}\right). \tag{25}$$

Competitive inhibition is easily recognized if kinetic data are presented in the form of a set of Lineweaver-Burk plots, each plot for a particular value of [I]. This is shown in Figure 12-4(a). Note that all lines intersect at a common y intercept, $1/V_m$. The slope, whose value is $(K_m/V_m)(1 + [I]/K_I)$, differs for each value of [I] because K_m varies with [I]. The x intercept also yields the value of K_I.

FIGURE 12-4   Lineweaver-Burk plots for three types of inhibition. The heavy curve is for [I] = 0.

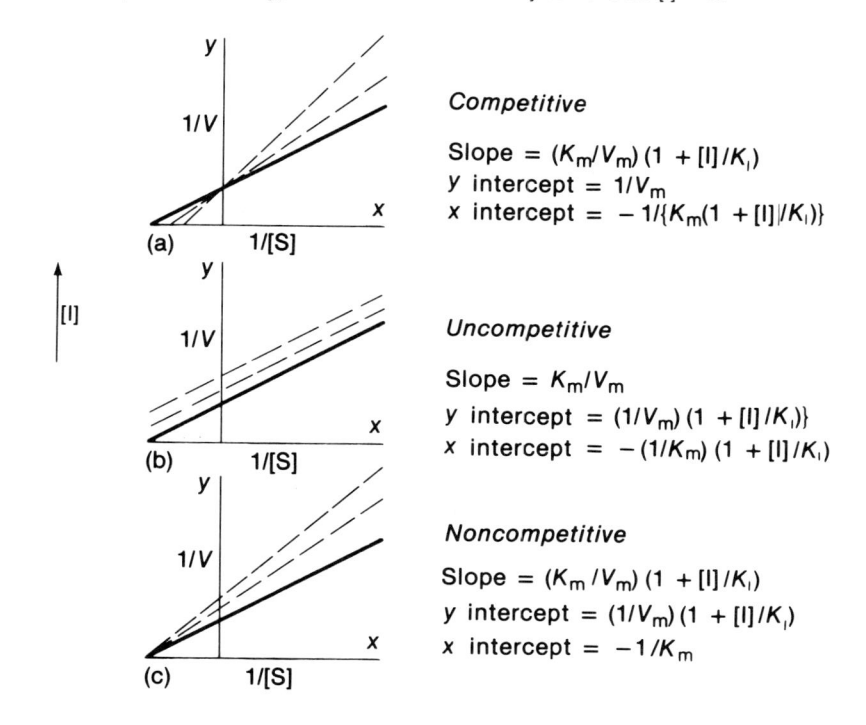

Competitive

Slope = $(K_m/V_m)(1 + [I]/K_I)$
y intercept = $1/V_m$
x intercept = $-1/\{K_m(1 + [I]/K_I)\}$

Uncompetitive

Slope = K_m/V_m
y intercept = $(1/V_m)(1 + [I]/K_I)$
x intercept = $-(1/K_m)(1 + [I]/K_I)$

Noncompetitive

Slope = $(K_m/V_m)(1 + [I]/K_I)$
y intercept = $(1/V_m)(1 + [I]/K_I)$
x intercept = $-1/K_m$

Studies of competitive inhibitors indicate the kind of molecule that is recognized by the binding site. A classic example of competitive inhibition occurs with the succinic dehydrogenase reaction:

$$\begin{array}{c} COO^- \\ | \\ CH_2 \\ | \\ CH_2 \\ | \\ COO^- \end{array} \quad \xrightarrow{\text{succinic dehydrogenase}} \quad \begin{array}{c} COO^- \\ | \\ CH \\ \| \\ CH \\ | \\ COO^- \end{array}$$

Succinate   Fumarate

This reaction is inhibited by

$$\begin{array}{c} COO^- \\ | \\ CH_2 \\ | \\ COO^- \end{array} \quad \text{or} \quad \begin{array}{c} COO^- \\ | \\ COO^- \end{array} \quad \text{or} \quad \begin{array}{c} COO^- \\ | \\ CH_2 \\ | \\ C=O \\ | \\ COO^- \end{array}$$

Malonate   Oxalate   Oxaloacetate

Since each of these has two negatively charged groups at the pH of the reaction, one might guess that the active site of the enzyme contains two positively charged groups.

2. Uncompetitive Inhibition

In uncompetitive inhibition, the inhibitor I combines with ES to form an enzyme-substrate-inhibitor complex:

$$ES + I \rightleftarrows ESI.$$

Increasing the value of [S] does not restore the reaction to an uninhibited form, so that V_m decreases as [I] increases. Uncompetitive inhibition is also recognized by Lineweaver-Burk plots obtained at several values of [I], as shown in Figure 12-4(b). Note that the curves for each value of [I] are parallel and that V_m decreases as [I] increases.

3. Noncompetitive Inhibition

Some enzymes contain a second binding site other than that which binds the substrate. A molecule I that can bind to this site on E and prevent formation of ES, and that can bind to the same site on ES to form ESI, is called a noncompetitive inhibitor. Thus there are two inactive forms, EI and ESI, produced by the reaction,

$$E + I \rightleftarrows EI$$

and

$$ES + I \rightleftarrows ESI,$$

SEC. 12–2 ENZYMATIC REACTIONS

and two inhibition constants,

$$K_{I,EI} = \frac{[E][I]}{[EI]} \quad \text{and} \quad K_{I,ESI} = \frac{[ES][I]}{[ESI]},$$

which are not necessarily equal. The value of V_m is reduced by the inhibitor but, because the affinity of E and S is unchanged, K_m is unaffected. This is again indicated by a family of Lineweaver-Burk plots, as shown in Figure 12-4(c). Note the constancy of the x intercept, which is $-1/K_m$, and the increasing values of $1/V_m$.

Entropic Advantage of Enzyme Catalysis

Equation 65 of Chapter 11 shows that the rate constant is proportional to $e^{\Delta S^*/R}$, in which ΔS^* is the entropy change in the formation of the transition state. Remembering that entropy is related to disorder and that entropy increases as disorder increases, we should realize that the intermolecular reaction between two freely moving molecules to form a single molecular intermediate is a process in which order increases. Thus, $\Delta S^* < 0$, and $e^{\Delta S^*/R}$ is small. We will see that enzymatic catalysis is often based on increasing the magnitude of this term.

Consider a reaction between two components A and B, which proceeds through a transition state (AB)* to form a compound C. Assuming that A and B are not bound to any other molecules (such as solvent molecules), entropy must be lost when A and B are brought together, because the A-B pair is more ordered than the separated molecules. However, to reach the transition state, the two molecules must not only be in contact but they must be oriented correctly with respect to one another—that is, the regions that are ultimately to react must be in contact. Thus, to achieve the transition state, both the rotational and translational freedoms of each molecule are restricted, so that the loss is one of translational and rotational entropy. The atoms within each molecule rotate and vibrate with respect to one another so that each molecule possesses vibrational and internal rotational entropy. These entropies may also change in the formation of the transition state—in fact, they may either increase or decrease, depending on the structure of the transition state. However, the major effect is the change in the translational and rotational entropies, and this typically amounts to about -230 J K^{-1} mol^{-1}, yielding a value of $e^{\Delta S^*/R}$ of approximately e^{-22}. If reaching the transition state were the rate-limiting step and if this bimolecular interaction could be totally avoided, the reaction rate would increase by a factor of e^{22} or about 10^9-fold. This means that if the concentration of the components were increased 10^9-fold, the bimolecular reaction would have the same rate as would occur at low reactant concentration with the bimolecular reaction being bypassed.

The following simple reactions show the enormous effect on the reaction rate caused by joining the reactants together. Let us consider two reactions in which a carboxylate group attacks an ester. The first reaction shown below is a bimolecular one:

(I) $CH_3COOAr + CH_3COO^- \xrightarrow{k_2} $ [acetic anhydride] $\xrightarrow{fast} CH_3COOH + CH_3COO^- + HOAr$

An acetate attacks an acetate ester to produce acetic anhydride and an alcohol. In the second reaction, the acetate and the ester are part of the same molecule, so that collision is vastly facilitated. In this reaction there is an *intra*molecular, nucleophilic attack of the carboxylate on the ester of a succinate half-ester to form succinic anhydride. There is a striking difference in the rates of the two reactions—the second reaction proceeds 10^5-fold faster than the first reaction:*

(II) [succinate half-ester with COOAr and COO$^-$] $\xrightarrow{k_1}$ [succinic anhydride] $\xrightarrow{fast}$ [succinate with COOH and COO$^-$] + HOAr

Presumably this is because there is not much loss of translational entropy (which typically amounts to 125 J K^{-1} mol^{-1}), which would alone result in a rate increase of about $e^{125/8.314} \approx 10^6$. The similarity in the experimental value and this crude calculation is spurious and is not always so good; the main point is that the effect is of the right magnitude.

It is currently believed that one way that enzymes catalyze reactions is by providing a way to achieve the transition state without a large entropy loss. Clearly if A and B are to form the transition state, the molecules must be brought together, but if the probability that this will occur is increased, the entropy loss will be less. Since an enzyme and a substrate molecule have a large binding energy, collision between the enzyme and the substrate will very often be effective—that is, the probability of an effective collision is high; thus, we can think of the entropy loss entailed in forming a bimolecular complex as being "paid for" in binding energy. To understand how this is done, one needs to have a picture of the molecular basis of the large binding energy between the enzyme and the substrate.

---

*To an organic chemist, reaction I seems bizarre, since no one would use a catalyst as poor as acetate to hydrolyze an ester. The point is that cells do not always have the catalyst that would be chosen by a good organic chemist, and often acetate is all that is available. The example shown gives an idea of how an enzyme can manage even with such a poor catalyst.

SEC. 12–2 ENZYMATIC REACTIONS

An enzyme molecule possesses an active site that consists of both the binding site and a place at which the chemical reaction occurs. In a normal (uncatalyzed) bimolecular reaction, two molecules collide and then usually separate because, in general, their orientation is not correct for bond formation; thus, the probability of their staying together is low. In an enzyme, the binding site consists of many amino acids presenting a collection of chemical groups and arrangement of groups. A substrate molecule approaching the site can engage in a large number of attractive interactions (van der Waals and London attractions, hydrogen bonding, ionic attraction). A particular enzyme is designed to make the probability very high for the binding of its substrate. The substrate molecule then tends to stay in place, rotating and vibrating (or trapped in a crevice when the active site is recessed in the enzyme) until it has become oriented for maximal binding to the enzyme. The substrate can, of course, come off with a certain probability, since binding is an equilibrium reaction; the probability of this occurring is reflected in the dissociation constant of the enzyme and the substrate. However, the enzyme is usually designed to have fairly tight binding, so that once it is bound, the substrate can be thought of as locked in place or anchored.* At this point the transition state has not yet formed, because the second reactant is not yet present. However, this reactant will bind to the enzyme with high probability for the same reasons that the first reactant binds. Thus, the binding energy has been utilized to increase the probability that the reactants are in contact. A final, important step is also taken by the enzyme. If the binding sites on the enzyme for the two reactants are so arranged that they are in the correct orientation for reaching the transition state, the probability of producing this intermediate is vastly increased. Thus, the binding energy has also paid for the loss of rotational energy. The net effect of all of this has been to buy negative entropy with binding energy so that ΔS^* is a small rather than a large negative number. Note that one way in which this has occurred is that the enzyme has utilized the energy of interaction with nonreacting portions of the reactants, whereas in the uncatalyzed reaction, only the potential reacting regions of the molecule are involved.

Many enzyme reactions are not bimolecular in the sense of having two interacting substrates; examples are reactions in which A is converted to C. In such cases the reaction is in fact bimolecular, because it usually involves a reaction between a site on the substrate and a particular group on the enzyme. The catalytic advantage is in this case also obtained from the binding energy; that is, the substrate binds in such a way that its reactive site is precisely oriented with respect to the reactive group on the enzyme.

---

*It is necessary, of course, that the binding not be excessively tight, because if it is, the enzyme might be unable to release the product.

REFERENCES

Bender, M.L., and L.J. Brubacher. 1973. *Catalysis and Enzyme Action*. McGraw-Hill. New York.

Bernhard, S. 1968. *The Structure and Function of Enzymes*. W.A. Benjamin. Menlo Park, Calif.

Boyer, P.D. (ed.). 1970. *The Enzymes*. Academic Press. New York.

Castellan, G.W. 1971. *Physical Chemistry*. Addison-Wesley. Reading, Mass.

Fersht, A. 1977. *Enzyme Structure and Mechanism*. W.H. Freeman and Company. San Francisco.

Glasstone, S., K.J. Laidler, and H. Eyring. 1941. *The Theory of Rate Processes*. McGraw-Hill. New York.

Gutfreund, H. 1965. *An Introduction to the Study of Enzymes*. Blackwell. Oxford, England.

Jencks, W.P. 1972. *Catalysis in Chemistry and Enzymology*. McGraw-Hill. New York.

Laidler, K.J. 1973. *Chemical Kinetics of Enzymatic Activity*. McGraw-Hill. New York.

Reiner, J.M. 1968. *Behavior of Enzyme Systems*. Reinhold. New York.

Walsh, C. 1979. *Enzymatic Reaction Mechanisms*. W.H. Freeman. San Francisco.

Westley, J. 1969. *Enzyme Catalysis*. Harper & Row. New York.

PROBLEMS

1. A particular catalyst reduces the free energy of activation of a reaction at 25°C from 75 kJ mol^{-1} to 30 kJ mol^{-1}. By what factor does the reaction rate change?

2. When a metal or a metal-metal oxide catalyst is used, the substance is usually deposited as a thin layer (called the *carrier*). Why is this done?

3. The initial velocity (moles per second) of an enzymatic reaction is studied as a function of temperature. The data obtained are the following: 20°C, 0.095; 25°, 0.0017; 30°, 0.0018; 35°, 0.0022; 40°, 0.00253; 45°, 0.00251; 50°, 0.0015; 55°, 0.0010. Explain the *decrease* in reaction velocity at higher temperature.

4. Since V_m is the maximal initial velocity achieved at a very high substrate concentration, and since these high concentrations are sometimes difficult to reach, it seems that V_m must be quite difficult to measure. How is this problem solved?

5. What are the units of K_m?

6. The initial velocity of an enzymatic reaction is measured at various substrate concentrations. The data are the following:

V	[S]
0.103	0.001 M
0.177	0.002 M
0.233	0.0033 M
0.310	0.005 M
0.362	0.007 M
0.382	0.008 M

What are the values of K_m and V_m?

7. Two enzymes A and B both catalyze reactions in which a substrate X is converted to products P_A and P_B, respectively. The value of K_m for the reactions with A and B are 0.001 M and 0.0036 M, respectively. To which enzyme does X bind more tightly?

8. Two milligrams of an enzyme whose mo-

lecular weight is 70,000 is dissolved in 100 ml of a solution containing 1.62, 3.18, or 10.0 grams of a substrate Q, whose molecular weight is 1500. The initial velocity of conversion of Q to the product is 0.39, 0.55, and 0.78 grams of Q per hour, respectively. What are the values of V_m and K_m?

9. An enzyme has a concentration of 10^{-5} M. The concentrations of the product at half-minute intervals starting at zero time are 0, 0.0312, 0.0602, 0.0934, and 0.125 M.

   (a) What is the turnover number?

   (b) How many substrate molecules are converted to product molecules each second per liter?

10. V_m for an enzymatic reaction is 5 μmol per minute when 2 μg of an enzyme whose molecular weight is 27,000 is present. What is the turnover number?

11. The molarity of a polymer solution is frequently expressed in terms of the molecular weight of the monomer. For example, the molar concentration of DNA at 1 μg per ml is expressed as 3×10^{-6} mol of nucleotide per liter and is independent of the molecular weight of the DNA. Thus, when an enzyme acts on a polymer, either to degrade it or synthesize it, the catalytic activity or turnover number is expressed as the number of bonds hydrolyzed or formed per unit time. An exonuclease at a concentration of 5 μg/ml and whose molecular weight is 23,000 attacks an excess of single-stranded DNA at the end of the molecule and releases nucleotides (molecular weight = 336) at a maximal initial velocity (V_m) of 6.4 μg per minute. What is the turnover number?

12. A regulator molecule R is known to reduce the velocity of an enzymatic reaction when it is bound to the enzyme. Propose a mechanism for regulation if

   (a) K_m is unaffected and V_m is reduced when R is bound.

   (b) K_m is decreased and V_m is unaffected when R is bound.

13. The following Lineweaver-Burk plot is observed.

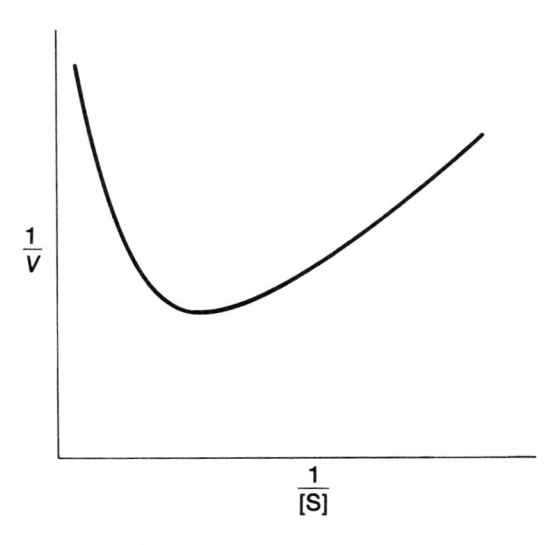

What might this suggest about the interaction between the substrate and the enzyme?

CHAPTER 13

SURFACES

Each living cell is separated from its surroundings by a surface. The cell surface has two main functions: to provide and maintain differences in concentrations between two regions and to give a means of distinguishing one type of cell from all other types. Cells also contain numerous internal surfaces. For example, the nucleus of a cell is enclosed in a nuclear membrane; cytoplasmic protein synthesis takes place on specialized membranes, and the generation of energy-rich compounds occurs on highly convoluted surfaces called mitochondria. These internal membranes also serve to separate cellular components but have another important function—namely, to facilitate certain chemical reactions that occur more effectively on a membrane than when free in solution. In order to understand how membranes perform these functions, it is necessary to understand the physical chemistry of surfaces.

13–1 SURFACE TENSION OF LIQUIDS

Molecular Origin of Surface Tension

The individual molecules of a liquid attract one another. However, the net attractive force experienced by any one molecule in the body of a liquid is zero, because an internal molecule is exposed to attractive forces from all

SEC. 13-1 SURFACE TENSION OF LIQUIDS

directions. The situation is quite different for a molecule on the surface; it is also attracted in all directions in the plane of the surface by other molecules within the surface, but in the body of the liquid it is attracted by molecules from only a single (under) side of the surface (Figure 13-1). Thus a molecule at the surface has a tendency to move from the surface into the body of the liquid. Clearly all surface molecules cannot move into the liquid. When such movement does occur, the liquid must change its shape. What this means is that the surface molecules of an irregularly shaped volume of liquid will move until a shape is achieved that has the minimal ratio of surface to volume.

FIGURE 13-1 The forces acting on two molecules in a liquid, one molecule totally within the body of the liquid, the other molecule on the surface. The forces that act on the molecule within the liquid cancel each other out.

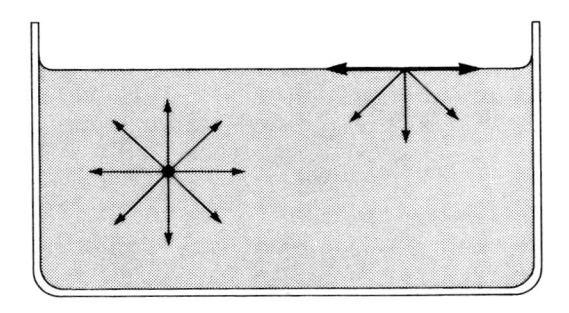

Motion from rest requires a force, so that we may say that there is *a force that acts to minimize the surface area*. Thus, if a liquid is not contained in a vessel (that is, if it is falling through air), the tendency to minimize the surface area imparts a spherical shape because, of all shapes, the sphere has the least ratio of surface to volume. Of course, if the liquid is in a container, it does not assume the spherical shape because gravity imposes an additional requirement, namely, that the air-liquid surface be uppermost and parallel to the plane of Earth's surface; this counteracts the tendency to form a sphere.

A process that results in an *increase* of the surface area of a liquid requires that molecules be moved *from* the body of the liquid *to* the surface. This means that work must be done against the cohesive forces within the liquid and that the free energy of the molecules in the surface must increase. Another way to recognize the increase in free energy when a surface expands is to note that a volume of liquid that is nonspherical spontaneously becomes spherical if there are no external constraints; a spontaneous change requires that ΔG is less than zero, so that the nonspherical shape must have greater free energy. This excess free energy of the molecules on the surface may also be described by stating that there is a force called the *surface tension* that acts *parallel* to the surface and opposes any attempt to expand the surface area. According to the usual definition of work, a small increment of work dW done in extending the surface by an increment in area dA is $-\gamma dA$, in which γ is the *surface tension*. If this work is done at constant temperature, $dW = -dG_\gamma$, in which G_γ is the surface free energy. Thus, the surface tension is

$$\gamma = \frac{dG_\gamma}{dA}. \tag{1}$$

The units of surface tension are erg cm^{-2} or dyn cm^{-1} or, in SI units, J m^{-2} and N m^{-1} (newtons per meter).

Surface tension is often misunderstood. The problem arises because of incorrect interpretations of certain phenomena, such as the following. If a needle is carefully placed on a water surface, the needle will float; this gives the appearance that surface tension acts perpendicular to the surface and thereby opposes gravity. This is the wrong way to think about surface tension, which acts parallel and not perpendicular to the surface; a more accurate statement is that some component of the surface tension opposes gravity. This component arises because the needle depresses the surface and causes the curvature shown in Figure 13-2. The surface tension only acts parallel to the surface, resisting expansion, but once the surface is forced by the needle to expand, neither the surface nor the surface tension remains perpendicular to the gravitational force in the vicinity of the needle. As long as the mass of the needle is small enough, its weight is not sufficient to overcome the component of the surface tension in the gravitational direction and break through the surface.

FIGURE 13-2 Depression of a liquid surface caused by a floating needle.

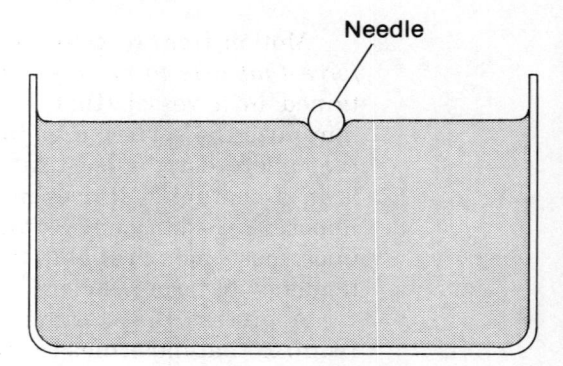

An important consequence of the existence of surface tension is that there is a difference in the pressure on the two sides of a surface. This can be understood most easily by thinking about a gas-filled rubber balloon that is stretched by the internal pressure. Since the stretched balloon is elastic, it experiences a force parallel to the surface that acts to eliminate the stretching. This force is analogous to the surface tension of a liquid; it also arises in the same way—namely, because the rubber molecules attract one another. The internal pressure of a balloon must be greater than the external pressure because the gas within the balloon is not only counteracting the external pressure but has also stretched the balloon. The same statement can be made about the pressure within a gas bubble in a liquid.

SEC. 13-1 SURFACE TENSION OF LIQUIDS

The internal pressure not only acts against the gas pressure on the body of the liquid but also against the surface tension, which tries to reduce the bubble to zero volume. This phenomenon is described quantitatively by the equation

$$\Delta P = P_1 - P_2 = 2\gamma/r , \qquad (2)$$

in which P_1 and P_2 are the internal and external pressures of a sphere of radius r bounded by a surface having a surface tension of γ. This equation indicates that of all of the properties of the substances forming the surface, only the surface tension determines the pressure difference. Furthermore, the pressure difference increases as the radius of the sphere decreases.

Surface tension is also responsible for capillary rise, as is described in the following section.

Capillary Rise

If a glass tube having a small radius (a capillary tube) is inserted in a pan of water, the water rises in the tube above the surface of the bulk liquid. This is called *capillary rise*. If the liquid is mercury instead of water, the level in the capillary tube is lower than the surface of the mercury in the pan; this is *capillary fall*. Here we are concerned with two different attractive forces: that between two molecules of the liquid and that between a molecule of the liquid and a molecule of the wall of the tube. These forces are called *cohesion* (attraction between like substances) and *adhesion* (attraction between unlike substances). Adhesion may be positive, as in the case of water wetting a glass surface, or may be negative (repulsive), as in the pulling away of mercury from glass. By convention cohesion is always positive.

Surface tension explains the phenomenon of capillarity. Molecules of water adhere to glass so that a thin film of liquid creeps up the walls. When the water creeps up, the surface of the top of the liquid expands considerably. This expansion is opposed by the surface tension at the air-liquid interface. The surface tension might prevent the rise from occurring, but the adhesive force is too great. Thus, in order to reduce the area of the air-liquid surface, liquid rises in the column. This rise is of course opposed by gravity. Thus the height at equilibrium is determined by the balance between the strength of adhesion, the surface tension, and gravity. As the strength of adhesion increases, the liquid creeps higher. In the standard mathematical description of capillary rise, the strength of adhesion is not mentioned explicitly but appears instead as the angle θ made by the surface with respect to the wall of the tube (Figure 13-3). This is called the contact angle θ. The equation describing capillary rise is then

$$h = \frac{2\gamma \cos \theta}{\Delta \rho g r} \qquad (3)$$

FIGURE 13-3 Capillary rise: θ is the contact angle; r is the radius of the tube; and **r** is the radius of curvature of the liquid surface. The dotted lines show the geometric construction that demonstrates that **r** = $r/\cos\theta$.

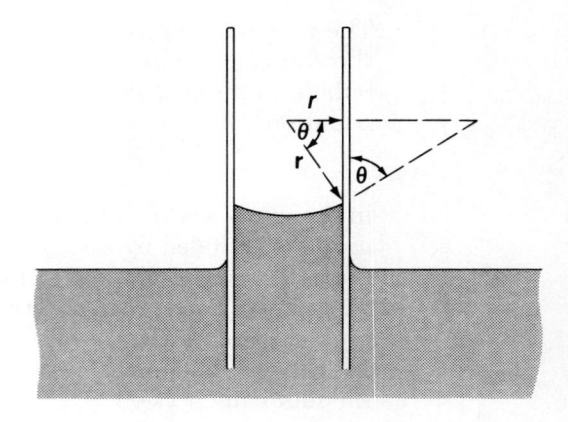

in which h is the height of the meniscus above the flat surface of the bulk of the liquid; $\Delta\rho$ is the difference between the density of the liquid and that of the surrounding fluid (usually air); r is the radius of the capillary; and g is the acceleration of gravity. In the case of capillary depression, which arises when $\cos\theta < 0$, the same equation can be obtained with h as the depth of the depression.

Measurement of capillary rise is one of the many ways to measure surface tension, although its accuracy depends upon a precise and relatively difficult measurement of θ and r; h is usually sufficiently large that it is easy to measure accurately.

An example of the use of Equation 3 follows.

EXAMPLE 11–A **A capillary tube 1 mm in internal diameter is immersed into 0.03 M sucrose (γ = 72.5 erg cm^{-2} and density = 1.0381 g cm^{-3}): how high above the surface of the liquid does the liquid in the capillary rise at 20°C and a pressure of 1 atm?**

The density of dry air is 0.0011 g cm^{-3} under these conditions. The contact angle is five degrees.

The value of $\Delta\rho$ will be $1.0381 - 0.0011 = 1.037$; $g = 980$ cm sec^{-2} and $r = 0.05$ cm. Thus, $h = 2(72.5)(0.996)/(1.037)(980)(0.05) = 2.84$ cm. Note that if we had not known the surface tension, a measured value of $h = 2.8$ cm would have provided the correct value of γ.

Capillary rise has been discussed for decades in terms of explaining the rise of sap in trees. The problem is that capillary rise cannot be the cause of the movement of the sap. The radii of the xylem tubes through which the sap rises are about 2×10^{-3} cm and the contact angle of sap in these tubes is found in the laboratory to be 0°. Using Equation 3 yields a value of h of about 60 cm, whereas trees grow to heights of 100 meters. The cause of the rise of the sap in the trees has not yet been satisfactorily explained, although there are numerous theories.

Surface Tension and Interfaces

Two phases in contact define an *interface* or a boundary between two substances. The phases may both be liquid, both solid, or be solid-liquid, solid-gas, or liquid-gas. Interfaces have properties that differ from the bulk of each phase and play an important role in many biological processes.

At the surface of any substance molecules are continually entering and leaving the material. For instance, at a liquid-gas interface, a liquid molecule might enter the vapor, return to the surface, enter the bulk liquid, and then return to the surface. When two different substances, for example, 1 and 2, are in contact, it is usually the case that the concentration of substance 1 at the interface is not the same as the concentration within the main body of substance 1. Similarly, the concentration of substance 2 at the surface differs from its concentration elsewhere. This variation of concentration at an interface is called *adsorption;* it results from cohesive forces between the molecules 1 and 2. Some examples are shown in Figure 13-4. Adsorption may either increase (*positive adsorption*) or decrease (*negative adsorption*) the concentration with respect to the concentration of some constituent of the system at the surface. We now examine this variation in concentration for a solution; this will enable us to understand adsorption in a quantitative way.

FIGURE 13-4 Two types of adsorption. In panel (a) the molecules represented by the solid circles are positively adsorbed at an air-liquid interface. The *x*'s are negatively adsorbed. In panel (b) the solid circles are positively adsorbed at a solid-liquid interface; the *x*'s are again negatively adsorbed.

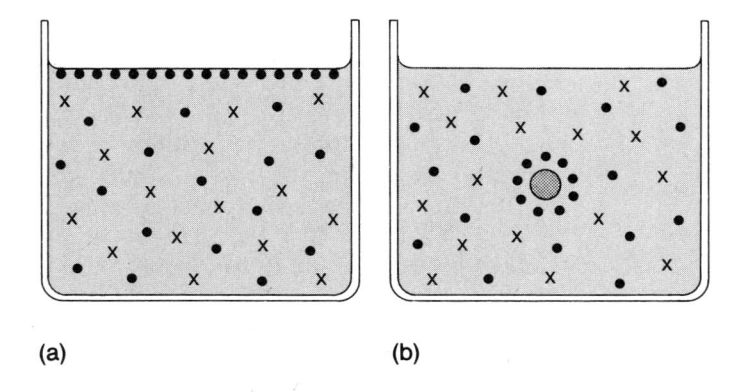

(a)   (b)

In a solution, the two components, the solvent and the solute molecules, are distributed uniformly throughout the body of the solution. However, at the surface the solute concentration is in general not the same as in the solution. The solvent concentration at the surface is of course also different from that in the solution. The relative concentration in the two regions depends on (1) the rate at which solute and solvent molecules enter and leave the surface layer; and (2) the cohesive forces between the solvent molecules, between the solute molecules, and between the solvent and the solute molecules. This concentration variation is described by adsorption at the surface of a solution. The following argument shows that such adsorption is a consequence of the effect of solution composition on the surface tension γ.

Consider two phases 1 and 2 that are in equilibrium and for which the pressure and the temperature are constant. We increase the surface area A between the two phases by an amount dA, which increases the surface free energy by an amount $dG = \gamma dA$, or

$$\left(\frac{\partial G}{\partial A}\right)_{T,P} = \gamma . \tag{4}$$

It is possible to change the surface tension in such a system by changing the composition of either phase—for instance, by increasing the number of moles n of some substance X by an amount dn. The free energy of the system then depends on the variables n and A, and

$$\left[\frac{\partial}{\partial n}\left(\frac{\partial G}{\partial A}\right)_n\right]_A = \left[\frac{\partial}{\partial A}\left(\frac{\partial G}{\partial n}\right)_A\right]_n = \left(\frac{\partial \bar{G}}{\partial A}\right)_n , \tag{5}$$

in which $\bar{G} = (\partial G/\partial n)_A$ is the partial molar free energy of substance X. By combining Equations 4 and 5, one obtains

$$\left(\frac{\partial \gamma}{\partial n}\right)_A = \left(\frac{\partial \bar{G}}{\partial A}\right)_n . \tag{6}$$

Applying a rule connecting partial derivatives to Equation 6, we obtain

$$\left(\frac{\partial \gamma}{\partial n}\right)_A = -\left(\frac{\partial \bar{G}}{\partial n}\right)_A \left(\frac{\partial n}{\partial A}\right)_{\bar{G}} = -\left(\frac{\partial \bar{G}}{\partial n}\right)_A \Gamma , \tag{7}$$

in which Γ is substituted for $(\partial n/\partial A)_{\bar{G}}$; Γ is defined more fully below.

Let us examine for a moment what Equation 7 means. Consider a system consisting of n_1 moles of solvent X_1 in equilibrium with its vapor (the vapor phase is the phase of constant composition) and a solution of n_2 moles of solute X_2. An increase in surface area by dA causes increased adsorption of X_1 at the interface, and this changes the composition of the bulk of the solution. This changes $\bar{G}$, since $\bar{G}$ depends on composition. If the addition of dn_2 moles of X_2 restores $\bar{G}$ to its original value, dn_2/dA is the amount of X_2 required to maintain the original composition of the solution. This is the molar excess of solute per unit area, or the *adsorption coefficient*, Γ.

Equation 7 can be rearranged to

$$\Gamma = -\left(\frac{\partial \gamma}{\partial n}\right)_A \bigg/ \left(\frac{\partial \bar{G}}{\partial n}\right)_A = -\left(\frac{\partial \gamma}{\partial \bar{G}}\right)_A , \tag{8}$$

the desired relation between surface tension, surface free energy, and adsorption. To make this equation more useful, we express $\bar{G}$ in terms of the activity a or the concentration c of the solute, using $d\bar{G} = RTd \ln a$, and write:

$$\Gamma = \frac{a}{RT}\frac{d\gamma}{da} \approx -\frac{c}{RT}\frac{d\gamma}{dc}. \tag{9}$$

This important equation is called the *Gibbs adsorption isotherm*.* It describes the number of molecules adsorbed to a unit area of the surface of a solution of concentration c whose surface tension varies with concentration by the factor $d\gamma/dc$. The Gibbs isotherm can be applied to the distribution of solvent and solute molecules in a solution because if the solute is nonvolatile, the constant phase is just the vapor above the solution. Equation 9 says that

1. A substance that lowers surface tension ($d\gamma/dc < 0$) is concentrated ($dn/dA > 0$) at the interface—that is, at the surface. Examples of such substances are soaps, fatty acids, alcohols, amines, and proteins.
2. A substance that raises the surface tension ($d\gamma/dc > 0$) is less concentrated ($dn/dA < 0$) at the surface. Some examples are sugars and polysaccharides.

An important point for biological systems is that *the surface concentration of such substances as fatty acids and proteins can be so high that there are virtually no solvent molecules in the surface layer*. That is, the surface is covered entirely by solute molecules—there is a *surface film*. In the next section we will see that when this occurs, the molecules in the film usually have a definite orientation with respect to one another.

How does a solute alter surface tension? Since the surface tension is a consequence of cohesive forces, the solute molecules must alter these forces. Thus, solute molecules that attract solvent molecules more strongly than solvent molecules attract one another will increase the surface tension. This is why sugars, which have multiple sites for hydrogen-bonding with water, increase the surface tension. Similarly, if the solute molecule either weakens the attraction between solvent molecules or is capable of attracting only a single solvent molecule, the surface tension will decrease. Thus fatty acids, which have bulky hydrophobic regions, decrease the surface tension. Substances that decrease surface tension are called *surfactants*.

The amount of liquid required to cover a surface also depends on the surface tension. This is because the surface tension acts to minimize the surface area by causing the liquid to round up. In the extreme case of zero surface tension, which could arise only if there were no cohesive force, a drop would spread over an infinitely large surface to form a monomolecular layer. In general, if a series of liquids are used to cover a surface, the least volume is required of the liquid having the lowest surface tension.

---

*Many such equations are called isotherms (Greek: *iso*, same; *thermē*, heat), because experimentally, measurements of adsorption are carried out at a single temperature.

Thus the addition of a solute that reduces the surface tension of water will cause a drop of water to spread on a surface. This effect is a familiar one—a drop of water forms a bead on a glass plate but a drop of soapy water spreads over the surface. The following examples show an important biological phenomenon based on this effect.

EXAMPLE 13-B   The role of surfactants in the lung.

Oxygen and carbon dioxide are exchanged in small sacs in the lungs called *alveoli*. The exchange occurs through a thin film of water on the surface of the alveolus. The rate of transfer clearly increases as the film becomes thinner. In order to cover the surface of one sac with a very thin film of liquid, the fluid contains a powerful surfactant called dipalmityl lecithin. This reduces the surface tension of water to nearly zero and allows the alveolus to be covered with a minimal amount of water. Premature human infants sometimes have an inadequate amount of dipalmityl lecithin and breathe only with great difficulty.

13-2 SURFACE FILMS AND MONOLAYERS

In the previous section we saw that solute molecules are often concentrated at an interface. It is a common phenomenon with asymmetric solute molecules that these molecules are oriented at an interface. This orientation, which is explored in this section, has important biological consequences.

Orientation of Solute Molecules at an Interface

Many organic and biological molecules have distinct polar and nonpolar regions. The fatty acids are an important example of molecules of this type in that they consist of a long nonpolar hydrocarbon chain and a terminal, charged carboxyl group. Another example is a normal alcohol, which has a single terminal hydroxyl group. Such molecules are called *amphipathic*. When such molecules are dissolved in or mixed with a polar solvent, such as water, it might be expected that the charged group would cohere more strongly to the polar water molecule than would the nonpolar hydrocarbon chain. Thus, a fatty acid molecule at an air-water interface should be oriented so that the carboxyl group is directed toward the water and the hydrocarbon chain toward the air. If there are many identical fatty acid molecules on the surface, this orientation should occur for each molecule. Hence, if a sufficiently small quantity of an insoluble fatty acid such as stearic acid is mixed with water, a monomolecular surface layer (a monolayer film) forms, in which all stearic acid molecules are oriented in the manner shown in Figure 13-5.

FIGURE 13-5 A monolayer of stearic acid molecules on water. The circles and the lines represent respectively the polar and nonpolar portions of the molecule.

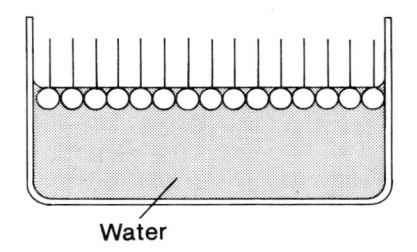

The reason for such orientation can also be understood in terms of surface tension, using the *principle of independent surface action* stated by Irving Langmuir. He proposed that each part of a molecule possesses a local surface free energy and that a monolayer or film is thermodynamically stable only when the free energy of the film is minimized. Interestingly, one can show by this principle that the molecules at the surface of a homogeneous liquid are also oriented if they have a polar and a nonpolar part. Let us consider the surface of pure ethanol. The two possible orientations of the molecule at the surface—that is, with either the OH group or the ethyl group on the air side—are shown in Figure 13-6. The surface

FIGURE 13-6 Two possible orientations of ethanol at an air-liquid surface. The dark and light portions represent the OH^- group and the ethyl group, respectively. In (a) the surface is strongly polar; in (b) it is a hydrocarbon surface.

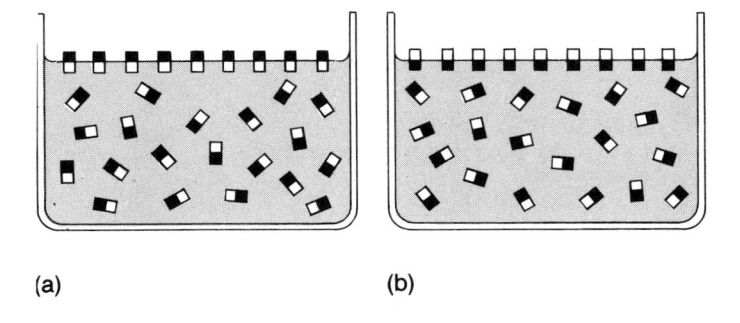

(a)   (b)

energy of water is 120 erg cm^{-2}, from which, by comparison to values for alcohols and hydrocarbons, a value of 190 erg cm^{-2} has been estimated for a layer consisting solely of the OH group. The surface energy of hydrocarbons ranges from 40 to 50 erg cm^{-2}. Therefore, the requirement for minimal energy says that the ethyl group will be exposed. This can be tested experimentally by measuring the surface tension γ of ethanol. If the OH group is exposed, γ should be greater than 72 erg cm^{-2}, the value for water. If the prediction is correct, the measured value of γ should be near that of pure short-chain liquid hydrocarbons, namely 20–22 erg cm^{-2}. The measured value is 22.75 erg cm^{-2}, in accord with the principle.

Formation of Insoluble Surface Films on Liquid Substrates

When small quantities of sparingly soluble molecules are put in contact with a liquid surface, they will spread on the surface to form a film one molecule thick. There are two principal requirements for this spreading phenomenon to occur: (1) that the surface of the liquid must be clean before the substance is added; and (2) that the cohesive force between the molecules applied must be low enough that the molecules will not attempt to maintain the initial three-dimensional structure of the solid; for example, insoluble crystalline material, such as sand, will not form a film.

The requirement for a clean surface can be met by sweeping a barrier rod across the surface. That is, a vessel is filled to the brim with a liquid so that the surface meniscus of the liquid will rise above the walls of a vessel (Figure 13-7). A rod longer than the diameter of the vessel is placed at one edge of the container and moved across to the other side. Surface material (for example, dust or surface films) is swept away and as long as the barrier rod is left in place, the swept surface will be clean. The second requirement is met with immiscible liquids, solids such as the fatty acids (which, because of the strong polar and nonpolar components, cohere poorly in the presence of a polar solvent) and certain soluble macromolecules, such as proteins. (How a soluble protein molecule can form an insoluble surface film will be explained shortly.)

FIGURE 13-7 Cleaning a surface of a liquid with a rod. Floating dust is pushed to the left.

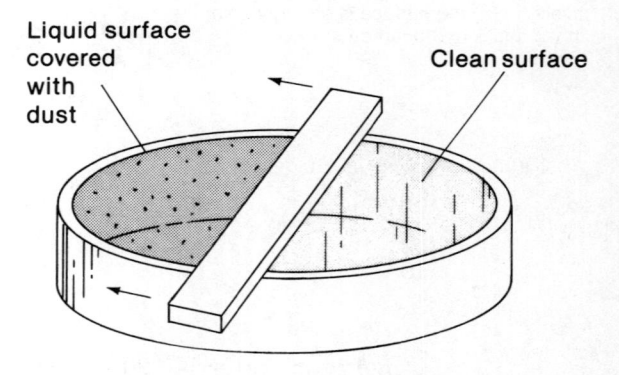

The physical properties of such films can be studied in a Langmuir trough (Figure 13-8). A barrier rod (#1) is used to prepare a clean surface and left in place. A barrier float F that is attached to a torsion device to measure force on the float is then put onto the clean surface and a second barrier (#2) is adjusted, as shown in the figure, in order to make a confined area in which a film can be formed. The film is then made in one of several ways:

SEC. 13-2 SURFACE FILMS AND MONOLAYERS

FIGURE 13-8 A Langmuir film balance. The trough is filled with water and rod #1 is swept across the surface to remove all surface material. The film is formed on the clean surface between rod #2 and the float F. Rod #2 is moved in the direction shown by the arrow and the film is pushed against the float F. A torsion device (not shown) measures the pressure on F. The area is determined by noting the separation of the barrier from the float as is indicated on a scale.

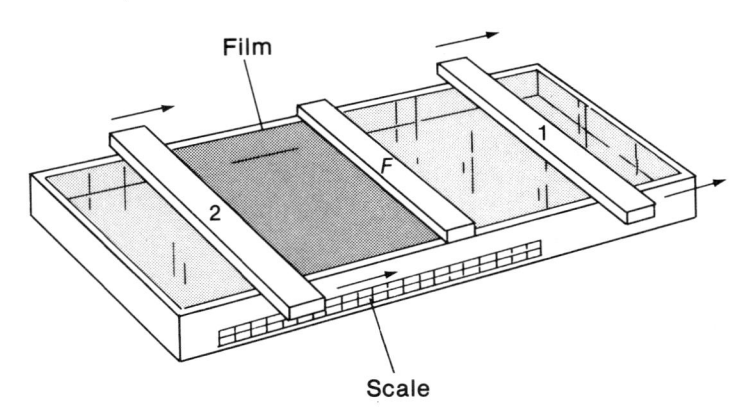

1. The substance of interest is dissolved in a low density, volatile solvent that is immiscible with the supporting liquid and placed on the liquid. For instance, stearic acid dissolved in benzene will float on water. The benzene evaporates, leaving the stearic acid molecules on the surface.
2. A needle is dipped into a solid or liquid of interest and then touched to the surface of a liquid. The particles come off the needle onto the surface, disperse, and move across the surface, forming a film. This method can also be used to apply a tiny amount of liquid onto the surface of a second liquid that is immiscible with the first.
3. A glass microscope slide is inserted into the supporting liquid at a low angle and a droplet of the substance of interest (or a solution of the substance) is allowed to slide down this glass ramp onto the surface. A film forms when the droplet reaches the surface. Because the film spreads out from the point of contact with the surface, this process is often called *spreading*. It is shown in Figure 13-9. This method is very useful for forming protein films, as we will soon see.

FIGURE 13-9 Spreading of a film on a liquid surface. A droplet is allowed to flow slowly down a clean glass slide (left). When it reaches the liquid surface (right), it spreads across the surface forming a film.

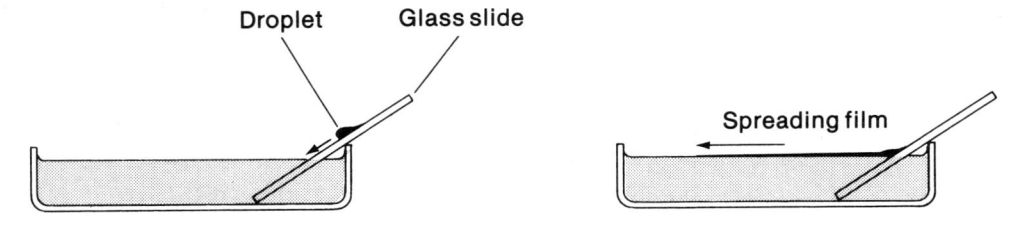

Once a film has formed on the Langmuir trough, barrier #2 is moved toward the float, and it pushes the film along the surface until the film meets the float. At this point the float will register a pressure due to the film. With continued movement of barrier #2, the pressure will increase, and if the area of the film is known, a force-area isotherm will be produced that is exactly analogous to the PV-isotherm for an ideal gas. If the number of molecules applied to the surface is known, it is convenient to plot the data as force F or pressure Π versus area A per molecule (Figure 13-10). As will be seen below, a force-area curve is a useful way to characterize a film and to understand its physical structure. The cross-sectional area of a long molecule can be determined from a force-area curve, as shown in the following example.

FIGURE 13-10 Force-area isotherm for a fatty acid. The Pockels point is approximately 21 Å^2 for all saturated fatty acids. This suggests that fatty acid molecules are oriented with the long axis of each perpendicular to the liquid surface. If this axis were parallel to the surface, the value of the Pockels point would increase with increasing molecular weight. It is important to realize that the area that is measured is that of the entire film, as enclosed by physical boundaries used in the experiments. The area per molecule is then calculated from some independent measurement that provides the number of molecules per film.

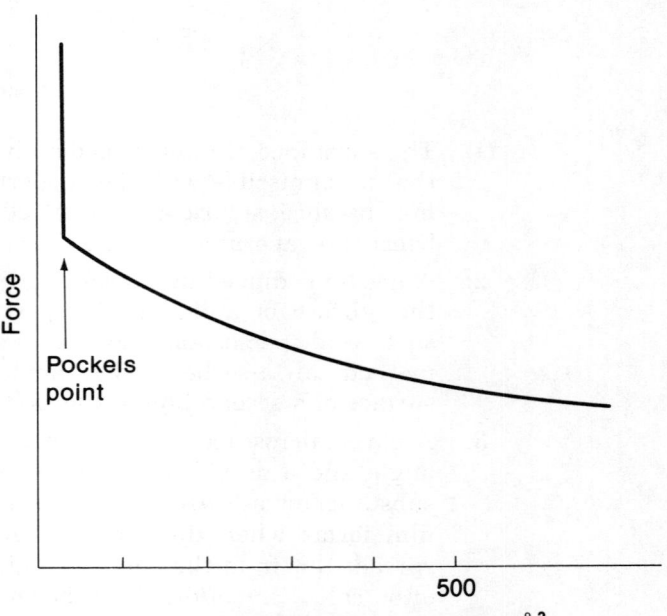

EXAMPLE 13-C Determination of the cross-sectional area of a molecule from the limit of compression of a pure film composed of the molecule.

It is usually found that a surface film is compressible. This indicates that the molecules in the film are not all aligned nor as near as possible to one another. As the film is compressed, an area will be reached at which all of the molecules are aligned, and close-packing (that is, contact in the smallest possible area) is achieved. At this point, there is an increase in the force on the float without a change in area. The area per molecule at which this occurs is called the *Pockels point;* it is considered to be the cross-sectional area of the molecule. For example, if 5 μg of a molecule whose molecular weight is 359 produces a film whose area at the Pockels point is 28.6 cm², the cross-sectional area of the molecules (assuming the area is a square so that there is no empty space between the molecules) is easily calculated. The number of molecules in the film is $(5 \times 10^{-6}/359) \times 6 \times 10^{23} = 8.36 \times 10^{15}$, so that the area is $28.6/(8.36 \times 10^{15}) = 3.42 \times 10^{-15}$ cm², or 34.2 Å². If the molecule is considered to be cylindrical, its radius is $\frac{1}{2}\sqrt{34.2} = 2.92$ Å and the area is $\pi(2.92)^2 = 26.8$ Å².

Molecular dimensions measured in this way are quite accurate, agreeing closely with values obtained by more complex techniques, such as X-ray diffraction.

Theory of Force-Area Curves

The compression of a surface monolayer is in many ways like the compression of a two-dimensional gas. In fact, if the molecules in the monolayer neither attract nor repel one another significantly, and if the pressure is sufficiently low that the individual molecules are on the average separated by distances greater than the molecular diameters, an equation having the form of the ideal gas law can be derived.

In a system consisting of a monolayer on a bulk liquid, there are two surface tensions—namely, γ_0, the surface tension of the pure bulk liquid at the liquid-film interface, and γ, that of the surface itself at the film-air interface. Since surface tension acts parallel to the plane of the film, the force f exerted upon one centimeter of a barrier is simply the difference between the two surface tensions, that is,

$$f = \gamma_0 - \gamma. \tag{10}$$

For any substance forming a surface film, there will always be some molecules that are forced into the bulk liquid by compression of the film. Denoting the concentration of these molecules by c and differentiating, there results

$$\frac{df}{dc} = -\frac{d\gamma}{dc}, \tag{11}$$

which can also be written as

$$\frac{f}{c}\frac{d \ln f}{d \ln c} = -\frac{d\gamma}{dc}. \tag{12}$$

Substituting the value of $d\gamma/dc$ from Equation 9 and using the expression for Γ in Equation 7, one obtains

$$-\frac{d\gamma}{dc} = \frac{\Gamma RT}{c} = \frac{RT}{A\mathcal{N}c} = \frac{kT}{Ac}, \tag{13}$$

in which A is the area in square centimeters per molecule of the film and N, R, and k are Avogadro's number, the gas constant, and the Boltzmann constant, respectively. Combining Equations 12 and 13 gives

$$fA = \frac{d \ln c}{d \ln f} kT . \tag{14}$$

Let us consider a substance for which the concentration of molecules "squeezed out" of the film is proportional to the force f exerted on the film; that is,

$$c = \text{constant} \times f . \tag{15}$$

When this is the case, $d \ln c = d \ln f$, or $d \ln c / d \ln f = 1$, so that Equation 14 becomes

$$fA = kT . \tag{16}$$

which is a relation similar to the ideal gas law. It is common to express f in dynes per centimeter and A in square ångströms, in which case k has the value 1.3805. As might be expected, Equation 16 can also be derived from the kinetic theory of gases, by assuming that the gas is two-dimensional.

As is the case with the ideal gas law, this equation is not usually obeyed for real monolayers—the principal reason is the failure of Equation 15, caused by the van der Waals attraction between the molecules and by their finite size. In fact, as will be seen shortly, many films behave not like gases but like liquids or solids and show phase transitions at particular combinations of pressure and temperature.

Use of Films to Determine Molecular Length

In Example 13–C we saw that the cross-sectional area of a molecule can be determined if one knows the Pockels point. With these surface films the length of the molecules can also be accurately measured in an extraordinarily simple way. If a clean glass slide is dipped into a closely packed film of a molecule such as a fatty acid (Figure 13-11) and then slowly withdrawn, the polar head of the molecule will adhere to the glass. The hydrocarbon tails are then exposed to the air. If the coated slide is again pushed through the film, the hydrocarbon tails in the film on the water surface will cohere to the exposed tails on the glass-bound molecules. Now the polar heads face the glass. By repeated dipping a layer of molecules is built up on the glass, one layer per dip. After 20 to 30 dippings there are enough layers that the thickness of the whole layer can be measured by shining a beam of monochromatic light on the slide and observing the interference pattern. The thickness of the layer can be expressed as a multiple of the wavelength of the illuminating light.* Thus, the thickness of the film on

---

*The optical method will not be described. It can be found in standard texts of physics and optics.

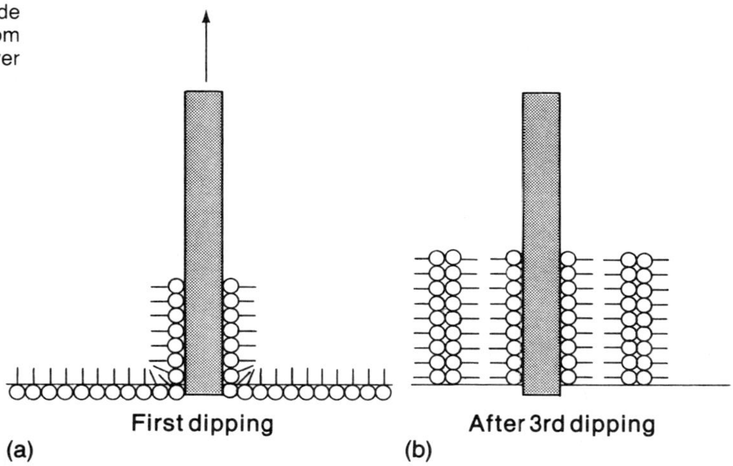

FIGURE 13-11 Coating of a glass slide with fatty acid molecules resulting from repeated dipping through a monolayer film.

(a) First dipping (b) After 3rd dipping

one side of the slide divided by the number of dips is the length of the molecule. This simple method has been applied to the determination of the dimensions of biologically important molecules, such as proteins, lipoproteins, lipids, and steroids.

EXAMPLE 13-D Determination of the length of a lipoprotein molecule and detection of shape changes at low pH.

A slide is dipped twenty-five times into a lipoprotein solution at pH 7. When the slide is illuminated with light, an interference pattern results indicating that the film is 4600 Å thick. Therefore the length of the molecule is $4600/25 = 184$ Å. The same thickness of film results if the solution is at pH 3.5. However, at pH 3.0, the film thickness is 3000 Å, so that at that pH, the molecule is only 120 Å long. Therefore, between pH 3.0 and 3.5, there is a change in the shape of the lipoprotein molecule. We can even make a guess as to how the shape change occurs. Since the length of the lipid portion of the molecule should not be affected by pH (because this part of the molecule is uncharged), it is probable that it is the protein component whose shape changes below pH 3.5.

Types of Monolayers

The gaseous, liquid, and solid states differ by the strength of the interaction between the individual molecules. In Chapter 3, in which deviations from the gas laws were examined, we learned that intermolecular attractions and repulsions cause departures from ideality that are primarily seen in the pressure-volume isotherms. We have also discussed force-area (fA) isotherms for films, and seen an equation, fA = constant, that describes the behavior of the film. The molecules in a film also frequently attract or repel one another, sometimes quite strongly, and departures from the constancy of the product of force and area occur. Some examples can be seen in Figure 13-12, which shows a variety of force-area curves. The existence of these different classes of curves has led to a classification of surface films as *gaslike, liquid-expanded, liquid-condensed,* and *solid,* as indicated in the figure. The four types of films have the following properties:

FIGURE 13-12 Force-area curves for different types of film. Curve G is for a gaseous film, one in which there is little or no molecular orientation and virtually no intermolecular contact. This curve is equivalent to Boyle's Law for gases. The L_1 curve is typical for a liquid expanded film. This curve is characterized by an inflection point that indicates a phase change as the pressure increases and by an extrapolated value of about 50 Å^2 at zero pressure. The L_2 curve for a liquid condensed film also extrapolates to about 50 Å^2 at zero pressure. At high pressure the film becomes nearly incompressible, and the P-A curve becomes linear, resembling that of solid films. This linear portion typically extrapolates to 22 Å^2 at zero pressure. The S curve describes a solid film. This curve is linear and, for hydrocarbon, extrapolates to 20.5 Å, the cross-sectional area of a hydrocarbon chain.

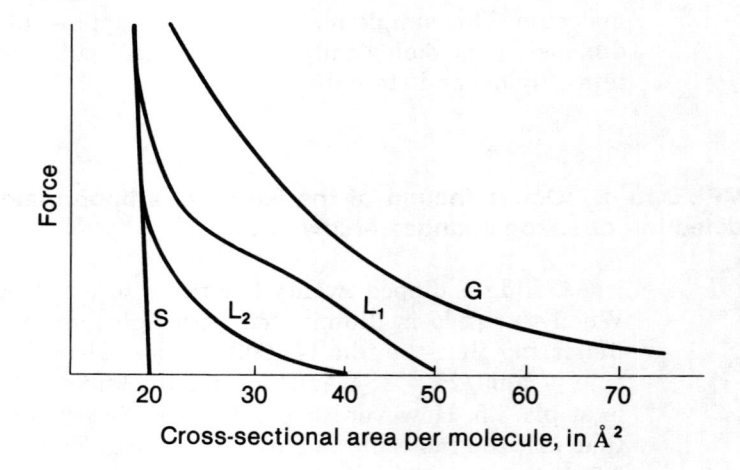

1. Gaseous film. In such a film the product fA is constant. In determining the Pockels point, the measured area of the molecule is found to be large compared to the actual area of the molecule. This type of film is obtained at high temperatures when the molecules are disoriented by thermal agitation and with molecules whose hydrocarbon chains are very short, so that there is not a very strong tendency for the molecules to be in a parallel array.

2. Liquid-expanded or L_1 film. At high pressure, an inflection point in the force-area curve is observed, as might be expected when the molecules of the film first come in contact. However, the area at which the inflec-

tion point occurs is again larger than the cross-sectional area of the molecule, suggesting that at that particular pressure the long axis of the molecule is not perpendicular to the surface. At higher pressure, the long axis of the molecule may be nearly perpendicular to the surface. Such a film is a continuous unit, which is known because the film is easily pushed around the surface. L_1 films are also formed at high temperatures or with short-chain molecules. As the temperature is reduced, L_1 films have a region of significant rise of A with increasing force known as the intermediate or I state.

3. Liquid-condensed or L_2 film. In a compressed L_1 film, the I region shows a linear decrease in A with force. At a lower temperature or with molecules having long hydrocarbon chains, A varies linearly with force, except for the region of high compression. A film with this characteristic is called an L_2 film and the molecules in the film are nearly perpendicular to the surface. The measured cross-sectional area is still greater than the true cross-section of a hydrocarbon film; this is usually a result of the presence of either C=C double-bonds (which produce kinks in the chain) or bulky, branched groups (which prevent alignment of the chains). A secondary weak polar group at a distance from the primary polar group will also produce an L_2 film, because at small distances there may be charge repulsion between the chains. (Note that the linear region is the rising region of the curve and it extrapolates to the cross-sectional area.)

4. Solid or S films. Solid films show very little compressibility, and a linear dependence of A on the force. At zero force, the area is the true cross-sectional area per molecule, because the molecules remain close-packed even when no external pressure is applied. Solid films result from long-chained, unbranched, fatty acids, or chains having a single, very polar terminus.

In the discussion of the four types of film just presented, it was pointed out that the structure of a film often depends on temperature. This is in some ways analogous to the phase transitions of bulk matter, such as melting and boiling. In fact, the same types of intramolecular forces that cause melting and boiling contribute to film structure. This can be seen by comparing the melting temperatures of bulk material and the *half-expansion temperature* of a film with the chemical structure of the molecule forming the film. The half-expansion temperature is a temperature at which the measured area of a molecule in the film is twice that of the true area. For fatty acids the melting temperature increases with the length of the hydrocarbon chain and the half-expansion temperature increases by roughly 8 to 10°C for each additional CH_2 group. Similarly, the introduction of a single *cis* double bond in the chain lowers the melting temperature (this accounts for the familiar observation that the unsaturated fats, such as cooking oil, are liquid at room temperature and the saturated fats, such as butter, are solid) and also lowers the half-expansion temperature. This effect must be steric and not due to a decreased van der Waals attraction when the double bond is present, because the *trans* form of the molecule, which is linear and not kinky, exhibits a higher half-expansion temperature than the *cis* form of the molecule, which is kinked.

Effects of Ionic Environment on Film Structure

In addition to the temperature, the physical properties of a film depend upon the pH and ionic composition of the aqueous liquid on which the film rests. This is for various reasons. For instance, in the case of a fatty acid at alkaline pH, the polar groups are ionized and mutually repelling, resulting in the formation of either a gaseous film or an L_1 film. If ions are present, they will affect the structure of the film by forming an ionic cloud around the charged portion of the fatty acid. Since different counterions (for examples, NH_4^+, Li^+, or Na^+) will interact to various extents, the identity of the ion will also affect the film structure (Figure 13-13).

FIGURE 13-13 A set of force-area curves for a fatty acid film formed on alkaline solutions containing the following ions: (a)Li^+; (b)K^+; (c)Na^+; and (d) $N(CH_3)_4^+$. At high pH, fatty acid molecules are charged and repel one another, thus producing a gaseous or a liquid expanded film. The dipole moment of each counterion alters the magnitude of the repulsion between the fatty acid molecules and thereby affects the film type.

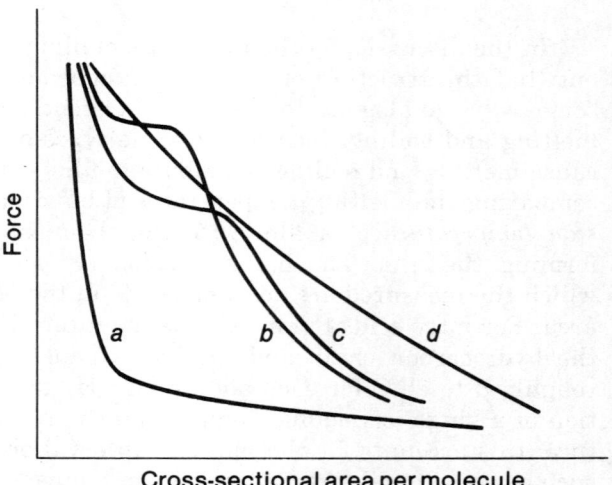

Protein Films

Protein molecules contain both charged regions (the polar side chains and the peptide bond groups) and numerous nonpolar side chains.* Unless these regions are randomly arranged, it should be expected that a typical protein molecule would be oriented at an air-water interface and could form a surface film. Furthermore, films might have somewhat unusual properties because proteins often form side-to-side complexes with one another.

Indeed, proteins form surface films with great ease. A standard procedure for the formation of such a film is to flow a drop of a protein solution down a glass ramp, as we have described in an earlier section. Protein films formed in this way are extremely stable. This may seem surprising, because most proteins are rather soluble in water and, because of this solubility, it might be expected that the protein molecules in the film would gradually break away from the film and dissolve in the bulk liquid. This does not occur and the explanation for the stability lies in the structural changes that occur in the molecule while the film is spreading and the decrease in solubility that accompanies these changes. There are various indications that such structural changes occur during formation of a film of protein. For instance, there is often a substantial decrease in enzymatic activity accompanying film formation and enzymatic activity is very sensitive to changes in the shape of the protein molecule. Furthermore, the optical properties of the films indicate that some unfolding of protein molecules frequently occurs. (See Chapter 15 for discussion of protein unfolding.) At one time it was thought that all of the orderly structure of a protein molecule is lost when it forms a film. The idea was that the polypeptide chain is totally unfolded, with the peptide backbone lying on the surface of the water, the polar side chains in the bulk liquid, and the nonpolar side chains projecting into the air. However, a great deal of evidence indicates that helical regions persist in the spread protein molecules, although the remainder of the molecule is probably unfolded, as just described. Thus, those protein molecules for which a large fraction of the amino acids are in a helical array suffer little or no change in structure when a film forms. The explanation of the unfolding of large parts of the nonhelical regions of the molecule caused by film formation may be the following. (We say "may be" because the mechanism has never been unambiguously identified.) The three-dimensional structure of a typical protein is maintained by polar bonds, such as hydrogen bonds and numerous nonpolar hydrophobic bonds. The hydrophobic bonds are a result of regions of the molecules attempting to escape from water (see Chapters 5 and 6). When a molecule is in a film, it will be oriented so that the nonpolar molecules are exposed to air. Since water is absent, there is nothing from which to escape, so that the strength of the hydrophobic bond approaches

---

*See Chapter 15 for the chemical formulae of side chains and peptide groups.

zero and the molecule partly unfolds. The unfolding process can also be thought of from the point of view of surface tension. Because a protein molecule is so large, it can be thought of as having a surface itself, and therefore, as having a surface tension. Its surface tension results from the existence of nonpolar amino acid side chains that cohere to one another better than they cohere to water molecules. Thus, the cohesive elements are drawn away from the surface of the molecule and into the interior; or, in other words, there is a pressure directed toward the molecule, tending to minimize its surface area. When a molecule is in a surface film, the side of the molecule in contact with the air does not experience a significant surface tension—that is, there is nothing to force it to hold onto itself, so it unfolds.

Protein Films at Oil-Water Interfaces

In biological systems, protein films usually occur at interfaces between the aqueous fluids and lipids rather than at an air-water interface. For this reason, protein films have been studied at the interface between water and various organic solvents, a common solvent being petroleum ether. At such an interface, a protein molecule has even less structure and is even more unfolded than at an air-water interface, because of the almost total breakdown of hydrophobic interactions on the organic side of the interface. This is because the polypeptide chain will tend to arrange itself with the charged regions in contact with the water phase and the nonpolar side chains in contact with the organic phase. Thus, the tendency for nonpolar side chains to cluster is absent.

An Important Use of Protein Films: Electron Microscopy of DNA by the Kleinschmidt Method

If two different types of molecules can adhere, spreading of a mixture of the two molecules on a liquid surface will usually yield a film containing both molecules. This is the basis of an important procedure, known as the Kleinschmidt method, for preparing DNA molecules for visualization by electron microscopy.

When the long strands of DNA molecules are dried on the supports used for holding a sample for viewing in an electron microscope, the DNA molecule collapses and forms what has been called a molecular puddle. Albrecht Kleinschmidt developed an extremely simple procedure by which the DNA molecule remains extended. A DNA solution is mixed with a solution of the protein cytochrome c. A droplet of the mixture is spread on a water surface. The DNA molecule interacts with the protein molecule and remains in the film. Thus, as the film spreads, the DNA molecule becomes extended. An electron microscope sample support is then touched to the surface of the spread film, which adheres to the surface of the support. When the support is dried, the DNA molecule remains extended, because it can resist the compacting caused by drying, since the protein film is rigid and strong. As an added bonus, the DNA molecule becomes coated with excess protein molecules so that it projects well above the surface of the protein film. Metal is then evaporated onto the film at a very low angle; the metal accumulates against the projecting protein-coated DNA molecule, producing a structure that strongly absorbs electrons and contrasts significantly with the film itself. This process is depicted in Figure 13-14.

A Practical Use of Monomolecular Films: Evaporation Retardation

When a liquid is totally covered with a stable film, the departure of molecules from the liquid to the vapor phase is mechanically retarded. This is of immense practical importance in arid areas, where water is stored in open reservoirs. For instance, in the western United States, approximately 16 million acre-feet of water are lost annually by evaporation. Hexadecyl alcohol films could retard this evaporation significantly, although this substance is not yet in widespread use. The problems involved in any such evaporation-prevention program are (1) that the film must heal ruptures caused by wind, waves, aquatic animals, boats (and so on); (2) it must not be biodegraded; and (3) it must not prevent adequate aeration of water and thereby interfere with aquatic life.

In the laboratory, the prevention of evaporation from water thermostats is important, since evaporation results in accelerated cooling of thermostats set above room temperature. In a thermostat in my own laboratory, temperature fluctuations were reduced from $\pm\ 0.006°C$ to $0.001°C$ when a thin film of silicone oil was placed on the water surface.

13-3 ADSORPTION

Adsorption to Solid Surfaces

Many substances, both liquid and gas, adsorb to solid surfaces. The amount of material adsorbed depends on (1) the cohesive forces between the adsorbent (the adsorbing substance) and the adsorbate (substance adsorbed); (2) the temperature; and (3) the concentration of the adsorbent and the adsorbate. The cohesive force depends upon both the chemical and physical

FIGURE 13-14 Preparation of DNA for electron microscopy by the Kleinschmidt method. [SOURCE: After *Physical Biochemistry*, First Edition, by David Freifelder. W.H. Freeman and Company Copyright © 1976.]

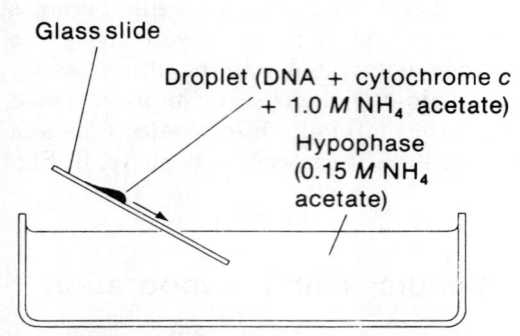

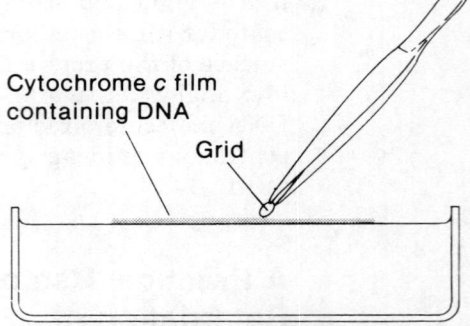

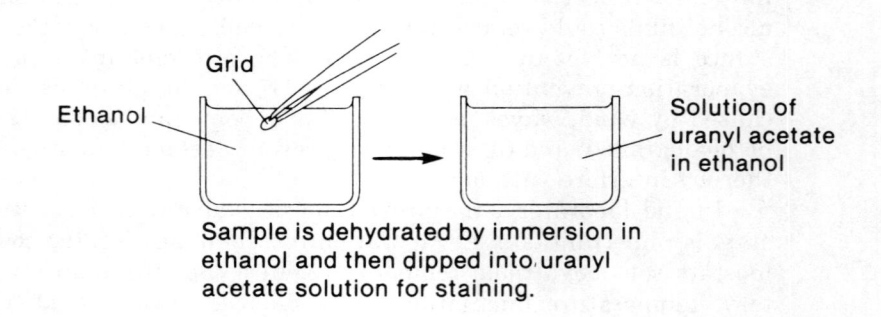

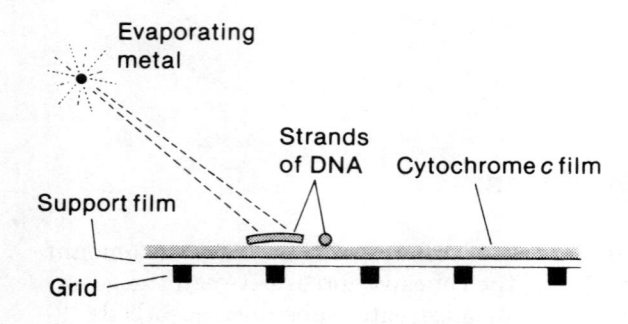

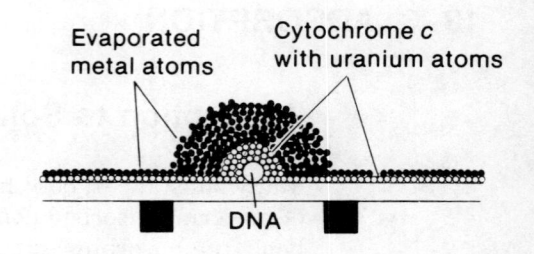

properties of the materials involved. For example, charcoal adsorbs ten times as much oxygen as it adsorbs helium at a pressure of one atmosphere and at 0°C. The number of molecules of a particular gas adsorbed onto a unit weight of a particular solid decreases with increasing temperature. This is because adsorption, like condensation, is an exothermic process ($\Delta H < 0$), so that raising the temperature favors the endothermic desorption process; alternatively, the effect of temperature can be thought of as the result of the fact that as temperature increases, the kinetic energy of an adsorbate molecule increases, and that this counteracts the cohesive forces. A related phenomenon is that if a large number of different gases is examined at a particular temperature and pressure, it is found that the fraction of molecules adsorbed rises with increasing boiling point of the gas. This simply indicates that the ability to adsorb is increased as the intermolecular cohesive force increases; and it also points out the similarity between adsorption and condensation.

As the concentration of the adsorbate molecules in the nonsolid phase increases, the fraction adsorbed also increases. For a gas, this means that a greater fraction of gas molecules adsorbs to a solid surface when the gas pressure increases. This results in part in the increased probability of a collision. An alternative and important explanation is that in many cases adsorption is *cooperative*. Two examples of cooperative binding are the following: (1) a free adsorbate molecule preferentially binds to a molecule that is already adsorbed, so that the former never comes in contact with the surface; and (2) an adsorbate molecule binds more tightly to an adsorption site adjacent to an adsorbed molecule than to a free site (see the section below on physisorption, in which these possibilities are examined in detail).

Types of Adsorption

There are two types of adsorption—physical adsorption or *physisorption* and chemical adsorption or *chemisorption*. These are depicted in Figure 13-15.

Physisorption is mainly the result of van der Waals forces between the molecules of the adsorbing surface (the adsorbent) and the adsorbate molecules. Because the adsorbate molecules themselves frequently cohere with one another by van der Waals forces, physisorption often results in the formation of multilayers of adsorbate molecules. Physisorption usually predominates at very low temperatures and occurs very rapidly. Furthermore, physisorption is reversible, in that the adsorbate molecules are easily removed by lowering the gas pressure or the concentration. Physisorption of gases presumably results from the same forces involved in condensation, and indeed the heat (enthalpy) of adsorption is usually near that of the heat of condensation.

In chemisorption a bond is usually formed that is stronger than a van

FIGURE 13-15 Examples of chemisorption (the adsorption of hydrogen by platinum) and physisorption (the adsorption of ammonia by copper). In chemisorption specific chemical bonds form at random sites on the surface. There is no tendency for adsorbed molecules to occupy adjacent sites, and adsorption is complete when the surface is covered. In physisorption, the binding is weaker and is frequently stabilized by lateral interactions. Since the adsorbing molecules can bind to one another, a second layer can begin to form before the first is complete. All surface sites may be filled but binding continues, owing to formation of a second and other layers. Squares represent molecules of the adsorbent; the circles are the molecules adsorbed.

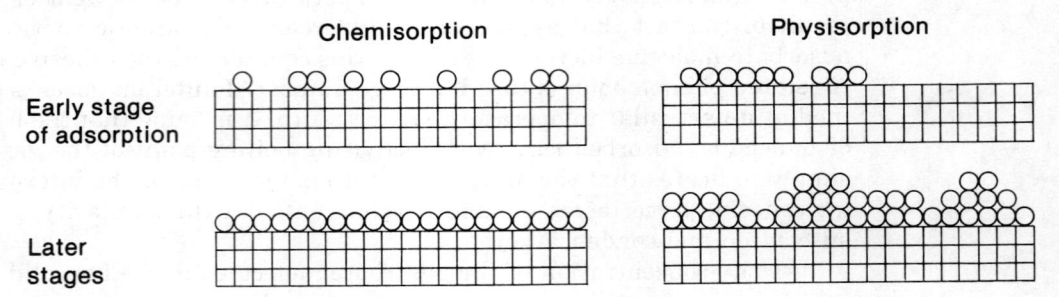

der Waals bond but it need not be a covalent bond. Chemisorption is distinguished qualitatively from physisorption in the following ways:

1. As is true of any chemical reaction, chemisorption can occur either rapidly or slowly. Since physisorption is a rapid process, slow adsorption suggests chemisorption. Rapid adsorption may be of either type.
2. The heat of adsorption for chemisorption is greater than that for physisorption and in the range found for chemical reactions.
3. There is often an activation energy for chemisorption.
4. In chemisorption, the binding is irreversible in the sense that desorption is difficult and often accompanied by chemical change. Quantitatively, chemisorption and physisorption can sometimes be distinguished by comparing experimental data to one of the common adsorption isotherms described in the following sections; for instance, if the data agree with the BET isotherm, binding is probably by means of physisorption, whereas agreement with the Langmuir isotherm often indicates chemisorption.

The line between physisorption and chemisorption is often not clear and sometimes a molecule might be adsorbed to a surface by both mechanisms. Also, it is not infrequent that a molecule adsorbs first by physisorption and then is chemisorbed.

A molecular distinction between the two types of adsorption that is frequently of value in thinking about the processes is that, because of the requirement for bonds between the adsorbent and the adsorbate molecules, chemisorption leads only to a monolayer. This is an important point to remember. Some examples of chemisorption are shown in Figure 13-16. The principal evidence supporting the covalent linkage of O_2 and C in panel (a) is that when oxygen is adsorbed to charcoal and then desorbed by

FIGURE 13-16 Adsorption of several gases to charcoal and metals. In panels (a), (b), and (d), covalent bonds are broken in the formation of the bonds to the surface; O_2 and H_2 molecules are cleaved in the binding to charcoal and platinum, respectively, and two hydrogen atoms are removed from ethylene. Often the C=C bond in ethylene is also broken and the addition of H_2 gas at high temperatures results in the release of methane. The bond between the C and W atoms in panel (c) is so strong that heating results in release of the compound WCO.

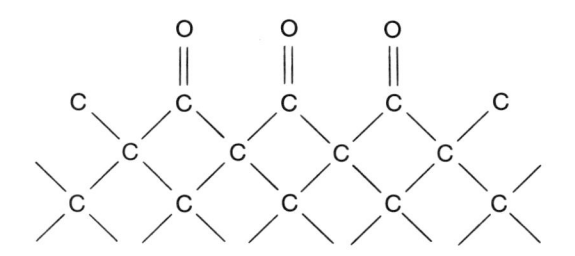

(a) Oxygen atoms adsorbed to charcoal

(b) Hydrogen atoms adsorbed to platinum

(c) Carbon monoxide adsorbed to tungsten

(d) Ethylene adsorbed to nickel

heating, both CO and CO_2 are obtained. Clearly, adsorption of O_2 has either required or led to the formation of a carbon-oxygen bond.

Adsorption Isotherms

Adsorption is usually measured in one of two complementary ways. In each an adsorbent surface is exposed either to a gas (the adsorbate) whose pressure is known or to a solution of known concentration of adsorbate. In one procedure, the number of moles of adsorbate bound per gram or per unit area of adsorbent is measured; in the other, the decrease in pressure or concentration is measured. The result of both methods is a relation between the amount adsorbed and the adsorbate concentration or pressure. The fraction adsorbed varies greatly with temperature, so that the measurement is always performed at constant temperature. A curve that describes the binding of adsorbate to adsorbent is called an *adsorption isotherm*.

Several mathematical forms have been proposed to describe the binding, some of which are a result only of empirical observations, and others which are in terms of specific models. We will examine some of these isotherms in the following sections.

A. The Langmuir Isotherm

The Langmuir isotherm is an important equation derived to explain the physisorption of a gas to a solid surface when only a monolayer forms.

In this description one assumes that the adsorption sites are independent in the sense that a molecule binds equally well to a site adjacent to a bound molecule as to a totally free site. Adsorption of a gas to a surface is then considered as a process in which condensation and evaporation occur simultaneously and at particular rates. The relative values of the rates define the degree of adsorption.

The original derivation by Irving Langmuir begins by considering a surface containing S sites of which S_1 are occupied and of which $S_0 = S - S_1$ are unoccupied. The rate of evaporation (the number of molecules leaving the unoccupied sites per unit time) is assumed to be proportional to S_1—that is, equal to $k_1 S_1$. The rate of condensation must be proportional to the number of empty sites and to the gas pressure P—that is, equal to $k_2 P S_0$. At equilibrium, the rates are equal, so that

$$k_1 S_1 = k_2 P S_0 = k_2 P (S - S_1) \ . \tag{17}$$

We then define

$$b = k_2 / k_1 \tag{18}$$

and the fraction θ of sites that are occupied,

$$\theta = S_1 / S \ , \tag{19}$$

which may also be thought of as the fraction of the surface area that is covered, if the surface is uniformly covered with binding sites.

Combining Equations 17, 18, and 19 yields

$$\theta = \frac{bP}{1 + bP} \ , \tag{20}$$

which is known as the *Langmuir adsorption isotherm*. This is a reasonable equation, in that the fraction of sites occupied approaches 1 as P becomes infinite (that is, as the concentration of gas molecules becomes infinite)

SEC. 13–3   ADSORPTION

and the fraction increases as the rate of condensation exceeds the rate of evaporation (that is, as b increases). An alternative formulation uses $\theta = V/V_m$, in which V is the volume adsorbed (at standard pressure and temperature) and V_m is the volume adsorbed when the surface is completely covered (that is, when the monolayer is complete). Equation 20 then becomes

$$V = V_m \left(\frac{bP}{1 + bP} \right), \qquad (21)$$

which is another form of the Langmuir adsorption isotherm. A variant of Equation 21 useful in data analysis is

$$\frac{P}{V} = \frac{1}{V_m b} + \frac{P}{V_m}, \qquad (22)$$

because a plot of P/V against P gives a straight line with slope $1/V_m$ and intercept $1/(V_m b)$, so that V_m and b can easily be measured. (See Chapter 16 for a discussion of the different ways to plot binding data.)

Some of Langmuir's own data for the adsorption of methane and nitrogen on mica are shown in Figure 13-17. If two gases adsorb to the same sites, as is thought to be the case for nitrogen and methane adsorbing to mica, the experimentally observed values of V_m for each gas should be the same. However, in the case of these two gases, the observed values differ

FIGURE 13-17 Adsorption of nitrogen and methane to mica. The values of V_m for nitrogen and methane are approximately 35 and 100 mm, respectively. [SOURCE: I. Langmuir, *J. Amer. Chem. Soc.* 40, 1361–1403 (1918). Copyright © 1918, The American Chemical Society.]

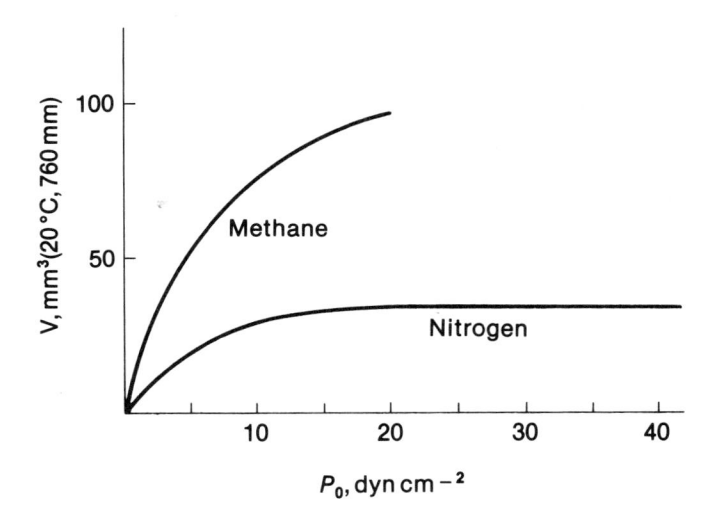

by a factor of 3, indicating either that there is physisorption in the system or that the Langmuir model does not apply. To explore this possibility, we examine the validity of five assumptions made in the derivation of Equation 20. These are the following:

1. The adsorbate behaves as an ideal gas in the gas phase. This is usually a good assumption at low pressure.

2. Adsorption is confined to a monolayer. This is true only if chemisorption occurs exclusively without any physisorption, which is often not the case.

3. All portions of the adsorbing surface have the same number of binding sites, and each of these has the same affinity for the adsorbate molecules. This is almost never true because of microscopic variation in surface structure (for example, pores and crystal planes often occur on surfaces).

4. There is no interaction between adsorbate molecules. This is rarely true, since virtually all molecules experience intermolecular attractive and repulsive forces.

5. The adsorbed molecules occupy fixed sites. This is true for chemisorbed molecules but physisorbed molecules drift around on the surface.

For the above reasons, the Langmuir isotherm is sometimes not a valid description of adsorption at a molecular level, so that a measured value of V_m is not always simply related to the number of adsorption sites per unit area. The principal problem is that a second layer starts to form before the first layer is completed. This phenomenon is dealt with in the BET theory described next.

B. The BET Isotherm

Adsorption such as that shown in Figure 13-18 is commonly observed. At high concentrations of adsorbate, this isotherm shows an increase in

FIGURE 13-18 Adsorption of nitrogen on a Al_2O_3-promoted iron catalyst. The solid circles are the experimental points. The solid curves are calculated from the BET isotherm, by using various values of n as indicated (see Equation 27) and values for V_m and c that allow the points obtained at low P to fall on the theoretical curve described by Equation 25. The experimental curve clearly indicates that multiple layers are present. [SOURCE: After S. Brunauer, P.H. Emmett, and E. Teller, *J. Amer. Chem. Soc.* 60, 314 (1938), with permission. Copyright © 1938, The American Chemical Society.]

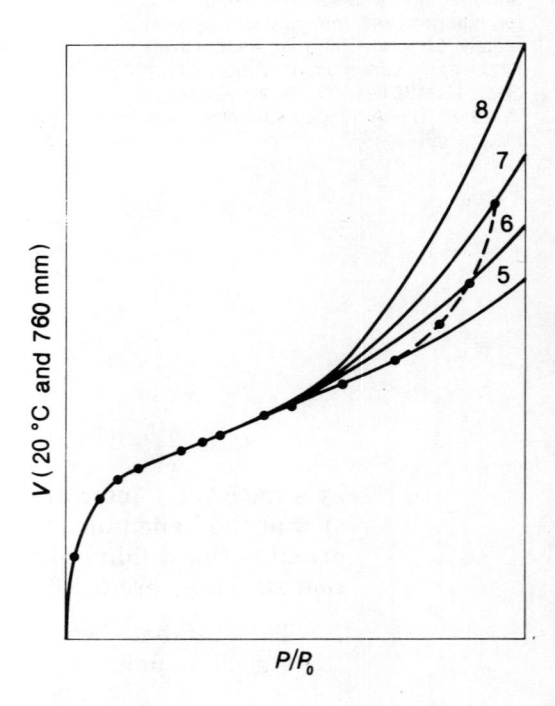

SEC. 13-3 ADSORPTION

binding; the rise is thought to represent the formation of a second layer of adsorbed molecules. This behavior is best explained by the model of *B*runauer, *E*mmett, and *T*eller, from which the name of the BET isotherm is derived.

The derivation begins by assuming that multiple layers form (and the number of layers may be infinite) and that the Langmuir model applies to each layer. Two further assumptions are made: (1) The heat of adsorption ΔH_{ads} for the first layer has a value determined by properties of the surface and of the adsorbate but for the second and all subsequent layers, it is just the heat of vaporization ΔH_{vap}; (2) Evaporation occurs only from an exposed surface. The physical picture of adsorption according to this model is shown in Figure 13-19. We shall not derive the BET equation but merely state it:

$$\frac{V}{V_m} = \frac{cx}{(1-x)[1+(c-1)x]}, \qquad (23)$$

in which $x = P/P_0$, P is the gas pressure, P_0 is the vapor pressure of the pure liquid adsorbate at the temperature of the measurement, and $c = e^{(\Delta H_{ads} - \Delta H_{vap})/RT}$. It should be noted that if $c \gg 1$, the equation becomes the Langmuir isotherm. This is expected, since when $\Delta H_{ads} \gg \Delta H_{vap}$, the attraction between the adsorbent surface and the adsorbate molecules greatly exceeds that between adsorbate molecules; that is, adsorption is primarily chemisorption and only a monolayer forms.

For the purpose of plotting data, Equation 23 is rearranged to

$$\frac{x}{V(1-x)} = \frac{1}{cV_m} + \frac{(c-1)x}{cV_m}, \qquad (24)$$

in which case c and V_m can be obtained from the intercept and slope of a plot of $x/[V(1-x)]$ versus x.

The BET isotherm has been successful in evaluating V_m. Values obtained are generally consistent with those obtained from other types of measurements.

FIGURE 13-19 Binding of molecules (circles) to a surface according to the BET theory. The surface is covered with patches of molecules one or more layers thick; in the early stages of binding, some regions remain uncovered even when multiple layers exist elsewhere. The theory assumes (1) that the Langmuir equation applies to each layer, (2) that evaporation and condensation can occur in any exposed layer, and (3) that the heat of adsorption is the heat of condensation, except for the layer bound directly to the surface for which there is a special value of the heat of adsorption.

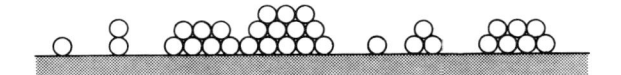

Experimental values of V_m are used to calculate the *specific surface area*, Σ—that is, the area per gram of adsorbent available for binding an adsorbate. This is

$$\Sigma = V_m \sigma° \rho, \tag{25}$$

in which $\sigma°$ is the area per adsorption site (cm² per molecule) and ρ is the number of gas molecules per cm³ at the temperature used in the measurement.* The value of $\sigma°$ is usually taken as the area of the adsorbate molecule. When this is done, it is found that the specific surface area for a particular adsorbent is independent of the nature of the adsorbate.

EXAMPLE 13–E A gas is allowed to adsorb to 6.6 g of a solid at 0°C. Calculate the volume required to form a monolayer, the enthalpy of adsorption, and the area of 1 g of the solid.

The volumes of the gas adsorbed at several pressures are the following:

P, mm Hg:	52.9	85.1	137.2	200.5	228
V, cm³:	2.93	3.83	4.85	5.90	8.08

At 0°C the vapor pressure of the gas is 778 mm Hg and $\Delta H_{vap} = 21.33$ kJ mol⁻¹. The area of each gas molecule is 32 Å².

We use Equation 24. First, $P/P_0 = x$ and $[x/(1-x)](1/V)$ are calculated. Each of these values is plotted against x. From the graph obtained, we measure the slope, which is 0.183, and the y intercept, which is 0.0129. Since $1/cV_m = 0.0129$ and $(c-1)/cV_m = 0.183$, then $c = 15.22$ and $V_m = 5.10$ cm³, which is the volume required to form a monolayer. Since $c = e^{(\Delta H_{ads} - \Delta H_{vap})/RT}$ and $\Delta H_{vap} = 21.33$ kJ mol⁻¹, $\Delta H_{ads} = RT \ln c + \Delta H_{vap} = (8.314 \times 10^{-3})(273) \ln 15.2 + 21.33 = 27.61$ kJ mol⁻¹. Equation 25 is then used to calculate the area. We convert V_m to the volume adsorbed per gram of adsorbent or $5.10/6.6 = 0.773$ cm³ g⁻¹ and multiply by the area per molecule (32.1 Å² = 3.21×10^{-15} cm²) and the number of gas molecules per cm³ at 0°C and at 1 atm pressure ($6 \times 10^{23}/22{,}400$) to obtain 6.65×10^4 cm².

At high pressures there are significant departures from the BET isotherm. This can be corrected by dropping the assumption that there is an infinite number of layers. If the equation is rederived for a maximum of n layers, one obtains

$$V = \frac{V_m c x}{1-x} \frac{1-(n+1)x^n + nx^{n+1}}{1+(c-1)x - cx^{n+1}}, \tag{26}$$

which gives a better fit to the data if an appropriate value of n is chosen.

---

*Usually the values of the pressure and the volume adsorbed are corrected to 1 atm and 0°C by using the ideal gas law before calculating V_m, even though another temperature is used in the measurement. Thus ρ is usually $6 \times 10^{23}/22{,}400 = 2.68 \times 10^{19}$ atom cm⁻³.

C. Other Isotherms

Models for adsorption other than the Langmuir and BET theories have been proposed. Each of these yields isotherms that agree satisfactorily with experimental data over a particular pressure range. The best known alternatives are the Polanyi potential theory and the polarization model. It must be emphasized, though, that agreement between an isotherm equation and experimental data is *not* sufficient proof of the validity of the model. In fact, detailed mathematical analyses of each of these theories lead to predictions that are definitely in violation of experimental fact. It is probably fair to say that at present a detailed molecular description of adsorption that applies to all cases has not yet been developed. It is felt, however, by some researchers in the field of surface physical chemistry, that the BET isotherm provides accurate values of V_m and of the heat of adsorption of the first layer.

We have so far been considering adsorbents having a single type of adsorption site. It is important to realize, though, that some solids are porous and some have different classes of adsorption sites (the so-called heterogeneous surface). For such solids, experimental data can depart drastically from the equations that have been presented. At the present time there is no quantitative theory available that can describe adsorbents of this type.

Adsorption of Materials in Solution to Solid Adsorbents: Langmuir and Freundlich Isotherms

In biochemical systems, one is usually more concerned with adsorption of solutes to an adsorbent that is in contact with a solution. This is of concern when using physicochemical techniques such as chromatography and when considering the properties of such important structures as cell membranes.

The physical picture of adsorption of solute molecules from a dilute solution is similar to that of chemisorption of gas molecules to a solid surface. That is, a monolayer of adsorbate molecules forms on the solid surface. This is because, in solution, the interactions between solute and solute are weaker than those between solute and solvent (a necessary condition for the solute to be in solution). That is, a multilayer is not likely, because solvent molecules will preferentially interact with adsorbed solute molecules, thus preventing binding of a second layer.* This type of monolayer adsorption differs from simple chemisorption of gases, though, in that the heat of adsorption from solution (which is the quantity evolved in transferring a solute molecule from the bulk of the solution to the surface) is small and comparable to the heat of solution rather than to the chemical bond

---

*This is actually too strong a statement because there are many instances known in which multilayers form.

enthalpies involved in chemisorption. Furthermore, if the concentration of the solute is decreased, desorption (loss of bound molecules from the adsorbent) occurs. It is likely that adsorption from solution is mainly physisorption, although usually without the formation of multiple layers.

In addition to the formalism describing binding of gases, Langmuir proposed a related model capable of accounting for some of the properties of adsorption from solution. The primary difference from gas adsorption is the proposal that the surface contains adsorption sites for both solvent and solute molecules and that these compete for the binding sites of the adsorbent. However, the resulting isotherm is similar to Equation 20 for the adsorption of a gas to a solid surface. That is, if n' is the number of moles of solute adsorbed per unit weight of adsorbent, and n is the number of moles of adsorption sites per unit weight, then

$$n' = nbc/(1 + bc), \tag{27}$$

in which c is the concentration of solute in solution and

$$b = \frac{1}{c} \frac{X_{solute}}{X_{solvent}}, \tag{28}$$

in which X_{solute} and $X_{solvent}$ are the mole fractions of the adsorbed solute and solvent, respectively. Thus, the magnitude of the product bc tells us whether the solute or the solvent adsorbs with greater affinity. The fraction θ of the surface occupied is

$$\theta = \frac{n'}{n}, \tag{29}$$

so that

$$\theta = \frac{bc}{1 + bc}, \tag{30}$$

which is identical in form to Equation 20.

A more practical form of Equation 27 is

$$\frac{c}{x} = \frac{1}{mb} + \frac{c}{m}, \tag{31}$$

in which c is the concentration of the solute, x is the mass of adsorbate, and m is the mass of adsorbent. A plot of c/x versus c gives a straight line of slope $1/m$ and intercept $1/mb$.

EXAMPLE 13–F A transparent solid sheet is immersed in a solution containing various concentrations of a red dye. After a fixed period of time the solid sheet is removed and from the intensity of the red color the mass of the dye adsorbed to the sheet is measured.

What fraction of the surface is covered in a 0.002-M solution and, at that concentration, what is the ratio of the amount of solute and solvent that is adsorbed?

SEC. 13-3 ADSORPTION

The data are the following:

Dye concentration, mM:	0.1	0.8	3.0	8.0	10.0
Mass of bound dye, μmol:	47.9	112.0	130.7	135.7	136.4

We calculate the ratio of the concentration c (in moles) of the solution and the number of moles x adsorbed—that is, c/x—and plot this against each value of x. By Equation 31, the slope of the resulting line is 7.19 = $1/m$, so that 139 μmol of surface material is available as an adsorbent. (Note that this is not the mass of the sheet itself but only the mass of material on the surface of the sheet.) The y intercept is 1.369 mol, so that b = 9.85×10^3. The fraction θ of the surface that is covered in a 0.002-M solution is calculated from Equation 30 with this value of b. Thus, θ = $(9.85 \times 10^3)(2 \times 10^{-3})/[1 + (9.85 \times 10^3)(2 \times 10^3)] = 0.95$. Equation 28 provides the relative amount of solute and solvent that is bound, or, bc = $(9.85 \times 10^3)(2 \times 10^{-3}) = 19.7$.

Equation 30 satisfactorily describes many adsorbent-adsorbate systems. However, there are numerous examples for which this equation does not apply. In many systems a strictly empirical equation called the *Freundlich isotherm* is applicable. This equation is

$$\log\left(\frac{x}{m}\right) = \frac{1}{n} \log c + a, \tag{32}$$

in which n is an empirical constant greater than 1, a is a constant that varies with temperature, and x and m are as in Equation 31. Since there is no physical model for adsorption that obeys Equation 31, the discovery that the data satisfy this equation is at present not particularly informative.

Let us consider at this point what can be learned from examining isotherms. Probably the main thing that is determined is whether binding occurs in a single layer or in a multilayer. For example, strict adherence to a Langmuir isotherm implies a monolayer, whereas if the data fit the BET isotherm better, a multilayer has certainly formed. This information usually enables us to distinguish chemisorption (which generally yields a monolayer) from physisorption, although the distinction is rarely completely unambiguous. When a Langmuir isotherm is followed, some information comes from the value of the parameter b—that is, it tells us whether the solvent or solute molecules adsorb more strongly to the surface in question. At the present time and development of the theory of adsorption, this is the extent of the knowledge gained from detailed examination of isotherms. Of greater significance are the many qualitative observations that have been made in binding studies. The more important findings are the following:

1. A polar adsorbent will preferentially adsorb the more polar component of a solution that is, in composition, either (polar solute : nonpolar

solvent) or (nonpolar solute : polar solvent). Similarly, a nonpolar adsorbent preferentially adsorbs the less polar component.

2. The degree of adsorption of a homologous series of organic substances in aqueous solution to a nonpolar adsorbent increases with the number of carbon atoms in the series. This is shown in Figure 13-20(a) and is known as Traube's rule. Similarly, adsorption of the organic molecules to a polar adsorbent decreases with increasing number of carbon atoms (Figure 13-20(b)). An important consequence of Traube's rule is that, as the number of carbon atoms increases, lower concentrations of solute are needed to get an equivalent fraction of molecules adsorbed.

FIGURE 13-20 Traube's rule. In panel (a) fatty acids in aqueous solution are adsorbed to charcoal, a nonpolar adsorbent. For a given concentration the amount of adsorption increases as the length of the nonpolar side chain increases. In panel (b) fatty acids in toluene are adsorbed to silica gel, a polar adsorbent. In this case, adsorption decreases as the length of the carbon chain increases. [SOURCE: Data from F.E. Bartell and Y. Fu, *J. Phys. Chem.* 33, 676 (1929).]

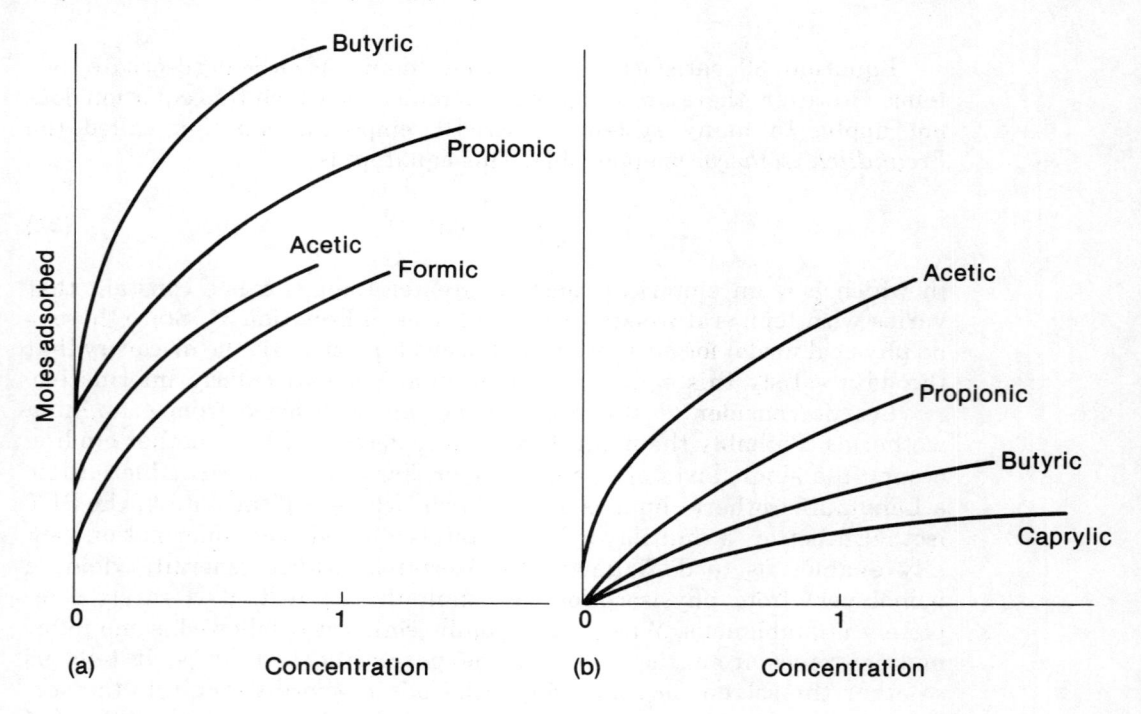

3. The less soluble a substance is, the more strongly it will adsorb. This is of course a result of the weak attraction between an insoluble substance and solvent molecules. In a homologous series, for example, of fatty acids, it is often found that if the number of grams adsorbed per unit mass of adsorbent is plotted against c/c_{sat} in which c is the concen-

tration used and c_{sat} is the concentration of a saturated solution, the curves obtained for each substance are superimposable. This effect is most easily thought of as a partitioning of the solute between the bulk solution phase and the adsorbed layer phase—that is, the adsorbate is found where it prefers to be.

Adsorption of Polymers

An extremely important topic in biochemistry and molecular biology is the adsorption of polymers (such as proteins) to solid surfaces or to interfaces. Unfortunately, adsorption of polymers is at present a poorly understood subject. There are two major complications. First, experimentally, the data for adsorption of polymers obeys a Langmuir isotherm. This usually means that the surface is saturated when a monolayer forms. However, several independent types of measurements indicate that multilayers of polymers form. A second peculiarity of the adsorption of polymers is that the fraction adsorbed often increases with rising temperature, which is not the case for small molecules.

Additional problems arise in understanding how biological polymers adsorb. The most important are the following: (1) biopolymers usually have a large number of different chemical groups that can engage in binding to a surface; (2) a change in the three-dimensional shape of the polymer often occurs upon binding; (3) the tendency to adsorb depends significantly on the nature of the solvent, the ionic strength, and the pH. Unfortunately, at the present time, there have been few investigations of the adsorption of biopolymers.

13-4 PRACTICAL ASPECTS OF ADSORPTION

Losses of Polymers by Adsorption to Solid Surfaces

In living cells adsorption is an important phenomenon that plays a role in enzyme reactions, macromolecular synthesis, and regulation of the expression of numerous genes, to name just a few processes. It seems that biopolymers, such as proteins and nucleic acids, and complex structures, such as membranes, have been designed to be powerful adsorbents—they are generally sticky. The tendency of these molecules to stick to surfaces plagues biochemists who often find that as certain polymers become more and more pure, they often seem to become lost; what in fact has happened is that they have adsorbed to the glassware used in the purification process or to the container used to store the purified molecules. Various procedures can be used to minimize or eliminate these losses but no general principles can be laid down because a particularly effective procedure frequently applies only to a few types of macromolecules. For example, treatment of the glassware with silicone compounds reduces adsorption of many protein molecules but increases adsorption of other proteins; DNA molecules adsorb particularly strongly to silicone-treated surfaces. The use of plastic containers is frequently of value, yet protein molecules having numerous hydrophobic regions may bind tightly to these surfaces.

In the study of nucleic acids there is an especially striking example of an artifact resulting from adsorption. In a series of experiments to determine whether both double- and single-stranded DNA can penetrate bacteria, a mixture of the two forms was separated by equilibrium centrifugation in a CsCl density gradient (see Chapter 15), using a plastic centrifuge tube. After centrifugation the material in the tube was divided into many samples that differed according to the density of the CsCl solution. This method was used because it was known that single- and double-stranded DNA molecules would be found in samples having different densities. The individual fractions of the centrifuge tube were collected into small glass tubes and then bacteria were added to each tube. Appropriate tests were carried out to detect uptake of the DNA by the bacteria and it was observed that only double-stranded DNA was found inside the bacteria. What was not noticed was that the tubes originally containing the single-stranded DNA no longer had any free single-stranded DNA molecules. Many years later it was found that if the tubes into which the various density fractions are collected contains the bacteria at the outset, both double- and single-stranded DNA are found inside the bacteria. This is because, in the first protocol, the single-stranded DNA molecules adsorbed to the glass collection tube so rapidly that the molecules never had a chance to come in contact with the bacteria.

Losses by adsorption are also serious when nitrocellulose filters are used. Problems arise when collecting bacteria from liquid growth media and when freeing suspensions of bacteriophages (bacterial viruses) from extraneous particles. In the first case a suspension of bacteria is filtered and the bacteria are retained on the filter. The bacteria are then collected by washing them off of the filter. When the number of bacteria filtered is less than 10^7 bacteria per square centimeter of filter, it is often found that nearly all of the bacteria do not wash off. In the second example, a suspension of bacteriophages whose cross-sectional area is much less than that of the pores of the filter is passed through the filter, yet most of the phages do not appear in the filtrate. In both cases the bacteria and the phages have adsorbed so tightly to the filter that they cannot be easily removed. Fortunately, the problem is alleviated simply by passing a solution of bovine serum albumin (a protein) through the filter before filtration of the bacteria or the phages. The molecules of bovine serum albumin also adsorb to the nitrocellulose filter and saturate the sites capable of adsorbing the bacteria and the phages.

Uses of Adsorption in the Laboratory

Adsorption is the basis of numerous analytical and separation techniques in common use in the laboratory. An important example is *adsorption chromatography,* in which various molecules are allowed to adsorb to an adsorbent under particular conditions of pH, ionic composition, solvent, temperature, and so forth. After the molecules have adsorbed, the adsorbent is washed ("eluted") while varying one of these factors, for example, Na^+ ion concentration. Often each molecular species desorbs at a different concentration in the elution procedure and the molecules are thereby separated. This method is discussed in greater detail in references given at the end of the chapter.

The ability of a molecule to adsorb specifically to nitrocellulose filters is frequently used as an analytical tool. For instance, DNA molecules usually do not bind to nitrocellulose filters, although many protein molecules do. This can be used to detect a protein molecule that binds to DNA. This is done as follows. Radioactive DNA is added to a mixture of proteins that possibly contains the protein of interest and the mixture is passed through a nitrocellulose filter. Free DNA is not retained in the filter, but if and only if the particular DNA-binding protein is present, the radioactive DNA molecules will bind to the filter. Thus, collection of radioactivity on the filter is an assay for the presence of a DNA-binding protein.

Utility of an Analysis of Adsorption of Gases: Catalysis and Gas Chromatography

Analysis of adsorption of gases to solid surfaces is essential to an understanding of the mechanism of catalysis—for example, the catalysis of hydrogenation by metallic platinum. This is described in some detail in Chapter 12.

Gas chromatography is a powerful technique for separating substances, which relies on the ability of gases to adsorb specifically to surfaces. The substances may be gaseous or, alternatively, solid or liquid, as long as volatilization is possible without degradation or chemical change. In this method, a sample of a gaseous mixture is passed slowly through a thin, very long tube (often several hundred meters long) that is coated internally with an adsorbent. The gas molecules adsorb, but each molecular type adsorbs with a characteristic strength. Desorption occurs behind the moving sample because of the equilibrium between bound and unbound molecules. The gas flows at a constant rate but the rate of movement of a particular chemical species is determined by how much its motion is impeded by adsorption. Thus, each molecular type is slowed down by a characteristic amount so that the different molecules emerge from the tube at different times. The composition of the emerging gas can be measured as a function of time so that the number of molecular species can be determined. Specific physical and chemical tests enable one to identify the species as they emerge. Alternatively, fixed volumes of gas can be collected successively as a function of time and allowed to condense so that the components of the mixture can be isolated. Gas chromatography is a very sensitive method and is widely used in the analysis of mixtures isolated from cells. It is especially valuable in the separation of lipids, fatty acids, and a variety of small molecules. It is currently being used to identify the hundreds of esters in wine that give each wine its characteristic flavor and bouquet.

13–5 MICELLES AND THE BILAYER MODEL OF MEMBRANE STRUCTURE

Long-chain, fatty acid molecules consist of a long hydrophobic portion and a small terminal polar region (thus they are amphipathic molecules). In the absence of the terminal carboxyl group, the hydrocarbon portion of the molecules would not be soluble in water at all. As we have mentioned earlier, if a very small amount of such a molecule is added to water, the molecules minimize their contact with water by forming a surface layer at the air-liquid interface with the polar region in contact with the water and the hydrophobic part facing the air. If larger amounts of fatty acids are added so that there are more molecules than can be contained in a monomolecular surface film, the excess molecules remain in the bulk of the solution.

When the nonpolar chain is a large fraction of the molecule, solutions of amphipathic molecules may have special properties owing to the great insolubility of the hydrophobic region. Clearly there must be a competition between the hydrophobic portion, which attempts to create a separate phase (that is, an immiscible liquid or a solid precipitate), and the polar region, which tends to make the molecule soluble. Consider two fatty acid molecules approaching one another in solution. The nonpolar regions of each molecule will attract one another by means of van der Waals forces in an attempt to escape from water. The polar regions will attract water molecules but because the polar regions of each have like charges, they will repel one another. With any pair of molecules the repulsive force is usually stronger than the attractive force, so that the pair does not remain together. As the number of molecules that chance to come together increases, the difference between the repulsive and attractive forces becomes smaller. Ultimately, a number is reached for which a closed surface can form (the necessary concentration is different for each type of amphipathic molecule). For a certain class of amphipathic molecules, the molecules form a spherical array as shown in Figure 13-21. Such a structure in which the hydrophobic region is encased in a polar shell is called a *micelle*. To be stable, a micelle must contain a particular and relatively large (50–100) number of amphipathic molecules. Furthermore, since the micelle is unstable below this number, the formation of the micelle is not a sequential

FIGURE 13-21 A typical spherical micelle shown in two dimensions. The inside of the sphere is basically a liquid hydrocarbon.

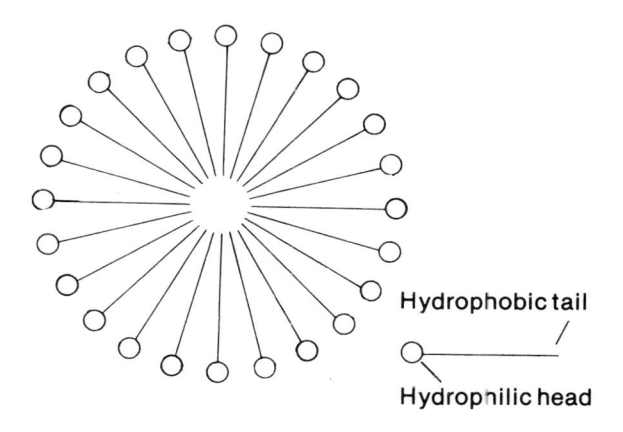

Hydrophobic tail

Hydrophilic head

process (addition of one molecule at a time) but a cooperative process (that is, all molecules join simultaneously). One might expect, then, that micelles only exist above a certain concentration of monomer, and indeed this is the case. This concentration is called the *critical micelle concentration* (cmc). At this concentration various properties of the solution change discontinuously as the solution changes from a uniform fluid to a liquid containing small nonpolar units that may be thought of as submicroscopic droplets of an organic liquid encased in a charged shell. Figure 13-22 shows the change in surface tension, osmotic pressure, and conductivity over a range of concentrations of sodium dodecyl sulfate (SDS).*

Note that each property changes dramatically near the cmc. The presence of micelles above this concentration range is indicated in two ways. First, above the cmc there are units that move rapidly if the sample is placed in an electric field; this is of course due to the charged groups on the surface of the micelle. Second, certain organic dyes that are relatively insoluble in water become very soluble above the cmc. This is because they dissolve in the hydrocarbon droplet within the micelle. For example, orange OT, a dye, is only sparingly soluble in water, yet if SDS is present at a concentration above the cmc of SDS, the solution turns deep red; furthermore, if the solution is placed in an electric field, the color moves with the micelles.**

---

*The decrease in conductivity occurs for the following reason. If there are 100 molecules of sodium dodecyl sulfate, there are 100 cations and 100 anions. If a micelle forms consisting of the 100 anions, the strong negative charge on the surface of the micellar sphere will bind many of the Na^+ ions. Suppose 60 Na^+ ions are bound. There are then 40 free Na^+ ions and one micellar unit having a charge of -40. The conductivity is determined by the *number* of ions and when the micelle forms, there are only 41 ions (40 + ions and 1 − ion) instead of the 200 that existed before formation of the micelle. The osmotic pressure also decreases because the effective molarity of the solution is decreased inasmuch as the micelle molarity is very low.

**See Chapter 15 for a discussion of this motion, which is called electrophoresis.

FIGURE 13-22 The effect of micelle formation in solutions of sodium dodecyl sulfate (SDS) on various physical properties. The shaded region is the range of concentration in which micelles begin to form.

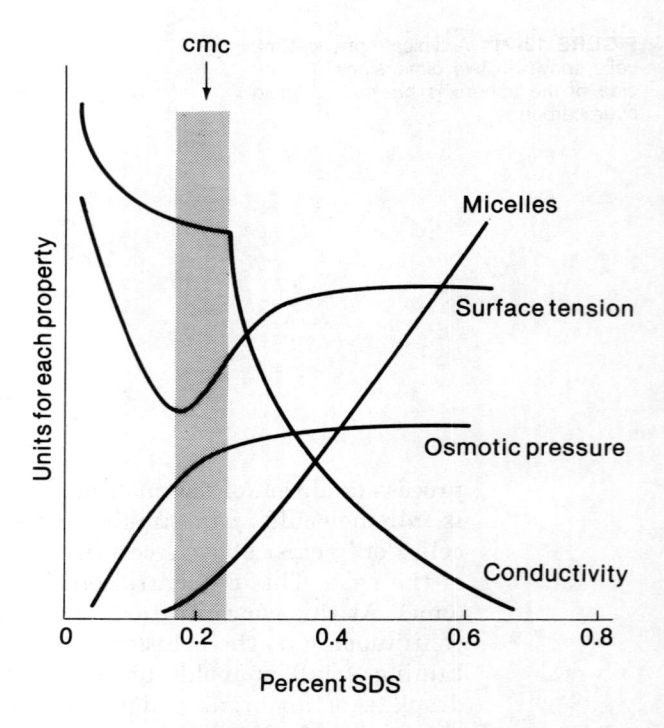

For the following discussion it is important to remember that

a requirement for formation of a micelle is not only an attractive force between the hydrophobic portion of each molecule but also a repulsive force between the polar regions.

This is because if the repulsive force were absent, the molecules would simply form a single separate phase.

We wish to consider how the size and properties of an amphipathic molecule determine the size and shape of a micelle. A spherical micelle consists of the array of molecules shown in Figure 13-21. The radius of the sphere is roughly equal to the length of the molecule. From the known dimensions of C-C bonds the length l of the extended molecule is

$$l = 1.5 + 1.265\, n_C, \qquad \text{in Å}, \tag{33}$$

in which n_C is the number of carbon atoms in the hydrocarbon chain. Simple geometric considerations that take into account the width of the chains indicate that the volume V of the sphere is

$$V = (27.4 + 26.9\, n_C)m, \qquad \text{in Å}^3, \tag{34}$$

in which m is the number of chains per micelle (usually 50 to 100). There is nothing in these simple geometric terms that tells anything about the stability of the micelle—that is, whether a spherical micelle can form for any value of n_C and m. To examine this question, one must remember that

SEC. 13-5   MICELLES AND BILAYER MEMBRANES

the polar heads must mutually repel if a micelle is to form and that the relative value of the repulsive and attractive forces are important for micelle stability. From simple geometry, as n_C increases and the sphere enlarges, the separation between the polar heads on the surface must increase. Since the repulsive force decreases with increasing distance, the stability of the spherical micelle must also decrease. Thus the size of the polar head must be a critical parameter in determining stability, so that in order to maintain a sphere, the polar head would have to increase in size as n_C increases. There is a second problem with large values of n_C—namely, that the outer regions of the hydrocarbon chain become farther apart and the very short-range van der Waals attractive force decreases. One might expect, then, that as n_C increases, either micelles have a lower probability of existence or the shape of the micelle must change to avoid the geometric problems just described. The first alternative seems unlikely since, if total disaggregation were to occur, the long hydrophobic chain would be forced to come in contact with water, which is exactly what it does not want to do. Thus the second alternative seems more likely. A geometric configuration that maintains close contact between the chains is a flattened ellipsoid (Figure 13-23(a)). Note that the geometric problem has been solved by no longer requiring that the tips of the hydrophobic tails meet at a single point (at the center of a sphere) but instead meet on the

FIGURE 13-23 Various forms of micelles. In panel (a) the hydrocarbon chain is so long that in order to maintain contact between the hydrophilic heads, an ellipsoid forms. Panel (b) is the limiting case of the ellipsoid— namely, the lipid bilayer.

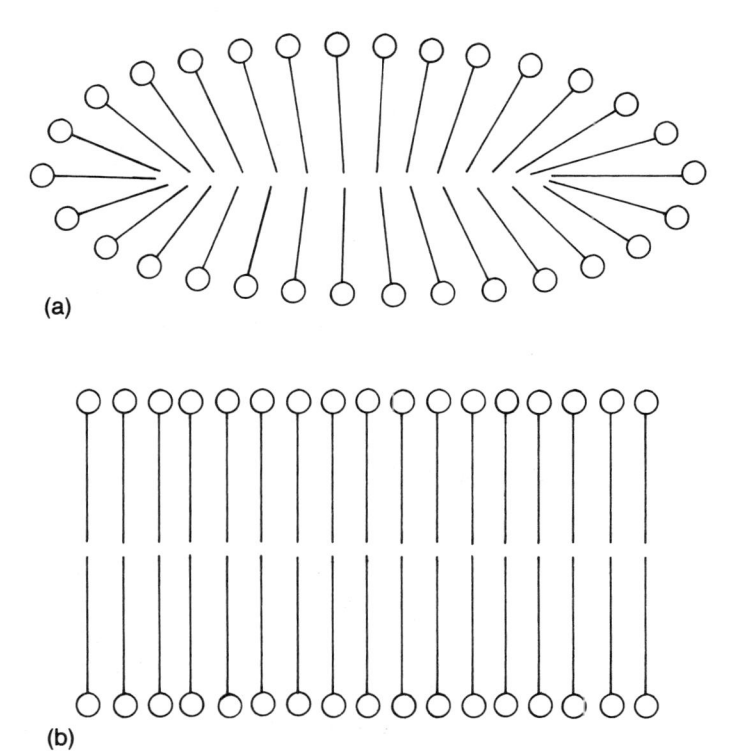

axis of the ellipsoid. As n_C gets increasingly larger, the long axis of the ellipsoid becomes greater but the thickness (i.e., the length of the short axis) is always twice the length of a single chain. The limit of this structure is the *lipid bilayer* (Figure 13-23(b)), the basic structure of biological membranes proposed by James Danielli and Hugh Davson.

This arrangement is of course not constrained to fatty acids but applies to any amphipathic molecule having a long hydrophobic portion and a terminal polar group, such as a detergent or the phospholipids found in biological membranes. It is important to realize that a lipid bilayer with external charged groups will not form at an air-liquid interface but only at the interface between two polar liquids such as two aqueous solutions.

Two variants of the lipid bilayer are of some interest—*soap bubbles* and *liposomes*. In a soap bubble, the nonpolar tails face the air and the polar groups are within the membrane (Figure 13-24). In a liposome, the lipid bilayer is not an infinite plane but folds back on itself to make a closed, hollow structure whose shell is the lipid bilayer and which contains an aqueous solution. Liposomes are models of cell membranes and are widely studied.

Although it is beyond the scope of this book, the current model of biological membranes, the Fluid Mosaic Model of Jonathan Singer and Garth Nicholson, deserves to be mentioned briefly (Figure 13-25). Biological membranes contain numerous protein molecules that are also amphipathic. Thus, they should be situated with the hydrophobic portion within the membrane and the hydrophilic portion projecting into the aqueous phase. These proteins make up the so-called active sites of the membranes. Some of the proteins contain transient pores through which small molecules can pass, some (for example, ion pumps) are capable of conformational changes that can result in the transport of ions, and some do not pass entirely through the membrane and are presumably binding sites for various regulatory compounds. Numerous studies have shown that the lipid bilayer remains highly fluid at ordinary temperatures, so that the proteins drift laterally in the membrane. It is of current interest that several recent experiments have suggested that the fluidity of the membrane is much greater in a cancer cell than in a normal cell.

13-6 DETERGENTS

A detergent or a soap is essentially no different from the kinds of molecules we have been discussing—that is, it consists of a hydrocarbon chain and a polar group. However, a molecule is usually not considered to be a detergent unless it has cleaning power. What this means is simply that it can remove a nonpolar substance from a solid surface that one wishes to clean. The detailed mechanism of detergency is very complicated and involves numerous factors, but can be thought of simply (but only partially

SEC. 13-6 DETERGENTS 461

FIGURE 13-24 Arrangement of amphipathic molecules in a soap bubble (polar groups within the film) and a liposome (hydrophobic groups within the film).

(a) Soap bubble

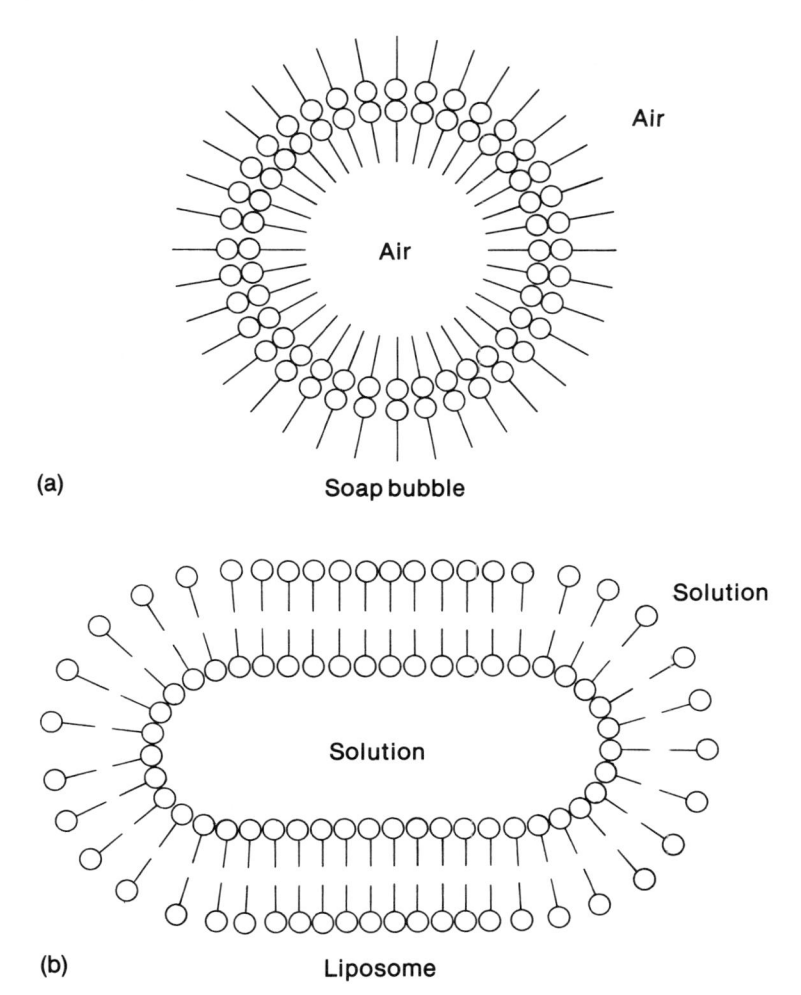

(b) Liposome

Figure 13-25 Membrane structure according to the Fluid Mosaic Model. The shaded portion of each protein represents the hydrophobic region.

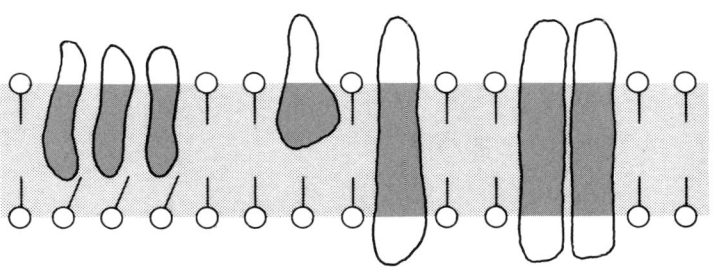

correctly) to occur in one of two ways: (1) solubilization, or (2) displacement of the unwanted substance. Solubilization occurs by an interaction between the nonpolar "dirt" and the nonpolar region of the detergent molecule. The dirt then acquires the charge of the detergent and becomes soluble. In displacement, the detergent binds more tightly to the surface than does the dirt. Both mechanisms are important in cleaning fabrics.

Detergents are of three types—anionic, cationic, and nonionic (neutral). The anionic detergents are the traditional soaps or salts of fatty acid. Cationic detergents are generally long-chain quaternary amines and amine salts; these are commonly used as emulsifying or dispersing agents. The nonionic detergents are also useful cleansing agents and are typically polyether alcohols, alkyl-aryl polyether alcohols, and amides.

The dispersing properties of detergents are of great use in biological experiments—in the solubilization of membranes and in the separation of complex proteins into the individual subunits. Many detergents are capable of breaking open bacteria. The detergents, when in great molar excess, render biological lipids so soluble that there is no longer any attractive force to stabilize the lipid bilayer, which then falls apart. The bacterial membrane is thereby removed and the cellular contents spill out into the suspending medium. Ionic detergents are the most disruptive, since they have an additional effect on the protein matrix of other layers of the bacterial cell wall. The neutral detergents tend to yield very large, completely permeable sacs that retain large molecules, such as the bacterial DNA, which cannot get through the newly created pores.

Protein aggregates are held together by a combination of ionic interactions and hydrophobic forces between nonpolar regions on the surface of the protein molecule. In the presence of a detergent several changes occur. First, the detergent molecules can bind to the nonpolar regions, thereby increasing their solubility and eliminating their tendency for mutual aggregation. This can occur since the protein aggregate (or oligomer) and the monomer are in equilibrium and the detergent simply stabilizes the monomer. Second, binding of detergent molecules can alter the shape of the monomer. The shape of most proteins depends in part on van der Waals attractions between nonpolar amino acid side chains buried within the three-dimensional structure. Since the configuration of a protein molecule fluctuates slightly in time (with an equilibrium that strongly favors a compact structure at ordinary temperatures), a detergent molecule can penetrate the structure and cause gross structural changes. Thus, the biological activity of most proteins is lost when detergents are present.

13–7 FLOTATION

If a liquid is beaten or stirred violently, air bubbles can form. Since creating the surface of the bubble requires an increase in surface free energy,

SEC. 13-7 FLOTATION

the formation of the bubble is a direct consequence of the work done on the liquid by the violent agitation. Since a liquid always tends to assume a shape that minimizes its surface area, the existence of bubbles that increase the surface area is an unstable situation. Many bubbles do collapse rapidly within the bulk of the solution, although owing to the buoyancy of the liquid, most bubbles rise rapidly to the liquid surface, where they disintegrate. Suppose a liquid in which bubbles are being formed also contains small particles whose dimensions are in general smaller than the dimensions of the bubble. If a particle were to become adsorbed at the air-liquid interface within the bubble, this would decrease the surface area of the bubble and thereby stabilize it somewhat.

The adsorption of a particle at such an air-liquid interface results in a loss of area of the solid-liquid interface of the particle and the formation of an air-solid interface. Thus, the ability of a particle to adsorb stably at the edge of a bubble should depend upon the relative values of the following surface tensions: solid-liquid ($\gamma_{s,l}$), solid-air ($\gamma_{s,a}$), and air-liquid ($\gamma_{a,l}$). The conditions for adsorption can be calculated by making reference to Figure 13-26. A spherical particle is at an air-liquid interface, and the contact angle θ is as shown in the figure. Let h be the distance of penetration into air. The area of the solid-air interface is $2\pi rh$, in which r is the radius of the particle, and the area of the solid-liquid interface is $4\pi r^2 - 2\pi rh$. The

FIGURE 13-26 Geometric construction for calculating the conditions for adsorption of a spherical solid at an air-liquid interface.

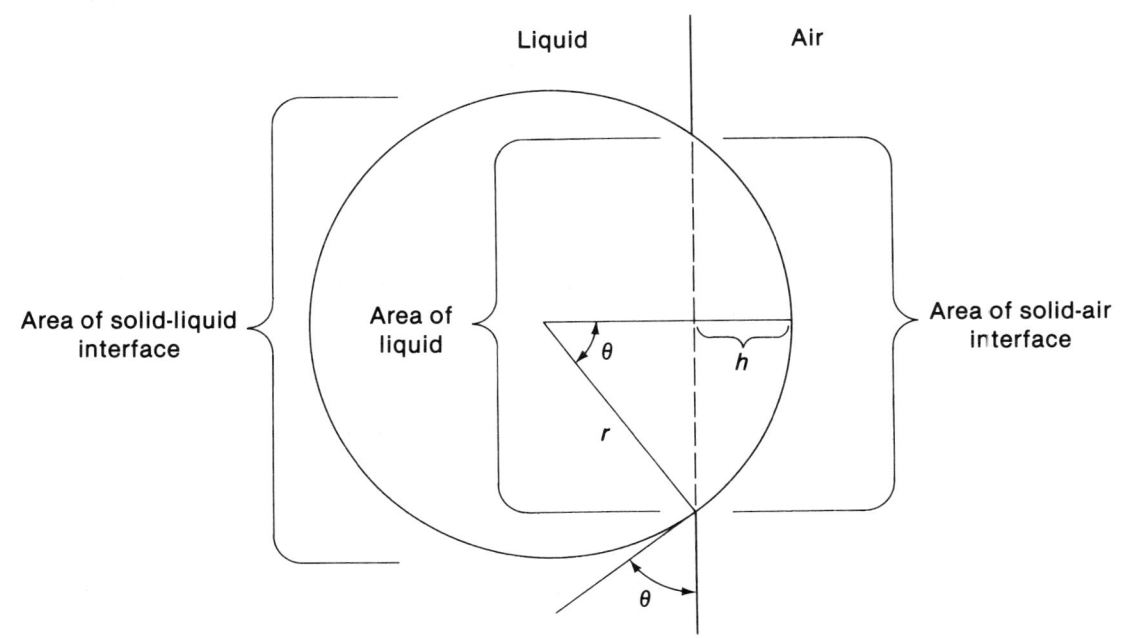

area of the air-liquid interface occupied by the particle is πl^2 in which $l = [r^2 - (r - h)^2]^{1/2}$, which reduces to $\pi(2\pi rh - h^2)$. Remembering the definition of surface free energy—namely, $\Delta G = \sum (\gamma \times \Delta \text{ area})$, and that at equilibrium a small displacement Δh causes zero change in the free energy (that is, $\Delta G = 0$), we have

$$0 = \Delta G = \gamma_{s,a}(2\pi r\Delta h) + \gamma_{s,l}(-2\pi r\Delta h) - \gamma_{a,l}\pi(2r - 2h)\Delta h , \quad (35)$$

or

$$\gamma_{s,a} - \gamma_{s,l} = (1 - h/r)\, \gamma_{a,l} = \gamma_{a,l} \cos \theta . \quad (36)$$

Therefore, as long as the values of $\gamma_{s,a}$, $\gamma_{s,l}$, and $\gamma_{a,l}$ are such that $\cos \theta$ is neither 0 nor 1, adsorption can occur. (If $\cos \theta = 0$, the particle is entirely within one phase.) For most particles this condition is satisfied; when it is not, it is usually possible to add some surfactant to make it so.

What all this means is that if a suspension of particles is agitated, so that bubbles form, particles will adsorb to the surface of the bubble and rise with the bubble to the surface of the bulk liquid. This process is called *froth flotation* and is widely used in the mineral industry to recover fine particles of essential minerals. If the liquid contains an agent that produces a stable surface foam, rapid bubbling of air through a suspension causes the particles to be carried from the body of the liquid to the foam. The foam can be skimmed off of the surface and with it will come the particles.

Froth flotation has no particular value in biological systems but is responsible for certain experimental difficulties. Bacteria and bacterial viruses are often grown in liquid nutrient media through which air is bubbled, because the growth rate and ultimate yield increases as the oxygen supply is improved. Sometimes at a high rate of bubbling there are enormous losses of the particles. This occurs because during the bubbling both bacteria and bacteriophages adsorb to the bubbles and are carried rapidly to the surface of the liquid. The bubbles burst at the surface of the liquid and the contents of the bubble, namely, the bacteria and the phage, are sprayed into the vapor phase above the liquid. The particles reach the walls of the vessel and adsorb there (and sometimes dry onto the surface). These particles are then, for all practical purposes, lost. The losses are accentuated by the fact that most microbiological growth media foam excessively; the particles are retained by the foam, which either froths over the side of the vessel or dries on the walls of the vessel. To reduce losses one generally uses antifoaming agents (silicone oils are in common use) and adjusts the air flow so that most bubbles collapse before reaching the liquid surface.

REFERENCES

Adamson, A.W. 1967. *Physical Chemistry of Surfaces*. Wiley-Interscience. New York.

Bond, G.C. 1962. *Catalysis by Metals*. Academic Press. New York.
Bull, H.B. 1964. *An Introduction to Physical Biochemistry*. Davis. Philadelphia.
Castellan, G.W. 1971. *Physical Chemistry*. Addison-Wesley. Reading, Mass.
Chapman, D. 1968. *Biological Membranes, Physical Fact and Function*. Academic Press. New York.
Drexhage, K.H. 1970. "Monomolecular Layers and Light." *Scientific American*, 222:108.
Haensel, V., and R.L. Burwell. 1971. "Catalysis." *Scientific American*. 225:46.
Harris, R., and G.G. Lunt. 1975. *Biological Membranes: Their Structure and Function*. Wiley. New York.
Kushner, L.M., and J.I. Hoffman. 1951. "Synthetic Detergents." *Scientific American*. October, pp. 26–43.
Ross, S., and J.P. Oliver. 1964. *On Physical Adsorption*. Interscience. New York.
Somorjal, G.A. 1972. *Principles of Surface Chemistry*. Prentice-Hall. Englewood Cliffs, N.J.

PROBLEMS

1. When water rises in a capillary tube, the contact angle on clean glass is 0°. The surface tension of water at 20°C is 72.75 dyn cm^{-1} and the density is 0.9982 g cm^{-3}. What is the capillary rise in a tube whose radius is 0.25 mm, if the water is in contact with air? Use $g = 980$ cm sec^{-2}.

2. An interesting way to determine the surface tension of a liquid is to place two capillaries having radii r_1 and r_2 into the liquid. The liquid rises to heights h_1 and h_2. If $\Delta h = h_1 - h_2$, show that the surface tension γ is given by

   $$\gamma = \tfrac{1}{2}\Delta h[r_1 r_2/(r_2 - r_1)]\rho g.$$

3. Two capillaries with radii 0.32 and 0.68 mm are inserted in a liquid whose density is 0.80 g cm^{-3} in contact with air of density 0.001 g cm^{-3}. The difference in the heights risen in each tube is 2.3 cm. What is the surface tension? Assume $\theta = 0$, and see problem 2.

4. The addition of almost all solutes to water decreases the surface tension—substances that increase the surface tension are uncommon. Explain why this is so.

5. A droplet of pure water is placed on a hydrophobic surface. A second droplet containing a solute that lowers the surface tension is placed on the same surface. Compare the shapes of the two droplets.

6. A test tube contains both water and an immiscible organic liquid. The densities of water and the liquid are 0.998 and 0.825 g cm^{-3}, so that the organic liquid floats on the water as a distinct phase. A capillary tube of radius 0.03 cm is filled with the organic liquid and then placed in the liquid, so that the lower end of the tube is in the water phase. Water enters the capillary tube and rises until it is 2.5 cm above the top surface of the organic phase. What is the surface tension of water when it is in contact with this organic liquid? Assume $\theta = 10°$.

7. A dense fog consists of droplets 10^{-3} cm in radius and the droplets are 1 mm apart. Calculate ΔG in producing a cubic kilometer of fog at constant temperature. The surface tension of water is 72.75 dyn cm^{-1}.

8. Butylamine, $CH_3(CH_2)_2CH_2NH_2$, is soluble in water and in benzene. How will the molecule be oriented at an air-water and an air-benzene interface?

9. Which part of the following molecules will face the air side of an air-water interface when each is separately dissolved in water?

(a) *n*-octanol.
(b) phenol.
(c) 1,4-diaminobutane.
(d) propionaldehyde.
(e) butyric acid.

10. A film is made by using 8 µg of a molecule whose molecular weight is 854. It is compressed to a minimal area of 18.5 cm². What is the cross-sectional area of the molecule if it is considered to be cylindrical?

11. A mixture of equal amounts of hydrogen (H_2) and deuterium (D_2) is allowed to adsorb to a metal surface. When desorbed, it is found that the gas mixture is 25 percent H_2, 25 percent D_2, and 50 percent HD. What is the probable mechanism of adsorption?

12. You are studying adsorption of CO_2 to metal oxides such as CaO. You suspect that they are chemisorbed and test this hypothesis by using CO_2 containing the heavy isotope ^{18}O—that is, using $C^{18}O_2$. What experimental result would support the idea that CO_2 is chemisorbed?

13. Gaseous H_2 is adsorbed separately to the surfaces of metal A and metal B. Desorption of H_2 from A occurs at 45°C whereas from metal B, 662°C is required. Without knowing the isotherm that describes the binding, you guess that physisorption occurs with one metal and chemisorption occurs with the other. To what metal is the H_2 probably physisorbed?

14. In preparing very dilute solutions in the laboratory, one must always worry about losses of the solute to the walls of the container, owing to adsorption. This is not a problem when the solute concentration is 0.01 *M* or greater because the amount of material that can adsorb is usually very small. However, when the concentration is 10^{-5} *M* or less, in some cases all of the solute molecules may be adsorbed. This often necessitates making a choice between using glass and plastic containers. Glass is polar and plastic is usually quite nonpolar.

(a) Which type of container would you choose for storing a dilute solution of DNA (an acidic polymer)?

(b) In a dilute solution containing both glycine, NH_2CH_2COOH, and leucine, $NH_2CH(C_4H_9)COOH$, which amino acid would preferentially be lost in a plastic test tube?

15. A carbon rod is placed separately in solutions of adenine at various concentrations and the amount of adenine (in micromoles) bound is measured. The data are: 0.0002 *M*, 46.8; 0.0005 *M*, 48.7; 0.0012 *M*, 49.4; 0.0025 *M*, 49.7. How many µmol of adenine would be adsorbed from a 0.008-*M* solution?

16. A protein dissolved in 0.01 *M* NaCl is adsorbed to a surface and the binding data agrees with the Langmuir isotherm. If the surface is exposed to protein-free 0.01 *M* NaCl, would you expect protein to appear gradually in the solution?

17. A gas adsorbs to 5 g of ZnO at 0°C. The volumes of adsorbed gas as a function of pressure are the following:

V (cm³): 1.02 1.35 1.52 1.61 1.65
P (atm): 2.9   8.2   13.7 21.1 27.5

(a) Show that the data agree with the Langmuir isotherm and calculate V_m.

(b) Assume that the gas molecule is a sphere whose radius is 3.5 Å and that the gas molecules are closed-packed. What is the area per gram of the sample of ZnO?

18. Gaseous N_2 is allowed to adsorb to 7.4 g of ZnO at −195°C. The volume in cm³ adsorbed at various pressures (mm Hg) are the following: 5.90 cm³, 56 mm Hg; 6.20, 94.8; 6.98, 145.2; 7.55, 183.4; 8.17, 222. (Each of these values has been converted to values at 0°C and 1 atm pressure by using the ideal gas law.) The vapor pressure of N_2 at this temperature is 760 mm, $\Delta H_{vap} = 5.56$ kJ mol^{-1}, and the area of a N_2 molecule is 16.2 Å². Calculate (a) the volume of N_2 needed to form a monolayer on the sample, (b) the enthalpy of adsorption, and (c) the area of 1 g of ZnO.

19. Enzymes are proteins that catalyze chemical reactions. The protein molecule is a highly folded polypeptide chain that has many internal interactions and a distinct three-dimensional structure. These structural features are characteristic of a particular enzyme and are necessary for its catalytic activity. Enzymes are often purified from cell extracts by adsorption to small particles that have a large surface area (*adsorption chromatography*). Usually the enzyme adsorbs and desorbs easily and the enzyme is recovered. However, sometimes the molecules obtained after adsorption have lost all enzymatic activity. If a different adsorbent is tried, this loss of activity may not occur. Explain this loss of activity.

CHAPTER 14

ELECTRICAL AND TRANSPORT PROPERTIES OF SURFACES AND MEMBRANES

A major role of membranes in biological systems is to define "compartments" within a system, each of which contains a solution of a particular composition and concentration. Membranes are never an absolute barrier, for they must allow many, although not all, substances to pass through. There are a variety of mechanisms for determining the substances to which a membrane is permeable as well as for determining the final concentrations on the two sides of the membrane. The transported molecules and the membrane components often carry an electrical charge, so that a study of transport and membrane structure requires studying the electrical properties of membranes and surfaces. This is the subject of this chapter.

Before starting the discussion, a few words are needed about electrical units to remind the reader of some of the points mentioned in Chapter 1.

14–1 A PRECAUTION ABOUT ELECTRICAL UNITS

There are three systems of units in use to describe the electrical properties of matter. These are called the SI, electrostatic, and electromagnetic units, and they differ primarily in how electrical charge, electrical potential, and distance are expressed. In chemistry and biology, only the SI and the electrostatic units are used. The names of the important units and the values of some constants are listed in Table 14-1. In this chapter we will use SI units.

TABLE 14-1 Comparison of the SI and electrostatic units for various measures and constants.

Measure and constant	SI	Electrostatic system
Measure		
Electrical charge	Coulomb (C)	Electrostatic unit (esu)
Electrical potential	Volt (V)	Statvolt (statV)
Distance	Meter (m)	Centimeter (cm)
Constant		
Charge of the electron	1.6×10^{-19} C	4.80×10^{-10} esu
Faraday constant, $\mathscr{F}$	96,485 C mol^{-1}	2.89×10^{14} esu mol^{-1}
Permittivity of vacuum, ϵ_0	8.85×10^{-12}	Does not appear

Conversion
1 statV = 3.33×10^{-3} V
1 V = 300 statV
1 C = 2.98×10^9 esu
1 esu = 3.35×10^{-10} C

Unfortunately, in laboratory practice, when studying surfaces and membranes, a mixture of the two systems is in use and this has given rise to great confusion. Charge and potential are normally stated in coulombs and volts; however, distances are usually expressed in centimeters, μm (10^{-6} m), or ångström units (1 Å = 10^{-8} cm). This practice arises because the distances encountered in laboratory systems are small (of the order of centimeters or much less) rather than of the magnitude of one meter. Thus, electric field strength is invariably stated as X volts per *centimeter* and the charge density on surface is almost always given as charge per cm^2, per μm^2, or per Å^2. We cannot state too strongly that

> *in order to use the equations developed in this chapter, all distances must be expressed in meters.*

To remind the reader of this possibility for confusion, the data given in all examples in this chapter will be presented first in common laboratory units of distance (cm, μm, or Å) and then converted to meters.

14-2 ELECTRICAL POTENTIALS ACROSS SURFACES

If two substances of different chemical composition are in contact, there exists a potential difference or a voltage across the interface. There are many ways in which this voltage difference can arise, a few of which are the following:

1. The surface of one of the substances may be electrically charged, as would be the case if one of the substances were a protein molecule.
2. Dipoles can be induced at the interface.

3. Ions having different signs of charge may have different rates of diffusion across the boundary.

In all cases the potential difference is the result of a separation of charges.

Electrical Potential across a Charged Film

Of particular interest is the potential difference that exists between two sides of a film of charged molecules as a result of the orientation of the molecules in the film. The magnitude of this quantity can be calculated from simple electrostatics as follows.

Two parallel conducting plates separated by a distance d (expressed in m) and enclosing a surface charge of density σ (expressed in C m^{-2}) in a medium of dielectric constant D have a potential difference $\Delta \phi$ (in volts, V) equal to

$$\Delta \phi = \frac{\sigma d}{D \epsilon_0}. \tag{1}$$

Use of this equation is demonstrated in the following example.

EXAMPLE 14–A A sheet of plastic whose dielectric constant is 5.2 is 1 mm thick and 10 cm^2 in area and has 10^9 univalent positive charges bound to one side and the same number of univalent negative charges bound to the other side. What is the potential difference across the sheet?

We first calculate the charge density on one side of the sheet. The total charge on one side of the sheet is $(10^9)(1.6 \times 10^{-19}) = 1.6 \times 10^{-10}$ C. Dividing by the area (in m^2), we obtain the charge density σ, namely, 1.6×10^{-7} C m^{-2}. The thickness is 0.001 m. Thus, by Equation 1,

$$\Delta \phi = \frac{(1.6 \times 10^{-7})(0.001)}{5.2(8.85 \times 10^{-12})} = 3.47 \text{ V}.$$

Equation 1 expresses the potential difference across a dielectric in terms of the density of charge on each surface. However, the charges on the two sides of a dielectric can be thought of as a collection of dipoles, so that the potential difference across the dielectric can be expressed in terms of their dipole moments, as follows. Let the charge density be due to a collection of dipoles of dipole moment μ oriented at an angle θ with respect to a line drawn perpendicular to the plates. If there are n dipoles per m^2, then $\sigma d = n \mu \cos \theta$, and if $D \approx 81$ (the value for water), then

$$\Delta \phi = 1.4 \times 10^{-9} \, n \mu \cos \theta. \tag{2}$$

SEC. 14-2 ELECTRICAL POTENTIALS ACROSS SURFACES

FIGURE 14-1 The polonium electrode method for measuring surface potential. The α particles emitted by the polonium (Po) ionize the air and make a conducting path between the polonium electrode and the surface, thereby completing the electrical current. E is an insulated reference electrode, R a variable resistance, B a battery, and G a galvanometer.

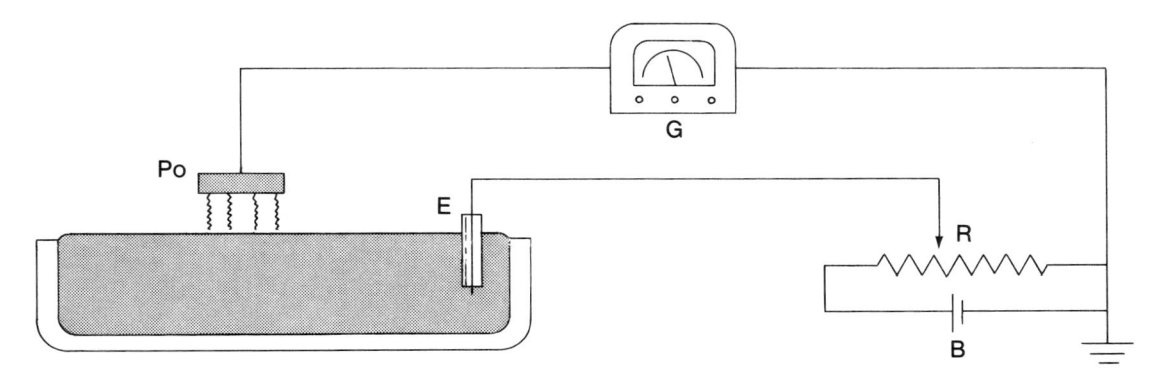

This equation is not totally correct, because we have not accounted for the fact that, in the case of water at least, the solvent molecules just below a film are also oriented by the film. However, Equation 2 is usually acceptable to the extent that $\Delta\phi$ is an indication of the degree of orientation of molecules in a film. As might be expected, if a film is compressed (as is done in obtaining a force-area curve), $\Delta\phi$ increases with rising pressure.

It is not possible to measure film surface potential by putting an electrode on a film surface because the film would then be disrupted. Measurement of the surface potential is usually carried out by what is known as the polonium electrode method (Figure 14-1). In this technique, the electrode is not put in physical contact with the film, although it is in electrical contact. An electrode coated with polonium is placed slightly above the surface; the α particles emitted by the polonium ionize the air between the electrode and the film, and thereby provide an air path of sufficient conductance to perform the measurement. This is an important and commonly used procedure.

A Solid Surface in Contact with a Solution of an Electrolyte—the Electrical Double Layer

Consider a positively charged surface of a solid within an ionic solution. Such a surface can attract the negative ions in the solution and if there are no repulsive forces, at the surface there will be a quantity of negative ions (*counterions*) sufficient to neutralize the positively charged solid surface. Thus, the object whose surface we are considering can be thought of as being covered by an electric double layer, one layer being the positive charge of the solid surface, the other being the bound negative charges of

FIGURE 14-2 Diagrams of two types of electrical multilayers. (a) The fixed or Helmholtz double layer. The positively charged surface of a solid binds negative charges in a solution, resulting in a negatively charged monolayer. The remainder of the solution is electrically neutral except for a small deficit of negative ions. (b) The Gouy-Chapman model. Many of the positive charges on the surface of the solid have bound negative ions from the solution, which in turn bind more positive ions. Some negative ions at a distance from the surface are also attracted toward the surface. A cloud of charges is found near the surface of the solution but the charge density decreases with increasing distance from the surface.

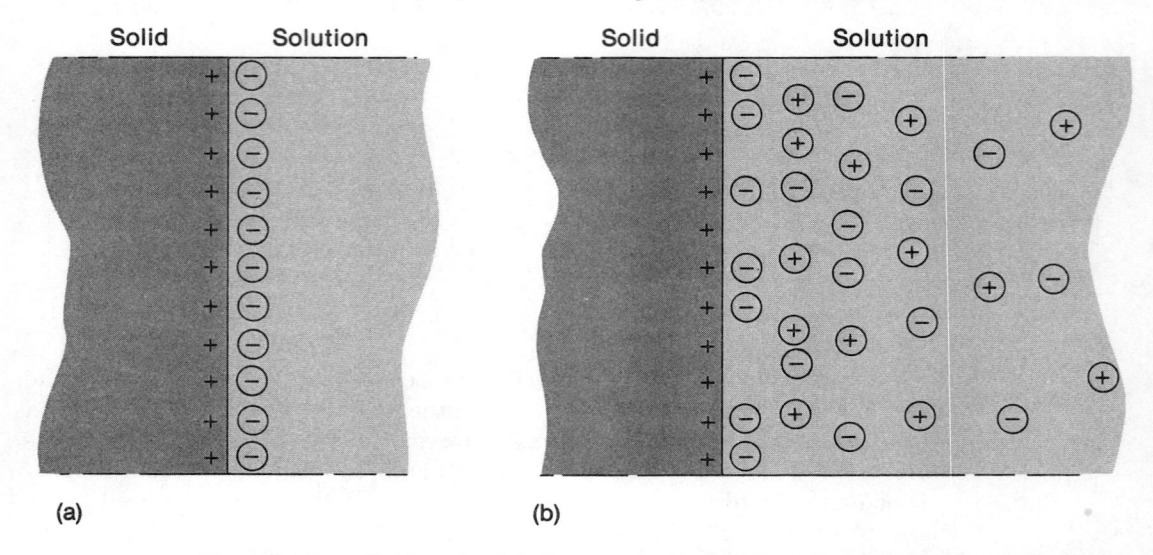

the solution. Such a double layer is called a *fixed* or *Helmholtz double layer* (Figure 14-2(a)). A plot of the electrical potential as a function of distance from the surface for four types of double layers is shown in Figure 14-3(a). The distance indicated in the figure by each of the vertical arrows is the boundary between the two layers. In general, an adsorbed negative layer will be loosely bound for two reasons: (1) thermal motion tends to disrupt the bound layer; (2) there must also be positive ions in solution that tend to draw the negative charges away from the positive layer. This loose collection of counterions near a surface is referred to as a *cloud* of charges. These negative ions will not only tend to attract positive ions from the solution but the positive ions will, in turn, attract negative ions. This results in a gradient of concentration of negative and positive charges in which the concentration of the negative charges increases and the concentration of the positive charges decreases as the surface is approached. Thus, the surface is said to be covered with a *diffuse layer* of charge known as a *Gouy layer* (Figure 14-2(b)). A plot of the electrical potential as a function of the distance from the surface for this arrangement is shown in Figure 14-3(b). The mathematical formulation of the charge distribution was analyzed by Gouy, who made the following assumptions:

1. If the electrical potential at a distance r from the surface is ϕ_r and the solutions contain positive and negative ions of charge $+ze$ and $-ze$, respectively, the number of ions of each type at a distance r from the

FIGURE 14-3 Variation of electrical potential in four different types of double layers in which each surface is positively charged. (a) Helmholtz layer; (b) Gouy layer; (c) Stern layer; and (d) Stern layer with a fixed layer having more bound negative charges than are necessary to neutralize the surface charge.

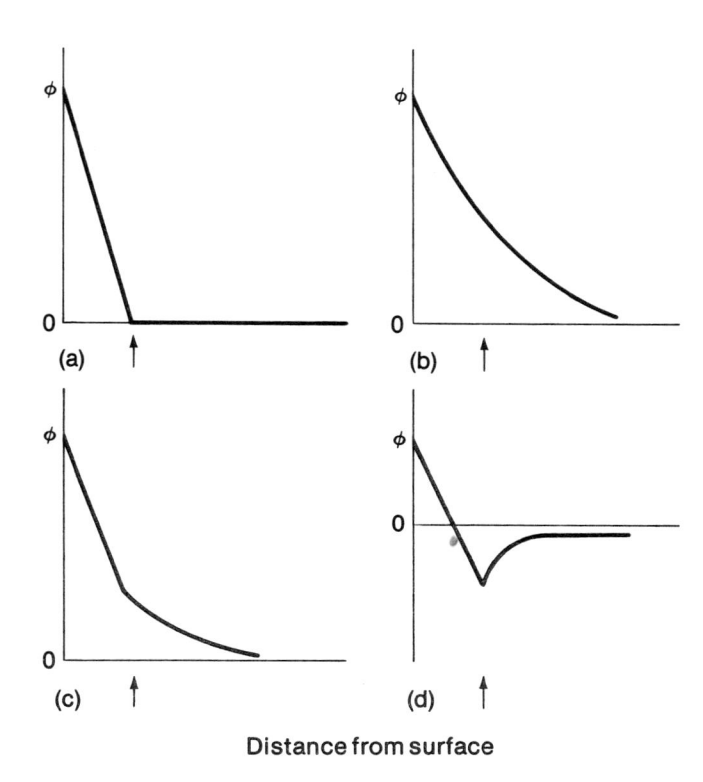

Distance from surface

surface will obey the Boltzmann distribution—namely, will be proportional to $\exp(\pm ze\phi_r/kT)$.

2. The system as a whole must be electrically neutral, so that, far from the surface, the concentrations of positive and negative charges must be equal.
3. All charges are assumed to be point charges—that is, to have no volume. (This third assumption is dropped in the Stern formulation, which we examine shortly.)

In many cases a more reasonable picture of conditions near a surface is one in which a layer of bound ions and a cloud of ions are combined. In such a case, some of the counterions are often tightly bound to the surface of a solid that is in a solution. Furthermore, it is possible that some of these negative ions will be bound by van der Waals forces and hydrogen bonds rather than only by simple charge attraction, so that there might be even more negative ions bound to the solid surface than there are positive charges on the solid surface responsible for the main part of the cloud. The electrical potential for these last two cases is plotted in Figures 14-3(c) and 14-3(d); panel (c) shows the effect of bound negative ions, the number of which is insufficient to neutralize the surface charge, and panel (d) shows the situation when there are excess negative charges. A mixture of a fixed layer (counterions) and a diffuse layer (panel c) is called a *Stern layer*. The counterions are always in equilibrium with the ions that are free in solution, so that the number of counterions on a surface must depend upon the ionic concentration of the solution. Thus, the situation in panels (c) or (d) in Figure 14-3 could be shifted to that of panels (a) and (d) by a suitable change in concentration.

The Stern treatment of adsorption is an attempt to explain certain surprising predictions of the Gouy-Chapman theory. For instance, surface potentials of 300 millivolts are not particularly unusual. If the concentration of ions in solution is 10^{-3} M, the Boltzmann distribution stated in the presentation of Gouy-Chapman theory yields an ion concentration of 160 M at the surface of the membrane, which is a highly unlikely value. This large value is a result of the assumption in the development of the Gouy-Chapman theory that the ions occupy no space. In the Stern theory, the size of the ions is taken into account, so that there is a limited number of ions that can bind to the membrane (and that also can take up space in the bulk solution). Thus one considers the membrane and the bulk solution to consist of a finite number of sites, and it is the fraction of sites that are occupied that is assumed to obey the Boltzmann distribution. We will not go through the theory but simply point out that it satisfactorily explains the charge distribution in many cases. It is important to realize, though, that there are cases, for instance, the charge distribution around a lipid bilayer, for which the Gouy-Chapman theory is a better description of the charge distribution. At the present time the theory that best applies to a particular system cannot usually be predicted.

The Liquid Junction Potential and the Contact Potential

We will see later that surface potentials are important in many biological systems. However, they can often be an annoyance in the laboratory. The liquid junction and contact potentials are two examples of surface potentials that frequently introduce difficulty in laboratory work (for example, in the design of pH meters and electrical cells—see Chapter 10).

Consider a dilute solution of HCl in contact with a more concentrated HCl solution. We will assume that the two solutions are prevented from mixing. This can be accomplished if one solution is considerably less dense than the other and the less dense one is carefully layered onto the more dense solution. Alternatively, they could be separated by a porous plug. Both H^+ ions and Cl^- ions will diffuse across the boundary between the two solutions until the solutions are ultimately at the same concentration of HCl. Let us instead consider the state of affairs before equilibrium is reached. The H^+ ion is much more mobile than the Cl^- ion, so that before equilibrium is achieved, charge separation occurs. That is, near the boundary there will be more H^+ ions and thus a net positive charge on the side of the boundary facing the solution having the lower initial concentration of HCl. Similarly, because of the H^+-ion deficit, a net negative charge will

FIGURE 14-4 Formation of a liquid junction potential. At zero time, two HCl solutions are placed in contact. The dashed line represents the junction or boundary between the two solutions. The net charge in each region is zero. At a later time, owing to the greater rate of movement of the proton with respect to the H$^+$ ion, there is a net positive charge in the upper solution and a negative charge in the lower solution. Of course, at equilibrium, the concentrations of all ions are the same in both regions.

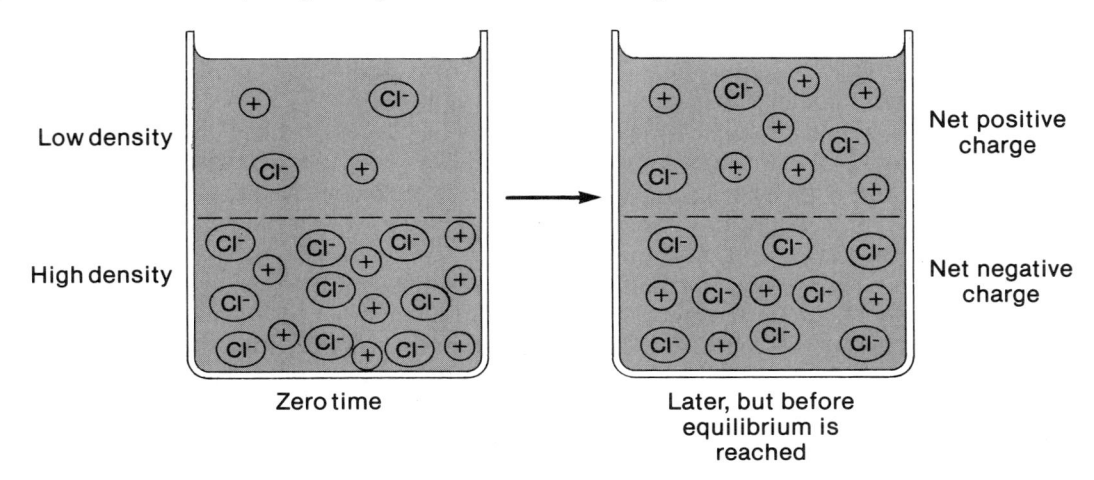

exist on the other side of the boundary (Figure 14-4). *Electroneutrality of the bulk solution is of course never violated, and there is only a local imbalance of charge in the immediate region of the boundary.* This local imbalance may only consist of a very small fraction (10^{-4} percent is not an unreasonable value) of the ions in the solution, so there is no significant violation of electroneutrality. Nonetheless, this small imbalance can give rise to a significant potential difference that is known as the *liquid junction potential*. The magnitude of this potential is based upon the relative charges and rates of diffusion of the ions. The two solutions need not contain the same ions initially—for instance, dilute HNO_3 layered on concentrated $MgCl_2$ will also produce a liquid junction potential.*

Note that the liquid junction potential exists only before equilibrium is achieved. It vanishes at equilibrium and its magnitude changes with time.

The operation of pH meters is affected by liquid junction potentials. A pH meter consists of two electrodes that are attached to a voltmeter and are placed in the liquid whose pH is to be measured (refer back to Figure 10-8). One electrode, the glass electrode, consists of a H$^+$-permeable glass tube containing 0.1 M HCl. A potential exists across the glass, the magnitude of which is proportional to the logarithm of the H$^+$-ion concentration of the sample. The other electrode, the reference electrode, is most commonly a Hg-Hg$_2$Cl$_2$ paste called *calomel* encased in a glass tube having a fine orifice. In order to make electrical contact between the calomel and the sample, the reference electrode is filled with saturated KCl. The K$^+$

---

*It is of course possible that the solutions in contact will not produce a liquid junction potential. However, for this to be the case, the mobilities of the ions would have to be equal.

and Cl⁻ diffuse through the orifice but at different rates. Thus, a charge is generated at the interface between the KCl solution and the sample—this charge separation causes a liquid junction potential. The magnitude of the potential, as we have just stated, changes with time. However, this change is very small during a single measurement. Furthermore, the value of this potential is independent of pH and ion concentration (in the range normally encountered), so that it is considered to be a constant. To avoid measuring this potential, which would ordinarily have to be accounted for in relating pH values to measured voltages, a pH meter is not used to measure absolute values of voltage. Instead, the electrodes are placed in a buffer having a known pH and the meter is adjusted to read that pH value. Thus, when the electrodes are placed in a sample to be measured, the correction for the liquid junction potential is already made. The operation of a pH meter is also discussed in Chapter 10.

When two metals are in contact, there is also a potential difference across the metal-metal interface. This results because the electrons of one metal usually find it energetically favorable to move into the other metal. When equilibrium is reached (which is very rapid because of the exceedingly great mobility of electrons in metals), *in the region of the interface* the electron-receiving metal has a *local* excess of negative charge and the electron donor has a local deficit; this creates a potential between the two metals that is called the *contact potential*. Typical values range from 0.1 to 2 volt. When measuring small voltages in the laboratory, the contact potential must be kept in mind because metal-metal contacts are almost always required in electrical circuits. The problem is eliminated if the same metal is used throughout the circuit. This is not always feasible though—for example, copper is the metal of choice for wires because of its high conductivity, battery terminals are generally made of zinc, and filaments are invariably tungsten wire. The contact potentials for various metal-metal combinations can be found in standard reference tables. It is important to understand that the values listed in such tables are for fresh metal surfaces. Many metals such as aluminum and lead have oxide coatings that significantly decrease the contact potential.

14-3 MEMBRANES

We define a membrane as any somewhat stable, material barrier that separates two components, which may be different substances or different concentrations of the same substance. A membrane may be a single molecular layer or a complex structure. Its composition and structure may change in time but is normally thought of as having a basic, defined structure. The membrane may be liquid, solid, or some combination of these phases. In biological systems, the substances separated by a membrane are usually two liquids, two gases, or a gas and a liquid.

The most important property of biological membranes is their *selective permeability*—that is, their ability to allow some, but not all, molecules to pass from one side of the membrane to the other. This gives rise to differences in concentration and in electrical potential. Furthermore, the passage of molecules across the membrane can occur (1) by *simple diffusion* through the membrane from the side having a higher concentration to that at a lower concentration (passive diffusion); (2) by means of a *carrier molecule,* also from a region of higher concentration to one of lower concentration; (3) by means of a *series of chemical reactions* (which may transport molecules against a concentration gradient); and (4) by *active transport* (that is, pumped from a region of lower to higher concentration with an expenditure of energy). Each of these mechanisms will be discussed in detail shortly.

In order to discuss transport of charged molecules, the potential differences created by such transport must be examined. This is done in the following section.

Membrane Potential

Consider a membrane separating two aqueous ionic solutions having identical compositions but different concentrations. The values of the Gibbs free energy or the chemical potential of each ion in the more concentrated solution are greater than the values in the less concentrated solution (see Chapter 6), so that the ions will flow from the region in which the free energy is higher to that of the lower free energy until at equilibrium the free energy is the same in both regions. At equilibrium the concentration of a particular ion will be the same on both sides of the membrane. In general the positive and negative ions will not have the same rate of movement, so that before equilibrium is reached, there will normally be a potential across the membrane similar to the liquid junction potential just described.

Let us consider two solutions of the ionic salt A^+X^-, each at a different concentration; the solutions are separated by a membrane that is permeable to the A^+ ion but not to the X^- ion. In this case A^+ ions will still tend to move from the region of high concentration to the region of low concentration, driven by the chemical potential gradient. However, there is a problem, because movement of A^+ ions without movement of the X^- ions creates a charge imbalance. That is, there would be a net positive charge in the solution containing the salt that was initially (and still is) at a lower concentration and a net negative charge in the other solution. Such a charge separation would immediately set up an electrical potential that would drive the ions back to the original arrangement. However, neither the chemical nor the electrical potential can "win," so that at equilibrium the two potentials must be in balance. What actually happens is the following. A very small number of A^+ ions moves across the membrane, driven by the chemical potential. To preserve electroneutrality in the two bulk solutions, the distribution of ions in each solution becomes nonrandom in the sense that A^+ ions accumulate at the surface of the membrane in the solution that was initially at the lower concentration. Similarly, X^- ions are concentrated on the other side of the membrane. Thus an electrical potential difference is established across the membrane; the value of this potential difference, $\Delta\phi$ (in V), is given by

$$\Delta\phi = -\frac{RT}{z\mathcal{F}}\ln\frac{[A^+]_{high}}{[A^+]_{low}} \tag{3}$$

in which z is the valence of the ion, $\mathcal{F}$ is Faraday's constant (in coulombs), R is the gas constant (which has the value 8.314 in these units), T is the absolute temperature, and the subscripts *high* and *low* refer to the initial concentrations in the high and low concentration solutions, respectively. This potential difference is called the *Donnan potential*. (Note that if $[A^+]_{high} = [A^+]_{low}$, then $\Delta\phi = 0$.) The value of $\Delta\phi$ for a tenfold difference in concentration is about -60 mV at room temperature.

The number of charges transferred depends on the thickness d and the dielectric constant D of the membrane. If the potential difference is $\Delta\phi$, the charge density σ on the surface of the membrane is defined by Equation 1. If we choose typical values for a cellular membrane—namely, $\Delta\phi = 50$ mV, $d = 10^{-8}$ m, and $D = 5$, then $\sigma = 2.21 \times 10^{-4}$ C m^{-2}. Since a monovalent ion carries the charge 1.6×10^{-19} C, then $\sigma = 1.38 \times 10^{15}$ charges per m^2 or 1.38×10^{-5} charges per Å^2. A simple calculation shows this to represent a rather small fraction of the total charge in a particular experimental situation. For instance, if a membrane whose surface area is 5 cm^2 (5×10^{-4} m^2) is in contact with 10 cm^3 (10^{-5} m^3) of an 0.01-M NaCl solution, the solution contains 1.2×10^{20} ions and the number of ions on the 5 cm^2 membrane is 6.9×10^{11} charges, or, 5 charges per 10^{10} total charges are on the membrane. If instead we consider a nerve cell 100 Å (10^{-8} m) in diameter and 1 cm (0.01 m) long, containing a solution whose concentration is about 0.15 M, the cell contains about 7×10^7 charges and the membrane will be covered by 4×10^5 charges; thus, about 0.006 of the charges are on the surface. Note that as the volume of the solution decreases, the fraction of the ions transferred increases. However, unless the volumes of the solutions are extremely small, the number of ions transferred is very small compared to the number of ions that are free in solution. Thus, the charge deficit in the bulk solution is very small and electroneutrality of the bulk solution is maintained.

EXAMPLE 14-B A membrane permeable to Cu^{2+} ions but not to NO_3^- ions separates 0.1 M $Cu(NO_3)_2$ from 0.0024 M $Cu(NO_3)_2$. What is the membrane potential difference at 25°C, and which side of the membrane is at a lower potential?

First, notice that $\Delta\phi = (RT/z\mathcal{F})\ln(0.1\ 0.0024)$. The Cu^{2+} ion is divalent, so that $z = 2$. In order to express the potential difference in volts,

SEC. 14-3 MEMBRANES

observe that $\mathscr{F}$ is 96,485 coulomb, and R is 8.314 J K^{-1} mol^{-1}. Thus, $\Delta\phi =$ [(8.314) (298)/(2)(96,485)]ln 41.67 = 0.0479 V = 47.9 mV. Cu^{2+} ions are transferred from the 0.1 M solution to the 0.0024 M solution; this creates an excess of negative charge in the concentrated solution, which must therefore be at the lower potential.

We now consider the situation in which both A$^+$ and X$^-$ ions can pass through the membrane. In this case, A$^+$ ions will again move from the region of high to low concentration. However, now, to preserve electroneutrality, one mole of X$^-$ ions will move with each mole of A$^+$ ions. The movement of the ions is accompanied by the free energy changes

$$\Delta G_{A^+} = RT \ln \frac{[A^+]_{II}}{[A^+]_I} \quad \text{and} \quad \Delta G_{X^-} = RT \ln \frac{[X^-]_{II}}{[X^-]_I},$$

in which I and II refer to the two sides of the membrane. At equilibrium, the total free energy change $\Delta G_{A^+} + \Delta G_{X^-}$ is zero, so that

$$\frac{[A^+]_{II}}{[A^+]_I} = \frac{[X^-]_I}{[X^-]_{II}}. \tag{4}$$

Of course, if no constraints are placed upon the concentrations of the ions, at equilibrium the concentrations on the two sides of the membrane will be the same; that is, $[A^+]_I = [A^+]_{II}$ and $[X^-]_I = [X^-]_{II}$. However, if there is some factor that prevents these concentrations from being equal (and we will see an example of this shortly), then, as we just saw, the statement that at equilibrium the *chemical* potential must be the same on both sides of the membrane must be replaced by the statement that the *electrochemical* potentials must be equal. The *electrochemical potential* $\bar{\mu}$ *is defined as the chemical free energy term plus the electrical term* $z\mathscr{F}\phi$, in which ϕ is the electrical potential of the solution. Thus, at equilibrium, we require that $\bar{\mu}_{A,I} = \bar{\mu}_{A,II}$ and $\bar{\mu}_{X,I} = \bar{\mu}_{X,II}$, or

$$\left. \begin{array}{l} RT \ln[A^+]_I + \mathscr{F}\phi_I = RT \ln[A^+]_{II} + \mathscr{F}\phi_{II} \\ RT \ln[X^-]_I - \mathscr{F}\phi_I = RT \ln[X^-]_{II} - \mathscr{F}\phi_{II} \end{array} \right\} \tag{5}$$

which can be rearranged to

$$\Delta\phi = \phi_I - \phi_{II} = \frac{RT}{\mathscr{F}} \ln \frac{[A^+]_{II}}{[A^+]_I} = \frac{RT}{\mathscr{F}} \ln \frac{[X^-]_I}{[X^-]_{II}}; \tag{6}$$

$\Delta\phi$ is the Donnan potential for this arrangement.

We now give an example of a kind of constraint that will produce a Donnan potential. The most important example found in biological systems is that of a charged protein molecule that cannot pass through a membrane. To see what happens in this case, we must analyze the situation with a membrane that is permeable to the ions A$^+$ and X$^-$ but not to a charged protein molecule P. The important point is that since the protein is charged, it will have counterions. We assume that these ions are A^+, if the protein carries a negative charge, or X^-, if it is positive. The arrangement is the following. The membrane separates one solution (I) that contains P, A^+, and X^- from a second solution (II) that contains only A^+ and X^-. (Note that if the protein were negative, it would be possible that, initially, solution I might contain no free X^- ions—that is, all of the A^+ could be associated with the protein. A similar statement can of course be made for a positive protein. However, for generality, we have assumed that the initial concentration of AX is such that all ions are present.) Let the protein have a net charge of z. Once again, $\bar{\mu}_{A,I} = \bar{\mu}_{A,II}$, from which, as we have just shown, $[A^+]_I[X^-]_I = [A^+]_{II}[X^-]_{II}$, because the membrane is assumed to be permeable to both A^+ and X^- ions. However, the condition of electroneutrality, that is,

$$z[P]_I + [A^+]_I = [X^-]_I , \qquad (7)$$

must also be satisfied, from which we obtain

$$[A^+]_I([A^+]_I + z[P]) = [A^+]_{II}[X^-]_{II} , \qquad (8)$$

which describes the ion distribution. Thus, the presence of a positively charged protein in compartment I results in $[A]_I < [A]_{II}$ and $[X]_{II} < [X]_I$. The converse is true for a negatively charged protein, namely, $[A^+]_I > [A^+]_{II}$ and $[X^-]_{II} > [X^-]_I$.

The presence of the protein results in the transfer of a particular amount of AX across the membrane. This amount is easily calculated, as follows. Let the initial concentration of the A^+ in the protein solution be c_I and that in the other solution be c_{II}. If the concentration of AX transferred is x, the final concentration in the AX solution (which has lost ions) is $(c_{II} - x)$. Equation 4 describes the equilibrium situation, namely,

$$\frac{c_{II} - x}{c_I + x} = \frac{x}{c_{II} - x} , \qquad (9)$$

which yields

$$x = \frac{c_{II}^2}{c_I + 2c_{II}} . \qquad (10)$$

We can express this also as the fraction x/c_{II} of the AX that is transferred, or

$$\frac{x}{c_{II}} = \frac{c_{II}}{c_I + 2c_{II}} . \qquad (11)$$

Note that if $c_I = c_{II}$, the fraction is 1/3. If c_{II} is very large compared to c_I, then $c_I \gg 2c_{II}$ and $x/c_{II} = \frac{1}{2}$, and the AX is uniformly distributed. This would of course also be the case if the protein were absent. The use of the equations is shown in the example that follows.

EXAMPLE 14-C Calculation of the ion distribution when a tetravalent, negatively charged protein is present.

We consider a vessel divided equally by a membrane that can pass the A^+ and X^- ions but not the protein molecule. The protein carries four negative charges. Compartment I contains only the protein with its counterions (that is, A_4P) at a concentration of 10^{-3} M; compartment II contains 10^{-2} M AX. Let the amount of AX that at equilibrium has moved from II to I be x. The concentrations at equilibrium will be

$$[A^+]_I = 4 \times 10^{-3} + x\ ;$$
$$[X^-]_I = x\ ;$$
$$[A^+]_{II} = 10^{-2} - x\ ;$$
$$[X^-]_{II} = 10^{-2} - x\ .$$

Note that $[A^+]_{II}$ must equal $[X^-]_{II}$ at equilibrium in order to maintain electroneutrality. Substituting these values into Equation 4 yields $x = 4.17 \times 10^{-3}$ M, so that the concentrations at equilibrium are $[A^+]_I = 8.17 \times 10^{-3}$ M, $[X^-]_I = 4.17 \times 10^{-3}$ M, $[A^+]_{II} = 5.83 \times 10^{-3}$ M, and $[X^-]_{II} = 5.83 \times 10^{-3}$ M. This situation is shown diagrammatically in Figure 14-5.

The Donnan potential for this arrangement with a protein can be written as follows:

$$\Delta\phi = \frac{RT}{\mathscr{F}}\ln\frac{[A^+]_{II}}{[A^+]_I} = \frac{RT}{\mathscr{F}}\ln\frac{[X^-]_I}{[X^-]_{II}} = \frac{RT}{\mathscr{F}}\ln\frac{[A^+]_I + z[P]}{[X^-]_{II}}\ . \qquad (12)$$

FIGURE 14-5 Formation of a membrane potential. At zero time, a membrane permeable to A^+ ions but not X^- ions separates two solutions. At equilibrium, there is a net positive charge in the solution at the right. The concentration of the A^+ ion to the right of the membrane never becomes as high as that to the left, because of the repulsion caused by this positive charge.

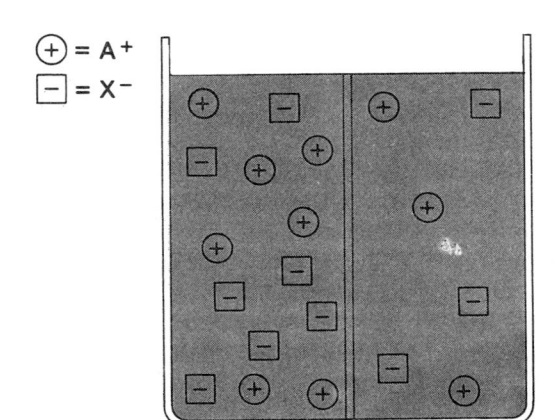

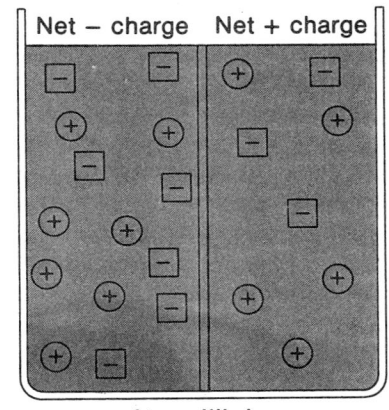

The form at the right in Equation (12) is especially useful. If [P] = 0, then at equilibrium, $[A]_I = [A]_{II} = [X]_I = [X]_{II}$, and $\Delta\phi = 0$. Thus, the potential difference is a result of the presence of a protein that cannot pass through a membrane. For a positively charged protein in compartment I, $\phi_I > \phi_{II}$, and for a negatively charged protein, $\phi_{II} > \phi_I$.

The actual cause of the potential should be clearly understood. In the example just given, the impermeant protein has created an imbalance in the concentration of AX on the two sides of the membrane. There is, as yet, no obvious cause of the electrical potential, since both sides of the membrane are, as we have so far described it, electrically neutral. However, we must remember our earlier discussion of the arrangement in which A, but not X, could pass through the membrane. That is, the concentration difference resulting from the presence of the protein results in a tendency for ions to move down the created concentration gradient. This would of course again create a charge imbalance so that a potential arises, as we saw before, because of the accumulation of charges on the two sides of the membrane, in the vicinity of the membrane. It is this potential that is described by Equation 12, in which the values of the concentrations are those unequal values produced by the presence of the protein.

Note that we have derived equations only for monovalent ions. To treat polyvalent ions, one must use a more general form of Equation 4, in which A^{z^+} would refer to a cation with charge z^+ and X^{q^-} would refer to an anion with charge q^-. This equation is

$$\left(\frac{[A^{z^+}]_{II}}{[A^{z^+}]_I}\right)^{1/z} = \left(\frac{[X^q]_I}{[X^q]_{II}}\right)^{1/q}. \tag{13}$$

It is also worth noting that the Donnan effect vanishes if a protein is at its isoelectric point—that is, if the protein is in a solution having a pH such that the protein has no net charge.

The significance of the Donnan effect is that most cellular potentials are approximated by the Donnan potential. The exceptions are those found in energy-transducing membranes, in which case ion pumps that are driven by the hydrolysis of high energy compounds produce potentials larger than could be the result of the Donnan effect.

Passive and Active Transport of Ions across Membranes

There are several ways that an ion can pass through a membrane. Most simply, it can diffuse freely through the membrane, either through *channels* or by using *carriers*. This type of transport is called *passive transport*. A relatively common mechanism of transport uses a series of chemical reactions, the components of which are contained in the membrane. By this scheme, molecules can move against a concentration gradient. What is done is that the molecule to be transported is altered chemically and undergoes a series of changes in which the original structure is regenerated. At each stage, the molecule is going down a concentration gradient. Energy is consumed in the process, so that even though the molecule itself has gained free energy, the set of reactions has lost sufficient free energy that the transport of the molecule is accompanied by a decrease in free energy. This kind of transport is known as *coupled transport,* because it is coupled to a set of chemical reactions.* A third mechanism is transport from one side of a membrane to another by an *ion pump*. In this case, there is not a series of reactions, but transport occurs as a *direct* consequence of a chemical reaction; this is called *active transport*. Among the best known examples of active transport are the following: the Na^+- and K^+-transporting ATPase found in all animal cell plasma membranes; the proton-transporting ATPase found in mitochondria, chloroplasts, and bacterial cell membranes; and the Ca^{2+}-transporting ATPase found in muscle tissue.

The passive and nonpassive mechanisms can be distinguished in another way; passive diffusion occurs by movement from a region of high concentration (or more precisely, high chemical or electrochemical potential) to a region of lower concentration, whereas nonpassive transport can occur against such a gradient. Both active and coupled transport require a significant expenditure of energy and both result in large concentration differences at equilibrium, which can be much greater than might result from the Donnan effect. When ions are moving freely through a membrane across which $\Delta\phi = 0$, at equilibrium the concentrations are identical and the ion flow from left to right is the same as that from right to left. However, when $\Delta\phi \neq 0$, and the final concentrations are unequal, the leftward and rightward motions are unequal. This is true for all three types of transport. Since the mechanisms of both coupled and active transport are of great interest in biology, one needs some criterion for distinguishing passive, coupled, and active transport. The following analysis provides an equation which to some extent serves this purpose.

We consider the flow J (in ions per second) of an ion across a membrane having an area A and a thickness dx. The flow J is proportional to the area, to the concentration c of the ions, and to the driving force (which is $d\bar{\mu}/dx$, in which $\bar{\mu}$ is the electrochemical potential), but is impeded by the frictional coefficient f of the ion; thus

$$J = -c\frac{A}{f}\frac{d\bar{\mu}}{dx} ; \tag{14}$$

J is evaluated in terms of the charge of the ion and the potential of the solution. An expression results that contains too many unknowns to be

---

*Some authors prefer to call coupled transport a kind of passive transport because the molecule is definitely moving passively down a chemical potential gradient; other authors call it a kind of active transport because it appears to be moving against a concentration gradient. The reader should keep this in mind when reading the scientific literature.

solvable; the unknowns can be eliminated by obtaining an expression for the ratio of the flow from left to right ($J_{l,r}$) to that in the reverse direction ($J_{r,l}$). The expression that results is

$$\frac{J_{l,r}}{J_{r,l}} = \left(\frac{\gamma_l}{\gamma_r}\right)\left(\frac{c_l}{c_r}\right)\exp\left[\frac{z\mathscr{F}}{RT}(\phi_l - \phi_r)\right], \tag{15}$$

in which γ is the activity coefficient of the ion, z is the charge, ϕ is the electrical potential of the solution, and the subscripts l and r refer to the left and right compartments. The values of $J_{l,r}$ and $J_{r,l}$ can be determined experimentally by adding radioactive ions to one side of the membrane and determining the rate of appearance of radioactivity on the other side; $\phi_l - \phi_r$ is just $\Delta\phi$, which is a measurable quantity. If a system satisfies this equation (known as the *Ussing equation*), the ion diffuses passively. However, departures from this equation are often observed that do not necessarily indicate that there is either active or coupled transport. The criterion for nonpassive transport is that $J_{l,r}/J_{r,l}$ is less than the right side of Equation 15. However, this is not a sufficient condition, because this inequality could have several other causes. One cause is an interaction between the ion transported and some component of the channel through which the ion passes. Another important cause of inequality is exchange diffusion, in which ions cross a membrane on a carrier molecule. That is, the membrane contains a molecule that can pick up ions from one side of the membrane and discharge them on the other. If the carriers are always saturated, and charging and discharging occurs with 100 percent efficiency, transport can occur against an electrochemical gradient without the consumption of energy that is characteristic of active or coupled transport. This seems to be a violation of our understanding of the meaning of potential gradients. The clue to understanding the phenomenon is the fact that even though transport may occur against a potential gradient, it only occurs with a chemical gradient. The explanation for the ability of the ions to ignore the electrical potential is that the ions are not being transported as charged units but as neutral particles, so that they are following the chemical rather than the electrochemical potential gradient. This type of transport is recognized by its lack of a requirement of an energy supply. Thus, in biological systems, the criterion for active or coupled transport is that the transport is stopped by the addition of inhibitors of oxidative phosphorylation, the primary source of energy. A common inhibitor is 2,4-dinitrophenol. If the Ussing equation is violated in the way just described, and transport continues in the presence of such an inhibitor, it is frequently the case that a carrier system is responsible for ion transport, although intrachannel interactions are not ruled out. There have also been observed cases of $J_{l,r}/J_{r,l}$ being greater than predicted and obeying a modified equation—namely,

$$\frac{J_{l,r}}{J_{r,l}} = \frac{\gamma_l c_l}{(\gamma_r c_r)^a}\exp\left[\frac{z\mathscr{F}}{RT}(\phi_l - a\phi_r)\right], \tag{16}$$

SEC. 14-3   MEMBRANES

in which a is a constant (usually a small positive integer). It has been shown that such an equation can result if there are defined channels through the membrane, through which ions pass in rows of a each.

Another deviation frequently observed occurs when there is solvent-flow across the membrane. This is a common effect in plant physiology.

The mechanism of passive transport is clear in that ions are simply following the electrochemical gradient—namely, the combined effect of concentration and potential differences. In active transport, the ions move against the gradient, using as a driving force the hydrolysis of high energy compounds, such as adenosine triphosphate. How the energy of hydrolysis is coupled to ion movement is at present unknown, and is a topic of current activity in biochemistry laboratories.

EXAMPLE 14-D   Identification of an ion that is actively transported.

The membrane of a particular cell is permeable to the Na^+ ion. Furthermore, the membrane potential is 80 mV at 20° C, with the inside of the cell at a lower potential than the outside. This potential is not a consequence of the distribution of Na^+ ions. The ratio of $[Na^+]_{inside}/[Na^+]_{outside}$ is 0.1. If the membrane were freely permeable to Na^+ ions, the Na^+ would accumulate on the negative side of the membrane—that is, $[Na^+]_{inside}$ would be greater than $[Na^+]_{outside}$. The opposite is true, which suggests that the higher external concentration is maintained by an active, energy-consuming process—that is, Na^+ ions must be pumped out of the cell. The Na^+ ion pump is in fact found in most cells.

We can calculate the free energy required to create the observed ratio of the Na^+ ion concentrations by using the electrochemical potential $\bar{\mu}$. For a single ion, this is $\bar{\mu} = \mu + z\mathcal{F}\phi = \mu° + RT \ln c + z\mathcal{F}\phi$. Calculating the difference between the electrochemical potential inside the cell, $\bar{\mu}_{inside}$, and outside the cell, $\bar{\mu}_{outside}$, we have

$$\bar{\mu}_{outside} - \bar{\mu}_{inside} = RT \ln\left[\frac{c_{outside}}{c_{inside}}\right] + z\mathcal{F}[\phi_{outside} - \phi_{inside}].$$

For the conditions stated above, $c_{outside}/c_{inside} = 10$,

$$\phi_{outside} - \phi_{inside}\ 80\ mV = 0.08\ V.$$

Thus,

$$\bar{\mu}_{outside} - \bar{\mu}_{inside} = (8.314)(293)(\ln 10) + 96{,}485(0.08)$$
$$= 13{,}328\ J\ mol^{-1}.$$

Thus, the electrochemical potential outside the cell is greater than inside the cell, indicating that if $[Na^+]_{outside} > [Na^+]_{inside}$, the external Na^+ ions must move against the electrochemical potential, and this requires an active, energy-consuming process.

Ion Transport across Membranes by Carrier Molecules

Films consisting solely of nonpolar molecules are generally impermeable to polar molecules. Thus, if a lipid film were formed between two aqueous solutions of KCl, each solution having a different concentration, neither K^+ nor Cl^- ions could pass through the film. Suppose there was added to one of the solutions a very small quantity of a molecule that has both a very large nonpolar part and a small charged region. Owing to hydrophobic interactions, this molecule will tend to escape from the aqueous environment by embedding its nonpolar portion in the lipid film and leaving its polar portion in the aqueous region. As a result, the film will acquire a charge. Let us assume the charge is negative. This negative region will bind the K^+ ions. However, once a K^+ ion is bound, the total charge on the molecule will be reduced and may even be zero. The molecule, which we call a *carrier,* can now be reoriented in the film by thermal agitation. This was not possible before the K^+ ion became bound, because the polar region would have been excluded from the liquid film. Given time, this reorientation might result in a 180° change in direction, so that the K^+ ion will have been carried across the membrane to the second solution. The probability that the carrier molecule will retain the K^+ ion is proportional to the concentration of the K^+ ion in that solution. Once a carrier molecule is reoriented, another K^+ ion might also be carried back. Note, however, that this mechanism cannot result in a net flow of K^+, because an empty carrier is charged and cannot reenter the film and flip back to the original position. Thus, to reenter the film, a K^+ ion must be carried back. Such a carrier molecule is called a *silent carrier*. If radioactive K^+ is on one side of the membrane and nonradioactive K^+ is on the other, the carrier will enable exchange to occur until the ratio of radioactive to nonradioactive K^+ ions is the same on both sides of the film. This enables one to demonstrate that a silent carrier is present. Such a carrier, however, has little utility to a living cell.

If the carrier could flip back in some way without binding an ion, net flow could occur. This is possible, and an example of such a carrier is *valinomycin,* a substance that transfers only the K^+ ion (a substance that can carry an ion through a nonpolar membrane is called an *ionophore*). Valinomycin is a flat nonpolar molecule in the center of which is a hole into which are directed the polar groups (Figure 14-6). It is readily incorporated into nonpolar films and moves freely through the film because its polar region is buried in the center of the molecule. While it is on the surface of a nonpolar film, a K^+ ion can slip into the polar hole, into which it just fits. By random motion, the nonpolar valinomycin molecule can move to the other side of the film, where its bound K^+ ion can be discharged. The probability of picking up and discharging a K^+ ion is proportional to the K^+ concentration on the two sides of the film. The important difference between a carrier such as valinomycin and a silent carrier is that valinomycin can move even if it does not contain a K^+ ion. Thus, it will shuttle K^+ ions from the region of high to low concentration. Note that

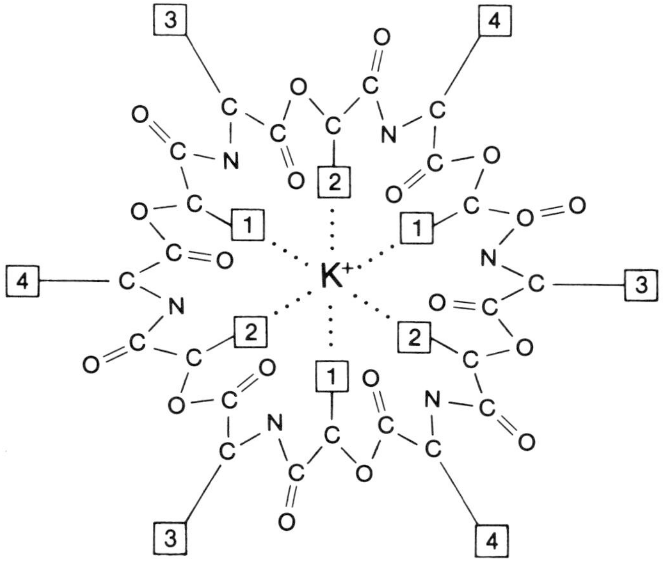

FIGURE 14-6 The structure of valinomycin. The numbers represent the following: 1, L-lactate; 2, D-hydroxyisovalerate; 3, L-valine; 4, D-valine. The K^+ ion makes coordination bonds within the hydrophilic interior of the molecule. The hydrophobic exterior, resulting from the hydrocarbon chain of valine, allows valinomycin to dissolve in the nonpolar regions of biological membranes.

since valinomycin carries only K^+ and cannot transport Cl^- at all, if the membrane is impermeable to Cl^-, at equilibrium, the K^+ concentration is not the same on both sides of the film. This will generate a membrane potential across the film and the ion distribution will be governed by the Donnan equilibrium.

A third and important type of ionophore is the so-called *channel*. This is usually a large molecule (often a macromolecule) that has a nonpolar exterior, which enables it to settle into a nonpolar film, and a pore that allows free passage through the film in both directions. Such carriers may transport only certain ions. This selectivity might occur in two ways: (1) the pore size might allow only ions of less than a certain diameter to pass through; or (2) the pore might contain either a positively or negatively charged group, which would exclude passage by either a positive or negative ion, respectively. Channels often have the important property that the pore can be opened or closed if either external agents or reagents are applied to the film. We will see an example of this shortly.

We have mentioned ion pumps several times without explanation. Although there are many kinds of ion pumps, and their mechanisms are poorly understood, the basic idea of one type of pump follows from the carrier model just described. Suppose a carrier molecule is in a membrane separating solutions of high and low K^+ concentration, and is oriented with its negatively charged region in the solution of low K^+ concentration. When a K^+ ion is bound, an energy-requiring system might either rotate the carrier by 180° or, if the carrier is a protein, induce a shape-change that transports the ion to the region of high K^+ concentration. However, in this movement, the carrier might be altered so that it does not bind K^+;

therefore the K^+ will be discharged into the region of high K^+ concentration. Reorientation to the original state could occur randomly, or the carrier could again be directed by an energy-consuming process. Thus, in this way, a K^+ ion could move against an electrochemical gradient.

Several kinds of bacteria synthesize antifungal agents that are actually Na^+ and K^+ carriers. These substances, which cause leakage of ions, act in the following way. Most living cells maintain a high internal K^+-ion concentration by means of a K^+-ion pump embedded in a membrane; the membrane is sufficiently impermeable to K^+ ions that outward leakage is small. The antifungal bacterial species excrete these carriers, which are unable to penetrate the membranes of other bacteria but are easily incorporated into the membranes of fungi. The carriers cause so much leakage that the ion pump is no longer able to maintain the concentration difference. The specificity toward fungi is usually a result of the fact that the antibiotic can only penetrate cholesterol-containing membranes, and cholesterol is present in fungal but not in bacterial membranes. Once the carrier has become embedded in the fungal membrane, Na^+ and K^+ ions are freely exchanged with the environment, so that if external Na^+ ions are present, the ion balance is disrupted and the fungal cell dies. In this way a bacterium producing such a substance can compete in nature with fungi.

14–4 EFFECT OF IONS ON THE CONDUCTIVITY OF CHARGED MEMBRANES

The concentration of the Ca^{2+} ion has a profound effect on the ability of some biological membranes to transport other ions. At one time, it was thought that this is caused exclusively by specific binding of the Ca^{2+} ion to sites on the membrane. However, it can be shown from Gouy-Chapman theory that with a solution of monovalent and divalent cations in contact with a negatively charged membrane, the contribution of the divalent cation to the effective electrical potential across the membrane is vastly greater than that of the univalent cation. This can be shown as follows.

Consider a membrane having a surface charge density in contact with a solution of NaCl and $CaCl_2$. Ignoring the Helmholtz layer,* we note that the concentration of the ions at a particular distance from the membrane is determined by the electrical potential ϕ (caused by the membrane) at that distance, according to the Boltzmann distribution. Thus

$$\left. \begin{array}{l} [K^+] = [K^+]_\infty \exp(-e\phi/kT) \\ [Cl^-] = [Cl^-]_\infty \exp(-e\phi/kT) \\ [Ca^{2+}] = [Ca^{2+}]_\infty \exp(-2e\phi/kT) \end{array} \right\} \quad (17)$$

---

*This is an important point. The Helmholtz layer consists of bound charges. We are discussing the Gouy layer, so that we are describing a *screening* effect of the ion cloud rather than an effect of bound charges.

SEC. 14-4 CONDUCTIVITY OF CHARGED MEMBRANES

in which e is the charge of the electron in coulombs, ϕ is expressed in volts, $[X]_\infty$ is the concentration of ion X far from the surface, and $k = 1.4 \times 10^{-23}$. We will see that it is the 2 in the exponent of the equation for $[Ca^{2+}]$ that produces the effects that have just been mentioned. The potential ϕ must also obey the Poisson equation,

$$\frac{d^2\phi}{dx^2} = -\frac{\rho}{D\epsilon_0}, \qquad (18)$$

in which D is the dielectric constant of water, and ρ, the charge density (in C m^{-3}) in the solution, is equal to $([K^+] + 2[Ca^{2+}] - [Cl^-])$. Combining these equations, integrating, using the facts that $\phi = 0$ and $d\phi/dx = 0$ at $x = \infty$, and that the double layer is electrically neutral (that is, $0 = \int_0^\infty \rho \, dx$), and applying the electroneutrality condition, $[K^+]_\infty + 2[Ca^{2+}]_\infty = [Cl^-]_\infty$, yields the relation between the charge density, ion concentration, and the potential ϕ_0 at the membrane surface—namely,

$$\frac{\sigma^2}{2\epsilon_0 DkT} = [K^+]_\infty\{\exp(e\phi_0/kT) + \exp(-e\phi/kT) - 2\}$$
$$+ [Ca^{2+}]_\infty\{2\exp(e\phi_0/kT) + \exp(-2e\phi_0/kT) - 3\}. \qquad (19)$$

At this point we restrict our discussion to those membranes for which the potential is relatively large and negative—that is, more negative than -5 millivolts. At commonly encountered temperatures, $\phi \ll -kT/e$, so we may approximate Equation 19 as

$$\frac{\sigma^2}{2\epsilon_0 DkT} = [K^+]_\infty \exp(-e\phi_0/kT) + [Ca^{2+}]_\infty \exp(-2e\phi_0/kT). \qquad (20)$$

Note that this is a quadratic equation with variable $\exp(-e\phi_0/kT)$, which can be solved by the quadratic formula to yield

$$\exp(-e\phi/kT) = \frac{-[K^+]_\infty + \sqrt{[K^+]_\infty^2 + 2[Ca^{2+}]_\infty \sigma^2/DkT\epsilon_0}}{2[Ca^{2+}]_\infty}. \qquad (21)$$

For biological systems, $[K^+]$ is between 0.01 M and 0.1 M, $[Ca^{2+}]$ is 0.0001–0.001 M, and σ ranges from 1 charge per 10 Å^2 (1 per 10^{-19} m^2) to 1 per 200 Å^2 (1 per 2×10^{-18} m^2). Under these conditions,

$$[K^+]_\infty^2 \ll [Ca^{2+}]_\infty \sigma^2/DkT\epsilon_0. \qquad (22)$$

Using this approximation and solving for ϕ_0 yields

$$\phi_0 = \frac{kT}{e} \ln\left(\frac{2[Ca^{2+}]_\infty DkT\epsilon_0}{\sigma^2}\right). \qquad (23)$$

Let us now examine what this equation means for a system consisting of a negatively charged membrane separating two solutions of NaCl, one of which also contains some Ca^{2+} ions. When the NaCl concentrations on both sides of the membrane are the same, the two sides will be at the same potential and the relation between the electrical potential and distance will be as shown in Figure 14-7(a). The significant points to be seen in this

FIGURE 14-7 Plots showing electrical potential as a function of distance from a negatively charged membrane separating solutions containing various concentrations of NaCl and CaCl$_2$. The central vertical rectangle represents the membrane in each case. The distance from the left to right side of each panel is about 100 Å. The horizontal dashed line within the membrane in panels (b)–(e) indicates the potential shown in panel (a), for comparison. The curvature in the plot for each potential is a result of the Gouy layer. Note that since, at distances greater than 100 Å, the potential is zero for all solutions, electrodes placed in the solutions will not detect the potential difference across the membrane; that difference is a local effect attributable solely to the Gouy layer.

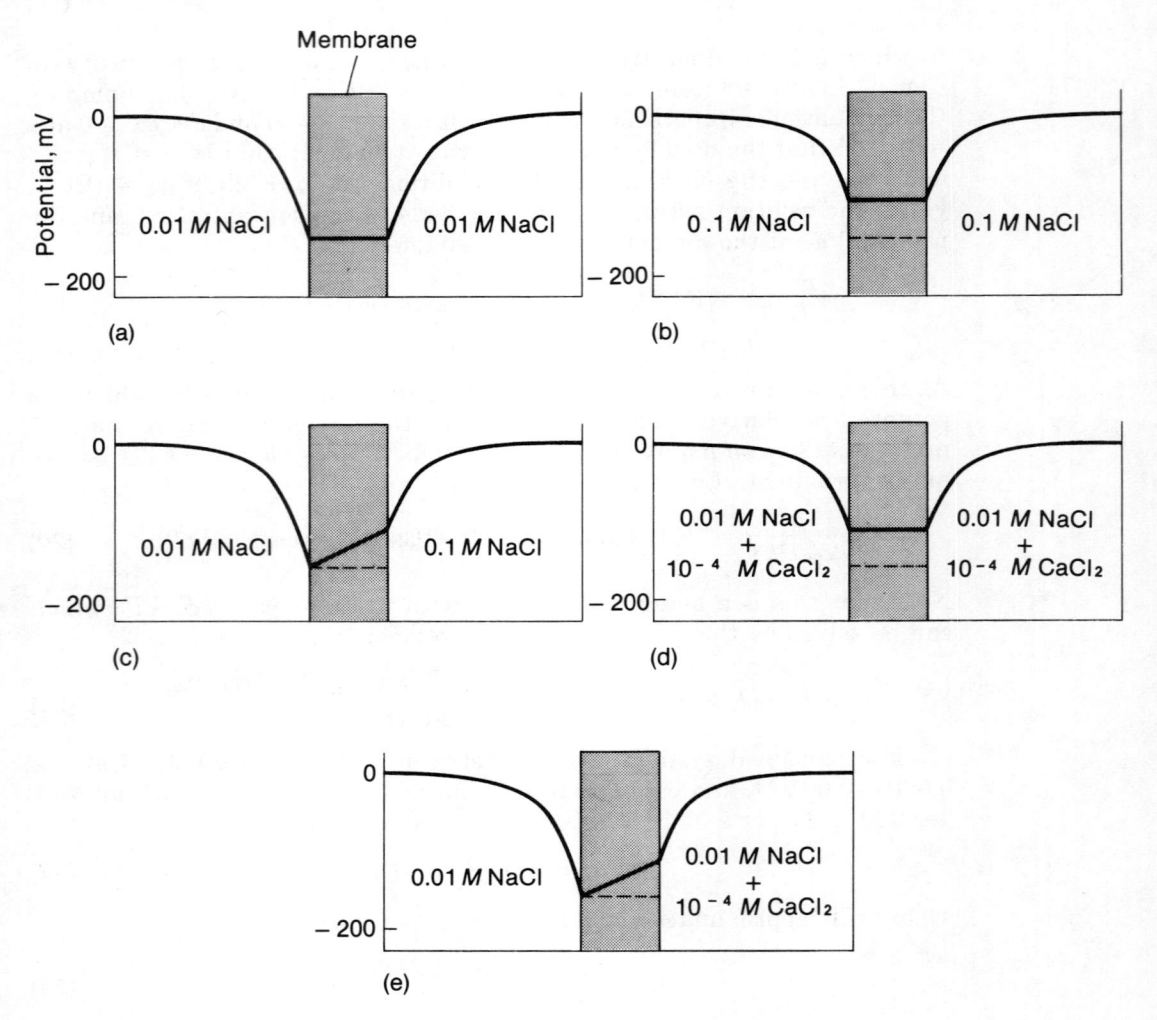

panel of the figure are that (1) the curvature of the plot near the membrane is caused by the Gouy layer, and (2) within the membrane the potential does not change with distance—that is, there is no electric field within the membrane. If the NaCl concentration is greater but still the same on both sides (as in Figure 14-7(b)), there is more screening by positive charges, so that the surface potential is less negative. However, as

SEC. 14-4 CONDUCTIVITY OF CHARGED MEMBRANES

long as the concentrations are the same on both sides of the membrane, there is still no electric field within the membrane. If the concentrations differ (as in Figure 14-7(c)), the potentials at a distance from the membrane remain the same but, owing to the different charge density on the two sides of the membrane, there is now an electric field within the membrane. Figure 14-7(d) shows the effect of adding a very low concentration of $CaCl_2$ to the solutions. Owing to the high valence of the Ca^{2+} ion compared to the Na^+ ion, the surface potential is again less negative, according to Gouy-Chapman theory. Figure 14-7(e) shows the effect of adding Ca^{2+} only to the right side. The total potential far from the membrane remains unchanged but now the potential on the right side of the membrane is less negative. As in Figure 14-7(c), this has the effect of increasing the strength of the electric field within the membrane. Note that the effect is the same as that produced by increasing the NaCl concentration on the right, but the concentration of NaCl needed is much greater than the very small amount of $CaCl_2$ that produces the identical effect.

What does this all mean? At a distance of more than 100 Å (10^{-8} m) the potential is zero for all solutions, so that electrodes placed in the two solutions will not detect the potential difference across the membrane, which is a local effect attributable solely to the Gouy layer. Let us now consider a membrane that is permeable to the Na^+ ion but not to the Ca^{2+} ion. What will be the effect of the Ca^{2+} ions on the movement of the Na^+ ions through the membrane? With no external voltage applied, there will be no current with or without Ca^{2+} because there is no thermodynamic driving force (that is, $[Na^+]$ is the same on both sides of the membrane). With an applied voltage, the effect of Ca^{2+} will depend on the particular mechanism of Na^+ transport by the membrane. If the membrane contains a Na^+ carrier, the Ca^{2+} ions will in general reduce the flow of Na^+ ions by lowering the $[Na^+]$ at the membrane surface.

The driving force for transport is of course just the electric field produced by the applied voltage. The situation is quite different if ion transport occurs by means of a voltage-sensitive channel. For instance, let us consider a voltage-sensitive transport system that is inoperative until an electric field of the appropriate polarity is applied—that is, the channels are closed when there is no electric field. When an external voltage is placed across the membrane and Ca^{2+} is not present, the voltage gradient occurring within the membrane might, if it is great enough, cause the channels to open. If Ca^{2+} ions are present, they create an electric field in the membrane, as we have just shown, so that a lower applied voltage is necessary to open the channels; this will manifest itself as an increase in electrical conductivity (that is, increased Na^+ ion transport), when Ca^{2+} ions are added.* In the more common biological situation, the voltage-sensitive channels have the property that the polarity of the potential that

---

*If $[Na^+]$ is not the same on both sides of the membrane, it is possible that the electric field induced by the Ca^{2+} ions may be great enough to cause the channels to open; in this case, one would observe transport of Na^+ in the absence of an applied field.

controls the opening of the channels is such that the applied voltage must counteract a potential that keeps the channels closed. Again the Ca^{2+} ions increase the internal electric field, but in this case, a higher applied voltage is needed to open the channels; thus the presence of Ca^{2+} ions *decreases* the electrical conductivity of the membrane. This type of channel will be discussed in a brief description of nerve transmission in the next example.

It is important to realize that the effect produced by the Ca^{2+} ions in the Gouy-Chapman theory depends only on valence, so that both Ca^{2+} and Mg^{2+} produce the same effect; hence, trivalent ions such as Al^{3+} produce a much larger effect. The effect of valence cannot be generalized to all biological membranes, because often the identity of the ion is important in the chemical reactions that are involved in ion transport. A second point worth repeating is that the general effect we have just described is a screening effect due to the clustering of ions in solution in the vicinity of the membrane rather than an effect of binding of ions.

We now consider a real biological example that can be understood in terms of the preceding discussion.

EXAMPLE 14–E Effect of Ca^{2+} on the properties of nerve cells.

Figure 14-8 shows a plot of the potential in the region around the membrane of a resting nerve cell. The left side of the membrane represents

FIGURE 14-8 Diagram of the electrical potentials in the neighborhood of a nerve cell membrane. The dashed line shows the effect of external Ca^{2+} ions.

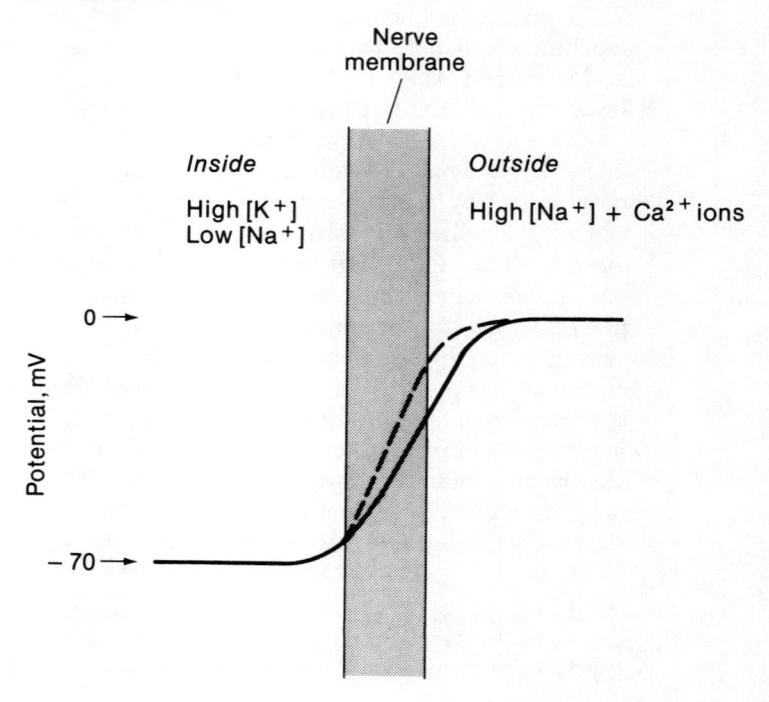

the inside of the cell; this is at a potential of −70 mV compared to the outside of the cell. The membrane contains two kinds of channels, one that can pass only Na$^+$ ions and another that can pass only K$^+$ ions. These channels are voltage-sensitive in the sense that each one opens when the membrane potential (that is, the electric field inside the membrane) is reduced below a critical threshold value (between zero and −70 mV). Transmission of an impulse along a nerve begins when an electric signal at the end of the nerve cell reduces the potential below the critical value and allows some of the channels to open. There are usually Ca^{2+} ions in addition to Na$^+$ ions in the fluid surrounding the cell, so that the electric field in the membrane is determined not only by the voltage differences across the membrane but also, as we saw above, by the Ca^{2+} concentration. It is known that increasing [Ca^{2+}] reduces the ability of the nerve to start propagation of an impulse. This is easily understood in terms of the Gouy-Chapman theory. When [Ca^{2+}] is increased, the electric field within the membrane increases, as was seen above. Thus an initial impulse sufficient to produce a particular reduction in potential does not decrease the field in the membrane to a value low enough to open the channels. Similarly, a lower-than-normal value of [Ca^{2+}] makes a nerve cell more active because, in that case, a smaller reduction in negative potential is needed to reduce the field within the membrane to a value that opens the Na channel.

Note that there is no need to require that Ca^{2+} bind to the cell membrane, since its mere presence in the Gouy layer has the observed effect on the efficiency of transmission of nerve impulses. However, there is evidence that specific binding of Ca^{2+} ions does play an important role in the behavior of the nerve cell—namely, that Mg^{2+} ions do not have the same effect that Ca^{2+} ions have—as is predicted from the Gouy model.

REFERENCES

Adamson, A.W. 1967. *Physical Chemistry of Surfaces*. Wiley-Interscience.
Chapman, D. 1968. *Biological Membranes, Physical Fact and Function*. Academic Press.
Harris, R., and G.G. Lunt. 1975. *Biological Membranes: Their Structure and Function*. Wiley.
Katz, B. 1966. *Nerve, Muscle, and Synapse*. McGraw-Hill.
Keynes, R.D. 1979. "Ion Channels in the Nerve-Cell Membrane." *Sci. Amer.* March.
Kotyk, A., and K. Janacek. 1975. *Cell Membrane Transport*. Plenum.
Rubinson, K.A. 1977. "Chemistry and Nerve Conduction." *J. Chem. Educ.* 54, 345.

PROBLEMS

1. A membrane that is permeable to small ions but not to protein molecules separates two compartments. Initially one compartment contains 1 M NaCl and the other contains a 0.1-M solution of the Na$^+$ form of a negatively charged univalent

protein (P⁻). What are the concentrations of Na^+, Cl^-, and P^- in each compartment at equilibrium?

2. A vessel is divided by a membrane permeable to A^+ and B^- ions but not to protein molecules. On the left side of the membrane is a solution containing a 10^{-4}-M solution of protein P to which six A^+ ions are bound. On the right side of the membrane is 10^{-2} M A^+X^-. What is the Donnan potential across the membrane at 25°C?

3. The Donnan equilibrium causes an imbalance in the concentrations of ions in solution. Consider a DNA solution at a concentration of 1 mg per cm³. Initially, this solution contains 0.01 M NaCl (left side); it is allowed to come to equilibrium with an equivalent volume of water (right side) separated by a membrane permeable to Na^+ and Cl^- ions but not to DNA. Calculate the osmotic pressure at equilibrium at 25°C owing to the concentration difference. The molecular weight of a nucleotide is 336, and there is one negative charge per nucleotide. The molecular weight of the DNA is 10^7.

4. A membrane permeable to Mg^{2+} ions but not to Cl^- ions separates solutions of 0.015 M $MgCl_2$ and 3×10^{-4} M $MgCl_2$. What is the potential difference at 20°C across the membrane? Is the potential higher or lower on the side facing the 0.015 M $MgCl_2$?

5. A membrane permeable to SO_4^{2-} ions but not to positively charged ions separates 0.1 M $ZnSO_4$ from 0.0035 M $CuSO_4$. What is the potential difference at 20°C across the membrane? Is the potential higher or lower on the side facing the $ZnSO_4$?

6. The electrical potential across the outer membrane of a cell is found to be 89 mV. The inside of the cell is negative with respect to the outside at 37°C. This potential is independent of the external K^+ concentration. However, for any value of the concentration of external K^+, the internal concentration is 28-fold greater than the external concentration.

(a) Is the K^+ ion concentrated by an active transport mechanism?

(b) What conclusion might you draw if the internal concentration were 40-fold greater than the external concentration?

(c) What conclusion would you draw if the internal concentration were 5-fold greater than the external concentration?

CHAPTER 15

THE PROPERTIES OF MACROMOLECULES

The word *macromolecule* can mean any large molecule but by convention the word denotes a polymer. The number of monomers in a polymer is the *degree of polymerization* of a polymer. The lower limit of the degree of polymerization that would require the use of the word macromolecule is undefined; most biochemists would not use the word to denote a tetramer but would use it to denote a molecule consisting of twenty monomers. The poorly defined word *oligomer* usually denotes a macromolecule that consists of more than one monomer but is not large enough to be called a polymer.

The range of macromolecular size is very great. For example, typical proteins contain 100 to 4000 amino acids, whereas the smallest naturally occurring RNA molecule has about 70 nucleotides and the largest known DNA molecule contains 5×10^7 nucleotides.

15–1 CHEMICAL AND PHYSICAL PROPERTIES OF MACROMOLECULES

Classes of Macromolecules

Macromolecules can be classified in several ways—by the number of their components, the kinds of their components, and the overall shape of the molecule. A polymer consisting of a single type of monomer is called a *homopolymer*. If the monomer is Q and there are n monomers per polymer, the polymer may be denoted poly Q or poly Q_n or $(Q)_n$. The first form, poly Q, is the most common notation. Sometimes the identity of a monomer is ambiguous. For example, polyethylene, H—(CH$_2$CH$_2$)$_n$—H, which is made by polymerizing ethylene (CH$_2$=CH$_2$), contains no double bonds and could just as well be called polyethane or polymethane. This confusion is especially prevalent in biochemistry because biochemical dimers often have special names. Thus the polysaccharide amylose can be viewed as a polymer of either α-D-glucose or α-D-maltose, since the latter is the name of the α-D-glucose dimer.

If two or more different monomers are found in a macromolecule, it is called a *copolymer*. Protein molecules (linear sequences of amino acids), nucleic acids (linear sequences of nucleotides), and polysaccharides (polymers of sugars) are all examples of biological copolymers. An *alternating copolymer* contains two monomers A and B and has the chemical sequence ABABAB . . . The formulas for the amino acids and nucleotides, and the basic structure of proteins and nucleic acids, are given in Figures 15-1 through 15-4.

FIGURE 15-1 The amino acids and their chemical structures.

Polar charged amino acids (tend to be on protein surface)

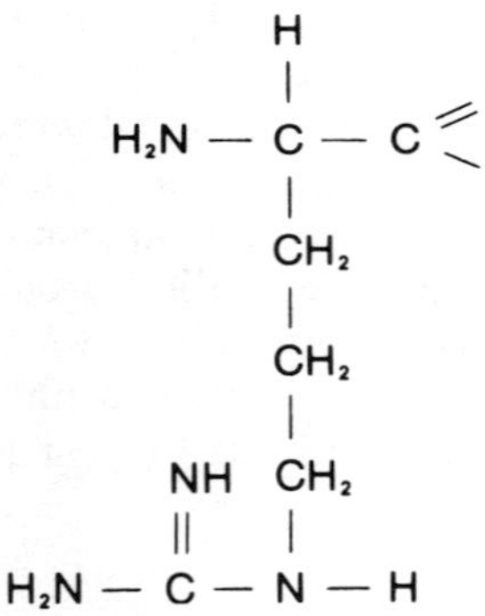

Arginine

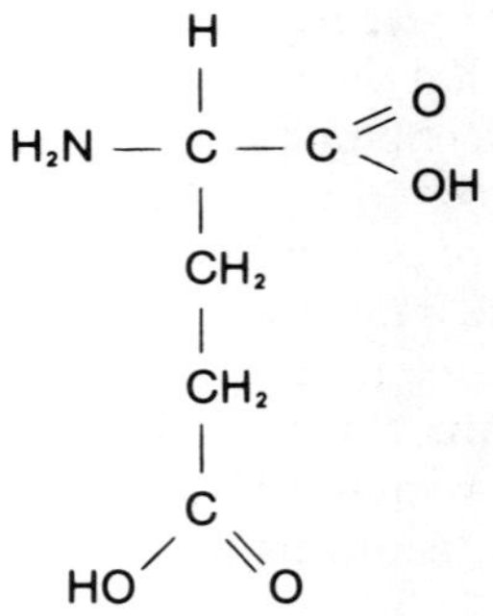

Glutamic acid

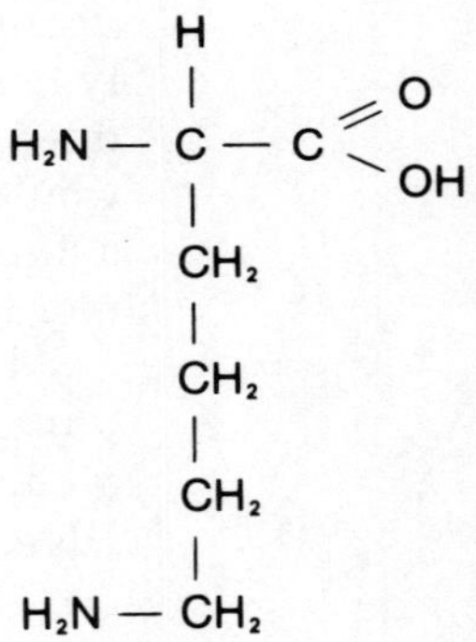

Lysine

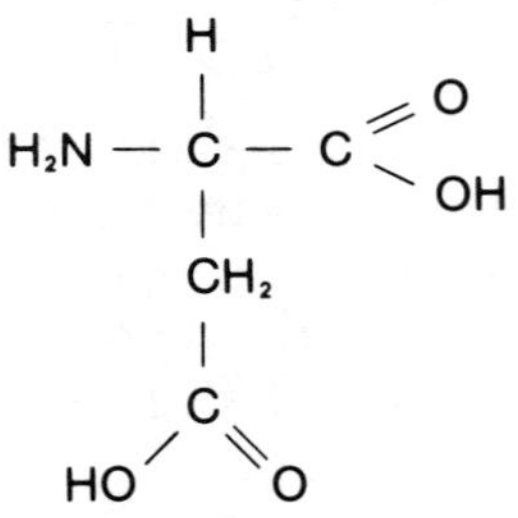

Aspartic acid

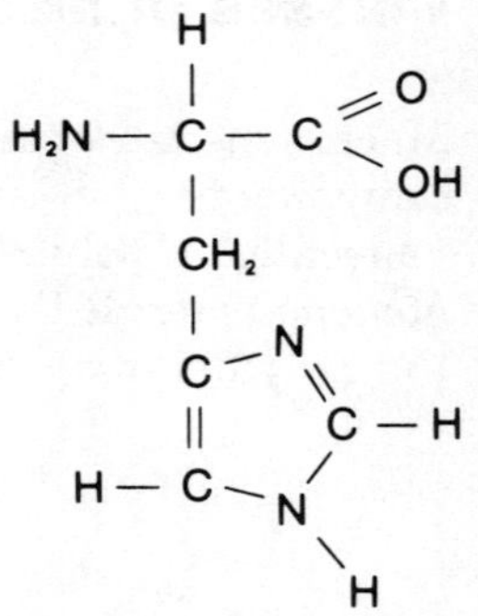

Histidine

FIGURE 15-1 (continued)

Polar uncharged amino acids (tend to be on protein surface)

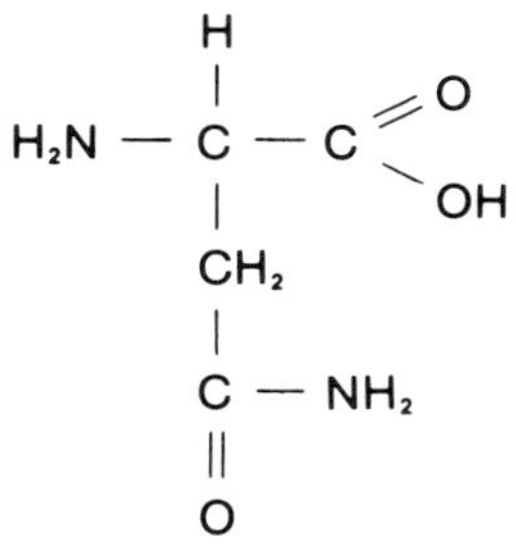

Asparagine 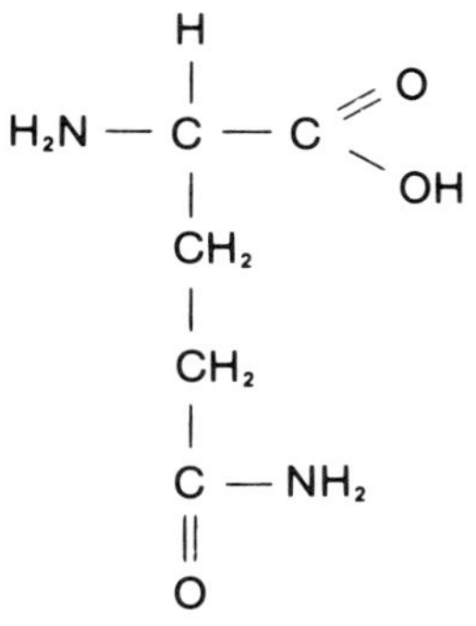   Glutamine 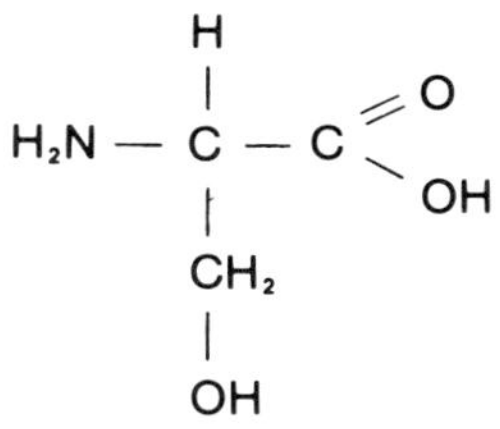   Serine 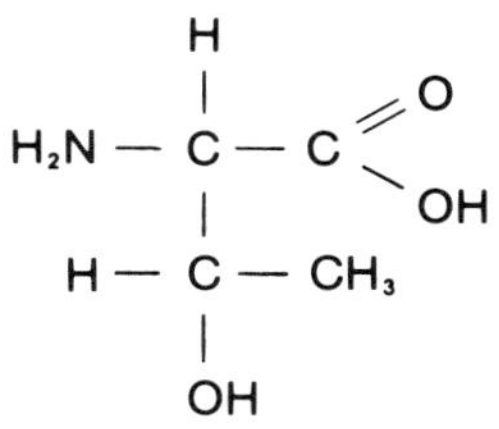   Threonine

Nonpolar amino acids (tend to be internal)

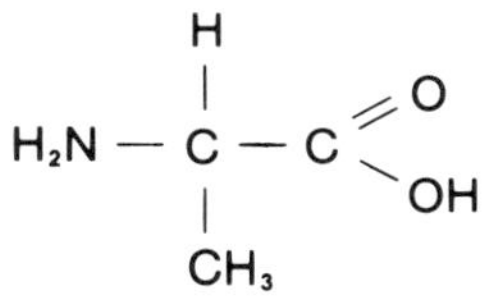

Alanine 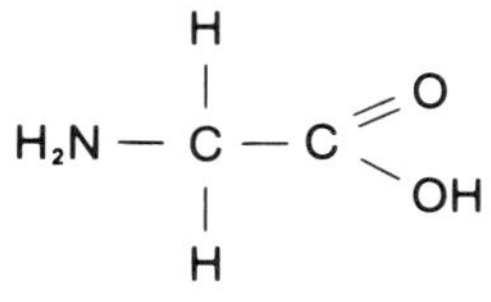   Glycine 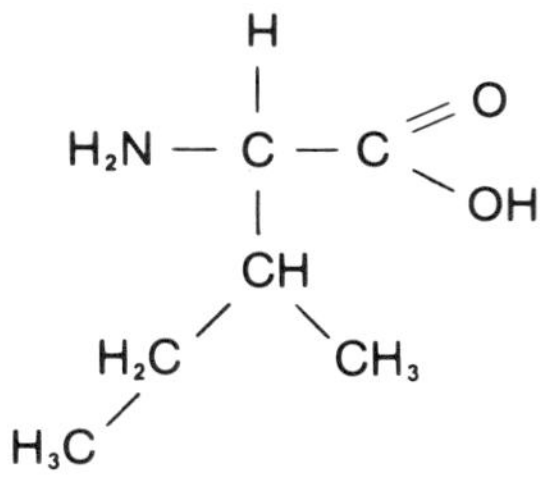   Isoleucine 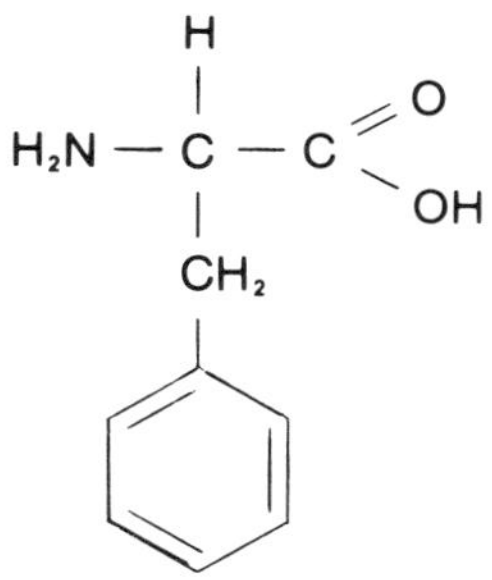   Phenylalanine

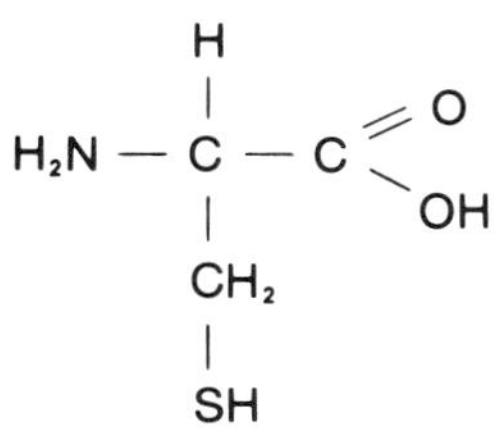

Cysteine 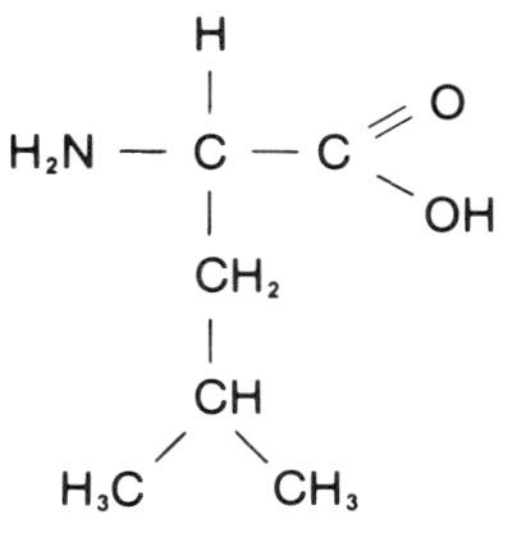   Leucine 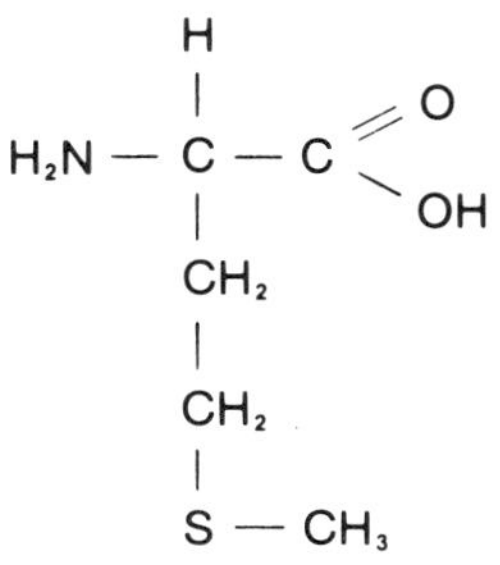   Methionine 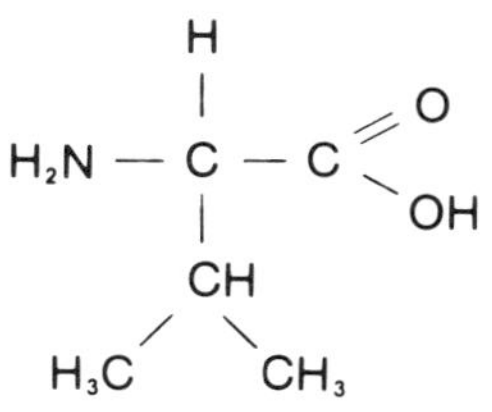   Valine

498 CHAP. 15 THE PROPERTIES OF MACROMOLECULES

FIGURE 15-1 (continued)

Amino acids equally frequently internal and external

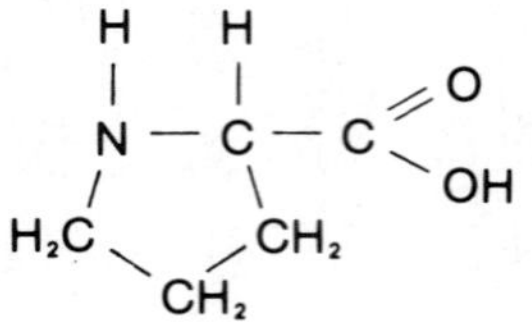

Proline

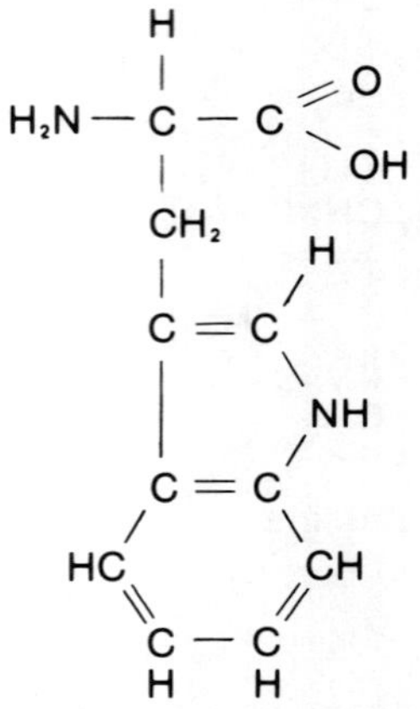

Tryptophan
(nonpolar)

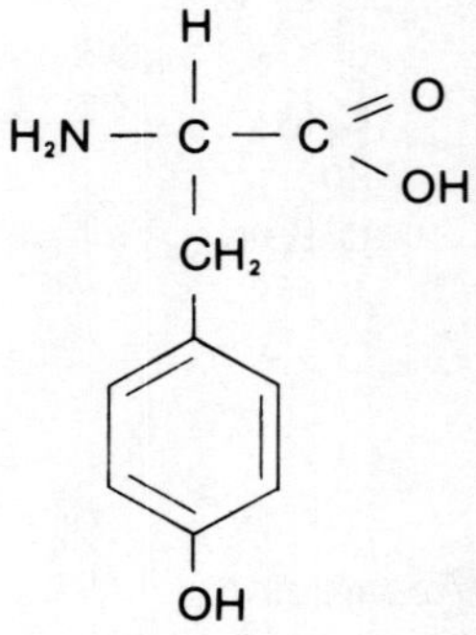

Tyrosine
(polar)

FIGURE 15-2 The four bases of DNA, showing the attachment of the sugar deoxyribose. If the encircled methyl group in thymine were replaced by a hydrogen, the result would be the RNA base, uracil. In RNA, the deoxyribose is replaced by ribose.

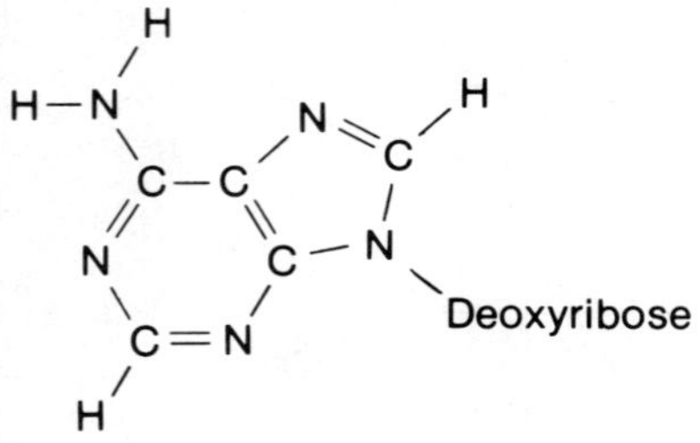

Adenine

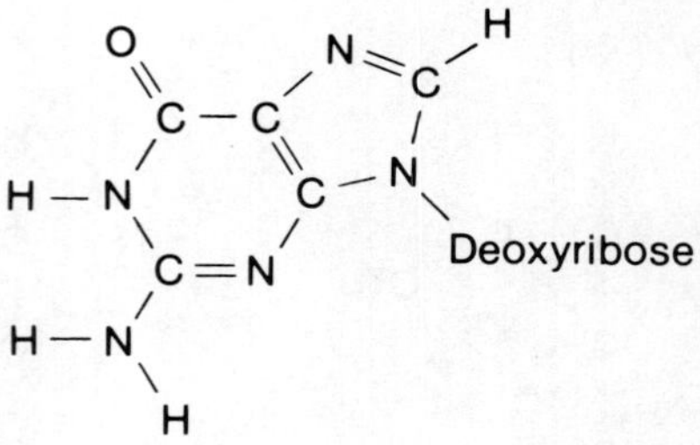

Guanine

FIGURE 15-2 (continued)

Thymine

Cytosine

FIGURE 15-3 Structure of a polypeptide chain, showing amino and carboxyl termini, peptide bonds, and the locations of the side chains (R_1, R_2, and R_3). If the polypeptide chain contains many amino acids, it is called a protein. The peptide bond has a partial double-bond character which makes each peptide group a planar unit.

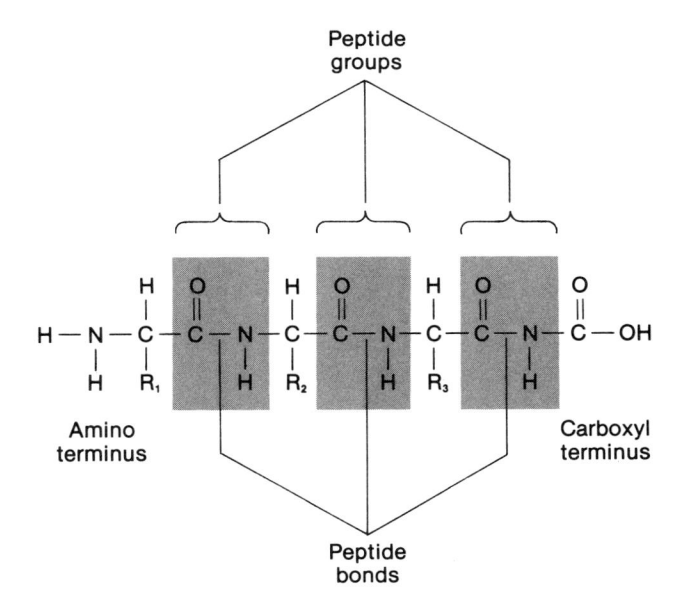

Macromolecules are also classified by the chemical nature of their components. Thus, there are simple polymers, such as proteins, nucleic acids, and polysaccharides, and complex polymers, such as glycoproteins (sugars attached to proteins), lipoproteins (lipids attached to proteins), and lipopolysaccharides (lipids linked to polysaccharides). There are also complex

FIGURE 15-4 Structure of a single polynucleotide chain. (a) The chemical structure. The sugar (ribose or deoxyribose) and phosphate moieties alternate, a phosphate always connecting the 3'- and 5'-carbon atoms. (b) A common schematic way of drawing a polynucleotide.

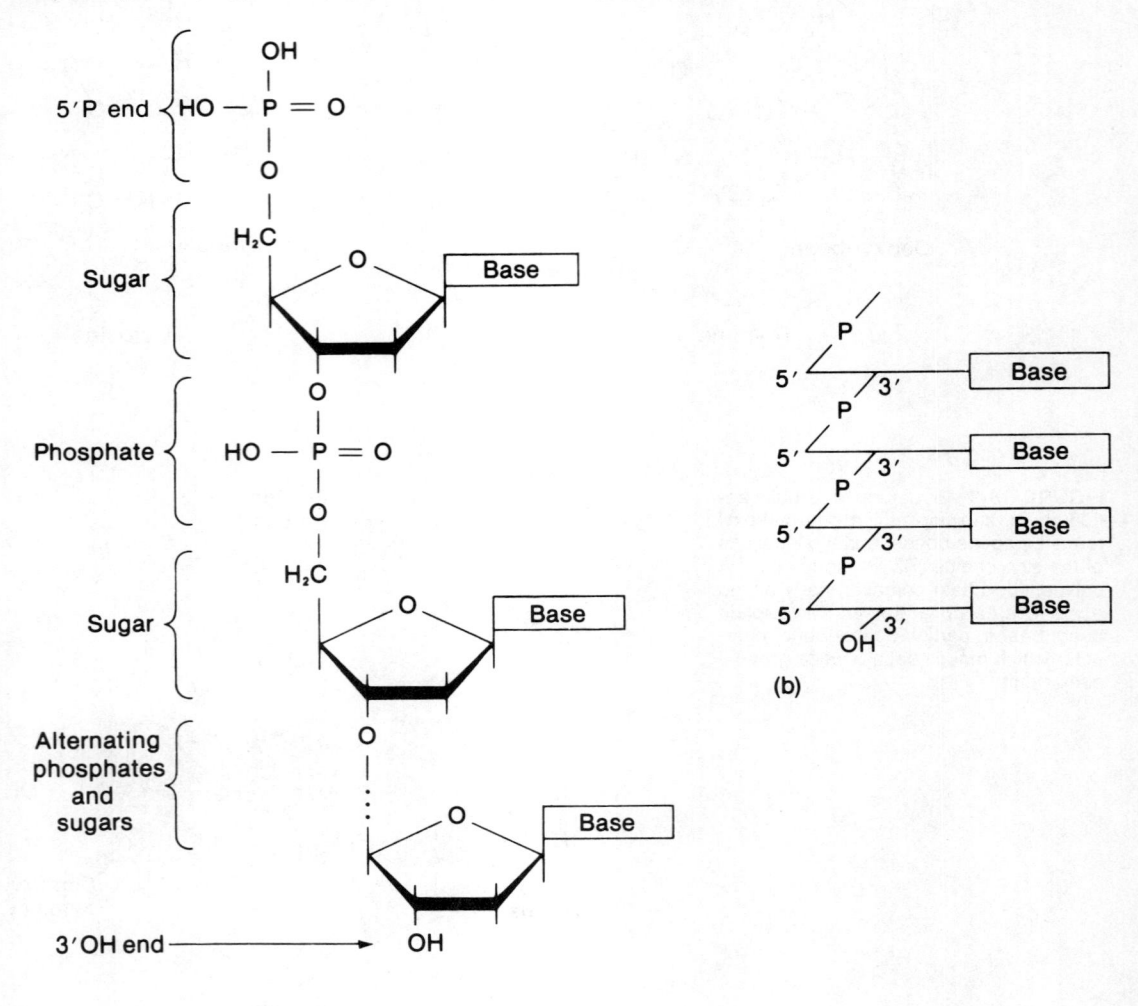

aggregates, such as nucleoproteins (nucleic acids and proteins held by noncovalent bonds), phages and viruses (a particular class of nucleoproteins), and membranes (a system of interacting proteins, lipids, lipoproteins, and lipopolysaccharides).

Polymers do not always have the same number of ends, and this gives rise to still another scheme of classification. For example, protein molecules and ribonucleic acid molecules have two ends and are *linear* polymers. Linear molecules are sometimes joined together to form helical

FIGURE 15-5 Several helical forms of macromolecules. [SOURCE: *Physical Biochemistry*, First Edition, by David Freifelder. W.H. Freeman and Company. Copyright © 1976.]

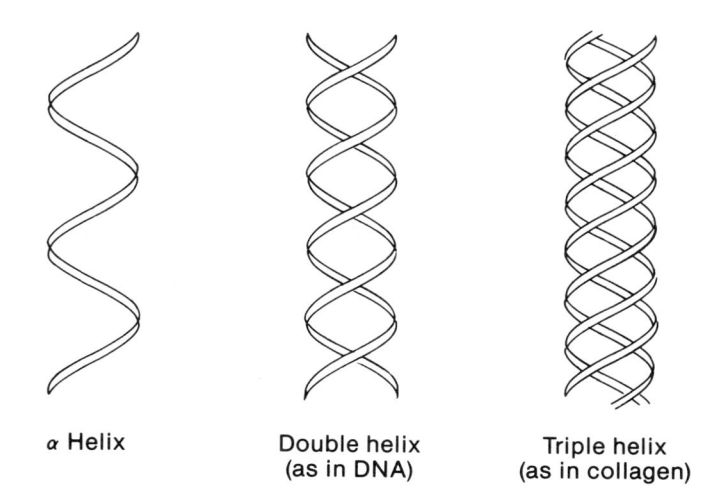

α Helix

Double helix (as in DNA)

Triple helix (as in collagen)

structures consisting of two or three molecules, as shown in Figure 15-5. Many polysaccharides (for example, glycogen) have many ends and are *branched* polymers (Figure 15-6). DNA molecules frequently have no ends and are *circular*. Circular DNA molecules are of three types: the *nicked circle,* in which one strand is continuous and the other is discontinuous; the *covalent circle,* in which both strands are continuous; and the *twisted circle* (also called the *supercoil* or *superhelix*), which can be thought of as the result of rotating one end of a linear DNA molecule with respect to the other end and then joining the ends with covalent bonds. Linked circles called *catenanes* have also been observed. These structures are shown in Figure 15-7.

Shapes of Linear Polymers

The function of a biological macromolecule is determined by its structure. Thus the biochemist goes to great pains to get as complete a description of the molecule as possible. The process usually begins by the determination of the general shape of the molecule and its molecular weight. In this

FIGURE 15-6 The kind of branched structure present in a sample of the polysaccharide glycogen.

FIGURE 15-7 Various conformations of DNA. [SOURCE: *Physical Biochemistry,* First Edition, by David Freifelder. W.H. Freeman and Company. Copyright © 1976.]

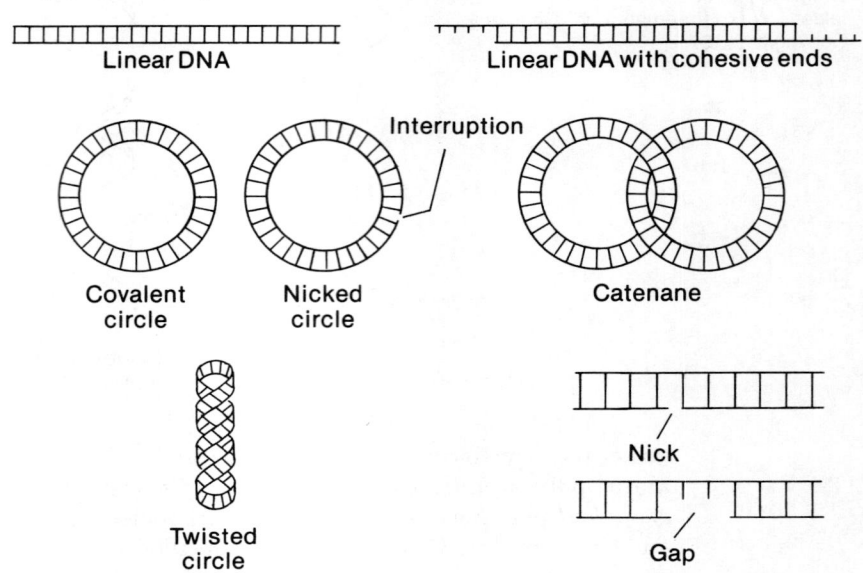

Linear DNA

Linear DNA with cohesive ends

Covalent circle

Nicked circle

Interruption

Catenane

Twisted circle

Nick

Gap

section we discuss the matter of shape; the determination of molecular weight by several methods is presented later in this chapter.

If the monomers of a macromolecule are connected by bonds that are easily bent, about which there can be unrestricted rotation, and if the monomers neither attract nor repel one another, the monomer chain is in a form known as the *random coil*. This is the structure that would exist if each monomer could be at *any* angle with respect to the adjacent monomer. Of course, such a perfectly random structure cannot exist because no bond is perfectly flexible and because no two distant monomers can occupy the same space. However, there is a tendency to consider a molecule to be a random coil if it is fairly flexible and if there are neither attractive nor repulsive forces (other than complete overlap) between any pairs of monomers. This is an extremely rare occurrence, if it happens at all, because in biological macromolecules there are usually very significant attractive interactions between the monomers.

Biological macromolecules are not usually random coils but have a complex orientation in space (they have *tertiary structure*), which is vital for activity of the molecule. The tertiary structure is governed by many forces and interactions, such as hydrogen-bonding, hydrophobic interactions, van der Waals forces, charge attraction (ionic bonds), and charge repulsion; these act between the backbone and the side-chain atoms. The term *side-chain interactions* is often used when referring to interactions that do not include the backbone. If the interactions maintain a generally spherical or ellipsoidal shape, the protein is said to be *globular;* if the molecule is extended and rodlike, it is said to be a *fibrous* protein. Often the interactions reduce the volume of the molecule to less than that of a random coil; in that case, the molecule is said to be *compact*.

Many polymers have a *helical structure* (Figure 15-5). A helix arises when monomeric units of the structure are related by a constant rotation about some axis plus a constant translation along that axis. Such a structure can take a cylindrical form. A simple analogy is the structure that would be obtained if a stack of playing cards were pinned together at one corner and then fanned so that each card is rotated at a fixed angle with respect to the one below. The α helix of proteins is one example of this structure. In an α helix a hydrogen bond is present between the C=O group of one peptide group and an NH group in a peptide group three residues farther along the chain (Figure 15-18). The NH group of the first peptide unit is also hydrogen-bonded with the C=O group three residues *back* along the chain. Thus, every peptide unit is engaged in *two* hydrogen bonds. Note that the side chains do not participate in formation of the α helix.

Other helical structures are also known. For instance, in polynucleotides, the planar nucleotide bases are stacked one above another but slightly rotated. In DNA, two polynucleotide strands, each extended by this stacking, hydrogen-bond to one another to form a *double-stranded helix;* some molecules (e.g., the protein collagen) can form a *triple-stranded helix;* these helical structures are shown in Figure 15-5. Helical molecules (single- and multiple-stranded) are examples of *extended* or *rodlike* molecules.

A common type of secondary structure in proteins is the β *structure,* in which two sections of a polypeptide chain (or in some cases two different chains) are aligned side by side and held together by hydrogen bonds. To maximize the number of hydrogen bonds, the polypeptide chains are pleated as shown in Figure 15-9. In this structure, the plane of the pleat contains the peptide group and the side chains are located alternately above and below the plane of the sheet. The figure shows only one of the

FIGURE 15-8 The α helix. (a) The dashed lines represent the two hydrogen bonds in which the peptide unit numbered 0 is engaged. These bonds join with peptide units three peptides upstream and downstream from unit 0. (b) The three-dimensional structure of the α helix, showing how the hydrogen bonds form a helix. The side chains project outward from the axis of the helix.

(a)

FIGURE 15-8 (continued)

(b)

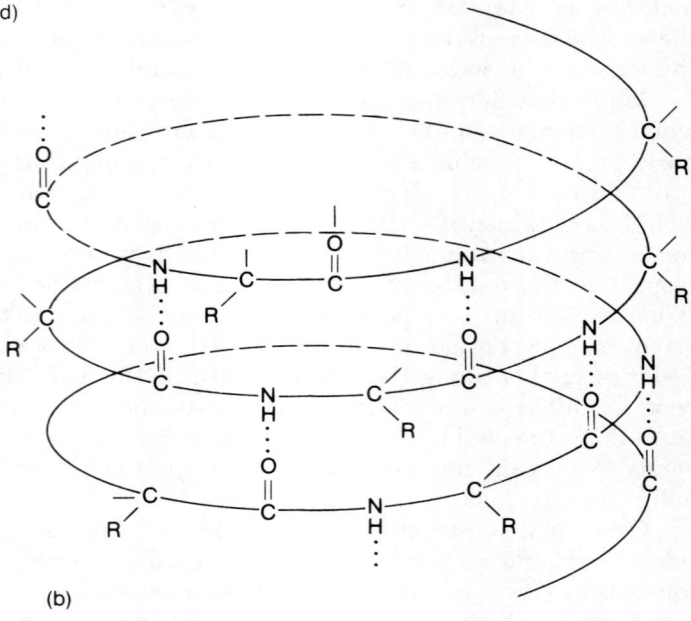

FIGURE 15-9 Several conformations of proteins. [SOURCE: *Physical Biochemistry*, First Edition, by David Freifelder. W.H. Freeman and Company. Copyright © 1976.]

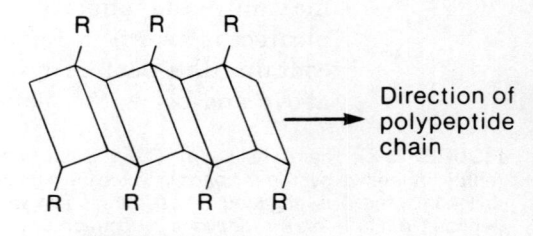

β Pleated structure

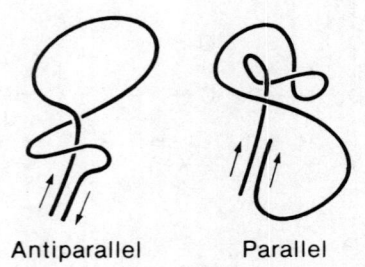

Antiparallel   Parallel

strands of the β structure; normally, a second strand would be adjacent to the one shown. The alignment of the two strands may be such that the adjacent chains are running either in the same direction (parallel) or in the opposite direction (antiparallel), as shown in Figure 15-9. Such regions of a protein are called parallel and antiparallel *pleated sheets,* respectively.

Parameters Used to Describe Macromolecules

Many parameters are used to describe macromolecules. One of the most important is its molecular weight, M. Often all of the molecules in a laboratory sample of a biological macromolecule do not have the same value of M. Such a sample is said to be *polydisperse;* it is *monodisperse* if the sample gives only a single value of M. When measuring M for a polydisperse sample, an average value is obtained. There are several types of statistical averages and not all techniques for measuring M yield the same kind of average value. The significance of these averages and the methods for measuring M are described in a later section.

When we discussed the van der Waals formulation of the gas law, we encountered the concept of the *excluded volume*—that is, a repulsive interaction that is simply a result of the fact that no two objects of finite size can occupy the same space. For a nonspherical particle, the excluded volume is greater than the real volume because, as a molecule rotates, it effectively takes up more space than its own volume. For a gas, this is one of the two causes for nonideality of real gases (the attractive van der Waals force is the other). However, the excluded volume effect is the major cause of nonideality of solutions of macromolecules because of the enormous size of a macromolecule compared to solvent molecules.

In the theory of solutions, the assumption is made that the solute and solvent molecules have roughly the same size, yet a macromolecule may be thousands (even hundreds of thousands) of times larger than a solvent molecule. The excluded volume effect is entropic in that it limits the number of possible arrangements of solvent and solute in a solution. The result is that the concentration at which a system obeys Raoult's Law and other relations pertaining to ideal conditions is much lower for a macromolecule than for a small molecule. At a concentration of 10^{-3} M, the activity and the concentration of a solution of a small molecule are approximately equal, yet with solutions of macromolecules, concentration is much greater than activity, down to concentrations of less than about 10^{-5} to 10^{-6} M. Since many equations that describe the properties of solutions—for examples, the colligative properties and sedimentation equilibrium equations—are derived thermodynamically, these equations require a determination of the activity, which is exceedingly difficult to measure for macromolecules. When these equations are used to determine the molecular weight of a macromolecule, and concentration is substituted for activity, the measured value of the molecular weight of a macromolecule is usually observed to depend on the concentration used in the measurement, a phenomenon that is not evident in the thermodynamic equations. The explanation is that the concentration is not necessarily proportional to activity. Even though these equations are based on activity yet are invariably written with concentration, they remain useful in the determination of the molecular weights, because to obtain the correct value, one need only measure the molecular weight at several concentrations and extrapolate the measured value to zero concentration. This extrapolation is discussed in a later section.

Dimensions of Macromolecules

A molecule with the shape of a rigid rod has two distinct dimensions, its length and its radius. However, most molecules are constantly being bombarded by solvent molecules so that their shape varies from one instant to the next. Thus the dimensions of a particular molecule are continually changing with time and some means of determining average dimension is required. Dimensions that are averaged over a long time interval could be defined, but because, experimentally, a large number of molecules are examined at once, it is more reasonable to take an average over the population at a particular instant; for a very large population such an average will be constant in time. The two parameters that are used are the *end-to-end distance* and the *radius of gyration,* for these averages result from several kinds of measurements; these will be discussed shortly.

The problem of defining dimensions also exists for a rigid molecule whose shape is not a simple geometric figure, as, for example, a greatly folded protein molecule. For such a molecule the radius of gyration is also a useful parameter. In addition, a molecule may be described in terms of the smallest sphere or ellipsoid that can contain the molecule.

Before proceeding, one should ask how the values of these average dimensions are of use to the experimenter. In fact, their information content is not very great, but they do enable one to do three things:

1. One can determine whether a molecule is either very long and thin, or is much more compact than a random coil, or is what might be termed "ordinary." A rough estimate of the ratio of the longest and shortest axes of the molecule can also be made. These facts are important in understanding the biological function of a macromolecule because, as we will see shortly, *the structure and function of a molecule are closely related.*

2. Changes in the shape of a molecule that accompany environmental changes can be detected; this is often a valuable aid in determining the detailed structure of a molecule.

3. The stiffness of a molecule can also be estimated and, again, this is useful in understanding structural details.

SEC. 15–1   PROPERTIES OF MACROMOLECULES

Detailed discussions of the end-to-end distance and the radius of gyration of a polymer are beyond the scope of this book; hence they will be described only in a limited way.

The end-to-end distance h is the average separation between the two ends of a linear molecule. For a rigid rod it equals the length of the molecule, but for a flexible molecule its value depends on the molecular weight and flexibility. For a true random coil that is perfectly flexible, h can easily be calculated from equations that describe a problem that in statistics is called "the random walk." In the one-dimensional random walk a person is constrained to walk in a straight line but the choice to take a step forward or backward is random. If N steps of length l are taken from the origin and m steps are made forward, the net displacement d is $(2m - N)l$. Since d can be greater or less than 0, if one wants to know the actual distance between the ends, it can be written either as the absolute value of d or, more conventionally, as $\sqrt{\langle d^2 \rangle}$, in which $\langle \rangle$ denotes the average of d^2. It can be shown that $\sqrt{\langle d^2 \rangle} = N^{1/2}l$, so that after N random steps, the walker is $\sqrt{N}$ paces from the origin. If the problem is extended to three dimensions, the same equation results. This analysis is equivalent to determining the end-to-end distance of a random coil as long as the bond angles are assumed to be capable of having any value. Thus we can write for the end-to-end distance h,

$$h = (Nl)^{1/2}l^{1/2} = L^{1/2}l^{1/2} , \qquad (1)$$

in which $Nl = L$, the total length, is called the *contour length* of the molecule. Since L is proportional to the molecular weight (M) of the molecule, h is also proportional to $M^{1/2}$.

Of course, the angle between two monomers is not random when they are linked in a polymer. However, with little error, we can assume that it is random and consider l to be the *effective segment length* of a hypothetical, truly random coil. This effective length can then be compared to the actual linear dimensions of the monomer. If they are similar, the macromolecule is flexible, and if they differ by a large value, the macromolecule must be rigid. If it is rigid, it will of course be very long and thin. The value of h can be determined (but not very accurately) from hydrodynamic measurements (principally sedimentation velocity), although how this is done will not be presented in this book. The value of N is just the degree of polymerization, so that $l = hN^{-1/2}$. For highly flexible organic polymers, such as polyethylene, l is 2 to 3 Å (the length of the CH_2 group), whereas for very rigid molecules, such as double-stranded DNA, $l \approx 1000$ Å, which is much greater than the length of the sugar-phosphate monomer. Thus, DNA is a very stiff molecule.

Another average molecular dimension is the radius of gyration, R_G. This is defined as the root mean square average of the distances of all parts of a molecule from its center of mass. That is,

$$R_G^2 = \frac{\sum m_i r_i^2}{\sum m_i} , \qquad (2)$$

in which m_i is the mass of the ith element at a distance r_i from the center of mass. The actual physical significance of R_G in a geometric sense is not obvious, but this average is chosen because its value appears in the analysis of the angular dependence of the scattering of light from a solution of macromolecules (discussed later in the chapter). For simple geometric shapes and for the random coil, R_G can be calculated. A few values are listed in Table 15-1.

TABLE 15-1 Calculations of R_G for various molecular shapes.

Sphere:	$(3/5)^{1/2}r$,	where r = radius of sphere
Rod:	$(1/12)^{1/2}L$,	where L = length of rod
Random coil:	$(N/6)^{1/2}l$,	where N = number of units of length l
Random coil:	$h/(6)^{1/2}$,	where h = end-to-end distance

The value of R_G can be used to estimate molecular shape by comparing the measured value with that expected for a sphere. This calculation requires writing r, the radius of a sphere, in terms of measurable quantities. Since the volume of a sphere is $\frac{4}{3}\pi r^3$ or $M\bar{v}/\mathcal{N}$, in which M is the molecular weight, $\bar{v}$ is the partial specific volume (a measurable quantity), and $\mathcal{N}$ is Avogadro's number, $r = (3M\bar{v}/4\pi\mathcal{N})^{1/3}$, so that $R_{G,\text{sphere}} = (3/5)^{1/2}(3M\bar{v}/4\pi\mathcal{N})^{1/3}$. Some values of $R_{G,\text{observed}}$ and $R_{G,\text{sphere}}$ for various macromolecules are listed in Table 15-2.

TABLE 15-2 Comparison of $R_{G,\text{observed}}$ and $R_{G,\text{sphere}}$ for several macromolecules.

Macromolecule	M	$\bar{v}$	$R_{G,\text{ observed}}$, in Å	$R_{G,\text{ sphere}}$, in Å	$\dfrac{R_{G,\text{observed}}}{R_{G,\text{sphere}}}$
Lysozyme	13,930	0.70	14.3	12.2	1.17
Myoglobin	16,890	0.74	16.0	13.2	1.21
tRNA	26,600	0.53	21.7	13.8	1.57
Myosin	493,000	0.73	468	45.2	10.4
DNA	4,000,000	0.55	1170	74	15.8

Table 15-2 indicates that the values of R_G for lysozyme and myoglobin are not much larger than for a sphere. In order to be at all larger, though, the molecule must have some extended regions such as would occur in an α helix or a β form. Myosin, which we know to be a relatively long, somewhat rigid molecule, has a value of R_G much greater than that of a sphere, as does DNA. We can also show that DNA must be somewhat flexible by calculating R_G for a rigid rod. The mass per unit length of DNA is, from X-ray diffraction studies, 200 molecular weight units per Å. Thus a molecule for which $M = 4 \times 10^6$ has a length of 2×10^4 Å and $R_{G,\text{rod}} = (1/12)^{1/2} \times (2 \times 10^4) = 5774$, roughly five times greater than the observed value.

Relation between the Shape of Molecules and Their Properties

We have seen that there are various types of structures in macromolecules—helices, β forms, random coils, and so forth. However, it is rare that a macromolecule would contain only a single type of structure—two notable exceptions are DNA and the protein collagen, both of which are totally helical. Thus it is common to distinguish fibrous molecules from globular molecules. Still another distinction is flexible versus nonflexible molecules.

The structure possessed by a macromolecule is always indicative of its biological function. For example, the so-called structural proteins—namely, those that are responsible for maintaining particular cellular structures—are usually fibrous and relatively inflexible. Examples are myosin, one of the two main proteins in muscle, and collagen, the protein of which tendon is composed.

Collagen is an especially good example of structure suited to function. It consists of a triple-stranded helix (called tropocollagen) in which one of the strands extends beyond both ends of the triple helix, resulting in two short, protruding peptide strands at both ends of the tropocollagen unit. The job of collagen in tendon is to provide tensile strength in a long fiber. In order to be an extended molecule, it has evolved to contain a large amount of the amino acid proline. This amino acid does not have an α-amino group (it is really an *imino* acid) and fails to form a typical peptide bond. The bond it forms is much less flexible than a standard peptide bond and thereby it helps to maintain the molecule in a more extended form than a random coil. Just as a bundle of sticks is stronger than a single stick, a side-to-side aggregate of molecules is stronger than a single molecule. The extended single strands are held together by intrastrand hydrogen bonds; the result, a triple-stranded helix, is the first stage in generating tensile strength. Additional strength is obtained by side-to-side aggregation of the triple helices. This is made possible by the numerous positive and negative amino acids arranged along the triple helix, so that a positive amino acid on one chain can interact via a simple electrical attraction with a negative amino acid on an adjacent chain. Having so many charges of both sign on a single chain, a single strand would tend to fold back on itself, but is prevented from doing this by the rigidity imposed by the triple-stranded helix. In order to form a fiber, the side-to-side aggregates must be able to aggregate end-to-end also; however, this does not occur. Instead the side-to-side aggregation involves a staggering of the ends of the triple helices by one-fourth the length of the molecule, which in effect causes longitudinal as well as lateral joining. This is shown in Figure 15-10. We can see how various structural features of the molecule combine to form a complex macromolecular array that has a great deal of tensile strength.

FIGURE 15-10 The pattern of aggregation of tropocollagen units.

The basic monomeric unit is helical

It is hydrogen-bonded with two other helical strands to form the triple-stranded tropocollagen helix

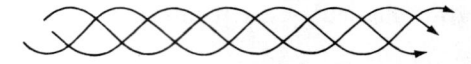

The tropocollagen joins other tropocollagen helices to form a side-to-side aggregate in which the constituent end-to-end lengths are displaced by one-fourth

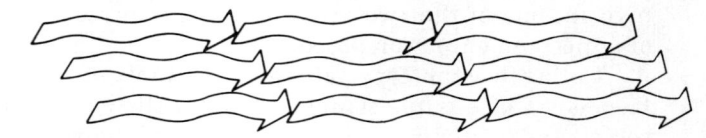

The DNA molecule is another example of a clear relation between structure and function. The double-stranded molecule provides the redundancy that is necessary if genetic information is not to be lost through the occasional attacks that are launched against this molecule, of which there is only one per cell. Furthermore, by having bases that are somewhat hydrophobic, the bases stack one upon another, minimizing contact with water and thereby decreasing the probability of molecular alteration by reactive chemicals in solution.

An enzyme molecule must exist in many copies per cell. Thus, in order to minimize the volume occupied by many such molecules and to enable them to move freely throughout a cell, molecules of an enzyme should not be extended but should be globular, as an enzyme usually is. However, the specificity of binding of a substrate molecule to an enzyme molecule requires that this binding site have a well-defined and constant shape. Thus, this part of the molecule must be relatively stiff. Regulation of the activity of an enzyme is most easily accomplished by altering the shape of the binding site. Thus, the enzyme must not be completely rigid. In accord with these assumptions about function, a typical enzyme is a highly folded, globular protein that contains a large number of hydrogen-bonded and interacting amino acids providing a relatively rigid, compact structure. The strengths of amino acid interactions are mutually dependent so that a single regulatory molecule can bind to a site on the enzyme molecule and alter the shape of the enzyme at the binding site. This alteration can, because of the dependent interactions, be propagated from one part of the molecule to another and thereby change the shape of the substrate binding site.

Intra- and Intermolecular Forces and Subunits

The three-dimensional structure of a macromolecule is determined by three factors—the allowable bond angles, the interactions between the components of the macromolecule (that is, the monomers), and the interaction between one component and another. The solvent interactions are of two types—*solvation* or solvent binding, which is an attraction between the components and the solvent molecules, and the *hydrophobic interaction*, which is a solute–solute interaction that is a consequence either of the inability of components to interact with the solvent or an avoidance interaction. The basic rule of the hydrophobic interaction is that

> *if a collection of molecules is unable to be solvated, they will instead stick close to one another in a way that minimizes contact with the solvent.*

This effect can be seen quite clearly in the behavior of nucleic acids and proteins. The nucleotide bases of nucleic acids interact weakly with water (that is, they are only sparingly soluble), and, to minimize contact with water, they stack one above another. This is one of the factors that makes even the single-stranded polynucleotides somewhat rigid. In the protein structure, hydrophobic forces may result in a stacking of the rings inside chains, such as in phenylalanine and, more significantly, a clustering of the very nonpolar amino acids, such as leucine, isoleucine, and valine. Hydrophobic forces are also responsible in part for the structures of biological membranes (see Chapters 5 and 12).

The principal positive interactions in a macromolecule are hydrogen bonds, ionic bonds, and the van der Waals attraction. Hydrogen bonds account for the double helical structure of DNA, for the α helix and β structure of proteins, and for certain other interactions. Ionic bonds are found in proteins, where they join pairs of oppositely charged amino acids, and in nucleoproteins, where they join the positively charged amino acids arginine and lysine with the negatively charged phosphates in nucleic acids.

All of these interactions serve to give every macromolecule a unique three-dimensional structure suited to its biological function. The surface of a macromolecule is also capable of interaction with the surface of another macromolecule (Figure 15-11). One example just given is that of a nucleoprotein. However, it is very common with proteins that there are hydrophobic patches on the surface. These are clearly not a result of a preference for interaction with the solvent but can be thought of as what is left over after the macromolecule has folded in order to minimize its free energy. Two molecules having these hydrophobic patches can produce a further decrease in free energy by pairing with one another in such a way that some of the patches are removed from solvent contact. There is a second cause of pairing of macromolecules. In the complex folding of a protein, for instance, there may be generated two surface regions whose three-dimensional shapes are complementary, as shown in Figure 15-11. The very weak van der Waals forces between two amino acids are quite insufficient to form a stable pair that, at ordinary temperatures, can resist thermal disruption. However, when the shapes are complementary, these weak forces can add up to make quite a strong attraction. This effect of shape is not only responsible for the pairing of proteins and other macromolecules but also for the powerful binding of small molecules to a macromolecule, such as in the specific binding of a substrate molecule to an enzyme.

These surface interactions are responsible for the widespread phenomenon of the existence of subunits of proteins. That is, most proteins whose molecular weights are above 50,000 consist of several polypeptide chains. Sometimes these chains are identical but they may also be of two or more types—for instance, hemoglobin consists of two α chains and two β chains. The large multifunctional proteins may contain several different chains, as for example, *E. coli* RNA polymerase, which consists of five different units.

The interactions we have just described are also responsible for the structure of large macromolecular aggregates, such as viruses, chromosomes, membranes, and so forth.

In the following section we examine how changes in the environment of a macromolecule that alter the strength of these interactions affect the structure of the macromolecule.

FIGURE 15-11 Interaction of two protein molecules having complementary shapes. This interaction would probably result mostly from many van der Waals interactions.

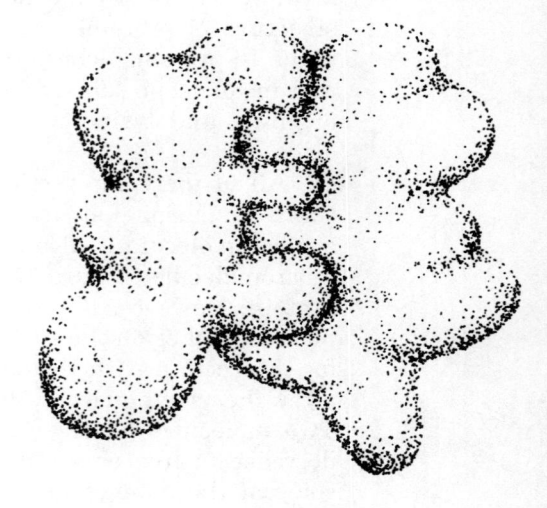

15–2 CHANGES IN THE STRUCTURE OF MACROMOLECULES

The Helix–Coil Transitions: Denaturation

A macromolecule can maintain a stable nonrandom, three-dimensional configuration only if the stabilizing forces exceed the disrupting influence of thermal motion. This will be true whether the molecule is completely helical, has helical regions, or has any other type of structure, such as the β structure of proteins, the stacked structure of the bases of nucleic acids, or the various hydrogen-bonded and ionic interactions of numerous macromolecules. If the temperature is raised to the point at which stability is lost, the molecule will assume some random configuration. The configuration that exists when all of these weak interactions are disrupted is, as we have stated earlier, not a completely random coil because there are geometric constraints imposed by the covalent bonds—nonetheless, it is common to call the disrupted molecule a coil. The structure that a macromolecule has in nature is called its *native* structure; once disrupted, it is said to be *denatured*. The process of converting a native molecule to a denatured molecule is called *denaturation* and the transition from the native to the denatured state is called a *helix–coil transition* (even if the native state is not helical!). A helix–coil transition can occur as a result of an increase in temperature, in which case the transition is called *melting,* or by varying any other parameter (for examples, the pH or the ionic strength) that would break down the interactions that maintain the native structure.

Helix–coil transitions can be detected in a variety of ways. For example, the viscosity of a solution of a nonspherical protein molecule or of a DNA molecule drops significantly when the molecule loses its structure, and the absorbance of DNA solutions at the wavelength of 260 nm increases substantially when the DNA is converted to the single-stranded form (Figure 15-12, solid curve).

A helix–coil transition can be either *noncooperative* or *cooperative,* a distinction best described by the following example. Consider a protein whose structure is determined by an ionic interaction between a single pair of amino acids A and B, one positively charged and one negatively charged. Owing to the statistical nature of matter, there is an equilibrium at any temperature between the number of protein molecules in a large population in which A and B are paired (the "helical" form) and the number in which A and B are unpaired (the "coil"). At a temperature at which the pairing is very stable, most of the molecules are native, whereas at some high temperature most, if not all, are denatured. There will be a

FIGURE 15-12 Changes in absorbance (at a wavelength of 260 nm) and viscosity for DNA solutions as a function of temperature. The melting temperature T_m is defined as the temperature at which the transition (vertical distance) is 50 percent complete. It is not necessarily the same for two different assays, as is indicated; T_1 is the value observed for the decrease in viscosity whereas T_2 is that for the increase in absorbance.

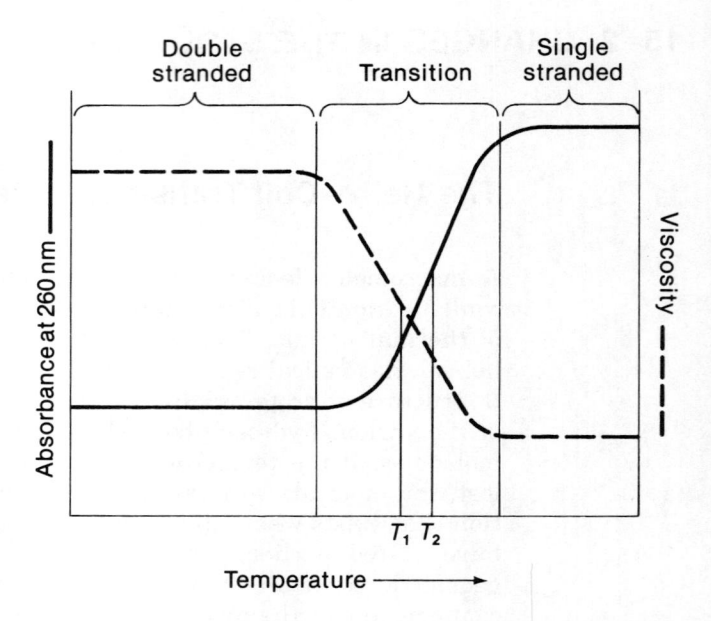

temperature range in which the equilibrium changes rapidly; that is, if a graph were to be made of the fraction of unpaired A and B versus temperature, it would show a rapid rise in this temperature range. Since there is only a single pair per molecule, the temperature range is narrow and the transition is said to be *sharp*.* Now let us consider a molecule whose shape is determined by a very large number of pairs of amino acids. We can now distinguish two different kinds of transitions. In one, the noncooperative transition, the probability of any pair existing is independent of the existence of any other pair; in the other type, the cooperative transition, the probability of a pair depends on the existence of other pairs. Let us assume also that all pairs do not have the same strength, because they are either chemically different or in different locations in the molecule. As the temperature is raised, pairs will be disrupted at different temperatures so that the helix–coil transition will not be sharp. In a cooperative transition, there is much greater difficulty in disrupting the first pair—because it is stabilized by the existence of all other pairs—than in breaking the last pair, which is stabilized only by its own intrinsic binding energy. Thus the first pair that is disrupted is broken at a higher temperature than if it were the only pair; this is also true of the second, third, . . . , etc., pairs, but not the last pair. However, once the first pair is finally disrupted, the second pair is less stable; when this one is disrupted then the third is weakened, and so forth. Therefore, *when there is cooperativity, a transition is much sharper than when there is no cooperativity.*

Noncooperative transitions are rare in biological macromolecules.

*We do not mean perfectly sharp because the interaction energies of all of the pairs in the population will not be exactly the same because of the statistical nature of the universe.

SEC. 15–2   CHANGES IN THE STRUCTURE OF MACROMOLECULES   515

Consider a protein, a large fraction of whose amino acids are hydrogen-bonded. At any particular temperature the molecule is being bombarded by solvent molecules, which has the effect that at any instant various regions of the molecule are moving with respect to one another. The motion is not very great because the regions are constrained by the many hydrogen bonds. As the temperature increases, the molecule is bombarded with greater force. The rigidity imposed by the hydrogen bonds still maintains the structure until a temperature is reached at which one hydrogen bond breaks. This allows the regions held together by this hydrogen bond to move more freely with respect to one another, which thereby strains nearby hydrogen bonds. A smaller increment in energy is needed to break these bonds, after which the molecule again increases in flexibility, making it even more susceptible to disruption. This process continues by addition of small increments of energy until the molecule is totally disrupted. Actually, this is somewhat idealized, in that the process is not necessarily completely progressive. That is, at a particular temperature *several* regions might be so weakened that the conversion to the disrupted form may be occurring *simultaneously* in several parts of the molecule.

We now examine a DNA molecule. The structure of the double-stranded helix is maintained by two factors: (1) the tendency for the bases within a single strand to stack one above the other with the planes of their rings parallel; and (2) the hydrogen bonds between the strands, which hold the two strands together. The hydrogen bond is weak unless the bases are stacked, because the stacking provides the orientation that is necessary for a large number of hydrogen bonds to form simultaneously. At a particular temperature, bombardment of the molecule by solvent molecules tends to break the hydrogen bonds and to alter the relative orientation of the bases. This is very difficult, though, because in order to break one hydrogen bond, the adjacent bonds would have to be strained in order to tip the plane of one base with respect to an adjacent base; the first would have to tip also with respect to its other neighbor. Thus, there is an enormous stabilization, and the huge DNA molecule typically undergoes a helix–coil transition at temperatures 30–40° above the values usually encountered for proteins. The bonds that are most susceptible to disruption are those at the ends of a DNA molecule, because the terminal base pair is stabilized only by one pair of stacked bases. Thus, as the temperature increases, the hydrogen bonds at the ends of the double helix are the first to break. This breakage of course destabilizes the next pair and denaturation proceeds progressively inward. The ends are not the only regions that denature early in the transition. Sequences rich in adenine and thymine (AT) pairs, which have two hydrogen bonds, are less stable than guanine-cytosine (GC) pairs, which contain three hydrogen bonds. An internal region at which base pairs are disrupted is called a *bubble*. Further breakage of hydrogen bonds occurs preferentially at the ends of AT-rich bubbles, because these regions are nearly equivalent to the ends of the molecule (Figure 15-13). Thus the bubbles enlarge, and denaturation of DNA

FIGURE 15-13 A DNA molecule in various stages of denaturation.

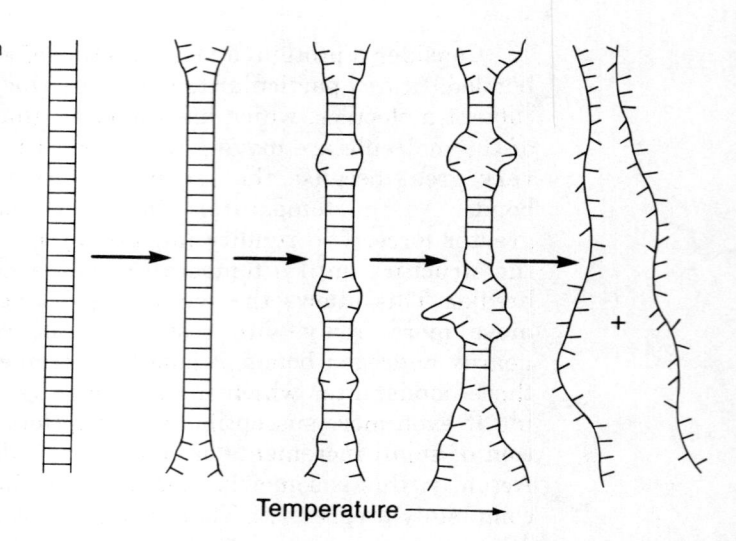

Temperature ⟶

proceeds both by growth of the bubbles and progressive opening of the helix from the ends.

Helix–coil transitions are frequently described by the temperature at which the transition (vertical distance) is 50 percent complete.* This temperature is designated T_m or the *melting temperature* (Figure 15-12). The value of T_m depends on the means of detecting the transition. For example, T_m for the change of absorbance of a DNA solution need not be the same as the value of T_m for the change of viscosity of the solution, as is shown in the figure. The melting temperature is sometimes called the midpoint of the transition; however, it is important to realize that this is not the temperature midway between the temperatures at which the transition stops and starts but the temperature at which the curve has risen to half of its maximal value. In fact, the melting curves are not usually symmetric on the temperature axis about the value of T_m.

We have been discussing the helix–coil transition as a conversion from an ordered to a disordered state. The opposite point of view—that of the coil–helix transition—can also be taken. In this way one can see that cooperativity strengthens the ordered state and provides a means for achieving it. Consider a protein in which many amino acids are paired. There may be alternate ways in which these amino acids can combine, and a molecule with these alternate pairings might be quite stable—it might also lack biological activity, though. If we assume that the molecule with the correct configuration has strong cooperative interactions and that one with the incorrect configuration does not, we can understand how cooperativity aids in the creation of the correct configuration. Let us examine the

---

*Other parameters that might be varied to cause helix–coil transitions—such as the pH, the ionic strength, and the concentration of a denaturant—are sometimes used as a basis for measurement, but temperature is by far the most common.

transition of a newly synthesized protein molecule to the correct configuration. Since all of the amino-acid side-chain interactions are weak and easily disrupted by thermal vibrations, the first amino acid pair that forms may have little stability and break spontaneously. Another pair may form a moment later, and this too may break. However, if the interactions that maintain the correct structure are cooperative, then even though the making and breaking of pairs will still occur, there will be a different outcome of the formation of the first *correct* pair. This is because the first correct pair, once formed, increases the probability of the formation of a second *correct* pair; furthermore, once the second correct pair has formed, the stability of the first correct pair will be increased. In this way, the probability of forming successive correct pairs increases and the stability of all pairs becomes greater. This effect of cooperativity is even easier to see in the formation of double-stranded DNA from two complementary single strands. That is, a base pair is stable only if the adjacent pair can form hydrogen-bonds, and the two pairs are stable only if the next-to-nearest neighbors can bond.

The examples just given demonstrate two important reasons why biological molecules have evolved with cooperative interactions. First, the stability of an ordered structure, in which there are cooperative interactions, is greater in the sense that it is more resistant to thermal disruption. If stability is of utmost importance, then why, one might ask, did the molecules evolve with weak bonds rather than strong bonds such as covalent bonds? The answer is the second reason for cooperativity—that is, in carrying out its biological function and in adapting to the varied conditions of nature, a molecule frequently must change its configuration. This would not be a simple matter if the structure was stabilized entirely by covalent bonds. For a molecule to be changeable, weak bonds are necessary, but in order to provide strength, several of the weak bonds must act together.

In the following, we give an example of a helix–coil transition that occurs in nature by reducing the effectiveness of the cooperative interactions that stabilize the native structure.

EXAMPLE 15–A   The effect of helix-destabilizing proteins on the DNA helix.

An essential step in the replication of DNA molecules within living cells is the separation of the two polynucleotide strands. This is accomplished in part by helix-destabilizing (HD) proteins. At any temperature, there will be temporary breakage of hydrogen bonds in small regions, owing to collisions with solvent molecules. (This phenomenon is called the *breathing* of DNA.) Sometimes an HD molecule, which binds tightly to several adjacent bases, enters the transient breakage "bubble," binds to one strand of the DNA molecule, and thereby prevents reformation of the hydrogen bonds. This distortion of the helix weakens the stacking interaction (the source of the cooperative stabilization) so that the bubble fluctuates in size. The HD protein that is bound undergoes a small change in shape that enables it to bind also to a second HD molecule. Thus another HD molecule, which is by itself capable of binding to the bases in the fluctuating bubble, now has a greater probability of binding because it can bind both to the bases and to the bound HD molecule. Furthermore, the second HD molecule binds *adjacent to the bound HD molecule*. This further enlarges the bubble by breaking down stacking and, if there are enough HD molecules, the process continues until the two polynucleotide strands are totally separate.

The principal reason for studying a helix–coil transition is that it provides information about the structure of a particular macromolecule. The rules used in these studies are the following:

Rule I. If a reagent lowers the T_m, it must reduce or eliminate some factor that stabilizes the native form. If T_m is increased, then the effectiveness of some stabilizing factor is enhanced.

Rule II. If a change in pH produces a helix–coil transition, the pH at which the transition is 50 percent complete is near the pK of a component of the molecule that is required for stability of the native structure. The interaction invariably alters the pK of the group.

Rule III-A. If T_m increases with increasing ionic strength, the molecule contains groups of like charge that repel one another and reduce stability, whereas if T_m decreases, there are unlike charges that attract one another and thereby enhance stability.

Rule III-B. If T_m increases with increasing ionic strength and there are no groups having like charge, then hydrophobic interactions are a major feature of the structure. This is because nonpolar groups are less soluble in a salt solution than in water, so that salt keeps the nonpolar groups together, thus requiring a higher temperature to disrupt structure. Rules III-A and III-B generally apply to nucleic acids and proteins, respectively.

The use of these rules can be seen in a few examples.

EXAMPLE 15–B   The effect of sodium trifluoroacetate on the stability of DNA.

Figure 15-14 shows helix–coil transitions (assayed by the increased absorbance of single-stranded DNA compared to double-stranded DNA) for DNA in 0.01 M, 0.1 M, and 4 M NaCF$_3$COO. The values of T_m for each solution are 75°, 82°, and 66°C, respectively. This is to be compared to similar curves for 0.01 M, 0.1 M, and 4 M NaCl, for which the values of T_m are 72, 85, and 99°C, respectively. For both salts, the value of T_m for a

SEC. 15–2   CHANGES IN THE STRUCTURE OF MACROMOLECULES   519

FIGURE 15-14 Melting curves for a DNA sample dissolved in various concentrations of sodium chloride (NaCl) and sodium trifluoroacetate. T_m increases as the NaCl concentration increases yet decreases above a concentration for sodium trifluoroacetate of 1.5 M, reaching a minimum at 4 M. (Data are not all shown, for the sake of clarity.)

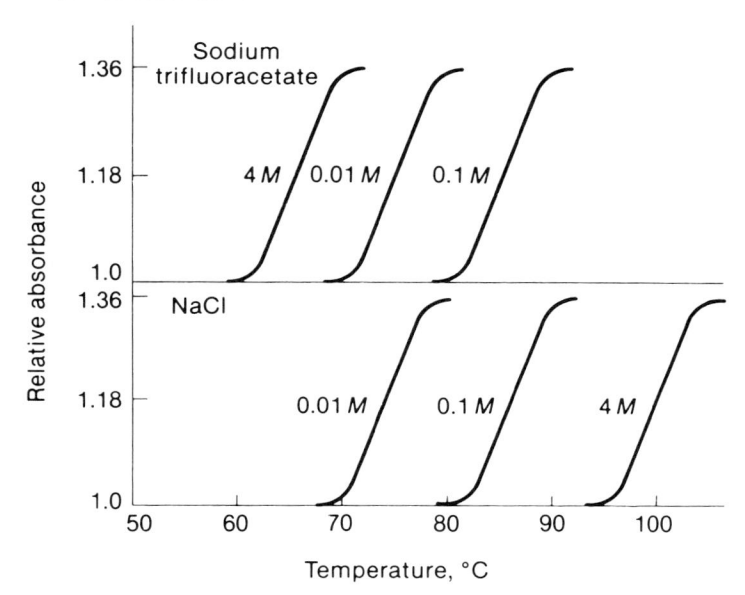

0.1-M solution is greater than that for a 0.01-M solution. This is because at low ionic strength, the negatively charged phosphate groups of the two strands repel one another and weaken the interaction, as we have discussed earlier. These charges are partly shielded in the 0.1-M solutions. (This exemplifies Rule III-A.) In 4 M NaCl, the T_m is greater than in 0.1 M NaCl because, as we have just said, the electrostatic repulsive force is incompletely eliminated at 0.1 M. However, in 4 M sodium trifluoroacetate, the T_m drops enormously, indicating that, between the concentrations of 0.1 M and 4 M, a stabilizing force has been eliminated. Sodium trifluoroacetate is one of a class of reagents that increases the solubility of weakly polar organic molecules in water and thereby reduces hydrophobic interactions (see Chapters 5 and 6 for a description of these interactions). This decreases the stacking tendency of the bases and thereby reduces the thermal stability of the DNA. Other reagents, such as $NaClO_4$, methanol, and ethylene glycol, have the same effect.

EXAMPLE 15–C   Effect of base composition of DNA on T_m.

In DNA the guanine-cytosine (GC) pair of bases has three hydrogen bonds, whereas the adenine-thymine (AT) pair has two. Thus, if hydrogen-bonding is important in the stability of DNA, a molecule with a large fraction of GC pairs should have greater thermal stability than one having fewer GC pairs. That this is the case is shown in Figure 15-15.

FIGURE 15-15 Melting temperature as a function of G+C content in 0.01 M sodium phosphate buffer.

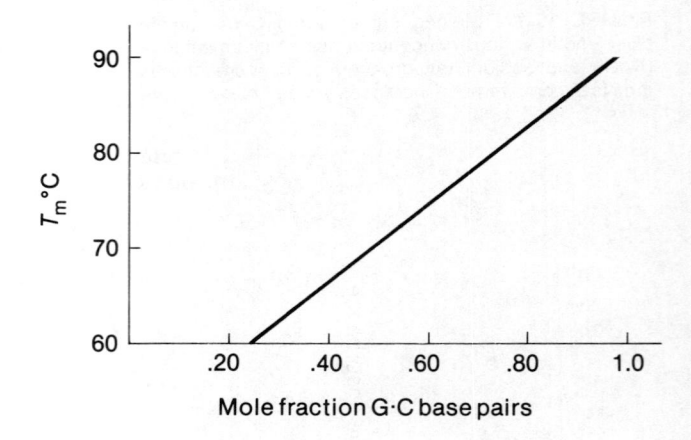

EXAMPLE 15-D Effects of various reagents on the viscosity of a protein.

The viscosity of a particular protein solution decreases significantly at 53°C, suggesting that the protein is a very nonspherical molecule that becomes a nonrandom coil at that temperature. At 25°C, the viscosity decreases by the same factor following addition of urea (NH_2CONH_2), to make a concentration of 7 M. Urea is a known breaker of hydrogen bonds because of its ability to combine with both amino and carbonyl groups. Therefore, the structure must in part be stabilized by hydrogen bonds. The dependence of viscosity on pH is shown in Figure 15-16. The decrease at a pH value of 10.2 is near enough to the pK value of the ϵ-amino group of lysine (pK = 10.5) to suggest that lysine is involved in stabilizing the structure.

Surface Denaturation of Proteins

In the preceding section, agents such as heat and reagents that break hydrogen bonds or disrupt hydrophobic interactions were shown to cause denaturation. However, some macromolecules are known to be highly labile (denaturing when you as much as *look* at them, it has been said). For example, many purified proteins lose biological activity during storage in solution. These molecules, as well as many other proteins, lose structure especially rapidly if the solution is stirred or shaken violently or if air is bubbled through the solution. This phenomenon, which is known as *surface denaturation*, has the following explanation. In aqueous solution protein molecules are diffusing throughout the water phase. By random motion each protein molecule will eventually reach the surface and become partially unfolded because of the reduction of the hydrophobic interactions that were just described. If the unfolding process is irreversible (as is frequently the case), the biological activity of each molecule reaching the

FIGURE 15-16 The change in viscosity of a hypothetical protein as a function of pH. The horizontal dashed line represents the viscosity when the transition is half complete. The vertical line indicates the pH when the transition is half complete.

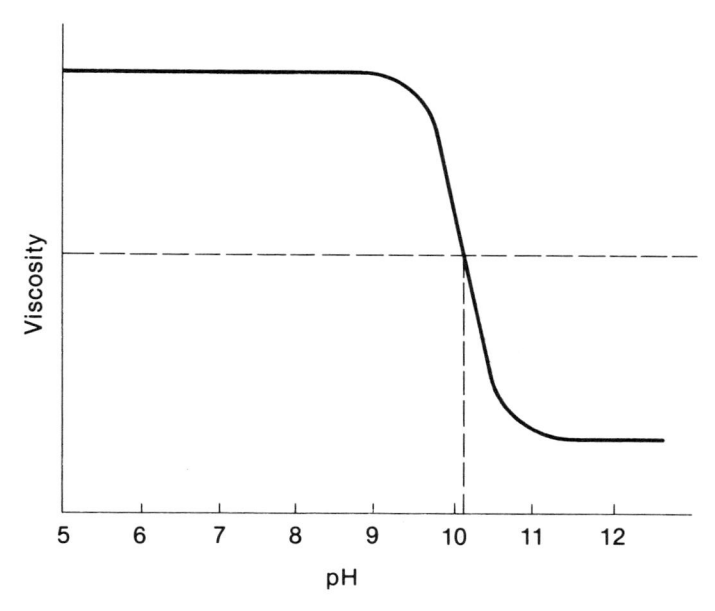

surface will be totally lost. Violent agitation increases the rate of loss of activity by increasing the surface-to-volume ratio of the solution and thereby increasing the probability that a molecule will reach the liquid-air surface. Many of the molecules that are unfolded remain on the surface and form a film (Chapter 13). If the protein concentration is high enough, the film will in time cover the surface and bring the surface denaturation process to a halt. Thus, the loss of activity of most proteins is much greater when the concentration of the protein solution is low than when the concentration is high. Also, solutions of unstable proteins can be stabilized by the addition of high concentrations of inert proteins, such as bovine serum albumin, because the inert protein forms a protecting surface film.

Lack of Intermediates in the Denaturation of Proteins

Denaturation of macromolecules has been studied for several reasons. First, a great deal of information about the identity of the interactions that stabilize a particular structure came from these studies. For instance, the existence of base stacking and hydrogen bonds in DNA was indicated by certain characteristics of melting curves. Second, it is believed that some information about the mechanism of folding of proteins can be obtained in this way. An important question in this regard is whether there are intermediate metastable states in either the folding or unfolding processes, as opposed to a more or less continuous transition through a very large number of states. One way that this question has been approached has been by calorimetric studies in which ΔH or C_P is measured as a function of temperature throughout the unfolding process. The rationale is that, just as in phase transitions, there are significant changes in enthalpy associated with each structural transition. That is, if protein-unfolding is a continuous process, a plot of ΔH versus T will give a straight line (refer to Figure 4-4) before and after denaturation and slope upward only during the unfolding process. However, if there is a metastable intermediate state, the ΔH-versus-T curve will show two distinct rises in ΔH in the temperature range in which unfolding occurs. In general, there will be $n + 1$ steps if there are n intermediate states. The technique of differential scanning calorimetry has made it possible to measure both enthalpy changes and the value of C_P for solutions of macromolecules undergoing thermal transitions. (This technique is explained in the final section of Chapter 4.) The results of application of this technique have been quite clear. For most proteins consisting of a single polypeptide chain, only a single transition is observed, so that there is no intermediate state in the unfolding process. For proteins containing several subunits, transitions representing dissociation of the subunits are often seen before the transition corresponding to the unfolding of each subunit.

Some proteins consisting of a single polypeptide chain do show several transitions. Such proteins usually contain regions of the chain that fold independently; these regions are called domains and each transition represents the unfolding of a particular domain. The domain substructure is a result of a tendency for amino acid residues in one polypeptide segment (one domain) to interact almost exclusively with other amino acids in the same segment. A polypeptide containing domains can be thought of as consisting of two or more subunits whose terminal amino acids are joined in peptide linkages. In general, if a protein known to consist of a single polypeptide chain has a discontinuity in a plot of ΔH versus T, it is reasonable to suspect that domains are present.

In contrast with proteins, many single-stranded nucleic acids (such as RNA) do pass through intermediate states when the molecule unfolds. These molecules contain numerous regions of hydrogen-bonded base pairs and, therefore, are highly folded compact molecules. Since the strength of a hydrogen-bonded region depends on base composition (Figure 15-15), the different base-paired segments will not in general have the same thermal stability and intermediate states will exist during thermal melting. A well-studied example is the *E. coli* transfer RNA molecule tRNA$_{val}$, one of a class of molecules containing 76–83 nucleotides and having extensive base-paired regions. The tRNA$_{val}$ contains 76 nucleotides, of which 21 are hydrogen-bonded, forming four distinct base-paired segments and yielding a cloverleaf configuration. Hydrogen bonds between groups in the sugar-phosphate chain are also present; these have the effect of bringing some of the branches of the cloverleaf together. As a tRNA molecule is heated, hydrogen bonds of both types are broken and the molecule is ultimately completely single-stranded. As unfolding occurs, several physical properties of the molecule change and the changes generally appear to occur continuously, giving no evidence for intermediate states. However, a plot of ΔH versus T obtained by differential scanning calorimetry shows, in certain ionic conditions, six distinct transitions. Four of these—those occurring at the higher temperatures—correspond to melting of the base-paired regions, and the remaining two transitions represent disruption of the weaker interactions that cause further unfolding of the structure. This kind of information is valuable in studying the location of molecules that bind to tRNA. Binding usually affects the stability of the region of the macromolecule in which binding occurs. Thus, if the temperature of a transition that involves a particular region of the macromolecule is changed by binding of another molecule, then binding probably occurs in that region.

The fact that there are intermediates in the unfolding of tRNA molecules (and also RNA molecules of other types) and that the base-paired regions contain nearby nucleotides suggests that the correct base pairs form in a segment of the molecule as soon as that segment is synthesized. The folding of a typical polypeptide chain that lacks domains probably occurs differently. In these cases, no intermediate states are seen and quite distant amino acids tend to interact. Thus, folding probably occurs after synthesis of the entire molecule is completed. With proteins having domains it seems likely that the correct folding of a domain occurs as soon as that domain is synthesized.

The Coil–Helix Transition: Renaturation

We have just seen that as the temperature of a solution of macromolecules is increased, the internal structure of each molecule is disrupted; this conversion of a structured molecule to a nonstructured one is termed a helix–coil transition, or denaturation. We now discuss the events that follow, when the temperature is lowered again. We will be concerned primarily with DNA because the process is best understood for this molecule.

We first consider a DNA sample dissolved in a solution of low ionic strength—for example, 0.01 M NaCl. In this solution, most of the negatively charged phosphate groups of the deoxyribose-phosphate chain are not shielded by an Na^+ counterion. Thus the phosphate groups repel one another and, as a result, a DNA molecule is less stable in 0.01 M NaCl than when it is in a solution in which almost all of the phosphates are shielded, as, for example, in 0.5 M NaCl. This causes the T_m to be lower in 0.01 M NaCl than in 0.5 M NaCl. If a DNA sample in 0.01 M NaCl is first heated to the temperature at which strand separation occurs and is then returned to room temperature, the mutual repulsion of the negatively charged phosphates keeps the strands separate; nor can hydrogen bonds form between different strands or parts of the same strand because the charge repulsion makes it impossible for nucleotides to approach one another.

If the DNA is in 0.5 M NaCl, the situation is quite different. Once a sample that contains strands that have been separated by heating is cooled, there is no charge repulsion, so that hydrogen bonds can re-form both within a single strand (in*tra*strand hydrogen-bonding) and between two different strands (in*ter*strand hydrogen-bonding). However, in a very long and complex molecule containing tens of thousands of nucleotides, it is not likely that hydrogen-bond formation will result in reconstitution of the original double-stranded helix: in a molecule of such length, even though many short complementary base sequences may form base pairs, adjacent noncomplementary bases will go unpaired. This is called *random hydrogen-bond formation*. The important point about these paired regions is that they are usually only a few bases long. Hence these short base-paired segments will not be very stable and will be disrupted at temperatures much lower than the T_m for double-stranded DNA. If a sample of DNA that contains these short random inter- or intramolecular hydrogen bonds is reheated to just a few degrees below T_m, all of the pairs in these short segments will break and the sample will consist entirely of separated single strands. However, at temperatures below T_m, the state in which single strands are separate is not a stable state; at these temperatures only a few regions of a double-stranded DNA molecule are unpaired—that is, double-stranded DNA is more stable than single-stranded DNA at temperatures that are a few degrees below T_m. Therefore, as long as no activation energy is required, the single strands ought to form perfect double-stranded DNA again. This process occurs in the following way; two strands collide and form short segments of randomly hydrogen-bonded base pairs; because short segments are not stable at the renaturation temperature, these paired bases then rapidly unpair. However, a collision sometimes leads to random base-pair formation at a position in the molecule where each *adjacent* pair of bases can also form hydrogen bonds, and the strands assume what might be called a "correct" position. In this event, a cooperative interaction occurs and the entire double-stranded molecule is "zipped up" again. Note that renaturation does not occur at room temperature because the process requires an activation energy, which is simply the energy required to break the random hydrogen bonds. A few degrees below T_m, there are no randomly hydrogen-bonded base pairs, so that there is no activation energy in that temperature range and renaturation can occur.

There are three requirements for renaturation to occur.

1. There must be complementary single strands in the sample.
2. The ionic strength of the solution of single-stranded DNA must be fairly high, so that the nucleotides can approach one another—operationally, this means the ionic strength must be greater than 0.2 M.
3. The DNA concentration must be high enough that intermolecular collisions occur at a measurable frequency.

The first two requirements determine whether renaturation *can* occur. However, the third requirement only affects the rate; that is, if the concentration is too low, the renaturation rate may be exceedingly low. In the

SEC. 15-2 CHANGES IN THE STRUCTURE OF MACROMOLECULES

remainder of this section, we examine the concentration dependence of the renaturation rate and how it is used in biological research.*

Renaturation is both a bimolecular and second-order reaction and is described by the rate law

$$\frac{d[A]}{dt} = -k[A]^2, \qquad (3)$$

in which [A] is the concentration of the single strands. (See Chapter 11 for a review of second-order reactions, if necessary.) If A_0 is the initial concentration, the rate law can be integrated to yield

$$\frac{1}{[A]} = \frac{1}{[A_0]} + kT. \qquad (4)$$

Experimentally it is convenient to measure the fraction f of the strands that remain dissociated. This is defined as

$$f = \frac{A}{A_0}. \qquad (5)$$

Combining this with Equation 4 yields

$$f = \frac{1}{1 + A_0 kt}. \qquad (6)$$

This equation tells us two things: (1) after a very long time the fraction that is dissociated approaches zero—that is, ultimately, renaturation is complete; (2) the rate of renaturation increases as the initial DNA concentration is increased, as is expected of a bimolecular reaction.

Concentration has a special meaning in the analysis of renaturation and requires some discussion. In ordinary chemical kinetics, concentration is expressed as the number of molecules per unit volume (or some form of this ratio). This is a useful measure because, in general, each molecule contains only a single site at which a particular reaction can occur. This is not the case for renaturation because any sequence of bases can appear several times in a particular DNA chain. We will see that the best measure of concentration is actually the number of copies of each sequence per unit volume. The effect of multiple copies of a particular sequence can be seen by examination of three simple types of DNA molecules, which we term types I, II, and III.

Consider a type I DNA molecule whose base sequence is ABC $\cdots$ IJ in one strand and A'B'C' $\cdots$ I'J' in the other strand. We use the prime to denote a complementary base. That is, according to the standard base-pairing scheme that guanine (G) pairs only with cytosine (C) and adenine (A) pairs only with thymine (T), we would write [G'] = [C], [A'] = [T], [C'] = [G], and [T'] = [A]. Thus we assume that AA' is a proper base pair, as are

---

*Proteins can also renature but because of the large number of potentially interacting groups, the analysis of renaturation and the conditions optimizing the process are not as well understood as for DNA.

the pairs BB′, CC′, and so forth. In order that a proper double helix forms, a base X in one strand must collide with the complementary base X′ in another strand. Since collisions are random, most of the collisions will not lead to base pairing. If the rate of renaturation of this single-stranded DNA, which has 10 bases, is compared to that of another molecule (a type II molecule), which has 20 base pairs and whose base sequences are ABC $\cdots$ ST and A′B′C′ $\cdots$ S′T′, the type II molecules having more bases and hence a higher molecular weight will renature more slowly for the same amount of DNA per cm^3, because so many more *in*effective collisions occur per unit time. Thus the initial concentration A_0 of DNA that appears in Equations 4 and 6 must somehow take into account the molecular weight or the number of base pairs in a complementary base sequence. If the type I molecules are cut in half so that a sample contains molecules having either the base sequence ABCDE or the sequence FGHIJ, the rate of renaturation will *not* be altered. (Be sure you understand this point.) This is because the number of ineffective collisions is unchanged—that is, B can initiate renaturation only by colliding with B′, and failure results whether B collides with an A′ on one strand or a G′ on the other strand. Let us now consider the rate of renaturation of the single strands of a type III molecule—that is, one whose sequences are ABCDEABCDE and A′B′C′D′E′A′B′C′D′E′. Such a molecule is said to have two copies of the *unique sequence* ABCDE. For a particular concentration (expressed as grams per cm^3), this molecule will renature faster than a type I molecule because, in a collection of type III molecules, more of the collisions with base B are effective than will be effective in a collection of type I molecules. As when the type I molecules were cut, cutting the type III molecules into fragments also does not affect the renaturation rate. Thus, if c_0 is the initial concentration now expressed in base pairs per cm^3* and X is the number of base pairs in each unique sequence, then

$$A_0 = \frac{c_0}{X}, \qquad (7)$$

and Equation 6 becomes

$$f = \frac{1}{1 + c_0 tk/X}. \qquad (8)$$

Let us now see how k can be evaluated. Consider a molecule containing only bases of type A in one strand and only of type A′ in the other. An example of this is poly(dA:dU), which consists of one strand of polydeoxyadenylic acid and one of polydeoxyuridylic acid. If this DNA is denatured

---

*Note that by expressing c as base pairs per cm^3, we are ignoring the molecular weight of the single strand and using instead the molarity of the bases. Thus if a DNA molecule whose molecular weight is 10^7 is at a concentration of 10 milligrams per liter, the molecular-weight-dependent molar concentration is $0.01/10^7 = 10^{-9}$ M, whereas expressed as base pairs (molecular weight = 660), the molecular-weight-independent concentration is $0.01/660 = 1.5 \times 10^{-5}$ M.

and renatured, *all* collisions between complementary strands are effective, so that $X = 1$. In general, for the time interval $t_{1/2}$, when $f = 1/2$, it is the case that $1 + (c_0 t_{1/2}/X) = 2$, so that

$$c_0 t_{1/2} = \frac{X}{k}.\tag{9}$$

Hence, for the DNA just described for which $X = 1$, $k = 1/c_0 t_{1/2}$, and a measurement of the time required for 50 percent renaturation for any initial concentration, c_0, yields the value of k. It is important to understand:

> k *is a constant characteristic of the physical process of renaturation and is independent of the molecular weight of the DNA and the value of* X.

Furthermore, since X is also a constant, $c_0 t_{1/2}$ is a constant but neither c_0 nor $t_{1/2}$ are constants. Thus, for any particular DNA-concentration c_0 that is chosen, $t_{1/2} = X/kc_0$. The important point is that the *product* $c_0 t_{1/2}$ *is proportional to* X, *the number of base pairs per unique sequence.*

The measurement of renaturation rates to obtain the value $c_0 t_{1/2}$ is called $c_0 t$ or "*cot*" analysis.

So far, we have not explained why anyone would carry out such an analysis. In order to understand this, we must consider the result of two more experiments.

In the first experiment, we compare the curves of f versus $c_0 t$ for a variety of DNA molecules whose molecular weights differ. It should be noted that in such an experiment we can use DNA of any concentration—that is, we can choose any value (or several) of c_0 because the curve of f versus $c_0 t$ is independent of c_0 (that is, $c_0 t_{1/2}$ is a constant). If each of these DNA molecules contains a single base sequence (that is, they have no repeated sequences), X is simply the number of base pairs in the unbroken DNA molecule. That is, $c_0 t_{1/2}$ *(not $c_0 t$) is proportional to* X *and hence to the molecular weight of the unbroken DNA.** This is the fundamental rule used to analyze experimental $c_0 t$ curves. The result of such an experiment is shown in Figure 15-17, in which it can be seen that $c_0 t_{1/2}$ is proportional to the number of base pairs per molecule. This means that we can use the measured value of $c_0 t_{1/2}$ to obtain the size of a particular sequence and the number of copies of the sequence in the organism from which the DNA was isolated.

Before proceeding to the second experiment, an important point must be made: the value of c_0 used in a $c_0 t$ curve is the *total* DNA concentration in the renaturing sample, rather than the concentration of each component. Thus, when analyzing multicomponent curves, the observed $c_0 t_{1/2}$ is $(c_0 t_{1/2})_{\text{true}}$ (the value if one component were present) divided by the fraction of the total

---

*For technical reasons the DNA molecules are broken into small fragments, but, as we saw when discussing molecules of types I, II, and III, such breakage does not alter the value of $c_0 t$.

CHAP. 15   THE PROPERTIES OF MACROMOLECULES

FIGURE 15-17 Kinetics of reassociation of various denatured double-stranded DNA molecules. The number of nucleotide base pairs in each DNA source is indicated by the arrows on the upper logarithmic scale. This is not the number of base pairs per individual DNA molecule used in the hybridization; for technical reasons, it is necessary to break the DNA molecules into small fragments prior to renaturation. Thus, the given values represent the number of base pairs per organism. Mouse satellite DNA is a short segment of mouse DNA that is repeated many times. Along the abscissa is shown the product of the initial concentration of the DNA sample (expressed as moles of nucleotide per liter), plotted on a logarithmic scale, times the number of seconds during which the molecules were subjected to renaturation conditions. The $c_0 t$ required for half-reassociation is proportional to the number of nucleotide base pairs in the DNA molecules. [SOURCE: After R. Britten, *Science* 161, 529(1968), by permission of the publisher.]

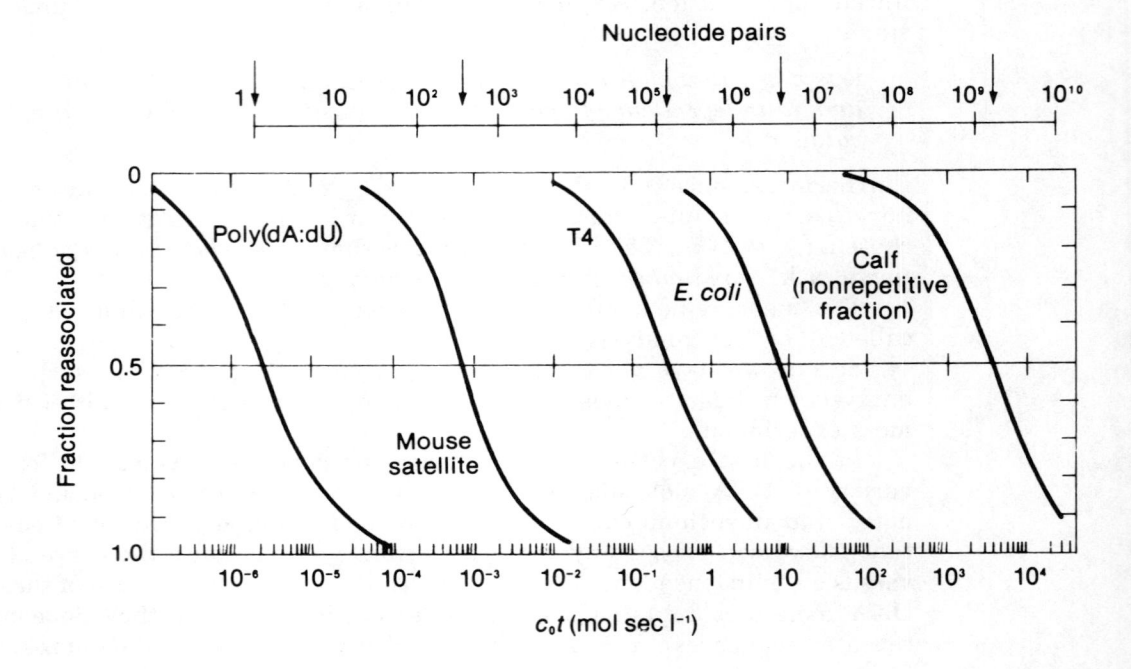

DNA represented by that component. Since the scale in Figure 15-17 relates size and $(c_0 t_{1/2})_{\text{true}}$, each $(c_0 t_{1/2})_{\text{obs}}$ must be converted to $(c_0 t_{1/2})_{\text{true}}$ before using such a scale.

In the second experiment, a molecule containing four copies of a sequence 5% the length of the intact molecule is examined (Figure 15-18(a)). If x is $(c_0 t_{1/2})_{\text{true}}$ for the unique sequence, $(c_0 t_{1/2})_{\text{obs}}$ will be $x/0.8 = 1.25x$ (80% of the total sequence is unique). For the redundant sequence $(c_0 t_{1/2})_{\text{true}}$ is $0.05x$ and $(c_0 t_{1/2})_{\text{obs}}$ is $0.05x/0.20 = 0.25x$. This is shown in Figure 15-18(b)—20% has $(c_0 t_{1/2})_{\text{obs}}$ of $q/5$ and 80% has $(c_0 t_{1/2})_{\text{obs}}$ of q.

The $c_0 t$ analysis is used as a means of detecting repetitive sequences in DNA, as is seen in the previous example. This is of great interest because in animal cells there are short sequences repeated up to one million times (such as mouse satellite DNA in Figure 15-17. An example of how one draws conclusions about repetitive sequences is given in the following.

SEC. 15-2   CHANGES IN THE STRUCTURE OF MACROMOLECULES   529

FIGURE 15-18 (a) A DNA molecule containing four repetitive sequences (heavy line) each equal to 5 percent of the length of the intact molecule. (b) The c_0t curve for the DNA of panel (a). Twenty percent of the DNA has a c_0t of q and 80 percent has a c_0t of $16q$.

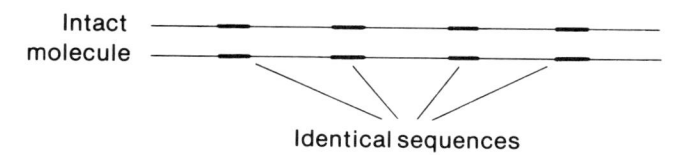

Intact molecule

Identical sequences

When this DNA molecule is fragmented to pieces that are 1% of the size of the intact molecule and then denatured, most of the single strands are ——, some are ——, and a few are ——.

(a)

(b)

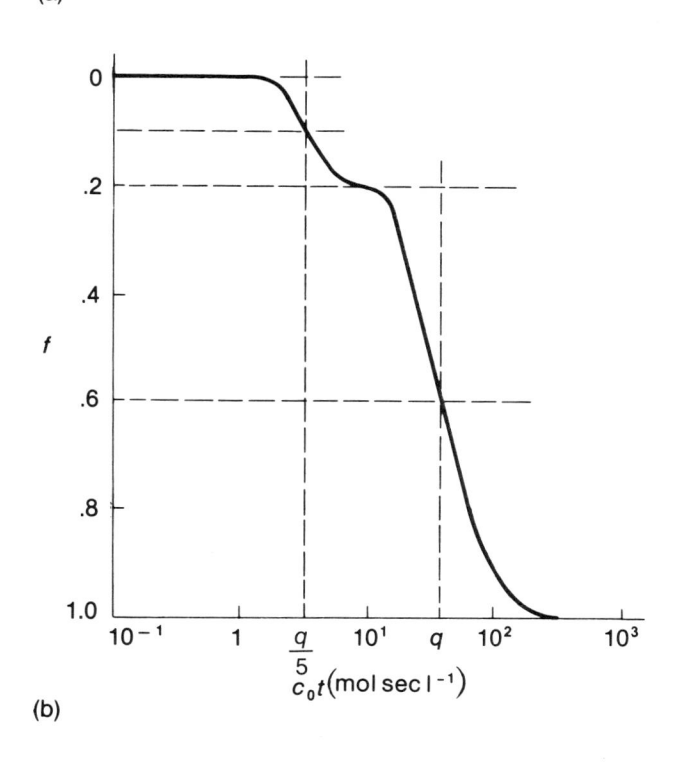

EXAMPLE 15-E Determination of the size and number of copies of various sequences in a DNA sample whose c_0t curve is shown in Figure 15-19.

The following $c_0t_{1/2}$ are observed for a DNA sample: 10^2, 53%; 1, 27%; 10^{-3}, 20%. The number of copies of each sequence is determined from the size scale in Figure 15-17. We first calculate $(c_0t_{1/2})_{\text{true}}$ for each component by

FIGURE 15-19 A hypothetical c_0t curve, analyzed in Example 15-E.

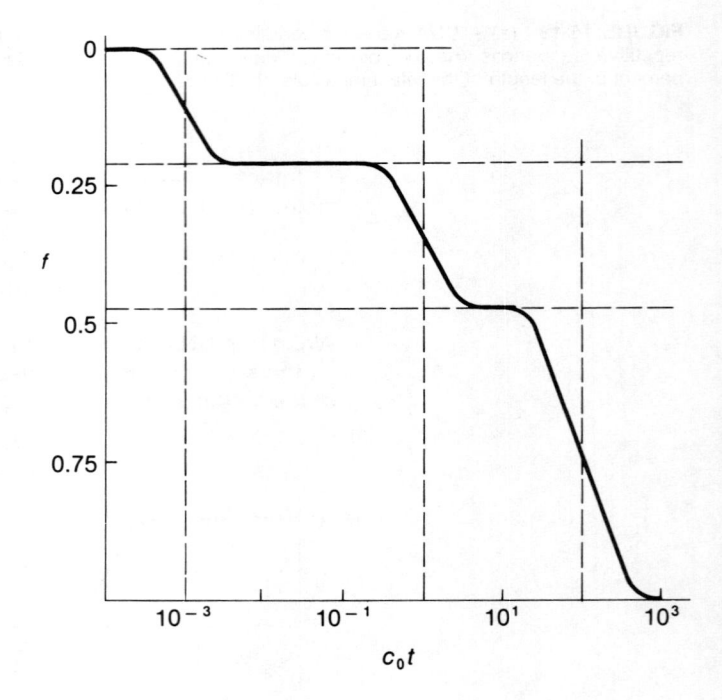

multiplying each $(c_0t_{1/2})_{obs}$ by the fraction of the total DNA that it represents. Thus, the $(c_0t_{1/2})_{true}$ are 0.53×10^2, 0.27×1, and 0.20×10^{-3}, which correspond to 4.2×10^7, 2.2×10^5, and 160 base pairs, respectively, for each component. The number of copies of each sequence is inversely proportional to $(c_0t_{1/2})_{obs}$; thus, if we assume that the genome contains one copy of the longest sequence, then the cell from which the DNA has been isolated would contain 1, 10^2, and 10^5 copies of sequences having 4.2×10^7, 2.2×10^5, and 160 base pairs, respectively. The total number of base pairs per genome would be $4.2 \times 10^7 + 100(2.2 \times 10^5) + 10^5(160) = 8 \times 10^7$, and the molecular weight of the total cellular DNA would be $(8 \times 10^7)(660) = 5.3 \times 10^{10}$. Note that if an independent measurement of the total molecular weight of the DNA were to yield the value $1 \times 10^{11} = 2(5.3 \times 10^{10})$, then there would be 2, 200, and 2×10^5 copies of these sequences, respectively.

The c_0t analysis is a powerful method for determining the number of copies of each base sequence, and this is its principal application at the present time.

15-3 MOVEMENT OF MACROMOLECULES

Diffusion

Macromolecules are in constant motion owing to their thermal energy and to bombardment by solvent molecules. This motion is random in direction, and if initially all of the macromolecules are in one region of a solution they will gradually move throughout the solution until the concentration is uniform everywhere. This process is called *diffusion*. Macromolecules can also be forced to move in a particular direction by a centrifugal force or, if they are electrically charged, by an electric field. These two processes are *sedimentation* and *electrophoresis*, respectively.

Diffusion, sedimentation, and electrophoresis also apply to small molecules but in this section we will consider only macromolecules and the information that can be obtained about them from studies of these transport processes.

It will help you to analyze diffusion if you imagine a rectangular prism (such as would be used in measuring stream flow) containing a solvent, in the center of which, initially, all of the solute molecules are confined to a thin vertical cross-section (by an unspecified means), as shown in Figure 15-20. The graph shows the concentration distribution within the prism at zero time (t_0). In time, solute molecules diffuse across the boundary in both directions out of the initial cross-section so that the concentration distribution changes as is shown in the lower part of the figure.

FIGURE 15-20 Diffusion from a thin layer of solute molecules initially at the center of a rectangular prism of length *l*. The graph shows the initial concentration distribution (t_0) and the distributions at two later times (t_1, t_2). The areas of the curves remain constant.

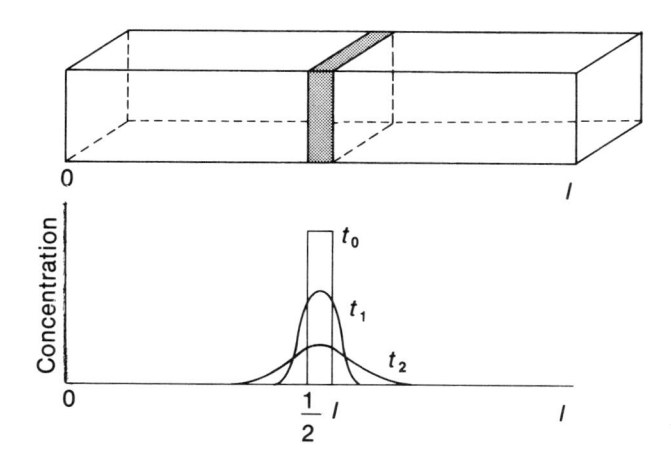

An analysis of diffusion shows that the rate of flow—that is, the movement of a number of molecules dn in a time t, across any cross-sectional area A, is proportional to the concentration (c) gradient dc/dx at that cross-section. This is expressed mathematically as

$$-\frac{dm}{dt}\frac{1}{A} = D\frac{dc}{dx}, \tag{10}$$

in which D, the constant of proportionality, is the *diffusion coefficient* and the negative sign is introduced purely so that D will be positive. This equation is called *Fick's first law of diffusion*. The units of D are $cm^2\ sec^{-1}$.

FIGURE 15-21 A diagram useful for a mathematical analysis of diffusion. See text for details.

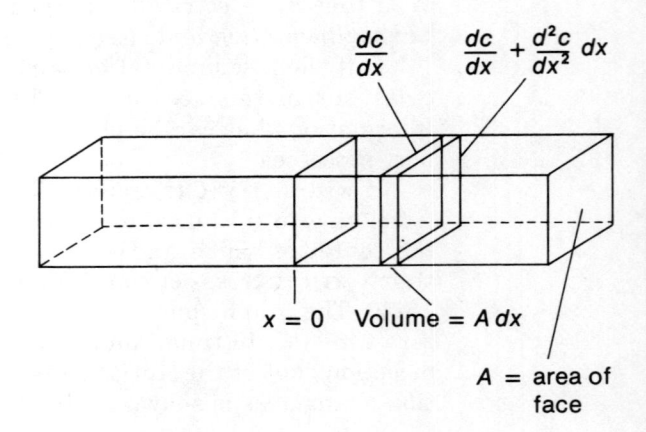

The diffusion process can be expressed in another way that is experimentally useful. Referring to Figure 15-21, we consider a volume element $A\,dx$ at a distance x from the origin and calculate the rate at which the diffusing solute accumulates in that volume element. This is done by calculating the difference between the rate at which molecules enter the element and the rate at which they leave the element. The concentration gradient dc/dx varies with x and the rate with which it varies is simply $(d/dx)(dc/dx) = d^2c/dx^2$. Thus, according to Fick's first law,

$$\text{Rate of entry of molecules at } x = -DA\frac{dc}{dx}, \tag{11}$$

and

$$\text{Rate molecules leave at } x + dx = -DA\left(\frac{dc}{dx} + \frac{d^2c}{dx^2}dx\right). \tag{12}$$

Subtracting Equation 12 from Equation 11 yields

$$\text{Rate of accumulation} = DA\frac{d^2c}{dx^2}dx. \tag{13}$$

SEC. 15-3 MOVEMENT OF MACROMOLECULES

To obtain dc/dt, we need only divide the rate of accumulation by the volume $A\,dx$, so that

$$\frac{dc}{dt} = D\frac{d^2c}{dx^2}, \qquad (14)$$

which is known as *Fick's second law*. The solution to this equation is

$$c = \frac{1}{\sqrt{4\pi Dt}}e^{-x^2/4Dt}, \qquad (15)$$

from which D can be calculated since c, x, and t are measurable.

In practice the arrangement in Figure 15-21 cannot be realized, so that D is measured instead by putting the solution in contact with the solvent, as is shown in Figure 15-22. The equations become more complex and D is obtained graphically as follows (see Figure 15-23).

FIGURE 15-22 A common experimental arrangement for studying diffusion.

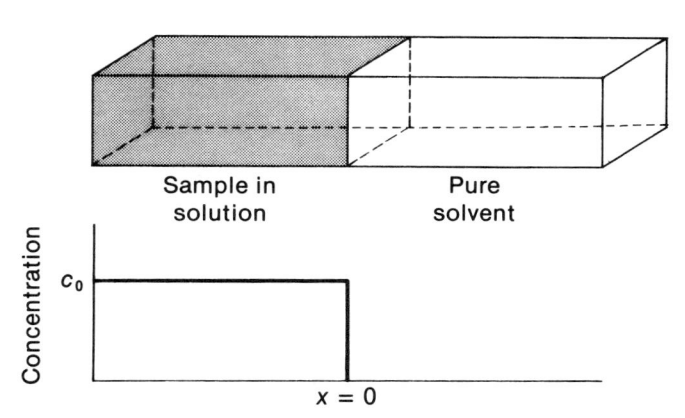

FIGURE 15-23 Measurement of the diffusion coefficient. (a) The concentration distribution at an early time shortly after the solvent and the solution are placed in contact; the curve c versus x shows the concentration of the solution across the cell and dc/dx versus x is the concentration gradient or the schlieren pattern; (b) the pattern at a later time. (c) A curve obtained from the data in parts (a) and (b). A and H are the area and height of the dc/dx curve. For small proteins, the time scale is of the order of several hours; for large viruses, it can be several days. Note that the curve does not pass through the origin. This is because the boundary is rarely perfect at zero time. The slope of the curve is $4\pi D$. [SOURCE: Based on *Physical Biochemistry*, First Edition, by David Freifelder. W.H. Freeman and Company. Copyright © 1976.]

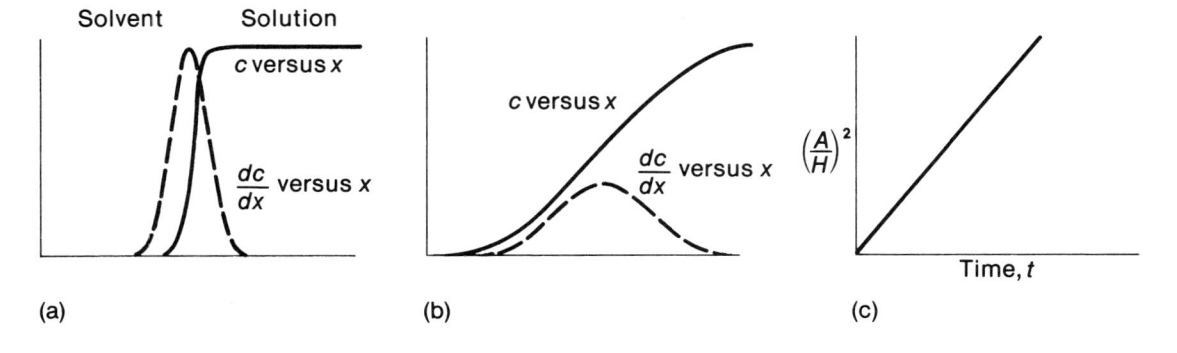

We measure c as a function of x at various times and calculate dc/dx as a function of x. Both the height H and the area A of the dc/dx versus x curve are calculated for each time t, and a graph of $(A/H)^2$ versus t is made. It can be shown that the slope of this graph is $4\pi D$. Actually, the work is made simpler by use of a particular optical system (schlieren optics) for viewing the tube, because this system gives a direct plot of dc/dx versus x at each time. Details of this method can be found in several references at the end of the chapter.

The utility of measurements of D is that it gives information about the size and shape of molecules. This can be seen by applying the following reasoning. We consider diffusion across an interval dx in which the concentration changes from c to $c - dc$. The force driving the molecules into the more dilute solution is the difference in the chemical potentials μ_c and μ_{c-dc}. From Equation 9, Chapter 7,

$$\mu_{c-dc} - \mu_c = kT \ln \frac{c - dc}{c},$$

which, by application of the approximation $\ln(1 - y) = -y$ for small y, yields

$$d\mu = kT \ln\left(1 - \frac{dc}{c}\right) = -kT \frac{dc}{c}. \tag{16}$$

This is equivalent to the free energy change, which is the work done in transferring one molecule a distance dx, so that $dG = F\,dx$, in which F is the driving force. Therefore,

$$F = \frac{dG}{dx} = -\frac{kT}{c} \frac{dc}{dx}. \tag{17}$$

The motion of any molecule is counteracted by a frictional force which, in the case of a sphere of radius a moving in a fluid of viscosity η at a velocity dx/dt, is

$$\text{Frictional force} = 6\pi a \eta \frac{dx}{dt}. \tag{18}$$

At constant velocity, the forces are equal, so that combining Equations 17 and 18 and rearranging yields

$$c \frac{dx}{dt} = -\frac{kT}{6\pi a \eta} \frac{dc}{dx}. \tag{19}$$

The product $c(dx/dt)$ is simply the rate at which the diffusing substance moves through the unit area at a position x, which is equivalent to the left side of Equation 19. Therefore,

$$D = \frac{kT}{6\pi a \eta}, \tag{20}$$

so that D is inversely proportional to the radius of the particle. Since the mass m of a sphere $= \frac{4}{3}\pi a^3 \rho$, in which ρ is the density, D varies inversely

SEC. 15-3 MOVEMENT OF MACROMOLECULES

with the cube root of the mass. Experimentally, this relation is not of much use. Of greater value, however, is the relation between D and the shape of the particle.

As a particle moves through a solvent, its motion is opposed by friction, as we have already stated. In the case of a sphere, the frictional force is $6\pi\eta a v$, but in the more general case it is fv, in which f is the frictional coefficient and v is the velocity. Thus, Equation 20 can be written

$$D = \frac{kT}{f}. \qquad (21)$$

Of all shapes, it is the sphere that encounters the least resistance to movement—that is, it has the lowest value of f. (This is easy to appreciate if one imagines the relative difficulty of walking through a swimming pool with arms drawn in versus arms extended—it is harder to move with the arms extended.) The more extended a molecule is, or, stated in another way, the greater its *axial ratio* (the ratio of the long axis to the short axis), the greater is its frictional coefficient. Thus, if we define the frictional coefficient of a sphere as f_0, comparison of Equations 20 and 21 shows that

$$f_0 = 6\pi a \eta.$$

This value can be compared to the measured value of f obtained from Equation 21, and the *frictional ratio* f/f_0 can be used as a measure of the departure from a spherical shape. To calculate f_0 requires knowledge of a, which is the radius of the sphere that the molecule would occupy *if* it were spherical. How this calculation is performed will be shown shortly.

Several quantitative theories have been developed for relating the value of f/f_0 to molecular dimensions. In each theory the value of f is calculated for simple, regular shapes, just as for the sphere. Of course molecules never have very regular shapes, so that the information provided by these theories is primarily suggestive. The two shapes examined in greatest detail are the oblate and prolate ellipsoids. An oblate ellipsoid is a disc-shaped ellipsoid generated by rotating an ellipse about its short axis. A prolate ellipsoid is rodlike (actually more like an American football) and is formed by rotating an ellipse about its long axis. The ratio of the long axis of either ellipsoid to its short axis is the axial ratio. A commonly used relation between the axial ratio and f/f_0 is shown in Table 15-3. Note that it is not necessary to distinguish the two kinds of ellipsoids until their axial ratio is greater than 10.

EXAMPLE 15–F **Estimation of the shape of a protein by measuring diffusion.**

Protein A has a molecular weight of 60,000, a diffusion coefficient of 7×10^{-7} cm^2 sec^{-1}, and a density* of 1.3 g/cm^3. The volume of the protein is $(6 \times 10^4)/(6 \times 10^{23} \times 1.3) = 7.7 \times 10^{-20}$ cm^3, so that, if it were a sphere, its

---

*The density of a protein is usually not measured. Instead, the partial specific volume, which is not very different from $1/\rho$, is measured. How this is done is explained in the references given at the end of the chapter.

CHAP. 15 THE PROPERTIES OF MACROMOLECULES

TABLE 15-3 Relation between axial ratio and f/f_0 for prolate and oblate ellipsoids.

	f/f_0	
Axial ratio	Oblate	Prolate
1	1.00	1.00
2	1.04	1.04
3	1.11	1.11
4	1.18	1.17
5	1.25	1.22
6	1.31	1.28
8	1.43	1.37
10	1.54	1.46
15	1.78	1.64
20	2.00	1.78
30	2.36	2.02
40	2.67	2.21
50	2.95	2.38
60	3.20	2.52
80	3.66	2.77
100	4.07	2.97

radius would be 2.64×10^{-7} cm. If D is measured in water at 20°C, $\eta = 0.01$ poise (the appropriate unit for this calculation). Thus $f_0 = 4.97 \times 10^{-8}$. From Equation 21, $f = kT/D = 1.38 \times 10^{-16} \times 293/(7 \times 10^{-7}) = 5.78 \times 10^{-8}$. Therefore the frictional ratio $f/f_0 = 1.16$. From Table 15-3, the axial ratio of this protein is about 3.8.

A similar calculation for myosin, a rodlike protein found in muscle, gives a value of $f/f_0 = 3.63$, and this corresponds to an axial ratio of about 80.

Diffusion measurements can also give information about the arrangement of the subunits in a protein that consists of several polypeptide chains. This can be seen in the following example.

EXAMPLE 15–G **The arrangement of the subunits in the protein hemoglobin.**

The blood protein hemoglobin consists of four subunits. We wish to know whether the four subunits are arranged in a row or are closely packed in a quasi-spherical cluster. The molecular weight of each subunit is about 16,000 and the value of D for a subunit in water is 11.3×10^{-7} at 0°C. Repeating the calculation in the preceding example, $f/f_0 = 1.1$, so that the axial ratio of each subunit is about 3 (Table 15-3). The values of the molecular weight and the diffusion coefficient of the hemoglobin tetramer are 64,500 and 6.9×10^{-7}, respectively, which yields $f/f_0 = 1.16$. Thus, the axial ratio of hemoglobin is about 3.5. This means that the subunits cannot be arranged linearly but must form a compact cluster; if they were arranged linearly, the axial ratio of hemoglobin would be about 12 and f/f_0 would be about 1.6.

Sedimentation of Macromolecules

If a particle is in a centrifugal field generated by a spinning rotor with angular velocity ω, it will experience a centrifugal force, $F_{\text{centrif}} = m\omega^2 r$, in which m is the mass of the particle and r the distance from the center of rotation. If this particle is not in vacuum but in a solvent,* the solvent molecules will, of course, be displaced by the motion of the particle. Their resistance to this displacement constitutes a buoyant force opposed to the centrifugal force. This *buoyancy* reduces the net force on the macromolecules by $\omega^2 r$ times the mass of the displaced solution; this mass is simply the volume of the particle multiplied by the density, ρ, of the solvent. The particle volume is $m\bar{v}$, in which $\bar{v}$ is the partial specific volume of the paticle, so that the buoyant force is $\omega^2 r m \bar{v} \rho$.

Clearly, the particles and the solvent molecules cannot slip by one another without experiencing friction. This frictional force—which opposes the motion of both—is proportional to the difference between the velocities of the particle and of the solvent molecules and is expressed as fv, in which f is the frictional coefficient and v is the velocity relative to the centrifuge cell, which holds the solvent. Because the velocity is constant when the net force is zero, the velocity of the particle is

$$v = \frac{\omega^2 m(1 - \bar{v}\rho)}{f}.$$ (22)

Equation 22 says that (all other things being equal):

1. A more massive particle (or molecule) tends to move faster than a less massive one.
2. A denser particle (that is, one having small partial specific volume) moves faster than a less dense one.
3. The denser the solution, the more slowly the particle will move.
4. The greater the frictional coefficient, the more slowly the particle will move.

These four statements constitute the basic rules of sedimentation and

---

*Solvent is used in the general sense of the suspending medium. That is, if the particle is in water, the solvent is water; if it is in 0.1 M NaCl, the NaCl solution is considered to be the solvent.

apply to all particles whether they are large structures, macromolecules, or small molecules.

Because the velocity of a molecule is proportional to the magnitude of the centrifugal field (that is, $\omega^2 r$), it is common to discuss sedimentation properties in terms of the velocity per unit field, or

$$s = \frac{v}{\omega^2 r} = \frac{m(1 - \bar{v}\rho)}{f}, \qquad (23)$$

in which s is the *sedimentation coefficient;* the determination of s is the immediate goal of most sedimentation velocity experiments.

Sedimentation of macromolecules is useful in the determination of the molecular weight and the shape of macromolecules. This can be done as follows.

The frictional coefficient f appearing in Equation 21 is the same as that in Equation 23. Therefore, we can combine these two equations to obtain

$$m = \frac{skT}{(1 - \bar{v}\rho)D}. \qquad (24)$$

This gives the mass in grams.* To obtain the molecular weight M, one need only multiply m by Avogadro's number $\mathcal{N}$ or rewrite Equation 24 as $M = sRT/(1 - \bar{v}\rho)D$.

The shape of a molecule can be estimated if M and s are known because, from Equation 21, f can be calculated and compared to f_0. This is shown in the following example.

EXAMPLE 15–H Determination of the shape of a protein from sedimentation measurements.

The molecular weight of the protein myosin is 525,000. The values of s and $\bar{v}$ are 6.4×10^{-13} sec and 0.73 cm^3 g^{-1}, respectively, in water (for which $\rho = 1.0$ g/ml) at 20°C. From Equation 22, $f = (525{,}000/6 \times 10^{23})[(1 - 0.73)/6.4 \times 10^{-13}] = 3.7 \times 10^{-7}$. From the molecular weight and $\bar{v} = 1/\rho$, we can calculate f_0 as in the preceding example to obtain a value of $f_0 = 1.02 \times 10^{-7}$. Therefore, $f/f_0 = 3.63$ and the protein is not spherical. By comparison with known proteins, such a large value of f/f_0 indicates that the molecule is at least 10 times as long as it is wide.

An important use of sedimentation data is to determine whether a sample of a macromolecule is homogeneous. If the sample contains molecules having several different shapes or molecular weights, one will typi-

---

*Remember to specify the value of k in erg—namely, 1.4×10^{-16}.

SEC. 15-3   MOVEMENT OF MACROMOLECULES   539

FIGURE 15-24 Sedimentation of a solution containing two types of molecules, represented by small solid circles (molecule 1) and large open circles (molecule 2). Panel (a) shows the distribution of molecules before centrifugation begins. Panel (b) shows the distribution some time after sedimentation begins. The open circles, which are assumed to have twice the molecular weight of the solid circles, also move more rapidly. The pellet represents the molecules that accumulate at the bottom of the centrifuge tube during sedimentation. Panel (c) shows the concentration distribution at the time in panel (a) (upper horizontal dashed line) and at the time shown in panel (b) (solid line). Note that the concentration after sedimentation does not exceed the initial concentration (c_0) except in the pellet. The lower dashed line shows c_1, the concentration of molecule 1. The steps represent the sedimentation boundaries for each sedimenting component. The rightward step is twice the height of that at the left because the molecules represented by the open circles have twice the mass (by assumption) and are equal in number to the smaller, more slowly moving solid circles. In general $c_1/(c_0 - c_1)$ is the ratio of the concentrations of the more slowly moving molecule and the faster molecule, respectively.

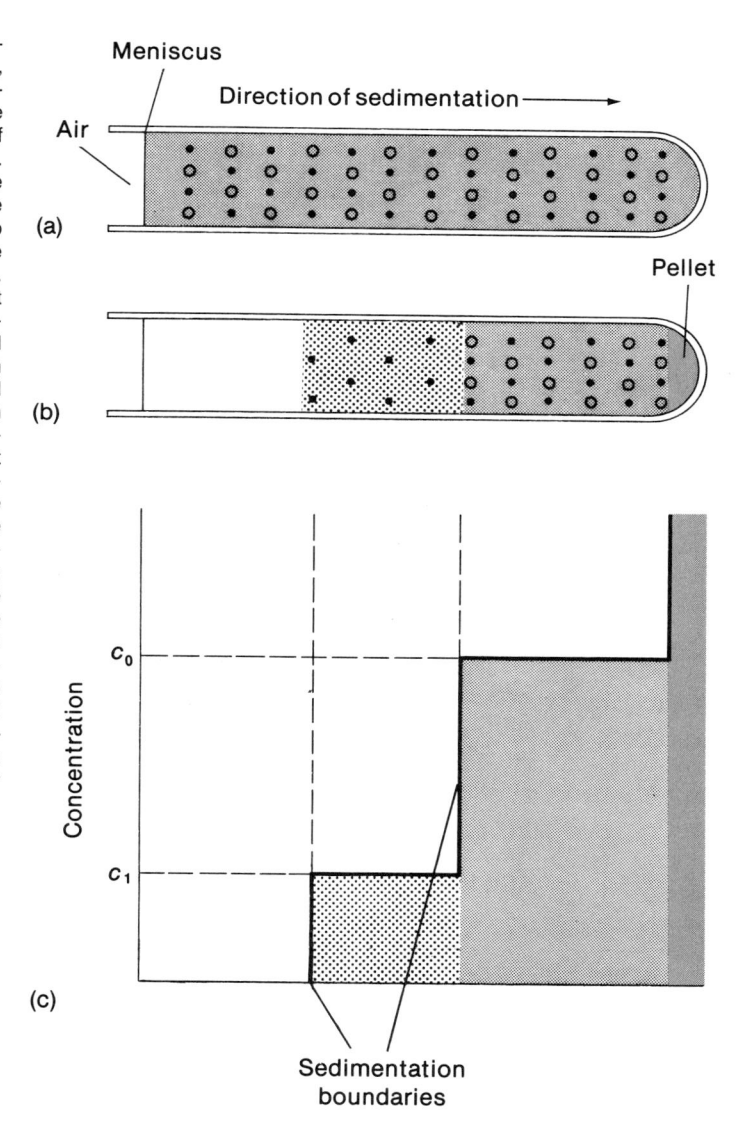

cally observe that the sample contains components having different sedimentation coefficients (Figure 15-24). This is a useful criterion for the purity of macromolecules.

When a solution of macromolecules whose density is greater than that of the solvent is sedimented, the macromolecules move to the bottom of the centrifuge tube. If the speed of rotation of the centrifuge is great enough, the molecules will form a dense solid mass at the bottom of the tube.

However, the motion of the molecules in the centrifugal direction is reduced by diffusion, which occurs in all directions. If the centrifugal force, or the rotational velocity, is sufficiently small, a distribution of the molecules can result in which the centrifugal motion is entirely balanced by back-diffusion and the concentration distribution is stationary. This is called *sedimentation equilibrium,* and it can be used to determine molecular weights with very great precision. How this is done will be described in a later section.

Electrophoresis

In the previous section we described the movement of molecules caused by a centrifugal force. If a molecule is electrically charged, it will also move under the influence of an electrical force. This motion is called *electrophoresis*. Electrophoresis is a useful method because it can be used both to determine whether a molecule is positively or negatively charged and as a means of separating molecules.

The basic theory of electrophoresis is simple. As in the case of sedimentation, a force (the electrical force) is applied and is countered by viscous drag. If a molecule with charge q is in an electric field E, the molecule experiences a force Eq. This force accelerates the molecule until the viscous drag fv, in which f is the frictional coefficient and v is the velocity, equals the force; when this occurs, the molecule no longer has a net force acting on it and thus moves at a constant velocity defined by

$$Eq = fv \,. \tag{25}$$

The mobility, u, is defined as the velocity per unit field, or

$$u = \frac{v}{E} = \frac{q}{f} \,. \tag{26}$$

Because the mobility of the molecule depends upon the frictional coefficient, which is in turn a function of the size and shape of the molecule, the value of u should give information about these parameters. However, a real charged molecule is surrounded by a cloud of charged ions (see Chapters 8 and 14) that shields it from the electric field; furthermore, many solvent molecules are bound to the molecule. These factors produce several effects. For example, the effective viscosity encountered by the molecule is different from the solution viscosity because the ions in the diffuse layer also move under the influence of the electric field. The net charge in the double layer is opposite in sign to the charge on the surface, and the ions move in a direction opposite to that of the molecule; these ions are solvated, so that there is also a net motion of the solvent opposite to that of the surface. The solvent flow retards the motion of the molecule and is called *electrophoretic retardation*. There is also another retarding effect resulting from the fact that the molecule must drag the diffuse layer along.

A final complication is a complex electrical effect due to the electrical conductivity of the double layer. The result of these complications is that it is not possible to calculate either the magnitude of the charge of a molecule or its frictional coefficient from its mobility. Thus electrophoretic measurements do not enable one to determine any of the structural parameters of a macromolecule from mobility measurements, although they can be used to determine the *sign* of the charge of a molecule.

The charge of an amino acid depends upon the pH of the suspending medium, so that the mobility of an amino acid will also be pH-dependent. If a protein contains both positively and negatively charged amino acids, there will usually be a pH value at which the number of positive and negative charges are the same and the net charge is zero. This is called the *isoelectric point,* and at this pH mobility is zero. Figure 15-25 shows the relation between u and pH for a typical protein; the isoelectric point is evident in the figure.

FIGURE 15-25 The mobility of a hypothetical protein as a function of pH. The isoelectric point—namely, the pH at which the mobility is zero—is 6.2.

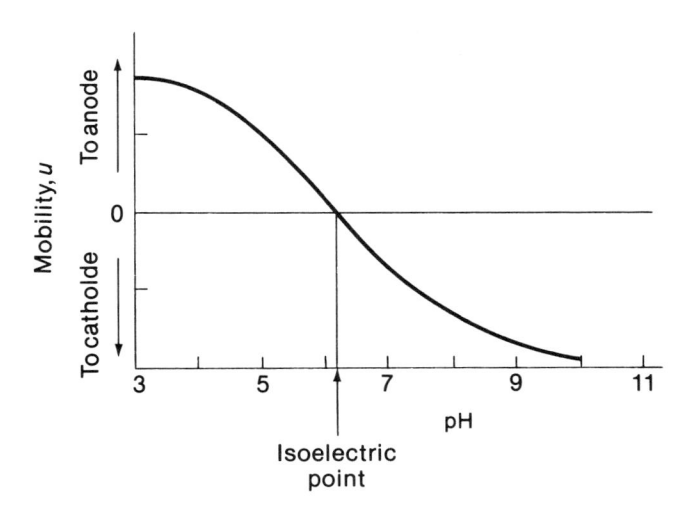

Electrophoresis has its greatest value as a separation technique. That is, if two particles have different mobilities, they will move at different rates in an electric field and will therefore separate from one another.*

Electrophoresis has been performed in the laboratory in three distinct ways.

---

*The reader might ask with good reason why we have bothered to discuss mobility, since it seems to be a useless parameter. Apart from the fact that numerous treatises, which are more or less understandable, have been written about the subject, we feel that it is important for the reader to appreciate the factors that affect mobility so as never to be tempted to draw simplistic conclusions from mobility data. In a later section we will describe an important electrophoretic method for determining the relative molecular weights of proteins and DNA molecules from their relative mobilities.

Free Electrophoresis. In this mode the molecules move through a liquid (invariably this is water) unimpeded by any physical barriers. The motion of the molecules is usually detected by optical means. Free electrophoresis yields a value for the mobility of a molecule but its principal value is that it gives some indication of the purity of a collection of molecules.

Figures 15-26 and 15-27 show the experimental arrangement for free electrophoresis and some typical results.

Surface Electrophoresis. With this method the molecules also move through liquid but the liquid is in intimate contact with a surface that can adsorb the molecules weakly. Thus, the rate of motion of a molecule is a function not only of its electrophoretic mobility but also of the extent to which it adsorbs to the surface. The stronger the adsorption, the greater is the fraction of time that a molecule is in contact with the surface and not moving. The most common adsorbents are paper and cellulose acetate. The molecules are usually detected either by their optical properties (absorption, fluorescence) or they are eluted from the supporting solid and analyzed in various ways. Figure 15-28 indicates a typical experimental arrangement.

Gel Electrophoresis. In this technique the liquid through which the molecules move is a gel. In such a system the motion of a molecule is again determined by its electrophoretic mobility and sometimes by adsorption but the principal factor is the ability of the molecule to pass through the pores of the gel. The manner in which pore size affects the mobility depends upon whether the gel is particulate or continuous (Figure 15-29). The principle is always the same however—that is, a smaller molecule is better able than a larger molecule to penetrate the pores and interstices of a gel. With a system consisting of gel particles, a molecule can either enter the particles or pass between the particles. The probability of entry is lower for larger molecules than for smaller molecules. Once inside the gel particle, a molecule is unable to move forward until it reemerges from the particle. Thus, a small molecule is impeded more than a large molecule is. This is the principle behind separation in starch gels and dextran gels. Examples of this technique are shown in Figure 15-30. In a continuous gel, the most common material being *agarose* (a component of agar), a molecule must move through the gel. However, the gel is a continuous network and presumably there are no blind alleys. In this system, the large molecules are impeded more than the smaller ones, because (1) the larger ones collide more often with the gel structure and (2) they must often be oriented end on for any motion to occur. This technique is commonly used to separate DNA molecules by molecular weight, as shown in Figure 15-31.

Figure 15-32 shows a simple experimental arrangement for the separation of proteins by electrophoresis. A flat slab of gel is prepared in which a small slot is made, into which a mixture of proteins is placed. The gel is made up in a dilute electrolyte at a fixed pH. A voltage is placed across the

SEC. 15-3 MOVEMENT OF MACROMOLECULES

FIGURE 15-26 A schematic diagram for moving-boundary electrophoresis. A U-shaped tube contains a mixture of a positively charged macromolecule A and a negatively charged macromolecule B. This mixture is overlaid with buffer. The initial position of the boundary between the solution and the buffer is called the origin. Under the influence of the applied electric field, A and B move in opposite directions. In the right arm, boundaries result between buffer and pure B (boundary 1) and between pure B and the mixture (boundary 2). Boundary 1 moves upward and boundary 2 moves downward, as is shown in the lower panels. The slight breadth of the boundaries is a result of diffusion. The vertical axes in the two right panels represent distance along the right arm of the U-tube.

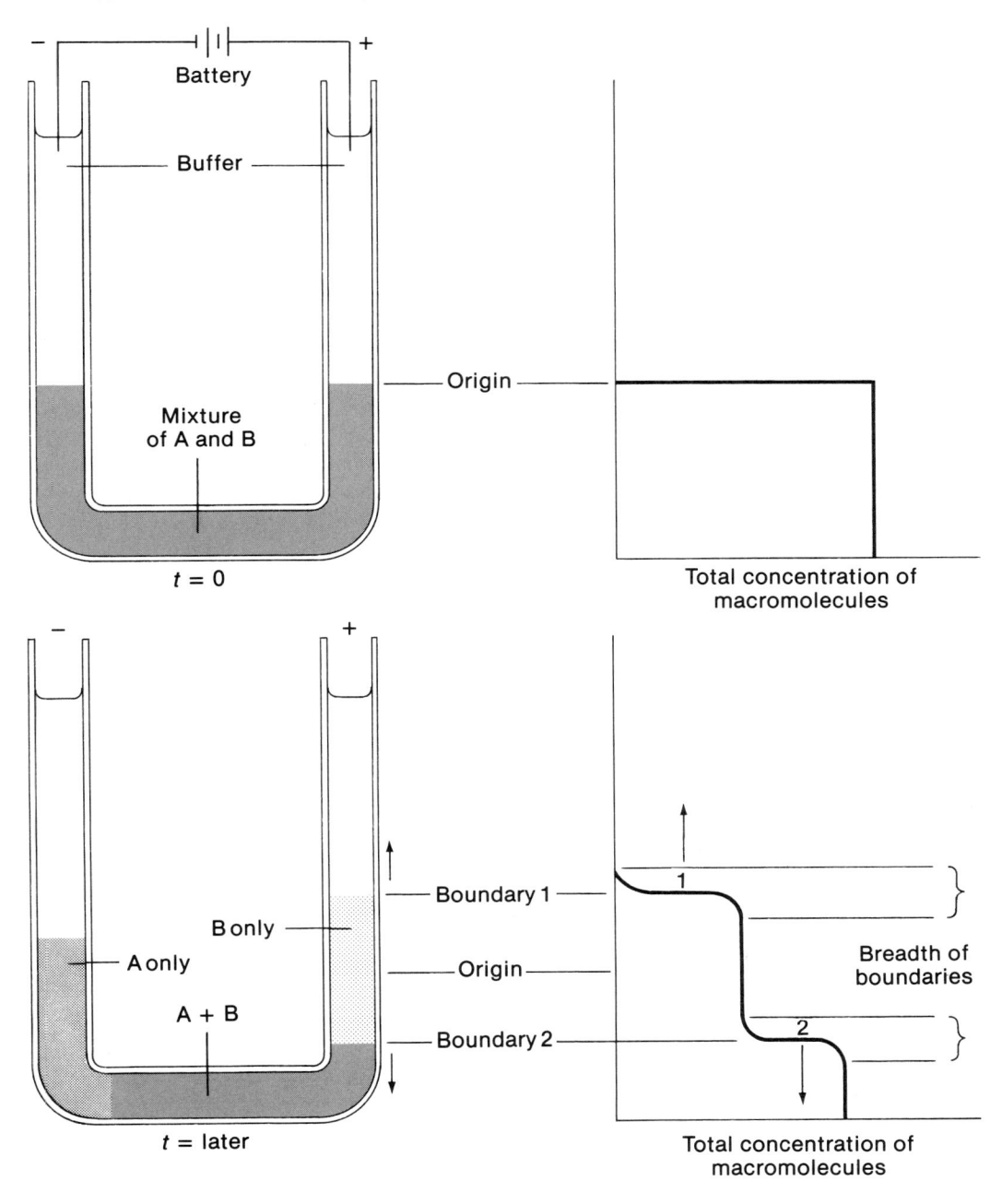

FIGURE 15-27 (a) Separation of various proteins in blood by moving-boundary electrophoresis. The graph shows the concentration gradient and not the concentration. The relation between the concentration and the concentration gradient is shown in panels (b) and (c). This representation is given because the schlieren optical system in the conventional electrophoresis instruments determines the concentration gradient. (b) A hypothetical concentration plot for two components separated by electrophoresis. The concentration of each component is proportional to the height of each plateau of the plot, as in Figure 15-24(c). (c) A concentration gradient (or schlieren) plot equivalent to the plot in panel (b). The area of each of the two peaks is proportional to the concentration of each component.

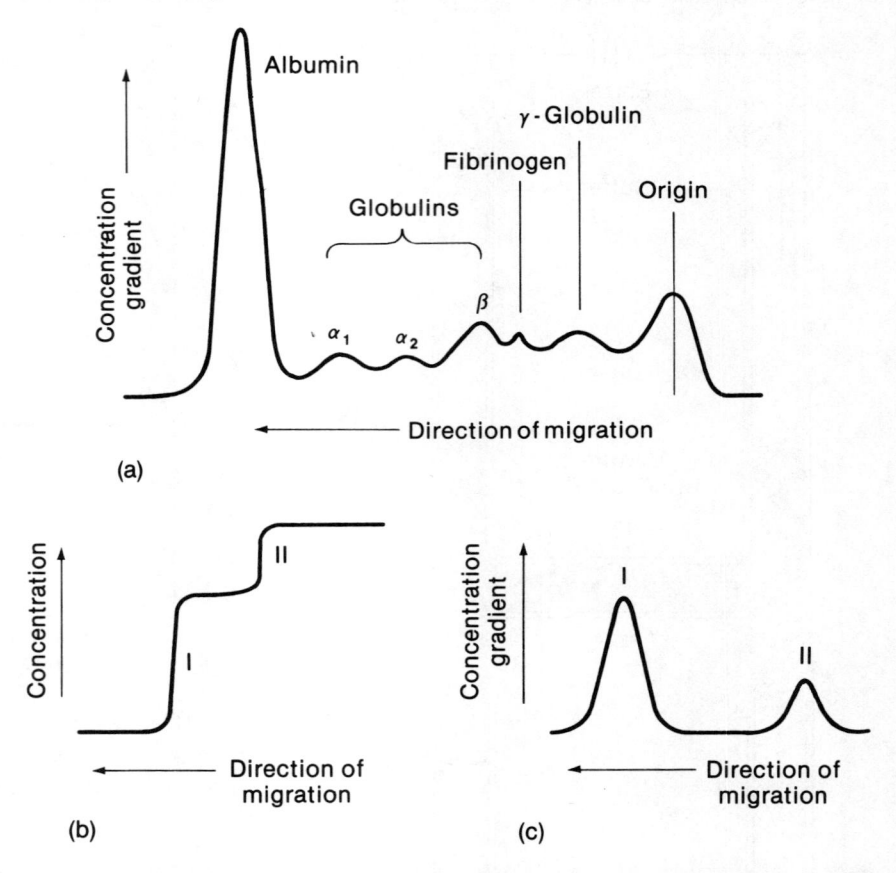

FIGURE 15-28 Experimental arrangement for low-voltage paper electrophoresis. A sample is spotted onto the paper near the middle of the paper. The paper is then saturated with buffer and both ends are placed in a buffer reservoir to make electrical contact with the power supply. The system is enclosed to prevent drying of the paper. At the end of a predetermined time, the paper is removed and dried. [SOURCE: *Physical Biochemistry*, First Edition, by David Freifelder. W.H. Freeman and Company. Copyright © 1976.]

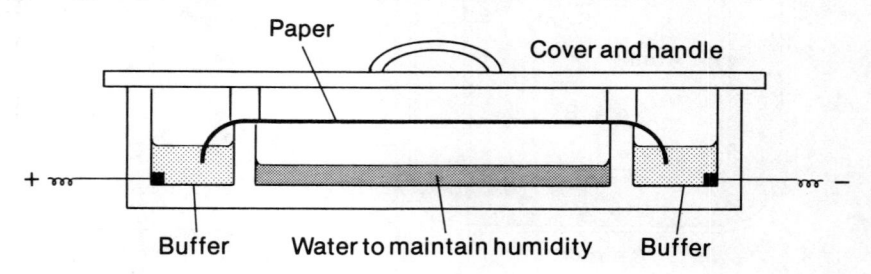

SEC. 15-3   MOVEMENT OF MACROMOLECULES

FIGURE 15-29 Separation of large molecules (open circles) from small molecules (solid circles) in particulate and contiguous gel systems. In the particulate gel we have assumed that the spacing between the particles is usually great enough for all molecules to pass through. However, the smaller molecules often enter the pores of the particles and are temporarily trapped. Therefore, on the average they move more slowly through the gel than do the larger molecules. In the contiguous gel the smaller molecules have available many narrow passages that the larger molecules cannot enter. Thus, on the average, the smaller molecules move more rapidly.

Particulate

Contiguous

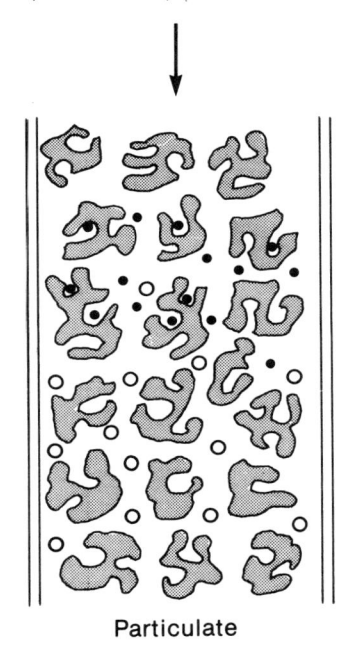

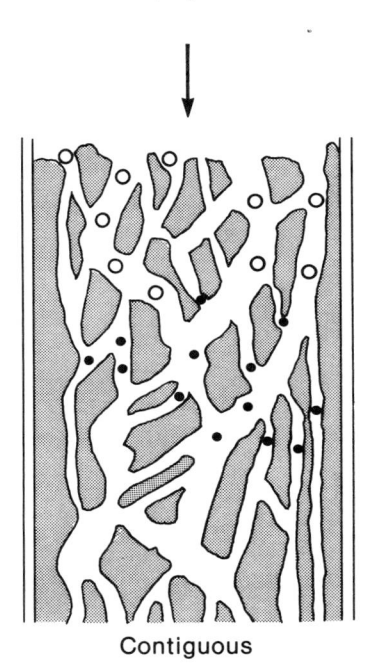

FIGURE 15-30 (a) Arrangement for starch-gel electrophoresis. The gel is formed in place by allowing the starch grains to swell in a buffer. The cut is made in the gel and the sample is placed in the cut. The system is covered with wax to prevent drying. (b) A starch-gel electrophoregram of separated proteins. When electrophoresis is finished, the gel is stained to make the proteins visible. Note that the proteins have migrated in both positive and negative directions, depending on their charge. [SOURCE: *Physical Biochemistry*, First Edition, by David Friefelder. W.H. Freeman and Company. Copyright © 1976.]

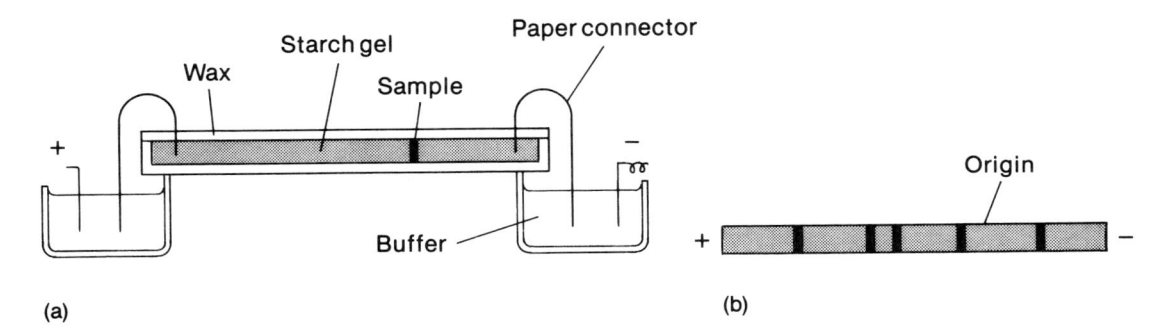

(a)

(b)

FIGURE 15-31 Separation of seven different DNA molecules according to molecular weight by agarose gel electrophoresis. The molecules have moved vertically in the figure. After electrophoresis was completed, the gel was soaked in a solution of the fluorescent dye ethidium bromide, which binds to DNA. The gel was then photographed by ultraviolet light. The white bands indicate the fluorescence and thus the position of the DNA molecules. The molecular weights of the molecules decrease from top to bottom. [SOURCE: Courtesy of Joel Loewenburg.]

FIGURE 15-32 An arrangement for electrophoresis in a slab gel. In (a) a mixture of four proteins is placed in a slot in the gel. Then a voltage is applied. After some time the four proteins have migrated to different positions in the gel, as shown in (b).

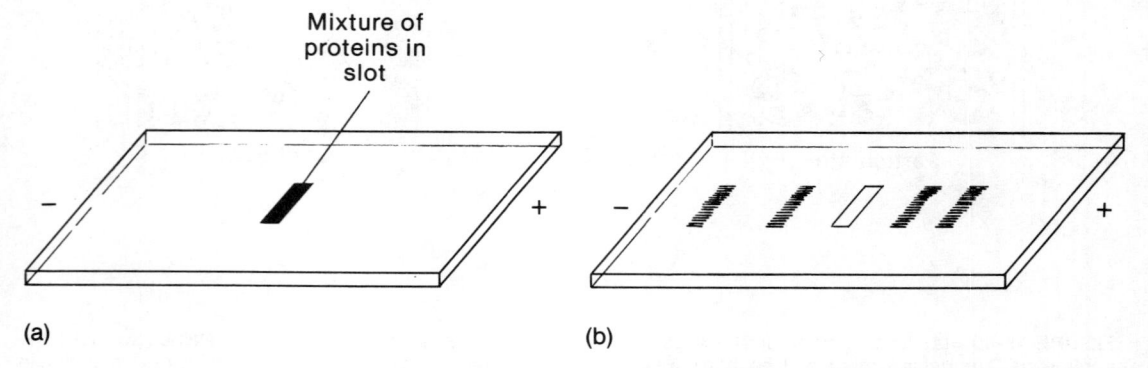

gel and the molecules move through the pores of the gel; the positively charged molecules move toward the cathode, the negative ones move toward the anode, and uncharged molecules remain in the slot. The figure shows a typical distribution of concentration after a particular time. The molecules that have moved the greatest distance do not have the highest charge, but the maximal mobility.

15–4 DETERMINATION OF MOLECULAR WEIGHTS

The measurement of molecular weight M is of great importance in the life sciences. For example, it is necessary to know M if one is to work out the structure of a macromolecule. In addition, M is often a useful parameter in a variety of biochemical studies—a molecule might, for example, be identified in a mixture purely by its molecular weight.

There are two classes of methods for measuring M—absolute methods and relative methods. An absolute method is one in which, by measuring well-defined parameters, such as the concentration or the concentration distribution, the angular dependence of the intensity of scattered light, the osmotic pressure, and so forth, one can calculate M without making any assumptions about molecular structure or specific properties of the molecules. A relative method is one in which M is calculated from the measured ratio of its molecular weight to that of a molecule whose molecular weight is known. Some of these relative methods are empirical in the sense that it is not known why they work. Table 15-4 lists methods of both types.

TABLE 15-4 Methods for measuring molecular weight.

Absolute Methods
Colligative properties
Light-scattering
Sedimentation diffusion
Sedimentation equilibrium
Sedimentation equilibrium in a density gradient
End-group labeling

Relative Methods
Sedimentation velocity
Viscosity
Electron microscopy
Gel electrophoresis
Gel permeation chromatography

Molecular Weight Averages

Ideally, M is determined by using a sample of material that is pure—that is, a sample in which no substance other than the solvent is present. However, since biological molecules are isolated from cells containing tens of thousands of different molecules, this can be a problem, although modern separation methods can give a very high degree of purity. Of equal importance is the fact that samples of large macromolecules are often contaminated with broken molecules. This is especially significant with large DNA molecules ($M > 10^8$), which are very fragile—for example, some of the molecules in a sample can be broken simply by passage of a solution through a pipette or by shaking the solution too violently. For this reason, it is necessary to consider the effect that molecular weight heterogeneity has on the measured value of M. Clearly the value will be an average of some sort and it is important that we recognize the kind of average that is obtained.

Some of the absolute methods count the number of molecules in a solution having a particular concentration (expressed as weight per unit volume). In this case, the average value of M is determined by the *number of molecules* whose molecular weights differ from that of the molecule of interest, but *not by their actual mass*. This average value is called a *number average*. However, other methods measure the weight of all particles having a particular mass, so that the average value is affected by the mass of the contaminating molecule rather than by their number. The resulting average is called a *weight average*. The number-average molecular weight M_n is defined as follows:

$$M_n = \frac{\sum n_i M_i}{\sum n_i} = \sum f_i M_i, \qquad (27)$$

in which n_i is the number of molecules and f_i is the fraction of the total number of molecules having molecular weight M_i.

The weight-average M_w is defined as

$$M_w = \frac{\sum w_i M_i}{\sum w_i}, \qquad (28)$$

in which w_i is the weight of all molecules having molecular weight M_i. Since $w_i = n_i M_i$, the equations can also be written

$$M_n = \frac{\sum w_i}{\sum \frac{w_i}{M_i}} \quad \text{and} \quad M_w = \frac{\sum n_i M_i^2}{\sum n_i M_i} = \frac{\sum f_i M_i^2}{\sum f_i M_i}.$$

The value of M_n is affected most by low-molecular-weight impurities, while M_w is affected most by large impurities, as we show in the following examples.

EXAMPLE 15–1 Calculation of M_w and M_n for a protein whose molecular weight is 100,000, and which is contaminated by a 1 percent-by-weight impurity—a molecule whose molecular weight is, in one case, 100, and in a second case, 10^6.

For the case of the contaminant for which $M = 100$,

$$M_n = \frac{\sum w_i}{\sum w_i/M_i} = \frac{0.99 + 0.01}{\left(\frac{0.99}{100,000}\right) + \left(\frac{0.01}{100}\right)} = 9.1 \times 10^3,$$

and

$$M_w = \frac{\sum w_i M_i}{\sum w_i} = \frac{(0.99)(100,000) + (0.01)(100)}{0.99 + 0.01} = 9.9 \times 10^4.$$

SEC. 15-4   DETERMINATION OF MOLECULAR WEIGHTS

Note that M_n is $[(9.1 \times 10^3)/100,000] \times 100 = 9.1$ percent of the true molecular weight, whereas M_w is hardly affected (99 percent of the actual value). For the case of the contaminant for which $M = 1,000,000$,

$$M_n = \frac{0.99 + 0.01}{\left(\frac{0.99}{100,000}\right) + \left(\frac{0.01}{1,000,000}\right)} = 10^5,$$

and

$$M_w = \frac{(0.99)(100,000) + (0.01)(1,000,000)}{0.99 + 0.01} = 1.09 \times 10^5.$$

It should be noted that M_n is more sensitive to contaminants of low molecular weight than is M_w, whereas M_w is more sensitive to high-molecular-weight contaminants.

In the following sections specific methods for determining M are described.

Colligative Properties

The measurement of M from the colligative properties—vapor pressure and freezing point depression, boiling point elevation, and osmotic pressure—are described in Chapter 7. The first three methods are useful only for small molecules ($M < 1000$), but osmometry is accurate up to $M \approx 30,000$.

Sedimentation Diffusion

This method requires a measurement of the sedimentation coefficient s and the diffusion coefficient D.* The relevant equation is obtained by modifying Equation 24, so that

$$M = \frac{sRT}{(1 - \bar{v}\rho)D}, \tag{29}$$

in which $\bar{v}$ is the partial specific volume and ρ is the solvent density. The principal limitation of the sedimentation-diffusion method is the accuracy in measuring D; for instance, for molecules having a large axial ratio (for example, DNA) D is too small to be measured accurately. This method is

---

*Usually the measurements are carried out independently, although there have been developed methods by which D can be measured from the shape of the curve relating concentration and distance during centrifugation. If $D = 0$, all molecules will move at exactly the same rate. However, if diffusion occurs, some molecules will appear to move more slowly and some faster than the bulk of the material. Thus there will appear to be a distribution of s values and the width of the distribution will increase with time. From this apparent distribution, D can be calculated. This method is described in the reference by Tanford listed at the end of the chapter.

useful for all proteins and for high-molecular-weight spherical particles, such as phages and viruses. In a heterogeneous system, the weight-average molecular weight M_w is obtained.

Sedimentation-Equilibrium in the Absence of a Density Gradient

The necessary formula for evaluating M can be simply derived by using the free energy concept. At equilibrium, the concentration of the molecules is not the same at all parts of a solution, yet the free energy must be the same. Thus we begin by calculating the free energies G_1 and G_2 at distances r_1 and r_2 from the axis of rotation of the centrifuge and set $G_1 = G_2$ or $G = G_1 - G_2 = 0$. The decrease in free energy owing to the centrifugal force for a molecule at r_1 and r_2 equals the work required to move a molecule from r_1 to r_2, or

$$G_{\text{centrif}} = -\int_{r_1}^{r_2} m'r\omega^2 \, dr$$
$$= -\frac{m'\omega^2}{2}(r_2^2 - r_1^2), \tag{30}$$

in which ω is the angular velocity in *radians per second* and m' is a mass that has been corrected for the buoyancy of the solute, or

$$m' = m(1 - \bar{v}\rho), \tag{31}$$

in which $\bar{v}$ is the partial specific volume of the molecule and ρ is the solvent density. Therefore

$$G_{\text{centrif}} = \tfrac{1}{2}m(1 - \bar{v}\rho)\omega^2(r_2^2 - r_1^2). \tag{32}$$

This free energy change is not zero but represents the tendency to move in the centrifugal direction. This tendency is balanced by diffusion, which is a result of the concentration difference produced by the centrifugal force. Thus, as the molecule moves from r_1, at which the concentration is c_1, to r_2, at which the concentration is c_2, there is an increase in free energy change, G_{diff}, equal to

$$G_{\text{diff}} = kT \ln(c_2/c_1). \tag{33}$$

But $G_{\text{centrif}} = G_{\text{diff}}$ at equilibrium, so that from Equations 32 and 33,

$$m = \frac{2kT \ln(c_2/c_1)}{(1 - \bar{v}\rho)\omega^2(r_2^2 - r_1^2)}. \tag{34}$$

Note that m is in grams, so that to get the molecular weight M, we must multiply by Avogadro's number, $\mathcal{N}$. Since $\mathcal{N}k = R$,

$$M = \frac{2RT \ln(c_2/c_1)}{(1 - \bar{v}\rho)\omega^2(r_2^2 - r_1^2)}. \tag{35}$$

SEC. 15-4 DETERMINATION OF MOLECULAR WEIGHTS

Thus, to determine M, one needs only to measure the concentrations at any two positions in the centrifuge tube. Commercially available centrifuges have optical systems that make measurement of c_1 and c_2 quite straightforward and accurate. The limiting element in measuring M in this way is the precision in evaluating $\bar{v}$, the measurement of which sometimes requires more material than may be available. How $\bar{v}$ is measured is described in several of the references at the end of this chapter. Two points should be made about this procedure: (1) the criterion for reaching equilibrium (which may take several days) is that c_1 and c_2 are the same at two different times separated by several hours; (2) if a sample of a macromolecule contains molecules having several molecular weights, the value obtained will be M_w, because the procedure measures the weight of each molecule rather than the total number of molecules. There is, however, a special way to plot the data that yields the value of M_n.

EXAMPLE 15–J Calculation of *M* for a macromolecule by sedimentation-equilibrium.

A pure protein sample is centrifuged to equilibrium at 20°C and 15,000 rpm in water. The concentration distribution is shown in Figure 15-33. The values of c_1 and c_2 at respective distances of 0.405 cm and 1.253 cm from the top of the tube are 2.98 and 15.84 g cm^{-3}. The top of the centrifuge tube

FIGURE 15-33 Concentration distribution of protein in a centrifuge cell at two times (t_1 and t_2) after starting to centrifuge and at equilibrium (when the curve no longer changes with time).

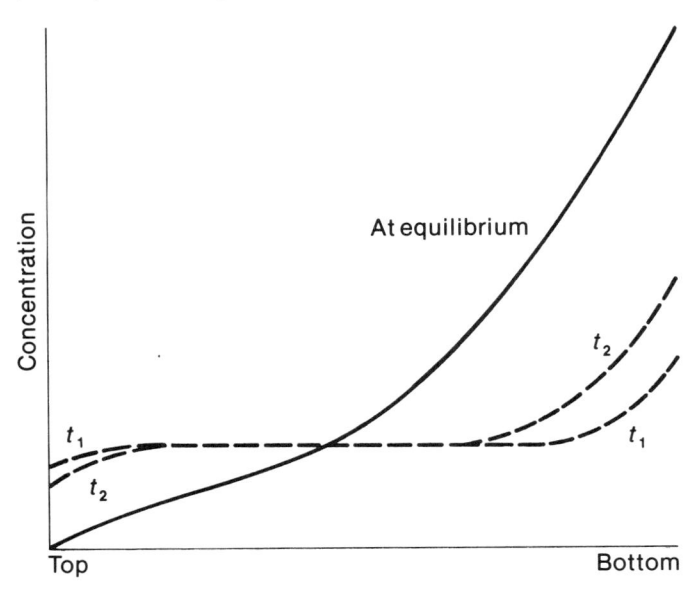

is 5.700 cm from the axis of rotation of the centrifuge so that r_1 and r_2 are 6.105 and 6.953 cm, respectively. The value of $\bar{v}$ is 0.752 cm^3 g^{-1} and the density of water at 20°C is 0.998 g cm^{-3}. Therefore $1 - \bar{v}\rho = 0.250$. The value of $\omega = (15{,}000 \text{ rpm}/60 \text{ sec min}^{-1}) \times 2\pi = 1.57 \times 10^3$ radians per second, and $\omega^2 = 2.47 \times 10^6$; $T = 293$ K and $R = 8.314 \times 10^7$,* so that

$$M = \frac{2(8.314 \times 10^7)(293)\ln(15.84/2.98)}{(0.250)(2.47 \times 10^6)(6.953^2 - 6.105^2)}$$

$$= 11{,}903 \ .$$

Sedimentation Equilibrium in a Density Gradient

Consider a centrifuge tube containing a concentrated salt solution on which is layered a dilute solution of a macromolecule (Figure 15-34). If the

FIGURE 15-34 Sedimentation of particles in solutions having various density distributions.

Particles are denser than the liquid

Particles are less dense than the liquid

Particles have the density of the solution, at point x

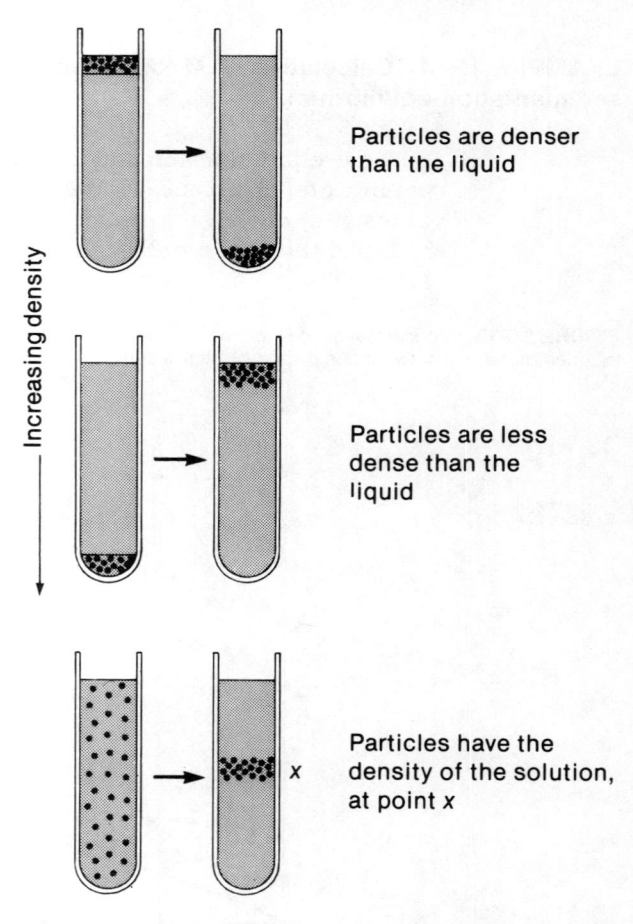

*The convention in sedimentation analysis is to measure all distances in centimeters. This necessitates using R in ergs—namely, 8.314×10^7.

SEC. 15-4 DETERMINATION OF MOLECULAR WEIGHTS

density of the macromolecule is greater than the density of the solution, the macromolecule will move to the bottom of the tube under the influence of a centrifugal force. If the solution density is greater than that of the macromolecule, the buoyancy effect will prevent sedimentation. Now consider the reverse situation, in which a macromolecule having density ρ_1 is in a salt solution of density ρ_2, and $\rho_2 > \rho_1$. When this solution is centrifuged, the macromolecules will *rise* to the top of the tube. As a final example, we consider a centrifuge tube containing a salt solution of continuously varying salt concentration (with, of course, the lowest density at the top of the tube); a macromolecule is present at a uniform and low concentration throughout the tube. Let the density of the macromolecule equal the density of the solution at some point x within the tube. When this solution is centrifuged, macromolecules above x will sediment downward toward x, while those below x will move upward in the tube toward x. Clearly, at equilibrium, all of the macromolecules will have formed a narrow band at x—that is, *they move to the point in the solution at which the density of the macromolecule equals the density of solution*. As in the case of sedimentation equilibrium, diffusion will prevent the zone containing the macromolecules from being infinitely narrow. Equation 35 indicated that from the concentration distribution in the absence of a density gradient, one can calculate M. This is true also when a density gradient is present, as we will now demonstrate.

Let us now consider a centrifuge tube filled with a concentrated solution of CsCl. This salt is chosen because solutions can be prepared having a density up to 1.85 g cm^{-3}. The solution, which has a density of about 1.7 g cm^{-3} (this is approximately a 7-M solution, so that it is far from being an ideal solution), also contains DNA molecules (at a concentration of about 10^{-5} M) whose density is also roughly equal to 1.7 g cm^{-3}. If this solution is centrifuged at high speed, the Cs$^+$ ions, which are much denser than water, will come to equilibrium with a concentration distribution described by Equation 35. To maintain electrical neutrality, the Cl$^-$ ions will have the same distribution at equilibrium. This produces a density gradient, because the density of a CsCl solution increases continuously with concentration. During the time it takes to reach equilibrium (about six hours at a typical centrifugal force of 100,000 × gravity), the DNA molecules barely move, because the density of a DNA molecule is very near that of the solution. However, ultimately (in 24–48 hours), the DNA molecules will have migrated to a position at which their density equals the density of the solvent. The distribution at equilibrium can be calculated as follows.

Equation 35 is first rearranged to exponential form and the derivative of the concentration (which depends on r) is taken with respect to r; this yields

$$\frac{1}{c}\frac{dc}{dr} = \frac{\omega^2 r M(1 - \bar{v}\rho)}{RT} \tag{36}$$

in which c is the DNA concentration at r. We now define r_0 as the position

FIGURE 15-35 (a) The position of molecules at equilibrium when centrifuged in a density gradient. (b) The concentration distribution of the macromolecule at equilibrium, showing the standard deviation σ.

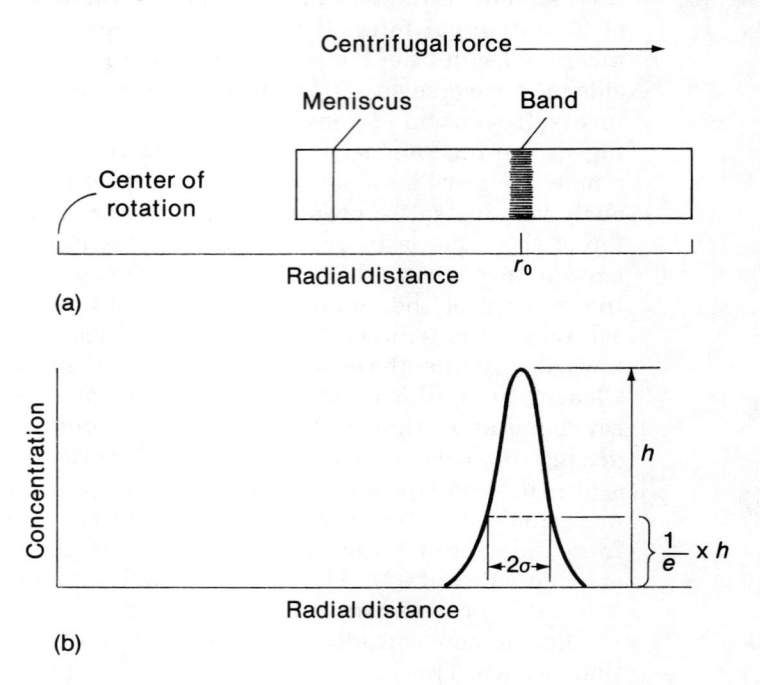

of the molecules at equilibrium (Figure 15-35(a)). We have ignored the fact that ρ is not constant but varies with r. An exact solution of the equation in which this is considered precisely is complex; however, since we are interested only in the distribution of the DNA molecules, which occupy a very small region near r_0, we make the assumption (which is experimentally verified) that in this small region, the density gradient is linear. That is,

$$\rho(r) = \frac{1}{\bar{v}} + (r - r_0)\frac{\Delta\rho}{\Delta r}, \qquad (37)$$

in which we recognize that the value of $\Delta\rho/\Delta r$ depends upon the particular salt used to generate the density gradient, the salt concentration, T, and ω. Thus, $\Delta\rho/\Delta r$ is not treated as a function of r but as an experimental constant (usually obtained from published tables) since in a particular sedimentation equilibrium experiment, the identity of the salt, its concentration, T, and ω are predetermined.

Combining Equations 36 and 37 yields

$$\frac{1}{c}\frac{dc}{dr} = -\frac{\omega^2 rM}{RT}\bar{v}(r - r_0)\frac{\Delta\rho}{\Delta r}. \qquad (38)$$

SEC. 15-4 DETERMINATION OF MOLECULAR WEIGHTS

We now rearrange this equation, substitute $(r - r_0)dr = \frac{1}{2}d(r - r_0)^2$ and make the approximation $r \approx r_0$ to obtain

$$d(\ln c) = -\frac{\omega^2 r_0 M \bar{v}}{2RT}\frac{\Delta\rho}{\Delta r}d(r - r_0)^2 \tag{39}$$

which, when integrated, becomes

$$c_r = c_{r_0}\exp\left[-\frac{\omega^2 r_0 M \bar{v}}{4RT}\frac{\Delta\rho}{\Delta r}(r - r_0)^2\right], \tag{40}$$

or

$$M = 4RT \ln(c_r/c_{r_0})\frac{1}{\omega^2 r_0 \bar{v}(\Delta\rho/\Delta r)(r - r_0)^2}. \tag{41}$$

In modern ultracentrifuges the concentrations c_r and c_{r_0} are measured by optical absorption, so that c_r/c_{r_0} is just the ratio of the optical densities at r and r_0. Equation 40 is the equation for a Gaussian band having a standard deviation σ, or the half-width at $1/e$ times the maximal height, as shown in Figure 15-35(b), in which

$$c_r = c_{r_0}\exp[-(r - r_0)^2/2\sigma^2]. \tag{42}$$

Thus, comparing Equations 40 and 42,

$$M = \frac{2RT}{\bar{v}(\Delta\rho/\Delta r)\omega^2 r_0 \sigma^2}. \tag{43}$$

The measurement of σ is the most common way of calculating M from the concentration distribution, although it is essential that the DNA sample be homogenous with respect to density. Any density heterogeneity will result in a distribution of values of r_0 that will widen the band. Since M is inversely proportional to the band width, density heterogeneity will give rise to an apparently low value of M.

It is important to note that if the macromolecule binds either the Cs^+ or the Cl^- ion, the molecular weight will be that of the molecule plus the bound ions. This is especially significant for DNA molecules, whose phosphate groups bind Cs^+ ions strongly. Thus, the molecular weight will be that of Cs-DNA, which will be higher than the values for Na-DNA that are commonly reported. However, to convert to Na-DNA, one need only multiply the measured value of M by the ratio of a Na nucleotide to a Cs nucleotide, or $(23 + 330)/(137 + 330) = 0.755$.

The density gradient method is in theory applicable to a wide range of substances, but it has not been used much for protein molecules, for two reasons. First, for low-molecular-weight molecules, such as proteins, the width of the band in the density gradient is so great that it may fill the entire centrifuge cell; second, most proteins are not very soluble in the high salt concentrations needed for this technique. Consequently the principal applications have been with DNA molecules, although with large DNA molecules there is an interesting problem. As we have pointed out, the band width decreases as M increases. For a DNA molecule whose M is 10^8 (a typical value for bacteriophage DNA molecules) the band width may be 0.05 cm. The length of such a DNA molecule is 0.005 cm but, owing to its flexibility, it is usually folded so that it fits in a sphere roughly 0.0001 cm in diameter. This has the effect that with large molecules, the excluded volume is an appreciable fraction of the volume of the band and the molecules are unable to get close enough together to produce a band as narrow as the one that calculations would seem to predict. Thus, the observed band width is greater than expected and results in a measured value of M that is too low. The discrepancy between the measured and true value of M should decrease with decreasing DNA concentration because the available space for the molecules then becomes greater. This just means that to obtain the correct value of M, one must measure M at various DNA concentrations and then extrapolate the measured values to zero concentration. We have encountered the necessity for such an extrapolation in other systems considered earlier.

The greatest utility of the density gradient technique is not in molecular weight determination but in separating molecules by density. For example, typically the density difference between the top and the bottom of a centrifuge tube is 0.100 g cm^{-3}. Thus, DNA that has a density of 1.7 g cm^{-3} can be easily separated from proteins whose density is 1.3 g cm^{-3} because if the initial density of CsCl is chosen to be 1.700 g cm^{-3} at equilibrium, the DNA will have formed a band and the protein will be floating on the surface. Densities are additive, so that a typical bacteriophage, which is 50 percent protein and 50 percent DNA, has a density of $(1.3 + 1.7)/2 = 1.5$ g cm^{-3}. Therefore, bacteriophage can be freed of contaminating bacterial DNA and protein by choosing an initial density of 1.50 g cm^{-3}; at equilibrium the phages will form a band, the protein will float on the surface of the liquid, and the DNA will be at the bottom of the tube. Separation by density is a very sensitive, highly effective method because of the narrow band widths and small values of the ratio $\Delta\rho/\Delta r$ that are produced. For example, the density of DNA isolated from an organism grown in the presence of the isotope ^{15}N is 1.714 compared to 1.700 g cm^{-3} for ^{14}N-DNA. At equilibrium they can be 0.3 cm apart, as shown in Figure 15-36. This method has been widely used in biochemistry and molecular biology and is described in greater detail in references given at the end of the chapter.

Light-Scattering

A measurement of the angular dependence of the light scattered by a solution of macromolecules yields the molecular weight and some information about the shape of the molecule. The theory is relatively complex and differs according to whether the dimensions of the molecule are nearly the same as or smaller than the wavelength of the incident light. For very large molecules the theory is inadequate. We begin with the theory for small molecules.

FIGURE 15-36 Separation in a CsCl density gradient of two DNA samples, one containing ^{14}N and the other containing ^{15}N.

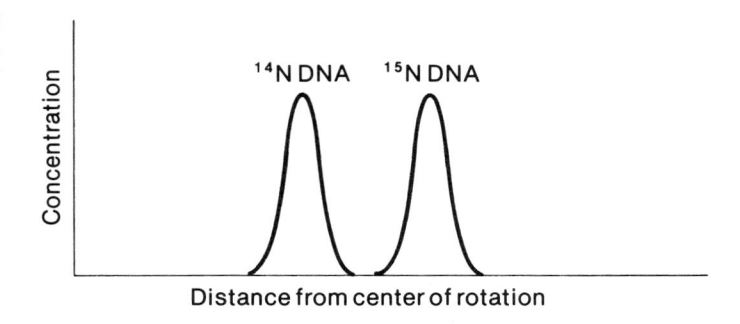

When a light wave falls on a molecule, the oscillating electric field of the wave sets up an oscillating dipole in the molecule which is itself a source of radiation. The ability of the dipole to oscillate is determined by how tightly the molecule holds on to its electrons or, alternatively, by the force required to move an electron. This is measured by the polarizability α of the molecule. Since the dipole has an orientation, the direction of emission of the light wave or the photon will be related to the component of the dipole vector seen by an observer at a particular observation angle with respect to the direction of the incident light. With a collection of molecules, the intensity of scattered light increases as the number of scattering centers (that is, the number of the molecules) increases. It is this point that enables one to determine the molecular weight. The equation describing the angular dependence of the scattered intensity is

$$\frac{I_\theta}{I_0} = \frac{8\pi^2 a \alpha^2}{\lambda^4 r^2}(1 + \cos^2\theta), \qquad (44)$$

in which I_0 is the intensity of the incident light and I_θ is the intensity of the light seen at angle θ with respect to the beam direction—both seen at a distance r (in centimeters) from the molecules—α is the polarizability, a is the number of particles per cubic centimeter and λ is the wavelength expressed in centimeters (Figure 15-37). This equation deserves several comments. First, the equation applies to unpolarized incident light—with polarized light the factor $(1 + \cos^2\theta)$ is replaced by $2\sin^2(90 - \theta)$, in which θ is expressed in degrees. Second, the equation describes scattering by a sample consisting of identical particles or, at least, particles having the same value of α. However, to determine the molecular weight of a macromolecule requires that the molecule be in solution. The solvent molecules and the macromolecules will rarely have the same polarizability and, in fact, it can be shown that if they do, the intensity of the scattered light from a solution will be identical to that of the solvent. In order to detect scattering by the macromolecule, the solvent molecules and the macromolecules must have different polarizabilities. The effect of this is that for a solution of macromolecules, the factor α^2 in Equation 44 is replaced by $(\alpha_1 - \alpha_0)^2$, in which α_1 and α_0 are the polarizabilities of the solute and the

FIGURE 15-37 Diagram of a light-scattering photometer showing the scattering angle θ. The photocell is mounted on a rotating arm centered on the sample so that the intensity can be measured at various values of θ.

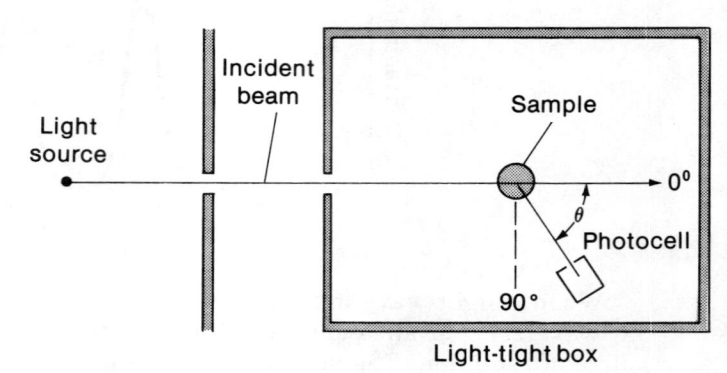

solvent molecules, respectively. The polarizability is not an easily measurable quantity, but fortunately it is related in a straightforward way to the index of refraction; thus we need only know the indices of refraction of the solvent and of the macromolecule. The value for the solvent is easy to obtain with a refractometer. However, the index of refraction of a macromolecule is not easily measurable unless a dried sample can somehow be made transparent; it can be expressed, though, as the product of the concentration c (in g cm^{-3}) of the macromolecule and the change in the refractive index of the solution with concentration, dn/dc, which can be measured in a straightforward way. When these factors are all taken into account, there results the equation

$$\frac{I_\theta}{I_0} = \frac{2\pi^2 n_0^2 (dn/dc)^2}{\mathcal{N}\lambda^4 r^2}(1 + \cos^2\theta)Mc \tag{45}$$

in which M is the molecular weight of the macromolecule, $\mathcal{N}$ is Avogadro's number, and n_0 is the index of refraction of the pure solvent. This equation indicates that the *intensity of the scattered light is proportional to the molecular weight,* and this is the basis for the determination of M by the light-scattering method. In an effort to simplify this equation, the definitions

$$R_\theta = \frac{r^2}{(1 + \cos^2\theta)}\frac{I_\theta}{I_0} \quad \text{and} \quad K = \frac{2\pi^2 n_0^2 (dn/dc)^2}{\mathcal{N}\lambda^4} \tag{46}$$

are usually made, so that Equation 46 simplifies to

$$R_\theta = KMc. \tag{47}$$

Experimentally the principal difficulty is that small particles of dust can scatter light significantly, increasing R_θ and thus yielding a value of M that is higher than the correct value. The removal of dust is probably the most important step in preparing a sample for a light-scattering measurement.

SEC. 15-4 DETERMINATION OF MOLECULAR WEIGHTS

In the derivation (which we have not given) of Equation 45 an assumption is made that the scattering by a particles is a times the scattering by a single particle. This is not the case, because, as the concentration increases, there are various contributions that break down the proportionality. Thus, we again encounter the problem with which we should by now be familiar—that *the observed value of* M *is concentration dependent*. It can be shown that this dependence is expressed by the equation

$$\frac{Kc}{R_\theta} = \frac{1}{M} + 2Bc + \cdots, \tag{48}$$

in which B is a constant. Thus, to determine the true value of M, one plots Kc/R_θ versus c and extrapolates to $c = 0$.

We mentioned earlier that the equations we have developed included an assumption about the relative size of the molecule and the wavelength λ of the incident light. The limit for the validity of these equations is that the size of the molecule must be about $\lambda/50$. For a typical wavelength of visible light (450 nm), this limit is about 9 nm (90 Å), which allows one to use the technique for all but the largest globular proteins and for small fibrous proteins. For most nucleic acids, however, and for macromolecular aggregates such as viruses, the equations are not valid and the theory must be extended. The problem with a large molecule is that different regions of the same molecule serve as scattering centers and the scattered waves can destructively or constructively interfere. The theory is made even more complicated by the rapid rotation and Brownian motion of the molecules. However, as long as the molecule is not so large that there is multiple intramolecular scattering—that is, a photon must not be scattered by one part of the molecule and then scattered again by another part of the *same* molecule—relatively simple equations can be developed. We will not go through the theory but simply state the result, which is

$$\frac{Kc}{R_\theta} = \left[1 + \frac{16\pi^2 R_G^2}{3\lambda^2}\sin^2\frac{\theta}{2}\right]\left[\frac{1}{M} + 2Bc\right], \tag{49}$$

in which R_G is the radius of gyration (which we discuss shortly). Now, to obtain M, one must perform an extrapolation to zero concentration and to zero angle. This extrapolation is carried out by an elegant procedure called a Zimm plot, which is described in references at the end of the chapter.

The value of the radius of gyration is a somewhat useful number that is derived from the light-scattering measurement and is related to the shape of the molecule. Unfortunately, there is no way to calculate the dimension of a molecule from the value of R_G, although R_G can be calculated if one knows the dimensions of simple molecular shapes, as we have seen earlier: once M is evaluated, one can calculate the dimensions that a molecule would have *if* it were a sphere, an ellipsoid, a long rod, and so forth, and then calculate R_G from the expressions given in Table 15-1. These values can be compared to the value of R_G obtained from light-scattering data analyzed according to Equation 49, and one will in this way have some idea of the shape of the molecule.

EXAMPLE 15–K Determination of the molecular weight of a protein Q by light scattering.

A series of solutions having different concentrations of a small protein Q in water are prepared. The index of refraction at each concentration is measured and a plot of n versus c is prepared. A straight line results whose slope, which is dn/dc, is 0.172. The index of refraction n_0 of water is 1.333. Light scattering is performed at a wavelength of 436 nm so that the constant K in Equation 47 is 4.76×10^{-7}. Samples at protein concentrations of 25, 50, and 75 mg per cm^3 are separately placed in the sample vessel and illuminated with light of intensity I_0. A photocell is placed 7 cm away from the sample; thus $r = 7$ cm. The light intensity is measured at an angle of 90° with respect to the incident light beam. The observed intensities for the three solutions are: 25 mg/cm^3, $I_{90} = 5 \times 10^{-6} I_0$; 50 mg/cm^3, $I_{90} = 10^{-5} I_0$; 75 mg/cm^3, $I_{90} = 1.49 \times 10^{-5} I_0$. The values of $R_{90} = (7^2/1 + \cos^2 90)/(I_{90}/I_0) = 49 I_{90}/I_0$ are 5.34×10^{-5}, 1.08×10^{-4}, and 1.61×10^{-4} for concentrations of 25, 50, and 75 mg/cm^3, respectively. From Equation 47, $R_{90}/Kc = 4600$. The same value is obtained for all three concentrations, which indicates that the protein molecule is sufficiently small that Equations 48 and 49 are unnecessary. Thus the molecular weight of the protein is 4600.

EXAMPLE 15–L Estimate of the shape of the protein myosin.

The values of M and R_G are 493,000 and 468 Å, respectively. The density of most proteins is about 1.3 g cm^{-3}, so that the volume of the protein is $4.93 \times 10^5/(6 \times 10^{23})(1.3) = 6.3 \times 10^{-19}$ cm^3. If it were a sphere, its radius would be 53 Å and R_G would be $(3/5)^{1/2} 53 = 41$ Å. We can guess, then, that the molecule is far from being spherical. If it were a long rod 10 Å in diameter, the length would be 2000 Å and R_G would be about $(1/12^{1/2}) 2000 = 577$ Å. This is larger than the observed value but sufficiently near that it suggests that the molecule is rodlike, although not a rigid rod.

It should be noted that R_G appeared in Equation 49 only when we included the contribution due to molecular size. This equation should of course apply to small molecules also, because there will always be *some* interference effects. However, with small molecules, R_G is usually not measurable because of experimental difficulties. The main problem is that as R_G becomes smaller, the scattered intensity becomes smaller also, and the amount of scattered light is difficult to measure accurately. To compensate, one can measure the intensity at very small values of θ, because I_θ increases as θ approaches zero. However, it is not possible to measure I_θ when θ is nearly zero, because the scattered light is superimposed on the transmitted light beam (refer again to Figure 15-37). There is another solution, though. The reduction in the intensity of the scattered light can be compensated for by decreasing the wavelength of the incident light, because the scattered intensity is proportional to $1/\lambda^4$. Thus, the smaller the wavelength, the greater is the reliability of the measurement of R_G. For this reason, scattering of X rays, for which λ is 1 to 5 Å, is more commonly used to determine the dimensions of small molecules.

We have just mentioned that if the molecule is too large, multiple intramolecular scattering becomes important and the theory must be revised. This becomes a problem with DNA molecules, which may be so large that multiple scattering occurs in every molecule. In practice, molecular weight cannot be measured by light-scattering in a DNA molecule that is larger than $M = 10^7$. Since most naturally occurring DNA molecules are much larger than this, the light-scattering method is not useful in the study of DNA. It remains a valuable method for the study of protein molecules, although usually equilibrium centrifugation is more convenient.

End-Group Labeling

If the concentration of a sample of macromolecules is known in g per cm^3 and the number of molecules per cm^3 can be evaluated, then M is known, because

$$M = \frac{\text{g per cm}^3}{\text{number of molecules per cm}^3} \times \text{Avogadro's number} . \qquad (50)$$

An important means of counting the number of molecules in a particular volume is by counting the number of ends of the molecule; this works as long as one knows whether one or both ends of each molecule are being counted. This procedure is carried out easily for both protein and DNA molecules because there are simple enzymatic procedures for linking a radioactive chemical to the terminus of the molecule. For proteins, a single terminus is labeled; for DNA molecules the 5' end of each polynucleotide chain is measured so that there are two labeled groups per double-stranded helix. In the following, we perform the calculation for a DNA molecule.

EXAMPLE 15-M **Molecular weight of a DNA molecule by end-group labeling.**

The enzyme *polynucleotide kinase* can transfer ^{32}P from [γ-^{32}P]-labeled adenosine triphosphate (ATP) to the 5' end of a DNA strand. A solution of DNA at 5 µg/ml is labeled with ^{32}P from [γ-^{32}P]-ATP, which is radioactive to the extent of three millicuries per micromole. After the reaction, the radioactivity in the DNA is found to be 10^{-3} microcuries. This is equivalent to 3.3×10^{-7} micromole of ATP or 1.98×10^{11} molecules of ATP. Since a DNA molecule contains two 5' ends, this radioactivity represents $1/2 \times 1.98 \times 10^{11} = 9.9 \times 10^{10}$ DNA molecules. The amount of DNA is 5 µg, so that $M = [(5 \times 10^{-6})/(9.9 \times 10^{10})] \times (6 \times 10^{23}) = 3 \times 10^{7}$.

In a mixture, the value obtained is the number-average molecular weight. Since DNA molecules are easily broken and may be contaminated with small oligonucleotides containing from 10 to 100 bases (compared to 5×10^4 bases for a molecule for which $M = 3 \times 10^7$), measured values of M_n can easily be very low compared to the correct value of M. This error is circumvented by carrying out the labeling reaction and then purifying the DNA by centrifugation—that is, selecting a fraction having the sedimentation coefficient of the unbroken molecule. In this way the labeled fragments are discarded before the radioactivity is counted.

Estimation of *M* from the Sedimentation Coefficient—a Relative Method

The sedimentation coefficient s of a molecule is a function of both the size and shape of the molecule. In no case so far has it been possible to calculate from first principles the relation between s and M for a real molecule—with sufficient precision, that is, to make this a useful absolute method. However, for a molecule whose overall shape is independent of M—for instance, DNA—s values can be determined for one kind of DNA on the basis of other DNA molecules for which M is known from other kinds of measurements, and an empirical curve relating s and M can then be obtained. Such curves have been obtained for both double- and single-stranded DNA as well as for circular double-stranded molecules, for specific conditions of pH and salt concentration. If the s value of a DNA molecule for which M is unknown is measured in these conditions, these curves can then be useful in estimating M for the particular kind of DNA that is being studied. It is difficult to obtain such an empirical curve for proteins, because of the great variation in the shape of proteins. Hence s measurements are rarely used to estimate M for proteins; however, sometimes this is done, and a rather shaky assumption can be made that the protein has the shape of a typical globular molecule.

Determination of *M* from Relative Electrophoretic Mobility

The motion of various molecules in an electric field could be related to M if all of the molecules were to have the same shape and if the charge per unit

SEC. 15-4 DETERMINATION OF MOLECULAR WEIGHTS

molecular weight were constant. This is the case for linear DNA molecules, which have one negatively charged phosphate per nucleotide. This state can be established for proteins. For example, if the ionic detergent sodium dodecyl sulfate (SDS) is added to a solution of protein molecules, the subunits of the protein dissociate but each polypeptide chain binds SDS at a constant weight ratio of 1.4 g SDS to 1 g of protein. The number of strongly charged sulfates bound to each protein molecule is so much greater than the number of charges on the protein molecule itself that the charge of the SDS-protein complex is proportional to the molecular weight. This has the effect that if either proteins or DNA molecules are electrophoresed through a gel (which allows smaller molecules to move through the pores of the gel more rapidly than a large molecule), the following relation is found to occur:

$$\log M = -ax + b, \tag{51}$$

in which x is the distance migrated and a and b are constants determined by properties of the gel. Interestingly, we do not yet fully understand the logarithmic dependence of M. A typical plot of the data for several proteins is shown in Figure 15-38.

FIGURE 15-38 Typical semilogarithmic plot of M versus distance migrated for determining molecular weight by SDS-polyacrylamide-gel electrophoresis. Proteins were reduced to eliminate disulfide bonds. The proteins are: (1) bovine serum albumin; (2) catalase; (3) ovalbumin; (4) carboxypeptidase A; (5) chymotrypsinogen; (6) lysozyme. [SOURCE: *Physical Biochemistry*, First Edition, by David Freifelder. W.H. Freeman and Company. Copyright © 1976.]

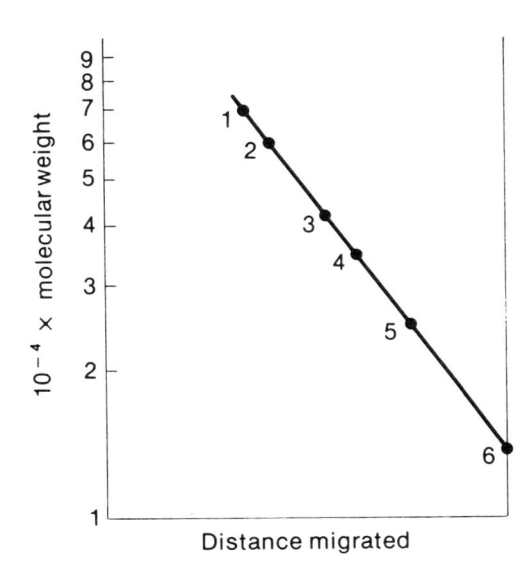

Thus, to determine M for a particular protein, this protein is mixed with several proteins for which M is already known, the mixture is electrophoresed, and the value of M is calculated from the curve determined by the known proteins. This is the same technique in which M is calculated for DNA molecules by using other DNA molecules for which M is known.

FIGURE 15-39 Electron micrograph containing a linear DNA molecule obtained from *E. coli* phage λ and several small circular molecules of *E. coli* phage φX174 DNA. The molecular weight of φX174 DNA is accurately known, so that the molecular weight of the λ DNA can be measured from the relative lengths of λ DNA and φX174 DNA. The two small dark structures are λ phage particles; a single λ DNA molecule is compactly contained within each phage particle. [SOURCE: Courtesy of Manuel Valenzuela.]

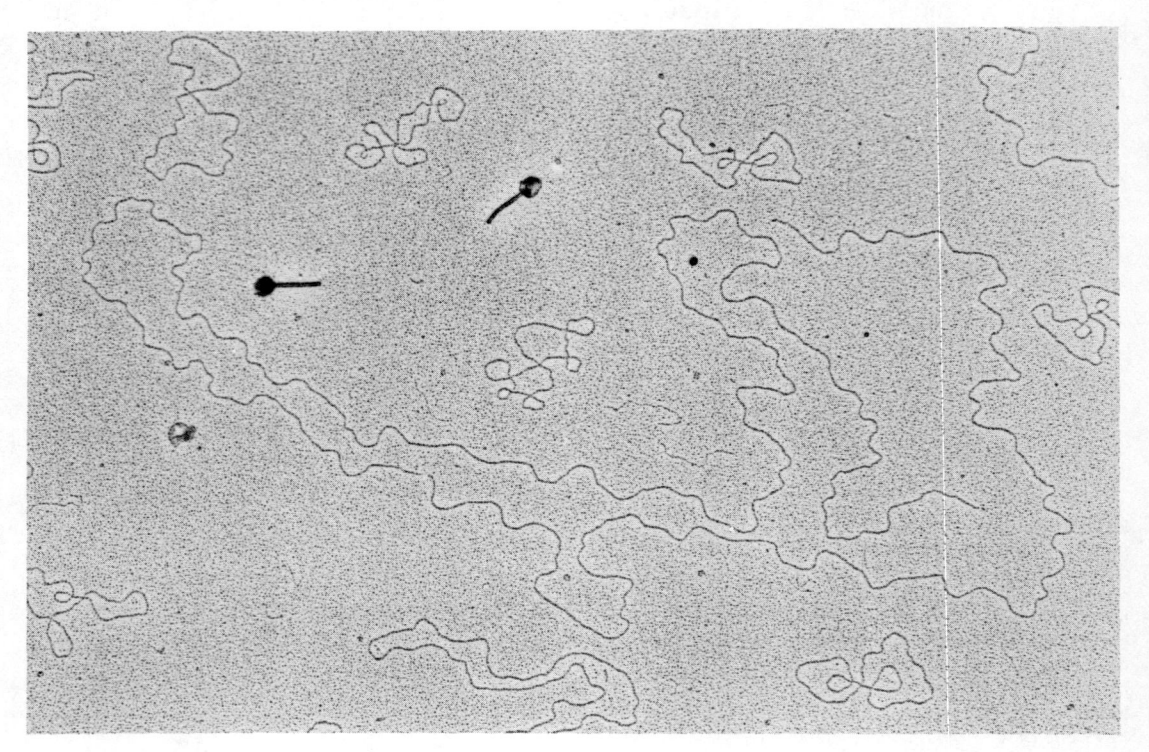

Electron Microscopy

Electron microscopy can, in principle, enable one to determine M as long as the magnification is known. However, for protein molecules this is not useful because details of the shape of these rather small molecules cannot be seen. Furthermore, for small spherical or rodlike particles, a small error in the dimensions can produce a large error in M. For instance, a 10 percent error in the diameter of a sphere (which is not an exceptionally large error in the dimensions of very small molecules) introduces an error of 34 percent in M. However, with large molecules, such as DNA, M is proportional to length and, because the molecule may be 10–100 μm long, an error of 50 Å is only a very small fraction of the length. Thus, by electron microscopy, M for a DNA molecule can be determined merely by adding another molecule of known molecular weight and measuring the relative length. A photograph and the kind of data obtained are shown in Figures 15-39 and 15-40.

FIGURE 15-40 Length distribution of single-stranded λ DNA molecules. A sample of DNA molecules, one of whose strands contained a broken phosphodiester bond, was denatured and the resulting DNA molecules were examined by electron microscopy. The largest peak is the unbroken strand. The fragments are of all sizes, which indicates that the broken bond was not in a unique position in the double-stranded λ DNA molecule. [From the author's laboratory.]

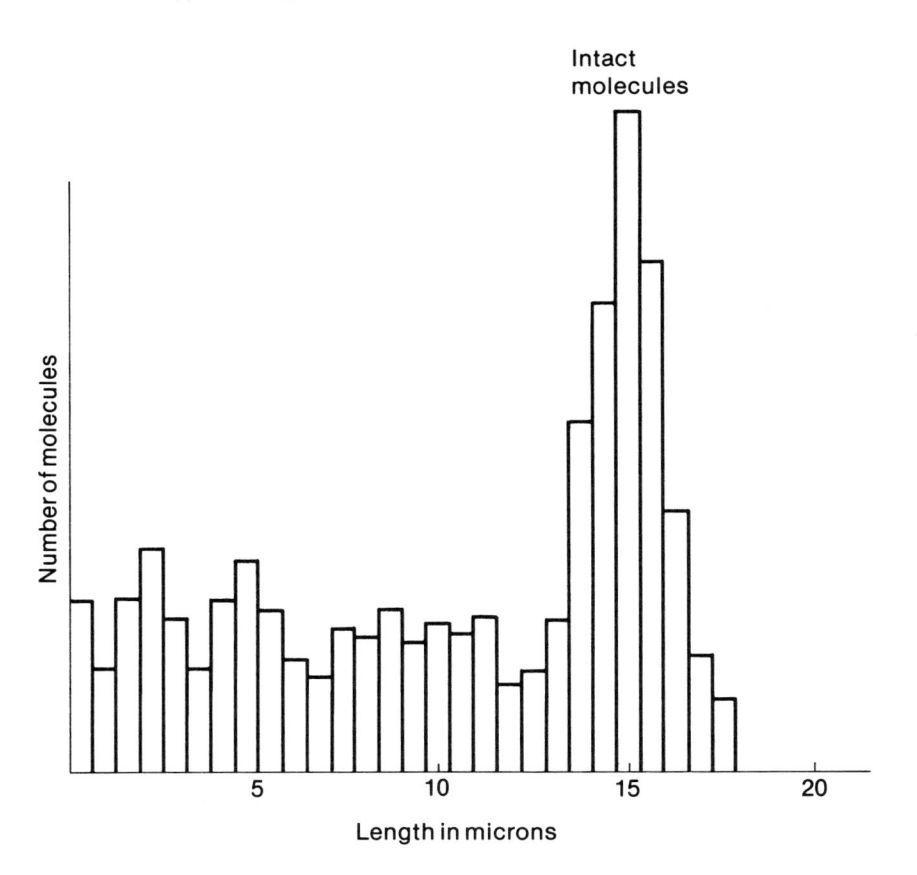

REFERENCES

Bovey, F. 1969. *Polymer Conformation and Configuration.* Academic Press. New York.

Cohen, E.J., and J.T. Edsall. 1965. *Proteins, Amino Acids, and Peptides.* Hafner. New York. An encyclopedia of information.

Dickerson, R.E., and I. Geis. 1969. *The Structure and Action of Proteins.* Harper & Row. New York.

Edsall, J.T., and J. Wyman. 1958. *Biophysical Chemistry.* Academic Press. New York.

Freifelder, D. 1976. *Physical Biochemistry.* W.H. Freeman and Co. San Francisco.

Freifelder, D. 1978. *The DNA Molecule: Structure and Function*. W.H. Freeman and Co. San Francisco.
Kaufman, M. 1968. *Giant Molecules*. Doubleday. Garden City, N.Y.
Morawetz, H. 1975. *Macromolecules in Solution*. Wiley. New York.
Steiner, R.F., and R.F. Beers. 1961. *Polynucleotides*. American Elsevier. New York.
Tanford, C. 1961. *Physical Chemistry of Macromolecules*. Wiley. New York.
Williams, V.R., W.L. Mattice, and H.B. Williams. 1978. *Basic Physical Chemistry for the Life Sciences*. W.H. Freeman and Co. San Francisco.

PROBLEMS

1. What is the degree of polymerization of polyglycine if the molecular weight is 10,600?

2. A particular polypeptide chain folds in such a way that there is a cluster of 4 leucines, 3 isoleucines, 6 valines, and 3 phenylalanines at one small region on the surface of the protein. What would you guess to be the state of this polypeptide chain in aqueous solution?

3. We have emphasized that the structure of biologically active proteins is determined by the sum of many weak interactions. Covalent disulfide bonds between cysteines are also common in most proteins. The biological activity of proteins—for example, an enzyme—is usually lost if the disulfide bonds are broken or if reagents are added that eliminate hydrophobic interactions. Thus, even in a disulfide bond-containing protein hydrophobic interactions are important. What is the likely role of the disulfide bond?

4. For every 100 molecules in a protein sample, there are 53 molecules having a molecular weight of 42,000, 38 whose molecular weight is 38,000, and 9 whose molecular weight is 1500. Calculate M_n and M_w.

5. You are determining the molecular weight of a protein sample. By osmometry you obtain a value of 18,520 and, by sedimentation equilibrium, a value of 18,580 is measured. What can you say about the degree of heterogeneity of the sample?

6. A sample of DNA molecules whose molecular weight is 22×10^6 contains some molecules whose molecular weight is 10^6; these smaller molecules amount to 2 percent by weight of the sample. What are the values of M_n and M_w for the sample?

7. A protein preparation gives a molecular weight of 16,500 by osmometry and 19,200 by sedimentation diffusion. Can you tell which number is probably nearer to the true value?

8. The molecular weight of a sample of a particular protein is measured by osmometry and found to be 25,000. However, by light scattering, the molecular weight is measured to be 27,000. If the sample is boiled and then cooled, the value of the molecular weight by both methods is 15,000. Furthermore, in a sedimentation velocity experiment, two components in equal amounts are found prior to boiling, but only one component is found after the boiling-cooling sequence. Interpret these results in terms of the structure of the molecule.

9. A protein molecule has a molecular weight of 24,340 and $\bar{v} = 0.72$ cm^3 g^{-1}. Sedimentation studies suggest that the end-to-end distance is 36 Å. What can you say about the shape of the molecule?

10. A protein molecule has a molecular weight of 37,650 and $\bar{v} = 0.70$ cm^3 g^{-1}. The observed radius of gyration is 64.5 Å. What can you say about the shape of the molecule?

11. What is the radius of gyration of four small spheres placed at the corners of a square 10 cm on a side? Two spheres weigh 5 g and are at opposite corners of the square. The other two weigh 1 g each.

12. Would you expect the end-to-end distance of a circular DNA molecule to be zero?

13. How would the end-to-end distance of each of the following vary with increasing temperature?

    (a) A random coil.

    (b) A rodlike molecule.

14. A sample of DNA is heated to a temperature sufficiently high that the two strands separate. It is then cooled, but the strands remain separate. Compare the end-to-end distance and the radius of gyration of the DNA before and after the heating-cooling cycle by stating which value is larger. Do this for two conditions—namely, 0.01 M NaCl and 1 M NaCl. In 0.01 M NaCl, the negatively charged phosphates in the sugar-phosphate chain repel one another. In 1 M NaCl each phosphate is neutralized by a Na^+ ion.

15. Consider a molecule whose structure is maintained by two noncooperative interactions. What determines the temperature at which the molecule undergoes a transition to the random coil configuration?

16. Two hypothetical protein molecules consist of 96 glycines, 2 lysines, and 2 aspartates. In molecule A the lysines are adjacent and the aspartates are adjacent; each lysine is ionically bonded to one aspartate. In molecule B the lysines are in positions 16 and 80 in the polypeptide chain and the aspartates are in positions 35 and 65. Ionic bonds exist between lysine-16 and aspartate-35 and between lysine-80 and aspartate-65. As the temperature is increased, both A and B undergo a helix–coil transition. For which molecule is the transition cooperative? Explain.

17. Are the two strands of DNA totally separated at the T_m?

18. How would you think the optical activity of a DNA solution would change when the DNA is denatured?

19. Formaldehyde reacts irreversibly with amino groups but not with carbonyl groups or hydroxyl groups. What effect would you expect formaldehyde to have on DNA?

20. What would be the density of a bacteriophage that is 2/3 by weight protein and 1/3 by weight DNA? The densities of the protein and the DNA molecules are 1.3 and 1.7 g cm^{-3}, respectively.

21. Answer the following.

    (a) What is the diffusion coefficient in water at 20°C of a virus particle with a molecular weight of 5×10^7, which is spherical? The particle is assumed to be 50 percent protein (density = 1.3 g cm^{-3}) and 50 percent DNA (density = 1.7 g cm^{-3}). The viscosity of water is 0.01 poise.

    (b) Many phages have long protein tails used for attachment to bacteria. If the phage described in (a) had a tail, would D be greater or smaller than the value calculated in (a)?

22. A macromolecule whose molecular weight is 22,600,000 and density is 1.79 g cm^{-3} has a value of D of 2.1×10^{-8} in water at 20°C. What may you conclude about the shape of the molecule?

23. A protein molecule has a molecular weight of 366,000 and a diffusion coefficient of 4.34×10^{-7} in water at 20°C. After heating to 75°C and restoring the temperature to 20°C, the molecular weight is found to be 61,000 and the diffusion coefficient is 6.2×10^{-7}. What can be said about the structure of the molecule?

24. The sedimentation coefficient s of a particular DNA molecule is 22×10^{-13}. How far will a molecule move at 40,000 rpm in 20 min at a distance 6.0 cm from the axis of rotation?

25. A rod and a sphere having the same density and same mass are centrifuged through a liquid of much lower density.

Which has a higher sedimentation coefficient?

26. A DNA sample consists of equal concentrations of linear and circular molecules, each having the same molecular weight. Which form will have the higher sedimentation coefficient?

27. The following questions concern the relation between shape, molecular weight, and the sedimentation coefficient.

    (a) If a protein sample gives a very sharp sedimentation boundary that clearly indicates that only a material having a single sedimentation coefficient is present, can one conclude that all molecules have the same molecular weight?

    (b) If, instead, two components are observed, can one conclude that material having two different molecular weights are present?

28. A particle having a molecular weight of 50,000 and a density of 1.35 g cm^{-3} has a sedimentation coefficient of 2.1×10^{-13} in water (density = 0.998 g cm^{-3}). Is the particle spherical?

29. The sedimentation coefficient and the diffusion coefficient for a particular protein are found to be (respectively) 18.3×10^{-13} sec and 4.62×10^{-7} cm^2 sec^{-1} at 20°C. The partial specific volume of the protein is 0.73. What is the molecular weight of the protein? At 20°C the density of water is 0.998 g cm^{-3}.

30. A protein is sedimented at 25°C in pure water (density = 0.998 g cm^{-3}). Its sedimentation coefficient is 8.6×10^{-13} and the diffusion coefficient is 6×10^{-7}. What is the molecular weight? Assume that $\bar{v} = 0.74$ cm^3 g^{-1}.

31. A protein solution is sedimented to equilibrium at 20°C at a speed of 20,000 rpm. The protein concentrations at 6.425 and 6.703 cm from the axis of rotation are 1.012 and 6.905 mg/ml, respectively. The partial specific volume of the protein is 0.71. The density of the solvent is 1.01 g cm^{-3}. What is the molecular weight of the protein?

32. A solution contains DNA and lipoproteins. The values of $\bar{v}$ for these molecules are 0.55 and 0.87, respectively. Sufficient sucrose and NaCl are added to the solution until the solvent density is 1.25 g cm^{-3} and the solution is centrifuged for several hours at 40,000 rpm. How are the DNA and lipoprotein molecules distributed in the centrifuge tube after this time?

33. Under particular conditions the band width of a DNA molecule whose molecular weight is 25×10^6 is 1.3 mm when centrifuged to equilibrium in CsCl. What would be the width of the band of a tRNA molecule whose molecular weight is 25,000 under the same conditions? The densities of DNA and RNA are 1.7 g cm^{-3} and 1.8 g cm^{-3}, respectively.

34. At a particular speed and temperature the density gradient in a CsCl solution at equilibrium is 0.1070 g cm^{-3} per cm in the region at which the density is 1.700 g cm^{-3}. Consider a DNA molecule whose density is 1.700 g cm^{-3} when all carbon atoms in the molecule are ^{12}C. What will be the separation in cm between this DNA and molecules containing ^{13}C?

35. The molecular weight of a small DNA molecule is determined, by centrifugation to equilibrium in CsCl, to be 3.2×10^6. What value of M will be determined by light scattering for a solution of DNA in 0.1 M KCl?

36. One way to evaluate M from data obtained from centrifugation to equilibrium in CsCl is to plot $\ln[c(r)/c(r_0)]$ against $(r - r_0)^2$, using Equation 41. This should give a straight line whose slope is $M\omega^2 r_0 \bar{v} (\Delta\rho/\Delta r)/2RT$. In such a plot it is sometimes observed that the graph is a concave upward curve, the departure from linearity being most noticeable as $(r - r_0)$ becomes large. Density heterogeneity does not explain this case if the data are obtained from both sides of the band and if the band is symmetric. Propose an explanation.

PROBLEMS

37. Will two molecules having the same molecular weight and charge have the same electrophoretic mobility? Explain.

38. Proteins A and B, having molecular weights of 16,500 and 35,400, respectively, move 1.3 and 4.6 cm, respectively, when electrophoresed through a gel. What is the molecular weight of protein C, which moves 2.8 cm in the same gel?

39. A protein sample whose concentration is 12 $\mu g/cm^3$ is labeled, at its amino terminus only, with a reagent that is radioactive to the extent of 1.5 millicuries per micromole. After the reaction the radioactivity of the protein sample is 0.1 microcuries. What is the molecular weight of the protein?

40. Some linear DNA molecules possess single-stranded termini that can join together. Would the viscosity of a solution of these molecules be higher or lower after joining than before joining?

41. Assuming that the denatured form of a macromolecule is always a random coil, is the viscosity of a denatured molecule always less than that of an undernatured molecule?

42. A sample of a circular DNA molecule is X-irradiated. This causes double-strand breaks—that is, the molecule becomes linear. Samples exposed to several doses of radiation are centrifuged and two species are observed. The ratio of the amount of slower-moving material to the amount of the faster-moving material is determined as a function of dose; the data are the following:

Dose, rad	Fraction fast	Fraction slow
0	1.00	0
1000	.61	.39
2000	.37	.63
3000	.23	.77
4000	.14	.86

On the average, how many rad are required to convert a circular molecule to a linear molecule. (Review Chapter 11, on first-order kinetics, if necessary.)

43. Many homopolymers of amino acids have been synthesized. One example is poly-L-tyrosine. At the isoelectric point (pH 10.6) measurements of s and D indicate that the molecule is a nearly spherical random coil. At low ionic strength (0.01 M), s decreases if the pH is adjusted to be more than two pH units from the isoelectric point. This does not occur if 0.5 M NaCl is added. Explain.

CHAPTER 16

LIGAND BINDING

A common event in living systems is the noncovalent binding of one or more molecules to a single macromolecule. A molecule bound to a macromolecule is called a *ligand*. An important example already discussed in Chapter 12 is the binding of a *substrate* molecule or a *cofactor* to an enzyme molecule. The ligand may itself be another macromolecule, as in the case of a protein binding to a proteolytic enzyme (a protease) or a regulatory molecule, as in the case of a repressor protein binding to DNA. When two macromolecules are bound together, one molecule is termed the ligand quite arbitrarily, because if molecule A binds to molecule B, then of course B also binds to A. Usually the particular molecule that is called the ligand is determined by convention. Thus, in the protein-protease interaction, the protein being attacked by the protease is the ligand; in the second case, the regulator molecule that binds to the DNA is the ligand.

The study of ligand binding is important for the following reasons: (1) to understand the mechanisms of action of enzymes and of regulatory systems often requires knowledge of the number of binding sites or the strength of the binding; and (2) there are probably as many intracellular reactions in which binding is an integral part as there are chemical reactions.

There are many types of binding. For example, a molecule may have only a single binding site, or there may be many sites. In the latter case the sites may be either identical or different, and they may be for only one kind of molecule or for several distinct molecules. Furthermore, multiple sites may be interdependent in the sense that occupation of one site by a ligand may affect the affinity other sites have for their ligands. In order to understand biochemical systems in which binding plays a role, it is necessary to determine the strength and number of binding sites and the magnitude of the interaction between multiple sites.

Experimentally, the measurement of binding is straightforward. However, interpretation of the data requires different modes of analysis that depend on various features of the binding process. These features can be incorporated into a mathematical theory of binding that is quite useful.

In this book mathematical analyses have been avoided in order to concentrate attention on an understanding of experimental techniques. However, it is not possible to describe binding equilibria in any useful fashion without resorting to equations. Fortunately, the mathematics does not go beyond simple algebra.

The analysis begins by first writing an equation for the dissociation constant of a binding reaction. The equation is then restated in terms of a particularly useful parameter—namely, the average number of ligand molecules bound to each macromolecule in a solution having a particular concentration of macromolecule and of added ligand. (For some types of binding the equation is very complex, so that simplifying assumptions must be made that apply to that mode of binding.) For convenience, the resulting equation is then rearranged to a form that enables the data to be presented usually in a straight-line plot; from this plot the number of binding sites and the dissociation constants can be obtained by measuring the slope and the intercepts. In the following sections we proceed from the simple case of a single binding site to more complex cases, such as of multiple, dependent binding sites. It is important to recognize, though, that a linear plot will occur only when the data are plotted with the equation that actually describes the mode of binding, and that the correct equation is not known in advance. Thus the usual procedure is to plot the data with various equations that are available; *the equation that is found to yield a linear plot is assumed to be a correct one.*

In the following sections the more common modes of binding and the relevant equations will be described.

16–1 MOLECULES WITH ONE BINDING SITE

Let us first consider the case in which a macromolecule P has a single binding site for a molecule A; that is, at saturation, one mole of P combines with one mole of A to form a complex, PA:

$$P + A \rightleftarrows PA . \tag{1}$$

As seen in Chapter 9, the equilibrium constant K for this association reaction is defined as

$$K = \frac{[PA]}{[P][A]} , \tag{2}$$

in which the brackets denote concentrations. In biochemistry and molecular biology it is common to use a *dissociation constant*, K_d, defined as $K_d = 1/K$. Thus

$$K_d = \frac{[P][A]}{[PA]}. \tag{3}$$

The equilibrium in Equation 1 is such that the fraction of ligands bound is affected by the value of [P] and [A]. Thus, we define r, the number of moles of ligand bound to one mole of macromolecules, as follows:

$$r = \frac{[A]_{bound}}{[P]_{total}} = \frac{[PA]}{[P] + [PA]}. \tag{4}$$

Most equations describing binding are written in terms of K_d and r. The value of r is usually determined in the following way. A particular concentration [A'] of ligand is added to a known concentration [P'] of the macromolecule. One then measures either the concentration of bound ligand [PA] or of unbound ligand [A]. Only one of these need be measured, because [A'] is known at the outset and [A] + [PA] = [A']. The value of [P] also need not be measured since [P] + [PA] = [P']. Methods for measuring [A] and [PA] will be described in a later section.

Combining Equations 3 and 4 yields

$$r = \frac{[A]}{K_d + [A]}, \tag{5}$$

which is known as the *Langmuir isotherm* (see also Chapter 13). This equation is the basis of evaluating K_d, but for precision in the evaluation of data (as will be seen below) it is usually rearranged either to yield

$$\frac{1}{r} = \frac{K_d}{[A]} + 1, \tag{6}$$

in which case a plot of $1/r$ versus $1/[A]$ (which is called a *double reciprocal plot*) gives a straight line whose slope is K_d, or

$$\frac{r}{[A]} = \frac{1}{K_d} - \frac{r}{K_d}, \tag{7}$$

in which case a plot of $r/[A]$ versus r gives a straight line having a slope of $-1/K_d$.* Equations 6 and 7 are both useful in handling data; however, in the example that follows we will see that, depending on the design of a particular experiment used to measure binding, one equation is usually preferable to the other.

---

*There is an advantage in using either Equation 6 or 7 rather than Equation 5 to determine K_d, although in both cases, the same parameters are measured. This will become clear when we consider a macromolecule with multiple binding sites; in this case the shape of the curves obtained by using Equation 6 or 7 provides, by simple inspection, information about the binding reaction and the effect of binding on the macromolecule itself.

SEC. 16-1 MOLECULES WITH ONE BINDING SITE

EXAMPLE 16-A Determination of K_d for binding of the Mg^{2+} ion to an enzyme having a single binding site.

In this experiment various amounts of Mg^{2+} are added to a 100-μM solution of an enzyme. The data obtained are plotted first according to Equation 6 and produce the graph shown in Figure 16-1(a). K_d is the slope of the straight line drawn through the points. Note that if the experimental points were more evenly spaced, the line could be drawn more accurately. If the same data are plotted with Equation 7, the graph shown in Figure 16-1(b) is obtained; here the points are more evenly spaced and the line can be drawn more precisely. Thus in this experiment Equation 7 is superior to Equation 6. This will not always be the case, because the spacing between the points is determined both by the values chosen by the experimenter for the amount of added Mg^{2+} and by K_d. Thus, one experimental design might give rise to clustered points when Equation 7 is used and to evenly spaced points when Equation 6 is used, and another experimental design might give the converse. However, the values of added ligand are not always freely chosen by the experimenter. For example, measurements at low concentration of added ligand (that is, at the right end of the curve in Figure 16-1(a)) are usually more difficult to make and have a larger experimental error than those at higher concentration; thus, the experimental points are frequently clustered as in Figure 16-1(b).

FIGURE 16-1 Two means of plotting the same binding data. In panel (a), Equation 6 is used; in panel (b), Equation 7 is used.

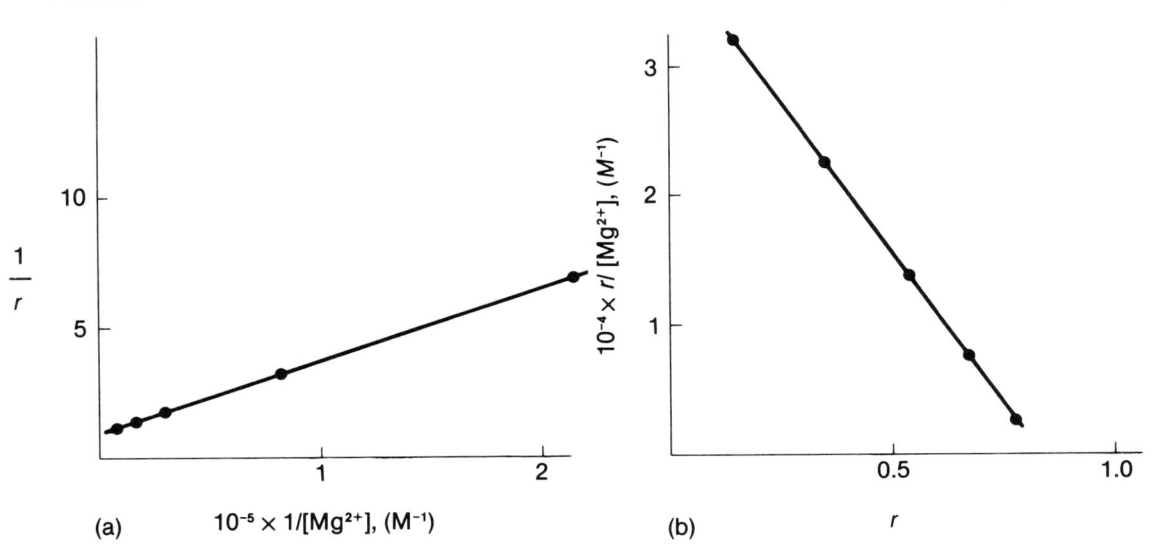

The preceding statements are not meant to imply that the means of displaying the data have determined the accuracy of evaluating K_d or any other binding parameter. Clearly the information content of a particular set of binding data is independent of the method used to interpret the data. The plots are visually different but a statistical treatment to obtain the best line through the points must yield the same value of K_d and the same error in its determination, independent of how the data are plotted. The point is that since a statistical analysis is not usually carried out, selection of appropriate graphical methods can help to show some aspects of the data more clearly.

16–2 MACROMOLECULES WITH SEVERAL BINDING SITES

Many biological macromolecules have more than one binding site. For example, a typical nucleic acid molecule can bind hundreds or even thousands of metal ions, and a typical protein molecule can sometimes bind several hundred H^+ ions. Furthermore the binding sites are not always identical; for example, an enzyme usually has distinct binding sites for a cofactor and a substrate molecule.

Multiple binding sites may be independent of one another, as in nucleic acid–metal ion binding, in which case the analysis of the binding is straightforward. However, it is a common occurrence that the binding sites are not independent (that is, occupation of one site influences binding at a second site) in which case the binding is said to be *interactive*. We will discuss independent binding sites first, distinguishing between the cases of tight binding and weak binding.

Tight Binding

Multiple binding equilibria are in general more difficult to analyze than the single binding site discussed in the preceding section. However, if the binding is very tight, a significant simplification occurs, because in this case, we assume that every ligand molecule is bound if the number of ligand molecules added to a solution of macromolecules is less than the total number of binding sites *in the solution*. Furthermore, once the amount of added ligand is greater than the number n of the binding sites in the solution, each of the n sites is occupied, so that additional ligand molecules remain unbound. From the concentration of added ligand at which some ligand remains unbound, the number of binding sites can be measured. This should be the same concentration at which the amount of bound ligand reaches a maximum. This simple method of determining n when binding is very tight is shown in the following example.

EXAMPLE 16-B Determination of the number of tight binding sites on a macromolecule.

Consider a protein with an unknown number of binding sites for some molecule X. To a solution of this protein at a concentration of 2 μM are added aliquots of X, and a plot is made of free versus added ligand. The data which are obtained are shown in Figure 16-2. We can see that binding continues linearly until the concentration of added ligand is 6 μM; at this point the concentration of free ligand increases markedly (but still linearly), so that saturation of the sites has occurred at 6 μM. Thus, the number of binding sites is $6/2 = 3$. Note that because of the very tight binding, it is not possible from this experiment to determine whether all binding sites have equivalent affinity for the ligand.

FIGURE 16-2 A binding curve illustrating tight binding. The initial protein concentration is 2 μM. There is no free X until the amount of added X exceeds 6 μM.

Therefore, 2 μM of protein bind 6 μM of X and there must be three binding sites for X per protein molecule.

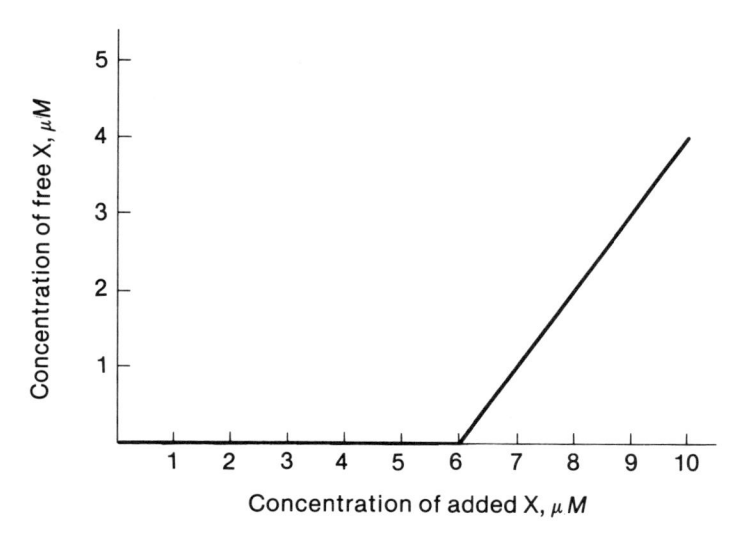

Weak Binding

Very tight binding of the type just described is not common; a more usual case is that of weak binding, the analysis of which is unfortunately considerably more complex than that of strong binding. Analysis of weak binding requires considering a set of equilibrium equations derived from the following set of reactions:

$$P + A \rightleftarrows PA_1, \quad PA + A \rightleftarrows PA_2, \ldots, \quad PA_{n-1} + A \rightleftarrows PA_n,$$

in which n is the maximal number of binding sites per macromolecule. The principal complication in this analysis is that usually the value of K_d is not the same for each of these reactions. Each value of K_d is defined by the equation

$$K_n = \frac{[PA_{n-1}][A]}{[PA_n]}, \tag{8}$$

which applies to the equilibrium

$$PA_{n-1} + A \rightleftarrows PA_n.$$

The general equation relating r, $[A]$, and the values $K_1, K_2, \ldots, K_n$ is derived in the usual way. The concentration of A that is bound is the sum of the amount of A in all species, or

$$[A]_{bound} = [PA] + 2[PA_2] + \cdots + n[PA_n].$$

The total concentration of P is the sum of the concentrations of all species of P, or

$$[P]_{total} = [P] + [PA] + [PA_2] + \cdots + [PA_n].$$

In order to evaluate $r = [A]_{bound}/[P]_{total}$ in terms of the dissociation constants, we note that

$$[PA] = \frac{[P][A]}{K_1}, \quad [PA_2] = \frac{[P][A]^2}{K_1 K_2},$$

and so forth, and obtain the general equation

$$r = \frac{\dfrac{[A]}{K_1} + \dfrac{2[A]^2}{K_1 K_2} + \cdots + \dfrac{n[A]^n}{K_1 K_2 \cdots K_n}}{1 + \dfrac{[A]}{K_1} + \dfrac{[A]^2}{K_1 K_2} + \cdots + \dfrac{[A]^n}{K_1 K_2 \cdots K_n}}. \tag{9}$$

This equation is complex and is not particularly useful in real experimental situations because of the virtually impossible task of fitting values of $K_1, K_2, \ldots$, to a plot of r versus $[A]$. We can, however, simplify the equation for several important cases.

Identical and Independent Weak Binding Sites

If all binding sites are identical and independent, Equation 9 can be shown to reduce to

$$r = \frac{n[A]}{K_d + [A]}, \tag{10}$$

in which K_d is an average dissociation constant.*

---

*If $K_1, K_2, \ldots, K_n$ are the dissociation constants for binding when $0, 1, \ldots, n-1$ sites are unfilled, K_d is the geometric mean of these constants or $(K_1 K_2 \ldots K_n)^{1/n}$.

SEC. 16-2 MACROMOLECULES WITH SEVERAL BINDING SITES

This equation is usually rearranged to the form of Equations 6 and 7, to yield

$$\frac{1}{r} = \frac{1}{n} + \frac{K_d}{n[A]} \qquad (11)$$

and

$$\frac{r}{[A]} = \frac{n}{K_d} - \frac{r}{K_d}, \qquad (12)$$

in which K_d is now an average dissociation constant. (Note that Equations 6 and 7 are special cases of Equations 11 and 12 with $n = 1$.) When Equation 11 is used, a plot of $1/r$ versus $1/[A]$ gives a straight line having a slope of K_d/n and a y intercept of $1/n$. With Equation 12 (often called the *Scatchard equation*) a plot of $r/[A]$ versus r gives a line having a slope of $-1/K_d$ and an x intercept of n/K_d.

A valuable aspect of this method of data analysis is that if the binding sites are not both identical and independent, neither of the plots, $1/r$ versus $1/[A]$ or $r/[A]$ versus r, will give straight lines. This is shown in Figure 16-3, which compares two macromolecules having, in one case, two identical sites and, in the other, two nonidentical sites. In a typical experimental situation, a smooth curve will result. If the sites are identical but dependent—that is, if binding is cooperative—a curved line will also result. These two cases will be treated in the following sections.

FIGURE 16-3 A Scatchard plot for a molecule with two independent binding sites, showing the difference between curves for identical and nonidentical sites.

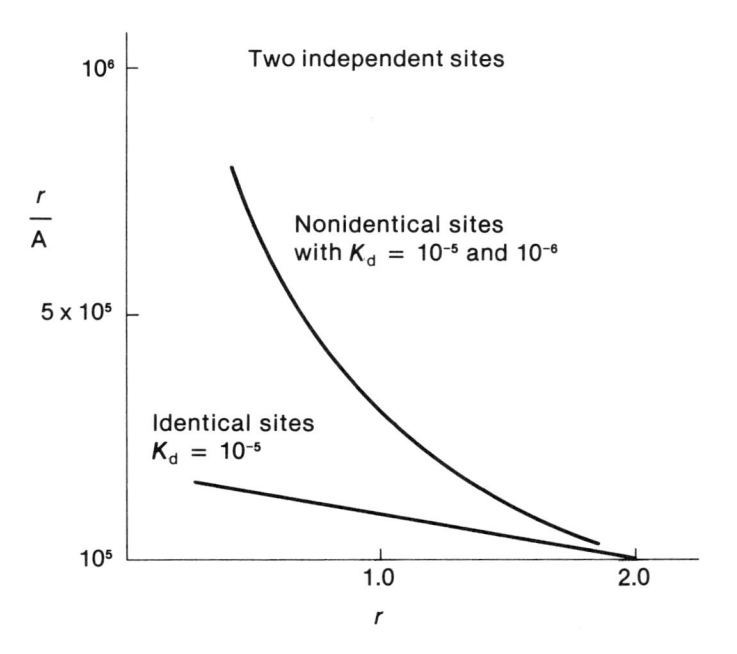

In analyzing binding data, it is a good practice to try first to apply Equations 6 and 7. If a linear graph results, there is no reason to test other equations; if a curved line results, other equations must be tried.

Macromolecules with Different but Independent Binding Sites for the Same Ligand

Up to this point we have discussed macromolecules with multiple and identical binding sites whose binding constants change as sites are filled. However, it is just as common for a macromolecule to have distinct types of sites, each having a different binding constant. An important biochemical example is the binding of H^+ ions by a protein, in which case the sites of binding are chemically different groups, such as amino, carboxyl, and hydroxyl groups. Other examples are the binding sites on DNA for RNA polymerase and the ribosome binding sites on messenger RNA molecules. If there is no cooperativity, a theoretical treatment is simple; unfortunately, extraction of information from experiments is difficult unless the number of classes of binding sites is small and the values of K_d differ greatly for each class.

Let us designate the different classes of binding sites as 1, 2, 3, . . . When ligand is added, binding to any site can occur but of course those sites with greater affinity for the ligand will be filled first. Thus for any particular concentration of ligand, each class will have a particular value of r that will be determined by the dissociation constant for that class. Thus, class 1 can be characterized by r_1, class 2 by r_2, and so forth. Experimentally one measures r, the *average* number of ligand molecules per macromolecule, which is

$$r = r_1 + r_2 + r_3 + \cdots.$$

If there are n_1 sites of type 1, n_2 of type 2, and so forth, Equation 10 may be written more generally as

$$r = \frac{n_1[A]}{K_{d_1} + [A]} + \frac{n_2[A]}{K_{d_2} + [A]} + \cdots. \tag{13}$$

A plot of r versus $[A]/K_d$ for two classes of sites having several ratios of K_d is shown in Figure 16-4. Note that only when the values of K_d differ greatly is it clear that there is more than a single class. In order to determine the number of classes unambiguously, it is necessary that the value of K_d be such that nearly all sites of one class are filled before there is appreciable filling of sites in a second class.

16-3 COOPERATIVE BINDING

When a curved double-reciprocal plot (Equation 12) is found, cooperativity is often the cause. The theory of cooperative binding is very complex and

FIGURE 16-4 Binding curves for a molecule having two classes of sites. Numbers adjacent to the curves denote ratios of the dissociation constants of the two classes. When the dissociation constants are in a 10 : 1 ratio, the curve remains smooth and there is no indication that there are two classes of sites, only when the ratio reaches 100 : 1 does the curve appear to have two components. Thus in this type of plot the values of K_d must differ substantially for two classes of sites to be evident.

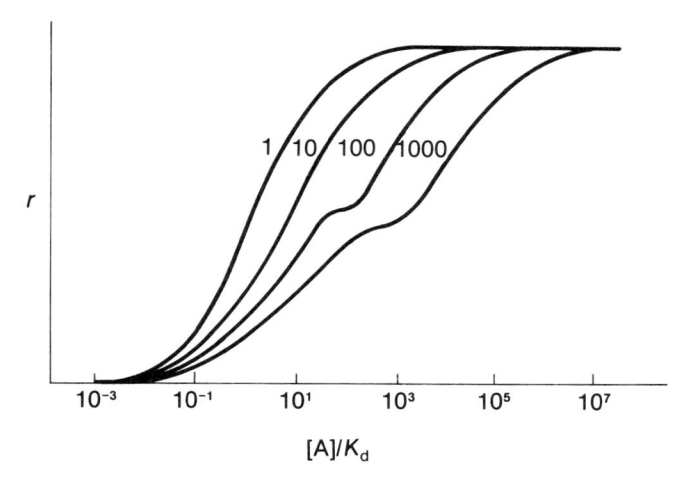

often not readily applicable to an experimental situation. The basis of the theory is the following. A molecule binds to one site of a macromolecule that has many weak binding sites. In response to this binding, one of two events occurs—either the affinity of all binding sites changes abruptly to a new value or one or more of the sites changes gradually. In the latter case, the affinity continues to change as more ligands are bound. At first we shall consider only *positive cooperativity*—that is, an increase of affinity as sites are filled. We will also proceed slowly through this relatively complex subject, beginning with highly cooperative binding of a single ligand, proceeding to weak cooperativity with a single ligand, and ultimately examine multisubunit macromolecules that bind several ligands.

Highly Cooperative Binding of a Single Ligand

In highly cooperative binding of a ligand at one site in a macromolecule, the affinity of the other binding sites is altered, so that if ligand molecules are available, each site is *immediately* filled. Thus, if there are n binding sites per macromolecule, then we need only consider the equation $P + nA \rightleftarrows PA_n$, for which

$$K_d = \frac{[P][A]^n}{[PA_n]}.$$

Therefore,

$$r = \frac{[A]_{bound}}{[P]_{total}} = \frac{n[PA_n]}{[P] + [PA_n]} = \frac{n\{[P][A]^n/K_d\}}{[P] + \{[P][A]^n/K_d\}},$$

or

$$r = \frac{n[A]^n}{K_d + [A]^n}. \tag{14}$$

Equation 14 is frequently rewritten as

$$\frac{r}{n-r} = \frac{[A]^n}{K_d} \qquad (15)$$

and is known as the *Hill equation*.

Let us now recall that r is the number of bound ligands per macromolecule. If there are n binding sites on each macromolecule, we may refer to Y, the fraction of binding sites that are filled, or

$$Y = \frac{r}{n}. \qquad (16)$$

In terms of Y,

$$\frac{r}{n-r} = \frac{Y}{1-Y} = \frac{\text{fraction of sites filled}}{\text{fraction of sites not filled}}.$$

Thus Equation 15 may also be written

$$\frac{Y}{1-Y} = \frac{[A]^n}{K_d}. \qquad (17)$$

In order to use Equation 17 to evaluate K_d or n, Y must be measured. Since Y is the fraction filled, it is necessary to determine the maximal amount of A that can be bound, the saturation value. Thus Y is sometimes called the *fractional saturation of sites*. Equation 17 should be tried whenever a plot of Equation 12 gives a curved line.

Highly cooperative binding is easily detected by a *Hill plot*, a graph of $\log(Y/1 - Y)$ versus $\log[A]$. According to Equation 17, a straight line with slope n and intercept $-\log K_d$ results.

Equation 17 is a statement of an idealized situation—namely, when cooperativity is so great that filling of each site is immediate. This is of course not a real situation and experimentally a value of n equal to the number of binding sites is never observed (when $n > 1$). For example, hemoglobin has four binding sites for O_2 yet n is found to be 2.8. Thus, the observed value of the slope of a Hill plot, which is called the Hill coefficient n_H, is a qualitative indicator of the degree of cooperativity. The value of n_H is always less than the actual number of binding sites. Note that a Hill plot is not as informative as a double-reciprocal plot in indicating whether cooperativity is present. A double-reciprocal plot indicates cooperativity merely by being curved, whereas to draw the same conclusion from a Hill plot requires that the total number of binding sites be known, which usually requires an independent determination. However, the Hill plot is useful because it clearly indicates when there is *no* cooperativity. For example, if the binding sites are independent, binding is described by Equation 12, which can be arranged to

$$\frac{n}{n-r} = \frac{1}{K_d}[A], \qquad (18)$$

so that a Hill plot would yield a slope of 1.

When cooperativity is not complete—that is, when $n_H < n$—the Hill plot is not linear because Equation 17 does not describe the real situation. Thus, if a Hill plot is made over a very wide range of values of [A], it is found that the curve is not a straight line for the entire range of values (Figure 16-5): at the extremes of [A] the line has a slope of approximately 1. This has the following explanation. When the ligand concentration is very low, there is no cooperativity. Thus, a Hill plot should represent single-site binding; that is, the binding of the first ligand molecule and the value of K_d for the first ligand should, in theory at least, be obtainable from extrapolation of this part of the curve. At high ligand concentration all sites are filled but one. Thus this region of the Hill plot should also indicate single-site binding with a value of K_d for the last ligand. As mentioned, an approximate slope of 1 is seen in the extremes of the Hill plot and the values of $-\log K_d$ should be obtainable by extrapolation to the y axis. However, binding data at very low and very high ligand concentrations are usually not very accurate, so that in practice these two values of K_d cannot be derived from a Hill plot.

FIGURE 16-5 A Hill plot for a hypothetical protein having three cooperative binding sites. The numbers indicate the slopes of various parts of the curve.

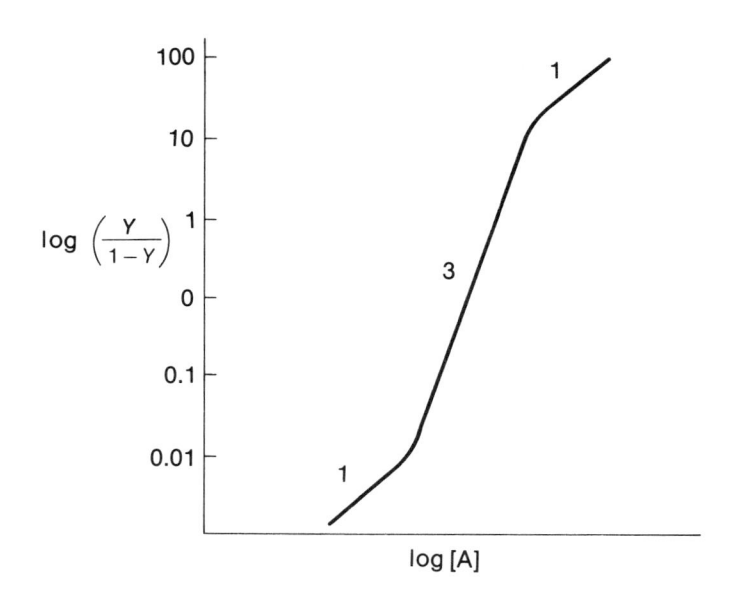

Macromolecules with Two Binding Sites, Each for a Different Ligand

In living systems there are many macromolecules that bind different ligands at different sites. The most commonly encountered example is that of an enzyme whose activity does not depend solely on the structure of the protein itself but on binding of a nonprotein ligand called a cofactor or coenzyme. These cofactors are usually metal ions, or small organic molecules often derived from vitamins. Usually when a cofactor is needed, the enzyme totally lacks activity when the cofactor is absent. Such an enzyme is called an *apoenzyme;* the term *holoenzyme* is used for the cofactor-apoenzyme complex. Usually the binding of the cofactor induces a *shape change* in the apoenzyme and this activates or creates a binding site for the substrate. An enzyme might also be subject to inhibition by a ligand that, when bound, prevents binding of a substrate molecule.

Analysis of the simple cooperative situation in which the enzyme must bind the cofactor C before it can bind a substrate S is not particularly informative since, if C is absent, r for the binding of S is zero for all values of [S], and if C is present and [S] is constant, the binding of [S] merely follows the r-versus-[C] curve; of course [C] and [S] could both be varied, but this would be overly complex experimentally.

The more complex analysis of a multisubunit macromolecule that binds two distinct types of ligand molecules will be discussed in a later section on allostery.

Parameters Used to Describe Cooperativity

In the preceding sections we have mostly been considering cooperativity in which the binding of the first ligand facilitates subsequent binding—namely, *positive cooperativity*. *Negative cooperativity*—that is, when a bound ligand reduces the binding of subsequent ligand molecules—also occurs, especially when ions and protons bind to proteins. Whenever Equations 11 or 12 yield a curved plot, cooperativity should be suspected, and a plot of r versus [A] (Equation 4) should be made. Whereas Equation 4 is not particularly useful for determining binding parameters, it gives a clear indication that cooperativity is occurring and also indicates whether it is positive or negative. Examples of binding curves for positive and negative cooperativity obtained in this way can be seen in Figure 16-6. Note that in positive cooperativity, a plot of r versus [A] is sigmoidal; that is, as [A] increases, r rises slowly at first, then rapidly, and finally levels off, becoming asymptotic to $r = n$. This sigmoidal shape is the most important criterion of cooperativity.

Three parameters are often used to discuss a cooperative binding curve. These are the values of r at 10, 50, and 90 percent of the maximal (saturation) value of r and are written r_{10}, r_{50}, and r_{90}, respectively. A curve can be described by stating $[A]_{10}$, $[A]_{50}$, and $[A]_{90}$, the values of [A] corresponding to r_{10}, r_{50}, and r_{90}, respectively. Positive cooperativity is greater as $([A]_{90} - [A]_{10})$ decreases.

Biological Significance of Cooperativity

The curves in Figure 16-6 give some indication of the utility of cooperativity in biological systems. For instance, consider a macromolecule that has

FIGURE 16-6 Binding data illustrating two kinds of cooperative binding and contrasting it with noncooperative binding. The scales of the x axes are given in arbitrary numbers for comparison of the three graphs; a concentration of 1 corresponds to $r = 0.1$ in the graph of no cooperativity.

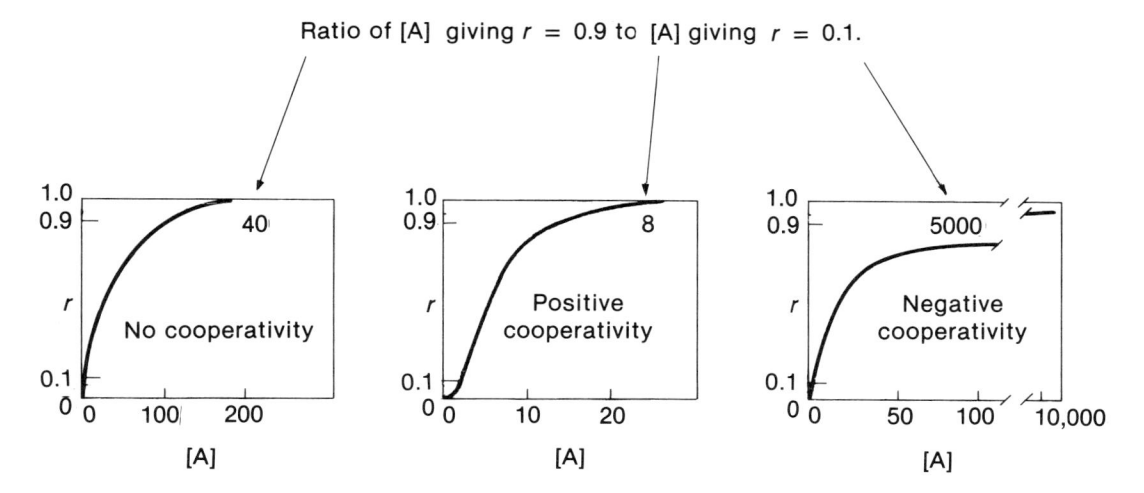

no biological activity when no ligand is bound and maximal activity only when *all* binding sites are filled. In this case the y axis of Figure 16-6 could be labeled the fraction of molecules having biological activity. Cooperativity then becomes a means of controlling activity of the protein by the concentration of ligand. Thus, if the system is positively cooperative, there is little activity at low ligand concentration and this activity will increase rapidly in a midrange of ligand concentration. If negative cooperativity is present, some activity exists at low ligand concentrations and the magnitude of this activity rapidly becomes insensitive to concentration changes as concentration increases. This modulating effect is even greater if a macromolecule binds two ligands A and B and if (1) B can be bound only if A is bound first (positive cooperativity) or (2) B cannot be bound easily if A is bound first (negative cooperativity). If the function of the macromolecule is to bind B (for example, an enzyme binding its substrate), the biological activity is regulated by the concentration of A. Cooperativity is one of the most important mechanisms for regulating biological activity and it has been suggested that the fact that a *macro*molecule can bind ligands cooperatively may be one reason why living systems have evolved with macromolecular components.

Very often cooperativity is a result of changes in the shape of a macromolecule induced by the binding of the first ligand molecule. A related mechanism, which is quite common, is one in which two or more macromolecules join to form an aggregate and the binding of a ligand to one of the macromolecules affects the ability of another macromolecule to bind the ligand. This is called *allostery* and is described in detail in the following section.

16-4 ALLOSTERY

In living systems many functional proteins are not single monomeric polypeptide chains but are aggregates of single-chain *subunits* (usually ranging from two to six) that may or may not be identical. In the case of a tetramer in which each subunit has a single binding site, the tetramer can be thought of as a macromolecule having four identical sites, and if these sites are independent, the binding properties can be analyzed by use of Equation 10. However, binding is usually cooperative, and the cooperativity is thought to result from structural changes in the polypeptide chains caused by or stabilized by binding of the first ligand molecule. In the case of a single macromolecule it is easy to see how binding of one ligand molecule can affect the binding of subsequent ligand molecules, because binding of the first can induce a change in the shape of the macromolecule that alters the shape of the remaining binding sites. It is a little more difficult to understand the mechanism of cooperativity in a multisubunit protein, because it is necessary to assume that a shape change in one subunit is somehow responsible for altering the shape of a binding site in other subunits. Such interactions can happen, though, because the subunits are in contact. Thus, we envision the following sequence of events. The binding of the first ligand molecule alters the shape of the subunit to which it is bound, and this results in changes in the sites on this subunit that are used by the subunit when it interacts with other subunits (Figure 16-7). If the subunits remain in contact, each subunit adjoining the first will undergo a shape change at its respective subunit interaction site, and this in turn alters the ligand binding site of each subunit. This may increase (positive cooperativity) or decrease (negative cooperativity) the affinity for the ligand molecules. Proteins undergoing such modifications are called *allosteric proteins*.

FIGURE 16-7 Schematic representation of the R and T forms of an allosteric enzyme. [SOURCE: From *Biochemistry, Second Edition*, by L. Stryer. W. H. Freeman and Company. Copyright © 1981.]

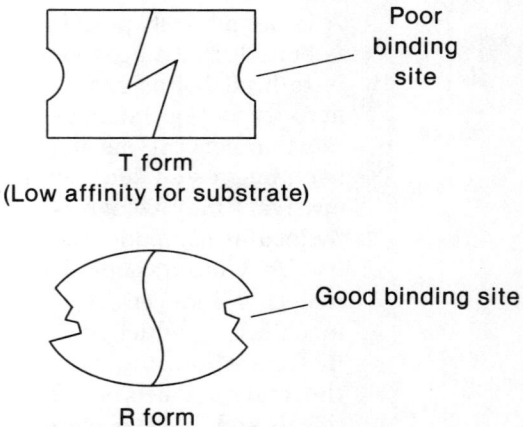

T form
(Low affinity for substrate)

R form
(High affinity for substrate)

Poor binding site

Good binding site

Models for Allostery

Two models have been presented to describe the mechanism by which the initial shape occurs and how this results in modification of the protein aggregate. These are called the *symmetry* or *concerted model* and the *sequential model*. In both models it is assumed that:

1. Each subunit exists in two forms, T and R, which bind the ligand with low and high affinity, respectively.
2. If a ligand molecule is firmly bound to a subunit, that subunit is very likely to be in the R form.

These models also differ in several ways. In the concerted model it is assumed that the T and R forms are in equilibrium, that significant binding *only* occurs to the R form (that is, rarely to the T form) and that this binding shifts the equilibrium strongly in the R-form direction. In contrast, in the sequential model it is assumed that binding tends to occur first to the T form and that this binding *induces* a transition to the R form. A second major difference between the two models concerns the number of possible forms of an allosteric protein whose binding sites are not all filled. Thus, in the concerted model the symmetry in the arrangement of the subunits is considered to be an essential component of the forces stabilizing the macromolecular aggregate. Thus, an R subunit can interact stably with another R form and two T subunits can aggregate, but an R and a T form cannot form a stable pair. Thus, as long as the subunits remain in contact, the stabilization of one R subunit causes conversion of *all* remaining T subunits to R subunits—that is, the symmetry of the protein aggregate is preserved. For instance, for a tetramer, only TTTT and RRRR are allowed forms of the complex; that is, TTTR, TTRR, and TRRR do not exist. In the sequential model, this notion of symmetry is discarded and it is assumed instead that the conversion of TTTT to RRRR occurs through such steps as

$$\text{TTTT} \to \text{TTTR} \to \text{TTRR} \to \text{TRRR} \to \text{RRRR},$$

in which each step occurs as another ligand molecule is bound. These two models are shown diagrammatically in Figure 16-8.

The concerted model can be described by a simple formalism. We consider a protein consisting of several R subunits each having one binding site, and define the high and low affinity forms of a subunit *having no bound ligand* as R_0 and T_0. These two states are in equilibrium and are related by an equilibrium constant L called the *allosteric constant*:

$$L = \frac{[T_0]}{[R_0]}. \tag{19}$$

The binding of a ligand molecule A to an R- or T-type subunit is described by the equilibrium constants K_R and K_T, respectively. It was stated earlier that in the concerted model one assumes that binding to the T form is very

FIGURE 16-8 (a) Sequential model for the cooperative binding of substrate in an allosteric enzyme. The empty active site in RT has a higher affinity for substrate than the sites in TT. (b) Concerted model for the cooperative binding of substrate in an allosteric enzyme. The low-affinity form, TT, switches to the high-affinity form, RR, upon binding the first substrate molecule. [SOURCE: From *Biochemistry*, Second Edition, by L. Stryer. W.H. Freeman and Company. Copyright © 1981.]

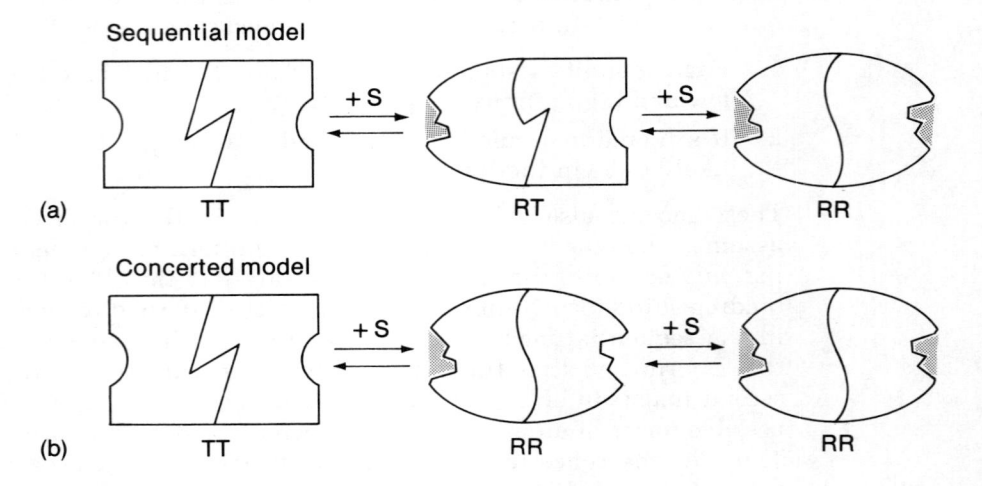

rare—that is, K_T is very large. This leads to a considerable simplification of the equation describing the binding. However, we will derive the general equation first.

We designate the complexes containing 0, 1, 2 ... n molecules of ligand as $R_0, R_1, \ldots, R_n$ and $T_0, T_1, \ldots, T_n$; the binding process then involves the following equilibria:

$$R_0 \rightleftarrows T_0,$$
$$R_0 + A \rightleftarrows R_1, \qquad T_0 + A \rightleftarrows T_1,$$
$$R_1 + A \rightleftarrows R_2, \qquad T_1 + A \rightleftarrows T_2,$$
$$\vdots \qquad\qquad \vdots$$
$$R_{n-1} + A \rightleftarrows R_n, \qquad T_{n-1} + A \rightleftarrows T_n.$$

In writing expressions for the dissociation constants for each reaction, we note that because it is assumed that once T is converted to R, all other subunits in the complex must be R, then the dissociation constants for the binding of each subsequent ligand molecule must be the same. However, in order to write expressions for the dissociation constants for *each reaction* one must take into account that there are $n!/(n-i)!i!$ ways of arranging i occupied and $n - i$ unoccupied sites (see Appendix IV at the end of this book). It can be shown that this results in the addition of a probability or weighting factor to each equilibrium equation so that

SEC. 16–4 ALLOSTERY

$$[R_1] = [R_0]n\frac{[A]}{K_R}, \qquad [T_1] = [T_0]n\frac{[A]}{K_T},$$
$$[R_2] = [R_1]\frac{n-1}{2}\frac{[A]}{K_R}, \qquad [T_2] = [T_1]\frac{n-1}{2}\frac{[A]}{K_T}, \qquad (20)$$
$$[R_n] = [R_{n-1}]\frac{1}{n}\frac{[A]}{K_R}, \qquad [T_n] = [T_{n-1}]\frac{1}{n}\frac{[A]}{K_T}.$$

We now follow the procedure used to derive Equation 9—that is, since R_n and T_n each have n bound ligands, the number of occupied sites is

$$\sum_{i=1}^{n} iR_i + \sum_{i=1}^{n} iT_i.$$

The total number of sites to be occupied is

$$n\left(\sum_{i=0}^{n} R_i + \sum_{i=0}^{n} T_i\right).$$

Thus, the fraction of sites occupied, Y, is

$$Y = \frac{\sum_{i=1}^{n} i[R_i] + \sum_{i=1}^{n} i[T_i]}{n\left(\sum_{i=0}^{n} [R_i] + \sum_{i=0}^{n} [T_i]\right)}, \qquad (21)$$

which, by substituting Equation 20, becomes

$$Y = A\frac{LK_R^n(K_T + [A])^{n-1} + K_T^n(K_R + [A])^{n-1}}{LK_R^{n-1}(K_T + [A])^n + K_T^n(K_R + [A])^n}. \qquad (22)$$

For simplicity, this equation is frequently rewritten, substituting the definitions

$$\alpha = \frac{[A]}{K_R} \qquad \text{and} \qquad C = \frac{K_R}{K_T}, \qquad (23)$$

and yielding

$$Y = \frac{LC\alpha(1 + C\alpha)^{n-1} + \alpha(1 + \alpha)^{n-1}}{L(1 + C\alpha)^n + (1 + \alpha)^n}. \qquad (24)$$

A plot of Y versus $[A]$ or α (Figure 16-9) has the sigmoidal shape characteristic of cooperativity. As the value of L increases, the curve broadens and shifts to higher values of α. Thus, cooperativity increases when the $R_0 \rightleftarrows T_0$ equilibrium leans heavily toward T_0. Note also the effect of the value of C. As C becomes very small, the sigmoidal shape of the curve becomes more obvious. Thus cooperativity requires that binding to the R form is much greater than to the T form. It should also be noted that when C approaches zero, Equation 24 simplifies to

$$Y = \frac{\alpha(1 + \alpha)^{n-1}}{L + (1 + \alpha)^n}. \qquad (25)$$

FIGURE 16-9 Plots of Y versus α for (a) several values of L and C, and (b) for two values of C when $L=1000$. The macromolecule is a tetramer (that is, $n=4$).

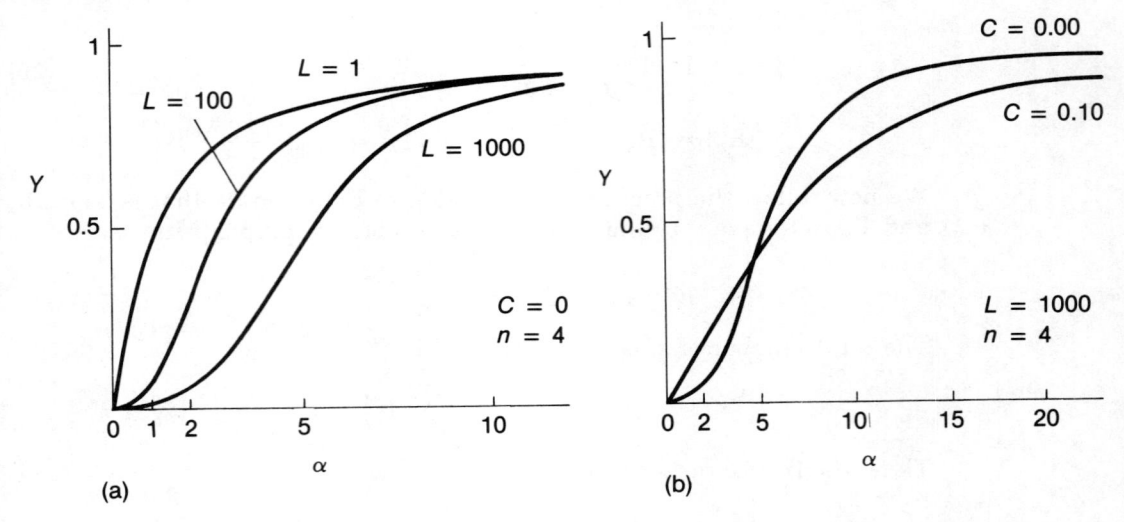

This is a useful equation and should be used whenever cooperativity is suspected.

Equations 24 and 25 enable us to understand the molecular basis of the cooperative binding of ligands to macromolecules consisting of subunits. For instance, the variation in the rate of reaction of some enzymes with substrate concentration can often be explained by the concerted model.

Allosteric Activation and Inhibition in Terms of the Concerted Model

In our discussion of allostery we have been thinking in terms of a macromolecule containing several *identical* binding sites. The cooperativity observed in such a case is called a *homotropic* effect, and the cooperativity is always positive. A macromolecule may have two distinct binding sites for different substances and filling one site with ligand A can affect the binding of B either positively or negatively. This is called a *heterotropic* effect. We have already explored this for the case of a single macromolecule; the interesting case in which there are subunits and the binding of A affects the course of allosteric changes will be examined now. Since allostery involves shape changes, the binding of A to an A site might either facilitate (activate) binding of B to its site or reduce (inhibit) the binding. Thus, we refer to *allosteric activation* (an example is the action of a cofactor on an enzyme) and *allosteric inhibition* (for example, end-product inhibition of an enzyme).

Allosteric activation and inhibition can easily be understood in terms of the concerted model. In activation the $R_0 \rightleftarrows T_0$ equilibrium shifts to the left since the activator increases the probability that a subunit is in the R form; similarly, the inhibitor shifts the equilibrium to the right by stabilizing the T form. Thus activation and inhibition are merely a decrease and increase, respectively, of L. It is important to appreciate two points. First, the binding of an activator need not always result in conversion of a T subunit to an R subunit, but may only increase the concentration of the R form. Of course, in so doing, it increases the number of proteins consisting only of R subunits. Second, the T and R terminology refers to the binding of B. Thus, the protein is allosteric only in the binding of B and the activator merely decreases the value of L in Equations 24 and 25.

We have not stated whether the binding of A is cooperative or noncooperative. In fact, it does not really matter, because whether it is or is not only determines the value of L (for binding of B) corresponding to a particular concentration of A.

EXAMPLE 16–C Activation and inhibition of a dimer for the binding of B.

Let us consider an allosteric dimer ($n = 2$) for which $L = 10^4$ and $K_R = 10^{-5}$ M. This value of L implies that at any time there is on the average one R-type molecule for every 10^4 T-type molecules. Let us also assume that binding to the T form is very poor ($C = 0$), so that Equation 25 may be used. Since the dimer is allosteric, the addition of the ligand B increases the number of R forms, since after it has bound to the R form, the complex cannot shift to the T form (Figure 16-10). Thus, the fraction of R-type molecules increases as more ligand is added. When all sites are occupied (that is, when the sites are saturated), all molecules are R forms, as shown in Figure 16-10. Applying Equation 25 with $n = 2$, $K_R = 10^{-5}$, and $L = 10^4$ as the state for which neither activator nor inhibitor is present, we can see,

FIGURE 16-10 In the concerted model, an allosteric inhibitor (represented by a hexagon) stabilizes the T state, whereas an allosteric activator (represented by a triangle) stabilizes the R state. [SOURCE: From *Biochemistry*, Second Edition, by L. Stryer. W.H. Freeman and Company. Copyright © 1981.]

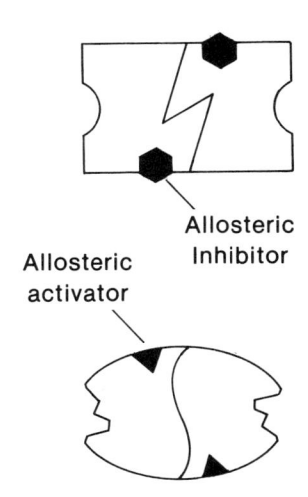

FIGURE 16-11 Saturation, Y, as a function of substrate concentration, according to the concerted model; the effects of an allosteric activator and inhibitor are shown. [SOURCE: From *Biochemistry*, Second Edition, by L. Stryer. W.H. Freeman and Company. Copyright © 1981.]

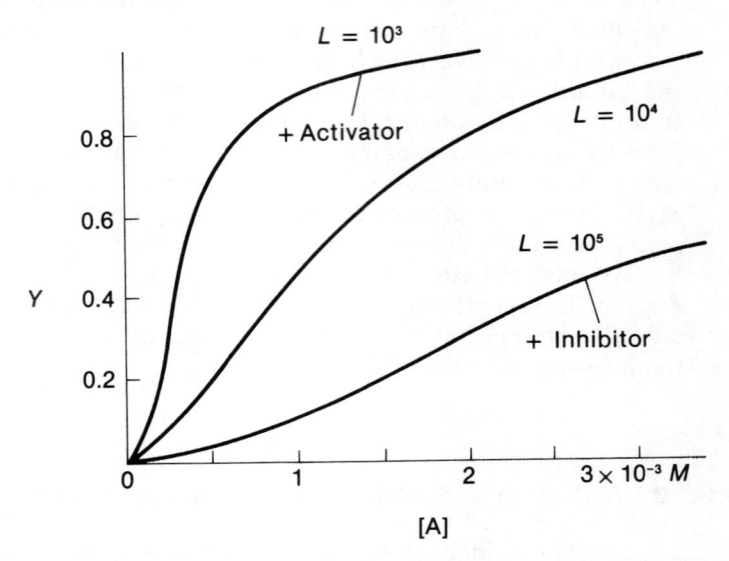

in Figure 16-11, the effect of a tenfold activation ($L = 10^3$) or inhibition ($L = 10^5$) on the plot of Y versus [A]. At any value of [B], Y is decreased by the inhibitor and increased by the activator.

In the sequential model it is assumed that when a single T subunit binds a ligand molecule, the T subunit is, with high probability, converted to the R form. As each additional ligand is bound, the T forms are sequentially converted to the R state. This process for a tetramer in a square array is depicted in Figure 16-12, which points out the major complication in the theory. Both the ligand-protein interactions and the subunit-subunit interactions are governed by equilibria, as shown in the figure; thus, the dissociation constants for these interactions must be included in the derivation. Furthermore, if a second ligand binding to the tetramer has the same probability of binding as the first ligand, there would be no cooperativity. Thus it is assumed that the existence of a single R form increases the probability that other subunits *in contact* with that R subunit will switch to an R form. This creates the next complexity, because the geometric arrangement of the subunits determines how many points of contact there are, as shown in Figure 16-13. With a tetramer, the array of subunits can be linear, rectangular, tetrahedral, or more complex in form. Note in the figure that when two subunits are in the R form there are one, two, and four possible arrangements for the tetrahedral, rectangular, and linear arrays, respectively. This means that not only must the dissociation constants for the equilibria shown in Figure 16-12 be included in the mathematical derivation but also weighting factors must be included to account for the number of arrays shown in Figure 16-13.

SEC. 16-4 ALLOSTERY 591

FIGURE 16-12 The possible intermediates in sequential binding of A to a tetramer. The circle and squares are the T and R forms, respectively.

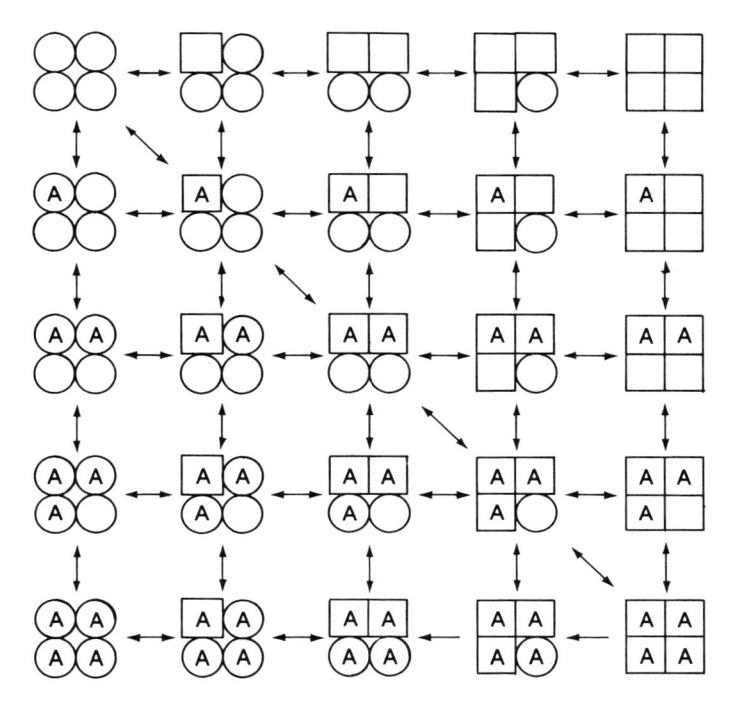

The derivation of the equations describing sequential binding of the ligand A is fairly involved, so that we will merely state the final equation for the rectangular case:

$$Y = \frac{K_{TR}^2 K_{TT}^4 Q + K_{TR}^4(1 + 2K_{TT}^4) + 3K_{TR}^2 K_{TT}^2 Q^3 + K_{TR}^4 Q^4}{K_{TR}^4 K_{TT}^4 + K_{TR}^2 K_{TT}^4 Q + 2K_{TT}^3(K_{TT} + 2K_{TR}^2)Q^2 + 4K_{TR}^2 K_{TT}^2 Q^3 + K_{TR}^4 Q^4} \cdot \quad (26)$$

The following substitutions are used in Equation 16-26:

$$Q = \frac{L[A]}{K_d}, \quad K_d = \frac{[T][A]}{[TA]}, \quad K_{TR} = \frac{[RR][T]}{[TR][R]}, \quad \text{and} \quad K_{TT} = \frac{[R][T]^2}{[TT][R]^2};$$

and [RR], [TT], and [TR] are the concentrations of the interacting subunits. (These concentrations are usually not easily measurable, so that K_{TT} and K_{TR} are rarely known.) This is clearly a very complex equation and since, in experiments dealing with binding, one usually has available only the values of Y and [A], the determination of the values of the other parameters is a formidable task. One possible method is to select arbitrary values for L, K_d, K_{TR}, and K_{TT}, and calculate curves of Y versus [A] and compare

FIGURE 16-13 Possible arrangements of two bound and two unbound subunits in a tetramer for three different arrays.

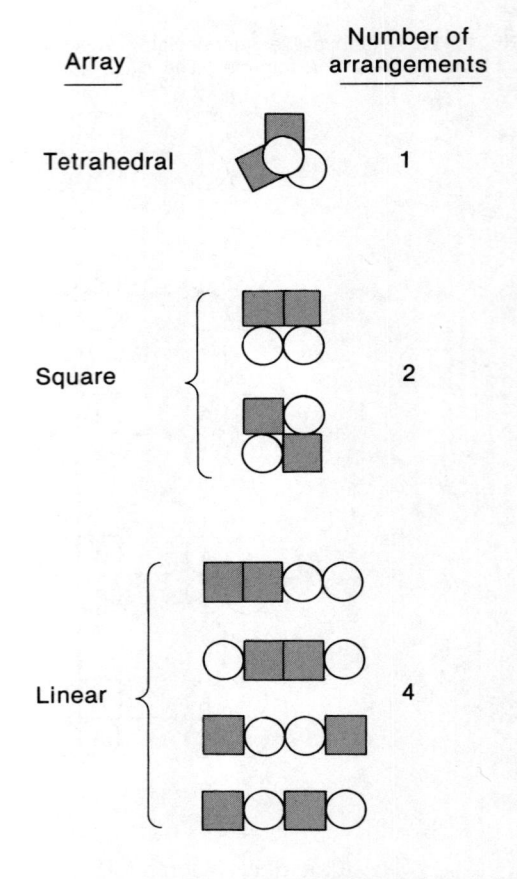

these to the experimental curves. This is not profitable, since with four varying parameters the number of possibilities is too great and often different combinations of values produce theoretical curves that are experimentally indistinguishable. Therefore, a different approach is taken. It can be shown that the basic shape of the saturation curves for the rectangular, tetrahedral, and linear cases are different; frequently one or two of these shapes can be eliminated by comparison to experiment. Alternatively, the arrangement of the subunits might be known from other data—for example, from X-ray diffraction analysis. At present, the use of the formalism of the sequential model is limited by the inability to measure or calculate the parameters needed to determine whether a particular experimental binding curve can be explained by sequential binding. For some macromolecules the sequential model may be ruled out. For instance, the binding curve may only be consistent with sequential binding to a linear array, but it might be known from other techniques that the array is tetrahedral; in this case it would be unlikely that binding would be sequential.

It is hoped that, in the future, means to evaluate the necessary parameters will become available.

Applicability of the Concerted and Sequential Models

The concerted and sequential models propose different mechanisms for allosteric changes and predict different relations between Y and [A]. If there is a single mechanism for allosteric changes, we would like to know which model is correct. Experiments show that the behavior of some macromolecules is best explained by the concerted model, while for others, the sequential model is better. There are, however, many molecules for which neither model is satisfactory. These molecules seem to exist in more than two states and complex models are needed to account for their properties. It would seem that there is not a single general mechanism, yet some investigators remain confident that a single model will some day be found.

Hemoglobin—An Example of an Allosteric Protein

Hemoglobin is the O_2-carrying protein of blood. It is a tetrameric protein containing two α and two β subunits, each of which binds O_2. A plot of the percent saturation (r) against the partial pressure of O_2 (P_{O_2}), which by Henry's Law is proportional to the concentration of dissolved O_2, shows a sigmoidal curve characteristic of cooperativity (Figure 16-14). As always, the cooperativity means that the affinity of the hemoglobin for binding the first O_2 molecule is low but once this is bound, the affinity increases. There is a tendency to interpret Y-versus-[A] curves in terms of starting with [A] = 0 and examining the results of increasing [A]. It is just as reasonable to do the reverse—that is, to begin with the sites that are saturated and study the effect of decreasing the ligand concentration. With this line of thought, cooperativity implies that once a single ligand molecule is removed, the remaining ones will follow quickly. Therefore, in the case of saturated hemoglobin, we may say that as P_{O_2} drops, the loss of one O_2 molecule from saturated hemoglobin causes rapid dissociation of the remaining ones.

We have mentioned that regulator molecules can affect cooperativity by altering the $R_0 \rightleftarrows T_0$ equilibrium. Although not usually thought of as a regulator, the H^+ ion can often cause very great shape changes by changing the interactions between individual amino acids. The pH can also regulate in another way if the binding of the ligand involves a simultaneous release or adsorption of a proton. This latter mechanism occurs with hemoglobin. The effect of pH on the O_2-binding power of hemoglobin (a phenomenon known as the Bohr effect*) is also shown in Figure 16-14. That is, the higher the pH, the greater is Y, and at lower pH values, Y decreases. This is a result of the following equilibrium:

$$HbH^+ + O_2 \rightleftarrows HbO_2 + H^+ ,$$

---

*Discovered by Christian Bohr, father of the Danish physicist Niels Bohr.

FIGURE 16-14 The binding of O_2 by hemoglobin at various pH values.

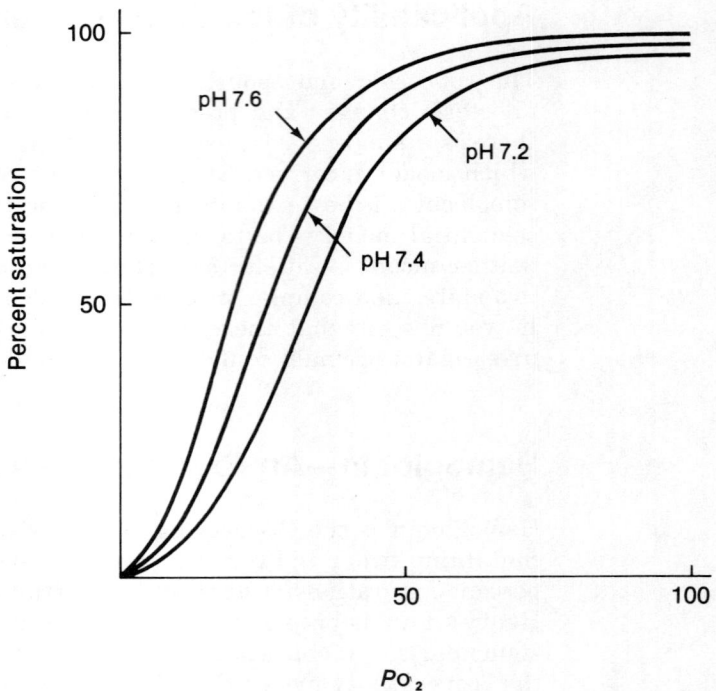

in which Hb^+ is a protonated subunit of hemoglobin (that is, it is the low-affinity or T form). The O_2-binding properties of hemoglobin can be successfully analyzed by assuming that release of the H^+ ion results in a transition to the R form. An X-ray diffraction study of the structure of hemoglobin, to which various amounts of O_2 are bound at several pH values, confirms this assumption, in that the study indicates that the first two oxygen molecules bind to the two α subunits of hemoglobin and release two H^+ ions. This results in a conversion of the two α subunits to R forms. The transition changes the interaction with the two β subunits, which then also undergo a transition to the R form. These subunits then bind O_2 with high affinity and release the remaining two H^+ ions. Thus, the H^+ ion can also be thought of as an allosteric inhibitor that decreases binding at low pH.

The Bohr effect explains how the P_{O_2} and the pH regulate the oxygen-carrying property of hemoglobin in circulating blood. In the lungs P_{O_2} is about 100 mm Hg, the pH is relatively high, and hemoglobin is 96 percent saturated with O_2. Within body tissues, owing to O_2 consumption, P_{O_2} is lower (approximately 45 mm Hg); furthermore, owing to the CO_2 formed as the end product of respiration, the pH is relatively low. Thus, hemoglobin binds O_2 less strongly and 96-percent-saturated hemoglobin releases O_2 to the tissue until the hemoglobin is 65 percent saturated. In this way, hemoglobin cycles between 65 and 96 percent saturation with O_2 and continually oxygenates the body tissues through which blood flows.

16-5 EXPERIMENTAL METHODS TO MEASURE BINDING

There are two principal experimental means of measuring binding. In the direct method, the macromolecules and the ligands are first mixed and then the macromolecules to which ligand molecules have bound are physically separated from the free ligand molecules. This is commonly done by *equilibrium dialysis* and *membrane filtration*. In the indirect method, binding is measured by observing whatever change in a physical property of either the macromolecule or ligand has been induced by the binding. Common observations are spectral changes or changes in sedimentation properties.

Equilibrium Dialysis

In equilibrium dialysis, a solution of macromolecules is placed in a dialysis bag—that is, within semipermeable membranes through which the ligand molecules but not the macromolecules can pass (Figure 16-15). The bag,

FIGURE 16-15 Equilibrium dialysis. A dialysis bag filled with macromolecules (shaded circles) is placed in a solution containing dialyzable small molecules (solid circles) that can bind to the macromolecules. At equilibrium, the concentration of free small molecules is the same inside and outside the bag. Because the macromolecules bind some of the small molecules, the total concentration of small molecules is greater inside the bag than outside. [SOURCE: From *Physical Biochemistry*, First Edition, by D. Freifelder. W.H. Freeman and Company. Copyright © 1976.]

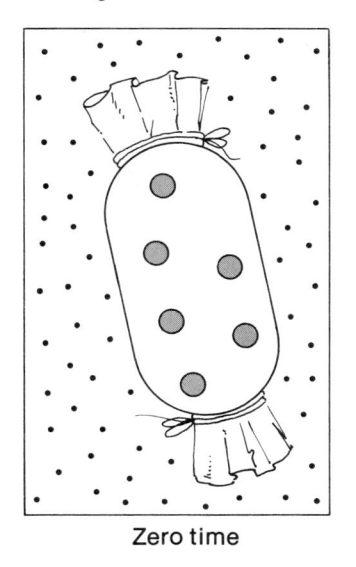

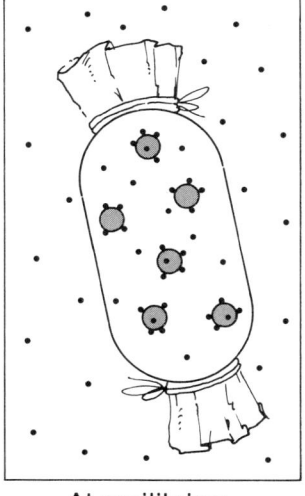

which contains a known concentration of the macromolecule, is suspended in a solution of ligand molecules. The ligand molecules then diffuse into the bag. If no macromolecules are present, the ligand concentrations inside and outside the bag are the same when equilibrium is achieved. However, when macromolecules that can bind the ligand are present, the total ligand concentration inside the bag is greater than that outside the bag by an amount determined by the number of bound molecules; this is because, at equilibrium, the concentration of *unbound* ligand molecules within the bag will always equal the concentration outside the bag.

The values needed to determine the concentration [A] of unbound ligands and the concentration [PA] of the complex are the following:

$$[\text{Unbound}] = [A] = [A]_{\text{outside}},$$

and

$$[\text{Complex}] = [PA] = [A]_{\text{inside}} - [A]_{\text{outside}}.$$

The concentration [P] of the unbound macromolecule is

$$[P]_{\text{initial}} - [PA] \quad \text{or} \quad ([P]_{\text{initial}} - [A]_{\text{inside}} + [A]_{\text{outside}}).$$

Thus, for the simple case of binding with a single site described by the reaction of $P + A \rightleftharpoons PA$ (Equations 1 and 2),

$$K = \frac{[A]_{\text{inside}} - [A]_{\text{outside}}}{[A]_{\text{outside}}([P]_{\text{initial}} - [A]_{\text{inside}} + [A]_{\text{outside}})}.$$

To obtain reliable data, certain conditions must be met: (1) the nonspecific binding of ligand and macromolecule to the dialysis bag itself must be small and preferably measurable; (2) binding must be strong, because of the following effect, which tends to indicate an artifactual negative binding (which, of course, is meaningless in the context of measuring binding). That is, if the macromolecule is very large and does not bind the small molecule at all, its great size takes up so much volume that it excludes the small molecule from the solution within the dialysis bag. Hence, binding must always be great enough that this negative effect is negligible. (3) If the macromolecule and the ligand are charged, which is often the case, there will be a Donnan effect. This effect, which is a result of the inability of the charged macromolecule to traverse the membrane and the requirement for electrical neutrality throughout the solution, causes an inequality in the concentrations of charged ligand molecules across the membrane. Since this inequality may have nothing to do with binding of the ligand to the macromolecule, the data can be misleading. To avoid the Donnan effect, one need only increase the salt concentration of the solution so that the charges causing the Donnan effect are shielded.

The use of equilibrium dialysis to evaluate K_d is shown by the following example:

SEC. 16-5 EXPERIMENTAL METHODS TO MEASURE BINDING

EXAMPLE 16-D Determination of K_d for the binding of Mg^{2+} to a protein having a single binding site.

Five milliliter of a solution containing a protein at 5×10^{-3} M is placed in a dialysis bag containing 10^{-3} M radioactive $^{24}MgCl_2$. At equilibrium, the bag contains 5.5 ml and 1653 counts of radioactivity per minute (cpm) per milliliter. The external fluid contains 1555 cpm per milliliter. One milliliter of $10^{-4} M$ $^{24}MgCl_2$ contains 1565 cpm. The bag is washed out and the radioactivity is again counted; only 35 cpm above the background level of radiation are found and this is negligible.

The internal and external concentrations of $^{24}MgCl_2$ are $(1653/1565) \times 10^{-4} = 1.056 \times 10^{-4}$ M and $(1555/1565) \times 10^{-4} = 9.936 \times 10^{-5}$ M, respectively. The volume in the bag has increased from 5.0 to 5.5 ml, so that the total protein concentration is $(5.0/5.5)(0.005) = 4.545 \times 10^{-3}$ M. Thus

$$K_d = \frac{(1.056 \times 10^{-4} - 9.936 \times 10^{-5})}{(1.056 \times 10^{-3})(4.545 \times 10^{-3} - 1.056 \times 10^{-4} + 9.936 \times 10^{-5})} = 1.3 \ .$$

Equilibrium analysis has also been used as a means of detecting macromolecules that are identifiable only by their ability to bind a particular ligand. An example of this procedure is the following.

EXAMPLE 16-E Measurement of the binding of the *E. coli* repressor for the lactose operator.

The *lac* (lactose) operon of the bacterium *E. coli* is regulated by a repressor protein that binds to the operator base sequence within the *lac* operon. When the repressor is bound to the operator, RNA polymerase is incapable of binding to the DNA and thus fails to initiate synthesis of *lac* mRNA. Several substances are known that are inducers of *lac* mRNA synthesis. Prior to purification of the *lac* repressor, genetic experiments had been performed that led to the hypothesis that an inducer binds to the repressor and causes a conformational change of the repressor that prevents the repressor from binding to the operator. Walter Gilbert developed a procedure for purifying the *lac* repressor by using repressor-inducer binding as a means of detecting the repressor. Cell extracts were fractionated by chromatography; then each fraction was tested by equilibrium dialysis for binding of the radioactive inducer isopropylthiomethyl galactoside (ITMG). That is, extracts containing [^{14}C]ITMG were dialyzed and the concentrations of ^{14}C inside and outside the bag were determined. The fraction having an internal ^{14}C-concentration greater than the outside concentration contained the repressor. By using separation procedures that continually maximized the extent of binding, the *lac* repressor was ultimately purified. The relative values of $^{14}C_{inside}$ and $^{14}C_{outside}$ were also used to determine the dissociation constant for ITMG-repressor binding.

Membrane Filtration

If a macromolecule can be collected on a filter through which unbound ligand molecules freely pass, binding of the ligand can be measured by filtering a solution containing both macromolecules and ligands and measuring the amount of ligand retained on the filter. This is a rapid and sensitive method that is applicable to the binding of ligand molecules to nucleic acids, proteins, and large particles such as membrane fragments. The filters most commonly used are made of either nitrocellulose or a mixture of nitrocellulose and cellulose acetate, both types of which usually carry a significant electrical charge. Surprisingly, nucleic acid and protein molecules can be collected on these filters even when the pores are larger than the macromolecule; *retention of these molecules is mainly by binding to charged sites on the filter*. Usually conditions can be found under which the macromolecule binds tightly to the filter but in which a ligand molecule that is not bound to the macromolecule passes through the pores unhindered.

Spectral Changes

Spectroscopic procedures are easily performed whenever the spectrum of either the macromolecule or the ligand changes significantly. The spectral change might be a shift in the absorption maximum, an increase or decrease of the absorbance, a change in the intensity of fluorescence, or other more complex changes, such as the band width in nuclear magnetic resonance spectra. If the spectrum in the absence of binding and the spectrum at saturation are both known and if the spectral change varies linearly with ligand concentration, the degree of binding at intermediate stages can be measured. An example is given in Figure 16-16, which shows the changes in the spectrum of the dye proflavin that accompanies the binding of proflavin to DNA. When unbound, proflavin has an absorption maximum at 450 nm; when bound, the maximum is at 468 nm. Note that the absorbance at 475 nm is 0 when no DNA is present but is substantial in the DNA-proflavin complex. Thus the absorbance at 475 nm can be used to measure binding.

Sedimentation Changes

An elegant method for detecting binding makes use of the shape changes of a macromolecule that sometimes occur when a ligand is bound. The shape change can be measured by observing changes in the sedimentation coefficient $s = $ (velocity/centrifugal force) when a solution containing macromolecules and ligand is centrifuged (see Chapter 15 to review macromolecular shape and the sedimentation coefficient).

FIGURE 16-16 Absorption spectra of proflavin and proflavin-saturated DNA.

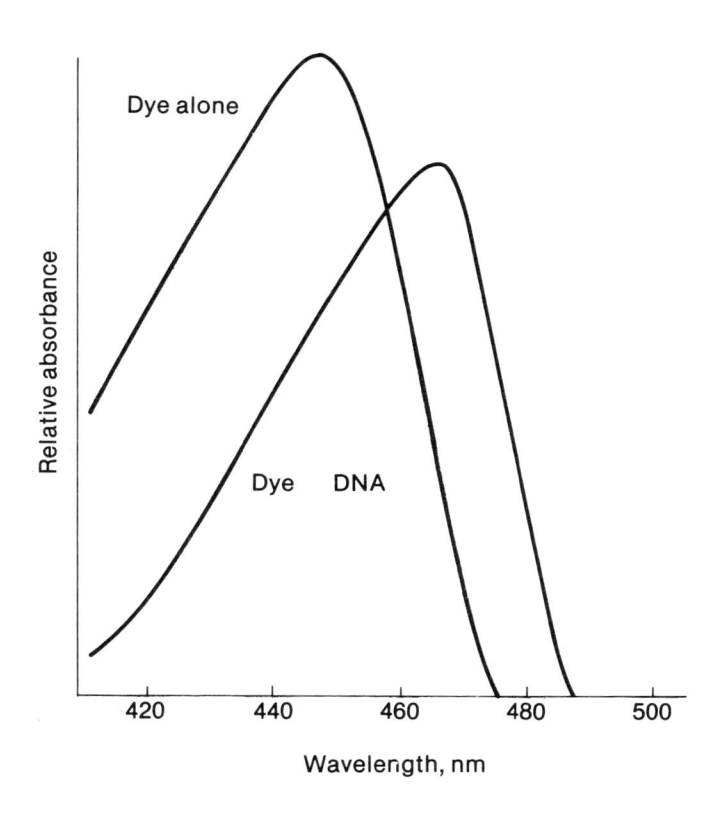

EXAMPLE 16-F Measurement of the binding of a ligand A by a protein P, by sedimentation analysis.

Suppose a protein P has an s value of 7.2 and that this value changes to 6.8 when a ligand A is bound; that is, the complex PA has an s value of 6.8. If a solution containing known concentrations of both P and A is centrifuged, the ratio f of the amount of material having s values of 6.8 and 7.2 is $f = [PA]/[P]$. If $[P_0]$ and $[A_0]$ are the starting concentrations,

$$[P] = \frac{1}{1+f}[P_0], \qquad \text{and} \qquad [PA] = \frac{f}{1+f}[P_0] \,;$$

thus r and K_d are easily calculated from equations presented earlier in this chapter.

In the case of a macromolecule that binds several ligand molecules at saturation, it is sometimes possible by the sedimentation method to determine the number of binding sites. This can be done if the addition of each ligand molecule produces a change in s so that as $1, 2, \ldots, n$ molecules are bound, $1, 2, \ldots, n$ distinct sedimenting species having different s values are observed. The number of binding sites determined in this way is always a minimal value because it might be the case that at some stage in the binding process there is either no change in s or a change that is too small to be observed. In an allosteric protein having several subunits, the existence of many sedimenting forms may be taken as evidence against the concerted model and should suggest analysis in terms of the sequential model.

In a particularly interesting variant of the sedimentation method, one observes a change in the s value of a ligand when a molecule is added. Usually a fraction f of the total ligand is observed to sediment at the same or nearly the same rate as the macromolecule. For example, there are many enzymes that use nicotinamide adenine nucleotide (NAD) as a cofactor. NAD absorbs ultraviolet (UV) light at wavelengths not absorbed by most protein molecules, so that the binding of the NAD to the protein can be detected by the appearance of UV-absorbing material that is sedimenting with the s value of the protein (s = 2 to 10, depending on the protein), inasmuch as NAD is too small to be sedimentable by itself. When this occurs, $[PA] = f[A_0]$ and $[A] = (1 - f)[A]$, and r and K_d can be calculated from the equations given earlier in the chapter.

REFERENCES

Koshland, D.E. 1970. "The Molecular Basis for Enzyme Regulation." In *The Enzymes,* edited by P. Boyer. Academic Press. New York.

Koshland, D.E. 1973. "Protein Shape and Biological Control." *Scientific American* 229: 52–65.

Monod, J., J. Wyman, and J.P. Changeux. 1965. "On the Nature of Allosteric Transitions: A Plausible Model." *J. Molecular Biology. 12:* 88–118.

Price, N.C., and R.A. Dwek. 1974. *Principles and Problems in Physical Chemistry for Biochemists.* Clarendon Press. Oxford.

Van Holde, K. 1971. *Physical Biochemistry.* Prentice-Hall. Englewood Cliffs, New Jersey.

PROBLEMS

1. The binding of a small molecule X to a protein is studied by holding the protein concentration constant and varying the concentration of the small molecule. The data obtained are the following:

Added X, mM	Bound X, mM
20	11.5
50	26.1
100	42.7
150	52.9
200	58.8
400	69.3

    Which method of plotting the data will give a more reliable value for K_d—Equation 6 or Equation 7?

2. The binding of a substrate to an enzyme increases if the ionic strength of the solution increases. Give two explanations for this increase.

3. The binding of a ligand A to a protein P is greater in a 50 percent methanol:50 percent water mixture than in water alone.

PROBLEMS

Propose two explanations for the difference.

4. A protein at a concentration of 10^{-3} M is placed on one side (I) of a semipermeable membrane. An excess volume of a solution of 10^{-4} M X is placed on the other side (II) of the membrane. The concentrations of X in compartment II have the values 10^{-4} M, 9.2×10^{-5} M, 8×10^{-5} M, 6.25×10^{-5} M, 6.12×10^{-5} M, and 6.12×10^{-5} M, at times 0, 1, 3, 7, 15, and 24 hours, respectively. After 24 hours the concentration of X in compartment I is 6.54×10^{-5} M.

(a) Does X bind to the protein, and, if so, what is the dissociation constant?

(b) A second molecule Y is added to compartment II. At equilibrium the two concentrations of X are 10^{-4} M (side I) and 10^{-4} M (side II) and the concentrations of Y are 10^{-3} M (side I) and 5×10^{-6} M (side II). What can you say about the binding sites for X and Y?

(c) Now 10^{-3} M Z is added instead of Y. At equilibrium, the concentration of Z is the same on both sides of the membrane but the concentrations of X are 9.72×10^{-5} M (side II) and 9.78×10^{-5} M (side I). What can you say about the interaction of Z and X?

5. Cytochrome c is a red protein that, under certain conditions, can bind to DNA molecules. The sedimentation coefficient s of cytochrome c is less than 5 (expressed in svedbergs) in a wide variety of aqueous solvents having different salt concentrations. The s-value of the DNA of bacteriophage T7 is 32 and is constant throughout the range of NaCl concentrations from 0.01 M to 2 M. If a particular concentration of DNA and of cytochrome c are mixed together in 0.01 M NaCl and the solution is centrifuged, the s-value of 10 percent of the cytochrome c is 32 and the remainder is less than 5. In 0.5 M NaCl no red material sediments that has an s-value of 32. What information do these experiments give you?

6. The antibody-antigen (Ab-Ag) reaction is an example of very tight binding. The following data were obtained in a binding experiment in which [Ab] = 1 μM:

Ag added, μM	Free Ag, μM
0.50	0.015
1.0	0.025
1.5	0.035
2.0	0.05
2.5	0.5
3.0	1.0

How many binding sites are there per antibody molecule?

7. Consider a protein P at a concentration of 10^{-4} M; substance X that binds to the protein is added at a concentration of 10^{-5} M. After ten minutes it is found that there is on the average one X bound per 100 molecules of P. If [P] = 10^{-3} M and [X] = 10^{-4} M, after ten minutes there is one molecule of X bound per 10 molecules of P. If this more-concentrated sample, in which binding has already occurred, is diluted tenfold (that is, back to 10^{-4} M P and 10^{-5} M X), it is found even after several hours that there is no unbound X—there is still one molecule of X per 10 molecules of P. What does this tell you about the binding of X to P?

8. A protein P has two identical binding sites for a molecule A. The shape of the protein is dependent on the pH. If the pH is varied, the affinity of P for X changes but the number of potential binding sites remains constant. How would the double-reciprocal plots obtained at various values of the pH differ from one another?

9. An enzyme E is suspected to consist of several subunits, but no one has measured the number nor determined whether they are identical or not. In the course of studying the binding of the substrate S to the enzyme, you obtain the data shown in the following Scatchard plot (page 602), in which $r = [S]_{bound}/[E]_{total}$.

(a) What can you say about the structure of the protein?

(b) What is K_d for the binding?

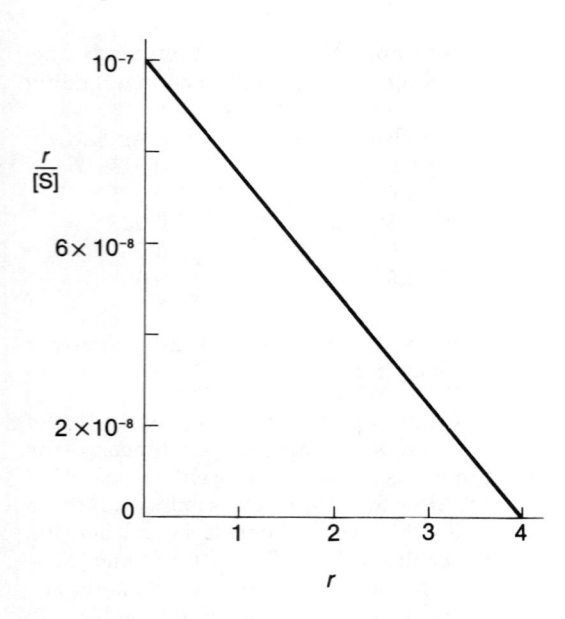

10. A titration curve of a weak acid is a binding curve in the sense that as the pH is decreased by the addition of a strong acid such as HCl, the dissociation of the weak acid will decrease. For such a system $r = $ [dissociated]/[undissociated], so that a double-reciprocal plot of $r/[H^+]$ versus r can be made. Describe qualitatively the differences between the double-reciprocal plot of a monoprotic acid and a triprotic acid.

11. A particular protein—that synthesized by gene 32 of the T4 phage (called the T4 gene-32 protein)—binds tightly to DNA. Its curved double-reciprocal plot suggests that the binding is cooperative. The gene-32 protein is readily visualized by electron microscopy when it is bound to a DNA molecule. Which of the following drawings depict what you would expect to see? The dots represent the protein molecules.

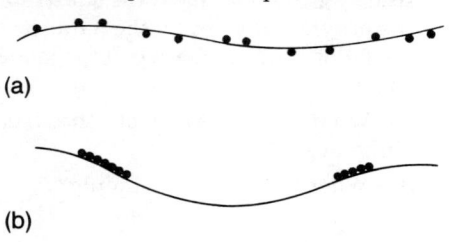

12. The binding of a ligand B to a protein is studied. The protein concentration is 10^{-3} M. The amounts of ligand bound, b, for each value of added B (that is, [B]) are the following:

[B], M	b, M
0.01	5×10^{-6}
0.02	1.55×10^{-4}
0.05	3.78×10^{-3}
0.07	4.72×10^{-3}
0.1	4.95×10^{-3}
0.2	5×10^{-3}
0.5	5×10^{-3}

Prepare a Hill plot and determine if the binding is cooperative, the number of binding sites, and the value of K_d.

13. A macromolecule H consists of four identical subunits M. The binding of a substance X for various values of [X] is shown in the figure below.

(a) Explain the different saturation values for M and H.

(b) Interpret the different shapes of the curves.

(c) What kind of plot might you make to test your interpretation in part b? If your interpretation is correct, what would the plot show?

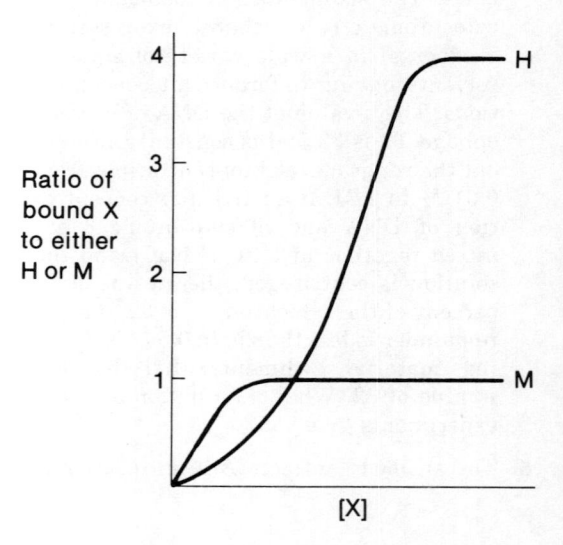

PROBLEMS

14. A Hill plot typically consists of three regions as shown below. The regions I and III invariably have a slope of 1. K_d can be determined for each of these regions by extrapolating to the y axis (the intercept is log K_d). If a Hill plot is made for an allosteric protein, what constants are determined that are necessary to perform an analysis of allostery?

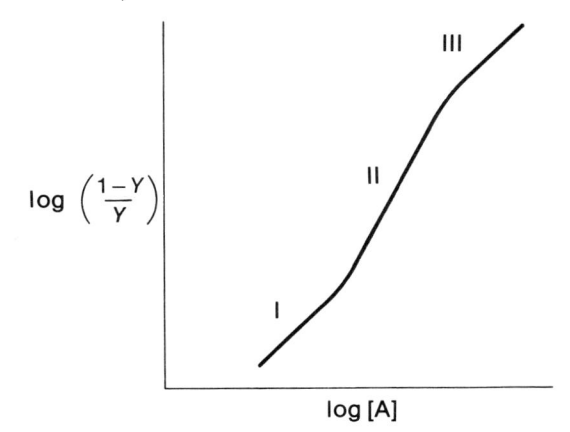

CHAPTER 17

QUANTUM MECHANICS AND SPECTROSCOPY

Between 1870 and 1910, a variety of phenomena were observed that could not be explained by the known laws of mechanics and electrodynamics. Ultimately it was found that each of these was understandable as long as the idea that a system can possess *any* amount of energy was rejected. A large number of these phenomena became understandable after the Planck hypothesis was introduced. We shall, however, only describe two of these phenomena—the photoelectric effect and the line spectra of atoms.

17-1 ORIGINS OF THE QUANTUM THEORY

In 1900, Max Planck introduced the then bizarre notion that the energy possessed by a vibrating particle must be an integral multiple of the product $h\nu$, in which ν is the frequency of oscillation and h is a constant. He called a package containing an amount of energy $h\nu$ a *quantum* of energy. This idea led to the additional hypothesis that energy can only be transferred as individual quanta. The value of h, later called *Planck's constant*, is so small, namely 6.625×10^{-27} erg sec^{-1}, that one does not detect this discreteness in the flow of energy when two macroscopic objects collide and transfer a large amount of energy, but it is evident when one considers the small energy changes accompanying either the collision of two atoms or the absorption and emission of light by a single atom.*

---

*In this chapter we retain the traditional cgs-esu units, so that the equations will be the same as in most physics books. Thus the charge of the electron is 4.7×10^{10} esu, distances are in centimeters, energy is in ergs, and the Planck constant h is 6.625×10^{-27} erg sec^{-1}.

The Photoelectric Effect

When certain metals are illuminated with visible or ultraviolet light, the metal atoms emit electrons. This was explained in classical physics* to be the absorption of the light wave by metal atoms; the energy of the light wave was thought to accelerate a freed electron away from a metal atom. In classical physics the energy of a beam of light would be related to the beam intensity. However, experimentally it was found that although the *number* of electrons is related to the incident intensity, the *kinetic energy* of the electrons, E, is not. Instead, it is a function of the frequency of the light; Einstein described this with the equation $E = h\nu - \phi$, in which h is a constant that is the same for all metals and, in fact, is just the Planck constant, and ϕ is a constant called the work function, which is specific for each metal and represents the energy necessary to overcome the attractive force holding the electron in the metal.

Einstein's use of the Planck constant in his equation describing the photoelectric effect led many other physicists to consider Planck's ideas in several optical phenomena. A highlight of this intellectual activity was Niels Bohr's explanation of atomic spectra, which we discuss in the following section.

Atomic Spectra and the Bohr Atom

A hot gas emits light: atoms of the gas absorb energy in interatomic collisions and emit energy as electromagnetic radiation. The spectrum (that is, the distribution of wavelengths) of the emitted light is not continuous but consists of discrete wavelength regions called spectral *lines*. Similarly, if white light—that is, a mixture of visible light of almost all wavelengths—passes through a gas, some of the light is absorbed. It is found that again the spectrum of the absorbed light consists of discrete wavelengths whose values for a particular atom are identical to the wavelengths in the emission spectrum.

In 1885, Johannes Balmer found empirically that the wavelengths of visible light in the spectrum of hydrogen could be described by a simple equation

$$1/\lambda = \mathcal{R}\left(\frac{1}{2^2} - \frac{1}{n_i^2}\right), \tag{1}$$

in which n_i has the integral values 3, 4, 5, ... The constant $\mathcal{R}$ is called the Rydberg constant and has the value 109,667.581 cm^{-1}.

---

*Classical physics is the term applied to the concepts of physics that existed before the quantum theory was developed. Another use of the term is to refer to those aspects of physics that can be explained without needing the quantum theory.

Examination of other lines in the ultraviolet region of the hydrogen spectrum suggested that a more general equation might be

$$1/\lambda = \mathcal{R}\left(\frac{1}{n_2^2} - \frac{1}{n_1^2}\right), \tag{2}$$

in which n_1 and n_2 are always integers. Indeed, the lines in the ultraviolet region of the spectrum obey the equation with $n_2 = 1$, and other collections of lines in the infrared region ($\lambda > 800$ nm) are found for $n_2 = 3$, 4, and 5. Similar sets of lines obeying Equation 2 have been observed for many of the elements.

The existence of line spectra requires a modification of the picture of an atom that was held in the early part of the twentieth century. According to the early model of the atom as formulated by Ernest Rutherford, electrons revolve around a positively charged nucleus, at a distance from the nucleus, and at a velocity such that the outward centrifugal force is exactly balanced by the attractive Coulomb electrical force. An immediate failure of this model is evident in the requirement of classical electrodynamics that an accelerated electrical charge continuously emit radiation. A revolving electron is necessarily accelerated—otherwise it would move in a straight line and leave the atom. However, if radiation is continuously emitted, this must mean that the kinetic energy—and, hence the velocity—of the electron are decreasing. Thus, the centrifugal force should also decrease; the electron should move nearer to the nucleus, and the electron should rapidly fall into the nucleus. However, atoms are stable, and continuous emission of radiation does not occur, so there is something inherently incorrect in the Rutherford model. A possible (and correct) explanation is that the electrons do not obey classical mechanics.

A new model for the atom was proposed in 1913 by Niels Bohr, who used Planck's notion that energy is quantized. Bohr suggested that electrons can revolve about a nucleus only in definite orbits, each orbit possessing a particular ("allowed") amount of energy. Radiation would be emitted only if an electron could move from one orbit to a second orbit possessing a lesser amount of energy; conversely, absorption of radiation would occur only if an electron could be raised from an orbit of lower energy to one of higher energy. He further insisted that the energy changes be quantized, as proposed by Planck, so that for a particular energy change $\Delta\epsilon$, radiation would have a unique frequency, ν, defined by

$$\nu = \frac{\Delta\epsilon}{h}. \tag{3}$$

This idea led to a prediction that spectra consist of lines, but additional hypotheses were needed to predict the values of the observed frequencies.

Bohr then postulated that the orbits are circular and that the angular momentum of each electron, mvr, is an integral multiple of $h/2\pi$; that is,

$$mvr = n\frac{h}{2\pi}, \tag{4}$$

in which m is the mass of the electron, v is its velocity in an orbit of radius r, h is Planck's constant, and n is an integer called a quantum number. This equation allows us to calculate the radii of electron orbits as follows.

We consider a stationary nucleus of charge ze and a particular orbiting electron of charge $-e$. The electrical attractive force F is given by Coulomb's law,

$$F = -\frac{ze^2}{r^2}, \tag{5}$$

in which F is given in dynes, e in electrostatic units (esu) with the value of 4.8×10^{-10} esu, and r in centimeters. In SI units, F is given in newtons, e is 1.6×10^{-19} coulombs, r is given in meters, and the equation is

$$F = -\frac{1}{4\pi\epsilon_0} \cdot \frac{ze^2}{r^2}$$

in which the constant ϵ_0, which is called the permittivity of vacuum, has the value 8.85×10^{-12}. The centrifugal force acting on an electron moving with velocity v in a circular orbit of radius r is mv^2/r. If the orbit is stable, the two forces are equal, or

$$\frac{mv^2}{r} = \frac{ze^2}{r^2}. \tag{6}$$

The total energy of an electron, $\mathscr{E}$, equals the sum of the kinetic energy ($\frac{1}{2}mv^2$) and the potential energy ($-e^2/r$), so that

$$\mathscr{E} = \frac{mv^2}{2} - \frac{ze^2}{r}. \tag{7}$$

From Equation 6,

$$\frac{1}{2} mv^2 = \frac{1}{2}\frac{ze^2}{r}$$

so that substitution into Equation 7 yields

$$\mathscr{E} = -\frac{ze^2}{2r}. \tag{8}$$

Equation 6 can be rewritten as $v = \sqrt{ze^2/mr}$, which, when combined with the quantum condition in Equation 4, yields

$$r = \frac{n^2h^2}{4\pi^2me^2z}. \tag{9}$$

The smallest integral value of n must be 1 because, if $n = 0$, then r would be 0, meaning that the electron would be inside the nucleus. Thus, setting $n = 1$ in Equation 9 and designating a_0 as the radius of the innermost orbit of the hydrogen atom ($z = 1$),

$$a_0 = \frac{h^2}{4\pi^2me^2} = 5.29 \text{ nm}. \tag{10}$$

This is called the radius of the first Bohr orbit or simply the *Bohr radius*. The energy of the electron in this orbit is obtained by combining Equations 8 and 9 to obtain, for any value of n,

$$\mathcal{E} = -\frac{2\pi^2 me^4}{n^2 h^2}, \tag{11}$$

so that for the innermost orbit of hydrogen ($z = 1$ and $n = 1$),

$$\mathcal{E} = -13.58 \text{ electron volts (eV)} = -2.179 \times 10^{-11} \text{ erg}.$$

This is called the *ground-state energy* of the hydrogen atom and represents the amount of energy needed to ionize a single hydrogen atom. Thus, it is sometimes called the *ionization energy*, $\mathcal{E}_I$, so that for any orbit

$$\mathcal{E} = \frac{\mathcal{E}_I}{n^2}. \tag{12}$$

To calculate the wavelengths of the lines in the spectrum of the hydrogen atom, Bohr postulated that when an electron moves from one orbit to another, a photon of light is emitted whose energy equals the difference in energy between the two orbits. Since the energy of a photon is $h\nu$, from Equation 12 we have

$$\nu = \frac{\mathcal{E}_I}{h}\left(\frac{1}{n_f^2} - \frac{1}{n_i^2}\right), \tag{13}$$

in which the subscripts i and f refer to the initial and final orbits. However,

$$\nu = \frac{c}{\lambda},$$

in which c is the velocity of light; substituting this in Equation 13 yields a form of Equation 2 with $\mathcal{R} = c\mathcal{E}_I$. Thus, a particular set of spectral lines, for example, the Balmer series, results from an electron jumping between a particular orbit (in this case, the innermost) and each of the other possible orbits. This is shown diagrammatically in Figure 17-1.

Let us examine for a moment the conditions that would lead to movement of an electron from one orbit to another. To do so, we must distinguish between absorption and emission spectra. First, consider an atom in the *ground state*—that is, *the state in which all electrons are in the lowest possible energy state*. This occurs, for example, when all inner orbits are filled and the first unfilled orbit contains all of the remaining electrons. If a collection of atoms is exposed to a beam of light having a range of frequencies, or—if we consider light to be particulate—exposed to a collection of photons having various energies, then a photon will be absorbed by an atom if the energy of the photon equals the difference between the energy an electron in the outermost orbit (for the ground state) would have and the energy the electron would have in an orbit of larger radius (that is,

SEC. 17-1 ORIGINS OF THE QUANTUM THEORY

FIGURE 17-1 Sets of transitions by which an electron moves between a particular Bohr orbit (the circles) and all orbits having larger radii. The wavelengths in each set are clustered and the sets are called series. Arrows directed toward a smaller circle represent emission; one directed toward a larger circle represents absorption.

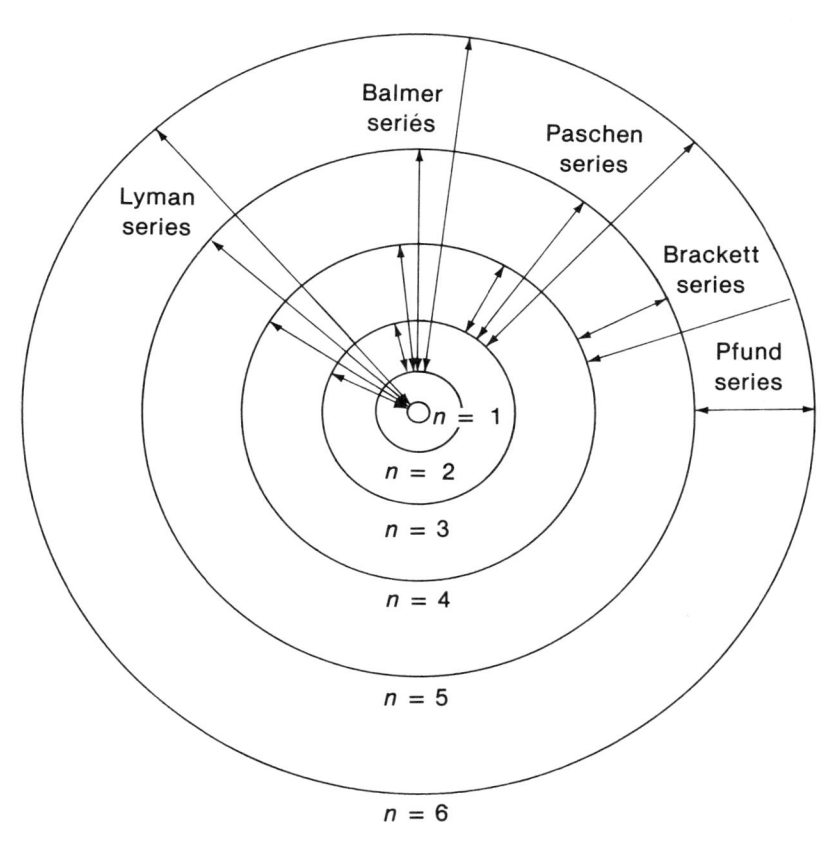

a frequency satisfying Equation 13). The energy of the photon is used to move the electron to a more distant orbit; the atom is then in an *excited state*. Thus, the absorption spectrum is just the collection of frequencies (or wavelengths) that are absorbed. The excited atom is not stable and will soon return to its ground state. To do so requires giving up the same amount of energy that was taken in. One possible way to do this (a rare one) is to reemit a photon having the same frequency as the absorbed photon. (This phenomenon will be considered later when we discuss fluorescence.) Another possibility is that by colliding with other atoms, which may or may not be excited, the quantum of energy can be converted to translational and vibrational energy—that is, to *heat*.

If a gas is heated to a very high temperature, it produces light (it becomes incandescent); the wavelength (or frequency) distribution of this light is the emission spectrum. This light production requires two steps, and light is produced in the second of these: (1) an excitation of the atoms, so that electrons have moved into orbits more energetic than that of the ground state, and (2) a drop to a lower energy state accompanied by emission of a photon. The source of the excitation energy is the kinetic energy of the atoms that, by collision, can transfer energy to an electron, thus raising the electron to a higher orbit and raising the atom to an excited state.

A more efficient way of producing emission of light is by bombarding a gas with electrons, in which case the source of energy is the kinetic energy of the electrons that can be transferred to the atom in a collision. The energy of the electron beam must be such that an incident electron that collides with the atom can totally remove an orbital electron from the atom. That is, the gas is ionized, and the electron is in a state characterized by $n = \infty$. The minimal energy of the incident electron producing this effect is called the ionization potential. For a hydrogen atom this energy is 13.58 electron volt, which is the value calculated from Equation 11 with $n = 1$. Once a gas atom is ionized (stripped of its electron), an electron can return from infinity to any orbit, emitting a photon in so doing; then, this electron can return to the ground state—either directly or in several steps—again emitting a photon in each transition. The set of all possible transitions for the gas is the emission spectra.

Electrons as Waves

In the discussion just presented, we saw that excitation of an orbital electron could result from absorption of a light wave, which we viewed as a particle called the photon. Similarly, excitation results from the absorption of the energy of an electron. A great many experiments (for example, diffraction of electrons) have indicated that a particle, the electron, can also be viewed as a wave. The wavelength of a moving electron and the momentum of a wave can be calculated as follows.

The momentum p of any particle equals its energy E divided by its speed. Thus, when light is viewed as a particle, its momentum is

$$p = \frac{E}{c} = \frac{h\nu}{c} = \frac{h}{\lambda}. \tag{14}$$

Louis de Broglie assumed that the momentum of an electron is also described by $p = h/\lambda$. However, an electron does not move with the velocity of light but with a velocity v, so that $p = mv$, and the wavelength of the electron is

$$\lambda = \frac{h}{mv}. \tag{15}$$

This concept can be related to the Bohr theory of the atom. If the electron is considered as a wave, an electron in orbit must be a wave that closes upon itself. This means that the circumference of the orbit must be an integral number of wavelengths, or

$$2\pi r = n\lambda ,\qquad(16)$$

in which r is the radius and n is an integer. Combining Equations 15 and 16 yields

$$mvr = \frac{nh}{2\pi},$$

which is identical to the Bohr assumption given in Equation 4. A use of the notion of an electron as a wave is given in the example that follows.

EXAMPLE 17-A   The limit of resolution of the electron microscope.

In an electron microscope an object is exposed to a beam of electrons, and an image is formed in the same way as in an ordinary microscope, because the electrons behave as waves. In the physical theory of optics, it is shown that the limit of resolution of a microscope (that is, the minimal distance between two objects that is allowable if the objects are to appear as two distinct objects rather than as a single larger object) is roughly equal to the wavelength of the illuminating light. Thus, the limit of resolution of an electron microscope is determined by the voltage V used to accelerate the electrons (because the voltage determines the velocity by the equation $\frac{1}{2}mv^2 = Ve$ and therefore determines the wavelength, according to Equation 15). At a typical operating voltage of 60,000 V, the wavelength of the electron is 0.015 Å. Actually, this limit of resolution is never achieved in a real electron microscope because of deficiencies in the design of electron lenses.

Modifications of the Bohr Theory

If a gas is placed in a strong magnetic field, some of the spectral lines disappear and are replaced by close pairs of lines. This is called the *Zeeman effect,* and to incorporate this finding into the Bohr model, an addition to the theory is necessary. When a moving electron is exposed to a magnetic field, the direction of motion of the electron changes. In the Bohr model, this can be thought of as a change in the plane of the orbit. We can then ask if all orientations of the orbits with respect to the magnetic field are allowable. Bohr proposed that they are not—that only certain orientations are possible—and that these are determined by the condition that *the component of the angular momentum* p_m *in the field direction must be an integral multiple of* $h/2\pi$, or

$$p_m = m\frac{h}{2\pi},\qquad(17)$$

in which m is an integer called the *magnetic quantum number*. The permissible values of m are ± 1, ± 2, and on, up to a limiting value related to the principal quantum number n, so that each electronic energy level is split into two levels when the atom is in a magnetic field. Thus, if transitions between states having different values of n are accompanied by a change in m of $+1$, then pairs of spectral lines will result. The differences in energy levels between a state having $m = +1$ and one with $m = -1$ is so small that the resulting pair of spectral lines is separated by only a small difference in wavelength. It can be shown that the separation is proportional to the magnitude of the magnetic field. (The reader should note that in this discussion the symbol m does not represent mass.)

17-2 THE NEW QUANTUM MECHANICS

The Bohr theory enabled one to understand the frequencies of the observed lines for atoms containing one electron in the outermost orbit, but failed to explain two phenomena—that the intensities of all of the lines are not equal, and that transitions occur between some but not all orbits. To approach this problem, as well as to explain other phenomena, a new quantum mechanics was developed by Erwin Schrödinger and Werner Heisenberg, who arrived at their findings independently. What they theorized is in principle somewhat at odds with our daily experience, and is mathematically very complex, but when the equations are solvable, they have enormous predictive value. We shall describe only the Schrödinger theory, which is based upon two things—the wave nature of electrons, and a new way of writing the law of conservation of energy. We shall present only a small part of the theory, with the single goal of indicating how the quantum numbers arise as a natural consequence of mathematics, rather than as a postulate, which was how these were introduced by Bohr. The quantum theory also takes a rather new point of view about the behavior of matter, and it is hoped that in the discussion of the theory, this will become clear.

The Schrödinger Theory

Classical mechanics is based upon Newton's laws, which are accepted without proof. The proofs of the validity of these postulates, such as that force equals mass times acceleration, and that energy and momentum are conserved, are only that they succeed in predicting the outcomes of innumerable experiments. Quantum mechanics is also based upon a set of postulates, but they differ from the seemingly reasonable laws of classical mechanics in that the quantum postulates seem to be peculiar, arbitrary, and unrelated to experience. Nonetheless, the predictions of quantum mechanics agree with experimental observations.

SEC. 17-2 THE NEW QUANTUM MECHANICS

Practically speaking, quantum mechanics and classical physics differ in two important ways. First, the mathematics of quantum mechanics becomes formidable for even relatively simple problems, and computer technology becomes essential for real situations (when the problem is solvable at all). Second, quantum mechanics is valid for systems of all sizes, but the so-called quantum effect—that is, those phenomena that are not explainable by classical physics—only appears when examining very small objects, such as electrons, atoms, and molecules.

The fundamental equation of the quantum theory is called the *Schrödinger equation*. It arises by writing an equation for the conservation of energy but with the expressions developed by Planck and de Broglie for energy and momentum, which include the constant h. A new function ψ is also introduced that incorporates the concept of the wave nature of particles. This is called the *wave function,* and knowledge of this function enables us to obtain a complete description of the behavior of the particle that it represents. This function contains the position x as a variable. In a classical wave equation used to describe a visible standing wave, such as the vibration of a string or the motion of a wave on water surface, there is also a wave function; this tells the location of a part of the string or a molecule on the water surface *as a function of time*. The basic Schrödinger wave equation does not contain the time variable* but is in general an oscillating function. In the absence of time as a variable, the significance of ψ becomes difficult to understand; however, a postulate of quantum theory tells us what ψ means.

This postulate is that, in considering a particle as a wave, we mean that *the probability of finding a particle in a given interval* dx *is* $\psi^2 dx$.† Mathematically, this means that if two particular coordinates are stated, for example, x_1 and x_2, the probability $P(x_1, x_2)$ of finding the particle between x_1 and x_2 is

$$P(x_1, x_2) = \int_{x_1}^{x_2} \psi^* \psi \, dx . \tag{18}$$

The meaning of this seemingly obscure equation can best be seen by an example. If ψ for the electron in a hydrogen atom were calculated, then a determination of the values of x_1 and x_2 for which $P(x_1, x_2)$ is maximal should tell the location of the innermost orbital‡ of the electron. Note, however, that *by talking about probabilities we have lost the idea of fixed*

---

*There is also a time-dependent form of the Schrödinger equation, but it will not be discussed in this book because its solutions are very complex.

†Since ψ often contains $i = \sqrt{-1}$, ψ^2 is more correctly written $\psi\psi^*$, in which ψ^* is called the complex conjugate of ψ—that is, the function that is obtained by changing the sign of $\sqrt{-1}$ wherever it occurs. For example, if $\psi = \cos x - i \sin x$, then $\psi^* = \cos x + i \sin x$.

‡Up to this point we have used the term *orbit* to mean the closed path of an electron in the Bohr theory. Henceforth we will use the term *orbital* to mean a three-dimensional region of space in which an electron may be found. Specifically, an orbital is a region defined by a solution to the Schrödinger equation.

positions for the orbit and also the ability to localize the electron precisely. What this means is that if an experiment were done repeatedly with a single electron, and if we had a means of determining its location with some degree of accuracy, repeated measurements would not necessarily yield the same value. If the measurement were done 1000 times, and we were looking for the electron in a given interval x_1 to x_2 (not at a given point), then the fraction of times we would detect it would be given by $P(x_1, x_2)$.

In calculating ψ from the Schrödinger equation, we usually obtain a function containing an arbitrary constant. This is evaluated from these probability notions because, if the particle exists, it must be somewhere—that is, the probability of its being in the universe must be 1. Mathematically this is expressed as

$$\int_{-\infty}^{\infty} \psi^*\psi \, dx = 1 \ . \tag{19}$$

In general, when making a measurement, one is not looking at a single particle—for example, a single hydrogen atom—but a large number of particles, and the value one obtains is an average. Time-independent quantum mechanics states that to obtain an average is the best one can ever do; it makes no sense to talk about the precise location of a particular electron because there is no experiment that can ever tell where it is, anyway. (In this sense, quantum mechanics is strictly the physics of observables.)

Another postulate of quantum mechanics tells us how to calculate the average value of some property f. If the property can be written as a function of x, that is, $f(x)$, the average value $\langle f \rangle$ (that is, the only value that is measurable) is given by

$$\langle f \rangle = \int_{-\infty}^{\infty} \psi^* f(x) \psi \, dx \ . \tag{20}$$

Thus, if we want to know the average radius $\langle r \rangle$ of the innermost orbital of the hydrogen atom, this would be

$$\langle r \rangle = \int_{-\infty}^{\infty} \psi^* r \psi \, dr \ , \tag{21}$$

in which ψ must be written as a function of r. One might ask why we wrote $\psi^* f(x) \psi$ rather than $\psi^* \psi f(x)$ in Equation 20. The reason is based upon a peculiar but necessary way in which a mathematical function must be expressed in quantum mechanics—namely, as an operator. That is, certain properties of a particle are obtained by carrying out a mathematical operation on ψ (for example, multiplication, or the taking of a derivative), and then multiplying the result by ψ^*. For an operation such as multiplication by x, the order of the operations is irrelevant, in that $\psi x \psi^* = \psi^* \psi x = x \psi^* \psi$, etc. However, in the case of taking a derivative, $\psi^* d(\psi)/dx$ is not necessarily the same as $d(\psi^*\psi)/dx$, and $\psi^*\psi(d/dx)$ makes no sense. Thus, in quantum mechanics, the momentum in the x direction (for example) of a particle having a wave function is obtained by use of the momentum operator $(h/2\pi i)(\partial/\partial x)$, in which $i = \sqrt{-1}$. Hence the average momentum p_x in the x direction is

$$\langle p_x \rangle = \frac{h}{2\pi i} \int_{-\infty}^{\infty} \psi^* \frac{\partial \psi}{\partial x} dx . \tag{22}$$

It should be noted that the formalism that has been introduced expresses a probabilistic view of the universe. This is an idea we have encountered before, in the concept of entropy in Chapter 5. However, there is a difference. In thermodynamics, there is no rule that governs the precise position of, for example, a particle in a box, but nonetheless the laws of classical physics enable one to describe the velocity and position of the particle at any time. That is, if the velocity and position of a particle were known precisely at one time, we would be able to calculate precisely its position at any other time. In classical mechanics, the bulk properties of matter are obtained by summing the properties of each atom. (Note we have said *"summing"* and not *"averaging"*.) That is, the pressure of a gas in a container is the sum of the forces exerted by each particle when it collides with the walls of the container. From the quantum mechanical viewpoint, such a sum is meaningless, because if there were one particle in a box, there is no way of knowing *both* its position and its momentum at a particular instant with perfect accuracy. The particle may be anywhere, although it has a certain probability of being in a particular volume.* It is important to appreciate, though, that quantum mechanics does provide a great deal of dynamic information about a particle—that is, one knows *roughly* where the particle is (the region of high probability) and how fast it is moving (the most probable velocity).

So far, we have described several properties of the wave function ψ but have not explained how to evaluate it. This is done by solving the Schrödinger equation. The fascinating consequence of these solutions is the appearance of quantum numbers without the necessity of postulating their existence.

To obtain the Schrödinger equation, we need only write an expression for the energy of a particle in terms of the wave function ψ. Schrödinger's postulate that ψ contains all the information that is available requires that ψ appear in the terms for the kinetic energy, the potential energy, and the total energy. This is expressed by the preliminary equation

Kinetic energy (T) + potential energy (U) = total energy (E),

---

*It is interesting to note that Albert Einstein never really liked these probabilistic ideas, saying "I shall never believe that God plays dice with the world," and for many years he debated with Niels Bohr about the validity of the quantum theory. However, disregarding the philosophical implications of the quantum theory, its justification as a viable mathematical expression of natural phenomena is that its predictions agree with experimental observation.

or
$$T\psi + U(x)\psi = E\psi, \qquad (23)$$

in which T, U, and E represent the *operations* that must be applied to ψ to obtain the different energies. T, U, and E are called *operators*. The kinetic energy is written in terms of the momentum, p, as $p^2/2m$, or, using the expression given in deriving Equation 22—that $p_x = (h/2\pi i)(d/dx)$—we can write Equation 23 as*

$$-\frac{h^2}{8\pi^2 m}\frac{d^2\psi}{dx^2} + U(x)\psi = E\psi. \qquad (24)$$

This is the one-dimensional Schrödinger equation for a particle of mass m. The solution of this equation is obtained after $U(x)$ is replaced by a function that describes the potential energy of the particle as a function of distance. We will give several examples of this shortly. For a particular expression for $U(x)$, the Schrödinger equation does not have a single solution, because E is not specified. Thus we obtain a collection of wave functions, one for each value of E, and have the problem of determining the ones that are valid for the system being examined. This can be done, because constraints are placed on ψ. For example, for any ψ, Equation 19 must hold, and this reduces the number of possible solutions tremendously. Second, the value of the potential energy is often known at particular positions, so that $U(x)\psi$ must have specific values at these positions. These values of ψ are called the *boundary conditions* of the system. After these two constraints are satisfied, there remains a collection of pairs of expressions for ψ and E. We can number these pairs (ψ_1, E_1) and (ψ_2, E_2), and so forth, or more generally, (ψ_n, E_n). It turns out that the value of n appears in the solution of the equation and, moreover, n is an integer. Thus, we obtain actual quantum numbers, *but not by postulate*, which was how these were obtained in the Bohr model. The fact that the values of E are limited to n different values indicates, as Planck proposed, that a system cannot have an arbitrary amount of energy, but is confined to a discrete number of values. In quantum theory the German word *eigen* (meaning inherent or particular) is used, and the values of ψ_n and E_n are called *eigenfunctions* and *eigenvalues*, respectively.

We have so far written the Schrödinger equation in a single dimension; real systems exist in three dimensions. Thus, a more correct statement of the equation for three dimensions is

$$-\frac{h^2}{8\pi^2 m}\left(\frac{\partial^2\psi}{\partial x^2} + \frac{\partial^2\psi}{\partial y^2} + \frac{\partial^2\psi}{\partial z^2}\right) + U(x,y,z) = E_{n_x n_y n_z}\psi. \qquad (25)$$

---

*The derivative d/dx is used instead of the partial derivative $\partial/\partial x$ because there is only one variable in this one-dimensional formulation.

Solutions to the Schrödinger Equation

There are three classes of solutions to the Schrödinger equation—the trivial, the difficult, and those that are so difficult in practice that they might as well be impossible. Unfortunately, most real systems are so complicated that the solution is almost always in the third category. For example, the hydrogen atom has been analyzed, as will soon be seen, but for no other atom or molecule has the equation been solved exactly. The major problem is either in knowing the correct expression for U or in tackling the formidable mathematics required to solve the equation once U is known. Thus, the solutions are usually approximations and invariably require modern computer technology. In the following, we present a few of the simple solvable systems, so that the reader can see how quantum numbers arise.

A. The Particle in a One-dimensional Box

We consider here a single particle that can move completely freely throughout a box having a width a. This freedom of motion means that $U(x) = 0$ in the box. We assume further that the walls of the box are so strong that the particle cannot pass through the walls. To do so would require infinite energy, so that $U = \infty$ when $x < 0$ and $x > a$ (Figure 17-2). We already know the value of ψ outside of the box; since the particle is never outside the box, ψ must be 0 outside the box. The Schrödinger equation for the particle within the box is

$$-\frac{h^2}{8\pi^2 m} \frac{d^2 \psi_n}{dx^2} = E_n \psi_n . \qquad (26)$$

The general solution of this equation is

$$\psi = A \sin\left(\frac{2\pi}{h} \sqrt{2mE}\, x\right), \qquad (27)$$

in which A is a constant. Remembering that $\psi = 0$ at $x = 0$ and $x = a$, we note that $\sin(2\pi/h)\sqrt{2mE}\, a = 0$, which is true whenever $(2\pi/h)\sqrt{2mE}\, a = n\pi$, in which n is an integer. (Note the appearance of the quantum number n.) Thus

$$E_n = \frac{n^2 h^2}{8ma^2}, \qquad (28)$$

which gives the permissible values of E for the particle. Substituting this into Equation 27 yields

$$\psi = A \sin(2\pi/h) \sqrt{\frac{2mn^2 h^2}{8ma^2}}\, x$$

$$= A \sin\frac{n\pi x}{a} . \qquad (29)$$

FIGURE 17-2 The potential energy of a particle in a box.

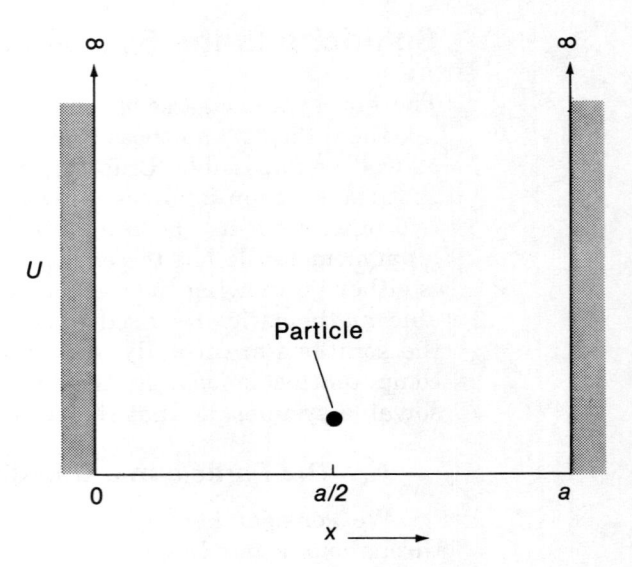

To calculate A, we turn to the boundary condition that the particle is definitely *somewhere* in the box; that is, from Equation 19

$$\int_{-\infty}^{\infty} \psi^*\psi \, dx = \int_0^a A^2 \sin^2\left(\frac{n\pi x}{a}\right) = 1 \, ,$$

which yields $A = \sqrt{2/a}$. Substituting this into Equation 29 yields

$$\psi = \sqrt{\frac{2}{a}} \sin\left(\frac{n\pi x}{a}\right) . \tag{30}$$

This is the solution of the Schrödinger equation for a particle in a one-dimensional box.

Let us see what Equation 30 tells us. First, we wish to determine the lowest value of n. We can see that n cannot ever be zero because then $\sin(n\pi x/a) = \sin 0 = 0$, so that $\psi = 0$ everywhere, which would mean that there is no particle anywhere. When $n = 1$, ψ is nonzero, so that 1 is the lowest value of n. This is interesting, because when n is 1, Equation 28 indicates that $E_1 \neq 0$. Thus, the lowest possible energy state is *not* one in which $E = 0$. Furthermore, by definition, the potential energy of the particle is 0 everywhere, so that the total energy is all kinetic energy. Thus, the kinetic energy is never zero; hence the particle can never be stationary. This minimal permissible amount of energy is called the *zero point energy*. The implication of this energy is discussed in the next section, which deals with the harmonic oscillator.

We now consider the information contained in the wave function

(Equation 30). This is a sine function that is zero at the walls and maximal at $x = a/2$. Thus, if we look at various positions in the box, we will find the particle most frequently in the middle. If somehow the particle were to obtain more energy, the least amount it could have above the *ground state* (the zero point energy) is that for which $n = 2$ in Equation 28. Having this energy, x would be zero at $x = a/2$ (see Equation 30) so that the particle would not be found at all in the center of the box. Instead, there would be two highly probable regions symmetrically placed about $a/2$ (the center). It is important that the reader does not attempt to rationalize this in terms of classical physics. For particles of atomic dimensions—that is, when m and a are small—one must accept the wave nature of particles. If, on the other hand, the particle were a golf ball contained in a macroscopic box, the spacing between the energy levels would be very small and as far as a real observation is concerned, there would appear to be a continuous range of possible energies. This is because, if the ball were moving visibly, n would have to be extremely large and the possible increments of energy would be so tiny (for example, n might change from 10^9 to $10^9 + 1$) that continuous energy states would be observed. Furthermore, with n so large, the sine wave in Equation 31 would be maximal at so many points that the probability of finding the ball at any one position would be the same as finding it at any other position. This agrees with our intuition that if the ball were bouncing back and forth in a box, it would have the same probability at any instant of being at any position. It will be found that whenever n is large (or, as a matter of fact, if h were 0) the quantum mechanical result always approaches the result expected by the classical theory.

It is important to note that the appearance of integral quantum numbers is a consequence of making the dimensions of the box finite. Returning to Equation 29, we note that if x could have any value, then x would be nonzero for *any* positive value of E. Thus the energy of a particle freely moving in the universe is not quantized. This agrees with the observation that once sufficient energy is put into an atom to ionize it (that is, to remove an electron to infinity), any additional energy yields a continuous absorption or emission spectrum, not a line spectrum.

B. The Harmonic Oscillator

An example of a harmonic oscillator is shown in Figure 17-3. This is the simplest example of a vibrating system and is a model for understanding molecular vibrations. The figure shows a particle of mass m attached by a spring to an immobile object (a wall). The coordinate system is defined so that $x = 0$ if the particle is at rest. If the particle is pulled out a distance x, there is a restoring force $F = -Kx$ in which K is the force constant of the spring. When released, the particle will oscillate about the equilibrium position, moving a distance x to the other side of the equilibrium position if the spring is a perfect one having no internal friction. Classically, ψ could have any value, so that the oscillation frequency could also have any

FIGURE 17-3 A diagram of a harmonic oscillator.

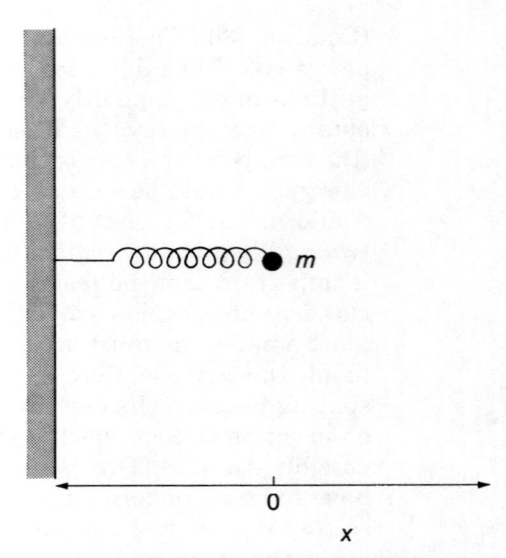

value. As we will see, this is not the case according to the view of quantum theory. The potential energy U of the particle is simply calculated from the well-known relation that $dU = -F\,dx$ for any force F. Thus, $dU = -Kx\,dx$, and integration yields

$$U = \frac{1}{2}Kx^2, \tag{31}$$

if $U = 0$ at $x = 0$. Therefore the Schrödinger equation for a harmonic oscillator is

$$-\frac{h^2}{8\pi^2 m}\frac{d^2\psi}{dx^2} + \frac{1}{2}Kx^2\psi = E\psi. \tag{32}$$

This equation also has a large number of solutions that arise as pairs (ψ_0, E_0), and (ψ_1, E_1), and so forth, which we simply state as

$$\begin{aligned}
\psi_0 &= \left(\frac{2a}{\pi}\right)^{1/4} e^{-ax^2}, & E_0 &= \frac{1}{2}h\nu_0; \\
\psi_1 &= \left(\frac{2a}{\pi}\right)^{1/4} a^{1/2} x e^{-ax^2}, & E_1 &= \frac{3}{2}h\nu_0; \\
\psi_2 &= \left(\frac{2a}{\pi}\right)^{1/4} (4ax^2 - 1)e^{-ax^2}, & E_2 &= \frac{5}{2}h\nu_0;
\end{aligned} \tag{33}$$

in which

$$a = \frac{\pi}{h}(Km)^{1/2} \quad \text{and} \quad \nu_0 = \frac{1}{2\pi}(K/m)^{1/2}.$$

There is an interesting difference between the values of ψ and E for a harmonic oscillator and those for a particle in a box. In the latter case, n

cannot be zero because then $\psi = 0$, meaning that there would be no particle in the box. For the harmonic oscillator, n can be zero because $\psi_0 \neq 0$. However, when $n = 0$, E_0 is still unequal to 0, indicating, as in the case of a particle in a box, that the minimal energy is not zero. If the energies are expressed in terms of a *vibrational quantum number*, v, a general expression for the energy levels is

$$E_v = \left(v + \frac{1}{2}\right)h\nu, \qquad v = 0, 1, 2, \ldots . \tag{34}$$

Note how once again the existence of discrete energy states is a natural consequence of using the Schrödinger equation, as is the constant value $h\nu_0$ for the difference in energy between adjacent energy levels.

The energy value for $v = 0$ is again called the zero point energy and it occurs repeatedly in molecular phenomena. What it means is that whereas classical mechanics allows the particle to come to rest, the quantum theory insists that the particle never stops moving—at least we can never tell if it has. Note again that if the particle were a golf ball, m would be very large, ν_0 would be very small, the separation of the levels would be tiny, and E_0 would approach 0. Thus, as our experience tells us, macroscopic objects can certainly stand still (at least they appear to) even if attached to a spring. When the golf ball vibrates, v must be enormous because ν_0 is so small.

Let us now examine the molecular significance of the zero point energy. In classical physics, a vibrating system is free *not* to vibrate. However, in quantum mechanics one must conclude that a particle attached by some springlike force to another object continually fluctuates about its equilibrium position—it never stops. This is the reason that the ψ function tells us the probability of the particle being found at a particular place. If the particle could actually be at rest, there would be a probability of 1 of finding it at its position of rest and zero elsewhere. However, in any given region (other than everywhere) the probability is always less than 1 because the particle never stops moving.

The harmonic oscillator is a reasonable model for intramolecular vibration, especially when examining a small functional group attached to a large molecule; in such a case, the small group and the large molecule are analogous to the particle and the immovable wall, respectively. An example is the vibration of a H atom in butane or a C—N bond in an amino acid. With this model, some information can be gained about the structure of the molecule in which the vibration occurs. First we note that since the spacing between the energy levels of a harmonic oscillator is always $h\nu_0$, then a photon of frequency $\nu = \nu_0$ can always be absorbed by a vibrating atom that behaves like a harmonic oscillator. Thus, in the spectra of molecules there should be frequencies absorbed whose values are not much different from one molecule to the next. That is, the frequency of vibration for a C=O double bond should be in the same range in a small amino acid as in a large molecule containing a keto group. Indeed, this is the case, although, as we will see shortly, the frequencies are not precisely the same (which proves to be useful) and there are numerous absorption bands in the infrared region of the spectrum that correspond to intramolecular vibrations. If we remember that

$$\nu_0 = \frac{1}{2\pi}\left(\frac{K}{m}\right)^{1/2},$$

the measurement of the frequency absorbed by a particular compound enables us to calculate the force constant K. This is valuable because K is a measure of bond strength. For example, if, in one compound, the C=O double bond vibrates at a particular frequency ν_1, and in a second compound the frequency is slightly different and has a value ν_2, and if $\nu_2 < \nu_1$, then $K_2 < K_1$, and we conclude that the strength of the C=O bond in the second compound is less than that in the first one. In this case, this difference should suggest that some of the double-bond character has been lost in the second compound; this is therefore a useful bit of information about the distribution of electrons in the two compounds.

Typical values for vibrational energy are 25 kJ or 6 kcal per mole, which is an appreciable amount of energy in chemical terms.

C. The Hydrogen Atom

One of the great triumphs of the Schrödinger theory was the calculation of the spectrum of the hydrogen atom. The Bohr theory required several postulates, as well as the idea of quantum numbers, in order to explain the spectrum. However, in solving the Schrödinger equation for the hydrogen atom, the quantum numbers appear as naturally as in the two preceding examples.

The hydrogen atom is a three-dimensional structure having spherical symmetry, so that the mathematics is vastly simplified by writing the Schrödinger equation in spherical coordinates r, θ, and ϕ, in which r is the radial distance, θ is the angle of the latitude, and ϕ is the angle of the longitude. The potential energy is given by Coulomb's law and is simply $-e^2/r$, in which e is the charge of the electron (it is $-e^2/4\pi\epsilon_0 r$ in SI units). We shall not go through the mathematics required to solve the formidable equation that one encounters when working with spherical coordinates. As in the cases we have just described, the equation has an infinite number of solutions involving a set of three integers or quantum numbers (one for each coordinate) and some arbitrary constants. These constants are evaluated in the usual way, by requiring that $\int \psi^*\psi dr\, d\phi\, d\theta = 1$ and by imposing the boundary conditions that at $r = 0$, $U = -\infty$, and at $r = \infty$, $U = 0$. The three quantum numbers that arise in the solution are the following:

n: the principal quantum number, which ranges from 1 to ∞.

l: the azimuthal quantum number, which has any value from 1 to $n - 1$. The angular momentum of the electron is $\sqrt{l(l+1)}(h/2\pi)$.

m: the magnetic quantum number, which, as in the Bohr model, is the

SEC. 17-2 THE NEW QUANTUM MECHANICS

component of the angular momentum in the direction of a magnetic field; it has the allowed values of $-l$ to $+l$, including zero.

It should be noted that, in the Schrödinger theory, n does not define the angular momentum, as it does in the Bohr theory. Its existence is a natural consequence of the use of the Schrödinger equation as a starting point, whereas, in the Bohr model, n was introduced expressly to define the acceptable values of the angular momentum. In the Schrödinger model, the angular momentum is defined in terms of l as $\sqrt{l(l+1)}(h/2\pi)$, which, because l can be zero, means that in the ground state, an electron has no angular momentum. If such a state were viewed in terms of classical physics, one would conclude that without angular momentum, there can be no motion except toward or directly away from the nucleus, so that the atom would collapse as the electron moved into the nucleus. This means, of course, that the classical view of the electron in orbit is just not the right way to think, and that we must return to the wave function and the probability concept—that is, in the ground state the electron has a certain probability of being in a particular region, and this is defined by ψ. This will become clearer through examination of the wave functions themselves.

Some of the wave functions are given in Table 17-1; they are quite complex. Graphs showing the radial distribution functions—that is, the probability of finding the electron at a particular distance from the nucleus—are more informative and are shown in Figure 17-4. The maximal probability for $n = 1$ corresponds to the radial distance for the Bohr first orbit. For the other sets of quantum numbers correspondence with the Bohr model is not the case, although the values are close. Note that for a particular value of n, the position of the maximum depends upon l, or the angular momentum.

TABLE 17-1 Some wave functions of the hydrogen atom.

n	l	m	Wave function
$n = 1$	$l = 0$	$m = 0$	$\psi = \dfrac{1}{\sqrt{\pi} a_0^{3/2}} e^{-r/a_0}$
$n = 2$	$l = 0$	$m = 0$	$\psi = \dfrac{1}{4\sqrt{2\pi} a_0^{3/2}} \left(2 - \dfrac{r}{a_0}\right) e^{-r/2a_0}$
$n = 2$	$l = 1$	$m = 0$	$\psi = \dfrac{1}{4\sqrt{2\pi} a_0^{3/2}} \dfrac{\cos\theta}{a_0} e^{-r/2a_0}$
$n = 2$	$l = 1$	$m = \pm 1$	$\psi = \dfrac{1}{4\sqrt{2\pi} a_0^{3/2}} \dfrac{r}{a_0} \sin\theta \cos\phi \, e^{-r/2a_0}$

Note: $a_0 = h^2/4\pi^2 me^2$ = the Bohr radius; r is the radial distance; θ and ϕ are the latitude and longitude angles, respectively, in spherical coordinates.

FIGURE 17-4 Some radial distribution functions for the hydrogen atom. These give the probability of finding an electron at a particular distance from the nucleus.

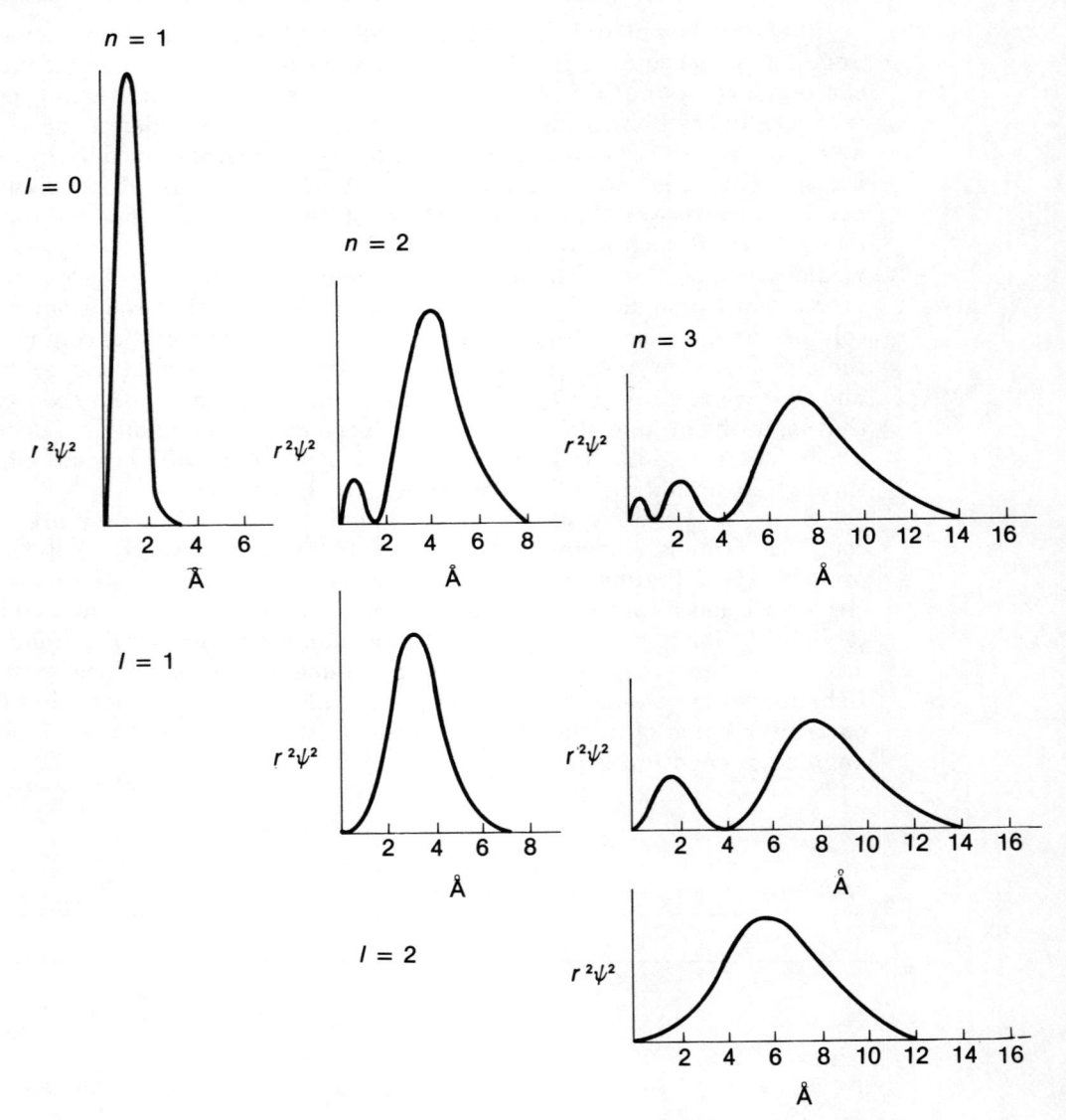

For interpreting spectra, the values of E_n are needed. These are

$$E_n = -\frac{2\mu e^4 \pi^2}{h^2 n^2},\qquad(35)$$

in which $\mu = mm'/(m + m')$, where m and m' are the masses of the electron and the nucleus respectively: μ is called the *reduced mass*. This may be written as

$$E_n = -\frac{\mathcal{R}}{n^2}, \qquad (36)$$

in which $\mathcal{R} = -2\mu e^4 \pi^2/h^2$; $\mathcal{R}$ is the Rydberg constant that appeared in Equation 2. The values of $\mathcal{R}$, expressed in the various units that chemists and biochemists use, are

$$\mathcal{R} = 2.179 \times 10^{-11} \text{ erg molecule}^{-1};$$
$$\mathcal{R} = 2.179 \times 10^{-18} \text{ J molecule}^{-1};$$
$$\mathcal{R} = 13.6 \text{ eV molecule}^{-1};$$
$$\mathcal{R} = 1.312 \times 10^6 \text{ J mol}^{-1}.$$

There is a remarkable feature about Equation 35, which is that the possible energies depend *only* on the quantum number n. Thus, differences in angular momentum (that is, $\sqrt{l(l+1)}(h/2\pi)$) affect the electron distribution function but not the energy. Another way of stating this is to say that there may be more than one state or electron configuration corresponding to a particular energy. When this occurs, the energy level is said to be *degenerate*. We will see that this plays a role in determining the intensity of spectral lines.

An exact solution of the Schrödinger equation has not been obtained for any atom other than the hydrogen atom, because the mathematics becomes too formidable as the number of electrons increase. The He$^+$ ion, which has lost one electron, is of course solvable, because it differs from the hydrogen atom only by the charge and mass of the nucleus.

The alkali metals (for example, lithium, sodium, and potassium) have been treated approximately by assuming that they are hydrogenlike in having a single outer electron. The wave functions for the inner electron have not been obtained, since it is not yet clear how to write the expression for the potential energy. The alkali metals differ from hydrogen in that their energies depend not only on n but also on l. The energy levels for lithium are shown in Figure 17-5. The various l values are denoted by the symbols *s, p, d,* and *f* for $l = 0, 1, 2,$ and 3, respectively, and each energy level is described by stating the n quantum number first and then the letter corresponding to l. Thus, the state for which $n = 2$ and $l = 1$ is called the 2*p* state.

We now turn to the prediction of spectra from the energy levels. The essence of the Planck theory and the Bohr model is that an electron can move from one energy level to another by absorption or emission of a quantum having energy $h\nu$, and this determines the frequency (and of course, the wavelength $\lambda = c/\nu$) of the light. The question is whether a transition can occur between *any* two energy levels. Because of the probabilistic nature of the quantum theory, a better way to phrase this question is: Is the probability of a transition between a *particular* pair of energy levels the same as for *any* pair of energy levels? The answer to this ques-

626   CHAP. 17   QUANTUM MECHANICS AND SPECTROSCOPY

FIGURE 17-5 The energy levels and the allowable transitions of the lithium atom. The numbers and letters next to each energy level tell the n and l quantum number respectively: for example, 3p means $n = 3$ and $l = 1$.

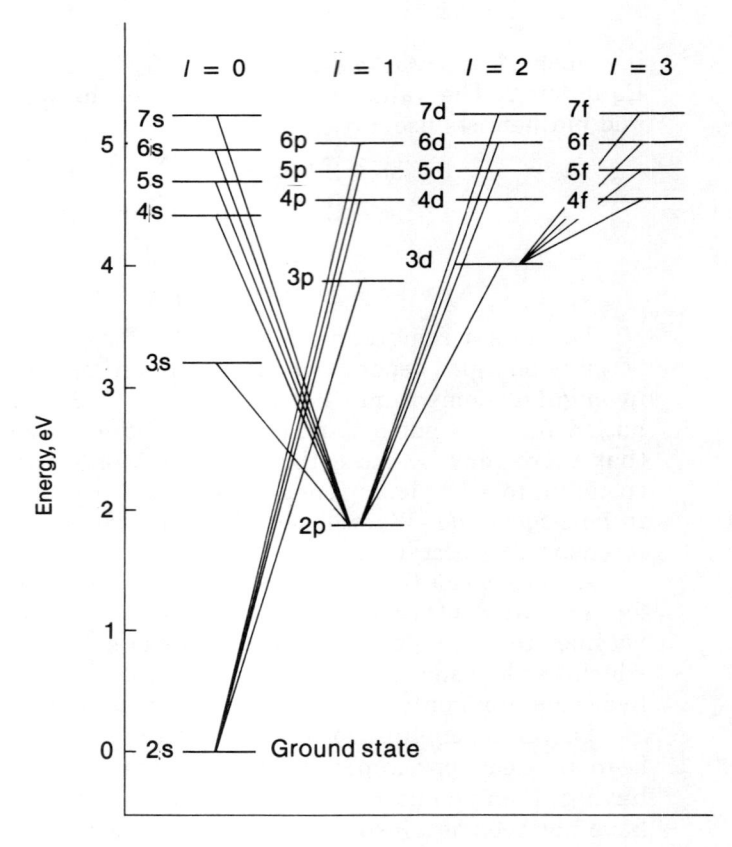

tion is an unequivocal *no*. This can easily be seen when the energy levels are calculated and compared to the observed spectrum. That is, we assume that the frequency of every spectral line must equal $\Delta E/h$, in which ΔE is the difference in energy between two energy levels. Study of the spectrum indicates immediately that there are not nearly enough spectral lines to account for a transition between each pair of energy levels. Furthermore, the frequencies that are observed correspond only to transitions for which $\Delta l = \pm 1$. This type of observation—that is, the presence of far fewer spectral lines than energy level differences—is seen in virtually every system examined by quantum mechanical reasoning. The transitions that occur are called *allowed* transitions and must satisfy particular requirements—for example, $\Delta l = \pm 1$, as in the case just described. Statements such as $\Delta l = \pm 1$ are called *selection rules*. Other selection rules exist for the magnetic quantum number, m. Transitions that do not satisfy the selection rules are called *forbidden*. An extension of the quantum theory allows the selection rules to be calculated. As might be expected, these rules define the highly probable transitions, and the forbidden ones are merely those having very low probability. An example of the allowed transitions of lithium are also shown in Figure 17-5.

Quantum-mechanical Tunneling

Our intuition tells us that if a ball is contained in a sealed box with all intact walls, there is no way for the ball to escape from the box without breaking through a wall. After studying both thermodynamics and quantum mechanics the reader should have recognized that "no way" is a term that should not be part of a scientific vocabulary; one must say instead that it is very improbable that the ball can leave the box and that the value of the probability depends on the potential barrier imposed by the walls. In part A of this section a particle in a box was examined and it was shown that if the walls comprise an infinite potential barrier, the wave function of the particle falls precisely to zero at the wall. However, if the Schrödinger equation is solved for a box bounded by finite (but very large) potential barriers, it will be observed that the wave function penetrates the walls slightly. Since the probability of a particle's being at a particular location is related to the magnitude of the wave function at that point, one must conclude that the particle might be found inside the material of the walls. Once in the wall, the wave function will fall off rapidly toward zero. However, if the walls are very thin, the wave function, which has some nonzero value on the inside edge, will have a nonzero (but smaller) value at the outside edge and will decrease to zero outside the box. This means that with a certain, albeit small, probability, the particle will be found outside the box, even though, according to classical physics, the energy of the particle is insufficient to surmount the potential barrier. This phenomenon—the passage of a particle through a classically forbidden zone—is called *tunneling* or the *tunnel effect*. A quantum mechanical description of the harmonic oscillator also shows the tunnel effect in that, with a finite probability, a particle can be found outside its classical limits of oscillation, where the classical potential energy of the particle is greater than the classical total energy. Tunneling is strongly dependent on the mass of the particle and, hence, is observed primarily for electrons and in certain cases with protons.

Tunneling explains in part the release of a beta particle by a nucleus in radioactive decay (the kinetic energy of the particle is not always sufficient to overcome the high potential barrier imposed by nuclear forces) and the properties of transistors and other solid-state electronic devices. In biochemistry, tunneling effects have been observed in analyses of chemical kinetics. For example, there are instances in which H atoms, H^+ ions, and electrons participate in simple reactions for which the potential barrier is so high that the reaction should be infrequent or, classically, should not occur, and in which the rate constant is 3–5 times higher than expected. Because of the strong influence of mass, the effects usually disappear if deuterium is substituted for hydrogen.

17–3 MOLECULAR SPECTROSCOPY

Molecules

The mathematical determination of the wave functions and energy levels of a many-electron atom surpasses present-day computer capabilities. The reader should therefore not be surprised to learn that such determinations for polyatomic molecules are all the more inaccessible. Various approximation methods have been used which have been more or less successful in that molecular spectra have sometimes been used to differentiate between two possible structures that an organic compound might have. As a trivial example, consider a compound of unknown structure having the formula C_6H_6O. Two possible structures are

$H_2C=CH-CH=CH-C\equiv C-OH$ and   [benzene ring with OH substituent]

Quantum theory is not sufficiently advanced to be able to calculate the wave functions and energy levels precisely, but the approximations are adequate to predict that the principal ultraviolet absorption bands of the second compound would be *about* 280 nm compared to less than 200 nm for the first.* Thus even a crude measurement of the spectrum should enable one to distinguish the two possible structures.

The calculation of spectra is not really of widespread use at the present time, because there exist tabulated spectra for so many compounds of known structure that structural determinations are usually made instead

---

*In general, ring compounds absorb at longer wavelengths in the ultraviolet-to-visible region of the spectrum than do similar but linear molecules. Also, for ring compounds having a similar number of atoms in their ring sections, a heterocyclic compound absorbs at longer wavelengths than a homocyclic compound. Substituents in either a ring or a linear molecule also tend to increase the wavelengths absorbed beyond what is found in the unsubstituted molecule.

by noting similarities between the spectrum of an unknown substance and that of a known molecule. This is certainly the point of view taken in biochemistry: there are well-known tabulations of spectra of the major cellular components, such as nucleic acids and proteins, and of many of the small molecules, such as amino acids, nucleotides, vitamins, and other cofactors, and metabolic intermediates. In a later section we will give some examples of spectral identification of various intracellular components.

Molecules possess several features that we have not yet discussed and that are understandable quantitatively only in terms of quantum mechanics. These are the vibration of the atoms of a molecule with respect to one another and rotation of the entire molecule. Vibrational and rotational energies (which are produced in conjunction with each other), are also quantized, and these energies give rise to a class of spectral lines that is of great value in the determination of structure. This is discussed in the following section.

Molecular Vibration and Rotation

Consider the molecule acetone

$$\begin{array}{c} \text{H} \quad \text{O} \quad \text{H} \\ | \quad \parallel \quad | \\ \text{H}-\text{C}-\text{C}-\text{C}-\text{H} \\ | \quad \quad | \\ \text{H} \quad \quad \text{H} \end{array}$$

If the chemical bonds are thought of as springs, the molecule will have three types of springs—those in the C—H bonds, in the C—C bonds, and in the C=O bond. Each of these springs can vibrate, and the simplest vibration is along the axis of the chemical bond; for example, a hydrogen atom moves toward and away from a carbon atom. This is like the harmonic oscillator that we analyzed by using quantum theory; the particular type of vibration along the axis is called *stretching*. We saw in the case of the harmonic oscillator that the separation of the energy levels is proportional to $(K/m)^{1/2}$ in which K is the force constant of the spring and m is the mass of the particle. In the case of two atoms having mass m_1 and m_2 that are vibrating with respect to one another, m must be replaced by the reduced mass, which equals $\mu = m_1 m_2/(m_1 + m_2)$. Thus, a transition from one vibrational level to another can occur by absorption of a photon whose frequency is proportional to $(K/\mu)^{1/2}$. Since the spring for each bond will not have the same strength (for example, a double bond is stronger than a single bond) and because bonds occur between atoms having many different masses, there will be characteristic vibration frequencies for each type of bond.

A second type of interatomic motion is called *bending*. For example, the O in acetone could vibrate along an axis parallel to the C—C—C chain. This would be called a *carbonyl bend:* the C—C and C—H bonds could also bend—for instance, a terminal C atom could move perpendicularly to the C—C—C axis. These motions are called bends because the bond angle changes as the bond bends with respect to its equilibrium direction. This type of vibration is, of course, also quantized, and gives rise to characteristic frequencies.

The characteristic vibrational frequency for a particular type of bond differs slightly from one molecule to the next. For example, the C=O bond does not vibrate at the same frequency in acetone as it does in methylethylketone,

$$CH_3-\overset{\overset{\displaystyle O}{\|}}{C}-CH_2CH_3 ,$$

because, in a vibrating system, two masses must move with respect to one another. Thus, in acetone, the oxygen is really vibrating with respect to the entire carbon chain. This has two effects. First, the molecular weight of the carbon chain of acetone is 42, whereas for methylethylketone it is 56. The fact that both masses are involved is expressed by the reduced mass μ, which appears in the expression for the frequency of vibration as $\mu^{-1/2}$. The values of $\mu^{-1/2}$ for the C=O vibration in acetone and methylethylketone are approximately 0.30 and 0.29, respectively, so that, based on this consideration alone, the frequency at which the C=O stretches would be expected to be about 3 percent lower in methylethylketone than in acetone. The second effect is that the C—C chains are not rigid rods but are flexible, and this flexibility produces some damping of the vibration. Since the two chains do not have the same length, the magnitude of the damping of each differs, so that the reduction of the frequency at which C=O vibrates is not the same. The important point is not that the frequencies are not the same but that the difference in frequency is actually quite small compared to the difference in vibrational frequencies of various bond types (for example, C=O and C—H), so that there is usually not much difficulty in identifying the bond that produces a particular vibrational frequency.

At ordinary temperatures, most molecules possess only ground state, zero point energy. This is because apart from the absorption of light, the only way a molecule has of gaining energy is by a collision with a second molecule. Since thermal energy is $\frac{1}{2}kT$ or 2×10^{-14} erg at room temperature and the spacing between the vibrational levels is $h\nu$ or, for a typical vibrational frequency, 4×10^{-13} erg, collisional energy is clearly insufficient to raise a molecule to a higher vibrational state. Of course, because of the range of translational energies of individual molecules (they are Boltzmann-distributed), a transition to a higher state will sometimes happen.

The acetone molecule can also rotate; for example, it could rotate about the C—C—C axis and about the C=O axis. Rotation is also quantized, in that only particular energy levels are allowed. This quantization depends upon certain properties of the molecule and can be understood most easily for a linear diatomic molecule. If the axis of the molecule is the z axis, there are two modes of rotation—about the x axis and about the y axis. For either mode, the moment of inertia I is

$$I = \sum m_i r_i^2 , \tag{37}$$

in which m_i is the mass of the ith atom and r_i is the perpendicular distance of that atom from the axis of rotation. Solution of the Schrödinger equation shows that the angular momentum M of the molecule can have the values defined by $M^2 = J(J + 1)h^2/4\pi^2$, in which J, the rotational quantum number, can have the values 0, 1, 2, ... The energy of a rotating system is $E = (1/2)(M^2/I)$, so that

$$E_J = J(J + 1)\frac{h^2}{8\pi^2 I} . \tag{38}$$

The separation between adjacent energy levels is

$$E_{J+1} - E_J = (J + 1)\frac{h^2}{4\pi^2 I} . \tag{39}$$

If a transition between a rotational level having the quantum number $J + 1$ and one having the number J could be accompanied by absorption or emission of a photon of frequency ν, then

$$\nu = \frac{Jh}{4\pi^2 I} . \tag{40}$$

Since J can have many values, there will be a set of frequencies. Furthermore, because $\Delta J = +1$, the difference $\Delta \nu = \nu_{J+1} - \nu_J$ between these frequencies is

$$\Delta \nu = \frac{h}{4\pi^2 I} . \tag{41}$$

Note that this difference is a constant, in contrast with the variable separations with the hydrogen atom. Thus, a measurement of this constant spacing of the spectral lines (which occur in the microwave region of the spectrum) yields the value of I and, from this value, the interatomic distance of a diatomic molecule. To perform this calculation, we need to relate the interatomic distance r to the values of r_i in Equation 37. This can be done as follows.

Consider two atoms of mass m_1 and m_2 at distance r_1 and r_2 from their center of rotation. The center of gravity is the center of rotation and is defined by

$$m_1 r_1 = m_2 r_2.$$

The interatomic distance, r, is $r_1 + r_2$, so that $r_2(m_2/m_1) + r_2 = r$, and $r_1(m_1/m_2) + r_1 = r$, or

$$r_1 = \left(\frac{m_2}{m_1 + m_2}\right)r \quad \text{and} \quad r_2 = \left(\frac{m_1}{m_1 + m_2}\right)r .$$

Substituting into Equation 37 yields

$$I = \left(\frac{m_1 m_2}{m_1 + m_2}\right) r^2 . \tag{42}$$

Thus, for carbon monoxide, for which $\Delta\nu$ is 1.15×10^{11} sec^{-1}, Equation 41 yields $I = 1.46 \times 10^{-39}$ g cm^2. Since $m_C = 1.92 \times 10^{-23}$ g and $m_O = 2.56 \times 10^{-23}$ g, one obtains $r = 1.15 \times 10^{-8}$ cm.

We will see shortly that measurement of I and of r can also be carried out in the infrared region of the spectrum from analysis of rotational frequencies that are superimposed on a vibrational transition.

There is an important difference in the transitions between electronic energy levels and those between vibrational and rotational energy levels. That is, in a change in an electronic energy level that is allowed by the selection rules, a photon can always be emitted or absorbed. This does not mean that emission or absorption necessarily occur, because excitation energy can also be removed as heat; rather, it means that every electronic transition can *potentially* be accompanied by absorption or emission of light. This is not the case for all vibrational and rotational energy changes—many of these changes are not accompanied by photon emission or absorption but are instead a result of intermolecular (and sometimes intramolecular) collisions. To understand why this is so, we must remember that *electromagnetic radiation is emitted only when a dipole oscillates, and conversely, radiation can be absorbed only if a dipole can be made to oscillate*. The quantum theory we have examined so far has only defined *how much* energy must be involved in a single transition—it says nothing about the *form* of the energy. In short, only molecules having a permanent dipole moment will emit light as a result of a vibrational or rotational transition. Thus, a homonuclear diatomic molecule, such as O_2 or Cl_2, has no dipole moment and does not produce a vibrational spectrum, whereas a heteronuclear molecule, such as HCl does. In triatomic molecules, a nonlinear molecule such as water usually has a vibrational spectrum, whereas a linear triatomic molecule such as CO_2, of structure O=C=O, has a vibrational spectrum for some, but not all, modes of vibration. For instance, as shown in Figure 17-6, in vibrational mode I, the dipole moment remains zero, whereas in modes II and III, the vibration causes a charge separation and hence a fluctuating dipole moment. Thus, only modes II and III produce a vibrational spectrum.

FIGURE 17-6 Three modes of vibration of CO_2, a linear triatomic molecule. The solid circle represents a carbon atom; the open circles represent oxygen.

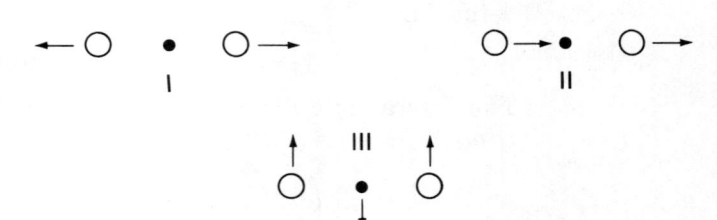

SEC. 17-3   MOLECULAR SPECTROSCOPY

FIGURE 17-7 Vibration-rotation spectrum of HCl. The frequency difference between all adjacent peaks, except for the central pairs is $\Delta \nu$. For the central pair, it is $2(\Delta \nu)$ because there is no peak where frequency is ν_0.

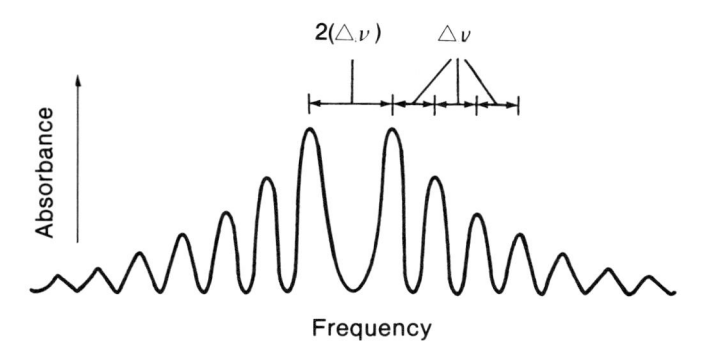

Molecules do not have pure vibrational spectra, because the selection rules for vibrational transitions require that a change in vibrational energy must be accompanied by a change in rotational energy. As a result, the vibrational spectrum consists of sets of closely spaced lines that result from superposition of the rotational levels on the vibrational levels. These are called *vibrational bands*. An example of such a band is shown in Figure 17-7. Bands of this type can be analyzed in a straightforward way for a diatomic molecule by assuming that the sum of the vibrational and rotational energy, E_{vr}, is

$$E_{vr} = \left(v + \frac{1}{2}\right)h\nu_0 + J(J+1)\frac{h^2}{8\pi^2 I}. \tag{43}$$

The selection rule for a vibrational change is $\Delta v = \pm 1$, so that the frequency ν corresponding to a shift from energy E_{vr} to E'_{vr} (for another vibrational and rotational energy level) is

$$\nu = \nu_0 + [J'(J'+1) - J(J+1)]\frac{h}{8\pi^2 I}. \tag{44}$$

The selection rule for a rotational change is $\Delta J = J' - J = \pm 1$; this produces two sets of frequencies, designated ν_R and ν_P. These are

$$\left. \begin{array}{ll} \nu_R = \nu_0 + 2(J+1)\dfrac{h}{8\pi^2 I}, & \text{if } J' = J+1 \text{ and } J = 0, 1, 2, \ldots \\[6pt] \nu_P = \nu_0 - 2J\dfrac{h}{8\pi^2 I}, & \text{if } J' = J-1 \text{ and } J = 1, 2, 3, \ldots \end{array} \right\} \tag{45}$$

These two equations define the leftward and rightward families of peaks shown in Figure 17-7. The value of J can never be zero, so that the equations simplify to

$$\left. \begin{array}{ll} \nu_R = \nu_0 + 2J\dfrac{h}{8\pi^2 I}, & J = 1, 2, 3, \ldots \\[6pt] \nu_P = \nu_0 - 2J\dfrac{h}{8\pi^2 I}, & J = 1, 2, 3, \ldots \end{array} \right\} \tag{46}$$

Note that because $J \neq 0$ the ν_0 never appears in the vibrational band (that is, there is no peak on the line of symmetry of the spectrum in the figure) and the spacing between the peaks is $h/4\pi^2 I$. Thus we could calculate the interatomic spacing r of HCl gas from the measured value of $\Delta \nu$, since

$$\Delta \nu = \frac{h}{4\pi^2 I}$$

and

$$I = \frac{m_H m_{Cl} r^2}{m_H + m_{Cl}},$$

so that

$$r = \left[\frac{h(m_H + m_{Cl})}{4\pi^2 m_H m_{Cl} \Delta \nu} \right]^{1/2}$$

In more complex molecules, the rotational levels are so near one another that they are not resolved and the vibrational spectral lines appear as very broad bands.

The photons that are absorbed or emitted by vibrating molecules have energies in the infrared region of the spectrum and have wavelengths greater than about 800 nm. Transitions arising from pure rotational changes yield wavelengths in the far infrared and microwave regions of the spectrum, which are difficult regions to study for technical reasons. Thus, pure rotational spectroscopy is rarely used in biochemical studies, and it will not be discussed further. For historical reasons, infrared spectral measurements are usually reported in terms of a wave number, which is denoted by the symbols k or $\bar{\nu}$ and is given a scale of cm^{-1}. The wave number is defined as

$$\bar{\nu} = \frac{1}{\lambda} = \frac{\nu}{c}. \tag{47}$$

We will say more about the use of these infrared spectra in a later section.

The Effect of Vibration on Electronic Energy Levels

Just as rotational energy levels are superimposed on vibrational energy levels, the vibrational levels are superimposed on electronic energy levels. That is, the total energy of a molecule in a particular state is the sum of the electronic and vibrational energies. This has an important effect on electronic spectra (that is, visible and ultraviolet spectra) because the vibrational levels need not be the same before and after a transition between two electronic levels. Thus the energy change and hence the frequency of the transition will have a spectrum of values depending upon the particular vibrational states. In considering vibrational spectra, we noted that the selection rule $\Delta v = \pm 1$ limits the number of transitions. However, in an electronic transition, Δv can have many values. Thus, in observing absorption of photons in the visible and ultraviolet ranges of the spectrum, we find that transitions are possible between so many different vibrational states that each electronic spectral line consists of a large number of closely spaced lines. Together these comprise what is called an *electronic band*. In molecules more complex than simple diatomic molecules, the lines are usually so close that they are not resolved and appear instead as a broad absorption band (see Figure 17-8, which compares the spectrum of a molecule with the line spectrum of an atom). *This is the most important cause of the width of absorption bands in the visible and ultraviolet regions of spectra produced by molecules.*

FIGURE 17-8 A comparison of the narrow line spectrum of atomic hydrogen (solid lines) with the broad spectrum of the visual pigment rhodopsin (dashed line).

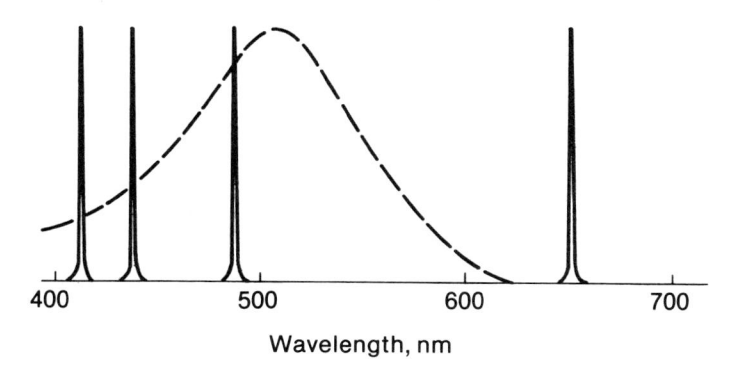

The superposition of vibrational energies on electronic levels is most conveniently described by a potential energy diagram. In a potential energy diagram we make use of the fact, which quantum mechanics tells us is so, that electrons are not localized; instead, we speak only of the probability of their being at a particular distance from the nucleus, a probability that is determined by the wave function. We may think of the potential energy an electron would have at a particular distance from the nucleus. This potential energy is minimal at the most probable location and increases as the electron both approaches and departs from the nucleus. If we now consider a diatomic molecule, we can talk about the potential energy as a function of the internuclear distance. Once again there will be a minimal potential energy at some equilibrium position and the potential energy will increase as the atoms approach one another because, in classical terms, the atoms cannot "overlap." Similarly, the potential energy would be greater as the distance increases. In the case of a diatomic molecule, it is meaningful to think about these changes because the individual atoms are vibrating and thus the potential energy is continually changing. A typical potential energy diagram for two electronic states of a diatomic molecule is shown in Figure 17-9(a). Curves 1 and 2 represent the ground

FIGURE 17-9 (a) A typical potential energy diagram showing two electronic levels (1 and 2) and many vibrational levels. (See text for details.) (b)(c) Two illustrations of the Franck-Condon principle. (See text for details.)

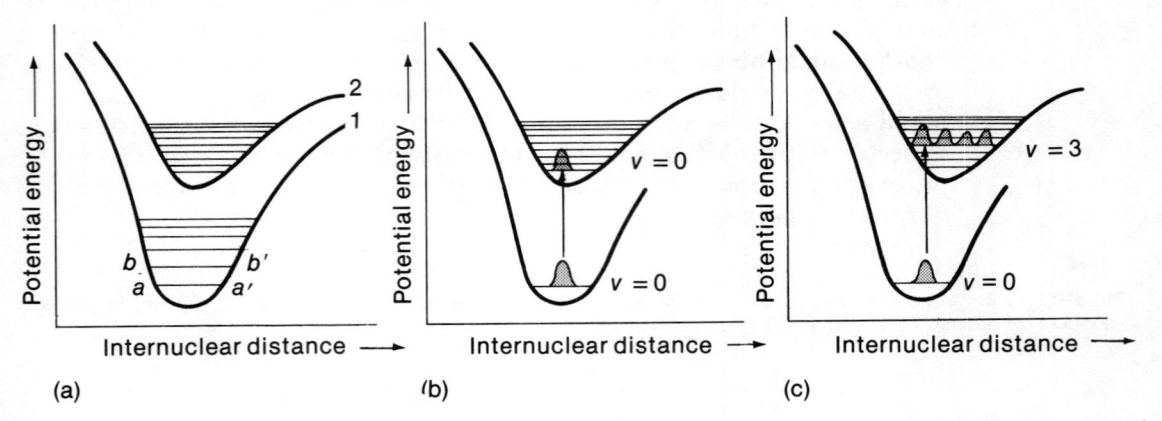

state and the first excited state, respectively. The horizontal lines represent the vibrational states and thus the *total energy* (not the potential energy). Thus in the electronic ground state the line aa' represents the energy of a molecule whose vibrational quantum number v is 0, and the line bb' corresponds to $v = 1$. Note that the total energy (kinetic + potential) is constant for any particular vibrational state as long as the internuclear distance stays within the bounds of the potential energy curve.

The vibrational structure of an electronic spectrum is accounted for in terms of the *Franck-Condon principle*. This states the following:

when electrons redistribute during absorption of energy, the nuclei, which are much more massive than electrons and hence move more sluggishly, tend to retain their initial internuclear distance.

What does this mean? When energy is absorbed and a higher electronic state is reached, the electrons move to a new distribution. If more energy is absorbed than is required to achieve this distribution, the excess energy is used to raise the molecule to a higher vibrational state. A nucleus may have many different maximally probable positions relative to other nuclei in different vibrational states. Since electronic excitation occurs very rapidly, and the more massive nucleus responds much more slowly, a massive nucleus is more likely to reach a vibrational state in which it need not move than to reach a vibrational state in which it must move. In quantum mechanical terms, this means that since the nucleus is, in general, in a position of high probability in the ground state, it should be in a position of high probability in the upper states also. The significance of this statement can be appreciated by examining panels (b) and (c) of Figure 17-9, in which the probability distribution of the nucleus is superimposed on the energy level diagrams. The maximum of this distribution is the most likely position of the nucleus. In both cases, we consider as an example the transition from the $v = 0$ state of the ground state electronic level. This is of course the most common transition, since most molecules are in this state. In the molecule shown in panel (b), in which the first electronic excited state is directly above the ground state (that is, internuclear distance is unchanged in the transition), the most probable transition is to the $v = 0$ level of the first excited state, because this is the first level that is reached that does not require that the nucleus move. However, in panel (c), in which the electronic distribution of the first electronic state is displaced rightward from the ground state (representing a changed internuclear distance), the maxima of the nuclear distributions coincide at the $v = 3$ energy level, as the straight vertical line shows. The most probable change in energy is from $v = 0$ to $v = 3$. Similar considerations apply to transitions initiating at other vibrational levels. This argument does not mean to imply that no other changes in energy are possible. In fact, the probability of a particular shift depends on the degree of overlap of the nuclear distribution functions for the two vibrational states. Because transitions are conveniently discussed in terms of potential energy diagrams such as those in Figure 17-9, the transitions are said to occur vertically. The important point is that whereas the selection rule for a vibrational transition within a single electronic energy level is $\Delta v = 1$, *when there is an electronic energy level change, this selection rule is superceded by the Franck-Condon principle.* If we remember that each vibrational level is really broken down into a collection of rotational levels, it is clear then that any electronic transition can occur with a large number of energy changes, and, as we have said before, the light absorbed or emitted covers a band of frequencies. Since the spectral lines will not usually be resolved, we can easily understand the great spectral width observed in absorption spectra of molecules.

It is important to note that whereas pure vibrational energy changes are not always associated with a spectral line (because these changes may not be accompanied by a change in dipole moment), all vertical transitions accompanying an electronic change are, because the electronic transition is always accompanied by a change in dipole moment. This means that for a molecule that does not have a pure vibrational spectrum, all of the information obtainable from a vibrational spectrum (for instance, determination of the bond strength by use of the force constant) can be calculated from the electronic spectrum of a molecule whose structure is sufficiently simple that the vibrational fine structure can be resolved.

Raman Spectroscopy

We have implied that a fundamental rule of quantum theory is that energy can be absorbed only if it precisely equals the difference between two energy levels. "Only" is a word that is inconsistent with quantum mechanical thinking; we will do much better to replace this word with the phrase "with high probability." In this section we describe a phenomenon that occurs with low probability but is nonetheless observable with appropriate instrumentation. Since it has low probability, the spectral lines we will describe have very low intensity.

Consider a population of molecules illuminated with monochromatic light of a frequency such that it is not absorbed—that is, $h\nu$ is not equivalent to the difference between any energy levels. The molecules are transparent to that frequency, so that the light passes right through the sample. Some of the incident light will be scattered, and this can be observed at an angle of 90° from the direction of the incident beam (see also Chapter 18). The frequency of the scattered light will be *predominantly* the same as that of the incident light—that is, usually no energy is lost. (This is called elastic scattering.) However, with very low probability, a small amount of energy of the incident photon can be absorbed by the molecules, which will be raised to higher vibrational states. The amount of energy the photon loses is exactly equal to the energy of a transition from one vibrational state to another, so that a very small fraction of the scattered light has a frequency that is decreased by the amount of the frequency associated with the vibrational state. Similarly, with equally low probability, if the energy of the molecule is not in the ground state, vibrational energy can be transferred from the molecule to the incident photon to raise the frequency of the light scattered by the molecule. Thus, at right angles to the incident beam there is observed a collection of faint spectral lines whose frequencies are determined by vibrational transitions. These lines constitute the *Raman spectrum*. The Raman method is an extremely valuable technique, because the Raman spectrum contains all the information contained in a vibrational spectrum yet *it is obtained at any convenient wavelength*. Furthermore, vibrational lines can be found that are not present in an infrared spectrum of a molecule for which the vibration is not accompanied by a change in the dipole moment—for example, homonuclear, diatomic molecules, such as O_2, produce Raman spectra but not an infrared spectrum. Second, transitions can be observed that correspond to frequencies that cannot be seen by infrared spectroscopy in aqueous solution because of the very strong absorption of certain infrared frequencies by water.

The difference between the frequency of the main Raman band (whose frequency is that of the incident light) and the frequencies of the other lines, is characteristic of particular groups in molecules (for example, C=O and CH groups), as we saw in infrared spectroscopy. The intensity of these lines is so faint that they are usually detectable only if high-intensity light sources, such as lasers, are used. An example of a Raman spectrum is shown in Figure 17-10.

FIGURE 17-10 (a) The Raman phenomenon. The center peak is the band of the incident light. Equally spaced on both sides are lesser peaks known as a Raman pair. The peak at the left with the lower frequency, which corresponds to a loss of energy, is stronger than the one at the right, which corresponds to an increase in energy, because energy can be lost to any molecule in the environment but can only be gained from the small fraction of the molecules that are above the ground state. (b) A portion of the Raman spectrum for DNA. The scale on the abscissa shows the *difference* between the frequency of the incident beam and the frequencies of the scattered beam. The Raman spectrum contains many more distinguishable lines than the infrared spectrum shown in panel (c). (c) A portion of the infrared spectrum for DNA; the frequencies on the *x* axis of (b) and (c) are identical but in (c) they are the actual frequencies of the incident light.

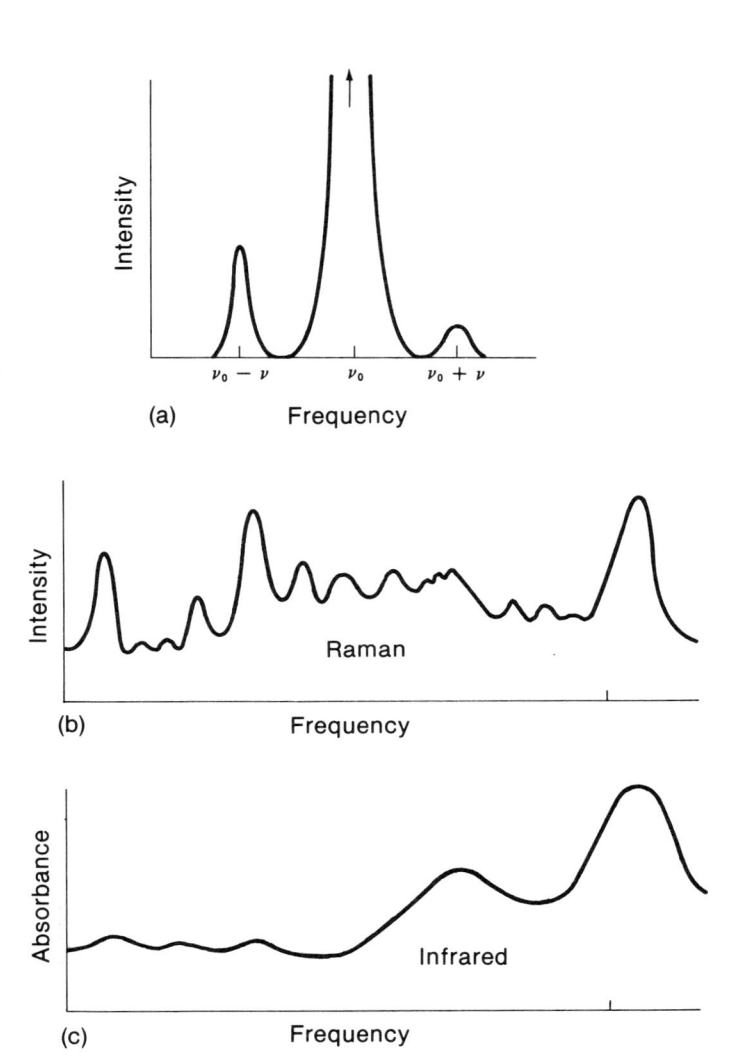

17–4 FLUORESCENCE

Properties of Fluorescence Spectra

Let us now consider a molecule that has absorbed a photon and been raised to a higher energy level, and ask what happens to this energy. The molecule is not in its most stable state and will attempt to return to that state, which it can accomplish only by discharging its excitation energy. This can occur either by loss of energy as heat, by chemical changes, or by reemission of a photon.

In an excited state, as in any other state, molecules are subject to numerous collisions. In a solution, for example, an excited solute molecule collides very frequently with solvent molecules. When an excited molecule collides with another molecule (assume that the second molecule is in a ground state, although this is not a necessary assumption), it can transfer some of its energy to the kinetic and vibrational energies of the second molecule. Its energy will be transferred in quanta, and the collisions will move the molecule down the ladder of vibrational energy by stepwise removal of energy so that this energy is converted to heat.

The result is that a molecule that has been excited to an upper vibrational level in the first excited state will very rapidly reach the lowest vibrational level of that state. At that point, several things can happen to the remaining electronic excitation energy. If upper vibrational levels of the electronic ground state are near the $v = 0$ level of the excited state, weak collisions can again remove energy stepwise. Strong collisions can, of course, reduce the molecule to the ground state in a single step. However, if the upper vibrational levels of the ground state are far from the bottom of the electronic excited state, *only* strong collisions can remove the energy. In the absence of such strong collisions, the molecule may remain excited long enough to emit a photon. This emitted light is called *fluorescence*. The emitted photons have a range of energies, because a photon may represent a transition from the $v = 0$ level of the excited state to a $v > 0$ level of the ground state. Figure 17-11(a) shows this process schematically. The lengths of the arrows in the figure are proportional to the energy change; it is clear that the range of energy of the fluorescent photons is less than that of the absorbed photons. That is, *the wavelengths of the fluorescence band are longer than those of the absorption band*. Figure 17-11(b) shows the absorption and fluorescence spectra corresponding to the diagram of panel (a). It should be noticed that as long as the vibrational structure is similar in the ground state and the excited state, the absorption band for transitions to the first excited state and the fluorescence band will have mirror image symmetry. Furthermore, the wavelengths of the fluorescence are independent of the wavelengths absorbed. The argument just given leads to the conclusion that the absorption and fluorescence spectra should not overlap and should be symmetric about the wavelength corresponding to the transition between the lowest vibrational levels. However, when spectra are obtained of molecules in solution, a small overlap is often observed. This has the following explanation. In solution, a solute molecule is not in isolation but is surrounded by solvent molecules. The energy levels of the solute molecule are determined both by the energy levels it would have in vacuum and its energy of interaction with the solvent molecules. When a solute molecule absorbs light and becomes excited, its electron distribution changes; shortly afterwards, the solvent molecules become reoriented so that their free energy is minimized. The vibrational energy levels of the excited molecules of solute among reoriented solvent molecules need not be the same as those reached by the solute molecules immediately after exci-

SEC. 17-4 FLUORESCENCE

FIGURE 17-11 (a) Energy level diagram showing absorption of light (solid upward arrows) from the ground state G to the first excited state E_1. The vertical wavy lines represent vibrational nonradiative losses. The dashed downward arrows represent fluorescence transitions. (b) Absorption (solid line) and fluorescence (dashed line) spectra of the transitions shown in panel (a).

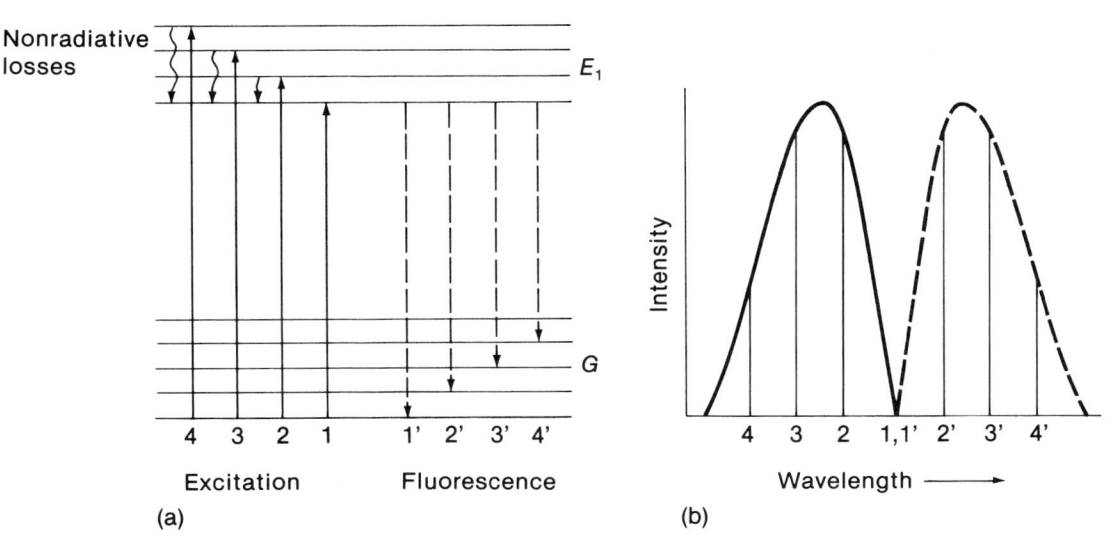

tation but before reorientation of the solvent molecules. If the vibrational energy levels happen to be far from the ground state, the vibrational energy will be slightly greater than the energy required for excitation between the lowest vibrational states, so that there will be a small overlap of the spectra. If they happen to be still further from the ground state, there will be a gap between the absorption and fluorescence bands.

It is sometimes found that the absorption spectrum is much broader than the fluorescence spectrum. This is because the absorption spectrum consists of several bands—for example, one from the ground state to the first excited state and another from the ground state to the second excited state. These bands may be distinct but more commonly they overlap, yielding a very broad band. As we have stated, fluorescence consists only of emission from the lowest vibrational level of the *first* excited state; the mirror image symmetry that has been described refers only to the first absorption band. This phenomenon is shown in Figure 17-12.

The intensity of fluorescence is determined by the probability with which deexcitation of a molecule by nonradiative means can be avoided until emission of a photon has occurred. Thus the intensity of the fluorescence does not depend only on the energy levels of the molecule that has been excited (the *fluor*), but also on the extent to which other nearby molecules (for example, a solvent or other solute molecules, if the fluor is in solution) can convert excitation energy to heat. Molecules with widely spaced vibrational levels (of which water is an important example) are

FIGURE 17-12 A common relation between the fluorescence spectrum (F) and the observed absorption spectrum (A). I and II represent the absorption bands for excitation to the first and second excited states, respectively. Curve A is the sum of the curves I and II. Curve F shows mirror image symmetry only with curve I.

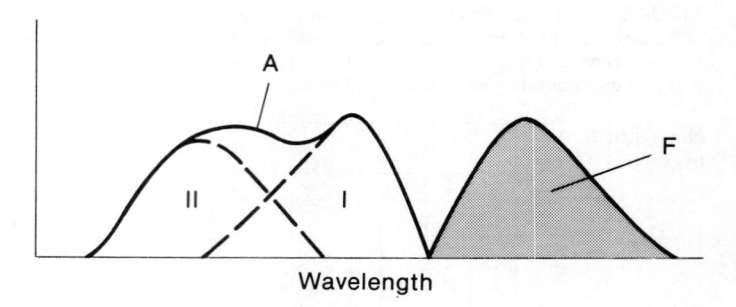

extremely efficient at removing excitation energy, and so, these diminish the intensity of fluorescence. A molecule that decreases the intensity of fluorescence of a molecule compared to that in the pure state is called a *quencher* of fluorescence. The iodide ion is a powerful quencher that is of great use in biochemical research, as will be seen in a later section. The O_2 molecule is also an efficient quenching agent.

The intensity of fluorescence is usually described in terms of the quantum yield Q, which is defined as

$$Q = \frac{\text{number of photons emitted}}{\text{number of photons absorbed}}. \tag{48}$$

Note that Q is not just an intensity ratio, because intensity is a measure of the amount of energy, whereas the number of photons carrying a particular amount of energy depends upon the frequency of the light. Thus, the intensity ratio is always less than 1 but, in a perfectly efficient system, Q can be 1. There are some, but not many, compounds for which $Q = 1$. The measurement of Q is very difficult, because fluorescence is emitted over a wide range of frequencies (even if only a single excitation frequency is used), so that it is necessary to know the intensity or energy emitted at *each* frequency in order to calculate the number of photons emitted. In most biochemical analyses the intensity of fluorescence—that is, the total energy produced by the whole fluorescent spectrum—is usually measured, because its value is sufficient for the kind of information that is desired in fluorescence spectroscopy. We will see some examples of the use of fluorescence in a later section.

When a molecule is excited, its excitation energy is lost very rapidly by the combined effect of collisions, photochemical changes, and fluorescence. The *fluorescent lifetime* is the time it takes for the intensity of fluorescence to drop to $1/e$ of an initial value. This can be measured by exciting a molecule with a nanosecond light pulse, and typical values are 3 to 20 nanoseconds. Thus, in an ordinary experiment, when the exciting light is turned off, fluorescence vanishes immediately. However, with some molecules, a long-lived emission of light called *phosphorescence* occurs. This has the following explanation. Sometimes an excited molecule will undergo a subtle rearrangement in which an electron pair in a particular chemical

SEC. 17-4 FLUORESCENCE

bond (a paired electron is said to be in a *singlet state*) is converted to two unpaired electrons (which are in a *triplet state*). This is not a very frequent occurrence because the selection rules for electronic transitions indicate that all singlet-to-triplet transitions are forbidden; *however, this only means that they occur with low probability*. Also, just as the triplet state is reached with low probability, the return from a triplet state to the ground state, which is a singlet state, is also an infrequent event. However, if all of the energy is not lost by collision, then when the transition occurs, the excess energy is emitted as a photon. A low probability of a transition merely means that, on the average, it takes a long time for it to occur, so that the half-life for the emission is not 10^{-8} seconds, as in the case of fluorescence, but much longer. Commonly encountered half-lives for phosphorescence are in the range of milliseconds to minutes. The conditions necessary for phosphorescence to occur are the following. The molecule must have a triplet state attainable by a small energy loss from a singlet state of *higher* energy, such as might occur in a collision. The vibrational levels of the molecule must not be too close—otherwise all of the energy will be lost by collisional losses. The vibrational energy levels of the triplet state must also be widely separated, so that its energy is not immediately degraded to heat. Of course, some collisional losses always do occur, so that the phosphorescent spectrum always has longer wavelengths than the exciting light, as is also true of fluorescence.

Phosphorescence is not often encountered in biochemical research. When it is, it is frequently the cause of experimental difficulties. For example, in the technique of liquid scintillation counting of radioisotopes, a radioactive decay excites fluorescent molecules that, as a consequence, emit photons that are counted by a suitable detector. The experimental arrangement is such that photons are produced only as a result of the radioactive decay, so that an optical measurement enables one to count the number of decays. Occasionally, trace quantities of materials (some acids, for example) are introduced into the experimental sample; chemically, and without radioactive decay, these trace materials cause the fluor to phosphoresce; since the samples are usually prepared in the presence of room light before being placed into the darkness of the liquid scintillation counter, phosphorescence is produced by this extraneous light source. This phosphorescence is independent of the amount of radioactive material, yet the detector counts this as if it *were* a result of radioactive decay. Phosphorescence is easily recognized by its lifetime, though. For example, suppose the phosphorescence has a half-life of one second and produces enough light to be counted as equivalent to 10^6 radioactive decays per second at the initiation of the counting period. In the first second of counting, it will contribute 5×10^5 decays, in the second 2.5×10^5, and so forth. Clearly, after 60 seconds, the total contribution will be about 10^6 decays, but the amount per second will be nearly zero. If the sample also contains radioactivity equal to 10^4 decays per minute, then a count of total decays in the time interval 0–1 minutes will yield about 10^6 decays of phosphorescence as well as of radioactivity. However, if it were counted in the interval 1–2 minutes, at the end of this time the phosphorescence would contribute nothing, and only the 10^4 true radioactive decays would be counted. Thus, *phosphorescence can be detected when it is noticed that a decay rate has decreased in two successive counting intervals.* To help insure that the observed rate is the true radioactive decay rate, one should count the sample in successive time intervals until the count rate has become constant.

Scintillation counting is an example of the production of fluorescence from excitation energy that is not a result of absorption of photons. When a rapidly moving atom or particle (for instance, an electron) passes through matter, a large amount of its kinetic energy can be released as excitation energy, as explained in Chapter 16. This energy can also appear as fluorescence. Some molecules in the solid state are especially efficient at fluorescing as a result of this type of excitation. Zinc sulfide is an example of such a substance, and it is used in electron microscopy to make the electron image visible. The electron beam passes through the sample being observed and makes a highly magnified image at some point. The electrons themselves are not visible, but if they impinge on a layer of zinc sulfide deposited evenly on a broad surface, causing the zinc sulfide to fluoresce, a fluorescent image of the sample appears that is easily visible to the observer. Another example is *bioluminescence,* as in the emission of light by fireflies, in which the energy generated in a chemical reaction is released as light.

Fluorescence and Energy Transfer

We have stated that quenching of fluorescence is caused by a draining away of the excitation energy through collisions of the fluor with either solvent molecules or certain solute molecules. In this section, a quenching mechanism is described that does not require contact between the fluor and the quencher; furthermore, the quencher itself fluoresces, although it does not directly absorb exciting light. This phenomenon is called *fluorescence energy transfer,* and it is very useful in determining intramolecular distances in macromolecules. It is often called a "spectroscopic ruler."

In discussing fluorescence energy transfer, it is necessary to distinguish between a *donor* molecule, which is the primary absorber, and the *acceptor* molecule, which quenches the fluorescence of the donor and, in so doing, fluoresces. The mechanism by which the transfer occurs has been explained by a complex quantum mechanical theory developed by Theodore Förster; sometimes energy transfer of this type is called Förster transfer. The theory, which is beyond the scope of this book, is based upon an analysis of the electronic interaction between the donor dipole and the acceptor dipole. The result of the theory is that the efficiency of transfer, $\mathscr{E}_T$, is

$$\mathscr{E}_T = \frac{r_0^6}{r_0^6 + r_6} , \qquad (49)$$

SEC. 17–4 FLUORESCENCE

in which r is the distance between the centers of the molecules and r_0 is a characteristic distance for the donor-acceptor pair; r_0 is also the distance for which $\mathscr{E}_T = 0.5$. The value of r_0 depends on the amount of overlap between the fluorescence spectrum of the donor and the absorption spectrum of the acceptor, and on the angle between the donor and acceptor dipole moments. That is, there must be a range of wavelengths of the donor fluorescence that is included in the absorption spectrum of the acceptor. This requirement suggests a trivial mechanism for energy transfer—namely, emission of a photon by the donor, followed by absorption of the same photon by the acceptor. However, such a mechanism would be incorrect, because it does not lead to the observed $1/r^6$-dependence of $\mathscr{E}_T$ that is indicated by Equation 49.

Two points should be noted about Equation 49: (1) the efficiency of transfer can be measured either by the degree of quenching of the donor or by the fluorescence of the acceptor; (2) the efficiency of transfer decreases very rapidly as the separation of the donor and the acceptor increases.

Equation 49 has been used as a means of measuring distances between two parts of a molecule. How this is done is best seen by examining a now classical experiment that proved the Förster theory. A series of molecules of the type dansyl-(L-prolyl)$_n$-α-naphthyl, in which n ranged from one to twelve, was prepared (Figure 17-13). Polyproline, a polyamino acid that lacks peptide bonds, is a rigid molecule, so that the dansyl and α-naphthyl groups are at fixed separations for various values of n. Solutions of these molecules were illuminated with a wavelength absorbed by the α-naphthyl group but not by the dansyl group, and the dansyl fluorescence was measured. The fluorescence was plotted as a function of the distance (in Å) between the dansyl and α-naphthyl groups. The data, which show the $1/r^6$-dependence and a value for r_0 of 34.6 Å, are presented in Figure 17-14.

Energy transfer is widely used to determine distances in macromolecules. For proteins, which contain the fluorescent amino acid tryptophan, the procedure is to couple an acceptor fluor to a known site in the protein

FIGURE 17-13 Formula for the compound poly-L-proline, which separates a dansyl (acceptor) group and an α-naphthyl (donor) group (in experiments confirming the Förster theory, n varied from 1 to 12). [SOURCE: *Physical Biochemistry*, First Edition, by David Freifelder, W.H. Freeman and Company. Copyright © 1976.]

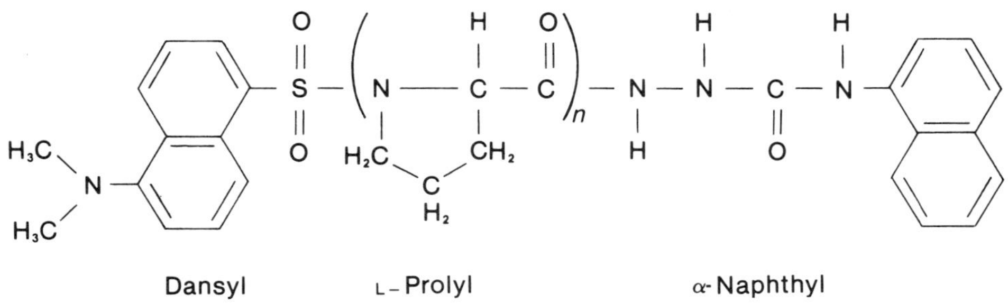

FIGURE 17-14 Efficiency of energy transfer as a function of distance in dansyl-(L-prolyl)$_n$-α-naphthyl, in which n varies from 1 to 12. The distances (in Å) for each value of n were independently determined from other types of measurements. The solid line corresponds to a $1/r^6$-dependence. [SOURCE: From L. Stryer and R.P. Haugland, *Proc. Nat. Acad. Sci.* 58 (1957):719–726.]

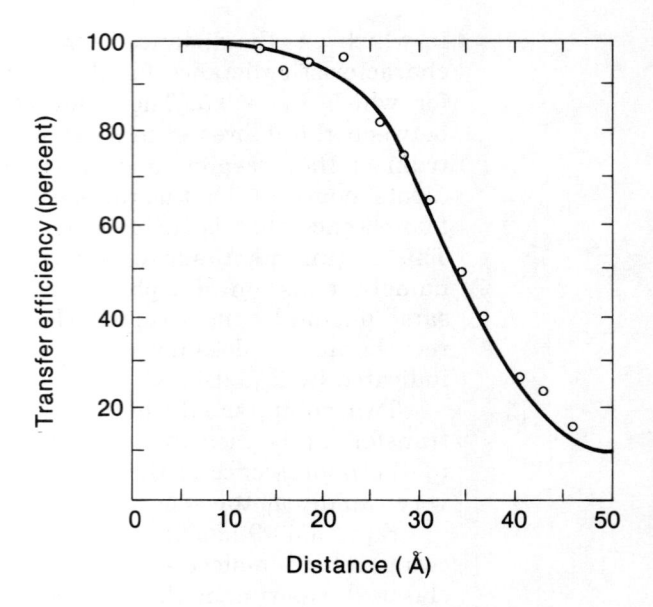

molecule. The tryptophan is then excited and the intensity of the fluorescence of the acceptor is measured. This tells the distance between tryptophan and the fluor, and hence, between tryptophan and the site to which the fluor is attached. This procedure is useful only if the protein contains a single tryptophan. Another procedure, which is very versatile, is to couple two fluors, an acceptor and a donor pair, to fixed sites on the macromolecule, and then to measure the efficiency of transfer. For example, the following method has been used to estimate the shape of small polypeptides. A donor fluor, trypaflavin, is coupled to the carboxylic end of the polypeptide, and an acceptor fluor, rhodamine B, is covalently linked to the amino terminus. The value of r_0 for this pair is well known. If measurement of $\mathscr{E}_T$ yields a value for r of 8 Å, the two termini of the polypeptides are very near one another. If the polypeptide consisted of twenty amino acids, the contour length of the polypeptide chain would be 100 Å. Thus, one would have learned that such a molecule is not extended in solution but has its ends very near one another.

17–5 PROTEINS AND NUCLEIC ACIDS

Ultraviolet Spectra of Proteins and Nucleic Acids

There are few biological macromolecules that absorb visible light. However, as a result of their containing aromatic rings, most of them absorb ultraviolet (UV) light in a range of wavelengths that is easily measurable.

SEC. 17-5 PROTEINS AND NUCLEIC ACIDS

FIGURE 17-15 Spectra of several amino acids. The molar absorption coefficient, ϵ, is a measure of the fraction of the incident light absorbed by passage through 1 cm of a 1 M solution; ϵ is discussed in detail in Chapter 18.

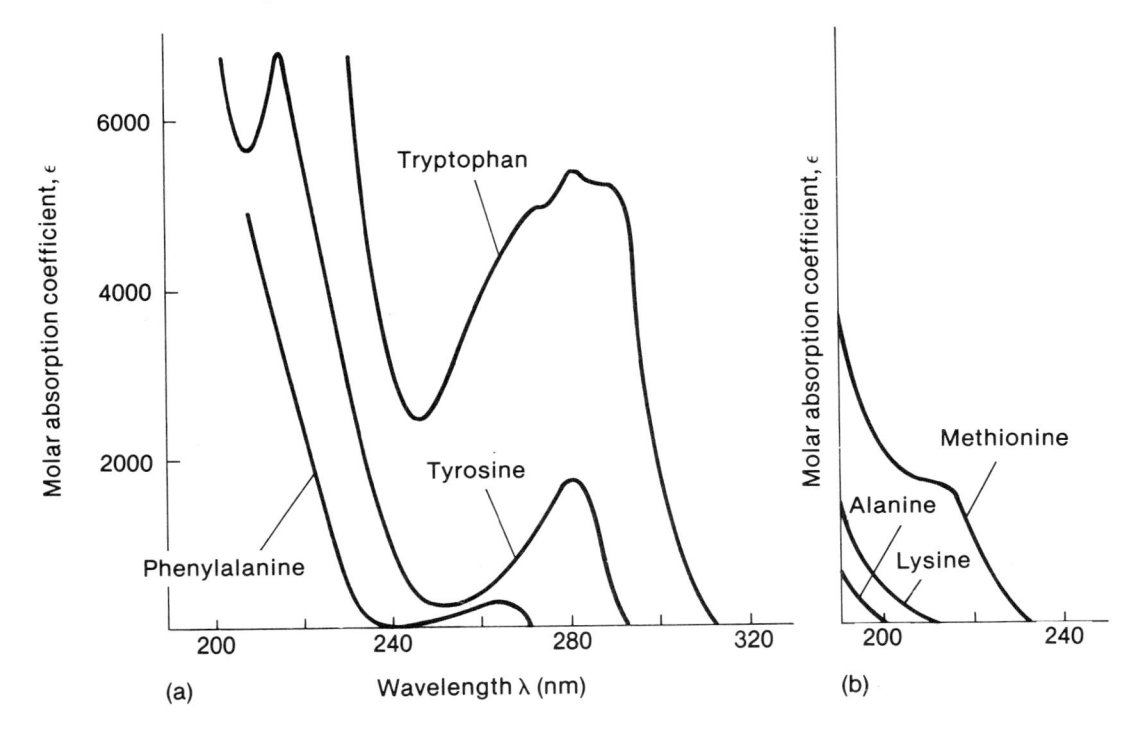

The absorption spectra of some of the amino acids and of the nucleotide bases in nucleic acids have been well studied and are of great use both in identifying substances and in determining the structure of proteins and nucleic acids.

The UV spectra of several amino acids are shown in Figure 17-15 (a) and (b). Only three aromatic amino acids, tryptophan, tyrosine, and phenylalanine, show strong absorption at wavelengths greater than 230 nm. The linear amino acids shown in (b) absorb significantly at very low wavelengths, but this region is experimentally difficult to study. The other aromatic amino acids, proline and histidine, absorb less per mole than does phenylalanine, and contribute only slightly to the overall UV absorption of proteins.

In solution, the structure of solute molecules depends upon certain features of the solvent, and thus it should be expected that the spectra are affected by the solvent. For example, the spectrum of the amino acid tyrosine in water is significantly altered by pH, because the pH value determines whether the phenolic OH is charged or uncharged. Furthermore, the spectrum of tyrosine in a 80-percent water–20-percent ethylene glycol solvent also differs from that in pure water because tyrosine is strongly

FIGURE 17-16 (a) Absorption spectrum of tyrosine at pH 6 and pH 13. Note that both λ_{max} and ϵ are increased when the phenolic OH is dissociated; ϵ is the molar absorption coefficient. (b) Effect of solvent polarity on the spectrum of tyrosine. Solvents: H_2O (solid line) and 20 percent ethylene glycol (dashed line). Notice the increase in λ_{max} in the less polar solvent [SOURCE: *Physical Biochemistry*, First Edition, by David Freifelder, W.H. Freeman and Company. Copyright © 1976.]

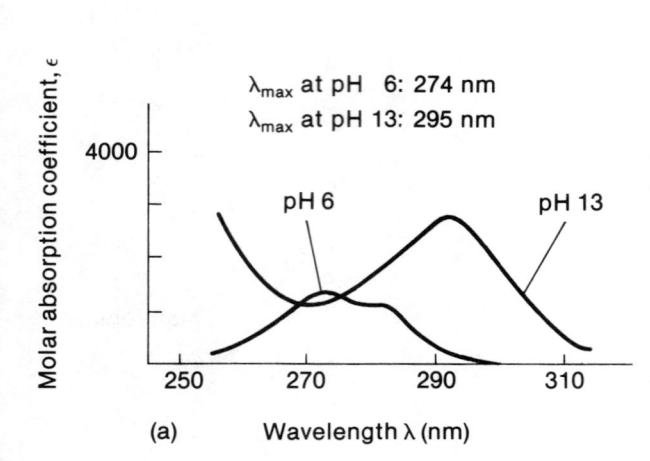

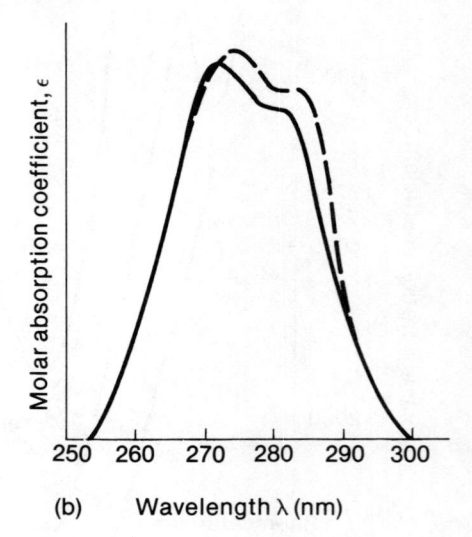

hydrated, and the degree of hydration differs in the two solvents—the spectrum is thus not really of pure tyrosine but of tyrosine plus the other molecules with which it interacts. Another way of stating this is that *the energy levels of the molecules are altered by the solvent*. These spectral changes are shown in Figure 17-16. Complete tables of absorption of amino acids in various conditions can be found in references given at the end of the chapter.*

The spectrum of a protein in a polypeptide chain contains contributions by the individual amino acids, although the spectrum of each amino acid is changed slightly, owing to interactions with one another, as might be expected from the changes just described and seen in Figure 17-16. However, in the formation of a protein, a new chemical bond, the peptide bond (Figure 17-17) is formed. The amide linkage in the peptide bond absorbs

FIGURE 17-17 The peptide unit in a protein. Arrows indicate the peptide bonds. *R* represent an amino acid side chain.

*In the following we use the terms absorbance and molar extinction coefficient as a measure of the absorption of light. The precise meaning of these terms is given at the beginning of Chapter 18.

significantly in the range 190–200 nm. Thus, the absorption spectrum of a typical protein containing most of the amino acids has two distinct maxima—one near 200 nm and the other near 280 nm, as shown in Figure 17-18. Absorption of light by the peptide bond is always much greater than that of the aromatic groups.

FIGURE 17-18 Ultraviolet spectrum of a typical protein. The dashed curve is drawn with a tenfold expanded scale.

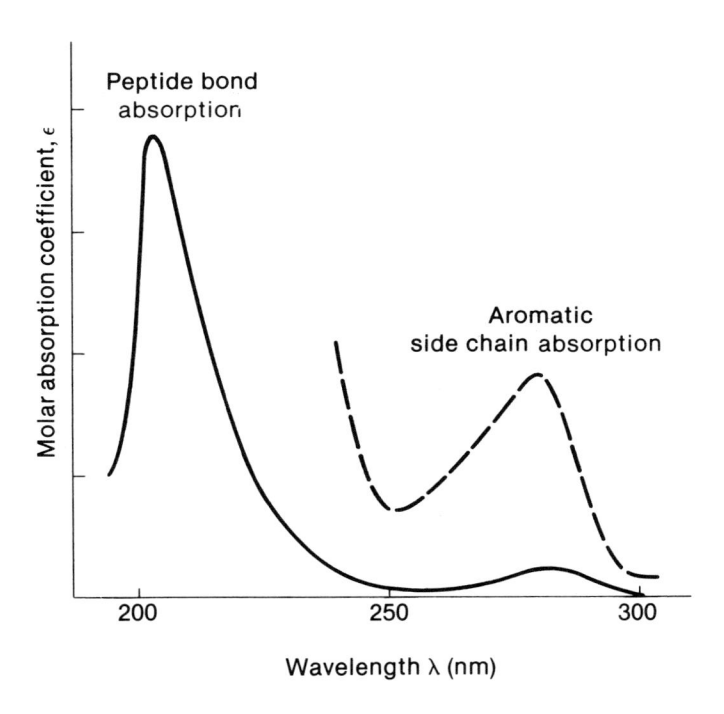

Proteins have a variety of three-dimensional configurations resulting from particular arrangements of the peptide bonds. For instance, in the *α-helical structure* the peptide units are hydrogen-bonded to one another and this produces significant spectral alterations in the 190–200 nm range. This can be seen dramatically in the polypeptide poly-L-lysine hydrochloride, which lacks an aromatic group, so that the UV absorption is almost entirely a result of the peptide bond. Figure 17-19 compares the spectra of pure lysine with that of poly-L-lysine when in the α-helical configuration and in the random coil configuration (in which there are no intrachain interactions and each amino acid may be oriented at random with respect to its neighbor). The absorbance of the α helix is reduced, compared to that of the random coil, and a subtle peak appears at 205 nm.

The absorbance of a protein in the 190–200 nm range can be used to determine whether the protein has a fixed three-dimensional structure. There are two methods that seem straightforward—to determine the absorbance at 195 nm, or to look at the shape of the spectrum. However, neither

FIGURE 17-19 The spectra of pure L-lysine and of poly-L-lysine in the α helix, β sheet, and random coil configuration; ε is the molar absorption coefficient, discussed in detail in Chapter 18.

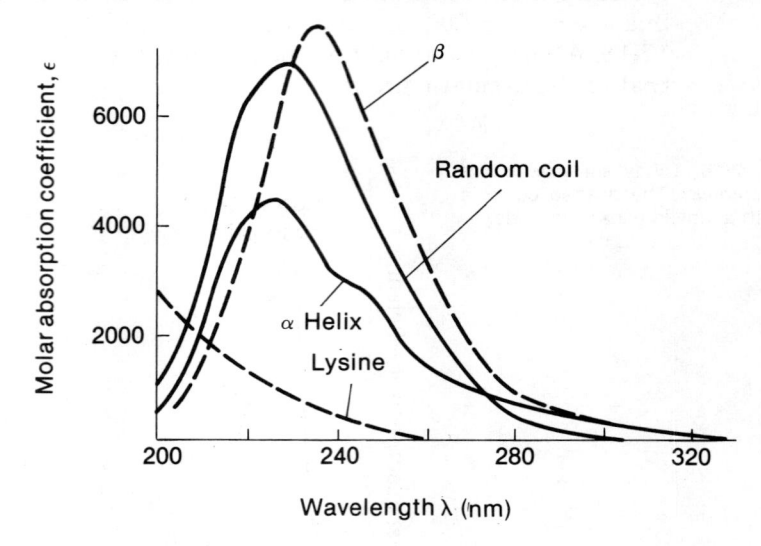

of these methods are useful for several reasons: the fraction of peptide bonds in the α-helical configuration varies from one type of protein to the next; the 205-nm peak is not usually visible in a protein containing many amino acids; and, most important of all, there are other factors that affect the absorbance.

However, what is almost always true is that if there is a fixed structure, the bonds that maintain the structure are usually so weak that they are disrupted by heat. Thus, if the absorbance of a protein at 195 nm is measured as a function of temperature, it will suddenly increase at the particular temperature at which the structure is disrupted, because the absorbance at 195 nm is about 70 percent greater for a random coil than for an α helix. An example of such a measurement is shown in Figure 17-20. The magnitude of the increase in absorbance would reflect the degree of helicity—the less the helicity, the smaller the change. Unfortunately, this method is not as useful as it would seem to be, because there is another structure called the β *structure* whose absorbance is greater than that of the random coil (Figure 15-19). Thus, if a protein possesses β structure, its absorbance will decrease with increasing temperature. One can easily imagine a protein in which there is both α-helical structure and β structure, in which there would be no change at all in the absorbance when the structure is disrupted. However, such a precise matching of structures is unlikely, so that for a typical protein, there is almost always a change in absorbance with temperature. If the protein has more α helix than β structure, the absorbance will increase, but if there is less α helix, it will decrease. There are at present more sophisticated techniques for assessing

FIGURE 17-20 The transition of a protein molecule from the α-helical form to the random coil as seen by an absorbance measurement. The absorbance of a solution is logarithmically related to the amount of incident light removed when the beam passes through 1 cm of solution; this is explained in detail in Chapter 18.

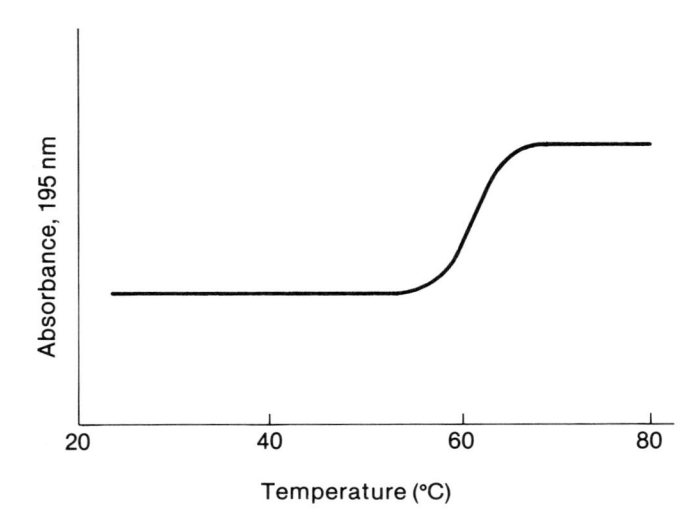

the fraction of amino acids in each configuration but, nonetheless, this example illustrates the kind of information that can be gained from simple measurements.

Absorbance measurements can also be used to determine whether a particular amino acid is on the surface or buried deep within a protein molecule. Referring back to Figure 17-16, note the effect of solvent composition on the spectrum of tyrosine. If a protein contains a single tyrosine and the tyrosine spectrum is measured first with water as a solvent and then with H_2O-ethylene glycol as a solvent, the spectrum will be identical if the tyrosine is buried in the folded molecule and thus inaccessible to the solvent; it will differ if the tyrosine is on the surface. Furthermore, if the protein contains two tyrosine groups, one inside the molecule and one on the surface, then the spectrum in H_2O-ethylene glycol will be an average of the two spectra. In this way, the fraction of tyrosine on the surface can be measured. This method is known as the *solvent-perturbation method*. There are many complexities with it, and the discussion here has been idealized; a detailed description of the method can be found in references given at the end of the chapter.

The absorption spectra of the five bases contained in DNA and RNA are rather similar (though distinguishable), with λ_{max} ranging from 250 to 275 nm. All DNA molecules have the same λ_{max} (259 nm) and have nearly indistinguishable spectra, unless a particular base pair is present in great excess. The most important aspect of nucleic acid absorption spectroscopy is the decrease in the absorbance of the nucleotide bases that occurs when an oligonucleotide forms. For example, the ratio of the absorbance of the oligonucleotide pentacytidylic acid (expressed as absorbance per mole of mononucleotide—*not* per mole of oligonucleotide)—to that of cytidylic acid monomer is 0.74. There is a further decrease in the absorbance per nucleo-

FIGURE 17-21 Spectra of phage T7 DNA, as a double-stranded DNA (solid line), as a single-stranded DNA (dashed line), or after hydrolysis to free nucleotides (dotted line), showing the decrease in absorbance (hypochromicity) that accompanies the formation of a more ordered structure. All spectra were obtained at the same concentration. The absorbance of a solution is logarithmically related to the amount of incident light removed when the beam passes through 1 cm of solution; this is explained in detail in Chapter 18. [SOURCE: *Physical Biochemistry,* First Edition, by David Freifelder, W.H. Freeman and Company. Copyright © 1976.]

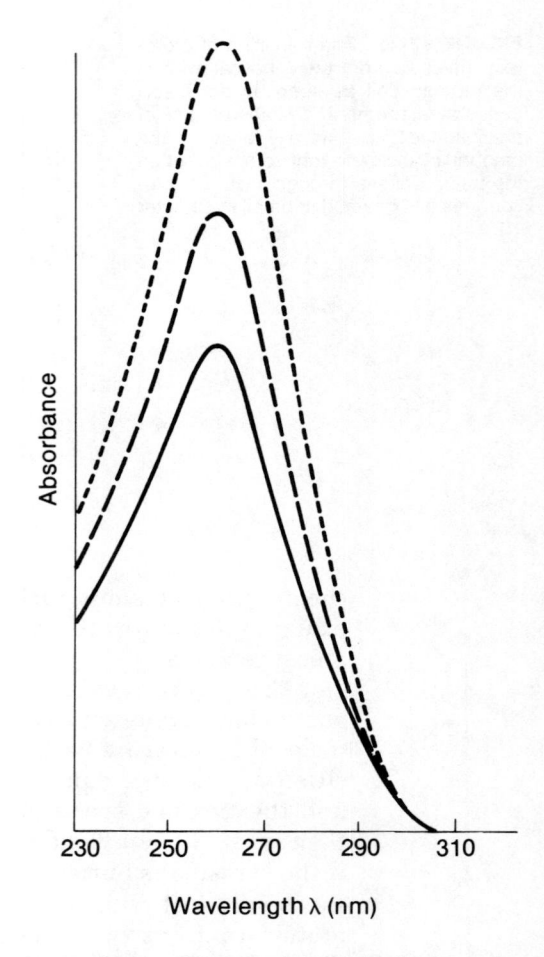

tide when a double-stranded structure forms (Figure 17-21), which compares the spectrum of a sample of DNA to that of a mixture of nucleotides having the same molar concentration. This phenomenon is described by the terms *hypochromicity* and *hyperchromicity*. DNA, whose absorbance is less than that of the free nucleotides, is said to be hypochromic. The terminology is curious, though, because if a double-stranded polynucleotide is converted to either a single-stranded form or to free nucleotides, it is said to become hyperchromic and the percent increase in absorbance is called the percent hyperchromicity. This confusion in terms is clarified in Table 17-2. Note that the structures of both DNA and RNA as isolated from nature are considered to be the standards. This is not meant to imply that these molecules have the same absorbance per molecule.

The origin of these absorbance differences is in the interactions between the dipole moments of the molecules and a neighboring molecule. The probability of absorption decreases as the molecules become nearer and as the planes of the bases become more nearly parallel to one another.

SEC. 17-5 PROTEINS AND NUCLEIC ACIDS

TABLE 17-2 Absorbance properties of nucleotides and nucleic acids.

Molecule	Arbitrary value of absorbance per mole of nucleotide	Percent hyperchromicity
Double-stranded DNA	1.00	0
Single-stranded DNA	1.37	37
Deoxyribonucleotides	1.60	60
RNA (partially double-stranded)	1.00	0
RNA (pure single-stranded)	1.17	17
Ribonucleotides	1.36	36

Thus there is a decrease in absorbance in a single-stranded polymer compared to the free nucleotide because the bases of the single-stranded polymer are constrained to be near one another; the planes of the bases are not all parallel, however, because of thermal disruption. In a double-stranded molecule, such as DNA, because of the hydrogen bonds between the two chains, the molecule is so rigid that the bases become parallel to one another and the absorbance reaches its minimal value.

These spectral changes are extremely valuable in detecting structural changes in nucleic acids. As we have just described for proteins, the three-dimensional structure of DNA is disrupted with increasing temperature. Thus, the absorbance of DNA should increase at the particular temperature at which the rigid double-stranded structure is broken down. This does occur and is shown in Figure 17-22. Such a curve is a general example of a *helix–coil transition,* which we discuss in detail in Chapter 18, and in the case of nucleic acids it is called a *melting curve.* These have been very useful in determining the factors responsible for maintaining the structure of DNA, as can be seen in several examples in Chapter 18.

Infrared Spectra of Proteins and Nucleic Acids

As we have pointed out before, each chemical bond vibrates with a characteristic frequency, and if the vibration is accompanied by a fluctuating dipole moment, light can be absorbed or emitted. The characteristic frequency for a particular bond, for example, varies according to the molecule containing the bond, because of the reduced mass term, as explained in the earlier section entitled "Molecular Vibration and Rotation." However, the range of frequencies is usually not very great, so that the observed absorption bands can be identified. The infrared spectra for each of the amino acids and each of the nucleotides are well characterized and are explained by their structures in that they consist of the absorption bands expected from the many bonds that they contain.

These spectra also contain information that is not obvious from their chemical formulae. For example, the frequencies found in the spectrum of

FIGURE 17-22 A melting curve for a DNA molecule. The molecular structure begins to be disrupted at T_1; is half disrupted at T_m, and the strands usually separate at T_2.

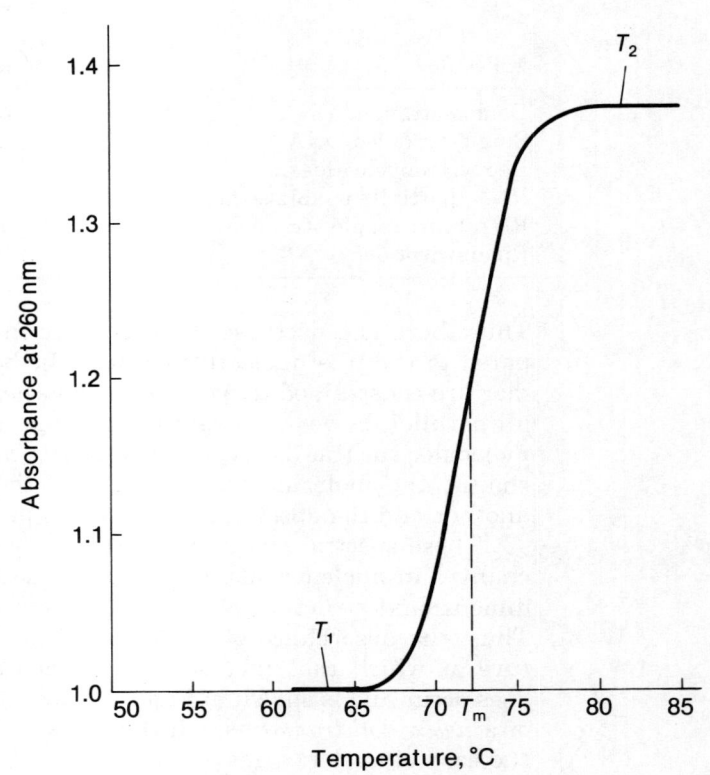

the DNA base cytosine are those expected from the chemical formula except for a single low-frequency band between 3 and 3.5×10^{13} sec^{-1}. All other bands expected from the formula have a frequency greater than 4.8×10^{13} sec^{-1}, so that cytosine must contain some bond other than those seen in the structural formula. Inspection of reference tables listing the characteristic frequencies for various bonds shows that of those bonds containing C, N, H, and O, one corresponding to the structure

$$\begin{array}{c} \text{OH} \\ | \\ =\text{C}- \end{array}$$

Enol

has just about the right frequency. This suggests that the C=O bond in cytosine sometimes takes an enol form. This agrees with the observation that the vibrational band resulting from stretching of the C=O bond at about 5×10^{13} sec^{-1} is weaker than expected. Thus, infrared spectra can indicate when tautomerism exists and can tell the relative amount of the keto and enol forms in particular conditions.

SEC. 17–5 PROTEINS AND NUCLEIC ACIDS

Another bond that is made evident by infrared spectroscopy is the hydrogen bond. This is shown in the spectra of Figure 17-23. The bases adenine and uracil should be capable of forming a hydrogen bond between the C=O group (in uracil) and the NH₂ group (in adenine), so that if a molecule containing both adenine and uracil were hydrogen-bonded, shifts in the frequencies corresponding to the C=O and NH₂ groups should be expected. This expectation is borne out by the data in the figure. In panel (a) we see a small part of the spectrum of the polymer polyuridylic acid (poly U). The principal peak is due to the C=O group. Panel (b) shows a portion of the spectrum of polyadenylic acid (poly A) in which the peak is that of the C—N bond. The monomers uridylic acid (UMP) and adenylic acid (AMP) cannot maintain hydrogen bonds at room temperature, so that the spectrum of a mixture of equimolar quantities of UMP and AMP is simply the sum of the spectra of each, as shown in panel (c). The polymers poly A and poly U, when mixed, form a double-stranded structure, presumably held together by hydrogen bonds. Panel (d) compares the measured spectrum of the mixture with a spectrum that is just the sum of the spectra of poly A and poly U. The C=O band has moved to a higher frequency, as the greater inertia of the oxygen atom that is a result of its being constrained to the hydrogen bond might lead us to expect. The C—N vibration is also shifted to a higher frequency, but to a lesser extent than the C=O shift. Thus, the spectrum gives supporting evidence for hydrogen-bonding. Although not shown in the figure, the two C=O bonds in uracil, that from C-2 and that from C-4, have different absorption frequencies. The band that is shown in the figure is for the C-2=O bond, so the spectrum indicates that this is the position at which the hydrogen bond forms.

Hydrogen-bonding could also have been studied in other regions of the spectrum—for instance, the band corresponding to the stretching of the N—H bond in the adenine amino group. A shift is also seen here, but to lower frequencies, indicating that the covalent N—H bond is weakened when the H is in a hydrogen bond. This fact has also been confirmed by other measurements.

The infrared spectra of proteins are extraordinarily complicated, because there are many different bonds engaged in numerous interactions. However, proteins contain a bond that is not present in the amino acids— the amide or peptide bond that is produced when two amino acids join to one another. There are many modes of vibration and at least ten different bands result, which are designated amide I, amide II, and so forth. The amide I band has been widely studied, because its frequency depends upon the environment of the bond. A polypeptide chain can have various three-dimensional configurations (for example, the α helix, the β sheet, the β turn, and the random coil, to name just a few), and the frequency of the amide I band depends upon the configurations. For instance, its frequency for the α helix is 1650 cm^{-1} but for the random coil is 1658 cm^{-1}. Thus, simply by measuring the intensity of absorption at these two frequencies, one can determine the relative proportion of the amino acids in the two configurations.

FIGURE 17-23 Infrared spectra giving supporting evidence for hydrogen-bonding. Panel (a) is the polyuridylic acid spectrum. Panel (b) shows the spectrum of polyadenylic acid (poly A). Panel (c) shows an equimolar mixture of uridylic acid and adenylic acid (solid line) and the expected curve (dashed line) that is the sum of the curves for pure adenylic and guanylic acids. Panel (d) shows the spectrum for a mixture of poly A and poly U (solid line) and the sum of the curves in (a) and (b) (dashed line). [From H.T. Miles, *Biochim. Biophys. Acta* 30, 324, 1958.]

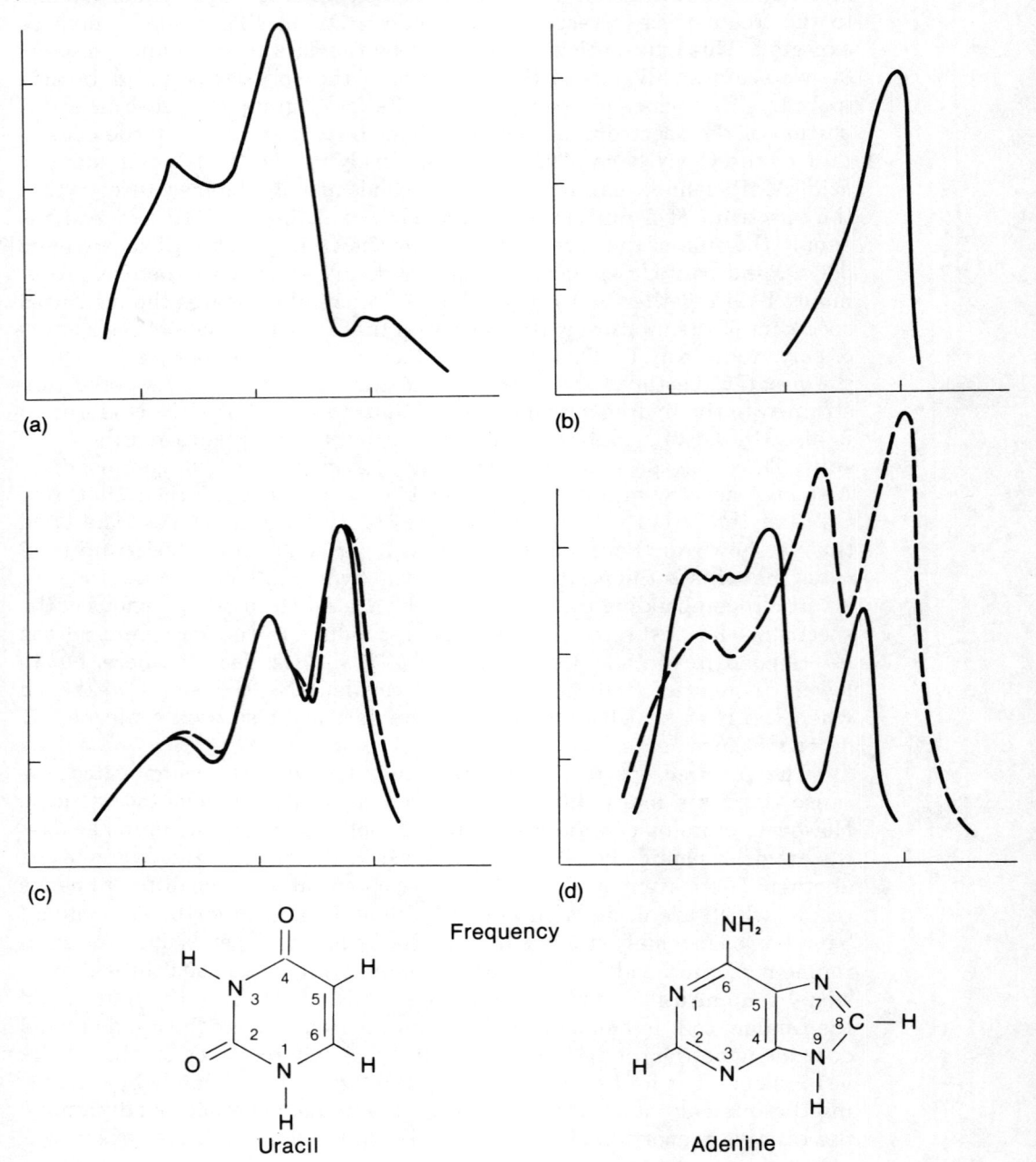

SEC. 17-5 PROTEINS AND NUCLEIC ACIDS

Proteins frequently contain bound molecules whose structure can be identified by infrared spectroscopy. For instance, the protein hemoglobin, found in blood, carries oxygen. If the spectra of hemoglobin and of oxyhemoglobin (hemoglobin carrying oxygen) are compared, it is found that there is a strong absorption band at a frequency of 3.321×10^{13} sec^{-1} when oxygen is present. O_2 itself does not have a vibrational spectrum, because it has no dipole moment, but its absorption frequency can be determined by Raman spectroscopy. The frequencies corresponding to other oxygen-containing groups such as superoxide, O_2^-, and peroxide, O_2^{2-}, have also been determined by either Raman or infrared spectroscopy of compounds containing these groups. The absorption frequency of the superoxide group, O_2^-, is also 3.321×10^{13} sec^{-1}, indicating that this is the form of oxygen in oxyhemoglobin. The protein hemerythrin is the oxygen carrier in many invertebrates. The spectrum of the oxygenated protein shows absorption at 1.532×10^{13} sec^{-1}, corresponding to peroxide, which is presumably the form of oxygen that is bound to this molecule.

Fluorescence Spectroscopy of Proteins and Nucleic Acids

The quantum yield for fluorescence of the amino acid tryptophan exceeds that of the other naturally occurring amino acids, so that fluorescence studies with proteins usually involve only tryptophan. The fluorescence and emission spectra of tryptophan are shown in Figure 17-24. The value of fluorescence in the study of proteins is based upon several facts, of which the most important are the following.

1. The maximal wavelength λ_{max} of the fluorescence spectrum of tryptophan is determined by the polarity of the environment of the particular tryptophan molecule that is fluorescing.

2. The intensity at λ_{max} is also determined by the polarity of the environment.

3. The fluorescence of tryptophan can be quenched—that is, the intensity at λ_{max} can be vastly reduced—by certain ions (for example, iodide), if the ion can collide with the tryptophan.

These three points (and others not mentioned) enable one to obtain significant information about the structure of a protein. How this is done can be seen in the following idealized examples.

EXAMPLE 17-B   The location of tryptophan in a protein containing a single tryptophan.

A solution of a particular protein shows a fluorescence spectrum characteristic of tryptophan. If the solution also contains 0.1 M NaI, no fluores-

FIGURE 17-24 Absorption and fluorescence spectra of tryptophan. Solid line: molar absorption coefficient as a function of wavelength. Dashed line: emission spectrum in arbitrary units. [SOURCE: *Physical Biochemistry*, First Edition, by David Freifelder, W.H. Freeman and Company. Copyright © 1976.]

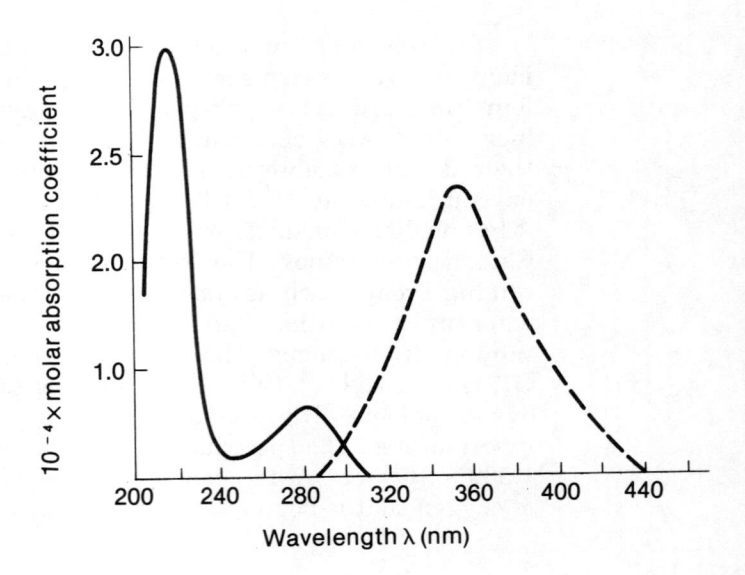

cence is observed. Since the I⁻ ion must collide with tryptophan to eliminate the fluorescence, the tryptophan must be on the surface of the protein. If the intensity were unaffected, the tryptophan would be internal.

EXAMPLE 17-C Locating tryptophan molecules in a protein that contains two tryptophan molecules.

Addition of 0.1 M NaI to a protein solution decreases the fluorescence intensity of tryptophan to one-half of its initial value. Since the protein contains two tryptophan molecules and the I⁻ ion must collide with tryptophan to quench the fluorescence, one tryptophan must be on the surface and one must be internal. When iodide is added and quenching occurs, λ_{max} of the remaining fluorescence is at a shorter wavelength than before addition of I⁻. It is known that λ_{max} has lower values if the tryptophan is in a nonpolar environment. Thus the residual fluorescence after quenching must come from a tryptophan in such a region, and hence the internal tryptophan must be surrounded by amino acids with nonpolar side chains.

EXAMPLE 17-D Locating a tryptophan in an enzyme.

In the presence of I⁻, the tryptophan fluorescence is quenched; therefore the tryptophan must be on the surface of the enzyme. In the absence of iodide, λ_{max} is smaller than that of free tryptophan; therefore the tryptophan must be surrounded by nonpolar amino acids. If the enzyme substrate, which is a nonpolar substance, is added, λ_{max} shifts to lower values,

FIGURE 17-25 Structures of two common extrinsic fluors.

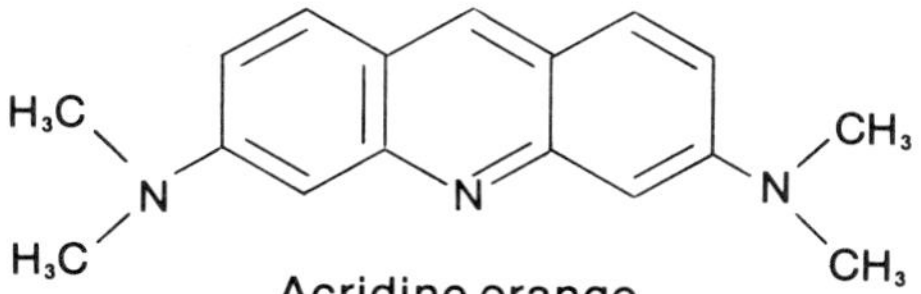

Acridine orange

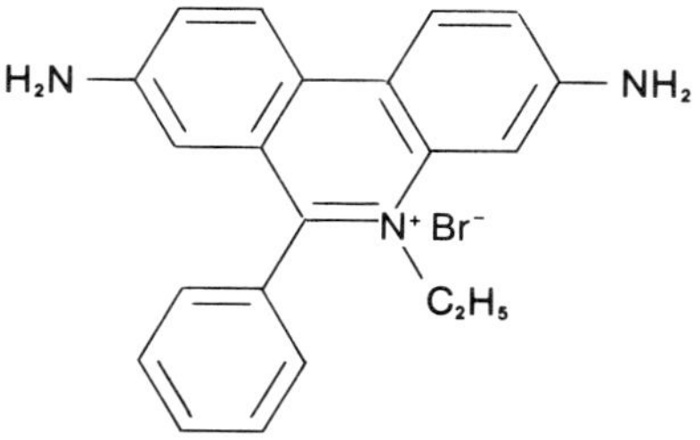

Ethidium bromide

suggesting that the tryptophan is in the binding site and that the nonpolar substrate has prevented the tryptophan from contacting the polar water molecules. The addition of iodide after the substrate is added does not quench the fluorescence, confirming that the tryptophan is no longer accessible to the solvent. Note that changes in fluorescence can also be used as an assay for substrate binding.

Other examples of the use of fluorescence in the study of proteins can be found in the references given at the end of the chapter.

The nucleic acids possess little fluorescence of their own. However, highly fluorescent compounds bind to the nucleic acids and their spectra give information about structure.

The two most useful fluors in the study of nucleic acids are acridine orange and ethidium bromide (Figure 17-25). Both substances bind strongly to double-stranded polynucleotides by intercalating between the stacked base pairs and, in so doing, the quantum yield for the fluorescence of these substances is increased significantly. For this reason they are very useful in the detection of DNA. For instance, the absorbance of DNA has a value that sets a lower limit of about 5 μg/ml for the quantitative measurement of the amount of DNA.* However, at a concentration of ethidium bromide of 1 μg/ml, the addition of 0.01 μg of DNA produces an enhancement of the fluorescence that is easily measured. Thus, fluorescence vastly extends the range of measurement of DNA concentration. It should be noted that if DNA and free nucleotides were mixed, the amount of DNA

---

*For a review of the relation between absorbance and concentration, refer to the beginning of Chapter 18.

would not be measurable by absorption, because the absorption spectra of both are nearly identical. However, there is no fluorescence enhancement by the nucleotides, so that the amount of DNA is measurable by the fluorescence method without any interference by the nucleotides.

The binding of both acridine orange and ethidium bromide to single-stranded polynucleotides is relatively weak. However, in the case of acridine orange there are changes in λ_{max} that are useful. When acridine orange is bound to DNA, there is a strong green fluorescence. However, if bound either to a single-stranded DNA or to RNA, the fluorescence is red-orange. Thus, just visually, single- and double-stranded polynucleotides can be distinguished by looking at the color of the fluorescence. If a sample contains both single- and double-stranded material, the fluorescence spectrum has two peaks (in the green and in the red) and by measuring the relative intensities of each, the ratio of single- to double-stranded material can be determined.

REFERENCES

Atkins, P.W. 1970. *Molecular Quantum Mechanics*. Clarendon Press. Oxford.
Barrow, G. 1964. *The Structure of Molecules*. W.A. Benjamin. Menlo Park, Calif.
Castellan, G.W. 1971. *Physical Chemistry*. Addison-Wesley. Reading, Mass.
Christodelas, N.D. 1975. "Particles, Waves, and the Interpretation of Quantum Mechanics." *J. Chem. Educ.* 40, 262.
Hochstrasser, R. 1964. *Behavior of Electrons in Atoms*. W.A. Benjamin. Menlo Park, Calif.
Hoffman, B. 1959. *The Strange Story of the Quantum*. Dover. New York.
Jammer, M. 1966. *The Conceptual Development of Quantum Mechanics*. McGraw-Hill. New York.
Karplus, M., and R.N. Porter. 1970. *Atoms and Molecules*. Benjamin. Menlo Park, Calif.
Semat, H. 1975. *Introduction to Atomic and Nuclear Physics*. Chapman and Hall. London.

PROBLEMS

1. A molecule has a spectral line at 2989 cm^{-1}.

    (a) What are its frequency and its wavelength?

    (b) In what part of the spectrum is it located?

2. A particular system has four energy levels that are characterized by the quantum numbers n, which have the values 1, 2, 3, and 4. The energies of these levels are: $E_0 = 10^{-12}$, $E_1 = 5 \times 10^{-12}$, $E_2 = 8 \times 10^{-12}$, and $E_3 = 10^{-11}$ erg. If the selection rule is $n = \pm 1$ or $n = \pm 2$, what are the wavelengths of the spectral lines associated with this system?

3. If NaCl is heated in the flame of a bunsen burner, the flame turns yellow. The wavelength of this yellow light is that of a strong spectral line of *atomic* sodium. Propose a molecular explanation for the occurrence of this emission.

4. A two-ton automobile moving 30 miles per hour possesses a wavelength, according to the de Broglie theory. Do you think the wavelength could be detected?

5. The maximal wavelength that will result in the emission of a photoelectron is 660 nm. What is the kinetic energy of an electron ejected by a photon whose wavelength is 450 nm?

6. From the point of view of the Bohr theory, where is the outermost orbital electron of an ionized atom?

7. At room temperature, would you expect most atoms to be in the electronic ground state, or some atoms to be excited simply by picking up a quantum of energy in an intermolecular collision?

8. What is the wavelength, in nm, of the fourth line in the so-called Pfund series ($n_1 = 5$) in the spectrum of atomic hydrogen?

9. How many spectral lines of atomic hydrogen are there if we consider transitions only between the first five energy levels?

10. The ionization energy of the hydrogen atom is 13.58 eV. What is the value of the ionization energy for removal of the second electron from helium?

11. In the spectrum of a hydrogen atom, there are several "series" of lines—namely, those for which $n_1 = 1$ (Balmer series), $n_1 = 2$ (Lyman series), $n_1 = 3$ (Paschen series), $n_1 = 4$ (Brackett series), and $n_1 = 5$ (Pfund series). The intensities of the lines in the absorption spectrum are not the same for all series. If the temperature of gaseous hydrogen is increased, the intensity of the higher-order series (for example, Pfund) becomes greater compared to the intensity of the lower-order series (for example, Balmer). Explain this phenomenon.

12. If one wanted to "see" atoms by using some kind of electromagnetic radiation, what region of the spectrum would be needed?

13. X-ray diffraction is the most precise method for determining the structure of protein molecules. Its single disadvantage in structure determination is that it is carried out with dry material, whereas one is interested in observing the structure of a protein in its natural, biological environment—namely, in solution. Whether the structure of the dry substance is the same as that in solution can be determined by Raman spectroscopy, which can be performed on a molecule both in the dry state and in solution. A protein is a macromolecule that consists of amino acids linked by peptide bonds (which contain an amide group). Each amino acid has a so-called side chain, which sticks out as a branch from the main linear polypeptide chain (see Chapter 15).

What would you conclude if the Raman spectral lines produced by the amide group were identical in wavelength and intensity for the dry protein and the protein solution, yet there were small changes in the spectrum derived from one amino acid side chain (for example, that of tyrosine)?

14. You have just prepared two concentrated solutions of the amino acids tyrosine and isoleucine (see Chapter 15 for the chemical structures of these). You neglect to label the bottles immediately and become confused about which is which. Short of making fresh solutions, which you hope not to have to do, what spectral measurement might you make to distinguish the solutions?

15. Calculate the three lowest frequencies expected in a rotational energy-band for a molecule whose moment of inertia is 10^{-39} g cm^2.

16. The spectrum of gaseous NaCl has a band in the microwave region. The frequency difference between the individual lines is 1.357×10^{10} sec^{-1}. What is the interatomic distance between the Na and Cl atoms?

17. Would you expect helium to have spectral lines in the microwave region?

18. We have seen that diatomic gases absorb microwave radiation. Radar is microwave radiation, and our atmosphere consists mostly of diatomic gases. If you were designing a radar generator, how would you avoid problems caused by atmos-

pheric interference, and which gases would not concern you?

19. We have so far only discussed microwave spectra for diatomic molecules and have implied that diatomic molecules produce only a single rotational band. How many bands will there be for a polyatomic molecule? (*Hint:* review the section in Chapter 2 about equipartition of energy.)

20. Draw the modes of vibration of carbon dioxide (a linear molecule) and indicate the modes that would give rise to infrared spectral lines.

21. The vibrational wave number of gaseous HCl is 2886 cm^{-1} and that of $^{12}C^{16}O$ is 2170 cm^{-1}. In which molecule are the atoms bonded more tightly?

22. Would the vibrational frequency of a hydrogen-containing compound be affected by substituting deuterium for hydrogen?

23. A great deal of information can be obtained from the study of infrared spectra. However, an important limitation occurs when the substance of interest must be studied in aqueous solution. The problem is that the stretching motion of the O-H bonds of water results in water absorbing very strongly in the infrared region at a wavelength of about 29,400 Å; this absorption often obscures many spectral bands of substances under study. The experimental solution to this problem is to dissolve the substance in D_2O rather than H_2O. Why does this solve the problem?

24. Consider the following pair of spectra. One is an absorption spectrum and the other is a fluorescence spectrum. Which is which?

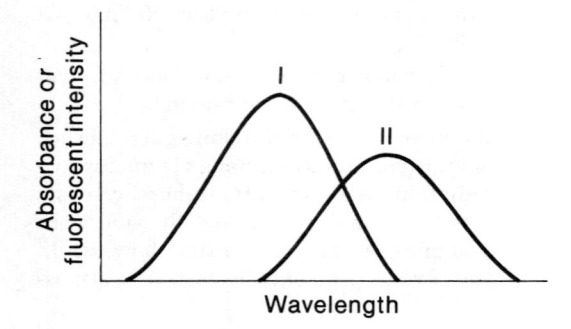

25. Two fluorescent compounds, A and B, are chemically attached to the end of the polyamino acid, poly-L-proline. A is at one end and B is at the other end of the molecule. This particular polyamino acid differs from other polyamino acids in that the bond joining the prolines is not a peptide bond but a more rigid bond. The result is that poly-L-proline is a much more extended molecule in solution than is polyalanine. Let us use the notation A-(proline)$_n$-B to designate the former polymer, in which n denotes the number of prolines in the chain. A and B have the property that if A absorbs exciting light, B frequently fluoresces.

(a) If solutions of A-(proline)$_8$-B and of A-(proline)$_{11}$-B are separately illuminated with light that excites A, for which solution will the fluorescence of B be brightest?

(b) If A-(proline)$_7$-B and A-(alanine)$_7$-B are separately illuminated as in part (a), for which compound will B fluoresce more intensely?

26. Iodide-quenching of tryptophan fluorescence can be used to determine whether tryptophans are exposed to a solvent.

(a) If a protein is known to contain only one tryptophan and iodide fails to quench fluorescence, what possible explanations might be given to account for the lack of quenching?

(b) If the protein contains eight tryptophans and iodide quenches 24 percent of the fluorescence, it is tempting to assume that two tryptophans are accessible to the solvent. State several factors that could make this conclusion invalid.

27. A protein containing ten tryptophans shows fairly strong fluorescence of tryptophan. A small molecule known to bind tightly to the protein produces virtually no change in the fluorescence, even though it is known that there are two tryptophans in the binding site. Give several possible explanations for this.

28. In the presence of a quencher of fluorescence, will there be a change in the shape of either the absorption spectrum or the fluorescent spectrum?

CHAPTER 18

PHOTOCHEMISTRY, RADIATION CHEMISTRY, AND RADIOBIOLOGY

The interactions of light and nuclear radiation that produce chemical and physical changes in matter are important topics for study, for two reasons: some of these interactions, such as vision, photosynthesis, and bioluminescence, are biologically crucial; others provide the basis for experimental techniques that are valuable in biological research. In this chapter we introduce the essential concepts of these topics and show their applications with several biologically relevant examples.

18–1 LIGHT

Light is a form of electromagnetic radiation that consists of perpendicular, sinusoidally oscillating electric and magnetic fields that move through space at a velocity known as the speed of light, denoted c (Figure 18-1). The value of c is 2.9979×10^{10} cm sec^{-1} or, in SI units, 2.9979×10^{8} m sec^{-1}. Every light wave is characterized by a *frequency*, ν, which is the number of oscillations per second, and a *wavelength*, λ, which is the distance between the peaks of the sine curve. The wavelength and frequency are related by the simple equation

$$\lambda \nu = c. \tag{1}$$

The units of frequency are cycles per second or hertz (Hz). Several units are used to describe wavelength. In the SI system, the unit of measure-

FIGURE 18-1 Propagation of an electromagnetic wave through space. The E and H vectors are mutually perpendicular at all times.

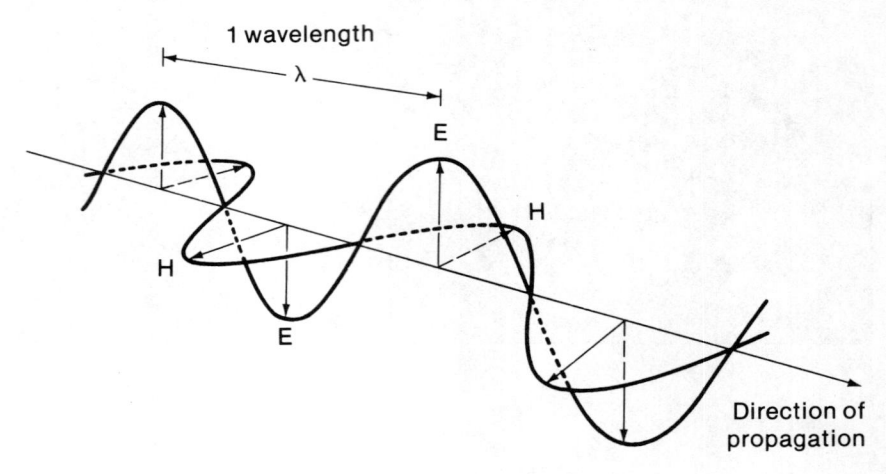

ment is the meter; the wavelengths that we commonly encounter are expressed in nanometers, nm, equivalent to 10^{-9} m. Other units frequently used are the millimicron, mμ, also equivalent to 10^{-9} m, and the ångström, Å, or 10^{-8} cm. Note that 1 nm = 1 mμ = 10 Å. Modern optical equipment is labeled in nm or mμ, but when electron microscopists and biochemists describe molecular size they generally use ångström units. In some instances, such as infrared spectroscopy, the *wave number,* or the number of wavelengths per unit of distance, is used to describe a light wave. The wave number is usually denoted as $\bar{\nu}$, or as k, and equals $1/\lambda$. The wave number is measured in cm^{-1}. Thus an infrared wave whose wavelength is 2000 nm would be described as having a wave number $\bar{\nu}$ = 5000 cm^{-1}.

There is an enormous range in the wavelengths of electromagnetic radiation, but only a small part of the spectrum of wavelengths is visible light, as is shown in Figure 18-2.

A beam of light can also be thought of as a collection of particles called *photons*. Each photon propagates through space at the speed of light and carries a unit of energy called a *quantum* of light (plural, *quanta*). The energy of such a quantum, E_λ is

$$E_\lambda = h\nu = \frac{hc}{\lambda}, \tag{2}$$

in which h is the Planck constant; h = 6.625 × 10^{-34} J sec^{-1} or 6.625 × 10^{-27} erg sec^{-1}. The product $h\nu$ is sometimes called a quantum of radiation. Note that *the energy of a quantum increases as the frequency increases.*

The intensity of a light beam is the energy carried by the light beam; it is measured as the energy falling on a unit area in one second, so that its units are J cm^{-2} sec^{-1}. If light is thought of as a collection of photons, the

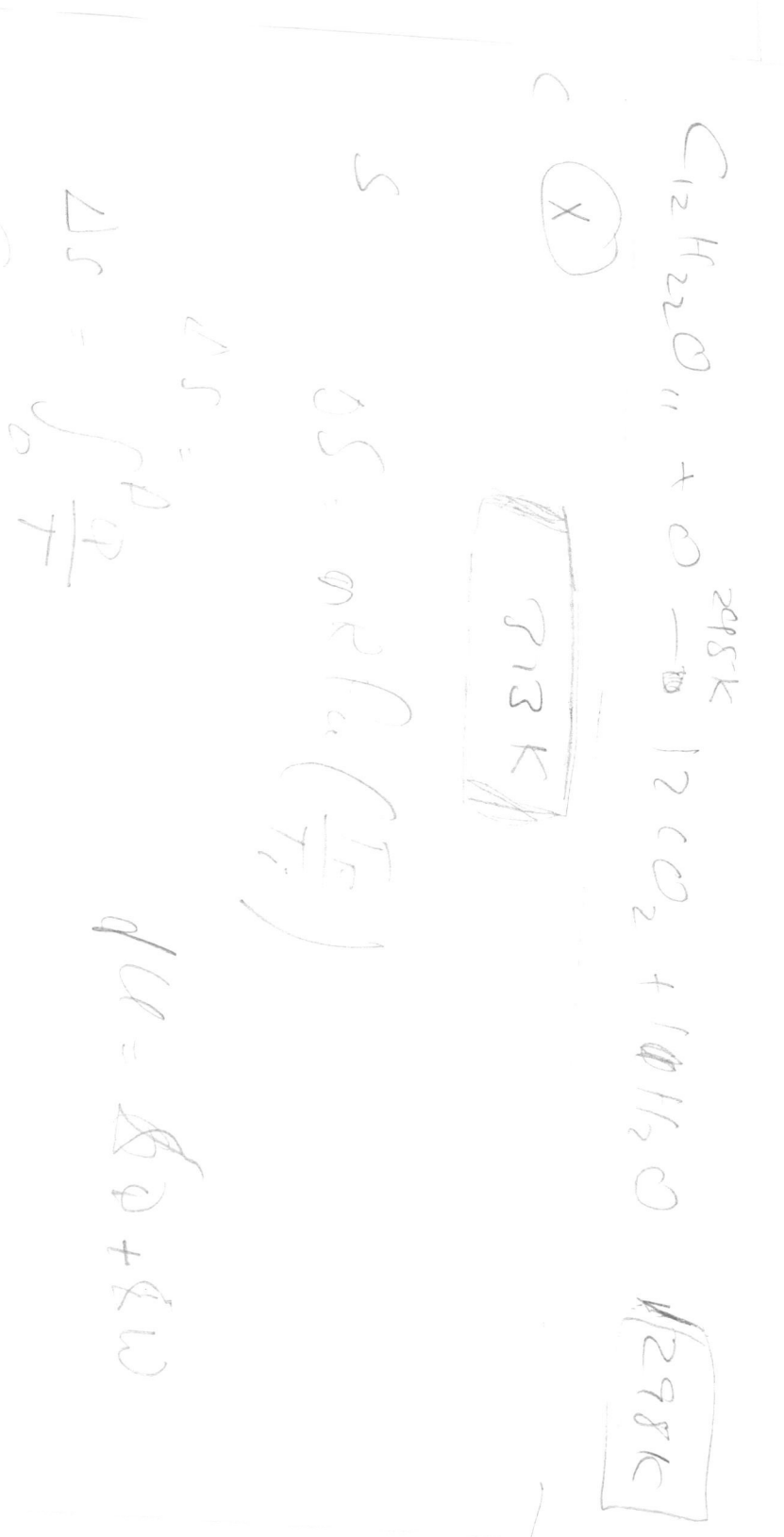

$C_{12}H_{22}O_{11} + O_2 \xrightarrow{298K} 12CO_2 + 11H_2O$ ΔH 298K

(x)

? 3 K

5

$\int_{0}^{5} p \, dp = 50$

$\left(\frac{1}{T_2}\right) \int dp = 50$

$Cp + \beta p = dp$

$5p - 2p = 25$

$\int_{0}^{T} \frac{dp}{p} = 25$

$\int_{0}^{P} dp = 25$

25

FIGURE 18-2 The part of the electromagnetic spectrum that is relevant to biochemistry.

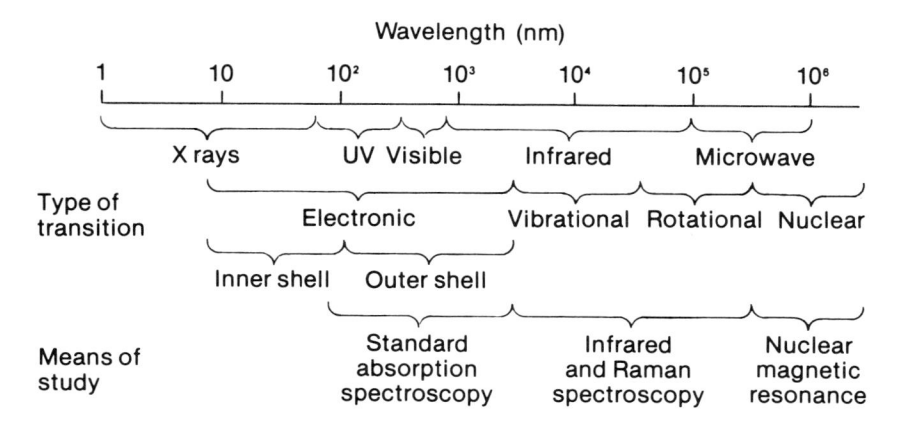

intensity of a beam of light equals the number of photons per unit area per second.

18–2 ABSORPTION OF LIGHT

When a light wave encounters matter, it may be *reflected, absorbed, transmitted,* or *refracted,* singly or in combination. (We will not discuss reflection or refraction in this book.) If the transmitted light moves in any direction other than that of the original light beam, it is said to be *scattered.* For many purposes, absorption of light by a sample substance is simply detected by measuring the difference between the intensity of the incident light beam and the intensity of the light that passes through the sample and continues onward in the same direction; hence, operationally, light that may have been scattered rather than absorbed is still counted as absorbed light. The reader should keep this in mind, because *in the analysis of photochemical reactions, we must consider only true absorption.*

The Beer-Lambert Law

Owing to the quantum nature of light, only an entire photon can be absorbed, never part of one. That is, a photon is either absorbed totally or not at all. Thus, the fraction of a light beam absorbed by a particular substance indicates the fraction of the photons that are absorbed. If 60 percent of a light beam is absorbed, then each of the following is true:

1. The intensity of the transmitted light beam is $100 - 60 = 40$ percent of the intensity of the incident beam.
2. 60 percent of the photons falling on the sample do not emerge from the sample.

3. The probability that a single photon will not be transmitted is 0.6.

The probability of absorption of photons by a sample of matter is a function of two factors—the thickness of the sample and the probability that a single molecule will absorb a single photon. This can be expressed quantitatively in the following way. Consider a beam of light of intensity I passing through an absorber of thickness dx and emerging with intensity $I + dI$. If N is the number of quanta in the beam and dN is the number of quanta absorbed, then the fraction absorbed, or, the probability of absorption, is dN/N. This is proportional to the number of molecules in the layer or $c\,dv$, in which c is the concentration and dv the volume of the absorber. If we let the area of the absorber be 1 cm², then $dv = dx$, and

$$\frac{dN}{N} = ac\,dx, \tag{3}$$

in which a is the proportionality constant. The intensity I is just the number N of quanta, so that $I + dI = N - dN$; or $dN = -dI$; Equation 3 then becomes

$$\frac{-dI}{I} = ac\,dx, \tag{4}$$

which can be integrated to yield

$$\ln I = -acx + \text{constant}. \tag{5}$$

When $x = 0$, there is no absorption, and $I = I_0$, the intensity of incident light, so that the constant is $\ln I_0$; thus

$$\ln\left(\frac{I}{I_0}\right) = -acx, \tag{6}$$

or,

$$I = I_0 e^{-acx}. \tag{7}$$

Equations 6 and 7 indicate that there is an exponential decrease in the intensity of transmitted light with increasing concentration and sample thickness. The equations are valid (1) for homogeneous samples, in which case the concentration c is related to the density and molecular weight M of the substance by the equation $c = \rho/M$; and (2) for solutions, in which case c is the molar concentration of the absorber.

By convention, Equation 6 is rewritten with a base-10 logarithm as $2.303 \log (I/I_0) = -acx$, and the constant a is replaced by $\epsilon = a/2.303$, so that Equation 6 becomes

$$\log_{10}\left(\frac{I}{I_0}\right) = -\epsilon c x, \tag{8}$$

which is called the Beer-Lambert Law.

Absorption is usually described quantitatively in one of two ways: as (1) the percent transmission, or 100 (I/I_0), or, more commonly, as (2) the *absorbance*,

$$A = -\log(I/I_0) = +\epsilon cx . \tag{9}$$

When the sample thickness is standardized as 1 cm, the absorbance is called the *optical density*, abbreviated as "OD."* Thus

$$OD = \epsilon c .$$

Usually, calculating absorbance is more useful than calculating the percent transmission because absorbance is proportional to concentration. Since absorption measurements may be carried out at various wavelengths, one always states the wavelength, λ, that is used, by writing either A_λ or OD_λ. In this notation, λ is universally expressed in nanometers, so that the absorbance or optical density at 260 nm = 2600 Å is written A_{260} or OD_{260}, respectively.

What is the significance of the constant ϵ? From Equation 3 we see that $\epsilon = a/2.303$ must be related to the probability that a single molecule will absorb a photon. Equations 6 through 8 are written in terms of concentrations rather than molecules, so that if c is expressed in molarity and x is 1 cm, then ϵ is the optical density of a sample 1 cm thick having a concentration of 1 molar—ϵ is called either the *molar extinction coefficient* or the *molar absorption coefficient*.

EXAMPLE 18–A **Calculation of the optical density of a solution at concentration c_1 if the optical density at another concentration c_2 is known.**

A DNA solution at a concentration of 20 μg/ml has an OD_{260} of 0.4. At a concentration of 37 μg/ml, according to the formula given above for OD, OD(20 μg) = 0.4 = 20ϵ and OD(37 μg) = 37ϵ, so that OD(37μg) = 37(0.4/20) = 0.74.

EXAMPLE 18–B **Calculation of the molar absorption coefficient of DNA from the data of the preceding example.**

Since DNA molecules can have many molecular weights, it is conventional to express the molar concentration as moles of nucleotide monomer per liter. The average molecular weight of a nucleotide is 336. Therefore, 20 μg/ml corresponds to

$$\frac{20 \times 10^{-6} \times 1000}{336} = 5.95 \times 10^{-5} M .$$

*Increasingly, optical density measurements are based on sample thicknesses other than the standard of 1 cm. Although both OD and A will be used in this book, A will be used more often.

From the formula for OD,

$$0.4 = 5.95 \times 10^{-5} \epsilon, \quad \text{or} \quad \epsilon = 6722.7.$$

EXAMPLE 18–C Calculation of the molar absorption coefficient of a pure liquid.

At a wavelength of 182 nm the percent transmission of a liquid 1 mm thick is 4.2 percent. The liquid has a density of 0.9 gm/ml and a molecular weight of 78. What is the molar absorption coefficient?

$I/I_0 = 0.042$, so that $\log(I/I_0) = -1.377$. Since the sample thickness is 1 mm, the optical density is 13.77. The molarity of the liquid is $900/78 = 11.54$, so that $\epsilon = 13.76/11.54 = 1.19$.

The Beer-Lambert Law is not always obeyed. That is, when a solution has a high molar concentration, solute molecules sometimes dimerize or interact in other ways and the molar absorption coefficient of the dimer may be either greater than or less than that of the solute at low concentration. In these two cases, there is a positive or a negative deviation from the Beer-Lambert Law, respectively. The deviation may usually be eliminated by appropriate choice of the wavelength because a change in the absorption coefficient is frequently accompanied by a change in the dependence of ϵ on λ (such a change is called a spectral shift). Figure 18-3 gives an example of this. Note that if λ_1 is used, there will be a positive deviation, and for λ_2 the deviation will be negative, but at λ_3, the *isosbestic point*, Equation 8 will be obeyed at all concentrations.

EXAMPLE 18–D Calculation of the molar absorption coefficient of a substance X showing positive deviation from the Beer-Lambert Law.

The OD of X at a concentration of $0.5\ M$ is 1.5. At $0.25\ M$ the OD = 0.9 rather than the expected value of 0.75. Successive dilutions yield the following values: at $0.15\ M$, OD = 0.54; at $0.075\ M$, OD = 0.27; at $0.05\ M$, OD = 0.18. Note that below $0.25\ M$, the OD is proportional to the concentration, so that any of these values can be used to calculate ϵ, which is $0.18/0.05 = 3.6$.

The equations just derived are applicable when true absorption has occurred. We pointed out, though, that the scattering of light will also reduce the intensity of the transmitted light and, indeed, Equation 8 can be applied when a sample at a low concentration has caused light to scatter. The proportionality of OD to concentration results from the fact that,

SEC. 18-2 ABSORPTION OF LIGHT

FIGURE 18-3 Positive and negative deviation from the Beer-Lambert Law and the causes. At the left is a spectral shift associated with increasing concentration—often a result of polymerization. Note that at one wavelength, λ_2, there is no change in the molar absorption coefficient with change in concentration. This wavelength is called the isosbestic point. At the right is a curve showing deviation from the Beer-Lambert law. At λ_1 the deviation is positive and at λ_3 it is negative. At the isosbestic point, λ_2, it is obeyed at all concentrations.

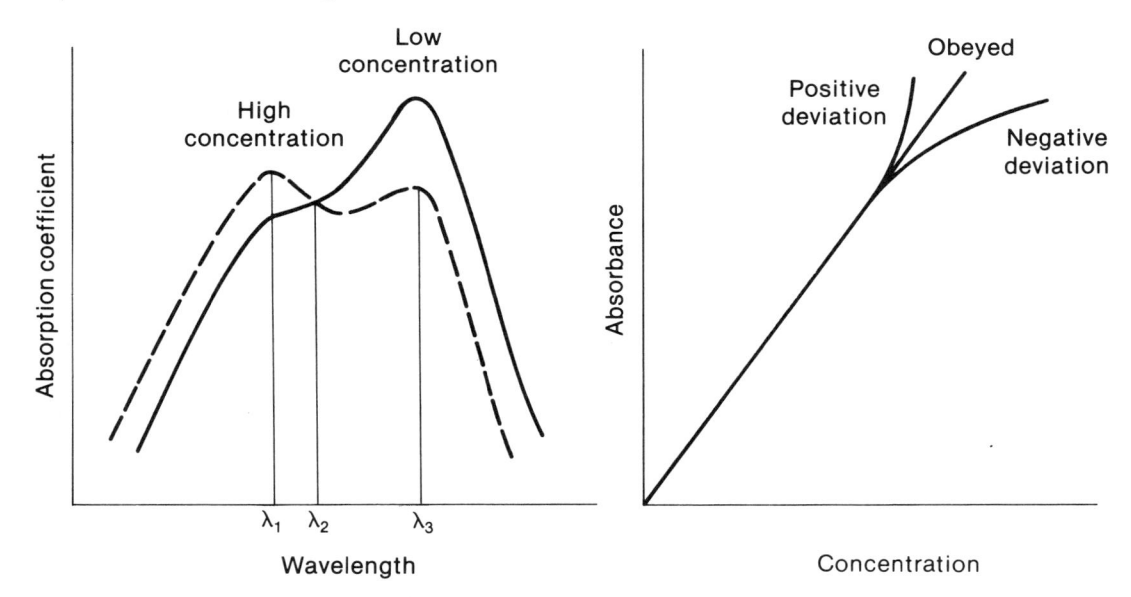

just as a single encounter between a photon and a molecule can, with a certain probability, result in absorption, a single encounter can also result in scattering. When light has been scattered, it is more difficult to apply the Beer-Lambert Law, though, because we must account for the fact that photons are sometimes scattered through various angles. Nonetheless, the Beer-Lambert Law is useful for determining the concentration of small particles, such as bacteria. The molar absorption coefficient has no real meaning in this case.

There may also be deviations from Equation 8 at high concentration because of multiple scattering. That is, a photon can be deflected away from the path of the incident beam at one scattering center and then be further deflected by a second particle, so that the photon is registered by a detector. This gives rise to negative deviation from the Beer-Lambert Law.

EXAMPLE 18-E Determination of the concentration of bacteria in a sample.

Absorption measurements of various bacterial suspensions yield the following data: for 10^8 cells per ml, $A_{650} = 0.24$; for 2×10^8 cells, $A_{650} = 0.48$; for 4×10^8 cells, $A_{650} = 0.96$; for 6×10^8 cells, $A_{650} = 1.12$; for 10^9

cells, $A_{650} = 1.4$. What are the cell concentrations of two suspensions having A_{650} values of 0.32 and 1.35?

Equation 8 is obeyed up to $A_{650} = 0.96$. Therefore, the cell density corresponding to $A_{650} = 0.32$ is $(0.32/0.24) \times 10^8 = 1.33 \times 10^8$ cells per ml. The cell density at $A_{650} = 1.35$ could be obtained from a graph of cell density versus A_{650} but such a graph would be inaccurate because a large change in cell density produces a small change of A_{650} in this region. The method of choice is to dilute the sample—for example, fourfold. If the A_{650} after the dilution were 0.53, the cell density would be

$$4 \times \frac{0.53}{0.24} \times 10^8 = 8.8 \times 10^8 \text{ ml} .$$

One type of positive deviation from Equation 8 is of special interest. Suppose a molecule has a low molar absorption coefficient for a particular wavelength but at a critical concentration the molecules aggregate to produce such large structures that the light is scattered significantly. Although there will be no real change in the amount of absorption, the amount of transmitted light will decrease, so that there will be an apparent increase in optical density. This phenomenon occurs with fatty acids and detergents that, at high concentrations, form *micelles*—large spherical aggregates in which the nonpolar portion of the molecule is inside the sphere and the polar portion is on the surface of the sphere in contact with the solvent. Micelles are discussed in detail in Chapter 12. The optical change that accompanies micelle formation is used to determine the critical micelle concentration, as is shown in Figure 18-4.

FIGURE 18-4 Increase in apparent absorbance by formation of a micelle at a particular concentration of a fatty acid, detergent, or other micelle former. Since this is a scattering rather than an absorption phenomenon, short wavelength light (ultraviolet) is usually employed in order to maximize the scattering; this is because the intensity of scattered light is proportional to λ^{-4}, in which λ is the wavelength, as shown by Equation 14 in this chapter.

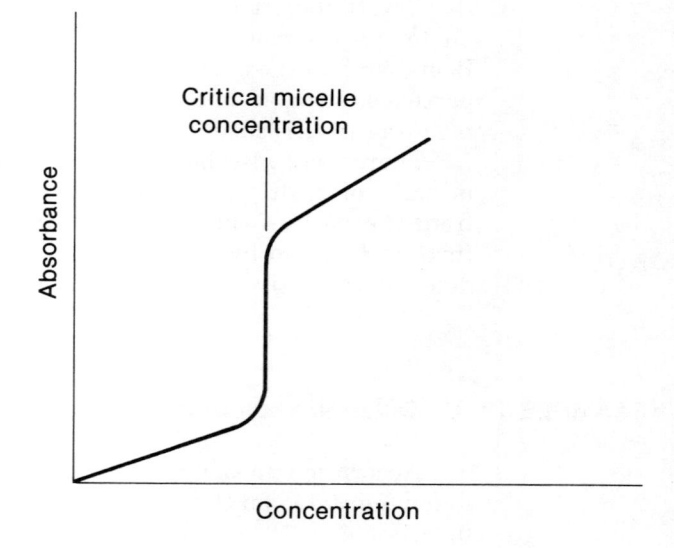

A similar optical change can be used to detect a change in a state of aggregation that is not a result of concentration changes. For example, most proteins are least soluble at their isoelectric point, that is—at the pH value for which the protein has no net charge. Sometimes the solubility is so low that a very fine precipitate occurs if the pH of a protein solution is adjusted to the isoelectric point. The precipitate usually has so little mass that it does not settle out and, instead, the initially clear solution becomes slightly turbid. This turbidity is easily detected by absorbance measurements, as is shown in Figure 18-5.

FIGURE 18-5 The absorbance at 350 nm of a solution of β-lactoglobulin at various pH values. The maximum absorbance occurs at pH 5.2, the isoelectric point.

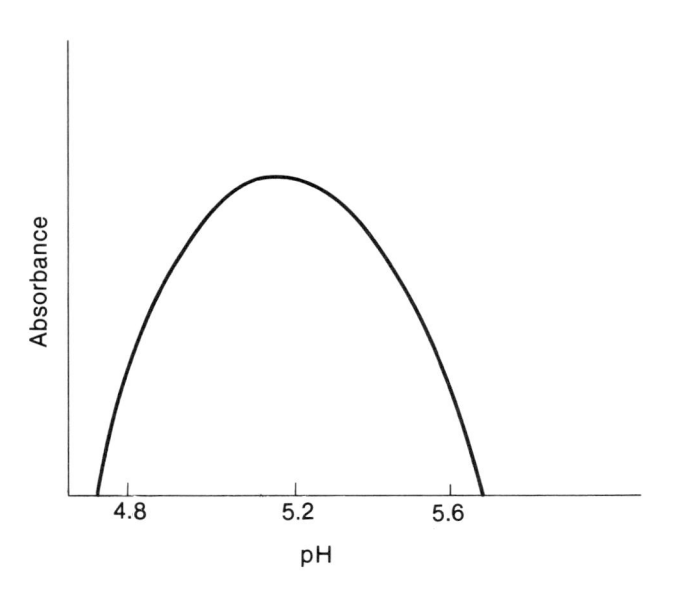

Application of the Beer-Lambert Law to Mixtures

We have just seen how Equation 8 can be used to determine the concentration of a substance in solution. If two substances are present and if their spectra—that is, the dependence of ϵ on wavelength λ—differ, the concentration of each can be determined by measuring the absorbance of two values of λ.

The absorbance of a mixture is simply the sum of the absorbance of its components, so that for a solution containing two substances X and Y, the total absorbance, A_{tot}, is

$$A_{\text{tot}} = A^X + A^Y = \epsilon^X[X] + \epsilon^Y[Y] . \tag{10}$$

If the measurement is now made at two wavelengths, λ_1 and λ_2,

$$A_1 = \epsilon_1^X[X] + \epsilon_1^Y[Y] \qquad \text{and} \qquad A_2 = \epsilon_2^X[X] + \epsilon_2^Y[Y] ,$$

which can be combined to yield

$$\left. \begin{array}{l} [X] = \dfrac{\epsilon_2^Y A_1 - \epsilon_1^Y A_2}{\epsilon_1^X \epsilon_2^Y - \epsilon_2^X \epsilon_1^Y} \,; \\[1em] [Y] = \dfrac{\epsilon_1^X A_2 - \epsilon_2^X A_1}{\epsilon_1^X \epsilon_2^Y - \epsilon_2^X \epsilon_1^Y} \,. \end{array} \right\} \quad (11)$$

If instead of the concentration of each component, one wants to know the total concentration of both, a measurement at an isosbestic point can provide this. An isosbestic point λ_{iso} for two substances is a wavelength at which the extinction coefficients of both are equal. That is, at λ_{iso}, $\epsilon_\lambda^X = \epsilon_\lambda^Y = \epsilon_{iso}$. At λ_{iso}, Equation 10 then becomes

$$A_{iso} = \epsilon_{iso}[X] + \epsilon_{iso}[Y] = \epsilon_{iso}([X] + [Y]) = \epsilon_{iso}[\text{total}] \,. \quad (12)$$

Isosbestic points are not present for all types of substances, however. Figure 18-6 shows a pair of spectra having three isosbestic points and a pair that has no isosbestic points.

FIGURE 18-6 (a) Spectra of two compounds for which there are three isosbestic points (arrows). (b) Spectra of two compounds for which there is no isosbestic point.

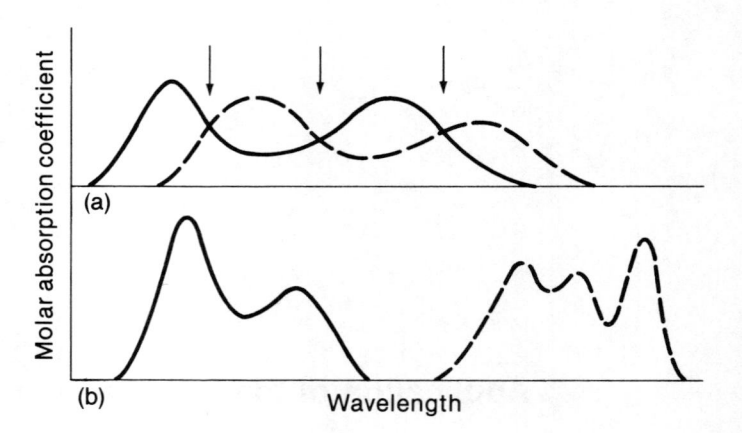

A particularly simple method of determining concentrations in multicomponent systems arises if there exist values of λ at which one substance absorbs and the other does not. For instance, in Figure 18-7, a component I can be measured at λ_1 and component II at λ_2. This is even possible if there is only one wavelength that is not common to the spectra of both substances. Thus, in Figure 18-7(b) the amount of component II is determined at λ_2 by subtracting from the measured value at λ_2 the absorbance at λ_2 that has resulted from component I. This is shown in the following example.

FIGURE 18-7 Examples of wavelengths useful for determining the concentrations of two components in a mixture. In (a) measurements at λ_1 and λ_2 allow independent determination of the concentrations of I and II. In (b), the amount of I is determined by measurement at λ_1. Knowledge of the spectrum of I enables one to calculate the absorbance A_2^I of I at λ_2. Measurement at λ_2 gives A_2 which is the sum of A_2^I and A_2^{II}. Thus, A_2^{II} is calculated as $A - A_2^I$.

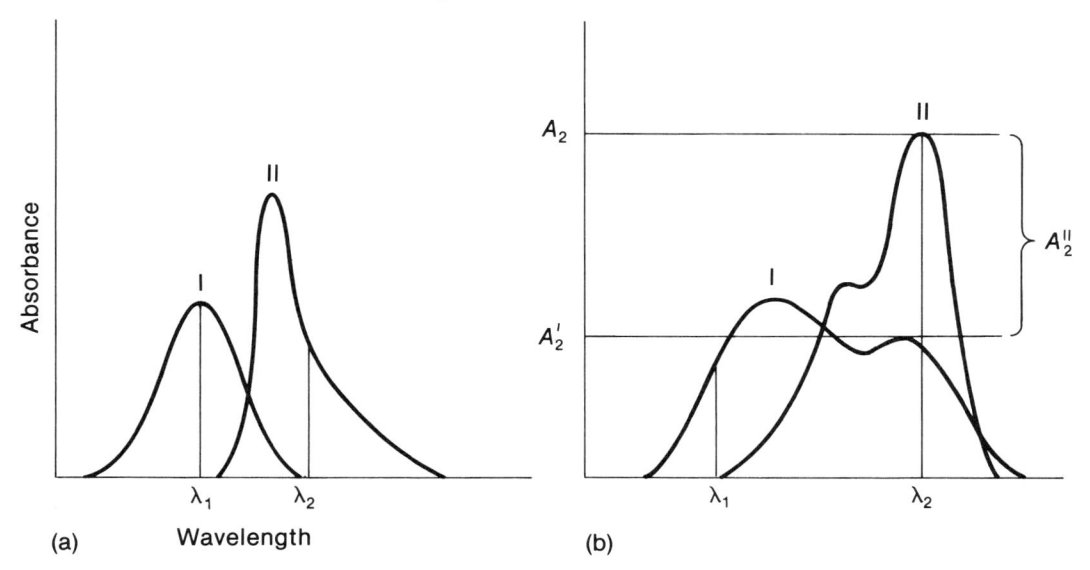

EXAMPLE 18-F **Determination of the amount of a contaminant in a DNA sample.**

The optical density of a DNA solution is found to be 1.32 at 260 nm and 1.15 at 280 nm. For DNA, $\epsilon_{260} = 6722$ and $\epsilon_{280} = 3908$. A contaminating substance is known to have $\epsilon_{260} = 1.3$ and $\epsilon_{280} = 1620$. The absorbance of this substance at 260 nm is so small that it can be neglected. Therefore, the DNA concentration is $1.32/6722 = 1.96 \times 10^{-4}\ M$, so that its contribution to the absorption at 280 nm is $(1.96 \times 10^{-4})(3908) = 0.77$. Therefore, the absorbance at 280 nm that is due to the contaminant is $1.15 - 0.77 = 0.38$, and its concentration is $0.38/1620 = 2.35 \times 10^{-4}\ M$.

18-3 SCATTERING OF LIGHT

In our discussion of the interaction of light and matter, we have so far considered only the absorption of light by molecules. However, light is also scattered. That is, a light beam falling on matter can be deflected so that all of it (or a fraction of it) is no longer moving in the original direction. To understand this phenomenon we must remember three things: (1) light consists of a sinusoidally oscillating electric and magnetic field; (2) atoms contain charged species—namely, protons and electrons; and (3) the charged particles are acted on by the force of the electric field. Therefore, if we imagine that, for an instant, the electric field of a light ray is not oscillating, then at that instant the electrons and protons will be slightly displaced in opposite directions, because the field has a direction and an electron and a proton have opposite charges. When two charges $+q$ and $-q$ are separated by a distance d, these charges constitute an electric dipole having a dipole moment $\mu = qd$. Thus, the charge separation means that the atom or molecule possesses an instantaneous dipole moment. The magnitude of μ is proportional to the field strength E at that instant. The constant of proportionality is called the polarizability, α (explained in Chapter 3), and its value is a complicated function of the spatial distribution of electrons and the energy of each electron. In reality, the electric field of the light oscillates sinusoidally between the values $+E$ and $-E$, in which the sign determines the direction of the electric field, so that the dipole also oscillates sinusoidally in magnitude and direction. Since the direction of the electric field of light is perpendicular to the direction of propagation, the direction of μ is also perpendicular to the direction of propagation.

The importance of these considerations rests on the fact that an oscillating dipole is an emitter of electromagnetic radiation because it produces an oscillating electric field. Furthermore, this light is propagated in all directions. Therefore, the energy of an incident light wave can induce a dipole to oscillate (that is, the incident energy is converted to the energy of oscillation) and the oscillating dipole produces light. The incident light is not absorbed and reemitted; rather, it is dispersed, so that its energy is no longer propagating in a single direction but in all directions. The probability of this scattering—that is, the fraction of the incident light that is scattered—depends upon the wavelength of the incident light: the more energetic the incident light, the greater is the probability that a dipole will be induced, since more energy will be available to separate the charges. Thus the intensity of the scattered light (in other words, the fraction of the light that is scattered) increases as the wavelength of the incident light decreases. Furthermore, the induced dipole is not spherically symmetric, so that light is not emitted with the same intensity in all directions.

We will not derive the scattering equation (the equation relating scattered intensity and incident wavelength) but simply state it with reference to Figure 18-8. This diagram shows the geometry of the scattering of light by a molecule or particle whose maximal dimensions are small compared to the wavelength of the incident beam. This restriction is important, because rather different and far more complex equations are needed for large particles. The equation describing the scattering of incident *un*polarized light (the most common experimental arrangement) by particles whose dimensions are small compared to the wavelength of the incident light is

SEC. 18-3   SCATTERING OF LIGHT

$$I = I_0 \frac{8\pi^4 \alpha^2 a(1 + \cos^2\theta)}{r^2 \lambda^4}, \tag{13}$$

in which I_0 is the incident intensity and I is the intensity measured at an angle θ and distance r from the scattering center. (See Equation 44, Chapter 15.) Two points should be noted. First, the scattered light intensity is maximal at 90° from the direction of the incident beam. Second, the wavelength dependence is $1/\lambda^4$. This second point is one of the most distinctive features of the scattering of light.

Light scattering accounts for several well-known phenomena. At sunset, because of the low angle of the sun's rays with respect to the surface of the earth, the white light of the sun passes through about 200 km of atmosphere. The molecules of air scatter the light. Blue light has a shorter wavelength than red and is scattered at right angles from our line of sight to the sun. The red light is scattered less efficiently, so that more red light is transmitted without deflection; in the line of sight the sun appears red. If the air is especially dusty, an even greater fraction of the blue light is scattered and the sun appears redder.

A related phenomenon is seen when samples of concentrated suspensions of macromolecules, such as proteins, and of small particles, such as viruses, are observed. These suspensions appear turbid, because only a relatively small amount of light is transmitted, and also bluish, because one is usually viewing the sample at an angle (hence is not in a position to receive the transmitted light) and mainly sees the scattered light, which is the shorter-wavelength blue light. When the transmitted light is viewed by looking through the suspension at the light source, it appears to be a pale reddish brown.

Scattering interferes with measurements of the absorbance of mole-

FIGURE 18-8 The scattering of light by a particle that is small compared to the wavelength of the incident light. The incident beam induces a dipole moment in an otherwise isotropic (directionless) particle. The magnitude of the dipole moment changes as the electric field of the absorbed light oscillates, and this oscillating dipole moment is responsible for the emission of the scattered light.

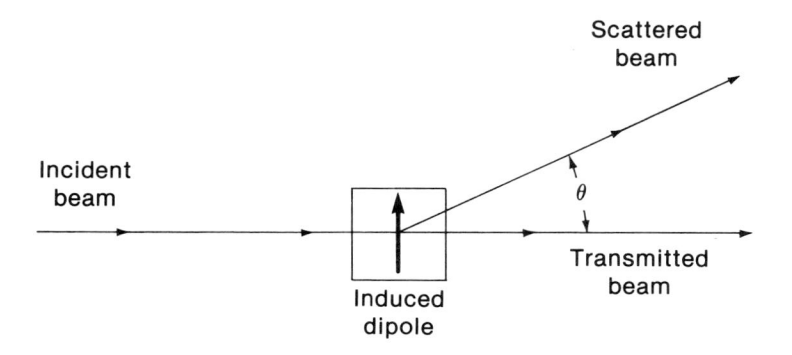

FIGURE 18-9 Spectrum of *E. coli* phage T7 showing the λ^{-4} scattering correction. [SOURCE: *Physical Biochemistry*, First Edition, by David Freifelder. W.H. Freeman and Company. Copyright © 1976.]

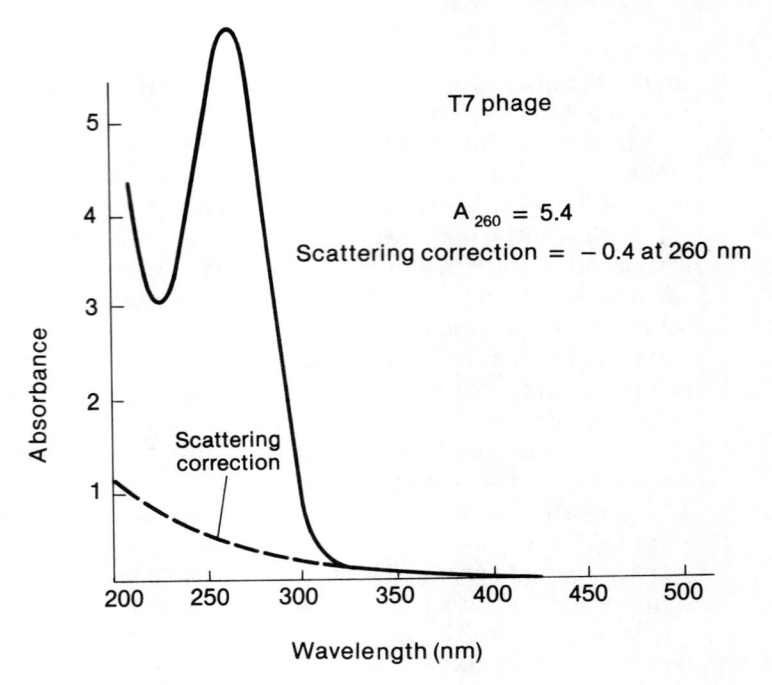

T7 phage

$A_{260} = 5.4$
Scattering correction $= -0.4$ at 260 nm

cules, because the transmitted light intensity equals the incident intensity minus the *sum* of the absorbed and scattered light intensities. In order to determine the true absorbance, the intensity loss resulting from scattering must be subtracted. This can be done in a simple way, shown in Figure 18-9. The apparent absorbance is measured (in the direction of the incident beam) at wavelengths that are not absorbed. A graph is made of the absorbance versus $1/\lambda^4$. By extrapolation to a wavelength that is absorbed, the reduction in intensity owing to scattering at that wavelength is obtained, and this is subtracted from the measured absorbance.

EXAMPLE 18–G Determination of the amount of a colored product P produced in an enzymatic reaction.

Enzyme reactions are frequently studied by measuring the conversion of a colorless substrate to a colored product. However, crude enzyme preparations are often turbid, and this results in errors in absorbance measurements. For one such enzyme, the product P has an absorption maximum at 350 nm and a molar absorption coefficient of 2133. It has no detectable absorption above 450 nm. In an enzymatic reaction, the A_{350} is found to be

0.782. To correct for scattering, the values $A_{500} = 0.085$ and $A_{600} = 0.052$ are obtained. Since the absorbance is proportional to the scattered-light intensity, we plot A versus $1/\lambda^4$. The slope of the line connecting the two points is

$$\frac{0.085 - 0.052}{500^4 - 600^4},$$

which must equal

$$\frac{x - 0.085}{350^4 - 500^4},$$

in which x is the absorbance at 350 nm resulting from scattering only. Solving for x, we have $x = 0.108$. Therefore, the true absorbance is the measured absorbance minus the apparent absorbance owing to scattering, or $0.782 - 0.108 = 0.674$, and the concentration of P must be $0.674/2133 = 3.16 \times 10^{-4}\ M$.

18–4 PHOTOCHEMICAL MECHANISMS

Three results are possible when a substance absorbs a photon. (1) The temperature of the absorbing substance may increase because the energy of the photon is converted to heat. (2) There may be fluorescence, in which a photon of the same or lower energy is emitted. (3) Chemical changes may result as the energy of the photon alters the bonding structures. Any combination of these three results may also occur. In this section, we will consider the third result.

The Energy of Absorbed Light: The Einstein

The absorption of light by matter is a process by which the absorbing molecule takes up the energy of a single quantum. This energy is equal to hc/λ or $h\nu$, according to Equation 2. When considering light-stimulated chemical reactions, it is convenient to use molar quantities; in these terms, the energy E taken up by one mole of molecules, each molecule absorbing one quantum, is

$$E = \mathcal{N}h\nu = \mathcal{N}hc/\lambda$$

in which $\mathcal{N}$ is Avogadro's number. This quantity of light is called an *einstein*. Knowing the amount of energy in an einstein makes it possible to know which chemical bonds in a substance can be broken by light of a particular frequency, since a bond can be broken only if the energy of a quantum exceeds the bond energy.

EXAMPLE 18–H Determination of the ability of green light to break a C-C bond.

One einstein of green light, for which $\lambda \approx 550$ nm, has energy $E = \mathcal{N} hc/\lambda$, or

$$E = \frac{(6.02 \times 10^{23})(6.62 \times 10^{-34})(3 \times 10^8)}{5.50 \times 10^{-7}}$$

$$= 2.17 \times 10^5 \text{ J} = 217 \text{ kJ}.$$

The energy of a C-C bond is about 340 kJ mol^{-1}, so that green light is insufficient to break this bond. The maximal wavelength that could break a C-C bond is

$$\frac{217}{340} \times 550 = 351 \text{ nm}$$

or, light that is in the near ultraviolet.

Quantum Yield

In our discussion of the Beer-Lambert Law, it was evident that when light falls on matter, all of the incident quanta are not absorbed. The energy of those that are absorbed can, as was said previously, follow three paths: (1) it can be converted to heat; (2) it can reappear as fluorescence; or (3) it can cause a chemical reaction. The fraction of the absorbed photons that follow either path 2 or path 3 is called the *quantum yield*, Q, for that path. For photochemical processes, it is defined as

$$Q = \frac{\text{number of molecules reacted}}{\text{number of quanta absorbed}}$$

$$= \frac{\text{number of moles reacted}}{\text{number of einsteins absorbed}}. \quad (14)$$

We will discuss the quantum yield for fluorescence in a later section.

EXAMPLE 18–I Determination of the quantum yield for the reaction A → B + C when 1.52 kJ of energy derived from light of wavelength 465 nm decomposes 7×10^{-4} mole of A.

One einstein of light of this wavelength is

$$\frac{\mathcal{N}hc}{\lambda} = \frac{(6.02 \times 10^{23})(6.62 \times 10^{-34})(3 \times 10^8)}{4.65 \times 10^{-7}} = 2.57 \times 10^5 \text{ J}$$

$$= 257 \text{ kJ}.$$

The number of einsteins absorbed is $1.52/257 = 0.0059$. The quantum yield is $(7 \times 10^{-4})/0.0059 = 0.12$.

Actinometry—The Measurement of Light Energy

In the preceding example, we calculated the amount of absorbed energy, but we have not yet explained how that number is obtained. The Beer-Lambert Law enables us to measure the fraction of the incident energy that is absorbed; to determine the amount we need only know the energy of the incident radiation. The measurement of the energy content of light is called *actinometry*. Actinometry can be accomplished through both direct and indirect methods. The direct method uses a device called a *thermopile*, an instrument based upon the simple physical principle that a truly black surface absorbs all light and that the light energy is converted totally to heat. A thermopile consists of a blackened metallic surface contained in an evacuated chamber. Under the surface are thermocouples, which convert the increase in temperature to a voltage that can be read on a meter. To measure the energy of a light beam, the thermopile (the area of which is accurately known) is placed in the beam and the temperature that is reached after a period of time is measured. (The reason the measurement is conducted in an evacuated chamber is to reduce heat loss.) In this way, the *flux* or *incident energy per unit area per unit time* is determined. It should be realized that because of the divergence of light from a light source, it is necessary to remove the sample temporarily and measure the flux at the precise position that would be occupied by the sample being illuminated. Thus, if the flux is 3.42×10^{-4} J per cm^2 per second, the total energy incident on a sample contained in a beaker 10 cm (0.05 m) in diameter after a period of one minute is $(3.42 \times 10^{-4}) \times 0.05^2 \times \pi \times 60 = 1.61$ J.

It is sometimes inconvenient to use a thermopile, and in fact, many laboratories do not even bother to acquire this instrument. This is because the indirect method of *chemical actinometry* is so extremely simple. In chemical actinometry, a photochemical reaction whose quantum yield is accurately known (having once been determined with a thermopile) is carried out in the vessel used to contain the sample to be illuminated. The simplicity of this method is shown in the following example; when studying this example, it should be kept in mind that a chemical actinometer can be used only in a limited range of wavelengths (those absorbed by the primary absorber) whereas a thermopile can be used at any wavelength, because it is a black body.

EXAMPLE 18–J Determination of the flux of light using uranyl oxalate actinometry.

A solution is prepared containing a uranyl salt (usually the acetate or nitrate). An oxalate salt is added, and the solution is illuminated. If the wavelength of the light is between 250 and 450 nm, the light will be absorbed by the uranyl ion (UO_2^{2+}), which thereby becomes excited (UO_2^{2+})*. The excited uranyl group then reacts with the oxalate ion. The scheme is the following:

$$UO_2^{2+} + h\nu \rightarrow (UO_2^{2+})^* \text{ ;}$$

$$(UO_2^{2+})^* + (COOH)_2 \rightarrow UO_2^{2+} + CO_2 + CO + H_2O \text{ .}$$

The concentration of oxalate is determined after illumination by adding $KMnO_4$ (a purple salt) and observing its reaction with oxalate to form a colorless solution. The quantum yield of the uranyl-stimulated photodecomposition is known to be 0.5, so that if a given flux of light decomposes 10^{-3} mole of oxalate, the number of einsteins absorbed must be $10^{-3}/0.5 = 2 \times 10^{-3}$. If the optical density of the uranyl oxalate solution is such that 25 percent of the incident light is absorbed, then the flux of incident light is

$$\frac{2 \times 10^{-3}}{0.25} = 8 \times 10^{-3} \text{ einstein .}$$

Since the value of one einstein is dependent on wavelength, the calculated value of the flux must be made for a particular wavelength. The way in which the wavelength enters this calculation is in the fraction of the light that is absorbed—that is, the absorption of 25 percent that has been used is for a particular wavelength.

Other chemical reactions, whose quantum yields are also known, are used for wavelengths outside of the range 250–450 nm.

Types of Photochemical Reactions

Photochemical reactions are of two sorts: (1) The absorption of a quantum causes a chemical change (for example, dissociation) of the absorber; and (2) the absorber is excited, collides with different molecules, and thereby initiates a chemical change in the second molecule. The second process is called *photosensitization;* in this process the absorber is sometimes regenerated but may ultimately be changed itself. Both types of photochemical reactions occur frequently in nature and are used in the laboratory.

An example of the first type of reaction is the well-studied photodecomposition of hydrogen iodide, which occurs by the following scheme:

$$HI + h\nu \longrightarrow H + I \text{ ,}$$

$$HI + H \xrightarrow{k_2} H_2 + I \text{ ,}$$

and

$$I + I \xrightarrow{k_3} I_2 \text{ ,}$$

for which the complete balanced equation is $2HI + h\nu \rightarrow H_2 + I_2$. The rate of a primary reaction such as this one is proportional to the intensity of the

SEC. 18-4 PHOTOCHEMICAL MECHANISMS

light absorbed by the absorber. This is to be expected, because a photochemical reaction is initiated by only a single photon.* The proportionality can be demonstrated as follows.

We apply the steady-state approximation (see Chapter 11) to the concentration of H and of I to obtain

$$\frac{d[\mathrm{H}]}{dt} = I - k_2[\mathrm{H}][\mathrm{HI}] = 0 \tag{15}$$

and

$$\frac{d[\mathrm{I}]}{dt} = I + k_2[\mathrm{H}][\mathrm{HI}] - k_3[\mathrm{I}]^2 = 0, \tag{16}$$

in which I is the intensity of the light absorbed. The overall rate is

$$-\frac{d[\mathrm{HI}]}{dt} = I + k_2[\mathrm{H}][\mathrm{HI}], \tag{17}$$

which, when combined with Equation 15 rearranged as $[\mathrm{H}] = I/k_2[\mathrm{HI}]$, yields

$$-\frac{d[\mathrm{HI}]}{dt} = I + I = 2I. \tag{18}$$

This expression indicates that the quantum yield Q is 2, because

$$Q = \frac{\text{number of HI molecules reacted}}{\text{number of quanta absorbed}}$$

$$= \frac{\text{number of HI molecules reacted per second}}{\text{number of quanta absorbed per second}}$$

$$= -\frac{d[\mathrm{HI}]/dt}{I} = \frac{2I}{I} = 2.$$

This value of $Q > 1$ occurs because not only has the absorber decomposed but the products of the decomposition of the absorber have led to decomposition of a second molecule. Equation 18, which is written in terms of the starting substance, is quite simple, but if the rate is expressed in terms of the concentration of the product, more complicated equations result. The calculation of $d[\mathrm{I}_2]/dt$ is left to the reader.

It should be noted that if the monatomic iodine produced in the first two steps were able to react with HI also, the reaction would have continued and the quantum yield would have been greater than 2—that is, the reaction would have been a chain reaction (see Chapter 11). In these more complicated reactions, the rate expressions can also become quite complex.

In the reaction just described, light was present throughout the reaction. In the analysis of some reactions, this leads to considerable complexity, and a great simplification results if the first photolytic step of a series

---

*This statement is sometimes called the Einstein-Stark Law.

of reactions is completed very rapidly. It is possible to do this with a technique called *flash photolysis,* in which a powerful flash of light lasting about 10^{-5} to 10^{-4} seconds is applied to the reactants, and the concentration of the product is then followed as a function of time. Some reactions are also conveniently studied by using periodic flashes of light. Further information about these techniques can be found in the references at the end of this chapter.

A well-understood and simple example of photosensitization is the photodecomposition of H_2, a process that is catalyzed by Hg vapor and light of 254 nm, a wavelength strongly absorbed by Hg atoms. An Hg atom is activated by absorbing a photon and, upon colliding with an H_2 molecule, the activated Hg atom gives up its energy to the H_2, which then dissociates. The complete scheme is the following:

$$Hg + h\nu \rightarrow Hg^*$$

$$Hg^* + H_2 \rightarrow Hg + 2H\cdot$$

The activated Hg atom is designated with an asterisk. If only H_2 were present, the H atoms would recombine. However, in the presence of other substances, a continued reaction can occur. For example, Hg can photosensitize the synthesis of formaldehyde (HCHO) from H_2 and CO, by the reaction $H_2 + CO \rightarrow HCHO$, according to the following scheme:

$$Hg + h\nu \rightarrow Hg^*$$

$$Hg^* + H_2 \rightarrow Hg + 2H\cdot$$

$$H\cdot + CO \rightarrow HCO\cdot$$

$$HCO\cdot + H_2 \rightarrow HCHO + H\cdot$$

or

$$2\ HCO\cdot \rightarrow HCHO + CO\ .$$

The decomposition of oxalic acid used in the chemical actinometer described earlier is an example of photosensitization, with the absorber being the one uranyl ion.

Photochemical Reactions in Nature

As we saw in Chapter 11, many chemical reactions require an activation step—that is, an input of energy. In the case of photosynthesis, this energy is provided by the absorption of light.

The overall reaction of photosynthesis is

$$6CO_2 + 6H_2O \rightarrow C_6H_{12}O_6 + 6O_2\ ,$$

for which $\Delta G = 2.87 \times 10^6$ J mol^{-1}. Since $\Delta G > 0$, the reaction requires an input of energy, and this is provided by light. We can calculate the wavelength of light needed if a single photon were to provide all of this energy.

SEC. 18-4 PHOTOCHEMICAL MECHANISMS

Since the energy of a mole of photons is Nhc/λ, then $\lambda = Nhc/\Delta G =$

$$\frac{(6 \times 10^{23} \text{ mol}^{-1})(6.625 \times 10^{-34} \text{ J sec})(3 \times 10^8 \text{ m sec}^{-1})}{(2.87 \times 10^6 \text{ J mol}^{-1})} = 41.5 \text{ nm}.$$

If, instead, one photon were needed for each molecule of CO_2, the wavelength would be $6 \times 41.5 = 249$ nm. Light having a wavelength as short as this (in the ultraviolet) is highly destructive to important biological molecules such as DNA and is screened out by the ozone layer of the earth's atmosphere. Hence, if light is to be the source of the energy for the photosynthetic reaction, longer wavelengths will be required and the reaction, if it is to be totally dependent on light, must result in the absorption of several photons.

This is in fact what occurs—the reaction of one CO_2 molecule with one H_2O molecule requires about 8 photons. This would seem to violate the Einstein-Stark Law, if it were necessary for 8 photons to be absorbed simultaneously. There are two mechanisms by which the energy of several photons is summed, however. In the first, the reaction proceeds through many steps and the absorption of each photon allows a particular step to proceed. This is in fact part of what happens, in that there are two reactions in the photosynthetic sequence that are light-activated. However, if this mechanism were not supplemented, the Einstein-Stark Law would require that there be eight photochemical reactions. By the second mechanism, the energy of an absorbed photon is not used immediately but is transferred to an energy storage system. Then, a second photon is absorbed and its energy is likewise transferred to the same storage system. This continues until the storage system discharges its energy and thereby initiates a single chemical reaction.

The second of these two mechanisms occurs in photosynthesis in the following way. The principal absorber *chlorophyll* is situated within plant cells in large clusters of hundreds of molecules. The chlorophyll molecules are linked together and are capable of transferring the energy of an absorbed photon from one molecule to another. In this cluster there are a few chlorophyll molecules, known as a reaction center, that are associated with other molecules capable of initiating chemical reactions. The energy of several photons is gathered together in the reaction center and is discharged when the first chemical step begins. The entire pathway is very complex and not yet thoroughly understood. Details about photosynthesis can be found in references at the end of this chapter.

Vision is another biological example of a photochemical process. The reaction that produces vision differs in principle from photosynthesis in that the free energy change for the total reaction is negative, so that the reaction should proceed spontaneously. However, in Chapter 11 we described processes that require an activation energy. An example of such a process is one in which there are a series of reactions whose net free energy change, ΔG, is less than zero but for which the first step has a positive free energy change. Another example is a simple reaction in which one component must undergo a structural rearrangement before the reaction can proceed, and this rearrangement likewise requires an input of energy—this is in fact what occurs in the process of vision. The visual pigment *rhodopsin* is a complex made up of a protein, opsin, and a colored substance, 11-*cis*-retinal. A single photon is absorbed by rhodopsin and, with a quantum efficiency of 1, the energy of the photon causes an isomerization of the colored substance to all-*trans*-retinal, which then initiates the sequence of reactions that results in the sending of an electrical signal to the brain. This type of reaction, in which the energy of a photon is used to provide activation energy, is the most common type of photochemical reaction in nature.

The emission of light by fireflies is another biological example of a reaction involving light, except that in this reaction light is produced rather than absorbed. A substance called *luciferin* reacts with an adenosine triphosphate (ATP) molecule to form luciferin adenylate, which contains adenosine monophosphate (AMP) instead of ATP. The energy produced by cleavage of ATP to form AMP and pyrophosphate is used to initiate a reaction between luciferin and molecular O_2; this reaction yields water and the molecule oxyluciferin, which is in an excited state. The oxyluciferin discharges its excess energy by emitting a photon of visible light.

Photochemical smog, common in large cities in the United States, is the product of a reaction involving various nitrogen oxides and hydrocarbons produced in automobile exhaust and activated by absorption of near ultraviolet light by the nitrogen oxides. The principal reaction is thought to be the following. The pollutant gas NO_2 absorbs near-ultraviolet light and dissociates to NO and free oxygen atoms. The highly reactive oxygen atoms attack the gaseous hydrocarbon pollutants that are released in automobile exhaust and emitted by some industrial processes, to yield a variety of noxious compounds, such as aldehydes (which irritate the eyes and respiratory passages) and peroxynitrates. Some of the free oxygen atoms combine with molecular oxygen to produce ozone, O_3, which can participate in a large number of other reactions and can by itself produce significant biological damage.*

Relation between the Rate of a Photochemical Reaction and the Molar Absorption Coefficient

We have just seen that the reaction rate of a photochemical reaction is related to the intensity of incident light. Since light must be absorbed, the reaction rate should be related to the molar absorption coefficient. This

---

*The ozone is harmful because it occurs near the surface of the earth—namely, in the life zone. There is also an upper atmospheric ozone layer, which protects all living things from ultraviolet radiation by absorbing all light-wavelengths shorter than 340 nm. This ozone layer would of course also be harmful if it were near the ground, but the functions of the two locations of ozone should not be confused.

relation is derived as follows. We consider a simple reaction in which light converts a reactant X to a product B. The rate $d[\text{B}]/dt$ is proportional to the intensity of the absorbed light, I_{abs}. The constant of proportionality is the quantum yield Q; that is,

$$-\frac{d[\text{B}]}{dt} = QI_{\text{abs}} . \tag{19}$$

Equation 7 relates the incident intensity I_0, the absorbed intensity I, a constant a, the path length x, and the concentration c with the formula $I = I_0 e^{-axc}$. Since $I_{\text{abs}} = I_0 - I$, we may write

$$-\frac{d[\text{B}]}{dt} = Q(I_0 - I) = QI_0(1 - e^{-axc}) . \tag{20}$$

The exponential can be expanded as $e^{-axc} = 1 - axc + (axc)^2/2 + \cdots$, and if the solution is dilute, c is small and we may substitute $1 - e^{-axc} \approx axc = ax[\text{B}]$, to obtain

$$-\frac{d[\text{B}]}{dt} = QI_0 ax[\text{B}] . \tag{21}$$

This can be integrated from $t = 0$ and $[\text{B}]_0$ to obtain

$$\ln\frac{[\text{B}]}{[\text{B}_0]} = -QI_0 axt , \tag{22}$$

which, in terms of the molar absorption coefficient $\epsilon \; (= a/2.303)$, is

$$[\text{B}] = [\text{B}_0]e^{-2.303 I_0 \epsilon x t} , \tag{23}$$

which enables one to calculate Q if I_0, ϵ, and x are known. These equations also demonstrate the frequently forgotten fact that the reaction rate depends upon the thickness of the solution being illuminated (that is, the value of x).

18–5 PHOTOCHEMICAL DAMAGE TO BIOLOGICAL MOLECULES AND TO LIVING ORGANISMS

The principal type of photochemical damage sustained by living organisms is a result of chemical changes produced in the genetic material deoxyribonucleic acid (DNA). These changes are caused either by direct absorption of ultraviolet light or by photosensitized oxidation.

So far we have mainly discussed the effect of light on the decomposition of a substance or on the reaction of one substance with another. A pure substance can also be stimulated by light to react with itself. An important reaction of this type—the dimerization of the base thymine—

occurs in DNA that has been irradiated with light of wavelength of 255–265 nm. The reaction is the following:

$$2\left[\begin{array}{c}\text{thymine}\end{array}\right] \xrightarrow{h\nu} \text{thymine dimer}$$

The compound on the right is called a *thymine dimer* and is frequently denoted T̂T. Dimerization of cytosine also occurs, but this has not been extensively studied. Thymine dimers are of importance in living cells because they inhibit DNA replication and the copying of ribonucleic acid (RNA) from the DNA.

In the early years of the earth but after life forms evolved, the composition of the earth's atmosphere allowed a great deal of ultraviolet light to reach the surface of the earth. This might have destroyed all living things were it not for the fact that thymine dimers are not only created but can also be converted back to monomers by ultraviolet light. For such conversion, the amount of energy needed to break the new bonds in the dimer should be at least as great as that needed to form the bonds initially. It is possible for this condition to be met, since the absorption spectrum of the dimer is shifted to shorter wavelengths than that of the monomer. The dimer has an absorption maximum of 240 nm, and ϵ_{240} (dimer) is roughly equal to ϵ_{260} (monomer). A 240-nm photon has more energy than a 260-nm photon, so it does possess the necessary energy to monomerize the dimer. It is important to realize that this is a necessary but not a sufficient condition for monomerization. That is, in general, a bond formed as a result of absorption of light is not always capable of being broken later also by absorption of light; not all bonds are subject to photochemical change, owing to the numerous ways in which the energy of a photon can be utilized or lost as heat.

As life forms became more complex, another system for restoring monomers, known as *photoreactivation,* arose. Photoreactivation was discovered by Albert Kelner, who observed that the ability of bacteria to survive the lethal effect of ultraviolet radiation is enhanced if the cells are exposed to sunlight after the initial irradiation. Photoreactivation is a light-stimulated enzymatic process: a photoreactivating (PR) enzyme is associated with a small cofactor molecule that strongly absorbs light of a wavelength of 325–375 nm. Absorption of a 350-nm photon alone cannot provide the energy for monomerization but, as we learned in Chapter 11, when binding its substrate molecule, an enzyme can utilize the binding energy to stimulate a chemical reaction. Thus, the cofactor of the PR enzyme absorbs a photon and the activated enzyme binds to a thymine dimer. The energy of the photon plus that of binding sum to provide sufficient

SEC. 18-5 PHOTOCHEMICAL DAMAGE TO BIOLOGICAL MOLECULES 687

energy to break the dimer bonds. The detailed mechanism of this reaction and the identity of the light-absorbing cofactor are not yet known.

A third mechanism for removing thymine dimers has also evolved and is widespread in nature. By a series of enzymatic reactions, this repair system can remove an entire thymine dimer (and the sugar-phosphates to which each thymine is linked) from an irradiated DNA molecule and restore the DNA molecule to its initial unirradiated state.

The three repair processes that have just been described are shown diagrammatically in Figure 18-10.

In a later section we shall show how one detects both irradiation damage and repair in living organisms and will describe the kind of information that can be derived from experiments in which cells are irradiated.

The type of damage that has just been described is a result of direct absorption of a photon by the molecule that is altered. Another type of photochemical reaction can also occur, in which a molecule A binds an absorber B and a photon absorbed by B causes an alteration of A. This is called photosensitization. The most common reactions of this class in biological systems require molecular O_2 and the excited molecule B causes oxidation of A; hence, the reaction is called *photosensitized oxidation*. When the reaction occurs within a living cell, it is called *photodynamic action*. The best-understood example is the dye acridine orange (AO), which has the following structural formula:

$(CH_3)_2N$ ⟨structure⟩ $N(CH_3)_2$

When a cell or virus to which AO has bound is illuminated with light of wavelength of 400–500 nm, the organism is killed. The mechanism by which this occurs is the following. When AO binds to DNA, it is positioned between the bases of DNA, as shown in Figure 18-11. (Molecules that bind in this position are called *intercalating agents*.) When the DNA-AO complex is illuminated, an AO molecule absorbs a photon and normally gives up this energy as fluorescence (reemission of light of a longer wavelength). If an O_2 molecule collides with the energy-rich AO molecule before fluorescence occurs, the energy is transmitted to the O_2 molecule, instead, and, in a subsequent collision with a guanine in the DNA, the O_2 molecule initiates an oxidation reaction that results in breakage of the ring of guanine. Since the process requires that O_2 collide twice (once with AO and once with the guanine) before it loses its excitation energy, the reaction requires that the AO molecule be bound to the DNA; if AO were not bound but floated freely in the solvent, an excited O_2 molecule would have a far greater opportunity to lose its energy by collision with molecules other than guanine (for instance, solvent molecules) than with guanine itself. When AO is intercalated in DNA, an impinging O_2 molecule will be so near the DNA bases that with high probability it will find a guanine ring. The detailed chemical reaction resulting in ring cleavage is not known.

688 CHAP. 18 PHOTOCHEMISTRY, RADIATION CHEMISTRY, AND RADIOBIOLOGY

FIGURE 18-10 Three ways to repair thymine dimers (T̂T) in DNA. In path I, UV light cleaves the dimer. In path II, the photoreactivating enzyme is activated by absorbing light and it cleaves the dimer. Path III is the dark-repair or "cut and patch" method. The dimer is removed from the DNA and new DNA is synthesized at the site of the removal.

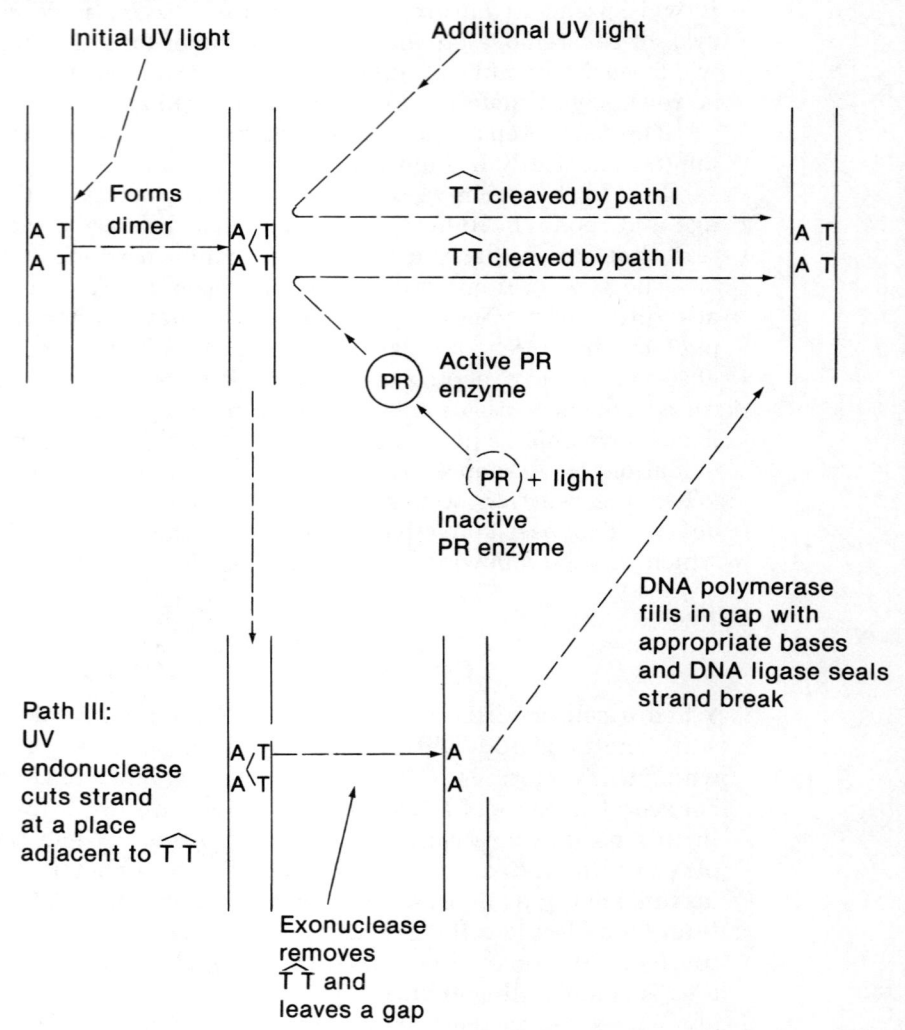

There are many substances known that cause photosensitized oxidation. The best-studied are the acridine compounds, the dye methylene blue, and the phenanthridium compounds (for example, ethidium bromide). In complex systems, such as viruses, which contain proteins as well as nucleic acids, these substances cause photosensitized oxidative reactions that link nucleic acids with proteins and that link various amino acids. In addition

FIGURE 18-11 Intercalation of acridine orange between the bases of DNA. [SOURCE: *Physical Biochemistry*, First Edition, by David Freifelder. W.H. Freeman and Company. Copyright © 1976.]

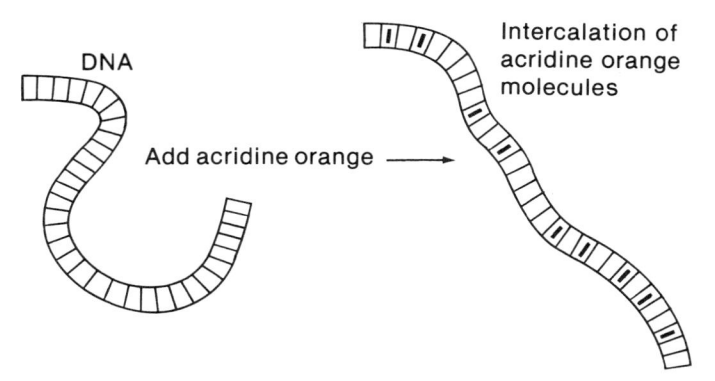

AO can also induce light-stimulated depolymerization of cellular polysaccharides, such as hyaluronic acid.

Photodynamic effects also play a role in determining the lifetime of dyed fabrics. Many commercial dyes in use in the textile industry are photosensitizers and cause depolymerization of cellular fibers such as in cotton and thereby decrease the mechanical strength of cotton fabrics exposed to light.

Photosensitizing agents have been used in the laboratory to cause deliberate damage to macromolecules—in particular, to DNA; however, the main use of these substances, principally AO and ethidium bromide, has been in various techniques for analysis or measurement: for instance, DNA can be localized in living cells by adding AO, illuminating the complex with light that is absorbed by the AO, and then observing the position of the fluorescence within the cell. Another important use of photosensitization derives from the fact that different forms of DNA molecules (linear molecules, twisted circular molecules, and so forth) do not bind the same amount of an intercalating compound; in the presence of these substances—for example, ethidium bromide—the various DNA forms have different densities; separation techniques are based on these density differences. It is essential to remember that the intercalating agents are also photosensitizers, and that the very act of looking at the cells or molecules that contain them may introduce chemical damage if the light used in the observation is of a wavelength that can be absorbed by the sensitizer.

18–6 IONIZING RADIATION

Types of Radioactivity

For the purpose of the present discussion we need to consider three parts of the atom: the positively charged proton and the uncharged neutron, both of which have the same mass and are present in the nuclei of most atoms, and the very light, negatively charged electrons found in the orbitals around the nucleus. An atom is defined by its atomic number—that is, by the number of its protons; to preserve electrical neutrality, these equal the number of its electrons. Atoms having the same atomic number can exist in different forms called *isotopes*, which are characterized by differing numbers of neutrons.

All isotopes of a particular atom are not equally stable. With a definite probability that is characteristic of each unstable nucleus, transition to a stable form occurs. This transition is accomplished for most of the lighter atoms by emission of an electron from the nucleus. There is no significant change in atomic weight but the process is invariably accompanied by an increase in the proton number by 1, so that, in the simplest view, it may be said that disintegration of a neutron yields a proton and an electron. The shift from an energetic, unstable nucleus to a low-energy stable nucleus is called radioactive decay, and an atom having an unstable nucleus is called a radioisotope. Radioactive decay does not always proceed by emission of a particle; in some cases the excess energy is released as a photon.

There are many mechanisms of radioactive decay, but only three, each of which results in emission of either a particle or a photon, are of biological relevance. These are the following:

1. Beta (β) Decay

In β decay, an electron called a β particle is emitted from the nucleus of an atom. The energies of the individual electrons produced by a sample of a radioisotope range from zero to some maximal value and result in a characteristic energy spectrum. Spectra of a few of the radioisotopes in common use in biochemistry are shown in Figure 18-12. The energy of the particles is usually expressed in millions of electron volts or with the unit MeV; 1 MeV = 1.6×10^{-5} erg. Each spectrum may be characterized by a maximal energy, E_{max}, and a mean energy, E_{mean}, both of which vary markedly from one isotope to the next. Of the radioisotopes in common use in biological work, the least-energetic β particle is that of tritium (^{3}H), for which the E_{mean} is 0.006 MeV, and the most energetic is that of ^{32}P, for which E_{mean} = 0.7 MeV. The high-energy β particles have a large momentum which, owing to the law of conservation of momentum, means that the nucleus *recoils* (that is, it moves in the opposite direction) when a highly energetic β particle is emitted. The significance of this will be discussed later.

2. Alpha (α) Decay

The heavy elements—for example, radium, thorium, and uranium—would not be made stable simply by β decay: they simply possess too much potential energy. These heavy nuclei decay by releasing a larger unit, one consisting of four tightly associated particles—two protons and two

FIGURE 18-12 Beta spectra for ^{3}H, ^{14}C, and ^{32}P: E refers to the energy of the particles and $N(E)$ is the number of β particles emitted with energy E. This is in fact a measure of the probability of emission of a β particle at that energy. [SOURCE: *Physical Biochemistry*, First Edition, by David Freifelder. W.H. Freeman and Company. Copyright © 1976.]

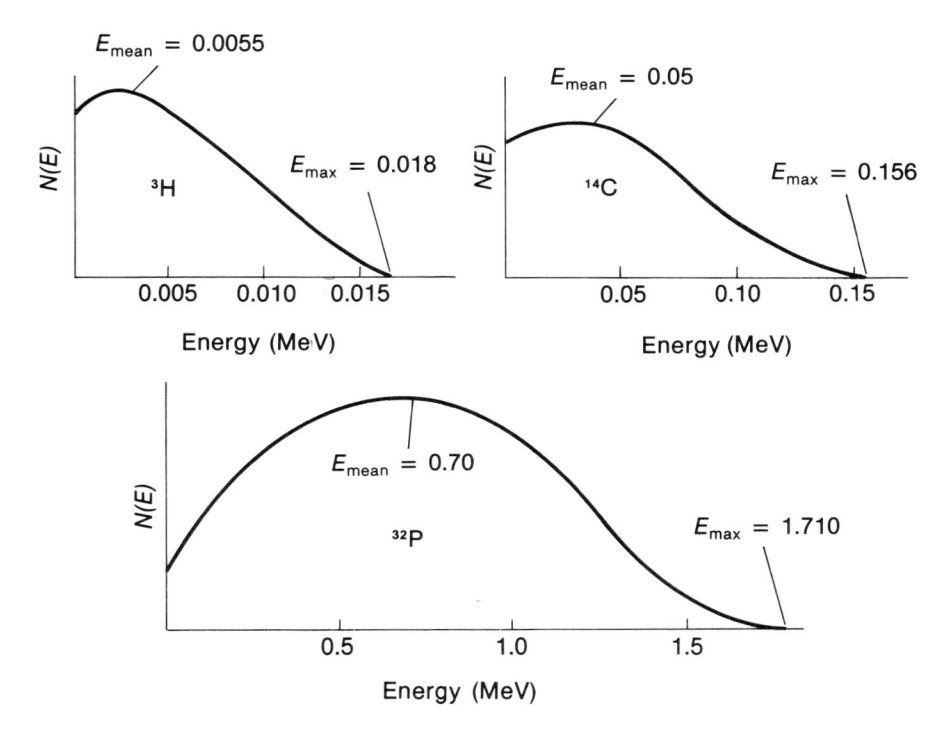

neutrons. This unit is called an α particle; it is equivalent to the nucleus of a helium atom. When an α particle is emitted, it has a very large momentum, owing to its great mass and, as in the case of a high energy β particle, the nucleus recoils considerably. Note that whereas β emission results in an increase in the atomic number by 1 and no significant change in the atomic weight, the emission of an α particle decreases the atomic number by 2 and the atomic weight by approximately 4.

3. Gamma (γ) Ray Emission

In many nuclei there is neither a deficit nor an excess of neutrons, and instability is caused by the way the nuclear particles are arranged in the nucleus. In this case, radioactive decay occurs by emitting the excess energy as short-wavelength (< 0.1 nm) electromagnetic radiation called a gamma (γ) ray or a γ photon. The emitted γ ray has a single wavelength. On rare occasions, a γ ray knocks an orbital electron from its orbit; what is termed an internal conversion electron is then emitted.

Electrons can also be emitted from matter without radioactive decay. One mechanism, called the *photoelectric effect*, is as follows. A photon, often of visible light, is absorbed by a solid; the energy of the photon reappears as the kinetic energy of an emitted electron called a photoelectron. This process occurs only in certain solids whose electrons are very mobile and loosely bound. The photoelectric effect is the basis of the operation of electric-eye door-openers. A second important mechanism is the *thermionic effect,* in which a metal is heated; the electrons, which are highly mobile and freely moving in any conducting metal, acquire so much kinetic energy from the heating that they overcome the cohesive forces in the metal. This process is often described as "boiling" the electrons out of the metal. The thermionic effect is the source of electrons in the vacuum tubes formerly used in radios, in the electron microscope, and in television picture tubes. Electrons produced in this way are called thermoelectrons. The value of thermoelectrons is that they can be used to produce an electron beam; one way to do this is shown in Figure 18-13.

Rate of Radioactive Decay

A radioactive atom decays in random fashion. When one observes a particular atom capable of decay, there is no way of predicting when it will decay. However, in a collection of a very large number, N, of radioactive atoms present at a time t, the number dN of atoms that decay in a time interval dt is proportional to the number of atoms present. This is reasonable—twice as many atoms should have twice as many decays per unit time. This statement can be expressed as

$$-\frac{dN}{dt} = \lambda N, \qquad (25)$$

in which λ is the constant of proportionality or the decay constant.* If N_0 is the number of atoms present at $t = 0$, integration of Equation 25 yields

FIGURE 18-13 Production of an electron beam. A heated filament produces thermoelectrons. The anode attracts the electrons and accelerates them to high velocity. Some of the electrons pass through a small opening in the anode. Their velocity is so great that they quickly escape the attractive force of the anode. The electrons that pass through the opening form an electron beam.

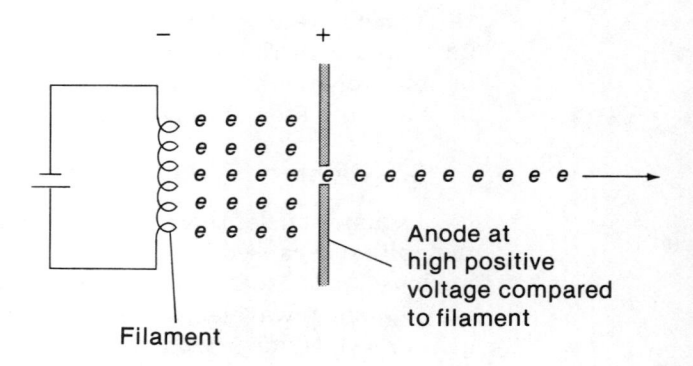

*The symbol λ does not here refer to wavelength; it is also commonly used as the decay constant. That is how it is used in Equations 25 and 26.

FIGURE 18-14 Kinetics of radioactive decay.

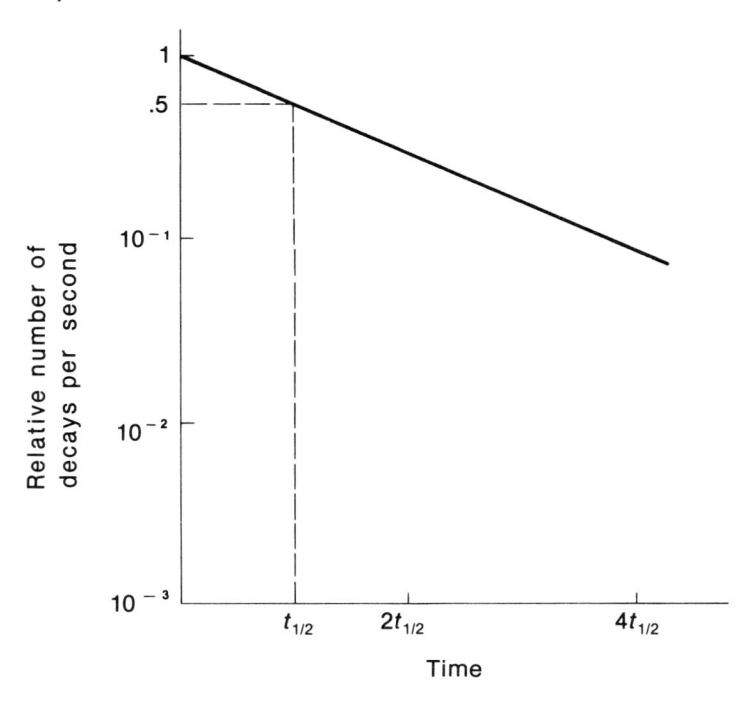

the number of atoms present at $t = 0$, integration of Equation 25 yields

$$N = N_0 e^{-\lambda t}. \tag{26}$$

An exponential decay equation states that in any particular time interval, the disintegration rate (also called the decay rate or simply the radioactivity) will decrease by the same fraction; thus, it is convenient to express the decay constant λ in terms of the half-life $t_{1/2}$, the time required for the decay rate to decrease by one-half. Hence, $t_{1/2} = -(\log_e 1/2)/\lambda = 0.693/\lambda$, and Equation 26 becomes

$$\frac{N}{N_0} = e^{-0.693 t/t_{1/2}}. \tag{27}$$

This simply means that after one half-life, half of the initial radioactivity remains, after a second half-life, one quarter of the radioactivity, and so forth. A simple plot demonstrating radioactive decay is shown in Figure 18-14.

The half-life of known radioisotopes ranges from nanoseconds to millions of years. Isotopes with excessively long or short half-lives are of little use in biological studies, because the former have decay rates that are too low, and the latter decay so rapidly that they are almost gone before one begins an experiment. A list of those that are commonly used is shown in Table 18-1.

TABLE 18-1 Characteristics of commonly used isotopes.

Isotope	Particle emitted	E_{max} (MeV)	Half-life
^{3}H	β	0.018	12.3 years
^{14}C	β	0.155	5,568 years
^{24}Na	β	1.39	14.97 hours
	γ	1.7, 2.75	
^{32}P	β	1.71	14.2 days
^{35}S	β	0.167	87 days
^{40}K	β	1.33, 1.46	1.25×10^{9} years
^{45}Ca	β	0.254	164 days
^{131}I	β	0.335, 0.608	8.1 days
	γ	0.284, 0.364, 0.637	

Radioactivity is expressed in units of *curies*. One curie (C) is defined as the number of disintegrations per second per gram of radium and equals 3.70×10^{10} disintegrations per second. For most biological applications, quantities much less than one curie are normally used and the millicurie (mC) or microcurie (μC) are the more common units. Furthermore, in biological applications, a minute is the standard time unit—hence

1 μC = 2.22 $\times$ 10^6 disintegrations per minute (dpm).

Transmutations and Their Consequences

With the exception of γ emitters, decay of an element by emission of a particle results in a change of the atomic number of the element. In the case of α decay, the atomic number decreases by 2; in β decay it increases by 1. A change in atomic number means that the element is converted to another element having different chemical properties. This is called *transmutation*. The α emitters are generally heavy metals—for examples,

$$^{238}_{92}U \rightarrow {}^{234}_{90}Th + \alpha,$$

or

$$^{210}_{84}Po \rightarrow {}^{206}_{82}Pb + \alpha.$$

Heavy metals are not present in biological systems; consequently, they will not be discussed further.

The β emitters used in biological systems are usually contained in chemical compounds, so that transmutation is of great significance. Consider the following transmutations of radioisotopes in common use in laboratory work:

SEC. 18-6 IONIZING RADIATION

$${}^{3}_{1}\text{H (tritium)} \rightarrow {}^{3}_{2}\text{He} + \beta\,;$$
$${}^{14}_{6}\text{C} \rightarrow {}^{14}_{7}\text{N} + \beta\,;$$
$${}^{32}_{15}\text{P} \rightarrow {}^{32}_{16}\text{S} + \beta\,;$$
$${}^{35}_{16}\text{S} \rightarrow {}^{35}_{17}\text{Cl} + \beta\,;$$
$${}^{42}_{19}\text{K} \rightarrow {}^{42}_{20}\text{Ca} + \beta\,.$$

In each case, a distinct chemical change must accompany β emission. For example, if ^{32}P is present in a compound of the form $^{32}\text{PO}_4^{3-}$, a likely conversion is to $^{32}\text{SO}_4^{2-}$, although other changes are possible depending upon the other substances that might be present. A drastic change always accompanies ^{3}H decay, since the ^{3}H converts to He, of valence 0.

A different type of chemical alteration results from recoil. We have already pointed out that when a particle is emitted in a particular direction, the nucleus must, because of conservation of momentum, move in the opposite direction. This recoil can result in the breakage of chemical bonds by two different mechanisms.

1. Breakage of the Bond Containing the Radioisotope

Let us consider a hypothetical compound of the form X—A—Y in which X and Y are any chemical groups and A is an isotope having a valence of +2. If A emits a β particle to become the trivalent isotope B, some chemical rearrangement must occur in accord with the valence change. If this occurs in aqueous solution, one possible alteration would create the compound

$$\begin{array}{c} \text{X—B—Y} \\ | \\ \text{OH} \end{array}$$

in which X and Y are still joined by chemical bonds. However, if the β-emission results in strong recoil, the bonds from B might be stretched to their limits and one or both might break. If this were to happen, there would result two distinct compounds,

$$\text{X—H} \quad \text{and} \quad \begin{array}{c} \text{H—B—Y} \\ | \\ \text{OH} \end{array}$$

in which the H and OH have come from a water molecule. This kind of process is an example of bond breakage by recoil. It is important in polymers containing radioisotopes because it can result in a substantial decrease in molecular weight.

2. Breakage of a Bond That Does Not Contain a Radioisotope

Consider a hypothetical compound whose three-dimensional structure is such that we should write the formula as

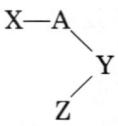

$$\begin{array}{c} X-A \\ \diagdown \\ Y \\ \diagup \\ Z \end{array}$$

As A is converted to B, all of the considerations just mentioned apply, with one addition. If the β particle is emitted in a direction away from the Z—Y bond, the recoil will be toward the Z—Y bond; if the recoil has sufficient momentum, the B nucleus might collide with Z or Y and stretch the Z—Y bond to the point of breakage. In this way, a bond that is not connected to the radioisotope can break. This cannot occur with every recoil and is strongly dependent on the energy of the emitted β particle and the geometry of the molecule.

EXAMPLE 18–K Structural changes of DNA caused by decay of ^{32}P.

A DNA molecule consists of two long strands entwined helically. Each strand is a polymer composed of alternate sugar and phosphate groups. By growth of cells in the presence of $H^{32}PO_4^{-2}$, DNA is obtained that contains ^{32}P in its sugar-phosphate backbone. When the ^{32}P decays, ^{32}S results, so that the phosphodiester bond becomes a sulfodiester bond. However, the recoil of the ^{32}P is so great that each decay results in bond breakage within the strand in which the decay occurs (Figure 18-15). This bond breakage results in single-strand breaks. Such a break has no effect on the molecular weight of the double-stranded DNA because the DNA molecule remains intact, as shown in the figure. However, in approximately one-tenth of the decays the recoil occurs with sufficient energy and in such a direction that the recoiling nucleus breaks a phosphodiester bond in the strand that does

FIGURE 18-15 Production of a single-strand break in DNA caused by β decay of a ^{32}P atom. The ^{32}P is converted to ^{32}S by emission of an electron. The ^{32}S atom recoils and stretches the bond holding it in the DNA chain and the bond breaks.

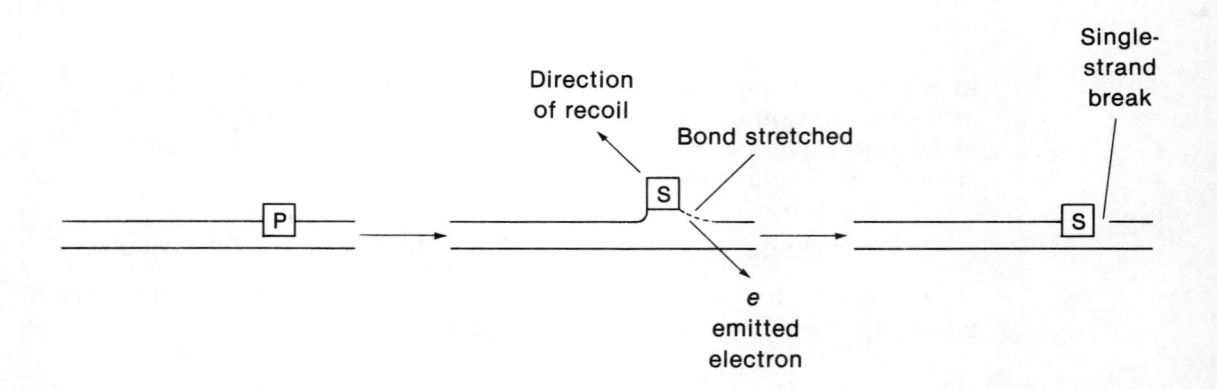

not contain the ^{32}P atom—that is, in the strand in which the decay has not occurred. When this recoil occurs, two bonds are broken—one is the single-strand break caused by the transmutation and accompanying cleavage, and the other break results from mechanical damage caused by the recoil. These two breaks are usually opposite one another in the two strands and thereby together comprise a double-strand break, which results in a decrease in the molecular weight of the DNA.

EXAMPLE 18–L   Damage to DNA caused by decay of ^{3}H.

A DNA molecule can be prepared in which the base thymine

contains a ^{3}H in the methyl group. When this ^{3}H decays, a molecular rearrangement occurs that results in breakage of the ring of thymine. The thymine is connected to the sugar deoxyribose in the backbone of DNA by a bond from the ring nitrogen (denoted in the structural formula above by the asterisk) and a carbon atom in the sugar. This is called an N-glycosylic bond, and it is highly susceptible to hydrolysis once the thymine ring is broken. Thus, shortly after ^{3}H-decay, the N-glycosylic bond breaks. Once this bond is broken, the phosphodiester bond becomes unstable, and it too hydrolyzes. Thus, the ^{3}H-decay in the thymine methyl group causes a single-strand break, which indicates how complex and indirect radiochemical alterations can be.

X Rays

X rays are a form of electromagnetic radiation having a wavelength in the range of 0.01–0.15 nm—that is, slightly longer than γ rays. X rays differ from γ rays in that they come from the inner electron shells rather than the nuclei of atoms. They are emitted when an electron moves from one inner orbit to another. For laboratory use, a beam of X rays is made in the following way (Figure 18-16). A metal filament is heated to a very high temperature, so that electrons are boiled off by the thermionic effect. A large piece of metal called a target is situated nearby. A large voltage difference, typically 10,000–200,000 volts, is created between the filament

FIGURE 18-16 Typical arrangement for the production of X rays. Electrons are boiled off of a filament and accelerated toward a metal target by the large voltage difference between the filament and the target. The entire unit is enclosed in a glass casing, which is evacuated, so that electrons are not deflected by air molecules. The X rays pass so efficiently through glass that little X-ray energy is lost.

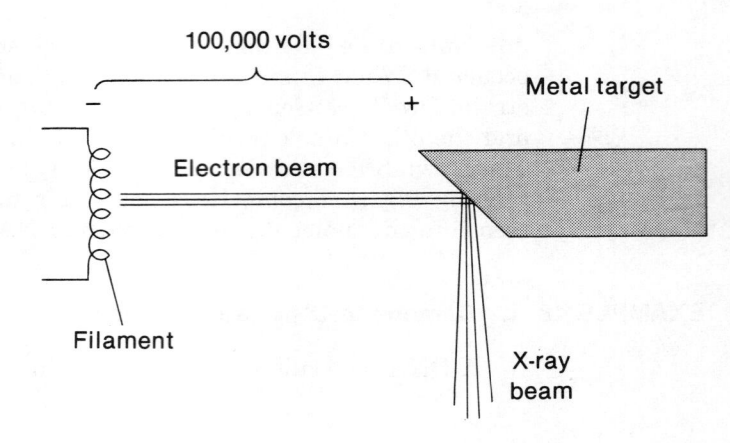

and the target, the target voltage being positive. The electrons, which are negatively charged, are accelerated toward the target and collide with it. Most of the electrons are stopped when they reach the target and their energy is converted to heat. (This necessitates using a metal that is a good heat conductor—Cu is very common—and cooling the target with circulating water; otherwise, the target melts.) About 0.2 percent of the accelerated electrons dislodge orbital electrons from the atoms of the target material, forming metal ions. These ions are unstable and they attract electrons to fill the gaps created in the orbitals. These electrons generally come from other orbitals; as an electron falls into a gap, its energy is reduced and an X-ray photon is emitted. Since the energy of an electron varies with its orbital (see Chapter 17), the released photon has an energy characteristic of the two orbitals between which the electron has moved. A particular metal will have several characteristic X-ray energies, since a metal atom has several orbitals. Note that when an electron moves from an outer orbital to an inner orbital, the outer orbital becomes electron-deficient; the missing electron is ultimately replaced by one of the many electrons bombarding the target and freely moving throughout the target. The X rays produced by jumping electrons comprise an X-ray line spectrum. However, the lines of the spectrum are always superimposed on a continuous spectrum, which arises as follows. Some of the electrons that bombard the target will penetrate an atom but will not dislodge an orbital electron. These electrons are deflected by the electrostatic attraction of the positively charged nucleus and the repulsion of the orbital electrons and therefore emerge from the atom with reduced energy. Thus emitted, these X-ray photons have a large and continuous range of energies, the values of which depend on how near each electron approached the nucleus. These photons yield the continuous or white radiation from an X-ray tube. An example of an X-ray spectrum is shown in Figure 18-17.

The use of X rays in biological research will be described in the following section.

FIGURE 18-17 Energy spectrum of an X-ray beam from a tube with a molybdenum target. The two peaks are called α and β and are sometimes referred to as "edges."

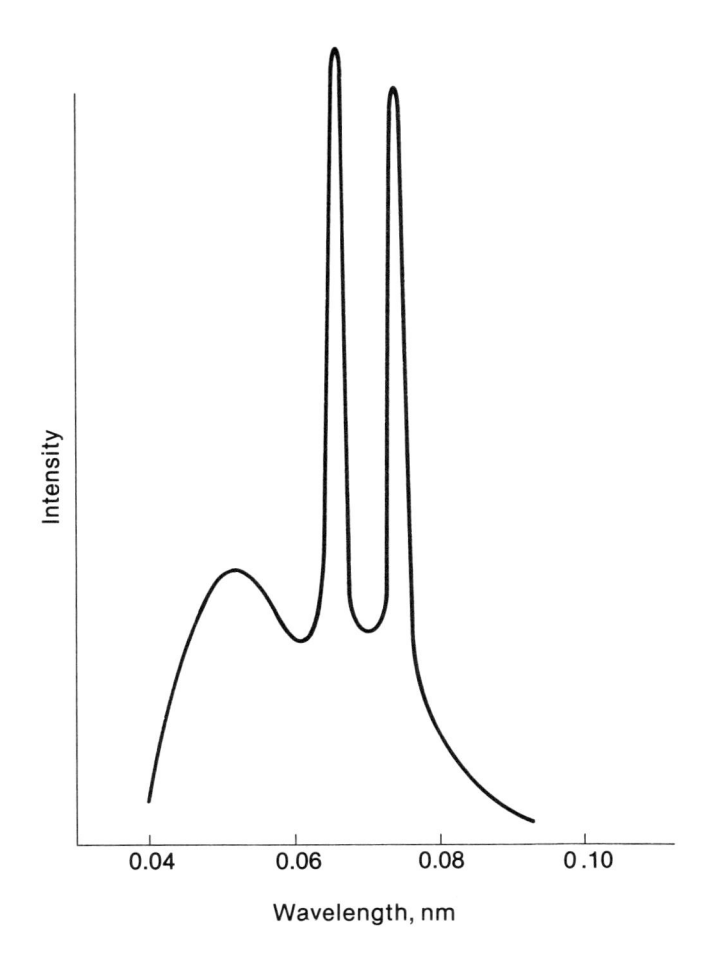

Interaction of Ionizing Radiation with Matter

When a fast-moving charged particle—for examples, an electron, proton, or α particle—passes through matter, ions are produced. How this occurs can be thought of in two ways that are physically equivalent. (1) A charged particle can collide with orbital electrons and knock one or more electrons away from the atom. If the moving particle is sufficiently energetic, it may even, when colliding with the nucleus (a rarer event than collision with electrons), dislodge nuclear particles. (2) In the second view, we regard the moving particle as carrying an electric field. If the particle passes sufficiently near the electrons, the electric field can deform the orbitals so much that an electron is driven out of its orbital. The result of this interaction is the production of two charged particles, an electron and a positively charged ion; usually the latter will be unstable and hence be chemically reactive.

A moving particle will not usually give up all of its energy in its first interaction and will go on to ionize many other atoms. The rate at which energy is lost has been estimated in a relatively complex way and is expressed by

$$\frac{dE}{dx} = -\frac{4\pi e^4 z^2 NZ}{mv^2} \ln \frac{2mv^2}{I_0} \qquad (28)$$

in which e is the charge of the electron in electrostatic units (esu), z is the charge of the fast particle, N is the number of atoms per cm^3 of the material traversed, m is the mass of the electron in grams, Z is the atomic number of the atom being bombarded, v is the velocity of the particle in cm/sec in the direction of motion of the particle, I_0 is the average ionization potential (expressed in ergs) of the atom being ionized, and $-dE/dx$ is the loss of particle energy E (in erg/cm) in the x direction. *This equation is not valid for electrons whose energy is greater than 10^5 eV;* this is because at higher energies the velocity is so great that certain effects explainable by the special theory of relativity become important. For high-energy electrons a much more complicated equation is needed. The equation is valid, however, for protons and α particles of energies up to 100 MeV.

Notice that the mass m' of the moving charged particle does not appear in the equation directly—the m in the equation is always the mass of the electron. However, the mass of the particle must be known to calculate the velocity of the particle from its kinetic energy. Since E is the kinetic energy, which equals $\tfrac{1}{2}mv^2$, then $v = (2E/m')^{1/2}$. We can replace v in Equation 28 by this expression to obtain

$$\frac{dE}{dx} \text{ (in erg/cm)} = -\frac{2\pi e^4 z^2 NZm'}{Em} \ln\left(\frac{4mE}{m'I_0}\right). \qquad (29)$$

As we have seen, it is common in the study of radiation to express energy in eV. To write Equation 29 in such units, we make use of the equivalence

$$1 \text{ erg} = 6.23 \times 10^{11} \text{ eV}$$

and rewrite Equation 29 as

$$\frac{dE}{dx} \text{ (in eV/cm)} = -\frac{1.29 \times 10^{-13} z^2 NZ}{E(\text{in eV})} \frac{m'}{m} \ln\left[\frac{4mE(\text{eV})}{m'I_0(\text{eV})}\right], \qquad (30)$$

in which now both E and I_0 are in eV and dE/dx is in eV/cm.

EXAMPLE 18–M The loss of energy of a 1-MeV proton passing through a protein molecule.

A typical protein is a polymer having the empirical formula $(C_4H_8NO_2)_n$, in which n is the degree of polymerization, and having a density of 1.3 gm/cm^3. From the mass of each atom, the value of N for each atom can be calculated. This is done as follows. The molecular weight of the monomer is 102. The number of atoms per cm^3 is $(6 \times 10^{23})(1.3)/102 = 7.67 \times 10^{21}$. The number of N, O, C, and H atoms per cm^3 are 1, 2, 4, and 8 times 7.67×10^{21}, respectively. The energy E is 1 MeV and the ratio of the proton mass to the electron mass, m'/m, is 1835. The value of z is 1. The atomic numbers Z are $Z_H = 1$, $Z_C = 6$, $Z_N = 7$, and $Z_O = 8$. The value of I_0 for each of these atoms is known to be $I_H = 16$ eV, $I_C = 64$ eV, $I_N = 81$ eV, and $I_O = 99$ eV. In an interaction of a charged particle with matter, the particle encounters each atom independently, so that the total energy loss is the sum of the losses due to each type of atom (this is called the Bragg additive law). Thus, for a 1-MeV proton, Equation 30 is

$$\frac{dE}{dx} = -\left(\frac{1.29 \times 10^{-13}}{1 \times 10^6}\right)(1835) \sum N_i Z_i \ln\left(\frac{4 \times 10^6}{1835 I_i}\right),$$

in which i refers to each type of atom (C, N, O, and H). Evaluation of the summation yields a value of $-dE/dx$ of 3.72×10^8 eV/cm or, in units useful for discussing molecules, 3.72 eV/Å.

A great many experiments have been performed with 1-MeV electrons, for which $-dE/dx$ is 0.025 eV/Å. A protein molecule of average size has a thickness of about 30 Å, so about 0.75 eV are lost by an electron passing through each protein molecule. A typical chemical bond has an energy of about 3 eV, so that very few bonds in a protein molecule would be broken by a 1-MeV electron. We will see shortly that the energy loss is not constant along the path of a particle and in fact an electron can do quite a lot of damage. For an α particle the value of $-dE/dx$ is 25.2 eV per Å, which could break 250 bonds when traversing a typical protein molecule.

An important point frequently misunderstood has to do with the relative efficiency of various particles at causing damage. It seems as if the efficiency should become greater as the particle energy increases. However, the logarithmic term in Equation 29 does not increase as rapidly with E as $1/E$ decreases, so that dE/dx decreases with increasing E. This is shown in Table 18-2, which lists the energy that is lost by α particles and protons of various energies as they pass through a typical protein molecule (with a thickness of about 30 Å). The same trend is not seen for high-energy electrons because of the relativistic effects mentioned earlier. Thus, the values for an electron of 1, 4, and 10 MeV are 0.75, 0.78, and 0.87 eV, respectively.

These considerations suggest that as the energy of a particle increases, the energy it deposits in a molecule per unit volume of the molecule decreases. However, the total amount of energy that is deposited does not necessarily decrease because the high-energy particle can penetrate farther into a particular mass of matter. Ultimately, if the matter is thick enough, the particle will lose all of its kinetic energy and stop moving, but the greater the initial energy of the particle, the greater is the thickness that is required for this to occur. The energy loss per unit distance is not

FIGURE 18-18 Energy loss by an ionizing particle such as a fast-moving electron passing through matter. Each dot represents a site at which energy is lost. This is a random process, but on the average the dots get closer together. When by chance there are several closely spaced dots (arrows), the trajectory of the particle changes. Near the end of the path the dots are very close, the trajectory is tortuous, and the particle stops.

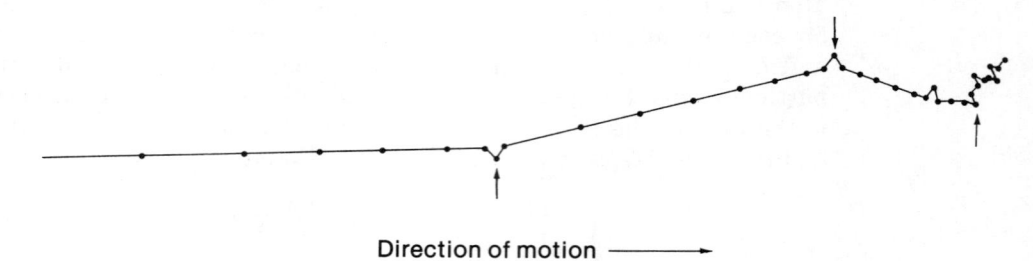

Direction of motion ⟶

constant because as a particle penetrates matter, v decreases and thus dE/dx increases. This is shown in Figure 18-18, in which we denote the position at which energy is lost by a dot. The dots are closer together as the particle penetrates more deeply; we assume the amount of energy discharged at each dot is the same; where the dots are densest near the end of the path of the moving particle, a large fraction of the initial energy has been given up. This phenomenon is of great value in tumor radiotherapy because, if one knows how far below the skin a tumor is located, the energy of the particles used can be selected such that the particles stop moving within the tumor; this has the effect that little tissue damage is done except in the tumor (Figure 18-19).

In the discussion just given of the rate at which a particle loses energy, we have made two assumptions that are not totally reasonable because of the statistical nature of the behavior of matter. These are (1) that the spacing between two ionizations at a particular depth is calculable from the value of dE/dx, and (2) that the energy discharged per event is

FIGURE 18-19 A principle of radiotherapy. A tumor is present in the human body. Radiation of such an energy is used that little energy is lost in the outlying tissue and most is deposited in the tumor. Dots represent sites of discharge of energy.

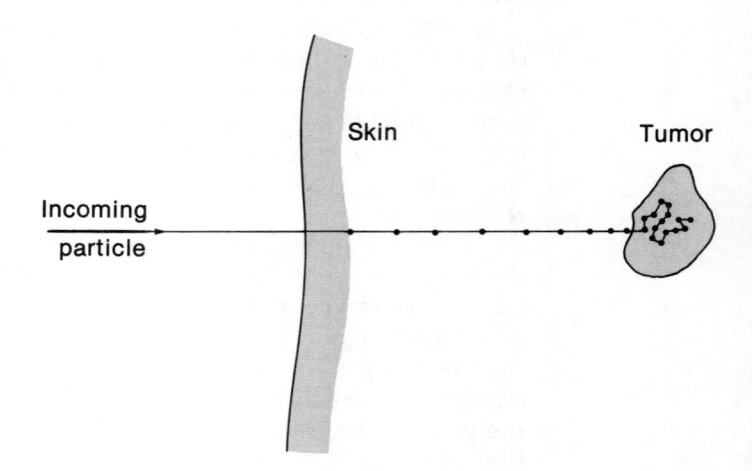

constant. Assumption 1 is not correct, because we must instead also consider the probability that the particle will interact with matter; thus, the value of dE/dx is an average per event. Assumption 2 has also been shown to be incorrect. Ionization results in the formation of a pair of ions (one positive and one negative). The amount of energy given up by a fast-moving particle in a particular volume is reflected in the *number* of ion pairs formed in that volume. Experimentally, it has been observed that the ion pairs usually occur in clusters and the size of the cluster varies; the average number is three ion pairs per cluster and a single ion pair is found in only half of the events. This enables us to account for the fact that from the value of dE/dx a 1-MeV electron apparently does not discharge enough energy in a protein molecule to break bonds, yet in fact, bond breakage does occur in a protein. This is because, in many of the interactions, the energy is greater than the average values given in Table 18-2. In the example, we showed that the average energy deposited per 100 Å by a 1-MeV electron is 2.5 eV. However, what we really need to know is the value of the energy discharged in a single ionization event in which a cluster is produced. These ionizations tend to occur wholly within a sphere having the dimensions of an atom—that is, about 2.5 Å in diameter. The energy of a single ionization is often called the *energy of a primary ionization*, and measurements show 110 eV to be a typical value in biological materials. Thus, in a particular event, there is sufficient energy to break numerous chemical bonds, which accounts for the extensive damage produced in biological systems by radiation. Note that the value of ΔE for various particles is a reflection of the different distances between successive ionization events.

The passage of X rays and γ rays through matter also results in the production of ion pairs but by two mechanisms that differ from the way charged particles produce ion pairs. These two processes are *photoelectric absorption* and *Compton recoil*. In photoelectric absorption, the energy $h\nu$ of the photon is totally absorbed by an atom and this results in ejection of an orbital electron with velocity v and kinetic energy $1/2\ mv^2$, and in the excitation of the atom to an energy W such that $h\nu = 1/2\ mv^2 + W$. Since absorption is complete, the reduction in intensity of an X-ray beam or a

TABLE 18-2 Loss of energy ΔE by α particles and protons having various initial energies when passing through a protein of thickness 30 Å.

Particle energy (MeV)	Particle	ΔE/protein molecule (eV)
1	α	756
4	α	390
10	α	195
1	Proton	96
4	Proton	35
10	Proton	17

γ-ray beam should obey an equation similar to the Beer-Lambert Law. The appropriate equation is

$$I = I_0 e^{-n\sigma x}, \tag{31}$$

in which I and I_0 are the transmitted and incident intensities, respectively, n is the number of atoms per unit volume, x is the thickness of the absorber, and σ is a parameter, the *absorption cross-section,* that expresses the probability of absorption and is equal to

$$\sigma = (2.04 \times 10^{-30})\frac{Z^3}{E^3}(1 + 0.008Z), \tag{32}$$

in which Z is the atomic number of the absorber and E is the photon energy in MeV. This equation indicates that, as we saw with the interaction of charged particles with matter, absorption is more probable as the photon energy decreases and the atomic number of the absorber increases.

An atom can only absorb energy of discrete values, and usually a single energy predominates in the energy range of X rays, so that electrons of a single energy are produced.

If the photon energy is greater than 0.3 MeV, Compton recoil occurs. In this process, a photon collides with an electron and rebounds. In this collision, which is an elastic collision, energy is lost by the photon and an equivalent amount is gained by the electron. The energy gained is often sufficient to enable the electron to leave the atom. If the photon energy is great enough, the rebounding photon may collide with several atoms so that several electrons can be produced. The process is ultimately terminated when the photon energy is sufficiently low that only photoelectrons are produced.

An X or γ photon can directly produce an ion pair by the mechanism just described. However, ions produced directly account for only a small fraction of the total number of ions that are created when this type of radiation traverses matter. Instead, it is the electron ejected from the atom by the X-ray photon that produces most of the ion pairs. This is because an ejected electron has fairly low energy and therefore produces a large number of ion clusters in a very small volume by the same mechanism by which electrons produce ions during passage through matter.

The biological importance of the ions that are produced is a result of their great chemical reactivity. That is, they cause substantial chemical alteration in their immediate environments, creating significant damage to macromolecules.

Measurement of Radiation

The amount of radiation in an ionizing beam is customarily measured in terms of either the number of ion pairs formed or the energy absorbed by a

SEC. 18–6 IONIZING RADIATION

particular amount of material. The principal unit of radiation is the *roentgen*, which is the amount of radiation that produces one electrostatic unit of separated charge in 1 cm³ of air at 0°C and at a pressure of 1 atm (that is in 0.001293 g of air). Another common unit is the *rad*, which is the amount of radiation that would result in an absorption of 100 erg/g of biological tissue—not a very precise unit! Since 1 g of tissue exposed to one roentgen absorbs approximately 93 erg, the absorption of exactly 93 erg is defined as the *roentgen equivalent physical* or the *rep*. One rad is approximately 1.07 roentgen; it is commonplace to use the two units interchangeably. For various biological molecules the number of ionizations per gram of material is from 6 to 7×10^{13}.

As might be expected, dosimetry, the measurement of a quantity of radiation, is complex; in fact, it is rarely carried out directly. The two most common methods of determining doses are by ferrous sulfate dosimetry and by use of an ionization chamber. In the first method, a solution of ferrous sulfate in H_2SO_4 is irradiated and the radiation-stimulated oxidation of the colorless Fe^{2+} ion to the yellow Fe^{3+} ion is measured as absorbance at 305 nm. This method has been accurately calibrated with radiation sources whose dose rates have been measured directly and yields the dose in rads. The ionization chamber consists of a charged rod in a chamber filled with air. The unit is placed in the radiation beam and the ionization of the air renders the air electrically conducting and allows the charge in the rod to leak off. The charge is measured by an electrometer. The amount of charge that is lost is proportional to the total dose received by the meter.

Most sources of radiation used in biological experiments—for example, X-ray machines and ^{60}Co (a source of γ rays)—emit radiation at a constant rate in time. Therefore, since in most experiments one is comparing the relative effect of two doses whose ratio is known, it is common to express effects in terms of time of irradiation rather than dose.

The most important effects of radiation on molecules are breakage of chemical bonds and molecular rearrangement. The number of changes of a particular class of bonds per 100 eV of energy absorbed is called the *G value* for that class. Typical values of G are from 0.1 to 4. This may seem odd since the bond energies are approximately 2–3 eV, so that it might be expected that G values would be from 30 to 40. This is not the case for two reasons. First, a substantial fraction of the absorbed energy is converted to heat. This happens, for example, when two ions simply recombine and their excess energy appears as vibrational energy that by collision is converted to kinetic energy. Second, there are usually several types of chemical changes that occur simultaneously and under the same conditions, and that compete with the reaction of interest for the absorbed energy. Since the G value is measured in terms of the *total* absorbed energy, it expresses the probability of production of a particular chemical change.

Chemical Effects of Ionizing Radiation on Biological Molecules

Since in biological systems the major component is water, let us examine for a moment the effect of ionizing radiation on water. The first step is the ionization of water,

$$\text{Radiation} + H_2O \rightarrow H_2O^+ + e^-$$

This is followed by two processes:

$$H_2O^+ \rightarrow OH\cdot + H^+$$

and

$$e^- + H_2O \rightarrow OH^- + H\cdot$$

The H· and the OH· are free radicals; they may recombine to yield water again; they may self-react, or, if O_2 is present, react with the O_2—that is,

$$H\cdot + H\cdot \rightarrow H_2 ,$$
$$OH\cdot + OH\cdot \rightarrow H_2O_2 ,$$

and

$$H\cdot + O_2 \rightarrow O_2H\cdot$$

The second and third reactions predominate, in one case producing the highly reactive compound H_2O_2 and, in the other, a reactive free radical. If solutes are present, many other reactions are possible.

When a solute that may be altered is present, one must distinguish between *direct* and *indirect effects* of ionizing radiation in that solute, because this information is valuable in determining the mechanism of radiation damage. A direct effect is a consequence of ionization of the solute itself. With an indirect effect, ionization of the *solvent* is the first step in a series of reactions that leads to damage of the solute molecule. In the case of a macromolecular solute, the definition of a direct effect is modified to include the consequence of ionizations that occur within the volume occupied by the molecule. That is, a water ion produced deep in a crevice of a highly folded protein molecule has such a high probability of reacting with some part of the protein that, practically speaking, it should be thought of as contributing to a direct effect. An operational distinction between direct and indirect effects that has sometimes been used is to determine if the effect depends in any way on diffusion of ions through a solvent. If damage to a molecule of solute results solely as a direct effect, diffusion of reactive ions does not play a role in producing the damage, and experimental changes in the diffusion rate do not alter the probability of damage; however, if an ionization occurs several hundred ångström units away from a solute molecule, the reactive species must diffuse to the solute, if there is to be a reaction involving solute molecules. Thus, if the viscosity of the solvent is substantially increased—for instance, by addition of glycerol—indirect effects should decrease, because the reactive species will either have decomposed or self-reacted before reaching the solute molecule. An example of this will be given shortly.

In a biological system that is irradiated, many chemical changes occur. However, each molecule in a cell exists in so many copies that most of the damage has little significance to the cell. DNA molecules are an exception, of course, in that there is usually only one or two copies of each type of DNA molecule per cell. For this reason, DNA is the principal target molecule for radiobiological damage, and we shall direct our attention henceforth entirely to radiation effects on DNA. To facilitate this discussion, we need some way to analyze the data obtained. This is provided by *hit theory,* which we describe in the following sections.

18–7 HIT THEORY

In hit theory, we consider a population of N identical organisms or molecules. This population is exposed to a dose D of some radiation that causes damage. We assume that there are one or more radiosensitive sites in each organism or molecule, and that inactivation or damage requires that n sites be damaged. The effect on the population is expressed in terms of the fraction of the population that has not sustained some particular type of damage, as a function of the dose. As will be seen, the data are most conveniently plotted as $\ln S$ versus the dose D, in which S is the surviving fraction. The resulting curve is called a *survival curve* or a *dose-response curve*. Hit theory enables us to calculate these curves for certain idealized situations.

Consider a population of N organisms (that is, the number when $D = 0$) exposed to a dose D of radiation. Let there be only a single site in each organism that, if damaged or "hit" by a radiation photon, will kill the organism. (Note that simple absorption of the photon is not sufficient—the photon must cause damage. However, the fraction of incident photons that cause damage will certainly be proportional to dose.) The number dN damaged by a dose dD is proportional to the number N that existed prior to receiving that dose, or

$$-\frac{dN}{dD} = kN, \tag{33}$$

in which the constant k is a measurement of the effectiveness of the dose and is proportional to the fraction of the incident photons that cause an inactivating hit—in other words, the probability that a single photon can cause such a hit. This equation can be integrated from $N = N_0$ at $D = 0$ to yield

$$N = N_0 e^{-kD}. \tag{34}$$

FIGURE 18-20 Survival curves for two irradiated populations, one of the bacterium *E. coli,* and the other of the phage T4.

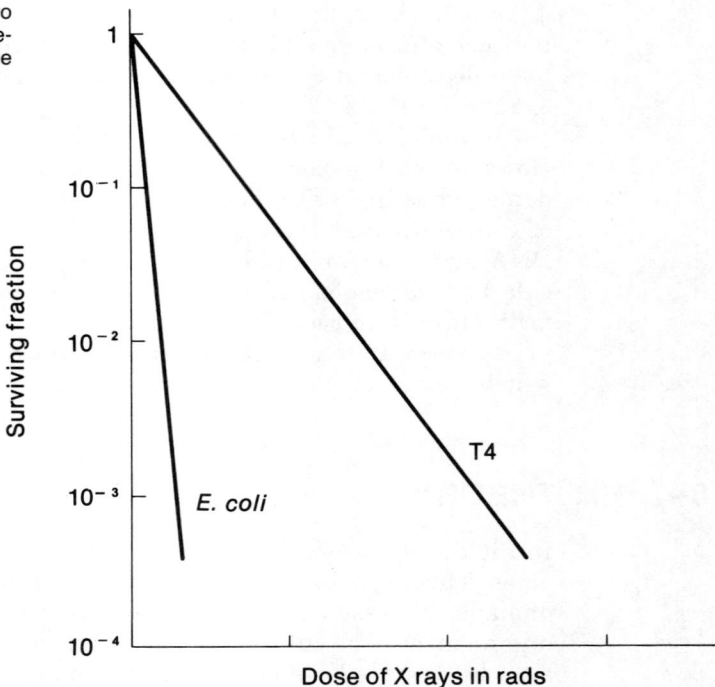

(Note how this is identical to the change in concentration of a reactant in a first-order chemical reaction, as described in Chapter 11.)

The surviving fraction $S = N/N_0$ is

$$S = \frac{N}{N_0} = e^{-kD} \:. \tag{35}$$

Clearly a plot of ln S versus D gives a straight line having a slope of $-k$. Curves of this type are called *single-hit* curves and are commonly observed with irradiated biological systems (Figure 18-20).

Let us now consider a population of organisms that differs from the one just described only in that each organism contains n units, *each* of which must be hit if the organism is to be inactivated. Thus, inactivation requires at least n hits—"at least," because it is assumed that two hits on a single unit are no more effective than one hit. The probability of one unit being hit by a dose D is $1 - e^{-kD}$, so that the probability P_n that all n units become inactivated is

$$P_n = (1 - e^{-kD})^n \:. \tag{36}$$

Thus, the surviving fraction S of the population is $1 - P_n$ or

$$S = 1 - (1 - e^{-kD})^n \:. \tag{37}$$

SEC. 18-7 HIT THEORY

FIGURE 18-21 Survival curves for various values of n showing that at high dose each curve becomes linear and that extrapolation to the y axis yields n as the intercept.

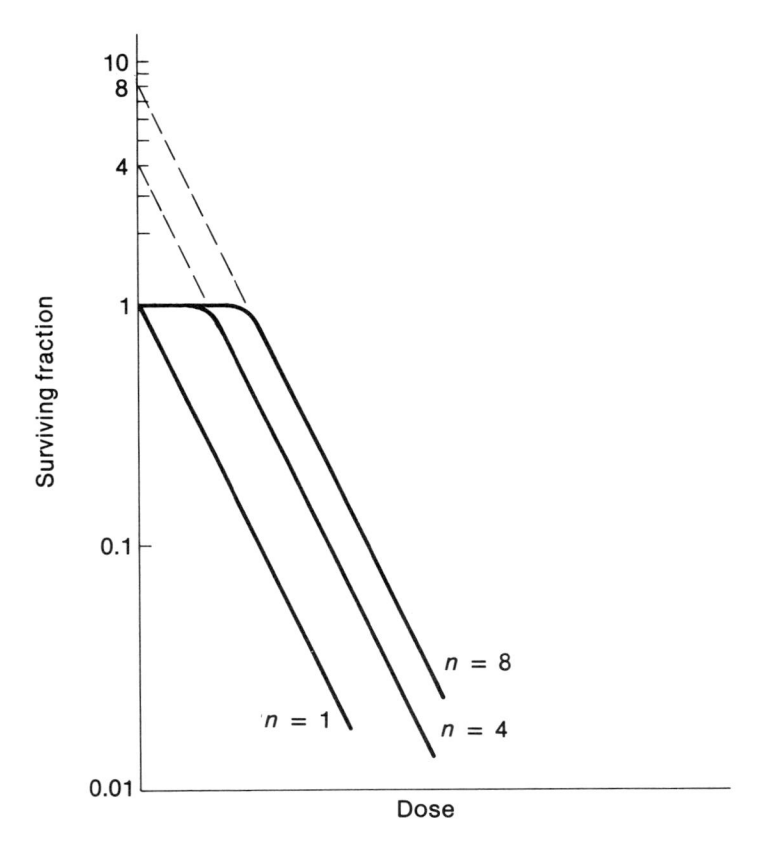

This equation can be expanded to yield

$$S = 1 - (1 - ne^{-kD} + \cdots + e^{-nkD}) . \tag{38}$$

As D increases, the terms containing e^{-2kD}, e^{-3kD}, and so forth, become negligible compared to ne^{-kD}, so that at high dose, $S = ne^{-kD}$, or

$$\ln S = \ln n - kD . \tag{39}$$

A plot of Equation 37 for $k = 1$ and various values of n show that for small values of D, $\ln S$ changes slowly (Figure 18-21). At large D, Equation 39 takes over and the curve becomes linear. Extrapolation of this linear part to $D = 0$ gives $S = n$ as the y intercept. Thus, as long as experimental data are obtained for high enough values of D, n can be determined. Curves of this sort are called n-hit curves and a system showing such a curve is said to have n-hit inactivation kinetics. An example of a system that shows this effect of hit number is the inactivation of baker's yeast by X rays. This yeast exists in both a haploid (one copy of each chromosome) and a diploid (two copies of each) state. Damage to any chromosome is sufficient to kill

FIGURE 18-22 Survival curves for haploid, diploid, and tetraploid yeast cells, showing the effect of ploidy on radiosensitivity. The two multihit curves extrapolate to approximately 2 and 4, respectively, at zero dose. Data from the author's laboratory.

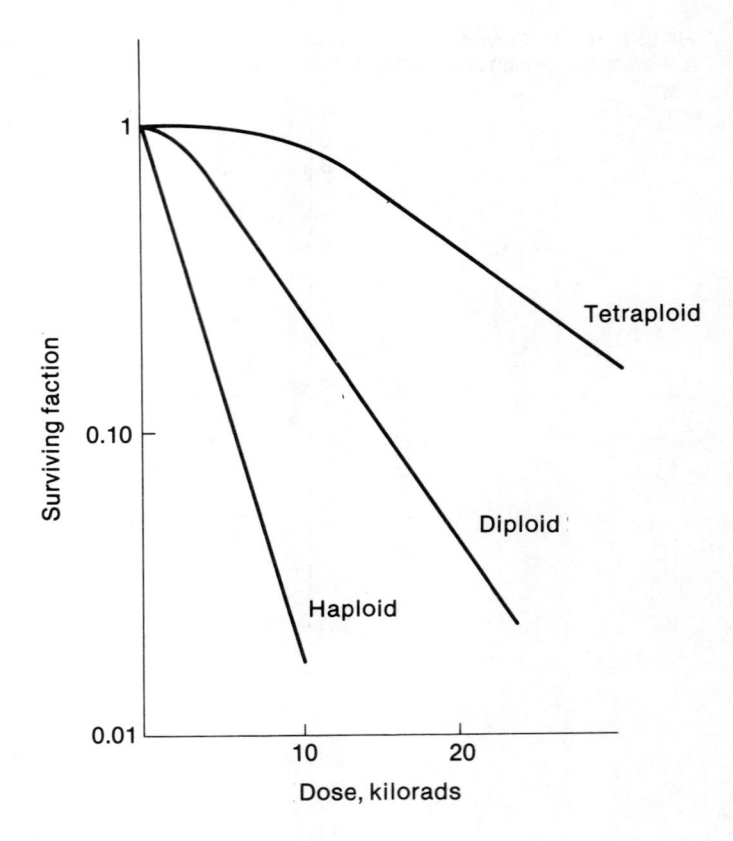

the haploid variety. However, inactivation of a diploid requires that both copies of any chromosome be inactivated. Therefore, as long as damage to a single chromosome follows single-hit kinetics, then the haploid and diploid strains should have single- and double-hit inactivation kinetics. That this is the case is shown in Figure 18-22.

More complicated situations can also be imagined. For instance, an organism might contain n targets, each of which might have to be hit m times. In this case the relevant equation (which we will not derive) is

$$S = [1 - (1 - e^{-kD})^n]^m , \qquad (40)$$

which, at high dose, is $S = n^m e^{-mkD}$, and at $D = 0$, the linear portion of the plot of ln S against D has a slope $-mk$ and extrapolates to $S = n^m$. We might also add to the complication by allowing each site to have a different radiosensitivity.

Real biological systems rarely follow any of these kinetics precisely. What can be said with certainty is only that if there is only one sensitive site, plotting ln S against D will always produce a straight line. It is of course true that if more than one hit is necessary for inactivation to occur, the survival curve will be multihit. However, the appearance of a curve

SEC. 18-7 HIT THEORY

FIGURE 18-23 Loss of colony-forming ability by ultraviolet irradiation of the bacterium *E. coli* and of a mutant that lacks a repair system. Data from the author's laboratory.

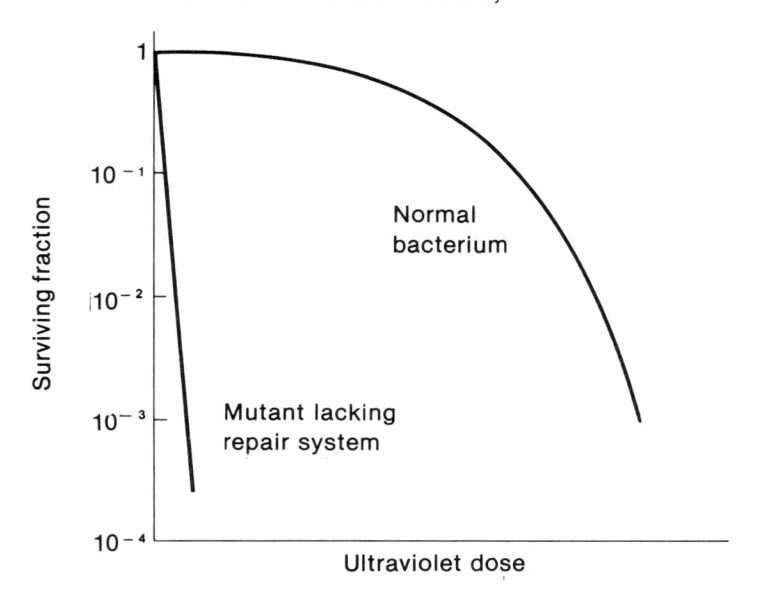

resembling a multihit curve—that is, having an initial slowly varying region (a shoulder) followed by a sharp drop in the value of ln S (such as in Figure 18-21)—is not sufficient to conclude that multihit inactivation is happening. In fact, the most common cause for inactivation kinetics of this sort is the existence of systems that repair damage. This is usually recognized by the fact that a linear region of the curve is never reached. (That is, careful examination of the experimental data indicates curvature in the apparently linear region.) Repair in a single-hit system would have the effect only of reducing the value of k, and the survival curve should, by that reasoning, remain single-hit. However, the repair systems are themselves radiosensitive and become increasingly less efficient at higher doses. Thus, k increases with dose and the curve resembles a multihit curve. An example of repair is shown in Figure 18-23, which shows the effect of ultraviolet radiation on a bacterium, *E. coli*, and on a mutant lacking the repair system. Note that only the mutant has single-hit kinetics.

It is not uncommon that a population is heterogeneous with respect to radiosensitivity. A simple example of this is a growing bacterial culture, because it contains bacteria whose content of DNA (the principal target molecule) per cell may range from a single DNA molecule that has not yet replicated to two just-completely-replicated molecules. Usually the effects of heterogeneity are relatively straightforward; some representative survival curves are shown in Figure 18-24. Panel (a) shows the curve for a population that is inactivated with single-hit kinetics even though it contains a fraction of totally resistant members. We denote this resistant

FIGURE 18-24 Different types of survival curves for mixed systems. See text for details. The fundamental rule in the interpretation of any survival curve is that if a portion of the curve is concave *upward*, the population must consist of at least two components having different radiosensitivity. In the calculation of the fraction f of a less-sensitive component, it is essential that the curve is extended to a sufficiently high dose that the curve becomes nearly linear. In panel (d), the evaluation of f_1 requires that f_2 is sufficiently small that the portion of the curve extrapolating to f_1 is linear.

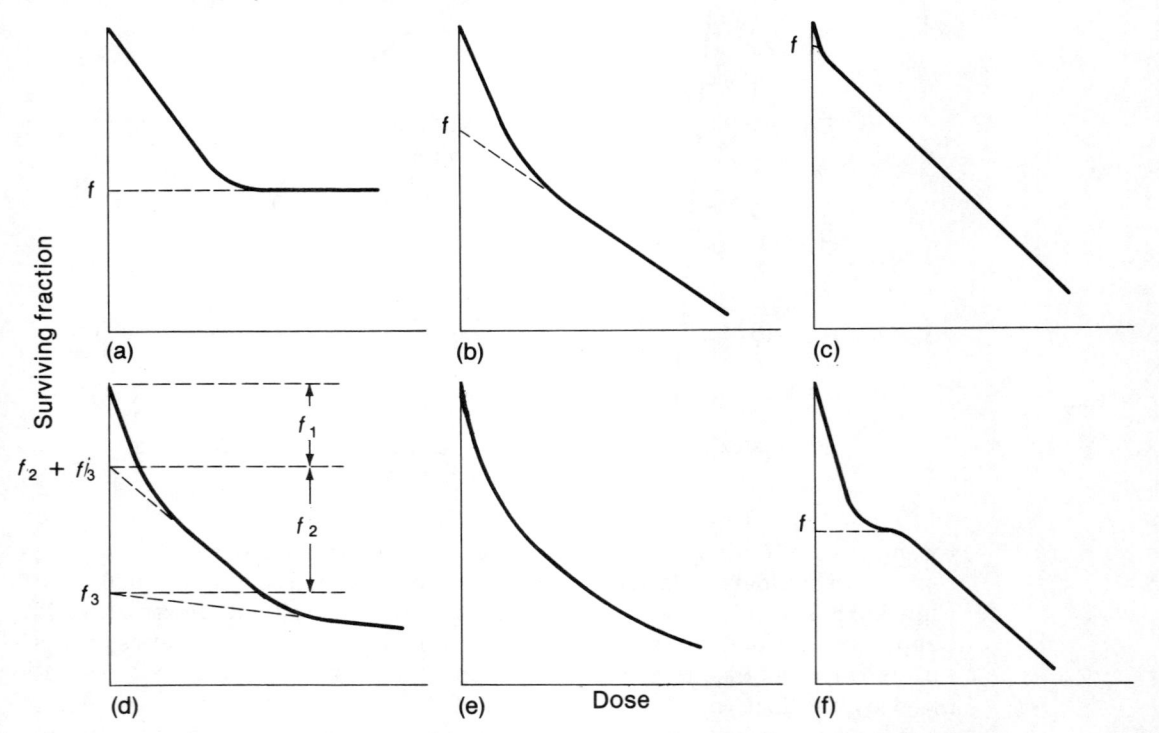

subpopulation as the fraction f; the surviving fraction can clearly never be less than f. In panel (b), a fraction f is present that is less sensitive than the major fraction but still is inactivated with single-hit kinetics; clearly the initial decrease in survival is due to inactivation of the more-sensitive fraction. A determination of f is now obtained by extrapolating the second component of the inactivation curve (that with the reduced slope) to $D = 0$. Panel (c) is similar to what is shown in (b) except that the major fraction is the less-sensitive one. This is in principle no different from (b) in that the more-sensitive fraction is still inactivated first. In panel (d) there are three components, denoted fractions f_1, f_2, and f_3. If the curves are sufficiently accurate to distinguish three linear regions, then f_1, f_2, and f_3 are measurable. In panel (e) the curve is concave upward, for which shape there are two indistinguishable reasons—either the number of components is so great that the linear regions are not separable, or the population has a continuous range of radiosensitivity. Panel (f) shows what appears to be at least a three-component system. However, this is not the case, because such a system would always have the curve shown in panel (d), as long as

SEC. 18–7 HIT THEORY

all components were inactivated through single-hit kinetics. Instead, panel (f) shows the result obtained if the minor component is inactivated through multihit kinetics. The fraction consisting of this minor component can again be determined by extrapolation to $D = 0$. Construction of curves for more complicated combinations is left to the reader.

In our derivation of Equation 35 we pointed out that the constant k is in some way a measure of the probability of a hit. The relation between k and this probability can be seen by looking at the effect of radiation in a slightly different way. First we note that the number of ionizations per unit volume is proportional to the dose; a fixed fraction of the ionizations produce a hit. Thus, if V is the volume of the sensitive target molecule or structure (V is called the *target volume* or the *sensitive volume*), then the average number of hits within the sensitive volume is cVD in which c is a proportionality constant relating the number of ionizations and the number of hits. If the hits are independent and random, the probability P_n that n hits occurs within the volume V is given by the Poisson distribution or

$$P_n = \frac{e^{-cVD}(cVD)^n}{n!} . \tag{41}$$

If x hits are required for inactivation, then all those target molecules receiving fewer than x hits survive and the survival S is

$$S = e^{-cVD} \sum_{n=0}^{x-1} \frac{(cVD)^n}{n!} . \tag{42}$$

If we consider a single-hit mechanism, this reduces to

$$S = e^{-cVD} . \tag{43}$$

Comparing this to Equation 35, we note that $k = cV$, or, k is proportional to the volume of the target. This is certainly not hard to appreciate—if one polymer A has twice the number of monomers (of the same kind) as a second polymer B, then the dose required to damage any one monomer in A will be half that needed to damage one in B. In other words, A is twice as sensitive as B. This is a commonly observed occurrence, as can be seen in the following example.

EXAMPLE 18–N Inactivation of bacteriophages by X rays in the absence of O_2.

Bacteriophages contain a single DNA molecule encased in a protein coat, and the DNA molecule is the radiosensitive site. When a phage suspension is X-irradiated in the absence of O_2, the principal damage to the phage is the production of double-strand breaks in the phage DNA. Only one double-strand break is necessary to destroy the ability of the phage to propagate itself, so that inactivation shows single-hit kinetics. All phages

FIGURE 18-25 Inactivation (loss of plaque-forming ability) of three bacteriophages irradiated with X rays. The molecular weights of the DNA molecule contained in each phage are: T7, 25 × 10⁶; λ, 31 × 10⁶; T5, 76 × 10⁶. Data from the author's laboratory.

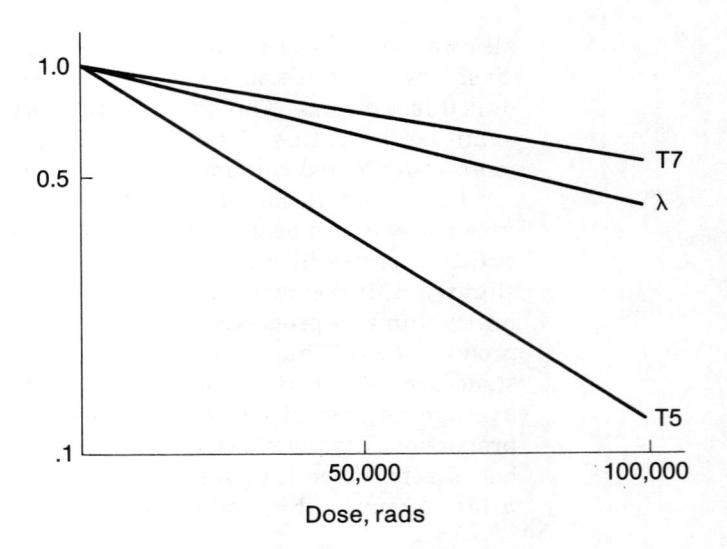

do not show the same radiosensitivity, though. Since a double-strand break can occur at any point in the DNA molecule, a phage having a larger DNA molecule should be more radiosensitive than one having a smaller DNA molecule. Furthermore, the ratio of the k values should be the same as the ratio of the DNA molecular weights. That this is the case is shown in Figure 18-25 for three phages—the larger the DNA molecule, the more rapidly is the phage inactivated. It is actually unnecessary to calculate the k values; as is shown in the figure, one need only select any survival value and read from the graph the dose required to produce that value. If D_1 and D_2 are the doses that produce a particular survival level for the molecular weights M_1 and M_2, then $D_1/D_2 = M_2/M_1$.

The dose producing a survival of 37 percent or $1/e$ has particular significance for a system having single-hit kinetics. From Equation 35 we can see that a survival of $1/e$ occurs when there is, on the average, 1 lethal hit per molecule. Since each lethal hit is a double-strand break, we can calculate from the data in Figure 18-25 the rate of production of double-strand breaks. For phage T7 the dose yielding $1/e$ survival is 1.49×10^5 rad, so that double-strand breaks are produced at a rate of 1 double-strand break per 1.49×10^5 rad per 25×10^6 molecular weight units or 2.7×10^{-7} break/rad/10^6 molecular weight units. This same value is of course also obtained from the λ and T5 curves, as the reader can verify.

EXAMPLE 18–O Estimating the molecular weight of a protein, which has not been isolated, by the radiosensitivity of the activity of the protein.

The proteins responsible for the transport of ions across biological membranes (see Chapter 14) are difficult to isolate, and some have not been isolated. An estimate of their molecular weight can be made from their radiosensitivity. Two assumptions are made in this analysis:

1. An ionization anywhere in the molecule causes loss of biological activity.
2. The radiosensitivity per unit molecular weight of the transport protein is the same as that for any other protein.

The second assumption is probably the weaker of the two. If a series of enzymes whose molecular weights are known is irradiated, all in the same suspending medium, it is generally found that enzyme activity decreases with single-hit kinetics and that the dose yielding $1/e$ survival of enzymatic activity per unit molecular weight is the same for all of the molecules. Within the same suspending medium, a biological membrane has been X-irradiated, and the loss of its ability to transport a particular ion has been measured. This ability was found to decrease exponentially, and the dose yielding $1/e$ survival corresponds to a molecular weight of 250,000, which is an estimate of the size of the transport protein. In the case that the protein has several subunits, the molecular weight cannot be taken as that of the individual subunit but only of the entire complex, because a single damaged subunit might result in total loss of the ability to transport the ion.

18-8 EFFECT OF THE ENVIRONMENT ON RADIOSENSITIVITY

Consider a molecule that is contained in a particular solvent A and that is broken by X rays with single-hit kinetics. In another solvent B the fraction of the ionizations leading to breakage may differ; either the radiation may produce different radiochemicals or the reaction of a free radical produced in the target molecule may re-form the original structure. The inactivation kinetics will of course remain the same—still single-hit kinetics—because the difference between the two solvents affects only the value of the constant c in Equation 43.

The solvent may exert yet another effect to which we alluded earlier when we distinguished between indirect and direct effects. Suppose many reactive chemicals are produced in the solvent and these are diffusible. If these can damage the target molecule, then there are two sources of damage—ionizations occurring within the molecule (the direct effect) and those occurring only in the solvent (the indirect effect). If there is a change in the solvent composition, such that the reactive chemicals in the solvent are reduced in concentration or even eliminated, then if the direct and indirect damage occurs by different mechanisms, the kinetics can change. This can be seen in the following example.

FIGURE 18-26 Breakage of DNA molecules by X-irradiation. The molecules were either in phosphate buffer (PB) or in PB containing histidine (PBH). Data from the author's laboratory.

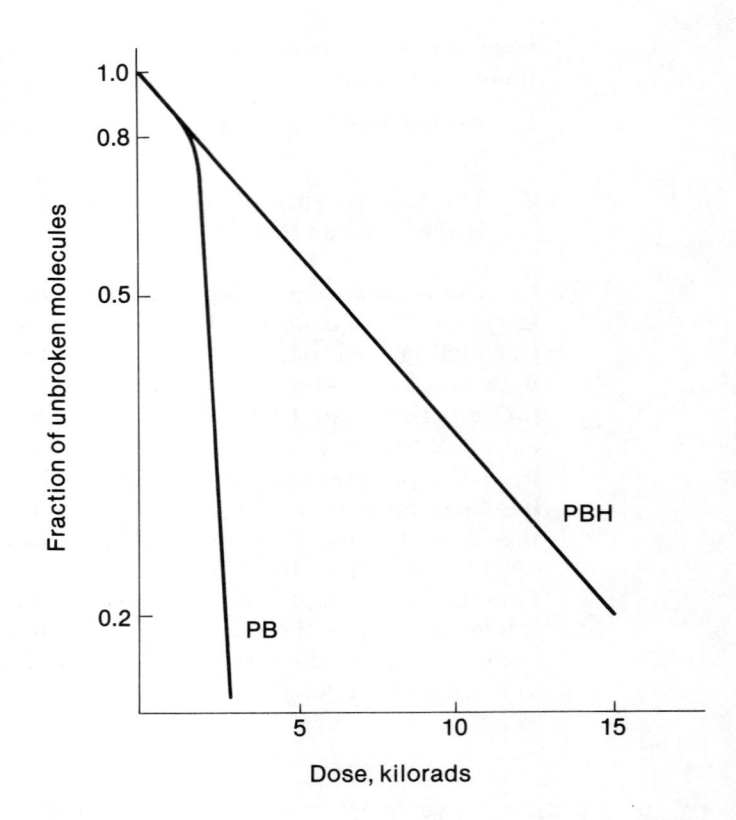

EXAMPLE 18–P   Effect of the amino acid histidine on the breakage of DNA molecules by X rays.

Here we consider a sample of DNA molecules, each of which has a molecular weight of 20×10^6. These molecules are suspended in either phosphate buffer (PB) or buffer containing $10^{-2}\ M$ histidine (PBH).

The following striking observation can be made. If PB is X-irradiated and then the DNA molecules are added, the molecular weight of the DNA begins to drop and continues to do so for about 48 hours. There is no such decrease if PBH is irradiated and DNA molecules are then added. The decrease in molecular weight with irradiated PB is an example of a *radiation after-effect;* clearly the responsible chemicals are present in irradiated PB but not in irradiated PBH.

We now consider experiments in which the DNA is irradiated in either PB or PBH. The data obtained from such experiments are shown in Figure 18-26 in which there is a clear difference in the kinetics of breakage. When the DNA is irradiated in PBH, breakage shows single-hit kinetics, whereas in PB, the breakage is much more rapid and shows multihit kinetics. It is clear that when histidine is present, a very rapid breakage mechanism has been eliminated. To understand how this occurs, we must realize that there are two ways to produce a double-strand break. When ionizations occur within a DNA molecule as a direct effect of irradiation, the amount of energy released in a small volume is so great that bonds are broken in both DNA strands—that is, a double-strand break occurs. However, a double-strand break might also occur if the DNA molecule slowly accumulates single-strand breaks until by chance two single-strand breaks in different strands are so near that they behave as a double-strand break. In this case, if single-strand breaks were made at a constant rate, a substantial dose would have to be received before opposite breaks matched. This is precisely what we have described as multihit kinetics—that is, many single-strand breaks or hits are required to inactivate (or break) the molecule. This is what is happening in the PB—many reactive chemicals are produced that cause single-strand breakage. Even though matching must occur, this happens at a greater rate than direct double-strand breakage, because the effective target volume is the entire solution rather than just the DNA molecule. When histidine is present, only the slower, direct effect is observed. This is because histidine reacts with and thereby eliminates the diffusible reactive chemicals. Actually, histidine is not unique in its ability to trap reactive molecules—tryptophan and cysteine, of the amino acids, have this capability, and so do a large number of reducing agents.

It is worth noting that when the DNA is irradiated in PB, there is also an after-effect—the molecular weight continues to drop, so that the precise curve that is obtained depends upon the elapsed time between the irradiation and the molecular weight measurement. In the experiment shown in Figure 18-26, histidine was added immediately after the irradiation to prevent this after-effect.

EXAMPLE 18–Q Effect of glycerol on X-ray sensitivity.

When direct effects occur, the reactive chemicals must diffuse to the target molecules before they self-react or react with something else. If diffusion is slowed down by an increase of viscosity of the solvent, indirect effects tend to disappear. This has been observed in a modification of the experiments just described. That is, if DNA is suspended in PB containing 30 percent glycerol, which yields a highly viscous solution in which diffusion is markedly inhibited, the survival curve shifts toward the PBH curve, as expected.

REFERENCES

Atwood, K.C., and A. Norman. 1949. "On the Interpretation of Multi-hit Survival Curves." *Proc. Nat. Acad. Sci.* 35: 696–709.

Castellan, G.W. 1971. *Physical Chemistry*. Addison-Wesley. Reading, Mass.

Clayton, R.K. 1971. *Light and Living Matter*. McGraw-Hill. New York.
Freifelder, D. 1976. *Physical Biochemistry*. W.H. Freeman and Company. San Francisco.
Setlow, R.B., and E. Pollard. 1962. *Molecular Biophysics*. Addison-Wesley. Reading, Mass.
Smith, K.C., and P.C. Hanawalt. 1969. *Molecular Photobiology, Inactivation, and Recovery*. Academic Press. New York.
Swallow, A. 1973. *Radiation Chemistry*. Halstead Press. New York.

PROBLEMS

1. The wavelength of the Hg spectral line used in many ultraviolet lamps is 2537 Å. What is its frequency?

2. The wavelength of the green line in the spectrum of Hg is 5438 Å. What is its value in nanometers?

3. A molecule absorbs infrared radiation having a wavelength of 2500 nm. What are the frequency and wave number of this wavelength?

4. Which has more energy—a photon whose wavelength is 5400 Å or one whose wavelength is 6250 Å?

5. Classify the following wavelengths as far ultraviolet, near ultraviolet, visible, and infrared: 196 nm, 540 nm, 3550 Å, 28,000 Å.

6. Two samples have OD values of 0.53 and 1.8. Which one allows less light to be transmitted?

7. A solution has an OD = 0.5. What is the percent transmission?

8. The absorbance A of a 0.5-M solution 5 cm thick is 0.62 at 260 nm.
   (a) What is A for this solution?
   (b) What is the molar absorption coefficient?

9. The absorbance of a 1-M solution of a substance at 540 nm is 0.52. If the Beer-Lambert Law is obeyed, what is the absorbance of a 0.25-M solution?

10. The A_{275} of a sample is 0.35. What can be said about the A_{450} for this sample?

11. The absorbance at 450 nm of four solutions of a substance X—0.1 M, 0.3 M, 0.5 M, and 0.8 M—are 0.15, 0.45, 0.75, and 1.0, respectively. What does this tell you about the optical properties of solutions of X?

12. Aqueous solutions of a particular substance show agreement with the Beer-Lambert Law up to a concentration of 1 M, above which there is negative deviation. A solution of the same substance in ethanol obeys the Beer-Lambert Law at least until 3 M. Propose several molecular explanations for this difference.

13. A 1-M solution of X has an A_{260} of 0.82, and a 1-M solution of Y has an A_{260} of 1.3. What is the A of a mixture that is prepared by mixing together equal volumes of 0.8 M X and 0.3 M Y?

14. A mixture of X and Y of unknown concentrations has an $A_{240} = 1.58$ and an $A_{450} = 0.82$. The molar extinction coefficients of X are $\epsilon_{240} = 5900$ and $\epsilon_{450} = 650$, and of Y are $\epsilon_{240} = 1520$ and $\epsilon_{450} = 8200$. What are the concentrations of X and Y?

15. DNA is frequently isolated from virus particles by shaking a virus suspension with phenol. In determining the final concentration of the DNA by absorption, it is essential to remove trace amounts of phenol, because of its large extinction coefficient in the spectral region in which DNA absorbs. For DNA, $\epsilon_{260} = 6722$ and $\epsilon_{280} = 3908$. Phenol also absorbs at these wavelengths and has $\epsilon_{260} \approx 500$ and $\epsilon_{280} \approx 1200$. The purity of the DNA can be assayed by measuring the ratio of A_{260}/A_{280}. What will be the minimum

PROBLEMS

ratio if the molar concentration of phenol is to be less than 1 percent that of the DNA?

16. Many bacterial enzymes can be assayed by exposing the cells to toluene, which releases the enzymes. The substrate is then added and the enzymatic activity is measured by a change in the absorbance of the substrate. The reaction mixture is frequently turbid because of the presence of bacterial debris and even of intact bacteria. The absorbance is then artificially high because of the scattering of light by these fine particles. Consider such a reaction, the product of which has $\epsilon_{350} = 4520$. If the reaction yields $A = 0.64$, what product concentration does this represent if the values of the absorbance at 550 and 600 nm are 0.085 and 0.062, respectively? The pure product has no detectable absorbance at 550 and 600 nm.

17. You are attempting to cleave a C-C bond of a particular molecule by illumination with light whose wavelength is the absorption maximum of the molecule. You do not succeed, and upon calculation, you discover that the energy of the photon is about 30 percent too small. To solve this problem, you choose a light source having a wavelength 40 percent shorter than the one you have been using. Whereas the energy of this light beam is adequate for what you wish to accomplish, there is no guarantee that you will be successful, because an essential condition for success will not always be met. What is this condition?

18. A photochemical reaction is being carried out in a beaker in an attempt to obtain a specific reaction product. After some time of illumination the yield of the reaction is sufficiently low that you decide to increase the volume of the reaction mixture. You use three times as much volume in the same beaker but discover you have not tripled the yield.

    (a) Why not?

    (b) Name two things you could do that would increase the yield.

19. The dissociation of HX, that is, $2HX \rightarrow H_2 + X_2$, is a photochemical decomposition with a quantum yield of 2. Show that this quantum yield agrees with the reaction sequence (a) $HX + h\nu \rightarrow H + X$. (b) $H + HX \rightarrow H_2 + X$. (c) $X + X \rightarrow X_2$.

    (*Hint:* Use the steady-state approximation.)

20. What is the energy content of one einstein of light whose wavelength is 260 nm?

21. The values of the energy required to dissociate H_2 and O_2 (in gaseous form) are 431.8 kJ mol^{-1} and 490.4 kJ mol^{-1}, respectively. If Hg vapor is mixed with some gases and the mixture is exposed to light having a wavelength of 2537 Å, the gas dissociates. Will this be the case for Hg vapor mixed with H_2 or O_2?

22. A particular chemical is exposed for one minute to ultraviolet light having a wavelength of 260 nm. The flux of light is 5850 joule per cm^2 per second. In a 1-cm^2 sample of the chemical, 10 percent of the ultraviolet light is transmitted, and 5×10^{-2} moles of the chemical are destroyed. What is the quantum yield for the destructive process?

23. What fraction of the initial amount of a radioisotope exists after 24 hours if the half-life is

    (a) 90 minutes,

    (b) 11 days, or

    (c) 5 years?

24. How many curies are represented by one microgram of ^{32}P? The half-life is 14.2 days.

25. Fill in the blanks.
    (a) $^{14}C \rightarrow \square + e^-$.
    (b) $^{9}Be + \alpha \rightarrow \square + $ neutron.
    (c) $^{2}H + \gamma \rightarrow \square + $ neutron.
    (d) $^{238}U \rightarrow \square + \alpha$.
    (e) $^{35}S \rightarrow \square + e^-$.

26. The radioactive isotope ^{14}C is continually formed from ^{14}N in the atmosphere owing to bombardment by extraterrestrial neutrons. Since during life all living things

take up and give off carbon, the ratio of nonradioactive carbon to radioactive carbon in live organisms is in equilibrium with the ratio present in the atmosphere. On the death of a living organism, however, carbon ceases to be exchanged with the atmosphere. Although the ^{14}C within a dead organism decreases with time owing to nuclear decay to ^{14}N, the ^{12}C content of the organism remains constant, since ^{12}C is nonradioactive. Thus the amount of ^{14}C that remains in a dead organism can indicate the number of years since death occurred. The half-life of ^{14}C is 5730 years and the decay rate of ^{14}C is about 14.5 decays per minute per gram of carbon, as it has been for tens of thousands of years.

(a) Derive an equation for the age of a previously living object in terms of both the ^{14}C when it was alive and its current ^{14}C content. By age one means the time elapsed since death.

(b) What is the age of a wooden post found in a prehistoric Indian ruin, if the radioactivity of the wood is 5.2 disintegrations per minute per gram of carbon? By age one means here the time elapsed since the tree was cut down.

27. This problem expresses the age of an object in a slightly different form from that derived in problem 26. In radiochemical dating, the age of a sample is determined from the ratio of the amount of a nonradioactive decay product D to the amount that remains of the original isotope I.

(a) Show that the age t is given by $t = (1/\lambda) \ln(1 + D/I)$, in which λ is the decay constant for I.

(b) What assumption is made about the origin of D?

(c) A 50-gram sample of a rock contains 120 mg of K and 0.5 mg Ca. The K is 0.119 percent ^{40}K, for which $\lambda = 5 \times 10^{-10}$ year^{-1}, and the Ca is 96.92 percent ^{40}Ca. How many years ago did the rock form? (Hint: $^{40}K \rightarrow {}^{40}Ca + e^-$.)

28. The energy loss of a 1-MeV electron passing through a protein molecule is about 0.75 eV; this is insufficient to break a typical bond whose ionization potential is about 3 eV. However, it has been found repeatedly that if an electron beam whose average energy is 1 MeV is used to irradiate (a) dry proteins, or (b) solutions of proteins, damage to the protein molecules does occur. The damage is greater in case (b). Explain this damage for the two cases.

29. A tumor has been irradiated with a proton beam but no damage to the tumor has occurred. In a second attempt, would you increase or decrease the energy of the protons?

30. You are using $FeSO_4$ to determine the dose rate, in rad, of an X-ray beam. You irradiate for various times and measure the following values of A_{305}: 1 min, 0.252; 5 min, 1.26; 8 min, 2.02; 10 min, 2.40; 12 min, 2.40; 20 min, 2.40. If $A_{305} = 1$ corresponds to 50,000 rad,

(a) Calculate the dose rate in rad/min.

(b) Explain the lack of proportionality between A_{305} and the irradiation time, assuming the Beer-Lambert Law remains valid at least up to an A_{305} of 10.

31. A beam of red light (650 nm) has an energy of 10^{-2} J/sec. What is the beam intensity in quanta per second?

32. Gaseous acetone can be converted to ethane and carbon monoxide by light. If a sample of gaseous acetone is irradiated with a beam of light having an intensity of 3.15×10^{16} quanta per second, and 0.005 mole of acetone is destroyed in 10 hour, what is the quantum yield? The sample absorbs 75 percent of the incident beam.

33. Light of 350 nm passes through five milliliters of a solution, and initially, 10 percent of the light is transmitted. The incident intensity on the solution is 10^{-2} J sec^{-1}. After one hour, the concentration of the solution has changed from 0.02 M to 0.015 M. What additional information is needed to calculate the quantum yield?

PROBLEMS

34. At a wavelength of 350 nm it is found that 6.2×10^{-5} mole of a compound X is decomposed by absorption of 8.25 joule of light. What is the quantum yield of the reaction?

35. Most naturally occurring radioisotopes are heavy metals; ^{40}K, a beta emitter whose half-life is 4.5×10^{10} years, is an exception. Potassium is normally 0.012 percent ^{40}K; furthermore, potassium constitutes 0.35 percent of the adult body weight. What will be the radioactivity in disintegrations per second of a 70 kg adult?

36. ^{32}P normally decays to ^{32}S by emitting a β particle. Some atoms have several modes of decay. Could ^{32}P ever decay to ^{32}Si? The atomic weights of these isotopes are: ^{32}Si, 31.9740; ^{32}P, 31.9739; ^{32}S, 31.9721.

37. A bacteriophage population is irradiated with gamma rays. The survival curve is shown below. What dose produces, on the average, one lethal event per phage?

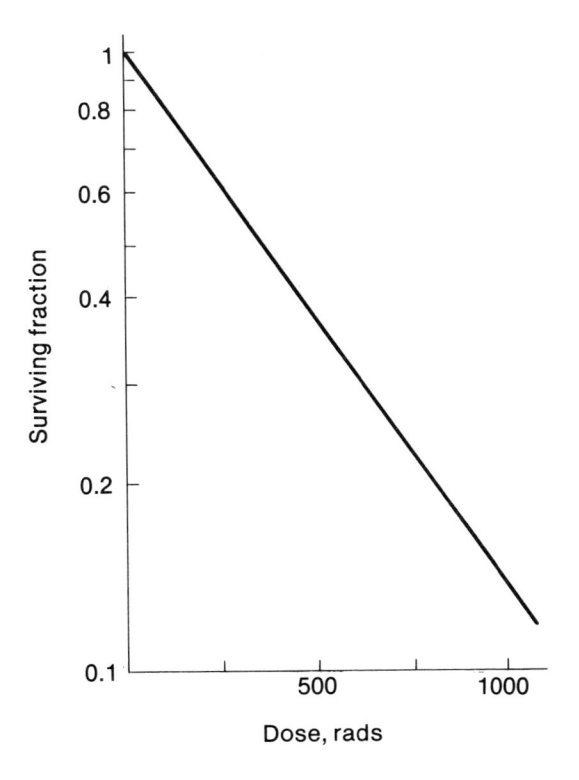

38. A chemical reaction has an activation energy of 314 kJ mol^{-1}. If this is to be provided by absorption of light, what is the maximal wavelength of the light that could be used?

39. Many years ago, in a study of substances that might prevent radiation damage, adenine was tested. It was found that a solution of 10^{-4} M adenine reduced the inactivation rate by ultraviolet radiation (254 nm) considerably. Adenine is not a chemical inhibitor of damage by ultraviolet light. Explain the protective effect.

40. Consider the after-effect described in Example 18–P. Would you expect the effect to have the same magnitude if the DNA were in 50 percent glycerol?

41. A single bacterium is capable of forming a single colony on a nutrient agar surface. A particular bacterium has the X-ray survival curve (#1) shown in the figure below. The curve is quite reproducible. However, one day an experiment is done that necessitates collecting the cells by centrifugation and then resuspending them in a fresh buffer prior to X irradiation. This sample gives the survival curve labeled #2. Several hours later the sample is irradiated again and curve #3 is obtained. The next day the irradiation is done again and curve #1 results. What has been happening?

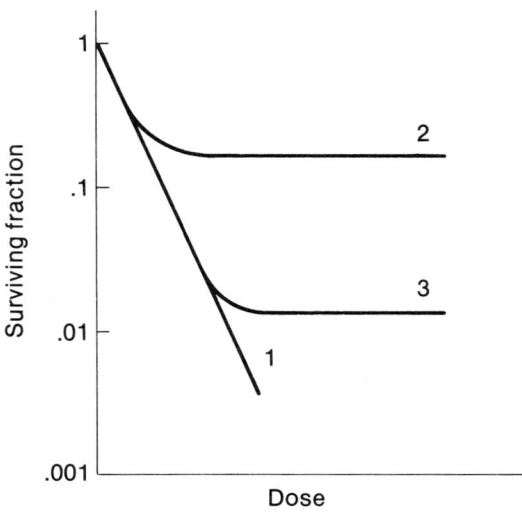

42. Which of the following statements, a or b, is true about two viral populations. If virus 1 is twice as sensitive as virus 2,

(a) a dose yielding 60 percent survival of a population of virus 2 will yield 30 percent survival of a population of virus 1.

(b) the dose yielding 60 percent survival of a population of virus 2 is twice the dose yielding 60 percent survival of a population of virus 1.

43. A phage sample is suspected of containing phages having two different kinds of DNA. In all known cases, the X-ray survival curve for phages is single-hit, and the inactivation rate depends on the DNA content. Thus, to test the suspicion the phage sample is X-irradiated and the survival curve shown below is obtained,

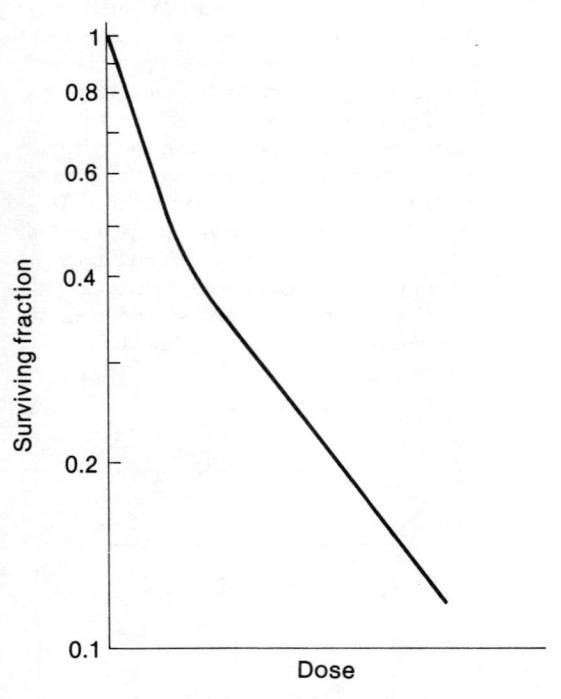

(a) What fraction of the population is the smaller phage?

(b) If the molecular weight of the DNA of the smaller phage is 18×10^6, what is the molecular weight of the DNA of the larger phage?

CHAPTER 19

SOLIDS AND LIQUIDS

The theories of the structure of pure liquids and of solids are in general beyond the scope of this book. Both are very mathematical and require a deep understanding of quantum mechanics. In this chapter these subjects are presented in the form of a survey. In particular, some aspects of crystal structure that are basic to the use of X-ray crystallography in determining the physical structure of macromolecules are discussed, and a few properties of liquid crystals that may have a bearing on membrane function are presented.

19–1 CRYSTALS

The main body of this book deals with gases and liquids. In this brief unit we consider some of the properties of crystals. In general, crystals are not important in biological systems, though there are notable exceptions such as the crystals of bone, teeth, and molluscan shells, and crystalline inclusions in a malfunctioning urinary tract. However, a great deal of information about the structure of macromolecules (such as nucleic acids and proteins) and about molecular interactions has been obtained from X-ray crystallographic analyses. This topic is generally for more advanced textbooks, as X-ray diffraction analysis is a highly mathematical subject. However, discussions of X-ray analysis in the scientific literature may seem a little less mysterious if the vocabulary of crystallography is understood. This section is designed simply to introduce some of the terminology and elementary concepts.

The Basis of a Crystal

A crystal is a solid with a repeating structure. The structural unit, called the *basis* or motif, generates the crystal structure. The basis may be a single atom or molecule, or a group of atoms, molecules, or ions. Each repeated basis group has the same structure and orientation in space as every other basis group in the crystal. The number of particles in the basis can vary. For Cu, the basis is a single Cu atom, for Zn it is two Zn atoms, and for diamond the basis is two C atoms (each surrounded by four carbon atoms in a tetrahedral array). For NaCl the basis consists of one Na^+ ion and one Cl^- ion.

The Space Lattice

If a point is placed at the same position in each basis group (centered on the atom, if the basis is a single atom), the set of points that results is called the space lattice of the crystal. The space lattice is a geometric abstraction rather than a physical structure; by placing the basis group at each lattice point (but not necessarily centering an atom on a lattice point), the crystal structure is generated. The distinction between the basis, the lattice, and the crystal structure is illustrated for a two-dimensional crystal in Figure 19-1. (Most crystals are three-dimensional but the concepts are visualized most easily in two dimensions. Also, two-dimensional periodic arrays do exist in biological systems—for example, membrane monolayers.) It should be noticed that in this hypothetical crystal only one of the two types of atoms lies on the lattice points. The choice is arbitrary in that the other atom could have been placed on the lattice points or the atoms could be placed such that each lattice point is at a particular point between the atoms.

FIGURE 19-1 The distinction between lattice, basis, and crystal structure. In this example, the basis contains two atoms; the atom represented by the open circle is centered on the lattice point.

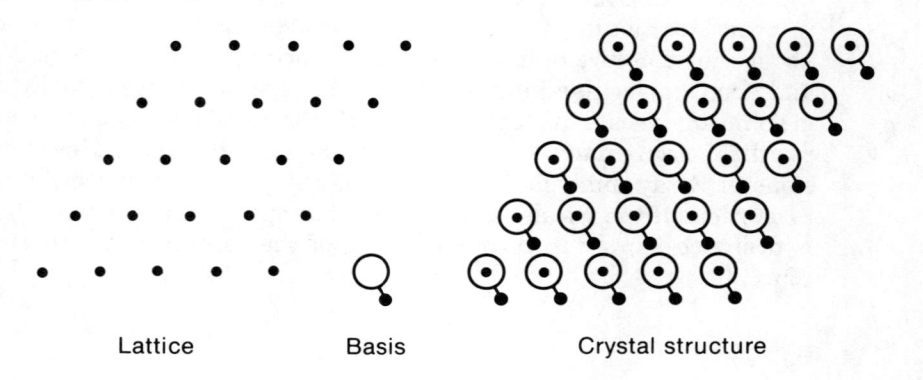

Lattice   Basis   Crystal structure

The Unit Cell

The space lattice of any two-dimensional crystal can be divided into identical parallelograms by joining lattice points by straight lines. For a three-dimensional crystal, parallelpipeds, rather than parallelograms, would be formed. Each parallelogram is called a unit cell. A lattice can be subdivided into different kinds of unit cells, as shown in Figure 19-2, again for a two-dimensional system. To simplify crystallographic analysis one usually chooses the unit cell that has both maximal symmetry and the smallest volume consistent with maximal symmetry. A unit cell is described by stating the length of the sides and the angles between the sides.

FIGURE 19-2 Two arbitrarily selected ways of forming unit cells in the lattice drawn in Figure 19-1. Note that the primary requirement of a unit cell—that adjacent unit cells can fill all space without overlapping—is satisfied by both choices.

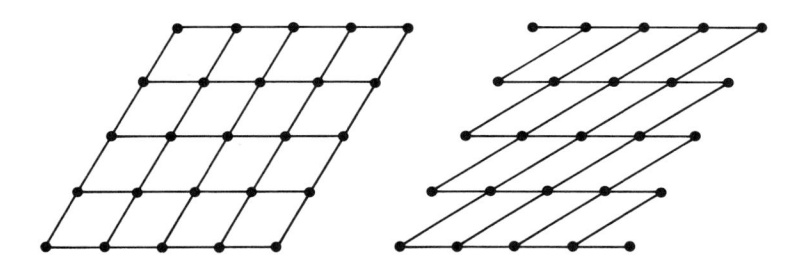

Three-dimensional Bravais Lattices and Unit Cells

In three dimensions fourteen different kinds of space lattices, called *Bravais lattices*, are possible. Each of these lattices has the property that if replicas of the unit cell are stacked side by side all space could be filled. The unit cells of the fourteen Bravais lattices are shown in Figure 19-3. These lattices can be grouped into seven crystal systems, based on the symmetry properties of the unit cell. The unit cells are subdivided into four classes:

1. *Primitive or simple.* These unit cells have lattice points only at their corners.
2. *Body-centered.* These unit cells have a lattice point at each corner and a lattice point centered within the unit cell.
3. *Face-centered.* These have a lattice point at each corner and a lattice point on each of the six unit-cell faces.
4. *End-centered.* These have a lattice point at each corner and one on each of the two faces at the opposite end of the parallelpiped on its longest axis.

FIGURE 19-3 Unit cell of the fourteen Bravais lattices.

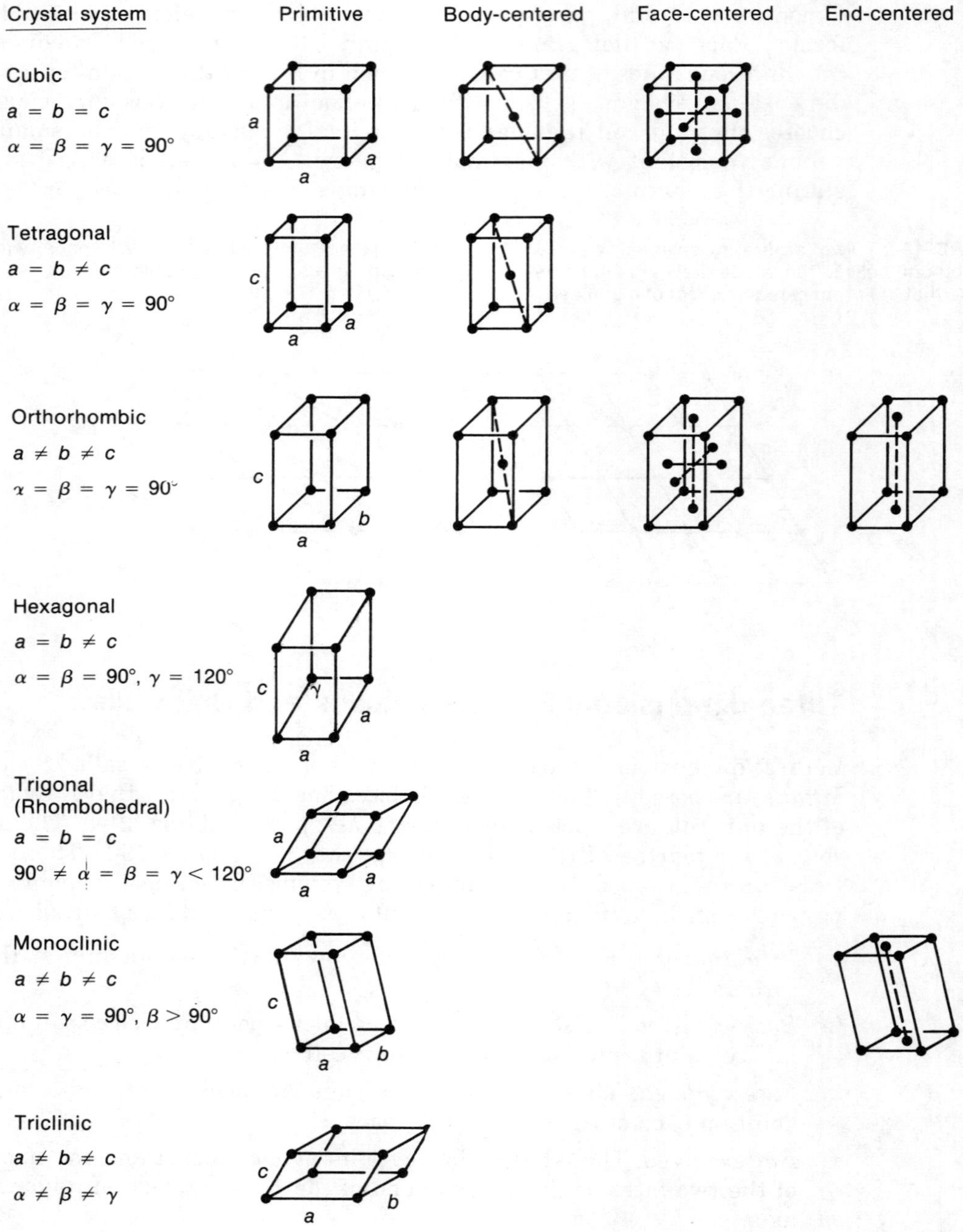

Crystal system	Primitive	Body-centered	Face-centered	End-centered
Cubic $a = b = c$ $\alpha = \beta = \gamma = 90°$				
Tetragonal $a = b \neq c$ $\alpha = \beta = \gamma = 90°$				
Orthorhombic $a \neq b \neq c$ $\alpha = \beta = \gamma = 90°$				
Hexagonal $a = b \neq c$ $\alpha = \beta = 90°, \gamma = 120°$				
Trigonal (Rhombohedral) $a = b = c$ $90° \neq \alpha = \beta = \gamma < 120°$				
Monoclinic $a \neq b \neq c$ $\alpha = \gamma = 90°, \beta > 90°$				
Triclinic $a \neq b \neq c$ $\alpha \neq \beta \neq \gamma$				

Except for central points, all lattice points are shared by other unit cells in the following way.

1. In any crystal each lattice point at the corner of a unit cell is shared by eight adjacent unit cells. Thus, for a primitive unit cell there is, on the average, one ($\frac{8}{8}$) lattice point and, therefore, one basis group per unit cell.
2. Each point on the face of a unit cell is shared by two unit cells. Hence, for a face-centered unit cell, there are, on the average, $\frac{8}{8} + \frac{6}{2} = 4$ lattice points and, therefore, 4 basis groups per unit cell. Similarly, for an end-centered system, there are $\frac{8}{8} + \frac{2}{2} = 2$ lattice points and 2 basis groups per unit cell.

These properties are summarized in Table 19-1.

TABLE 19-1 Some properties of unit cells.

Type of unit cell	Number of lattice points	Average number of lattice points per cell
Primitive	8	1
Body-centered	9	2
Face-centered	14	4
End-centered	10	2

Miller Indices

In X-ray diffraction analysis the angles at which X rays are deflected are determined by atoms that lie in particular planes in the crystal. These planes are not necessarily parallel to the crystal faces, as illustrated in Figure 19-4, which shows three different sets of planes in a rectangular parallelpiped. These planes are described by listing the coordinates of certain defining points in an arbitrary coordinate system. This is done in the following way.

A coordinate system is set up with the origin at one corner and axes parallel to the edges of the unit cell. (Note that the axes are perpendicular only for some types of unit cells.) The unit of length along each axis is the length (a, b, or c) of the side of the unit cell parallel to that axis. The position of any point in the unit cell is specified, as in any coordinate system, by giving the coordinates as a fraction of the unit lengths a, b, and c. The point at the origin is 0 0 0 (note that the commas are usually not written), and the point on the unit cell most distant from the origin is 1 1 1. The interior point of a body-centered lattice is designated $\frac{1}{2} \frac{1}{2} \frac{1}{2}$, and the uppermost point in an end-centered unit cell is $\frac{1}{2} \frac{1}{2} 1$.

The label of crystal planes is more complex. For reasons that derive from the mathematics of crystallographic analysis, planes are denoted by numbers—known as the *Miller indices hkl*—derived from reciprocal values of coordinates. These values are obtained as follows:

1. The intercepts of the plane on the *a, b,* and *c* axes are stated in terms of multiples of unit-cell lengths (that is, axis units). This step yields a coordinate *xyz*.
2. The reciprocal of each number in the coordinate is taken, yielding $\frac{1}{x}\frac{1}{y}\frac{1}{z}$. If these numbers are all integers, they are the Miller indices.
3. If some of the reciprocals are fractions, each number is multiplied by the smallest integer that will convert all of the fractions to whole numbers. How this is done is shown in Example 19-A, which refers to the planes in panel (a) of Figure 19-4.
4. If the coordinate is a negative number, the Miller index will also be negative. This is indicated by a bar over the number. Thus, the Miller indices 3 −2 −2 are written 3 $\overline{2}$ $\overline{2}$.

FIGURE 19-4 Several sets of planes in an orthorhombic system. (a) Five 220 planes. (b) Two 111 planes. (c) Three 100 planes. Two unit cells are shown in (a) and (c) and one and a portion of a second one are shown in (b).

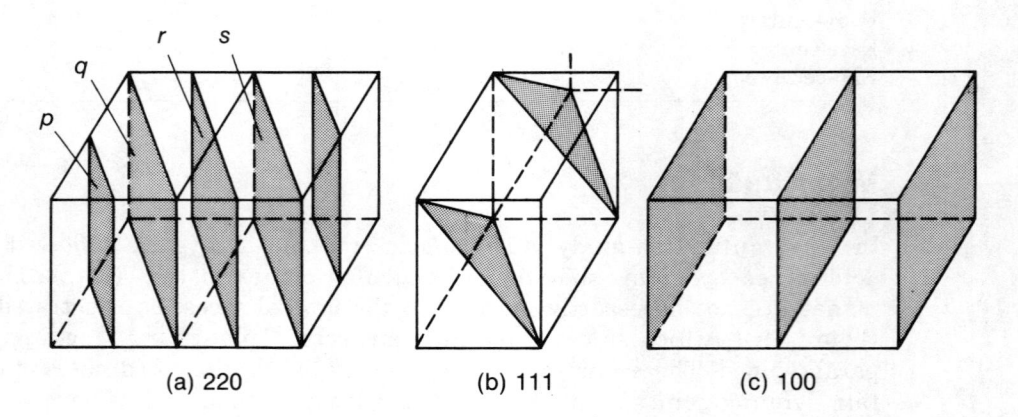

(a) 220   (b) 111   (c) 100

EXAMPLE 19–A Determination of the Miller indices of several planes shown in Figure 19-4.

The shaded plane labeled *p* in Figure 19-4 has intercepts $\frac{a}{2}$ and $\frac{b}{2}$ and ∞ (infinity) with the *a, b,* and *c* axes, respectively. Thus, step 1 yields the triplet $\frac{1}{2}\frac{1}{2}$ ∞. Taking the reciprocals in step 2 gives 2, 2, and 0, so the plane is defined by the Miller indices 220. Plane *q* has intercepts *a, b,* and ∞, and hence has Miller indices 110. To determine the indices for plane *r* requires the use of step 3. Step 1 gives $\frac{3}{2}a$, $\frac{3}{2}b$, and ∞; and step 2 yields $\frac{2}{3}\frac{2}{3}$ 0.

For step 3 each number is multiplied by 3, yielding the Miller indices 220, which are the same as those for plane *p*. Calculation of the Miller indices for plane *s* yields the values 110, the same as those for plane *q*. Two points should be noticed. First, a single set of Miller indices applies to a collection of parallel planes; in fact, all planes parallel to the 220 plane and separated by the distance between p and r would comprise the 220 set. Second, all planes that are parallel do not have the same Miller indices.

An advantage of the use of Miller indices in defining a crystal plane is that it is generally straightforward (though not always simple) to express the spacing between planes in terms of these indices. For example, a simple geometric argument shows that in a cubic crystal with lattice dimension a, the distance d_{hkl} between the parallel planes having Miller indices hkl is given by

$$d_{hkl} = a(h^2 + k^2 + l^2)^{1/2} \ . \tag{1}$$

Similarly, for an orthorhombic system with unit lengths a, b, and c the spacing between parallel hkl planes is

$$\frac{1}{d_{hkl}^2} = \left(\frac{h}{a}\right)^2 + \left(\frac{k}{b}\right)^2 + \left(\frac{l}{c}\right)^2 \ . \tag{2}$$

EXAMPLE 19–B What is the spacing between the 123 planes of an orthorhombic crystal in which a = 0.7 nm, b = 1 nm, and c = 1.2 nm?

$$\frac{1}{d_{123}^2} = \left(\frac{1}{0.7}\right)^2 + \left(\frac{2}{1}\right)^2 + \left(\frac{3}{1.2}\right)^2 = 12.29 \ .$$

Thus,

$$d_{123} = 0.29 \text{ nm} \ .$$

An Example of a Crystal Structure and Its Unit Cell

Crystal structures are formed from repetition of unit cells. It is beyond the scope of this book to discuss the numerous kinds of crystals and their symmetries and how they are formed from the unit cell. Instead, we will give a simple example—that of the formation of crystal whose external morphology is a right regular hexagonal prism in which the component atoms are packed as closely as possible (hexagonal close-packed—often written hcp). We start with parallel rows of touching spheres in which each sphere touches six other spheres (Figure 19-5(a)). This mode of packing is closer than if each sphere touched only four spheres. A second layer is then formed by placing spheres in the hollows marked by the dots in the figure. A third layer is formed by placing spheres in the hollows of the second layer. The process is continued until the structure shown in panel (b) is produced. The structure that results has a primitive hexagonal space lattice. The basis consists of two spheres associated with each lattice point; one sphere is on a lattice point and the second lies at the point $\frac{2}{3}\frac{1}{3}\frac{1}{2}$, which is not a lattice point. The unit cell is outlined by heavy lines in panel (b).

FIGURE 19-5 The positions of atoms in a hexagonal close-packed array. (a) Two layers of atoms (one white, the other shaded) showing the hexagonal array and the relative positions of the two layers. The open circles lie on the lattice points. (b) A hexagonal crystal. The shaded circles correspond to the shaded circles in (a). The unit cell is outlined by heavy lines. The shaded circles are not above the open circles but have the coordinates $\frac{2}{3}\frac{1}{3}\frac{1}{2}$ in a unit cell with an origin at the solid black circle.

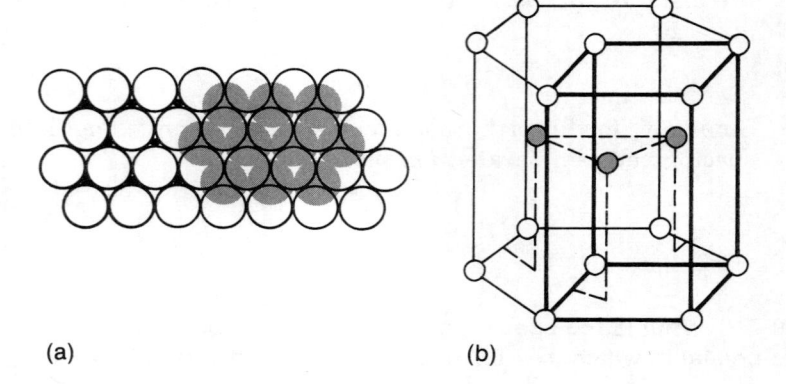

(a)   (b)

Types of Crystals

All atoms and molecules cannot form the same types of crystals. The prime requirement for crystal formation is that a cohesive force exists that is strong enough to maintain the components in a regular array. Geometric requirements must also be met; for example, the components must be arranged in such a way that they can fit side by side. Electrical charge is also important: for example, in a crystal of NaCl, two Na^+ ions cannot be next to one another. Quantum mechanical effects are also significant, but these will not be discussed.

There are basically four kinds of crystals—ionic, covalent, molecular, and metallic—the properties of which are described below.

1. An ionic crystal is usually hard and brittle, has a high melting point, and is a poor conductor of electricity. The inviolate rule of packing of the components of ionic crystals is that only unlike charges can be adjacent. Usually, the sizes of the ions (the ionic radii) also influence the geometric arrangement in that bulky ions tend to alternate with smaller ions. NaCl and CsCl exemplify the effect of size. NaCl forms a face-centered cubic lattice, whereas CsCl (the Cs^+ ion is much larger than the Na^+ ion) forms a body-centered cubic crystal (Figure 19-6). Note that in NaCl the Na^+ and Cl^- ions (which together form the basis

FIGURE 19-6 Arrangement of positive (dark circles) and negative (light circles) in (a,b) NaCl and (c,d) CsCl. The unit cell, which is outlined with heavy lines, is face-centered for NaCl and body-centered for CsCl. (b) A view down onto the shaded plane in (a). (d) A view of a face of the unit cell. In (b) and (d) the positive and negative ions are drawn with the correct relative radii. The scale is not correct in (a) and (d).

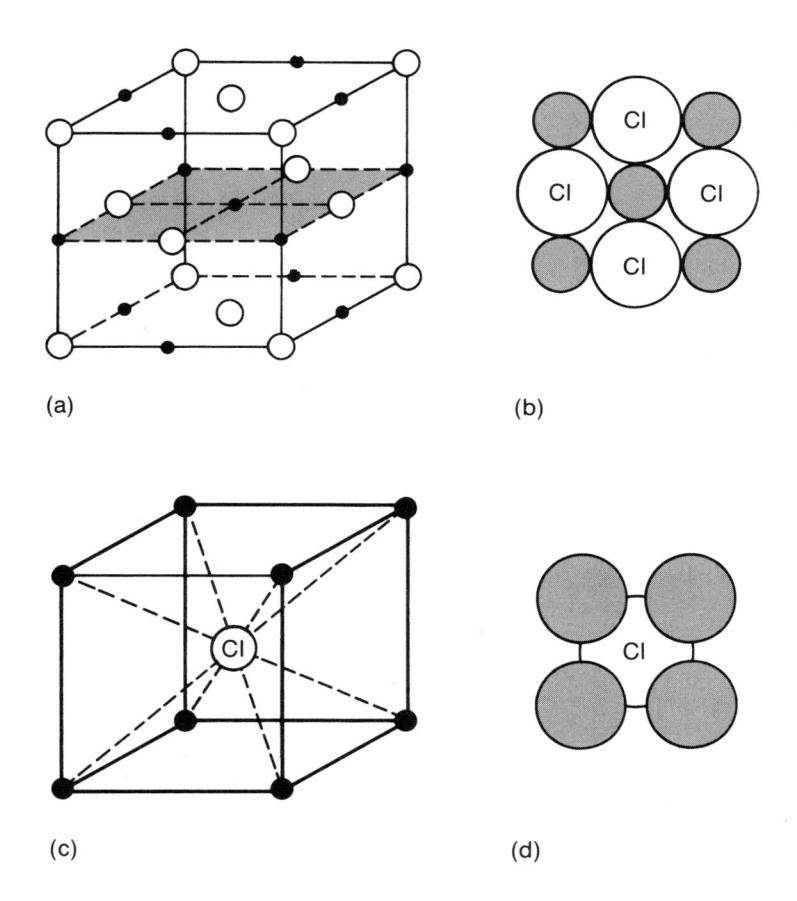

of the unit cell) both lie in the same plane perpendicular to the crystal axes, whereas in CsCl the Cs$^+$ and Cl$^-$ ions lie in parallel but different planes perpendicular to the crystal axes.

2. In a covalent crystal the components are held together by covalent bonds. Because these bonds are very strong, covalent crystals are generally hard solids possessing a very high melting point. A well-understood example is diamond, whose structure is based on a face-centered cubic lattice. Each carbon atom is covalently bonded to four other carbon atoms forming a tetrahedron (Figure 19-7). Since every possible bond is formed, a diamond crystal is, in essence, one gigantic single covalently bonded molecule. The complete network of bonded atoms is responsible for the extreme hardness of diamond. Since carbon atoms can also form

FIGURE 19-7 The covalent crystalline structure of diamond and graphite, both of which contain only covalently bonded carbon atoms. (a) The diamond face-centered cubic unit cell. The edges are drawn with heavy line. Note that these heavy lines are not chemical bonds; the thin lines are chemical bonds. The six atoms in the face are numbered. Five carbon atoms forming a tetrahedron are shown as solid circles. Note that three atoms are on faces and one is at the corner formed by intersection of the faces. The unit cell contains four internal carbon atoms; these have a c coordinate of either $\frac{1}{4}$ or $\frac{3}{4}$. (b) Two layers of the graphite array. The solid lines are covalent bonds. Each atom makes three bonds. Note how the central solid circle bonds to three other blackened atoms. The dashed lines are van der Waals bonds between the layers.

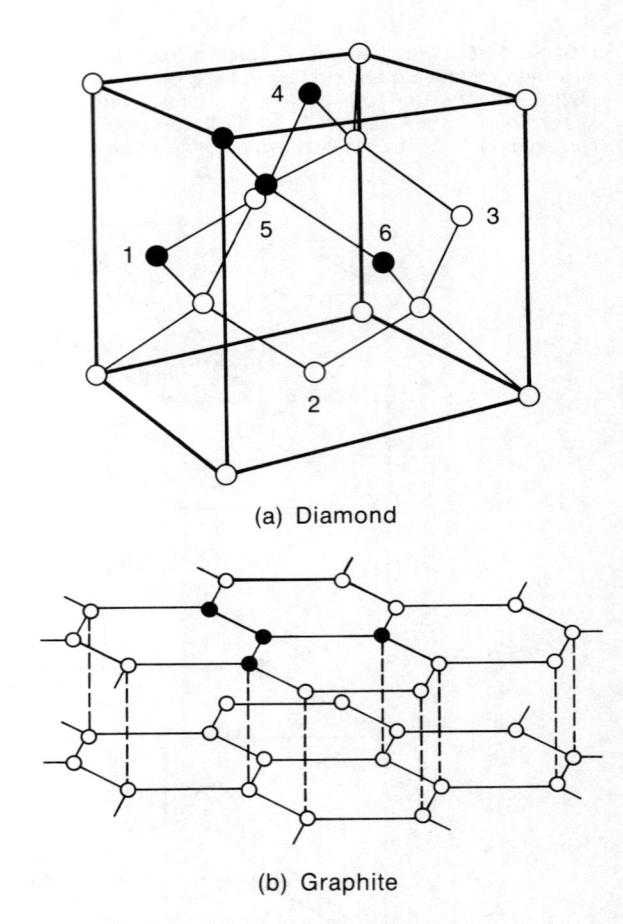

(a) Diamond

(b) Graphite

double and triple bonds, other kinds of carbon crystals exist. For example, in graphite, which also consists exclusively of carbon atoms, each atom is covalently bonded to only three other atoms—two by single bonds and one by a double bond, but resonating to give three bonds with two-thirds single-bond character and one-third double-bond character. These three bonds lie in a plane so that instead of a three-dimensional network the atoms form two-dimensional (planar) layers of tightly bonded atoms. The layers are held together only by weak van der Waals forces. The weak bonding allows the planes to slide along one another, so graphite is easily deformed in directions parallel to the planes. This ability to slide accounts both for the brittleness and the lubricating qualities of graphite. Different crystalline forms are also found with sulfur.

3. Molecular crystals are generally very soft solids, having a low metal point and conducting electricity poorly. The molecules cohere by van der Waals forces and are packed together in a way that is determined

primarily by their shapes. Many organic compounds and weakly ionizable inorganic salts form molecular crystals. In many cases the shapes of particular molecules prevent the molecules from approaching too closely; in such cases, the van der Waals forces are not strong enough to produce a stable periodic structure and crystals do not form. Such solids are called *amorphous*.

4. Metal atoms form crystals, whose structure can only be understood in quantum mechanical terms. The principal feature of these crystals is that the valence electrons are highly delocalized throughout the crystal and in fact can be thought of as freely moving. For complex (quantum mechanical) reasons, the atoms are very tightly bonded, which accounts for the high tensile strength of metals. The extraordinary mobility of the delocalized electrons is responsible for the high electrical conductivity of metals.

Determination of the Structure of Crystals

The arrangement of atoms in any crystal can, in theory, be determined by X-ray diffraction. However, in practice, the technique becomes more difficult and time-consuming as the basis becomes larger. For large molecules such as macromolecules the possibility of successful determination of structure at the atomic level is, at present, limited to the smaller globular proteins (molecular weight < 50,000) and such fibrous molecules as nucleic acids and collagen whose repeated unit contains a fairly small number of atoms. For simplicity, we will only show how, in principle, X-ray diffraction can yield information about molecular structure.

Consider a beam of X rays falling on the surface of a crystal at an angle θ (Figure 19-8). Some of the X rays will be scattered by the atoms on the surface, but because of the great penetrating power of X rays, most of the radiation will enter the crystal. A small fraction of the X rays will be scattered by the atoms within the crystal and the scattered waves will move away at all angles. It is convenient to think of the beam as being reflected by the various sets of parallel planes in the crystal and, in fact, one factor that determines the intensity of the light reflected at a particular angle is the density of scattering centers (electrons) in a particular plane. A second factor, and the one most relevant to the present discussion, is that waves reflected at various angles will be subject to optical interference, which can reduce the intensity to zero at some angles (destructive interference). However, some of the reflected waves will constructively interfere and produce detectable radiation. Constructive interference can occur only when the difference in the path lengths (AB + BC in the figure, in which it is assumed that the observation point is at infinity) of two beams reflected by two surfaces is an exact integral multiple of the wavelength. Only when that condition is met will the two waves be in phase. The condition for constructive interference is given by the *Bragg equation*:

$$n\lambda = 2d \sin\theta \tag{3}$$

in which n is the order of diffraction, λ is the wavelength, and d is the spacing between the planes. If the planes are defined by the Miller indices hkl, d is d_{hkl}.

In practice, the use of this equation for most crystals is not simple. Figure 19-8 shows an arrangement for a single reflecting plane, but of course many planes can give rise to detectable reflections; thus, a very large number of images of the incident beam may be observed. Furthermore, the first-order reflection by one set of planes may coincide in angle with a second-order reflection by another set. Also, the relative intensities of the different reflections depend on the relative density of scattering centers in each plane (the number and size of the atoms) and on the order of diffraction. Suffice it to say that for crystals containing a small number of different scattering centers (for example, crystals of inorganic salts containing a small number of different types of scattering centers—such as NaCl, diamond, and Na_2SO_4), straightforward calculations predict the pattern of reflections that is expected, and comparison between theory and experiment enables one to identify the crystal system, the planes that produce reflections at particular angles, and the spacing between many of these planes.

Before proceeding with a specific example, it is valuable to understand how the data about angles and intensities are obtained. This is shown in the following section.

FIGURE 19-8 The geometry of reflection of X rays impinging on a crystal surface at an angle θ.

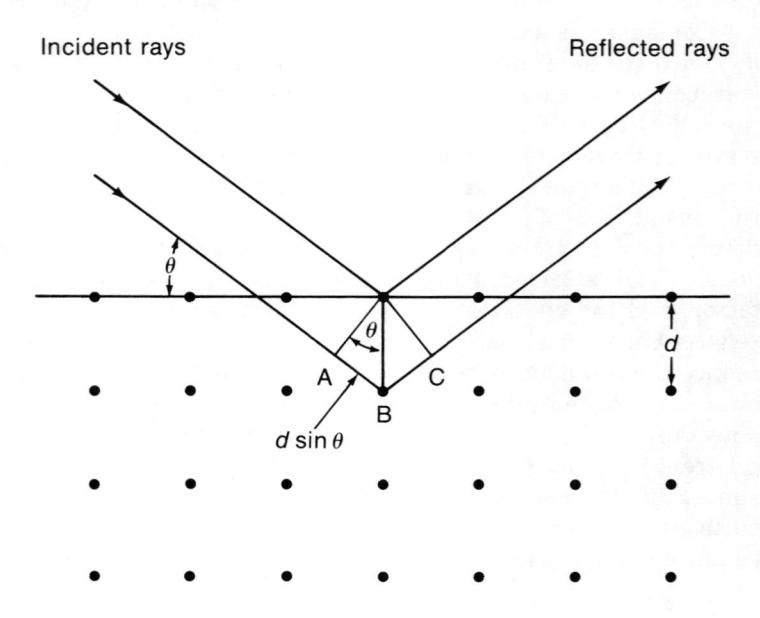

Obtaining Data

Figure 19-8 shows an X-ray beam that is incident on a single plane parallel to the crystal surface and that is producing a first-order ($n = 1$) reflection from that plane. To detect the reflected beam, an ionization chamber (various types will suffice) can be mounted at an angle 2θ (in which θ is the angle in the Bragg equation) with respect to the incident beam (Figure 19-9). However, with a detector at a fixed position, reflections emerging at other angles (for example, from other planes or with different values of n) would be missed. A possible solution to this problem is to rotate the detector around the crystal, but two major difficulties would remain: (1) some first-order reflections would be quite weak because they would emerge from the back of the crystal, and (2) it would be difficult to detect a reflection from a plane that is nearly parallel to the incident beam because the weaker reflection would be very near the much more intense transmitted beam. These experimental problems are avoided in the Bragg X-ray diffractometer by rotation of the crystal instead (Figure 19-9). By such rotation, the reflected beam for each set of planes is brought to the detector chamber, which is always located away from the transmitted beam. The diffraction data then consist of a set of intensities I of radiation observed at each angle of rotation 2θ (sometimes called the diffraction angle). An example of the kind of data obtained is shown in Figure 19-10, in which I is plotted against θ for crystalline tungsten. Note that the detection system used gives quantitative infor-

FIGURE 19-9 A diagram of a Bragg X-ray diffractometer. The detector is at a fixed angle with respect to the transmitted beam and the crystal is rotated.

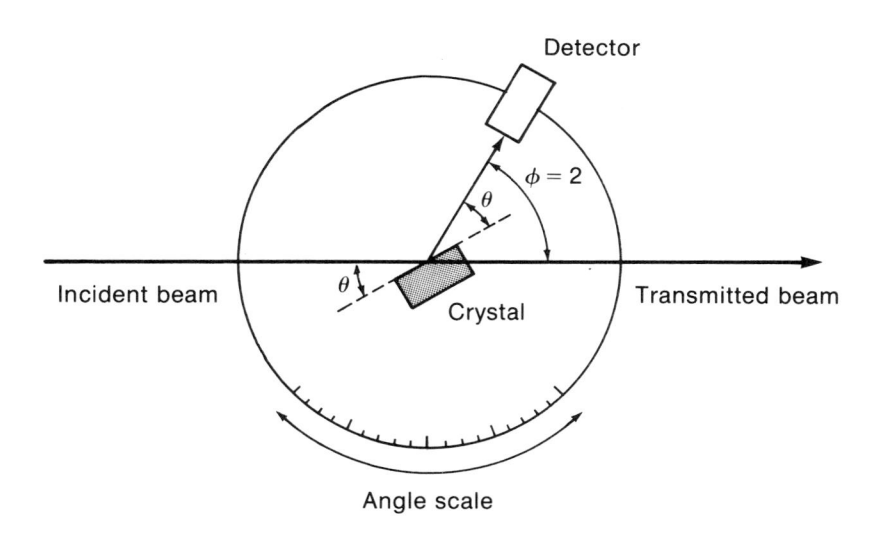

mation about the relative intensities of the reflected lines; these intensity differences are a result of the fact that all sets of planes do not have the same density of electrons (either because the number of atoms per plane or the atomic number of the atoms differ) and from interference between waves reflected from planes that are parallel but have different Miller indices.

FIGURE 19-10 A typical pattern of intensity of the diffracted beam at various angles, obtained with a Bragg diffractometer. This is a portion of the diffraction pattern of crystalline tungsten.

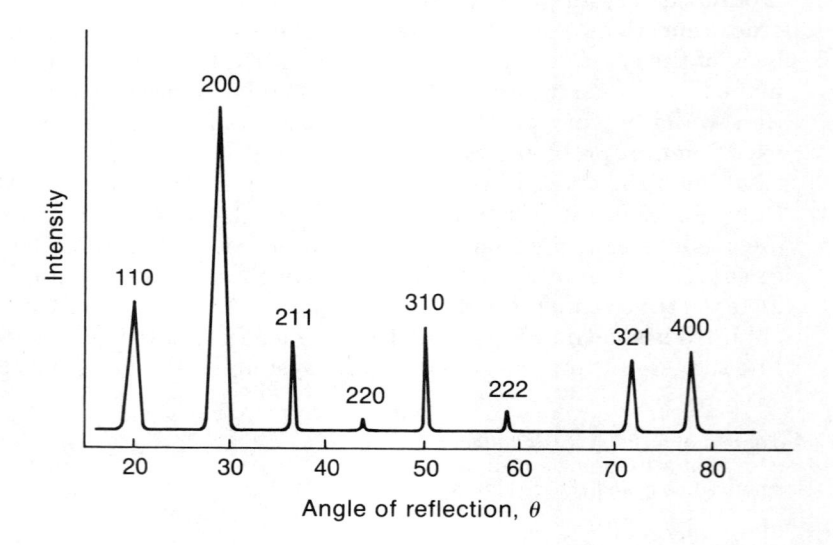

The Bragg technique is experimentally difficult because large crystals must be grown (which has not been easy for large organic molecules and has not yet been accomplished for some macromolecules) and the crystal must be accurately mounted with respect to the incident beam. For obtaining details of the structure of complex polyatomic molecules (for example, viruses, nucleic acids, proteins), the Bragg method is an exceedingly important, though tedious, technique. However, for simple molecules for which only the dimensions and symmetry properties of a unit cell are to be determined, a much less exacting technique—*the Debye-Scherrer powder method*—can be used.

In the powder method a sample of crystalline material is ground to a fine powder, placed in a thin-walled glass tube, and then exposed to a beam of monochromatic X rays. Since the orientations of the crystals with respect to the incident beam are random, some crystals will be oriented in a way that satisfies the Bragg equation for a particular set of planes and others for different sets of planes. In fact, constructively interfering reflections from all planes should be present (as they are), and they can be found in many places. The experimental problem, which is solved quite simply, is to find a suitable location for a detector. Consider a reflection from one set of planes of a single microcrystal; this reflection will be a beam emerging at a particular angle. If the crystal is rotated in a plane parallel to the incident beam (that is, either "forward" or "backward"), θ will change and the Bragg condition will not be satisfied. However, if the crystal is rotated in a plane *perpendicular* to the incident beam, θ will be unchanged and the reflected beam will rotate around the incident beam and trace out a cone. Thus, in a collection of *randomly* oriented microcrystals, the reflections from each plane will emerge as a cone of radiation, concentric about the incident beam (Figure 19-11). Thus, the reflections from planes with different Miller indices ob-

FIGURE 19-11 Diffraction cones of a powder diagram. For clarity, only the forward cones are shown. The sample is surrounded by a strip of film. The heavy lines show the intersection of the cones and the film.

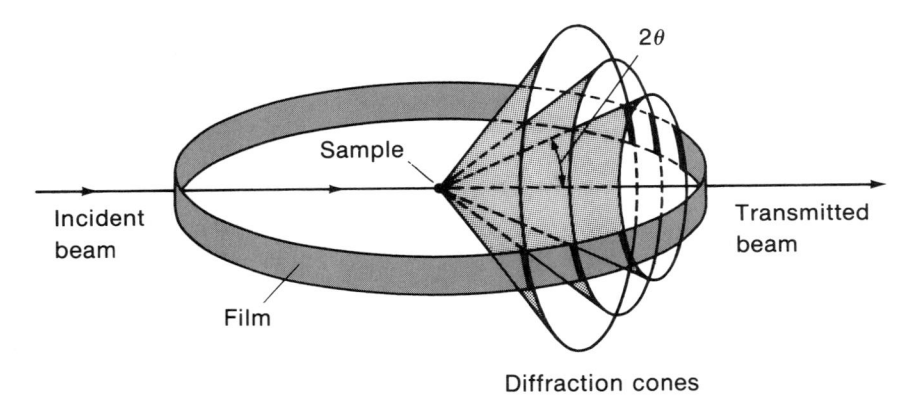

tained from a crystalline powder consist of a set of cones centered on the sample, each cone corresponding to a particular value of d and n.

The reflection cones are easily detected by surrounding the sample by a circular strip of film (Figure 19-11). The different cones will intersect the film in various places and produce slightly curved lines; the degree of curvature depends on the distance of the film from the sample. An example of the kind of data obtained is shown in Figure 19-12.

FIGURE 19-12 A portion of exposed and developed film showing the diffraction pattern for powdered crystals of KCl. The circles are holes in the film; these are the points (separated by 180°) through which the incident and transmitted beams pass. Note that the lines represent both forward and backward cones. Orders of diffraction >1 are shown.

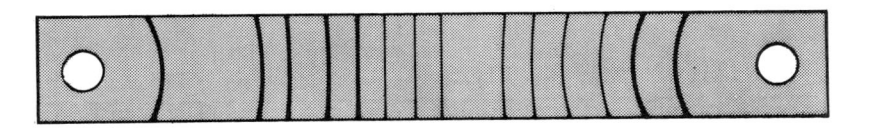

Analysis of a Cubic Crystal—An Example

Let us consider a simple example—that of a cubic lattice with lattice constant a.

Combining the expression relating d_{hkl} for cubic lattices and the Bragg condition, one obtains

$$\sin \theta_{hkl} = \frac{\lambda}{2d_{hkl}} = \frac{\lambda}{2a} (h^2 + k^2 + l^2)^{1/2} . \tag{4}$$

The quantity $(h^2 + k^2 + l^2)$ for planes defined by various Miller indices is shown in Table 19-2. Note that there is no value of $(h^2 + k^2 + l^2) = 7$, for no three integers exist whose sum is 7. Other numbers—for example, 15—are also missing.

TABLE 19-2   Values of $h^2 + k^2 + l^2$ for various planes having Miller indices hkl.

hkl	100	110	111	200	210	211	220	221
$h^2 + k^2 + l^2$	1	2	3	4	5	6	8	9
hkl	310	311	222	320	321	400	322	330
$h^2 + k^2 + l^2$	10	11	12	13	14	16	17	18

A plot can then be prepared that shows the positions of the reflections on a $\sin \theta$ scale for three different types of cubic lattices (Figure 19-13). The pattern for the primitive cubic lattice is straightforward with all lines present except those for $n = 7$ and 15 (see Table 19-2) and such a pattern is indicative of a primitive cubic lattice. The patterns for the body-centered (panel (b)) and face-centered (c) cubic lattices differ. In a body-centered cubic lattice some planes having different Miller indices are parallel—for example the 100 and 200 planes. The X rays reflected by the 100 planes constructively interfere with one another as do the X rays from the 200 planes. However, waves reflected from a 100 plane are out of phase by a half-wavelength (or an integral multiple of a half wavelength) with those reflected from a 200 plane and thus destructive interference (cancellation of intensity) will occur. The 100 planes are a subset of the 200 planes in that all 100 planes are also 200 planes, but all 200 planes are not 100 planes. Since the number of atoms in a 100 plane is the same as the number in a 200 plane, all light reflected from the 100 planes will be cancelled, though reflections from 200 planes will remain detectable. This destructive interference accounts for the fact that the 100 line is missing from the pattern for the body-centered cubic lattice. Similar arguments explain other missing lines in the body-centered and face-centered cubic lattices. Thus, the pattern of observed lines enables one to determine the Bravais lattice for a particular crystal; from the observed values of $\sin \theta$ the spacing between the planes can be calculated from Equation 4.

We now use what has been developed to determine the structure of a NaCl crystal. Table 19-3 shows representative data that might be obtained for an X-ray diffraction analysis using molybdenum X rays (obtained by an

FIGURE 19-13 Plots of $(\lambda/2a)(h^2 + k^2 + l^2)^{1/2}$ on the $\sin\theta$ scale for three cubic unit cells. The $(\lambda/2a)$ factor has arbitrarily been set equal to 0.14.

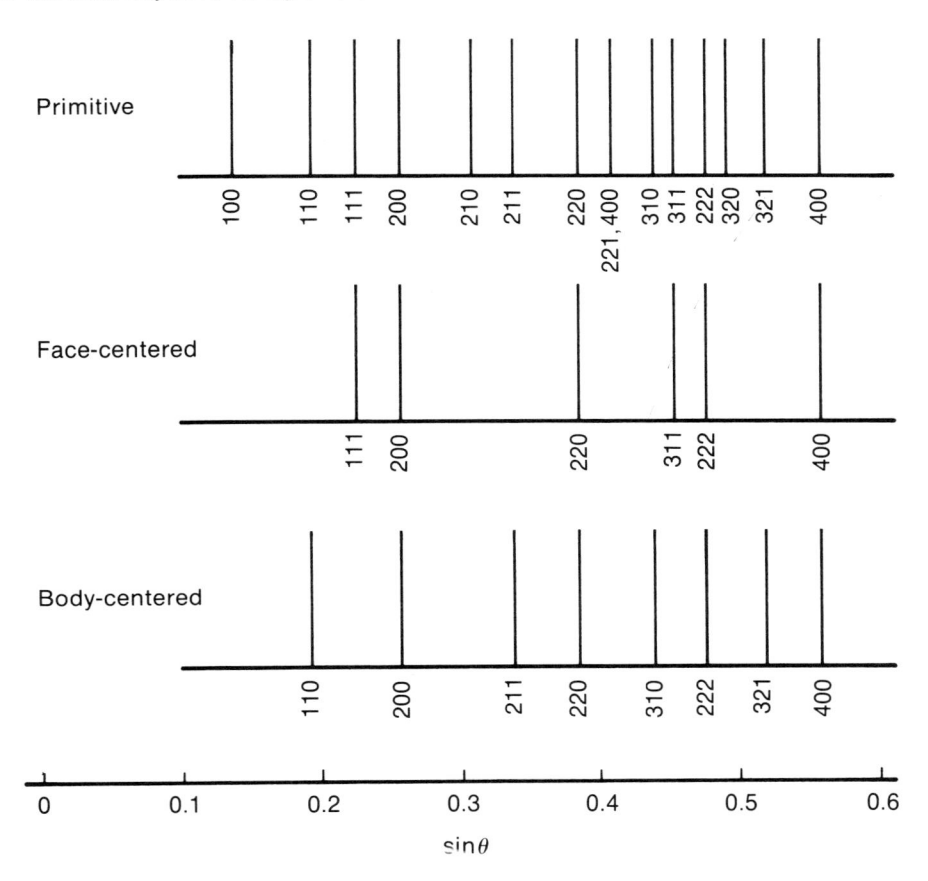

electron beam impinging on a metal surface), which have a wavelength of 0.0708 nm. The lines corresponding to the data in the table are positioned on a $\sin\theta_{hkl}$ scale, as shown in Figure 19-14. Visual observation of NaCl crystals indicates that the crystal is cubic; comparison of the pattern of line positions with the three patterns in Figure 19-13—in particular, noting the lines that are absent—indicates that NaCl has a face-centered cubic unit cell. Remembering that the sum of the squares of the Miller indices must be an integer and that $\sin^2\theta_{hkl}$ is related to these integers by the common factor of $\lambda^2/4a^2$, we need only seek the factor. As indicated in the table, this factor is 0.00396. Setting the factor equal to $\lambda^2/4a^2$ yields $a = 0.562$ nm, which is then the length of the side of the cube in the unit cell. We now proceed to determine the basis of the unit cell.

The number of ions per unit cell is determined from the density (2.16 g cm^{-3}) and the molar mass (58.44 g) of NaCl. These values provide the molar

CHAP. 19 SOLIDS AND LIQUIDS

TABLE 19-3 Data from an X-ray diffraction analysis of NaCl.

θ_{hkl}	$\sin^2\theta_{hkl}$	$\dfrac{\sin^2\theta_{hkl}}{0.00396}$	hkl
6°16′	0.0119	3	111
7°13′	0.0158	4	200
10°16′	0.0317	8	220
12°03′	0.0436	11	311
12°36′	0.0476	12	222
14°35′	0.0634	16	400

volume—namely, mass/density = 58.44/2.16 = 27.06 cm³. The molecular volume is obtained by dividing this value by Avogadro's number, yielding 44.9×10^{-24} cm³. From the value of a (the length of a side of the unit cell), the volume of the unit is a^3 or 177.8×10^{-24} cm³. Thus, the number of NaCl molecules per unit cell is 177.8/44.9 = 4. The structure of the crystal is now obtained by finding an arrangement of the Na⁺ and Cl⁻ ions such that (1) the ions form a face-centered cubic array, (2) positive and negative ions alternate, and (3) each unit cell contains, on the average, four Na⁺ ions and four Cl⁻ ions. One possible (and correct) arrangement is shown in Figure 19-6. To ensure that the number of ions is correct, the ones that are shared by other unit cells must be counted. Note that except for the center Na⁺ ion, each ion is shared by adjacent unit cells. Each of the eight Cl⁻ ions at the vertices is shared by eight unit cells, and each of the six Cl⁻ ions on the faces are shared by two unit cells. Thus, the average number of Cl⁻ ions per unit cell is $(8 \times 1/8) + (6 \times 1/2) = 4$. Each of the 12 Na⁺ ions on each edge is shared by 4 unit cells and the one in the center is unshared. Thus, the average number of Na⁺ ions per unit cell is $(12 \times 1/4) + 1 = 4$. The number of Na⁺ and Cl⁻ is four each, which agrees with the total number of NaCl molecules per unit cell calculated above.

The reader must not be misled into thinking that the analysis is always so easy. In fact, for those crystal classes in which the number of parameters (angles and dimensions of the unit cell) exceeds two, the initial process of

FIGURE 19-14 A plot of the data in Table 19-3 on a $\sin\theta$ scale.

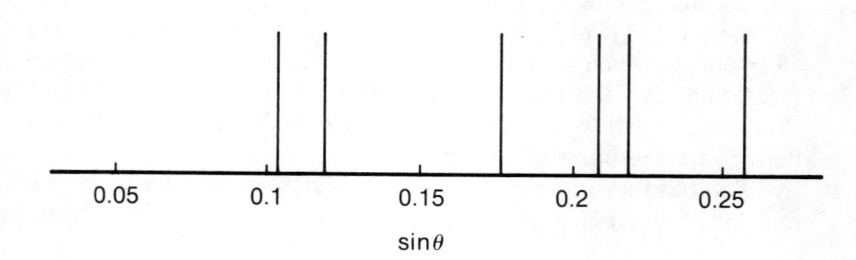

identifying the Miller indices that correspond to the observed lines is quite difficult. Furthermore, when the number of atoms in the basis exceeds two, the procedure becomes even more complex, because determination of Miller indices requires knowledge of the relative scattering powers of the individual components. The analysis (completed in 1955) of the first complex organic crystal, phthalocyanine, which contains 56 atoms, took several years, and the determination of the structure of myoglobin, a protein having several thousand atoms, originally took 18 years. Now, with the aid of high-speed computers, it is sometimes possible (using a battery of sophisticated tricks) to determine the structure of a protein or other biological macromolecule—as long as it is not too large—in a matter of a year or so.

19–2 LIQUIDS

The distinction between a crystalline solid and a gas is a simple one—the molecules of the solid are arranged in a highly ordered way, whereas those of a gas are randomly distributed. Furthermore, in a solid there are strong intermolecular cohesive forces, which are responsible for keeping the molecules in close proximity and for providing the solid with some degree of rigidity; in a gas the cohesive forces are quite weak and at best are responsible for deviations from the ideal gas law.

Classically the liquid state is defined in terms of how a vessel containing a substance is filled—and thereby contrasted with the solid or gaseous state. A gas fills its container totally and uniformly, whereas a solid retains its own shape, independent of the shape of the container. A liquid assumes the shape of a portion of the vessel, filling the container from the bottom upward, and maintaining an upper surface that is perpendicular to a radius of the earth (except for meniscus effects). Fairly strong intermolecular cohesive forces must be present in a liquid; otherwise, the molecules would freely drift around and the substance would be a gas. However, the cohesive forces must be weaker than in a solid. The phenomenon of viscosity—resistance to flow—is also indicative of the existence of cohesive forces in liquids.

Few substances appear to be as formless as a transparent liquid in a container.* In fact, if the temperature of liquid is the same throughout so that no differences in density exist, a liquid is, from a macroscopic viewpoint, totally lacking in internal structure. However, when the nature of intermolecular forces is examined, it seems likely that on a molecular scale, even pure liquids should show some structure; for example, in a hydrogen-bonded liquid (such as water) adjacent molecules should be situated at fixed distances and angles with respect to one another.

*In Chapter 8 it was pointed out that ionic solutions are not microscopically homogeneous in that the ions form clusters. Solutions of macromolecules also may show internal heterogeneity—for example, asymmetric macromolecules are oriented when the solution is flowing. However, in this section, we are referring only to pure liquids.

The best evidence for the existence of internal structure in liquids is obtained from the scattering of X rays by a liquid. When a beam of X rays passes through a gas, the beam is scattered in all directions, but diffraction lines are not seen because of the random arrangement of all molecules of the gas. In contrast, strong diffraction occurs with crystalline solids, as discussed in the last section. If the molecules of a liquid were located randomly, diffraction lines would not be seen, but this is not the case. Lines are observed, though they are very broad and blurry. The existence of lines indicates that liquids do possess some order; however, the blurriness and great line width says that the order is very short-range, in contrast with the long-range order of crystals. Quantitative analysis of the diffraction pattern of liquids (which is beyond the scope of this book) yields information about internal molecular order, though the information is much less detailed than with solids. For water it is clear that many molecules are organized in microunits that consist of a central water molecule hydrogen-bonded to four other molecules, forming a tetrahedral array resembling that present in ice. In general, the fraction of molecules of any liquid that are in an ordered array depends on the magnitude of the intermolecular cohesive forces and on geometric factors.

A characteristic of most liquids is that they are *anisotropic*—that is, their physical properties are independent of direction. This is not the case if the molecule is very long and thin because, then, the molecules tend to be locally oriented in a more-or-less parallel array. This orientation has dramatic effects on the optical properties of the liquid, as explained in the following section.

19-3 LIQUID CRYSTALS

A liquid crystal is a substance whose physical properties are between those of a solid and a true liquid. The material has the form, vessel-filling quality, and flow properties of a liquid, intermolecular spacings are irregular, and the molecules can move about; however, long-range order—similar to, but not the same as, that of a solid crystal—is present. This order results from most of the molecules having the same spatial orientation rather than in a fixed position with respect to one another, as in a solid crystal.

Multiple Transitions in the Melting of a Solid

Substances that form liquid crystals differ from those having a normal solid-to-liquid melting transition in that more than one visible transition in form occurs as the temperature is raised. At the lowest transition temperature the solid changes its form to that of a free-flowing liquid; however, the liquid is turbid. At a higher temperature the turbid liquid acquires the properties of a true liquid. Because of the existence of intermediate states, substances having multiple transitions are called *mesomorphic*. The original report of a mesomorphic system was the observation in 1888 that solid cholesteryl benzoate melted at 145.5°C to form a turbid liquid, which then became clear at 178.5°C.

Sometimes, more than one intermediate state (and multiple transitions) may occur. Each intermediate state is called a *mesophase* and is characterized by a particular transition temperature and enthalpy. The presence of intermediate states indicates that several different orientations are possible.

Types of Molecules That Form Liquid Crystals

In order for any substance to be liquid, the molecules must be able to approach one another to distances at which the cohesive forces are strong enough to overcome the disruptive effect of thermal motion. Molecules able to form liquid crystals always have a three-dimensional shape that tends to minimize the number of possible relative orientations of one molecule to another, when the molecules are in close proximity. For example, long molecules can pack most closely if their long axes are parallel, and flat molecules can pack most efficiently if their flat faces are parallel. The charge distribution also affects the possible orientation in that like charges cannot be nearby. The most common type of molecule that forms a liquid crystal is a long thin organic molecule. The liquid crystal state is enhanced and stable over a broad range of temperature if the molecule has a polar and a nonpolar end. Flat conjugated ring systems, which are very asymmetric, such as steroids, also tend to form liquid crystals. Figure 19-15 shows some well-studied molecules that form liquid crystals.

Types of Liquid Crystals

Liquid crystals can be divided into three classes—*nematic, smectic,* and *cholesteric* (Figure 19-16).

1. In nematic liquid crystals the long axes of the molecules lie generally in a particular direction. The short axes and centers of gravity are randomly situated.
2. In smectic liquid crystals, again the long axes are parallel and the short axes are randomly arranged, but both the centers of gravity and the ends of the molecules lie in parallel planes; the long axes are usually, but not necessarily, perpendicular to these planes.

FIGURE 19-15 Four of many types of organic molecules that form liquid crystals. Note that each contains a polar group and an extensive nonpolar region.

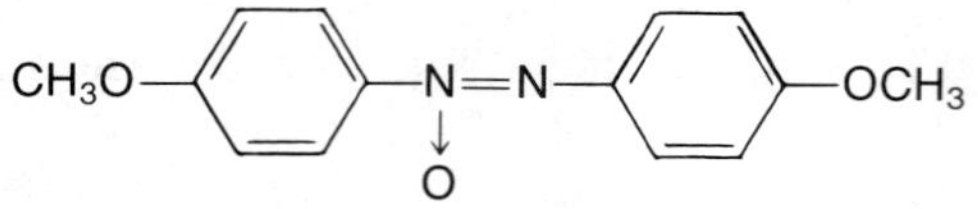

p-Azoxyanisole

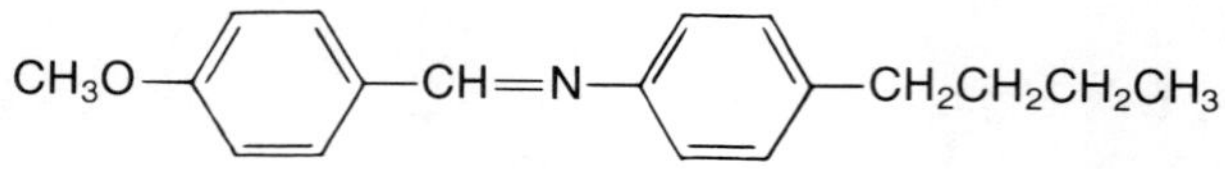

4-Methoxybenzylidene-4'-*n*-butylaniline

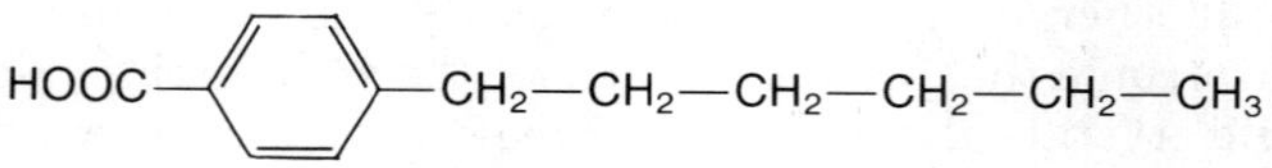

n-Hexylbenzoic acid

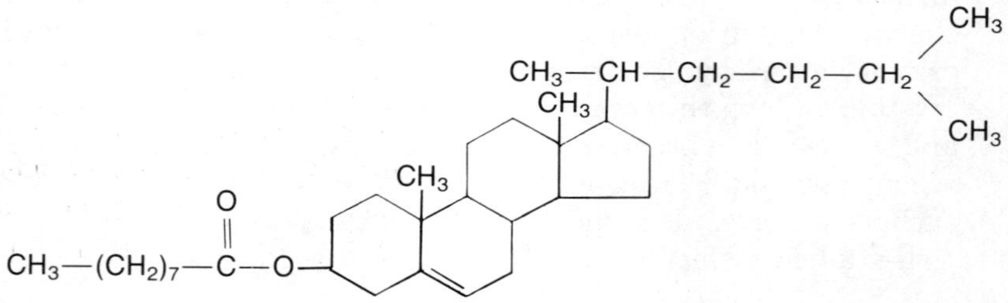

Cholesteryl nonanoate

3. Cholesteric liquid crystals also contain molecules in layers. Each layer consists of molecules organized as in the nematic phase—parallel long axes, random short axes, and randomly arranged centers of gravity. The layers are stacked one above the other and generally rotated with respect to the sheet above and below it, forming a roughly helical array. The angle of rotation ranges from about 3° to 90° per layer.

Long, thin, but not necessarily asymmetric, molecules can form nematic and smectic liquid crystals, whereas the molecule must be asymmetric (optically active) and usually flat to form a cholesteric array.

FIGURE 19-16 Arrangement of molecules in a true liquid and four types of liquid crystals.

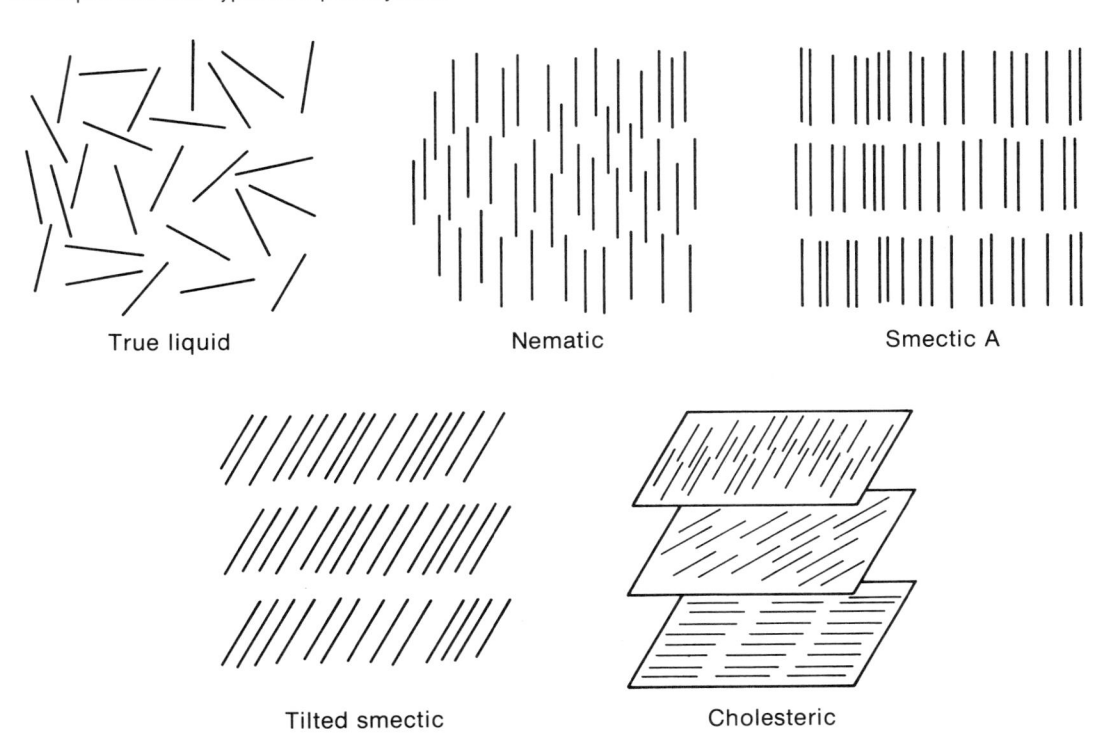

True liquid

Nematic

Smectic A

Tilted smectic

Cholesteric

The molecular orientation in a liquid crystal is not present on a macroscopic scale; instead, the liquid consists of ordered domains, which may be oriented at random with respect to one another. Also, transitions can occur between the three types of liquid crystals. For example, a liquid crystal in the smectic phase, which is generally more ordered than the nematic phase, may on heating undergo a transition to the nematic phase before becoming a true liquid. Furthermore, if the molecule is asymmetric, the transition from smectic to cholesteric will predominate rather than that to the nematic phase. The kinds of transitions that are commonly observed are shown in Figure 19-17. All types of molecules cannot undergo all of the transitions shown in the figure; also the transitions are not always reversible in that they may appear only in a cooling cycle or only in a heating cycle.

Physical Alteration of Liquid Crystals

High temperature produces disorientation of the molecules simply by increasing molecular motion. Other treatments can also change orientation. For example, the extent of the nematic phase can often be increased by

FIGURE 19-17 Relations between solid, liquid, and various liquid-crystalline phases. Several types of smectic phases are known and sequential transitions between them occur. However, usually only one is directly approached from the solid phase and another directly from the nematic and cholesteric phases. The transitions occur at particular temperatures except for the nematic-cholesteric transition, which only occurs without intermediate stages if an electric or magnetic field is applied.

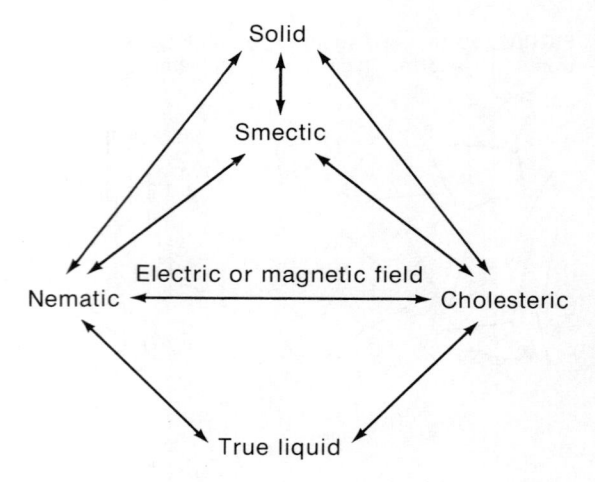

allowing the liquid to flow through a tube or across a surface, or by rotating the vessel containing the liquid. Application of an external electric or magnetic field can either decrease or increase orientation, depending on the structure of the molecule. For example, if a long thin molecule contains terminal polar groups, such that it is a dipole whose axis is parallel to the long axis of the molecule, then parallel orientation of a group of these molecules will be increased by an electric field. On the other hand, if charged groups are present on side chains, creating a dipolar axis that is very different from the long axis of the molecule, an electric field will induce disorder. The degree of orientation of the layers in a cholesteric liquid crystal can also be altered electromagnetically either clockwise or counterclockwise.

Uses of Liquid Crystals

Most liquid crystals possess useful optical properties. For example, since a liquid crystal is turbid, simply by raising the temperature the liquid can be clarified, and light will pass through it. Thus, a small container of a liquid crystal can be used as a temperature-mediated optical switch. Cholesteric liquid crystals are optically active and circularly dichroic. Thus, white light passing through a cholesteric liquid crystal is broken down into bright iridescent colors. The color depends on the angle of incidence, the angle of viewing, and the degree of rotation and spacing between the successive layers. Since this spacing is increased at higher temperatures, thin layers of liquid crystals can serve as a thermometer. An important application is the measurement of skin temperature. A broad sheet of a liquid crystal is placed on the skin and viewed with white light; the color indicates the temperature. Often the system is designed so that the colors allow particular colored numbers to appear so that the temperature can be read directly. These indicators are valuable in the detection of certain malignancies in that the increased metabolic activity of the tumor causes a measurable increase in local skin temperature ("hot spots"). The spacing between successive layers is also decreased by increasing pressure, so cholesteric liquid crystals have also been used to detect pressure variation by virtue of changes in color. For totally unknown reasons, the color of some cholesteric liquid crystals is changed by exposure of the liquid to certain chemical vapors, which can be detected by virtue of the induced color changes.

Liquid crystals are also used in a variety of electro-optical devices and are common in display devices in clocks and digital watches. The principle here is that an external electric field can alter the orientation of the molecules—either producing turbidity or clarifying the liquid, depending on the particular molecule, as explained in the preceding section. Therefore, a voltage applied in a particular region—for example, in the shape of a letter or number—can change either the ability to transmit or reflect light in that region and thereby form a visible image. In digital watches the numbers are usually formed by an applied voltage that produces turbidity that causes light to be reflected.

Biological Significance of Liquid Crystals

So far, we have discussed only pure liquids. However, certain mixtures—namely, some solutions of long chain nonpolar molecules with polar regions, dissolved in polar liquids—also have the attributes of liquid crystals. These solutions were discussed in Chapter 13 and are those in which the solute forms micelles, vesicles, or membranes. In these solutions the liquid-crystalline properties appear only above certain concentrations—for example, the critical micelle concentration; the most evident property is the appearance of turbidity when a concentration is reached at which molecular orientation occurs. When the solute concentration becomes even higher, close-packed structures form that may "come out of solution"; these structures are molecular aggregates that may be hexagonal or cubic arrays of molecules, or expansive lamellar structures. If formed under appropriate conditions, the lamella is a thin layer, which is simply a membrane.

Biological membranes are sometimes true liquid crystals: they contain almost no water, they are turbid, and can undergo a transition at a lower temperature to a solid or a higher temperature to a true liquid. At body temperature the lipid layers in the adrenal cortex and the ovaries, and the myelin layer of nerve cells are liquid crystals. Possibly, small changes in structure of these layers, caused by temperature or the penetration of potentially disrupting small molecules, may be biologically significant. Since the physical properties of liquid crystals near a transition temperature are exceedingly sensitive to small temperature changes, these substances may play an important role in temperature-sensing systems in animals. Also, the dramatic effects of certain vapors on cholesteric liquid crystals may be important in the sense of smell and taste.

The fatty deposits in atherosclerotic arteries consist mostly of cholesteryl oleate, which is a liquid crystal at body temperature. Since factors that can disrupt liquid crystals in the laboratory are known, an attempt is being made to use this knowledge to eliminate these deposits from diseased patients.

REFERENCES

Atkins, P.W. 1978. *Physical Chemistry*. W.H. Freeman. San Francisco.
Brown, G.H. 1972. "Liquid Crystals and Their Roles in Animate and Inanimate Systems." *Amer. Scient.*, 60, 64.
Chang, R. 1981. *Physical Chemistry with Application to Biological Systems*. Macmillan. New York.
Eliot, G. 1973. "Liquid Crystals for Electro-optical Displays." *Chem. Brit.*, 9, 213.
Holden, A. 1965. *The Nature of Solids*. Columbia University Press. New York.
Kapecki, J.A. "An Introduction to X-ray Structure Determination." *J. Chem. Educ.*, 49, 231.
Levine, I.N. 1978. *Physical Chemistry*. McGraw-Hill. New York.
MacIntyre, W.M. 1964. "X-ray Crystallography as a Tool for Structural Chemists." *J. Chem. Educ.*, 41, 526.
Wheatley, P.J. 1959. *The Determination of Molecular Structure*. Oxford University Press. Oxford, England.

PROBLEMS

1. What are the coordinates of the points on the six faces of a face-centered cubic unit cell?

2. Crystal planes intersect the crystal axes at the coordinates
    (a) 1, 1, 1
    (b) 2, 3, 1
    (c) 6, 3, 3
    (d) 2, -3, -3

   What are the Miller indices for each of these planes?

3. What are the Miller indices of the following planes?
    (a) The set of points parallel to a face of a primitive cubic lattice.
    (b) The set of points parallel to a face and passing through the central point of a body-centered cubic lattice.
    (c) The set of points passing through opposite pairs of corner points and the central point of a body-centered cubic lattice.
    (d) The set of points passing through opposite pairs of corner points and two face points of a face-centered cubic lattice.

4. An orthorhombic unit cell has dimensions $a = 5.2$ Å, $b = 10.8$ Å, and $c = 14.3$ Å. What is the spacing between the 322 planes?

5. Theoretically, what is the lower limit to the spacing between crystal planes that could be detected by molybdenum X rays

for which $\lambda = 0.708$ Å?

6. If you had available a source of heterochromatic X rays (that is, including a wide range of wavelengths) and a large NaCl crystal, and you needed a beam of monochromatic X rays, how would you go about obtaining such a beam?

7. Potassium fluoride has the same structure as NaCl. Its density is 2.48 g cm^{-3} at 20°C. Calculate the length of the unit cell at this temperature.

8. One crystalline form of TiO_2 has a tetragonal lattice with dimensions $a = 4.59$ Å and $c = 2.96$ Å at 25°C. There are two molecules per unit cell. What is the density of this form, assuming that the crystal contains no water?

9. The density of diamond is 3.51 g cm^{-3} at 25°C. What is the length of the C—C bond in this crystal? Remember that the tetrahedral angle is approximately 109°.

10. Crystalline iron exists in two forms, a body-centered cubic cell ($a = 2.90$ Å) and a face-centered cubic cell ($a = 3.68$ Å).

    (a) What is the ratio of the densities of the two forms?

    (b) One of these forms can be converted to the other by high pressure. Which one?

11. A crystal has a primitive cubic unit cell with $a = 4.20$ Å. What are the first five diffraction angles for

    (a) The 100 plane.

    (b) The 110 plane.

12. One way to check the accuracy of the determination of the dimensions of a unit cell is to use X-ray data to calculate molecular weight. A crystal of the protein lysozyme, whose molecular weight is known to be 13,900 (from direct analysis of the amino acid sequence) is being studied. The orthorhombic unit cell is found to contain an average of eight molecules and have dimensions $a = 79.1$ Å, $b = 79.1$ Å, and $c = 37.9$ Å. The density of the crystal is 1.242 g cm^{-3} but the crystal contains some water; by weighing crystals before and after vacuum drying, it is found that only 64.4 percent of the weight of the crystal is protein. Calculate the molecular weight and compare it to the known value.

13. A diffraction photograph of a crystal shows a set of lines having Miller indices 110, 200, 211, 220, 310, 222, 321, ... What type of cubic unit cell does this crystal have?

14. An X-ray diffraction diagram was obtained for a particular crystalline sample on a cold winter day on which the heating system of the building was not functioning and the temperature of the laboratory was 12°C. The same sample was used on a hot summer day when the temperature was 32°C. The two diffraction patterns are not the same. Why might this be the case? (Do not assume that the crystal melts or decomposes at the higher temperature.)

APPENDIX I

SOME NECESSARY MATHEMATICAL RELATIONS

Although the mathematics in this book is simple, the following is provided as a review.

Derivatives

We have on several occasions had to take derivatives of functions. The following rules should enable the reader to carry out this operation in all expressions in this book.

1. The derivative of a constant is 0.

2. $\dfrac{dx}{dx} = 1$.

3. $\dfrac{d}{dx}(ax) = a$, if a is a constant.

APPENDIX I

4. $\dfrac{d}{dx}(ax^n) = nax^{n-1}$.

   Example: $\dfrac{d}{dx}(3x^4 + 2x^3) = 12x^3 + 6x^2$.

5. $\dfrac{d}{dx}[f(x)]^n = nf(x)^{n-1}\dfrac{df}{dx}$.

   Example: $\dfrac{d}{dx}(3x^2 + 1)^3 = 3(3x^2 + 1)^2(6x) = 18x(3x^2 + 1)^2$.

6. $\dfrac{d}{dx}(fg) = f\dfrac{dg}{dx} + g\dfrac{df}{dx}$.

   Example: $\dfrac{d}{dx}[(3x^2 + 2)(2x^4 + x^3)] = (3x^2 + 2)(8x^3 + 3x^2) + 6x(2x^4 + x^3)$.

7. $\dfrac{d}{dx}\left(\dfrac{f}{g}\right) = \dfrac{1}{g^2}\left(g\dfrac{df}{dx} - f\dfrac{dg}{dx}\right)$.

   Example: $\dfrac{d}{dx}\left(\dfrac{x^2}{3x^2 + 1}\right) = \dfrac{(3x^2 + 1)2x - x^2(6x)}{(3x^2 + 1)^2} = \dfrac{2x}{(3x^2 + 1)^2}$.

8. $\dfrac{d}{dx}(e^f) = e^f\dfrac{df}{dx}$.

   Examples: $\dfrac{d}{dx}(e^{2x}) = 2e^{2x}$ and $\dfrac{d}{dx}(e^{-3x^2}) = -6xe^{-3x^2}$.

9. $\dfrac{d}{dx}(\ln f) = \dfrac{1}{f}\dfrac{df}{dx}$.

   Example: $\dfrac{d}{dx}[\ln(3x^2 + 1)] = \dfrac{6x}{3x^2 + 1}$.

Partial Derivatives

Partial derivatives appear frequently in thermodynamic equations. Their notation makes them appear complicated but they are really quite simple.

Consider a function $F(x,y)$ of two variables. We define the partial derivative with respect to x as the result of considering y to be a constant and then taking the derivative d/dx by using the rules of the preceding section. The notation for the partial derivative of F with respect to x is $(\partial F/\partial x)_y$. We could also assume that y is the variable and that x is constant; the partial derivative of F with respect to y would have the form $(\partial F/\partial y)_x$. If F is a function of three variables, x, y, and z, we could form

$$\left(\dfrac{\partial F}{\partial x}\right)_{y,z}, \quad \left(\dfrac{\partial F}{\partial y}\right)_{x,z}, \quad \text{and} \quad \left(\dfrac{\partial F}{\partial z}\right)_{x,y},$$

in which the subscript always lists the variables that are assumed to be constants.

EXAMPLE: What are the two partial derivatives of $F(x, y) = axy + bx^2y^3$?

They are

$$\left(\frac{\partial F}{\partial x}\right)_y = ay + 2bxy^3 \quad \text{and} \quad \left(\frac{\partial F}{\partial y}\right)_x = ax + 3bx^2y^2.$$

The *total differential* of $F(x,y)$ is defined as

$$dF = \left(\frac{\partial F}{\partial x}\right)_y dx + \left(\frac{\partial F}{\partial y}\right)_x dy.$$

This equation says that for small changes, the change in F—that is, dF—is the sum of how F changes with respect to x, when y is constant, and how F changes with y, when x is constant.

EXAMPLE: Imagine a rectangle with dimensions that are changing. The length x of a rectangle is increasing at a rate of 2 cm/sec and the width y is decreasing at a rate of 0.25 cm/sec. What is the rate of change of its area A when x = 10 cm and y = 18 cm?

Since $A = xy$, $dA = (\partial A/\partial x)_y dx + (\partial A/\partial y)_x dy = y\,dx + x\,dy$. Thus, the rate of change of A or dA/dt is

$$\frac{dA}{dt} = y\frac{dx}{dt} + x\frac{dy}{dt} = 18(2) + 10(-0.25) = 33.5 \text{ cm}^2/\text{sec}.$$

Several relations are often useful. The first is

$$\left[\frac{\partial}{\partial x}\left(\frac{\partial F}{\partial y}\right)_x\right]_y = \left[\frac{\partial}{\partial y}\left(\frac{\partial F}{\partial x}\right)_y\right]_x.$$

That this is true can be seen from the first example of this section in which the partial derivatives of $F(x,y) = axy + bx^2y^3$ were given. For the partial derivative $(\partial F/\partial y)_x = ax + 3bx^2y^2$,

$$\left[\frac{\partial}{\partial x}\left(\frac{\partial F}{\partial y}\right)_x\right]_y = \left[\frac{\partial}{\partial x}(ax + 3bx^2y^2)\right]_y = a + 6bxy^2.$$

For the partial derivative $(\partial F/\partial x)_y = ay + 2bxy^3$,

$$\left[\frac{\partial}{\partial y}\left(\frac{\partial F}{\partial x}\right)_y\right]_x = \left[\frac{\partial}{\partial y}(ay + 2bxy^3)\right]_x = a + 6bxy^2.$$

A second relation is

APPENDIX I

$$\left(\frac{\partial x}{\partial y}\right)_z = 1 \bigg/ \left(\frac{\partial x}{\partial y}\right)_z .$$

For example, if $xy^2z^3 = k$,

$$\left(\frac{\partial x}{\partial y}\right)_z = -\frac{2k}{y^3z^3} \quad \text{and} \quad \left(\frac{\partial y}{\partial x}\right)_z = -\frac{k^{1/2}}{2x^{3/2}z^{3/2}} = -\frac{y^3z^3}{2k} .$$

A third relation is

$$\left(\frac{\partial x}{\partial y}\right)_z = -\left(\frac{\partial x}{\partial z}\right)_y \left(\frac{\partial z}{\partial y}\right)_x .$$

Combining the above equations yields the Euler relation

$$\left(\frac{\partial x}{\partial y}\right)_z \left(\frac{\partial y}{\partial z}\right)_x \left(\frac{\partial z}{\partial x}\right)_y = -1 .$$

Exact and Inexact Differentials

In thermodynamics, an important distinction is that between exact and inexact differentials. Consider a function $F(x, y)$ and a differential $dF = M(x, y)\, dx + N(x, y)\, dy$. An exact differential has the property that

$$\Delta F = \int_{F_1}^{F_2} dF = F_2 - F_1 .$$

That is, the value of the definite integral is merely the difference between the initial and final values of F; ΔF is independent of path and F is a state function. For an inexact differential the equality would not hold and the value of ΔF would depend on the path. An exact differential can be recognized by the fact that

$$\left(\frac{\partial M}{\partial y}\right)_x = \left(\frac{\partial N}{\partial x}\right)_y .$$

Thus, $dF = (4xy + y^3)dx + (2x^2 + 2xy)dy$ is an exact differential because

$$\left[\frac{\partial(4xy + y^2)}{\partial y}\right]_x = 4x + 2y = \left[\frac{\partial(2x^2 + 2xy)}{\partial x}\right]_y .$$

$dF = (4xy + y^3)dx + (2x^2 + 2xy)dy$ is not.

Pressure is a state function so, for a van der Waals gas, dP must be an exact differential.

$$dP = \left[\frac{-RT}{(V-b)^2} + \frac{2a}{V^3}\right]dV + \left(\frac{R}{V-b}\right)dT$$

$$\left(\frac{\partial M}{\partial T}\right)_V = -\frac{R}{(V-b)^2} = \left(\frac{\partial N}{\partial V}\right)_T ,$$

so, as expected, dP is an exact differential.

APPENDIX I SOME NECESSARY MATHEMATICAL RELATIONS

Indefinite Integrals

The following integration formulas are useful:

1. $\int k\,dx = k\int dx,$ when k is a constant.

2. $\int (f + g)dx = \int f\,dx + \int g\,dx$.

3. $\int x^n dx = \dfrac{x^{n+1}}{n+1} + k,$ when $n \neq -1$.

Example: $\int x^4 dx = \dfrac{x^5}{5} + k$.

4. $\int \dfrac{1}{x} dx = \ln x + k$.

5. $\int e^x dx = e^x + k$.

The Definite Integral $\int_a^b f(x)\,dx$

To evaluate the definite integral, first determine the indefinite integral; then substitute $x = b$ and $x = a$ separately, and then subtract the expression containing a from the expression containing b.

EXAMPLE:

$$\int_a^b (x^2 + 2)dx = \left[\dfrac{x^3}{3} + 2x + k\right]_a^b = \left(\dfrac{b^3}{3} + 2b + k\right) - \left(\dfrac{a^3}{3} + 2a + k\right)$$

$$= \dfrac{1}{3}(b^3 - a^3) + 2(b - a).$$

Logarithms

The notation $\log_b a = p$ is equivalent to $b^p = a$.

Two kinds of logarithms are in use—$\log_{10}$, which in this book we denote simply as log, and the Napierian logarithm (to the base e), denoted herein by ln.

APPENDIX I

EXAMPLES:

$$\log 10 = 1.$$
$$\log 100 = 2.$$
$$\log 10^n = n, \quad \text{and} \quad \log 10^{-n} = -n.$$

To convert $\log x$ to $\ln x$, we make use of the fact that $\ln 10 = 2.303$, so that $\ln x = 2.303 \log x$.

The important properties of logarithms that hold for both log and ln are

1. $\log x + \log y = \log(xy)$.
2. $\log x - \log y = \log(x/y)$.
3. $k \log x = \log x^k$.

EXAMPLE:

$$\begin{aligned}\log 4462 &= \log(4.462 \times 10^3) \\ &= \log 4.462 + \log 10^3 \\ &= 0.650 + 3 \log 10 = 0.650 + 3 \\ &= 3.650 .\end{aligned}$$

Useful Trigonometric Relations

In some derivations within the text we have made use of trigonometric relations and their derivations. Following are a few that should be kept in mind.

$\cos n\pi = 1,$ when $n = 0$, or $n =$ an even integer.
$\cos n\pi = -1,$ when $n =$ an odd integer.
$\sin n\pi = 0,$ when $n =$ any integer.
$\cos(\theta - \tfrac{1}{2}\pi) = \sin \theta.$
$\sin(\theta + \tfrac{1}{2}\pi) = \cos \theta.$
$\sin^2\theta + \cos^2\theta = 1.$
$\tan \theta = \sin \theta / \cos \theta.$

For a right triangle

$$\sin \theta = a/c, \cos \theta = b/c, \tan \theta = a/b.$$

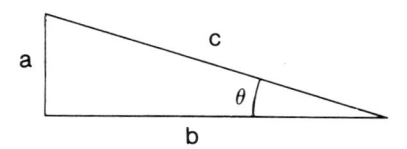

Also, when k is constant,

$$\frac{d}{d\theta}(\cos k\theta) = -k \sin k\theta,$$

and

$$\frac{d}{d\theta}(\sin k\theta) = k \cos k\theta.$$

APPENDIX II

SOME STATISTICAL RELATIONS

The number of ways of arranging n distinguishable objects in n boxes when placement of only one object per box is allowed is called the number of permutations of n or P_n. This is

$$P_n = n! = n(n-1)(n-2) \cdots 3 \cdot 2 \cdot 1.$$

Thus for three objects, a, b, and c, which can be arranged in different order in three boxes, there are $3! = 6$ permutations; namely,

$$abc, \ acb, \ bac, \ bca, \ cab, \ cba.$$

Suppose now we wish to take r objects at a time from a collection of n objects and form all the possible groups containing r elements. The order in which the objects are arranged is not considered. We wish to know the number of such groups; this is called the number of combinations C_r^n of n objects taken r at a time:*

$$C_r^n = \frac{n!}{r!(n-r)!}.$$

Thus there are

$$\frac{4!}{2!(4-2)!} = 6$$

---

*But keep in mind that the superscript n is not an exponent.

possible pairs of 4 objects, namely (1,2), (1,3), (1,4), (2,3), (2,4), and (3,4). Note that (1,2) is the same as (2,1) because order is irrelevant. C_r^n is also the number of ways of arranging r indistinguishable objects in boxes. In this case r boxes contain objects and $n - r$ boxes are empty. Thus, if we have three boxes and two objects,

$$C_r^n = \frac{3!}{2!(3-2)!} = 3.$$

The three arrangements are (1,1,0), (1,0,1), (0,1,1), in which 0 and 1 are the number of objects in each box. Thus, (1,1,0) means one object in each of the first two boxes and an empty third box. This is identical to the number of ways in which three pennies could be placed into groups of two tails and one head; namely, (T,T,H), (T,H,T), (H,T,T).

In the mathematical analysis usually encountered in physical chemistry, n and r are huge numbers (for example, 6×10^{23}) so that direct calculation of $n!$ is impossible. For large values of n, the Stirling approximation is valid. This is

$$\ln n! = n \ln n - n.$$

Another useful relation is the Poisson distribution. If there are N objects distributed randomly into r boxes, the Poisson distribution tells us the fraction of the boxes that have 0, 1, 2, or n objects per box ($n \leq N$). This is expressed as a probability, $P(n)$, in which

$$P(n) = \frac{a^n e^{-a}}{n!},$$

in which $a = N/r$ is the *average* number of objects per box. An important consequence of this distribution is that

$$P(0) = e^{-a}.$$

This is the so-called "Poisson zero" and enables one to determine a and, hence, N.

EXAMPLE: An unknown number N of balls have been thrown at 1000 boxes. Every ball lands in some box; some boxes contain one or more balls but others are empty. If 132 boxes are empty, how many balls were thrown?

The fraction of boxes that is empty is

$$\frac{132}{1000} = 0.132 = P(0).$$

Thus $a = -\ln(0.132) = 2.025 = N/1000$. Therefore, 2025 balls were thrown. This number, like all numbers calculated from statistical formulas, is an approximation. Statistical methods, which we will not describe, exist for evaluating the probability that it is the true value.

APPENDIX II 759

The Poisson distribution is especially useful in analyzing interaction between particles.

EXAMPLE: An experiment is done in which 10^9 bacteriophage particles are allowed to adsorb to 3×10^8 bacteria. What fraction of the bacteria are uninfected? What fraction are infected with more than one phage?

The bacteria may be thought of as the boxes and the phages are the balls. Thus $a = 10^9/(3 \times 10^8) = 3.33$. The fraction of uninfected bacteria is therefore $e^{-3.33} = 0.036$. The fraction infected with one phage is $P(1) = 3.33e^{-3.33} = 0.120$. Thus the fraction infected with more than one phage is $1 - (0.036 + 0.120) = 0.844$.

APPENDIX III

SOME USEFUL THERMODYNAMIC RELATIONS

In carrying out derivations it is often useful to be able to substitute thermodynamic state functions by differentials and to interchange differentials. Following is a list of some useful relations.

$$\left(\frac{\partial U}{\partial S}\right)_V = T \qquad \left(\frac{\partial U}{\partial V}\right)_S = -P$$

$$\left(\frac{\partial H}{\partial S}\right)_P = T \qquad \left(\frac{\partial H}{\partial P}\right)_S = V$$

$$\left(\frac{\partial G}{\partial T}\right)_P = -S \qquad \left(\frac{\partial G}{\partial P}\right)_T = V$$

$$\left(\frac{\partial U}{\partial S}\right)_V = \left(\frac{\partial H}{\partial S}\right)_P \qquad \left(\frac{\partial H}{\partial P}\right)_S = \left(\frac{\partial G}{\partial P}\right)_T$$

$$\left(\frac{\partial T}{\partial V}\right)_S = -\left(\frac{\partial P}{\partial S}\right)_V \qquad \left(\frac{\partial T}{\partial P}\right)_S = \left(\frac{\partial V}{\partial S}\right)_P$$

$$\left(\frac{\partial S}{\partial V}\right)_T = \left(\frac{\partial P}{\partial T}\right)_V \qquad \left(\frac{\partial S}{\partial P}\right)_T = -\left(\frac{\partial V}{\partial T}\right)_P$$

The last four equations are known as the Maxwell relations.

APPENDIX IV

A STATISTICAL DERIVATION OF RAOULT'S LAW

Consider N_A and N_B molecules of type A and B, respectively, for which the interaction energies between A and A, A and B, and B and B are identical. Let us think of a liquid as an array of sites each occupied by a single molecule. Thus pure liquid A has N_A sites occupied by N_A molecules. Since the sites are indistinguishable, the thermodynamic probability Ω_A of a liquid A is one. (See Chapter 5 for a definition of Ω.) Similarly, $\Omega_B = 1$ for substance B. The mixture, however, has a different probability, because in the mixture there are $N_A + N_B$ total sites, each of which can have either an A or a B. Let us construct the solution by first adding the A molecules one by one to the sites. For the first molecule, there are $N_A + N_B$ available sites; for the second there are $N_A + N_B - 1$; and for the ith there are $N_A + N_B - i + 1$. The total number of ways that N_A indistinguishable A's might be placed is

$$(N_A + N_B)(N_A + N_B - 1) \cdots (N_A + N_B - N_A - 1) = \frac{(N_A + N_B)!}{N_B!}.$$

Since the A's are indistinguishable, the number of distinguishable ways in which N_A molecules can be placed in $N_A + N_B$ sites is

$$[(N_A + N_B)!/N_B!](1/N_A!)(1/N_A!).$$

Now, because the number of B molecules equals the number of remaining sites, there is only one way to put the B molecules in the B sites, so that

761

the thermodynamic probability of the mixture Ω_{mix} is the previous expression multiplied by one, or,

$$\Omega_{mix} = \frac{(N_A + N_B)!}{N_A! N_B!}.$$

The change in entropy of mixing ΔS_{mix}, which is easily derived from Equation 35 of Chapter 5, is

$$\Delta S_{mix} = k \ln \frac{\Omega_{mix}}{\Omega_A \Omega_B} = k \ln \frac{(N_A + N_B)!}{N_A! N_B!},$$

in which k is the Boltzmann constant and which, by the Stirling approximation (that is, for large values of N, $\ln(N!) = N \ln N - N$), becomes

$$\Delta S_{mix} = -N_A k \ln \frac{N_A}{N_A + N_B} - N_B k \ln \frac{N_B}{N_A + N_B}.$$

From the definition of the mole fraction X, this is just

$$\Delta S_{mix} = -n_A R \ln X_A - n_B R \ln X_B.$$

Returning to the definition of an ideal solution—that is, a solution in which cohesive forces are all the same—it must be the case that there is no heat change in mixing; namely, $\Delta H_{mix} = 0$. Therefore, since $\Delta G = \Delta H - T\Delta S$,

$$\Delta G_{mix} = n_A RT \ln X_A + n_B RT \ln X_B = \sum n_i RT \ln X_i,$$

in which n_i is the number of moles of the ith component. We now also assume that the vapor above an ideal solution is an ideal gas. The free energy change per mole resulting from a change of pressure of an ideal gas from P_0 to P_1 is $\Delta G = RT \ln(P_1/P_0)$. The solution and its vapor are in equilibrium, so that ΔG_{mix} should equal the ΔG resulting from the change in vapor pressure from $P_{0,\,pure}$, the value for the pure solvent (that is, $X_A = 1$, $X_B = 0$) to P_1, that of the solution. Equating these expressions term by term, we obtain $\ln(P_1/P_{0,\,pure})$, or

$$P_1 = X_1 P_{0,\,pure},$$

which is Raoult's Law.

APPENDIX V

CALCULATION OF SOLUTE ACTIVITY FROM THE SOLVENT ACTIVITY

The necessary equation is obtained by starting with Equation 7 from Chapter 7, written as

$$X_A \frac{\partial \ln a_A}{\partial X_B} + X_B \frac{\partial \ln a_B}{\partial X_B} = 0 \,,$$

in which $X_A = 1 - X_B$. Subtracting the identity (which the reader can easily verify)

$$X_A \frac{\partial \ln X_A}{\partial X_B} + X_B \frac{\partial \ln X_B}{\partial X_B} = 0$$

from the previous equation, and rearranging terms, yields

$$\frac{\partial \ln (a_B/X_B)}{\partial X_B} = - \left(\frac{1 - X_B}{X_B} \right) \frac{\partial \ln (a_A/X_A)}{\partial X_B} \,,$$

which we integrate from $X_B = 0$ to X_B, to obtain

$$\ln \left(\frac{a_B}{X_B} \right) - \lim_{X_B \to 0} \ln \left(\frac{a_B}{X_B} \right) = - \int_0^{X_B} \left(\frac{1 - X_B}{X_B} \right) \frac{\partial \ln (a_A/X_A)}{\partial X_B} dX_B \,.$$

Since we choose the infinitely dilute state to be the reference state,

$$\lim_{X_B \to 0} \ln \frac{a_B}{X_B} = \ln 1 = 0 \,,$$

so the preceding equation becomes

$$\ln\left(\frac{a_B}{X_B}\right) = -\int_0^{X_B} \left(\frac{1-X_B}{X_B}\right)\frac{\partial \ln(a_A/X_A)}{\partial X_B} dX_B.$$

The value of a_A is known as a function of X_B and $X_A = 1 - X_B$, so that we can plot $(1 - X_B)/X_B$ against $\ln(a_A/X_A)$. The area under the curve between 0 and a particular value of X_B is equal to the numerical value of the integral. If this value is Q for a particular value of X_B, the value of a_B is evaluated from $\ln(a_B/X_B) = Q$.

This graphical method of determining the solute activity can always be carried through when one has experimental values of the activity of the solvent as a function of the mole fraction. The solvent values can be obtained from measurement of any of the colligative properties.

It should be noted that the solvent activity does not depart significantly from 1 because it is present in great molar excess.

APPENDIX VI

PHYSICAL PROPERTIES OF WATER

Water appears so frequently in chemical and biochemical calculations that a summary of its most important physical properties is given below:

Property	Value*
Molecular weight	18.016 g mol^{-1}
Freezing point	273.15 K
Boiling point	373.15 K
Density (0°C)	0.99987 g cm^{-3}
(ice, 0°C)	0.9167 g cm^{-3}
(4°C)	1.00000 g cm^{-3}
(20°C)	0.9982 g cm^{-3}
C_P	75.3 J K^{-1} mol^{-1}
Enthalpy of fusion	6.01 × 10^3 J mol^{-1}
Enthalpy of vaporization	4.079 × 10^4 J mol^{-1}
Absolute entropy (25°C)	69.96 J K^{-1} mol^{-1}
$\Delta G°_{form}$ (25°C)	−2.37 × 10^5 J mol^{-1}
Enthalpy of ionization (25°C)	5.73 × 10^4 J mol^{-1}
Dipole moment	1.82 debye, or 6.082 × 10^{-30} coulomb-meter
Dielectric constant (25°C)	78.54
Viscosity (20°C)	0.01 poise
Surface tension (20°C)	72.75 dyn cm^{-1}
Dissociation constant (25°C)	1.86 × 10^{-16} mol l^{-1}

*These values are for the mixture of ^{1}H$_2$O and ^{2}H$_2$O found in nature.

APPENDIX VII

DERIVATION OF THE DEBYE-HÜCKEL LIMITING LAW

The development of the theory is as follows. We consider spherical ions of radius a and charge q in a solvent of dielectric constant D. The final term in Equation 9 of Chapter 8 would be zero if $q = 0$, so that the energy represented by that term is the work done in creating such a charge; this is equivalent to bringing the charge q from infinity to the surface of the sphere. If we denote the electrical potential of the sphere by ψ, the work done in charging the ion sphere can be expressed as

$$W = \int_0^\infty \psi \, dq \, . \tag{1}$$

Thus, this is the energy possessed by an ion having charge q. If there are many ions present, this work W equals the sum of two terms; W_1, the work required to charge an isolated sphere; and W_2, the work done in bringing the charge to the sphere when other charges are present. The term W_1 is independent of concentration because it is the work done when only one charge is present; therefore it must be contained in the value of μ_0; hence W_2 is the interaction energy that we seek. Thus, since $W_1 + W_2 = W$,

$$W_2 = W - W_1 = kT \ln \gamma \, , \tag{2}$$

in which we have used the Boltzmann constant k instead of R since we are considering the work required to charge a *single* ion. The term W_1 is evaluated from the fact that the potential ψ of a sphere of radius a immersed in a dielectric medium is $\psi = q/4\pi\epsilon_0 Da$, so that

$$W_1 = \int_0^q \frac{q}{4\pi\epsilon_0 Da} \, dq = \frac{q^2}{8\pi\epsilon_0 Da}. \tag{3}$$

Our problem, then, is to calculate W, which requires obtaining an expression for ψ. This is done as follows.

We define the center of a positive ion (the choice of the sign of the charge is arbitrary) as the origin of a spherical coordinate system and consider a point P located at a distance r from the origin. It is at this stage that we make the assumption that the charge is spherical. If there is spherical symmetry, the potential ψ at P is related to the charge per unit volume ρ by the Poisson equation,

$$\frac{1}{r^2}\frac{d}{dr}\left(r^2\frac{d\psi}{dr}\right) = -\frac{\rho}{D\epsilon_0}. \tag{4}$$

The term ρ is then evaluated as follows. If z_+ and z_- are the valences of the ions whose molar concentrations are c_+ and c_-, the positive and negative charges in one cubic centimeter are $Nez_+c_+/1000$ and $Nez_-c_-/1000$, respectively, in which N is Avogadro's number and e is the charge of an electron. The charge per unit volume is the sum of these two charge densities, or

$$\rho = \frac{Ne}{1000}(c_+z_+ + c_-z_-). \tag{5}$$

Since the potential at P is defined to be ψ, the potential energies of the positive and negative ions at P are $ez_+\psi$ and $ez_-\psi$, respectively. The next step requires an important assumption. We first note that the entire solution, which is at a concentration c, is electrically neutral; that is, $\psi = 0$. At any other concentration, $\psi \neq 0$, and the ions in a region in which their arrangement is nonuniform will have a potential energy differing from that of the whole solution. There will of course be variations in potential energy and in local concentration, and we assume that the distribution of energies of the ions is a Boltzmann distribution; in other words,

$$c_+ = c_+^0 \exp(-ez_+\psi/kT) \quad \text{and} \quad c_- = c_-^0 \exp(-ez_-\psi/kT), \tag{6}$$

in which c_+^0 and c_-^0 are the molar concentrations in any region in which ψ might be 0. The relation between c_+^0 and c_-^0 comes from the fact just stated that in any region in which $\psi = 0$, the distribution will be uniform, the solution will be neutral, and ρ will equal 0. From Equation 5, if $\rho = 0$, then

$$c_+^0 z_+ + c_-^0 z_- = 0. \tag{7}$$

Combining Equations 5, 6, and 7 yields

$$\rho = \frac{Ne}{1000}[c_+^0 z_+ \exp(-ez_+\psi/kT) + c_-^0 z_- \exp(-ez_-\psi/kT)]. \tag{8}$$

For mathematical simplicity, it is common to expand exponentials in a power series and drop as many terms of the series as is possible. For very small values of x, e^{-x} can be replaced by $1 - x$; that is, *the exponential can be linearized.* But in order to make use of exponential linearization, it would have to be the case that $ez\psi/kT \ll 1$ or $ez\psi \ll kT$: these expressions say that the electrical potential energy must be much less than thermal energy for this linearization to be valid. This in fact is the case at commonly encountered temperatures if the concentration is sufficiently low that the ions are far apart. Debye and Hückel made this assumption, *which now limits our discussion to concentrations less than about 0.01* M. Equation 8 can now be simplified to

$$\rho = \frac{e}{1000}\left[c_0^+ z_+ + c_0^- z_- - \frac{e\psi}{kT}(c_+^0 z_+^2 + c_-^0 z_-^2)\right]. \tag{9}$$

From Equation 7, the first two terms sum to zero, so that we are left with an expression in which the valence is no longer in an exponential term but in a quadratic term. *It is for this reason that the ionic strength plays a major role in the Debye-Hückel theory.* When we introduce the ionic strength I from Equation 10 of Chapter 8, Equation 9 becomes

$$\rho = -\frac{2\mathcal{N}e^2 I \psi}{1000 kT}. \tag{10}$$

For convenience, a quantity κ^2 is defined as

$$\kappa^2 = \frac{2\mathcal{N}e^2 I}{1000 DkT\epsilon_0}, \tag{11}$$

so that Equation 4 becomes

$$\frac{1}{r^2}\frac{d}{dr}\left(r^2 \frac{d\psi}{dr}\right) - \kappa^2 \psi = 0, \tag{12}$$

whose general solution is

$$\psi = B_1 \frac{e^{-\kappa r}}{r} + B_2 \frac{e^{\kappa r}}{r}, \tag{13}$$

in which B_1 and B_2 are constants. The constant B_2 is evaluated as follows. Since $e^{\kappa r}/r$ is infinite as $r \to \infty$, and ψ is finite everywhere, then $B_2 = 0$, so that $\psi = B_1 e^{-\kappa r}/r$. To obtain an expression for B_1, we once again expand the exponential and retain only the linear terms to obtain $\psi = (B_1/r) - B_1\kappa$. Note that if there are no other charges but the one at the origin, then $I = 0$, and thus $\kappa = 0$, and the potential at P is $z_+ e/Dr$. However, when $\kappa = 0$, $\psi = 1/r$; therefore B_1 equals $z_+ e/4\pi\epsilon_0 D$, and Equation 13 will become

$$\psi = \frac{z_+ e}{4\pi\epsilon_0 Dr} - \frac{z_+ e \kappa}{4\pi\epsilon_0 D}. \tag{14}$$

APPENDIX VII

Thus, ψ_a, which is the potential at the surface of a sphere of radius $r = a$, is $(ez_+/4\pi\epsilon_0 Da) - (ez_+\kappa/4\pi\epsilon_0 D)$, and we can evaluate W_2. If all ions in solution are charged except for the ion at the origin, the work W required to give this ion a positive charge is

$$W = \int_0^\infty \psi_a \, dq = \frac{1}{8\pi\epsilon_0 D}\left[\frac{ez_+^2}{a} - (ez_+)^2\kappa\right]. \tag{15}$$

The first term $ez_+^2/8\pi\epsilon_0 Da$ is simply W_1, so that

$$W_2 = \frac{(ez_+)^2\kappa}{8\pi\epsilon_0 D}; \tag{16}$$

and from Equation 2,

$$kT \log \gamma_+ = -\frac{(ez_+)^2\kappa}{8\pi\epsilon_0 D}. \tag{17}$$

Similarly, for a central negative ion, $kT \log \gamma_- = -(ez_-)^2\kappa/8\pi\epsilon_0 D$. Now, to obtain $\gamma_\pm$, by using the relation $\gamma_\pm^\nu = \gamma_+^{\nu_+}\gamma_-^{\nu_-}$, it is easily shown that

$$\ln \gamma_\pm = \frac{e^2\kappa}{8\pi\epsilon_0 DkT}z_+z_-. \tag{18}$$

This expression is usually written with $\log_{10}$, as

$$\log_{10}\gamma_\pm = \left[\frac{(2\pi N)^{1/2}}{2.303}\left(\frac{e^2}{4\pi\epsilon_0 DkT}\right)^{3/2}\right]z_+z_-I^{1/2}, \tag{19}$$

or

$$\log_{10}\gamma_\pm = Az_+z_-I^{1/2}, \tag{20}$$

in which A is the expression in braces in Equation 19. Equation 20 is known as the *Debye-Hückel limiting law*. Since most studies of electrolytes are in aqueous solution and conducted at 25°C (it is convenient for technical reasons to perform experiments slightly above room temperature), the limiting law is frequently written with A evaluated for water at 25°C—namely, $A = 0.509$ so,

$$\boxed{\log_{10}\gamma_\pm = 0.509z_+z_-I^{1/2}.} \tag{21}$$

This is the equation we have sought—a relation between $\gamma_\pm$ and measured concentration, and remarkably, it is not simply the concentration but the ionic strength I that appears in the expression.

ANSWERS TO PROBLEMS

CHAPTER 1

1. (a) From the ideal gas law, $n = PV/RT = (1)(1)/(0.082)(308) = 0.04$ mol, or $(0.04)(32) = 1.28$ g.
   (b) 1.28 g of CO_2 is equivalent to $1.28/44 = 0.029$ mol. Thus $T = PV/nR = (1)(1)/(0.029)(0.082) = 419$ K $= 146°C$.

2. The number of moles of CO_2 and of ethane are 0.023 and 0.035, respectively. The pressure in the flask at 30°C $= nRT/V = (0.057)(0.082)(303)/(1) = 1.42$ atm. The partial pressures of CO_2 and of ethane are $(1.42)(0.023/0.057) = 0.57$ atm and $1.42 - 0.57 = 0.85$ atm, respectively. The mole percents are 40.1 percent and 59.9 percent for CO_2 and ethane, respectively.

3. From Boyle's law, $P = 2(1000/100) = 20$ atm.

4. Friction with the road has heated up the tire and the air contained in it.

5. (a) The pressure in atm is $742.3/760 = 0.9767$. The number of moles $= PV/RT = 0.004216$. Thus the molecular weight is $0.5262/0.004216 = 124.81$.
   (b) The formula of the gas is probably $CH_3\text{-}(CH_2)_x\text{-}CH_3$. If $x = 7$, the molecular weight should be 128, so the substance is probably C_9H_{20}.

6. At the boiling point the vapor pressure is 1 atm. Therefore the volume of 1 mol is $RT/P = 32.06$ l. One mole has a mass of 60 g; therefore the density is $60/32,060 = 0.00187$ g cm^{-3}.

7. $(1/V)(\partial V/\partial T)_P = R/PV = R/RT = 1/T$.

8. The weight of O_2 is 0.1374 g, so it contains 0.00429 mol. The volume of the flask $= nRT/P = (0.00429)(0.082)(298) = 0.105$ l. Let m and e represent the number of moles of methane and ethane. Since the gases are ideal, $m + e = 0.00429$. The number of grams of methane and ethane are $16m$ and $30e$. The flask contains 0.1167 g of the mixture, so $16m + 30e = 0.1167$. Solving for m and e yields $m = 0.000919$ mol and $e = 0.00343$ mol. Therefore the mixture is 21.4 percent (mol percent) methane.

9. Ninety grams of CO_2 = 2.045 mol. Thus $P = nRT/V = 24.59$ atm.

10. In atm, 10^{-6} torr $= 1.32 \times 10^{-9}$ atm. The molar volume is 22,400 cm^3 at 0°C or

770

ANSWERS TO PROBLEMS

24,041 cm³ at 20°C and a pressure of 1 atm. At 1.32×10^{-9} atm the number of molecules per cm³ is

$$\frac{(6 \times 10^{23})(1.32 \times 10^{-9})}{24,041} = 3.3 \times 10^{10}.$$

11. The hot air has a lower density than the cooler air around it.

12. (a) The sun heats the land, whereas its rays are reflected by the water. The air on the land surface becomes hot. Thus, it expands, which lowers its density, causing it to rise. The cooler and denser air over the sea moves in to replace the rising air. This movement is the sea breeze.

   (b) The land is usually colder than the sea and the sun fails to raise the land air to a higher temperature than the sea air. When snow covers the land, the rays of the sun are reflected more than those falling on the water.

CHAPTER 2

1. The mean free path is $\lambda = 1/(\sqrt{2}n\pi d^2) = (RT/\mathcal{N}P)(1/\sqrt{2}\pi d^2)$. We express d in cm; thus $d = 2.24 \times 10^{-8}$; $R = 0.0821$ l-atm $= 82.1$ cm³ atm. Thus $\lambda = 1.79 \times 10^{-5}$ cm at 1 atm and 0.179 cm at 10^{-4} atm.

2. dn/N
   $= 2\pi(1/\pi kT)^{3/2}(0.9kT)^{1/2}(e^{-0.9kT/kT})$
   $\times (0.2kT)$
   $= 0.087$.

3. From the density and molecular weight the number of molecules per cm³ is $(0.00278/35)(6 \times 10^{23}) = 4.77 \times 10^{19}$. Thus the volume available to each molecule is $1/(4.77 \times 10^{19}) = 2.1 \times 10^{-20}$ cm³. If each molecule were in the center of a cube, the side of the cube would be 2.76×10^{-7} cm, which is also the separation between the molecules.

4. $(3RT/M)^{1/2}$ = the root mean square velocity. If the desired units are cm sec⁻¹, R must be 8.314×10^7 erg. Thus, the velocity is 4.82×10^4 cm sec⁻¹.

5. $(3/2)RT = 3716$ J.

6. The number of collisions z by a single molecule is $\sqrt{2}\pi d^2 \langle c \rangle n$. The value of n is $(6 \times 10^{23})/22,400 = 2.68 \times 10^{19}$ molecules per cm³; $\langle c \rangle = (8RT/\pi M)^{1/2} = 4.5 \times 10^4$ cm sec⁻¹; and $d = 3.74 \times 10^{-8}$. Thus, $z = 7.5 \times 10^9$ per cm³ per sec. The total number of collisions is $\frac{1}{2}\sqrt{2}\pi d^2 \langle c \rangle n^2 = 1.0 \times 10^{29}$ per sec.

7. The mean free path is $1/(\sqrt{2}n\pi d^2)$. The value of n is 10^{-5} atom cm⁻³. Thus the mean free path is 5.64×10^{19} cm. One light year is 9.5×10^{17} cm; so that $\lambda = 59.4$ light years.

8. Each ball has a momentum of $(0.5)(500) = 250$ g cm sec⁻¹. The total momentum change per second is the number n of balls times $2(250)n = 500n$ g cm sec⁻¹. The downward force of the plate is mg or $(1000)(980)$. To keep the plate suspended, the upward force must equal the gravitational force, or $(1000)(980) = 500\ n$; or, 1960 balls would have to collide with the plate each second.

9. We assume that the pressure is low enough and the orifice small enough that the effusion law is obeyed. If so, the rate of depletion of air is independent of the pressure and, assuming an average molecular weight of air of 30, it is $(RT/2\pi M)^{1/2} = 1.15 \times 10^4$ cm³ per sec per cm². The area of the orifice is $\pi r^2 = \pi(0.01)^2 = 3.14 \times 10^{-4}$ cm; thus the rate of loss of air is 3.6 cm³ per sec. The volume of the ship is 2×10^7 cm³; thus the atmosphere will be gone in

$$\frac{2 \times 10^7}{3.6} = 5.6 \times 10^6 \text{ sec}$$

or 1543 hours. Therefore, it will take $(0.75)(1543) = 1157$ hours for the pressure to drop to one-fourth the initial value, and the ship will make it back to earth safely.

10. Use the barometric formula with $R = 8.413 \times 10^7$ and $g = 980$ cm sec⁻². $P = P_0 e^{-Mgx/RT} = 0.99977$ atm at the top of the container.

11. Use Equation 18. For P we must substitute the partial pressure of CO_2 or 0.00033 atm. Also, to keep the units correct, pressure must be in dyn cm⁻². Since 1 atm $= 10^6$ dyn cm⁻², $P = 3.3 \times 10^2$. Thus the number of collisions per cm² per sec is

$$\frac{3.3 \times 10^2}{2\pi(44)(1.6 \times 10^{-24})\ (1.38 \times 10^{-16})(298)^{1/2}}$$
$$= 7.7 \times 10^{19},$$

which is equivalent to 5.4×10^{-3} g of CO_2 per sec per cm². Thus $(200)(5.4 \times 10^{-3}) = 1.08$ g of CO_2 collides per second per leaf.

12. Use the equation $dW/dt = \rho(RT/2\pi M)^{1/2}$, which relates the weight of each gas effusing per unit time and the density of each gas. At a pressure of 0.1 atm = 10^5 dyn cm^{-2}, the number of moles in the container is $PV/RT = 0.41$, of which 0.406 mol are H_2 and 0.004 mol are D_2. Thus the density of each is

$$\frac{(0.406)(2)}{10^5} = 8.12 \times 10^{-6} \text{g cm}^{-3} \text{ for } H_2$$

and

$$\frac{(0.004)(4)}{10^5} = 1.6 \times 10^{-7} \text{g cm} \times 13 \text{ for } D_2.$$

The values of $(RT/2\pi M)^{1/2}$ are 4.40×10^4 for H_2 and 3.1×10^4 for D_2. Thus the weight loss for H_2 is 0.357 g per cm² per sec and for D_2 is 0.005 g per cm² per sec. Thus, through an area of $\pi(0.01)^2$ and after 30 minutes (1800 sec), the amount of H_2 lost is 0.201 g or 0.1005 mol. For D_2 this value is 0.00282 g or 0.0007 mol. The number of moles remaining are $0.406 - 0.1005 = 0.3055$ for H_2 and 0.0033 for D_2. Thus the mole percent of D_2 in the vessel is 1.07.

13. When moving horizontally, the atoms will be pulled downward by gravity. However the horizontal speed of the atoms is Boltzmann-distributed whereas the vertical speed is not (because it is determined by gravity). Thus the molecules moving horizontally faster will suffer less downward motion than a more slowly moving molecule. The spreading of the image will clearly be downward.

14. The total energy of the CH_4 is $(3N - 3)RT = 12RT = 3516R$; the total energy of the CO_2 is $(3N - \frac{5}{2})RT = 1937 R$. The total energy of the system is $(3516 + 1937)R = 5453R$. At an equilibrium temperature T' this energy is distributed over all of the molecules, so that $12RT' + 6.5RT' = 18.5RT' = 5453R$. Thus, $T' = 294.76$ K = 21.60°C. (We have used the more precise value for 0 K of −273.16°C.)

15. As the temperature increases, diffusion should occur more rapidly because the velocity of the particle increases. It should decrease with increasing pressure because the mean free path would be less; that is, the molecules travel less far at higher pressure before being deflected by a collision.

16. Consider a gas flowing through an orifice or between parallel plates. Owing to the friction with the edge of the orifice or with the plates, the layer of gas molecules adjacent to the surface moves more slowly than molecules away from the surface. We can then imagine that the viscosity is a result of friction between all of the adjacent layers of molecules. As the temperature increases, the collision frequency between molecules in adjacent layers increases and hence there is greater friction; this increases the viscosity.

17. Starting at the end indicated by the arrow, the horizontal distances are (in units of 10^{-4} cm), 1, 2, 1, $\frac{1}{2}$, $1\frac{1}{2}$, 0, 1, $1\frac{1}{2}$, $\frac{1}{2}$, $1\frac{1}{2}$, $\frac{1}{2}$, $\frac{1}{2}$, $2\frac{1}{2}$, 1, $1\frac{1}{2}$, $\frac{1}{2}$, $1\frac{1}{2}$, $1\frac{1}{2}$, $1\frac{1}{2}$, 1, and 1. The sum of the squares is 33.25×10^{-8} cm and $\langle(\Delta x)^2\rangle = 1.58 \times 10^{-8}$ cm. Using Equation 38 and solving for η yields 0.027 poise. Thus the viscosity is three times that of pure water.

CHAPTER 3

1. (d), (e), (f), and (h).
2. Yes, because it is not symmetric.
3. −80°C.
4. (a) For o-dichlorobenzene: $\mu_x = 1.69 \cos 30° + 1.69 \cos 90° = 1.46$; $\mu_y = 1.69 \sin 30° + 1.69 \sin 90° = 2.54$. The vector sum is $(\mu_x^2 + \mu_y^2)^{1/2} = 2.93$ debye. For m-dichlorobenzene the sum is 1.69 debye.
   (b) No; p-dichlorobenzene is symmetric, so that $\mu = 0$.
5. The dipole moments are antiparallel. Therefore $\mu = 4.22 - 1.69 = 2.53$ debye.

ANSWERS TO PROBLEMS

6. At the critical point, $(\partial P/\partial V)_T = 0$ and $(\partial^2 P/\partial V^2)_T = 0$. From $(\partial P/\partial V)_T = 0$, one obtains $RT/(\overline{V} - b)^2 = a/\overline{V}^2$. From $(\partial^2 P/\partial V^2)_T = 0$, one obtains $2RT/(\overline{V} - b)^3 = 2a/V^3$. These two equations can only be simultaneously true if $b = 0$. Thus, only a gas for which $b = 0$ can have a critical point.

7. In urea, an amino group of one molecule can hydrogen-bond to a carbonyl group of a second molecule, providing a significant intermolecular attractive force.

8. The OH group of phenol can hydrogen-bond to water.

9. Formaldehyde forms intermolecular hydrogen bonds resulting in a hydrogen-bonded trimer that is a ring, which can absorb short-wavelength ultraviolet light. Heating breaks the hydrogen bonds but in time these can reform.

10. This is an example of cooperativity. Once one hydrogen bond has formed, the adjacent molecule is able to hydrogen-bond more easily because it is in the correct orientation to form this bond. The second hydrogen bond makes formation of a third easier, and so forth.

11. Ethane is larger. Ethanol has a permanent dipole moment and can form internal hydrogen bonds.

12. A significant fraction of the molecules are hydrogen-bonded in the vapor phase.

13. Solubility would be good in ethanol, with which it can interact because of permanent dipoles and hydrogen bonds. Solubility would be poor with ether and chloroform.

14. The intense electric field of the divalent ion attracts water molecules resulting in an orientation of the water molecules, which take up less space than when they are randomly arranged.

15. Less, because the electron clouds overlap when a covalent bond forms.

16. Yes. The H $\cdots$ O bond distance ought to be less than the sum of 1.2 and 1.4 but not as small as 0.95. A length of 1.5 to 2.0 Å would not be an unreasonable guess. Since the O—H bond length is 0.95 Å, the O—H $\cdots$ O distance should be between 2.45 and 2.95 Å.

CHAPTER 4

1. 3.5×10^9 erg = 350 J = 83.7 cal.
2. $[(100)(80) + (50)(25)]/(100 + 50) = 61.7°C$.
3. (a) 0°C, because the ice will not melt.
   (b) Let T = the final temperature. Use C_V = 75.3 J K^{-1} mol^{-1} = 4.18 J K^{-1} g^{-1} and ΔH_{fus} = 6042 J mol^{-1} = 335 J g^{-1}. The number of joules released in cooling from 40°C to T is $4.18(40 - T)(50) = 8360 - 209T$. This equals the number of joules needed to raise the temperature of ice from 0°C to T or $\Delta H_{fus} + C_V\Delta T = (335 + 4.18T)(10) = 3350 + 41.8T$. Thus, T = 20.0°C.
   (c) Analysis such as in part (b) yields a negative value of T, which is not possible. Note that the cooling of 50 g of water releases only $(4.18)(50) = 209$ J, which is less than ΔH_{fus} per 50 g. Thus, all of the ice will not melt and the final temperature will be 0°C.
4. 427 m.
5. Positive for (a) and (d); negative for (b) and (c).
6. (a) $W = RT \ln(V_2/V_1) = RT \ln(P_1/P_2) = 5.7$ kJ.
   (b) T is constant, so $\Delta U = 0$ and $W = Q$. Thus 5.7 kJ of heat are evolved.
7. The volumes of the solid and the liquid are $150/2.09 = 71.77$ cm^3 and $150/2.00 = 75$ cm^3, respectively. Thus, $\Delta V = 3.23$ cm^3. The work is $P\Delta V = 0.00323$ l-atm = 0.328 J. The heat given off is 100 J, so $Q = -100$ J. Hence $\Delta U = -100 - 0.328 = -100.328$ J, and $\Delta H \approx -100$ J.
8. $W = RT \ln[(V_2 - aT + b)/(V_1 - aT + b)]$.
9. (a) The work done is $P\Delta V$, in which ΔV represents the change in the volume of 10 g of water (that is, 10 cm^3) to the volume of the steam at 100°C. The theoretical volume of one mole or 18 g of water vapor at 0°C is 22.4 l. By the ideal gas law, the volume of 10 g at 100°C is $(373/273)(10/18)(22.4) = 17$ l. Thus $\Delta V = 17 - 0.01 \approx 17$ l. The work is $P\Delta V = (1)(17) = 17$ l-atm. The conversion factor is 101.5 J = 1 l-atm. Therefore the work is $(101.5)(17) = 1726$ J.

(b) To vaporize 10 g yields $(10/18)(40.58)$ = 22.5 J. Thus $[(1725/22.5)](100) = 7.6$ percent.

10. (a) Work = force × distance. The force of a 50-kg mass = $(5 \times 10^4)(980)$ dyn so the work done on the gas is $(5 \times 10^4)(980)(3.5) = 1.715 \times 10^8$ erg = 17.15 J. Since the compression is adiabatic, $Q = 0$ and $\Delta U = W = 17.15$ J. From the kinetic theory, $\Delta U = \frac{3}{2}R\Delta T$ for an ideal monatomic gas, so ΔT is $+1.375°C$.
(b) For an ideal diatomic gas, $\Delta U = \frac{5}{2}R\Delta T$, so $\Delta T = 0.825°C$.

11. (a) None, because the process is adiabatic.
(b) In an adiabatic process, $Q = 0$, so $dU = -W$. Since $dU = C_V dT$, $W = -C_V(T_1 - T_2) = (12.6)(298 - 119) = 2254$ J.

12. The compression requires that at least one of the walls of the container move toward the molecules. As the wall moves, because it has momentum, by collision with the gas molecules the momentum is transferred to the molecules. Thus, their kinetic energy increases. By intermolecular collision this additional kinetic energy will be redistributed over the population of molecules with the result that both the vibrational and rotational energies of the molecules also increase. The total increase in kinetic energy is seen as a temperature increase.

13. $(\partial H/\partial P)_T = 0$ or $dP = 0$.

14. In a constant volume process all of the heat absorbed goes into kinetic, vibrational, and rotational energy, which causes a temperature increase. The heat capacity C_V is just a reflection of how much energy the molecule can hold. In a constant pressure process the system changes volume (for example, expands) against a resisting pressure. The heat put into the system is utilized in two ways—some is used to move whatever exerts the pressure and the remainder goes into the energy of the system. Thus, less of the heat is available to raise the temperature, so more heat is taken in per degree of temperature increase. This means that C_P > C_V. For a real gas, if there is a volume change, some heat is used to move the molecules apart against their intermolecular attractive forces, so that even more heat is required for a real gas to produce a change in temperature of one degree. Thus $C_P(\text{real gas}) > C_P(\text{ideal gas})$.

15. The heat capacity of a liquid is greater than that of a gas because, as explained in the answer to problem 4-14, energy must be put into the system to separate the molecules against the attractive forces. In a liquid the molecules are nearer to one another than in a gas. The attractive force increases as the separation decreases, as shown in Chapter 3, so that the heat capacity of a liquid is larger than that of a gas. By this same reasoning one would expect the heat capacity to be even greater for a solid. This is usually true, but not for water. Remember that the density of ice is less than that of water (ice floats in water); in other words, the molecules are farther apart. The reason for this is the geometrical constraint imposed by hydrogen bonds. In ice a smaller fraction of the molecules can hydrogen-bond than in water because fewer molecules are appropriately oriented for hydrogen-bond formation. Thus, compared to ice, the larger number of hydrogen bonds in water can take up a greater fraction of heat from the environment without raising the temperature. Thus the heat capacity of liquid water is greater than that of ice and of steam.

16. The heat given off is $Q = P(V_2 - V_1)$. The initial volume V_1 was 22.4 l. We must be sure to keep the units consistent. Hence we convert 2000 J to l-atm, using the relation 101.5 J = 1 l-atm. Thus $Q = 19.70$ l-atm. Hence $19.70 = 1(V_2 - 22.4)$ and $V_2 = 43.01$ l.

17. (a) At constant volume $\Delta U = nC_V\Delta T = 376$ J. To calculate ΔH we must first calculate ΔP. By the ideal gas law the initial pressure is $3RT_1/V$ and the final pressure is $3RT_2/V$. Thus $\Delta P = (3R/V)(T_2 - T_1)$ and $\Delta(PV) = 3R\Delta T = 3(8.314)(10) = 249$ J. Thus $\Delta H = 376 +$

ANSWERS TO PROBLEMS

249 = 625 J.

(b) The increase in ΔU is the increase in the kinetic, rotational, and vibrational energies of the molecules.

18. Zero, by definition.
19. 100 g = 5.56 mol. The heat required is $(\Delta H_{fus} + C_V \Delta T + \Delta H_{vap})(5.56) = 301$ kJ.
20. 50 g of water = 50/18 = 2.78 mol; C_V = 75.3 J K^{-1} mol^{-1}; thus the heat transferred is (17.8)(75.3)(2.78) = 3.726 kJ. The molecular weight of sulfur is 32; thus 0.4 g = 0.0125 mol. Therefore, ΔH_{comb} = 3.726/0.0125 = 298 kJ mol^{-1}.
21. (a) -395.4 kJ mol^{-1}.
    (b) -37.0 kJ mol^{-1}.
    (c) -193.0 kJ mol^{-1}.
22. The heat of reaction = ΔH_{comb}(reactants) $- \Delta H_{comb}$(products) = $-1787 - (-874.5)$ = -911.5 kJ mol^{-1}. However, in the combustion reaction, water is a gas and in this equation we are considering a temperature at which it is liquid, so we can add the heat generated by converting water vapor to liquid, namely, ΔH_{vap}. Thus, $\Delta H_{react} = -911.5 - 44.0 = -955.5$ kJ mol^{-1}.
23. The reactions are

    $2C_2H_2 + 5H_2O \rightarrow$
    $\quad\quad 4CO_2 + 2H_2O \quad (\Delta n = -1),$

    and

    $2C_6H_6 + 15O_2 \rightarrow$
    $\quad\quad 12CO_2 + 6H_2O \quad (\Delta n = -3).$

    Then

    $\Delta H_{comb}(C_2H_2)$
    $= -2(1303) - (8.314 \times 10^{-3})(293)$
    $= -2608$ kJ mol^{-1},

    and

    $\Delta H_{comb}(C_6H_6)$
    $= -2(3274) - 3(8.314 \times 10^{-3})(293)$
    $= -6555$ kJ mol^{-1}.

    The reaction for the synthesis of benzene is $3C_2H_2 \rightarrow C_6H_6$, which is obtained by subtracting half of the benzene combustion reaction from 3/2 of the acetylene combustion reaction, which yields 635 kJ mol^{-1}.

24. The reaction is

    $CH_2(COOH)_2(s) + 2O_2(g) \rightarrow$
    $\quad\quad 3CO_2(g) + 2H_2O(l).$

    The molecular weight of malonic acid is 104, so (104/10)(83.3) = 866 kJ is evolved per mole. In a calorimeter, $\Delta V = 0$, and no work is done, so $Q_V = \Delta U$. Thus $\Delta U = -866$ kJ mol^{-1}. To calculate the heat evolved at constant pressure, assume that the gases are ideal, so that $\Delta H_{comb} = \Delta U + P\Delta V = \Delta U + \Delta nRT$. In this reaction, $\Delta n_{gas} = 3 - 2 = 1$. Thus

    ΔH_{comb}
    $= -866 + (8.314 \times 10^{-3})(298)$
    $= -863$ kJ mol^{-1}.

25. Five moles of gaseous B_2 become two moles of gaseous A_2B_5. Thus there is a volume change, or, $\Delta(PV) = (2-5)RT = -7.5$ kJ. Hence $\Delta H = \Delta U + \Delta(PV) = 63 - 7.5 = 55.5$ kJ.
26. Using Trouton's rule, we calculate the ratio $\Delta H_{vap}/T_b$ and obtain 50.2, 90.0, 103.8, 89.1, 85.8 and 100.8 for liquids 1–6, respectively. When the ratio is > 88, as for liquids 3 and 6, there are excessively strong intermolecular forces, usually hydrogen bonds. For liquid 1 the ratio is much less than one, which indicates that the forces are so strong that they persist in the vapor phase. Thus the vapor of substance 1 is probably hydrogen-bonded.
27. The total number of joules produced in five minutes is (1000)(60)(5) = 300 kJ. The value of ΔH_{vap} for water is 40.58 kJ mol^{-1} so that you need 300/40.58 = 7.4 mol or (7.4)(18) = 133 g of water.
28. He needs (18)(150) = 2700 Calories per day to maintain his weight. His deficit is 700 Calories per day. The caloric value of burning one pound (= 454 g) of fat is (7)(454) = 3178 Calories. Thus to lose ten pounds he must eliminate (10)(3178) = 31,780 Calories from his diet or eat 2000 Calories for 31,780/700 ≈ 46 days. Note that if he eats 1300 Calories per day, it would take only 23 days.
29. (a) The air expands owing to decreased atmospheric pressure. If the expansion is adiabatic, the temperature decreases.
    (b) As it cools, the moisture condenses

and precipitates and gives off heat so that the temperature decrease is smaller.

(c) The air is compressed adiabatically by increasing atmospheric pressure and therefore warms. In southern Germany it is said that the hot foehn drives people crazy.

(d) No. It would be higher on the lee side. This is because the cooling on the windward side has been reduced by condensation and precipitation. Because of the precipitation, there is less water in the air to take up heat when the air descends the lee side. Thus the foehn is a hot, dry wind.

CHAPTER 5

1. Reversible: (b) and (e); irreversible: (a), (c), and (d).
2. No. A reversible process proceeds infinitely slowly but such slowness is not a sufficient criterion for reversibility. The process is irreversible because the heat flow cannot be reversed; this would require heat flow from a cold body to a hot body.
3. No, because the Earth is not an isolated system. Work could be done by the light that comes from the Sun and this work could decrease entropy.
4. No, because $S = Q_{rev}/T$, not Q/T. The adiabatic expansion is an irreversible process.
5. (a).
6. The royal flush, because it is attainable in only four ways and the "entropy" would be minimal for the least probable hand.
7. (a) $|\Delta S_{surr}| > |\Delta S_{cell}|$ so that $\Delta S_{univ} > 0$.
   (b) Release of waste products, release of oxygen to the atmosphere, and production of light by the Sun.
8. (a) Pentane;
   (b) Acetic acid;
   (c) Methylethyl ketone;
   (d) Glycine.
9. (a) The number of moles of gas in the box is $n = PV/RT = 41.57$, so that the number of molecules is $nN = 2.49 \times 10^{25}$. The probability, Ω, of one molecule permanently being in a particular half of the box is $\frac{1}{2}$, so that the probability of all of the molecules being in the same half of the box is $(\frac{1}{2})^{2.49 \times 10^{25}}$, which is exceedingly small.
   (b) $\Delta S = k \ln \Omega = (2.49 \times 10^{25}) k \ln(\frac{1}{2}) = -9.97 \times 10^9$ J K^{-1}.
10. $\Delta H = 459.0$ J; $\Delta S = 1.67$ J K^{-1}.
11. The process is irreversible but because S is a state function, one can use the expression for a reversible change. Thus $\Delta S = nR \ln(P_1/P_2) = 4.08$ J K^{-1} mol^{-1}.
12. The temperature at equilibrium is 36.67°C. Thus $\Delta S = 2.41$ J K^{-1} mol^{-1}.
13. If it cannot occur spontaneously, ΔS_{univ} will be less than zero. The heat of reaction is 2802 kJ mol^{-1}. This is heat taken in from the surroundings at 25°C. Thus $Q = -2802$ kJ mol^{-1} and $\Delta S_{surr} = -2802/298 = -9.403$ kJ K^{-1} mol^{-1}. For the reactants, $\Delta S = \Sigma S(\text{products}) - \Sigma S(\text{reactants}) = -238$ J K^{-1} mol^{-1}. Thus $\Delta S_{univ} = \Delta S_{surr} + \Delta S = -9.403 + (-0.238) = -9.641$ J K^{-1} mol^{-1}, which is less than zero. Thus the reaction cannot occur spontaneously.
14. (a) No. There are retarding forces, although small, and ultimately the Earth will move into the Sun.
    (b) It is an example of perpetual motion. It does not violate the Second Law for two reasons: (1) the Second Law does not apply to individual atoms and (2) perpetual motion is allowable as long as no work is being done.
15. Since the reaction does occur, ΔS must be > 0. However, $\Delta S°$ is not the same as ΔS but refers to 25°C. The reaction will proceed at 25°C, but the product is not water but steam at a very high temperature; $\Delta S > 0$ for this reaction. Thus the fact that $\Delta S° < 0$ means that it is very unlikely that liquid water at a temperature of 25°C would be the result. In other words, in an isolated and hence thermally insulated system, so much heat is produced that the steam so formed cannot spontaneously condense and cool to 25°C. Of course if the reaction occurred in a thermally conducting box, $\Delta S_{reaction}$ might

be less than zero but the heating of the environment results in $\Delta S_{univ} > 0$.

16. ΔS = S(leucylglycine) + $S(H_2O)$ − S(leucine) − S(glycine) = 281.2 + 69.91 − 109.2 − 103.5 = 138.4 J K^{-1} mol^{-1}.

17. ΔS = 26 ln(353/323) = 2.31 J K^{-1} mol^{-1}.

18. $\Delta S_{vap} = \Delta H_{vap}/T$ = 26,100/326 = 80.1 J K^{-1} mol^{-1}. However since we have used the normal boiling point, this value of ΔS_{vap} is based on a pressure of 1 atm. Thus if the evaporation occurs at 0.2 atm, we must include ΔS for the pressure change or $\Delta S = R \ln(P_1/P_2)$ = 13.4 J K^{-1} mol^{-1}. Thus the entropy change is 80.1 + 13.4 = 93.5 J K^{-1} mol^{-1}.

19. The ratio of work done to heat removed is $W/Q = (T_H/T_C) - 1$, in which H and C represent the high and low temperatures. T_H = 298 K, W = 100 watts = 100 J sec^{-1}, and Q = 418 J sec^{-1}. Thus T_C = 240.5 K = −32.5°C.

20. The total entropy change is $(Q_H/T_H) - (Q_C/T_C)$. Since the engine is cyclic, $W = -Q_H - Q_C$, so that $\Delta S_{tot} = (Q_H/T_H) - [(W + Q_H)/T_C]$, with T_H = 573, T_C = 298, W = 2.5 J, and Q_H = −4.18 kJ. Thus ΔS = −1.68 J per joule drawn from the high-temperature reservoir. Since ΔS must be > 0, the claim is impossible. Another way of looking at this is to calculate the engine's efficiency. The maximal efficiency would be that of a reversible engine, for which the efficiency would be 1 − (298/573) = 0.48. The new engine is claimed to have an efficiency of 0.6, which is greater than the maximal possible efficiency.

21. The maximal efficiency of an engine operating between 20° and 300°C is 1 − 293/573 = 0.49. The reactor operates at an efficiency of 50 percent of maximum or at 0.245. The efficiency is by definition the work produced divided by Q_H, the heat withdrawn from the high-temperature reservoir, which is the reactor. That is, Q_H = 10^6 kW/0.245 = 10^9 J sec^{-1}/0.245 = 4.08 × 10^9 J sec^{-1}. Of this, 1 − 0.245 = 0.755 is transferred to the river or 0.755 × 4.08 × 10^9 = 3.08 × 10^9 J sec^{-1}. Dividing by the flow rate of the river, the energy transferred is (3 × 10^9)/200 = 15.4 J cm^3 or approximately 15.45 J g^{-1}. The specific heat of water is 4.18 J g^{-1} K^{-1}, so that the increase in temperature of the river is 3.7°C.

22. Since $\Delta S > 0$, the molecule must become more disordered. Hence there must be some loss of the internal structure of the molecule, which then assumes a more random configuration. In fact most of the noncovalent bonds break and the molecule usually collapses into what is known as a random coil (see Chapter 15).

23. Separation is just the reverse of mixing. Thus one need only calculate the entropy of mixing and reverse the sign. Therefore $\Delta S_{mix} = -\sum n_i R \ln X_i$, in which n_i and X_i are the number of moles and the mole fraction of each component. The number of moles and the mole fractions of each component are: $^1H_2^{16}O$, 0.9974; $^1H_2^{17}O$, 0.0004; $^1H_2^{18}O$, 0.002; $^2H_2^{16}O$, 0.000156; $^2H_2^{17}O$, 6.24 × 10^{-8}; $^2H_2^{18}O$, 3.1 × 10^{-7}. Thus ΔS = −0.9 J K^{-1}.

CHAPTER 6

1. The value of ΔG is 0, because ice and water coexist at 0°C and 1 atm.

2. The volume of 100 g is 7.39 cm^3. Calculate ΔG in this way: $dG = -S\,dT + V\,dP$; in this case dT = 0, so that $\Delta G = \int_1^{100} V\,dP$ = 99 × 7.39 × 10^{-3} liter-atm = (99)(7.39 × 10^{-3})(101.3) = 74.11 J mol^{-1}.

3. Start with $dG = -S\,dT + V\,dP$. Thus $(\partial G/\partial P)_T = V$ and $(\partial G/\partial T)_P = -S$. Since $(\partial/\partial T)(\partial G/\partial P) = (\partial/\partial P)(\partial G/\partial T)$, then $(\partial S/\partial P)_T = -(\partial V/\partial T)_P$. Therefore $S = -(\partial V/\partial T)_P\,dP = -(\partial V/\partial T)_P \Delta P$.

4. Figure that $\Delta G_{blood} = \Delta G°_{form} + RT \ln c_{blood}$ and $\Delta G_{urine} = \Delta G°_{form} + RT \ln c_{urine}$. Then $\Delta G = \Delta G_{urine} - \Delta G_{blood} = RT \ln c_{urine} - RT \ln c_{blood} = RT \ln 75 = 11.13$ kJ mol^{-1}. Note that both the identity of the substance and the value of $\Delta G°_{form}$ are irrelevant.

5. It will be $\Delta G = \Delta G°_{form} + RT \ln c = -22.2 + (8.314 × 10^{-3})(298)\ln(0.01) = -33.6$ kJ mol^{-1}.

6. (a) Ice, because it melts at 5°C.

(b) 1 M NaCl, because if two solutions are separated by a semipermeable membrane, NaCl from the 1 M solution will flow into the other solution.
(c) At 10 atm, because if each is on opposite sides of a piston, the piston will move in such a way that the 1 atm sample is compressed.

7. Begin with $K = P_C^2/P_A P_B^2 = (0.54)^2/(0.25)(0.65)^2 = 2.76$ atm^{-1}. Then $\Delta G° = -RT \ln K = -2.5$ kJ mol^{-1}.

8. Use Equation 65. The value of $\Delta H° = -12.71$ kJ mol^{-1}.

9. Use $\ln K = -\Delta H°/RT + $ constant, and assume that $\Delta H°$ is independent of T. Thus, the value $K = 1.7 \times 10^{-10}$ at 25°C converts to $K = 7.6 \times 10^{-9}$ at 75°C. Since $K = [Ag^+][Cl^-]$, then $[Ag^+] = [Cl^-] = (7.6 \times 10^{-9})^{-1/2} = 8.7 \times 10^{-5}$ M.

10. The value $\Delta G°$ for the reaction is $124.5 - 3(209.2) = -503.1$ kJ mol^{-1}, so that by using $\Delta G° = -RT \ln K$, the value of K is 1.6×10^{88}. Thus the yield of benzene ought to be exceptionally good.

11. From $\Delta G° = -RT \ln K$, we get $K = 7.85 \times 10^{-35}$ at 25°C and 5.54×10^{-29} at 75°C. Assuming that $\Delta H°$ is independent of temperature, we use $\ln K = -\Delta H°/RT +$ constant, to obtain $K_{125°C} = 1.3 \times 10^{-24}$.

12. For the liquid-to-gas reaction, $\Delta G°$ is $-228.6 - (-237.2) = 8.6$ kJ mol^{-1}. Therefore $K = 3.11 \times 10^{-2}$ atm; and $P_{H_2O}/1$ atm $= 3.11 \times 10^{-2}$ atm^{-1}.

13. The values for ΔG_{form} of CO, H_2O, CO_2, and H_2 are -137.7, -228.6, -394.4, and 0 kJ mol^{-1}, respectively. Thus $\Delta G° = -28.1$ kJ mol^{-1} and $K = 8.43 \times 10^4$.

14. It must be > 0, because $[H^+] = [OH^-] = 10^{-7}$ M. Under standard conditions all concentrations are one molar which, if $\Delta G° < 0$, would mean that dissociation is very extensive, which is not true.

15. The value of $\Delta G° = -RT \ln K = -27.07$ kJ mol^{-1}.

16. The value of $\Delta G° = RT \ln K = +16.5$ kJ mol^{-1}. $\Delta G°' = \Delta G° + 16.118 \, RT$ (see Equation 66), so that $\Delta G°' = 56.4$ kJ mol^{-1}.

17. The partial pressure of each gas is reduced to 0.5 atm. Therefore for each gas $\Delta G = -RT \ln (1/0.5)$ and $\Delta S = R \ln(1/0.5)$. For both gases together, $\Delta G = -2RT \ln 2 = -3434.6$ J mol^{-1} and $\Delta S = 2R \ln 2 = 11.53$ J K^{-1}. Then $\Delta H = \Delta G + T\Delta S = -3434.6 + (298)(11.53) = +1.34$ J mol^{-1}. (Actually, if the more precise value of T, namely 298.16 K, is used, $\Delta H = 0$.)

18. The value of K corresponding to $\Delta G° = -28.45$ kJ mol^{-1} is 9.7×10^4. Thus [B] ought to be nearly 1 M. However this is at equilibrium, and clearly equilibrium has not been reached in the time allotted. Apparently, the conversion is slow.

19. (a) The value of $\Delta G° = -66.9$ kJ mol^{-1}.
(b) Yes.
(c) The reaction has an activation energy.

20. First we calculate ΔG for diluting glucose and oxygen from the interstitial fluid to the cells using $\Delta G = RT \ln(c_2/c_1)$. This gives -10.08 and -2.84 kJ mol^{-1} for glucose and oxygen, respectively. Then we calculate $\Delta G°$ for the reaction from the values of $\Delta G°_{form}$. This is $6(-386.2) + 6(-237.2) + 917.2 - 6(0) = -2823$ kJ mol^{-1}. In the cell $\Delta G = \Delta G° + RT \ln\{[CO_2]^6[H_2O]^6/[glucose][O_2]^6\} = -2707$ kJ mol^{-1}. Next we calculate the free energy of dilution of CO_2 from the cell to the blood; this is $\Delta G = -10.12$ kJ mol^{-1}. The total free energy change is the sum of these terms or $-10.08 + 6(-2.84) - 2707 + 6(-10.12) = -2794$ kJ mol^{-1}.

21. Let g_A and g_B represent the free energies of one mole. Using a symbol ° for the standard state, a symbol ' for the final state, and no superscript for the initial state, $\Delta G = G' - G = n_A g'_A + n_B g'_B - (n_A g_A + n_B g_B) = n_A(g'_A - g_A) + n_B(g'_B - g_B) = n_A(g°_A + RT \ln p_A - g°_A - RT \ln P) + n_B(g°_B + RT \ln p_B - g°_B - RT \ln P) = n_A RT \ln(p_A/P) + n_B RT \ln(p_B/P) = n_A RT \ln X_A + n_B RT \ln X_B$.

22. (a) Yes.
(b) No, because there is no common intermediate.

23. (a) 1.3×10^3.
(b) No. The enzyme is a catalyst and merely enables equilibrium to be achieved more rapidly.

24. Hydrolysis has a significant activation energy.

25. (a) Use Equation 64 to obtain $\Delta H° = 88.3$ kJ mol^{-1}.
(b) First calculate $\Delta G = -RT \ln K = -6083$ J mol^{-1}. Then $\Delta S = -(1/T)(\Delta H - \Delta G) = 247$ J K^{-1} mol^{-1}.
(c) An increase in entropy when the ordered helix forms is inconsistent with the fact that an increase in entropy implies a decrease in order. Thus there must be something else in the system for which a great deal of order is lost. A reasonable idea is that solvent molecules are tightly bound in an ordered way to the protein molecule. The absorption of heat ($\Delta H > 0$) at 60°C is consistent with the notion that the solvent molecules are driven from the chain by the increased kinetic energy when the helix forms.

26. The value of ΔH is greater than 0, because higher temperature favors denaturation. The value of ΔG is less than 0 because the reaction is spontaneous. Since $\Delta G = \Delta H - T\Delta S < 0$, then $T\Delta S > \Delta H$, which is positive, so that ΔS is positive at temperatures at which the process occurs.

27. It is exothermic; thus the temperature should be low. There is a volume decrease in the reaction and this will be aided by high pressure.

28. The fuel that runs the machine (that is, the source of free energy) is the difference in the free energy between the high-salt and low-salt solutions. As the machine runs, salt is picked up by the belt on the high-salt side and diluted into the low-salt side. Eventually the salt concentrations in A and B become equal, and the machine stops.

CHAPTER 7

1. There are $65/250 = 0.26$ moles of solute, so the molarity is 0.26. The total weight of the solution is 1065 g, so its volume is $1065/1.020 = 1044$ cm^3. Thus the molarity is $0.26/1.044 = 0.25$. One l of water is $1000/18 = 55.56$ mol, so the mole fraction of the solution is $0.26/(0.26 + 55.56) = 0.0047$.

2. In each cm^3 of solution there is $(0.575)(1.700) = 0.978$ g or 0.00581 mol. Thus, the concentration is 5.81 M. The water content is 0.722 g; therefore the molality is $(0.00581)(1000/0.722) = 8.04$ M. There are $0.722/18 = 0.04$ mol of water per cm^3. Thus, the mole fraction of CsCl is $0.00581/0.04581 = 0.126$.

3. (a) The molality is proportional to the number of moles of solvent, so $\bar{v} = \partial V/\partial \mathbf{M}$, which is $2(\mathbf{M} - 0.12)$. This is zero when $\mathbf{M} = 0.12$.
(b) At a concentration of 0.2 $\mathbf{M}$, $\bar{v}$ equals $2(\mathbf{M} - 0.12) = 0.16$.

4. The value of $V = \sum n_i \bar{v}_i = n_1 \bar{v}_1 + n_2 \bar{v}_2$. The partial molal volume $\bar{v}_1$ of water is $18/0.9986 = 18.0252$ cm^3 mol^{-1}. In 1000 cm^3, there is $(0.0013)(120) = 0.156$ g of A and $1000(0.9990) - 0.156 = 998.844$ g of water = 55.491 mol. Thus, V is equal to $1000 = (55.491 \times 18.0252) + 0.0013 \bar{v}_2$; and therefore $\bar{v}_2 = -181.825$ cm^3 mol^{-1}. The negative sign indicates that when A is added to water, the volume decreases. This is common for electrolytes containing divalent ions.

5. The volume of B in the solvent is $(120)(1.25) = 150$ cm^3. Thus there is $225 - 150 = 75$ cm^3 available for 1.5 mol of A, so the molar volume of A is $75/1.5 = 50$ cm^3 mol^{-1}.

6. (a) The number of moles of A and B are 0.738 and 0.034, respectively. The mole fraction of A is $0.738/(0.034 + 0.738) = 0.956$. The vapor pressure of A is $(0.956)(42.3) = 40.44$ mm Hg.
(b) No.

7. There is $200/352 = 0.57$ mol of S and $3.2/75 = 0.043$ mol of B. Thus the mole fraction of S is 0.93 and, according to Raoult's Law, the vapor pressure of the solvent is $(0.93)(6.5) = 6.04$ mm Hg.

8. The number of moles of M and N are 0.17 and 0.11, respectively. The mole fractions are 0.61 and 0.39, respectively. Thus the vapor pressure of M and N are $(0.61)(6) = 3.66$ mm Hg and $(0.39)(4.3) = 1.68$ mm Hg, respectively.

9. $P_A = 730 X_A$, $P_B = 410 X_B$, and $P_A + P_B = 620$. In addition, $X_A + X_B = 1$. Solving these equations for X_A yields $X_A = 0.66$ and $X_B = 0.34$.

10. The increased order of the water molecules causes a negative deviation from Raoult's Law.

11. The mole fraction of Q is 0.012. Thus the molarity is 0.665, which is equivalent to 60 g/l. Thus the molecular weight is 90.22.

12. Their boiling points.

13. Calculate P/c, in which c is expressed as g/l, and extrapolate to $c = 0$ to yield $(P/c)_0 = 1.80 \times 10^{-4}$ l atm g^{-1}. Then $M = RT(P/c)_0^{-1} = 1.41 \times 10^5$.

14. First, $P_{A,pure} - P_A = X_B P_{A,pure}$; thus $105 - 93 = X_B(105)$ or $X_B = 0.114$. Then, $X_B = (w_B/M_B)/[(w_A/M_A) + (w_B/M_B)] = (1/M_B)/[(10/M_A) + (1/M_B)]$, or $M_B/M_A = 0.78$.

15. (a) At a pressure of 10 atm, $(5)(3.28) = 16.4$ mg of H_2 will be dissolved per liter, or $0.0164/2.016 = 0.0081\ M$.
    (b) The mole fraction of the H_2 at 1 atm is 1.46×10^{-5}. Thus, $K_{H_2} = 1$ atm$/1.46 \times 10^{-5} = 6.85 \times 10^4$ atm.

16. At high pressure, more gas—especially N_2—dissolves in the blood. If the pressure is reduced rapidly, the blood becomes supersaturated and releases the gas in the form of bubbles. If the pressure is reduced slowly, the gas is slowly released into the lungs.

17. When the cold soda water reaches the stomach, it is warmed. Since the solubility of a gas decreases as the temperature of a solution is increased, the CO_2 comes out of solution.

18. 120 mm Hg.

19. There is no Cl_2 in the atmosphere, so that Cl_2 leaves the water and enters the atmosphere in an attempt to achieve the correct partial pressure. The atmosphere is so great in volume that this partial pressure is never reached and Cl_2 continues to leave the water until none remains in solution.

20. The concentration of CO_2 in soda water is higher than that which is in equilibrium with an atmosphere of pure CO_2 at a pressure of one atmosphere. Therefore it must be kept under high pressure to maintain this concentration.

21. The value of K_{frz} is proportional to the square of the freezing temperature and inversely proportional to the enthalpy of fusion, neither of which are the same for all molecules. The freezing point depends on the internal cohesion between the individual molecules, and this varies widely with the polarizability and dipolar structure, as described in Chapter 3. The value of ΔH_{fus} depends on the amount of internal energy (vibrational, rotational, and so forth) that can be taken up by a molecule, and this also varies from one molecule to the next.

22. (a) A = solid; B = liquid; C = gas.
    (b) X is the decrease in the freezing point.
    (c) Y is the increase in the boiling point.

23. 0.0114 moles of solute are dissolved in 0.194 kg of solvent, so the molality of the solute is $0.0114/0.194 = 0.0588$. The freezing point depression should be $(0.0588)(18.7) = 1.1°C$. Since the observed value is exactly half the expected value, a reasonable explanation is that the solute forms dimers in this particular solvent.

24. If the protein and the NaCl were acting independently, the freezing point depressions would be additive and T_{frz} would be $-0.074 - 0.09 = -0.164$. Note that $-0.12 \approx -0.074 + \frac{1}{2}(-0.09)$; that is, the protein is making only a half-contribution. Thus, the molarity of the protein must be lower than $0.02\ M$. Since the total concentration is the same but the molarity seems to be half, it is likely that the protein dimerizes in the presence of $0.02\ M$ NaCl.

25. (a) The molality is 0.27, so the freezing point depression should be $(0.27)(14.1) = 3.83°C$. It is less, suggesting that the effective molality is less than 0.27.
    (b) The effective molality is $2.374/14.1 = 0.168$. If m and d are the numbers of grams of monomer and dimer, respectively, then $(m/95) + (d/190) = 0.168$. However, $m + d$ must equal 25.8 g. Solving this pair of equations for m and d yields $m = 6.12$ g and $d = 19.68$ g. Thus the fraction of A that is a dimer is

19.68/25.8 = 0.76. This is a useful test for dimers.

26. (a) Three.
    (b) Two, because after choosing the concentration of one component, the concentration of the second component determines the concentration of the third.
27. (a) The total molar concentration will be halved, so that, assuming an ideal solution, the osmotic pressure will also be halved and the level of solution I will be 25 cm above that of solution II.
    (b) The molar concentration would be zero, so the levels would be equal.
28. Water passes through the membrane into the PEG compartment and the PEG dissolves. Since the membrane is impermeable to the PEG, the water continues to flow into the compartment until no water remains on the water side of the membrane.
29. Water passes through the membrane into the sucrose compartment. The sucrose dissolves. Because the membrane is permeable to sucrose, the sucrose molecules pass through to the water side. However, sucrose molecules are larger than the water molecules, so water will flow to the sucrose side more rapidly than the sucrose will pass to the water side. Thus the liquid level on the sucrose side will at first rise to a higher point than the water level, although at equilibrium the two levels will be equal.
30. (a) The total activity of all dissolved material is 0.3 M. Substituting this into Equation 23 yields a freezing point depression of $-0.54°C$ from $0°C$.
    (b) Use Equation 43. Note that if $R = 0.0821$, the molar volume $\bar{V}_A$ must be in liters, or, 0.018. Thus the activity of water is 0.995.
31. Each solution is diluted in half and contributes 1.2 atm (for A) and 2.3 atm (for B). The total osmotic pressure is a result of the total solute concentration so that it is the sum of 1.2 and 2.3, or 3.5 atm.
32. Use Equation 32 to calculate concentration; then, the molecular weight = 9622.
33. (a) 10^{-2} M $MgSO_4$ + 1 M KCl.
    (b) No, because there may be concentration dependence. The solution should be diluted several times with 10^{-2} M $MgSO_4$ + 1 M KCl.
34. From Equation 32, $\mathbf{M} = 0.0368$. Thus the molecular weight is 2273.
35. Calculate the molecular weight for each pair of numbers; the values are 5600, 5700, 5800, and 5900 for 100, 200, 300, and 400 mg, respectively. Extrapolating to zero concentration yields a molecular weight of 5500.
36. At equilibrium, for water, $\Pi = RT[\text{glucose}]$. The pressure exerted by a column of water of height h is $\rho g h$, in which ρ is the density of water. Equating the pressures yields $h = RT[\text{glucose}]/\rho g$. Being careful about units, we have [glucose] = 1.5×10^{-4} mol cm^{-3}, so that $h = 3.74 \times 10^3$ cm or 37.4 m.
37. From the solution to the preceding problem, $\Pi = \rho g h = RTc/M$ or $M = RTc/\rho g h$. Substituting $h = 0.6$ cm and $c = 10^{-3}$ g cm^{-3}, we have $M = 42{,}000$.
38. The addition of sugar reduces the mole fraction of water and therefore lowers the chemical potential. Water will therefore move down its chemical potential gradient from I to II, raising the level of the liquid in II, causing the floating object to rise. This flow continues until the osmotic pressure equalizes the chemical potential. Thus, in order to recover the object, one must be sure to add enough sugar.
39. Solution II is hypertonic with respect to solution I. Remember, it is only the concentrations of the impermeant substances that determine the tonicity.
40. In distilled water the vesicles rapidly swell. Pores in the membrane expand and X passes through the stretched pore. Salts and other substances leak out of the vesicle until a stable osmotic situation results and X cannot leak out. In 0.1 M NaCl the swelling does not occur.
41. If the final concentration of P in water is p, it is 18.6p in benzene. Thus there are 0.01p moles of P in the water and $\frac{1}{2}(18.6)p = 9.3p$ moles in the benzene. The initial number of moles was 0.0001. Since $9.3p + 0.01p = 0.0001$, the concentration of P is 1.07×10^{-5} M in water and 2×10^{-4} M in benzene.

42. Let [Y] and 3.62[Y] be the molar concentrations in water and benzene, respectively. The number of moles of Y in water and benzene are 0.057[Y] and 0.362[Y]; the sum of these values must equal the total number of moles of Y added, or, 5.205/243 = 0.0214. Thus, the concentrations are 0.511 M and 1.85 M in water and benzene, respectively, and the number of moles of Y in the benzene is 0.185.

43. (a) The concentration in $CHCl_3$ will be 0.0034 M. Since 0.0999 moles are removed, the volume of $CHCl_3$ must be 0.0999/0.0034 = 29.38 cm^3.

44. The molarity of X in water is (3.5/143)(1000/125) = 0.196 M. In $CHCl_3$, the molarity is 0.322 M. Thus, the partition coefficient is $[X]_{CHCl_3}/[X]_{H_2O}$ = 0.322/0.196 = 1.64.

45. The concentrations are so low that all activity coefficients are nearly 1, so there is no problem. The exception is the physical study of macromolecules, because their great size affects the value of even at low concentration.

46. Use Equation 50. Then the value of γ_c is (0.72)[1.16 − (0.001)(2)(356)/0.998] = 0.32. This value is much lower than γ_c, because the molality is much higher than 2.

47. (a) The value of a_A = 81.5/86.2 = 0.945.
    (b) The number of moles of the solvent is 1000/85 = 11.76. Thus the mole fraction is 11.76/11.96 = 0.983. Thus γ = 0.945/0.983 = 0.961.

48. The equilibrium constant K is a true constant only when activities and not concentrations are used.

49. No. There may be both positive and negative deviations from Raoult's Law, in which case γ will be > 1 and < 1, respectively.

CHAPTER 8

1.

	ν_+	ν_-	ν	c_+
Na acetate	1	1	2	$(1^1 \cdot 1^1)^{1/2}c = c$
$CuSO_4$	1	1	2	$(1^1 \cdot 1^1)^{1/2}c = c$
$Cu(NO_3)_2$	1	2	3	$(1^1 \cdot 2^2)^{1/3}c = 1.59\ c$
$Al_2(SO_4)_3$	2	3	5	$(2^2 \cdot 3^3)^{1/5}c = 2.55\ c$

2. There are three ions, so that $\nu_+ = 2$, $\nu_- = 1$, and $\nu = 3$; $R = 8.314 \times 10^{-3}$, so $\mu_e = \mu°$ −13.679 + 7.432 ln $\gamma_\pm$.

3. They are: for 0.08 M NaCl, I = 0.08; for 0.05 M $MgCl_2$, I = 0.15; for 0.02 M Na_2SO_4, I = 0.06; for 0.05 M $Fe_2(SO_4)_3$, I = 0.75.

4. (a) The concentrations of each ion are 0.016 M Na^+, 0.007 M Mg^{2+}, 0.024 M Cl^-, and 0.003 M SO_4^{2-}. Thus

   $I = \frac{1}{2}(0.016 \cdot 1^2 + 0.007 \cdot 2^2 + 0.024 \cdot 1^2 + 0.003 \cdot 2^2)$

   = 0.04 M.

   (b) Since $-\log \gamma_\pm = 0.509\ z^2\sqrt{I}$ and I = 0.04 from part (a), $\gamma_\pm$ = 0.80 and 0.41 for the monovalent and divalent ions, respectively.

5. Probably 0.5 M $CaCl_2$, because it has more ions per mole. However, one cannot say this for sure because, as the concentration increases, the activity coefficient becomes a characteristic of the individual ion and can no longer be calculated from the Debye-Hückel theory.

6. (a) The charge is surrounded by an ion cloud that reduces the external electric field.
   (b) Thermal motion disrupts the ion cloud.
   (c) The charged object attracts the ions more strongly.
   (d) There are more ions to form the ion cloud.
   (e) The value of κ increases with increasing ionic strength. Therefore it is smaller for MCl_2. This is because the increased charge of an ion implies tighter binding.

7. They are the same because the Debye length κ depends on ionic strength.

8. Remember that the constant 0.509 is for 25°C and is proportional to $(1/T)^{3/2}$. Therefore, for 30°C and 37°C the constant is 0.496 and 0.479, respectively.

9. From Equation 15, $\gamma'_\pm$ = 0.64.

10. (a) First, calculate $[H^+]$:

    $[H^+] = [(0.05)(1.8 \times 10^5)]^{1/2}$

    $= 9.4 \times 10^{-4} = I$.

    From the limiting law, $\gamma_\pm$ = 0.96. Thus

$\alpha = (1/0.96)(K/c)^{1/2} = 1.037(K/c)^{1/2} = 0.0197$ and $[H^+] = (0.0197)(0.05) = 9.85 \times 10^{-4}\, M$ and pH = 3.01.

(b) The only change from part (a) is that $I = 0.05$, so that $\gamma_\pm = 0.77$. Thus $\alpha = 1.30(K/c)^{1/2} = 0.025$, $[H^+] = 1.2 \times 10^{-3}\, M$, and pH = 2.91.

11. The coefficient is 0.67.
12. The solubility product is $(3 \times 10^{-4})^2 = 9 \times 10^{-8}$. To achieve a solubility of 6×10^{-4}, $\gamma_\pm$ must be $(1/c_{sat})K_{sp}^{1/2} = \frac{1}{2}$. From the limiting law, $\log \frac{1}{2} = -0.509 z^2 I^{1/2}$, so that I would have to be 0.35. Thus 0.35 M KCl should be added. Of course this value is not accurate because the limiting law does not apply for such large values of I.
13. It will increase because there is less shielding.
14. The protein structure is probably maintained in part by electrostatic interactions. These are neutralized in 1 M NaCl, so that the protein changes shape and the binding site is altered.
15. The undissociated insoluble material is in equilibrium with the dissolved ions. As the ionic strength increases by addition of an inert electrolyte, the ions of the sparingly soluble salt that are in solution attract one another less strongly, so that the equilibrium is shifted in the direction of increased solubility.
16. It should decrease as thermal motion disrupts the hydration shell.
17. The dielectric constant of the solvent will be lower, so the negative charge on the protein will attract the Na^+ ion more strongly and the average number of bound Na^+ ions will increase.
18. The radius of the hydrated Na^+ ion is greater than that of the less hydrated K^+ ion so that it has more difficulty passing through the pores of the membrane. This is an important consideration when one is removing ions from solutions of proteins and nucleic acids by their selective passage through a semipermeable membrane—a process called *dialysis*.
19. The RNA polymerase can bind to DNA by ionic and nonionic bonds. Only nonionic binding occurs within cells.
20. That they are electrostatic forces.
21. This polymer is highly hydrated and binds water more effectively than many proteins. At high concentrations, the polymer binds so much water that the molar concentration of water is reduced to the point that there is an insufficient amount of water to maintain the hydration layer of the proteins; the proteins need the hydration in order to remain in solution.

CHAPTER 9

1. Antilog$(-3.86) = 1.38 \times 10^{-4}$.
2. From Equation 15, $K_a' = (0.02)(10^{-4})^2/1 = 2 \times 10^{-10}$. Therefore $pK_a' = 9.7$.
3. $[H^+] = 10^{-3}$, $K_a' = 1.86 \times 10^{-5}$. Recall that $[H^+] = [Ac^-]$. Thus, if c is the desired molarity, $(10^{-3})(10^{-3})/(c - 10^{-3}) = 1.86 \times 10^{-5}$, and $c = 0.054$.
4. pH = 2, pOH = 14 − 2 = 12, and $[OH^-] = 10^{-12}\, M$. Alternatively, $[H^+][OH^-] = 10^{-14}$, so $[OH^-] = 10^{-14}/0.01 = 10^{-12}\, M$.
5. $pK_b' = 14 - pK_a' = 3.2$. Hence $K_b' = 6.3 \times 10^{-4}$.
6. Let $x = [H^+] = [Ac^-]$. $K_a' = 1.86 \times 10^{-5}$. Thus, $x^2/(0.1 - x) = K_a'$, so that $x = 1.36 \times 10^{-3}$, and pH = 2.86.
7. $[NH_4^+][OH^-]/[NH_3] = 1.79 \times 10^{-5}$. Thus $[OH^-] = 1.34 \times 10^{-3}\, M$. $[H^+] = 10^{-14}/[OH^-] = 7.5 \times 10^{-12}\, M$, so that pH = 11.1.
8. pH $= \frac{1}{2}(pK_a' - \log c_a) = \frac{1}{2}(9.24 + 1) = 5.12$.
9. $K_a' = 10^{-7}$, so that $[H^+] = 7.1 \times 10^{-5}\, M$. Therefore, $\alpha = 7.1 \times 10^{-5}/5 \times 10^{-2} = 0.0014$.
10. pH $= \frac{1}{2}(pK_a' + 14 + \log c_b) = \frac{1}{2}(4.73 + 14 - 2) = 8.34$.
11. The mixture contains 0.005 mole of formic acid and 0.001 mole of NaOH. Therefore there is 0.001 mole of sodium formate and 0.004 mole of formic acid. The final volume is 105 cm^3, so that the concentrations are 0.0095 M sodium formate and 0.038 M formic acid. Thus, from Equation 18, pH = 3.77 + log(0.0095/0.038) = 3.17.
12. $K_b' = 4.37 \times 10^{-4}$, and $K_a' = 10^{-14}/4.37 \times 10^{-4} = 2.29 \times 10^{-11}$, and $pK_a' = 10.64$. As in Example 9-E, the concentration of the

protonated and nonprotonated forms are 0.05 M and 0.1 M, respectively. Thus, pH = 10.64 + log(0.1/0.05) = 10.94.

13. Only Tris buffer. Acetate fails to buffer in the pH range desired; and the citrate and phosphate salts of calcium are poorly soluble and would precipitate.

14. The pK values are 2.35 and 9.78 so the pH at the isoelectric point is $\frac{1}{2}$(2.35 + 9.78) = 6.06.

15. At pH = 6, [H$^+$] = 1.0 × 10^{-6}, so that 1.0 × 10^{-6} equivalents are excreted.

16. They all require the same amount.

17. The relevant pK_a' is 4.602 for which $\Delta H° = -2.85$ kJ mol^{-1}. From Equation 36, ΔpK_a' = 0.02. From Equation 18 pH is 4.95 + 0.02 = 4.97.

18. (a) pK_a' = 3.77. From Equation 18 pH = 3.77 + log(0.05/0.1) = 3.47.
    (b) The buffering capacity is obtained from Equation 27, with c_A = 0.1 M, c_S = 0.05 M, K_a' = 1.7 × 10^{-4}, and [H$^+$] = 3.4 × 10^{-4} M; its value is 0.077.
    (c) One cm^3 of 1 M HCl contains 0.001 mole; this is equivalent to 0.01 mole per liter. The buffering capacity is the number of moles of added acid that would change the pH by one unit. Thus 0.077 = 0.01/ΔpH, or ΔpH = 0.01/0.077 = 0.13. The pH would then be 3.47 − 0.13 = 3.34.

19. The buffering capacity required is 0.01/0.2 = 0.05. The pH of the buffer is 4, so that [H$^+$] is 10^{-4} M; K_a' = 1.7 × 10^{-4}. From Equation 27, we can solve for c_A + c_S to obtain the value c_A + c_S = 0.093. Equation 18 gives us the ratio c_S/c_A, namely, 4.00 = 3.77 + log(c_S/c_A) or c_S/c_A = 1.70. Thus c_A = 0.034 M and c_S = 0.059 M.

20. As shown in Example 9-C, $I = c_S$. Therefore the maximum value of c_S is 0.02. From Equation 18, 5.0 = 4.8 + log(c_S/c_A), so that c_A = 0.013. The buffering capacity is obtained from Equation 27, by using K_a' = 1.8 × 10^{-5} and [H$^+$] = 10^{-5}, and is 0.017.

21. Prepare a solution that is 0.05 M HAc, 0.05 M KAc, and 0.95 M KCl.

22. The first dissociation gives [H$^+$] = [HSO$_4^-$] = 0.2 M. If x is the concentration of SO$_4^{2-}$ that is produced in the second dissociation step, [H$^+$] = 0.2 + x, [HSO$_4^-$] = 0.2 − x, and [SO$_4^{2-}$] = x. Since $(0.2 + x)x/(0.2 − x) = 1.3 × 10^{-2}$, x = [SO$_4^{2-}$] = 0.012 M, [H$^+$] = 0.212 M and [HSO$_4^-$] = 0.188 M.

23. Use $K_1' = 7.5 × 10^{-3}$. If x = [H$_2$PO$_4^{2-}$] = [H$^+$], then $x^2/(0.01 − x) = 7.5 × 10^{-3}$ and x = $5.7 × 10^{-3}$ M, so that pH = 2.25.

24. A pH value of 9.50 corresponds to [OH$^-$] = 10$^{-4.50}$. If V is the volume of added 0.1 M HCl, then [CH$_3$NH$_3^+$] + [CH$_3$NH$_2$] = 0.2{50/(50 + V)}, [NH$_4^+$] + [NH$_3$] = 0.1{50/(50 + V)}, and [Cl$^-$] = 0.1{V/(50 + V)}. We also know that [NH$_4^+$] [OH$^-$]/[NH$_3$] = 10$^{-4.75}$, so that [NH$_4^+$]/[NH$_3$] = 10$^{-4.75+4.50}$ = 0.56. By similar reasoning [CH$_3$NH$_3^+$]/[CH$_3$NH$_2$] = 10$^{-3.26+4.50}$ = 17.4. The final equation needed to solve the set of simultaneous equations is the one indicating the equality of charges, namely, [NH$_4^+$] + [H$^+$] + [CH$_3$NH$_3^+$] = [Cl$^-$] + [OH$^-$]. Solving these equations for V yields V = 112 cm^3.

25. Glycine is dipolar in the solid, and the dipoles attract one another strongly, bringing the molecules nearer to one another than they are in crystals of the much less polar substance glycolamide.

26. At the isoionic point. This is because at this pH there are two charges, each drawing water molecules near. This is the same phenomenon as the electrostriction of water by MgSO$_4$, described in Chapter 8.

27.

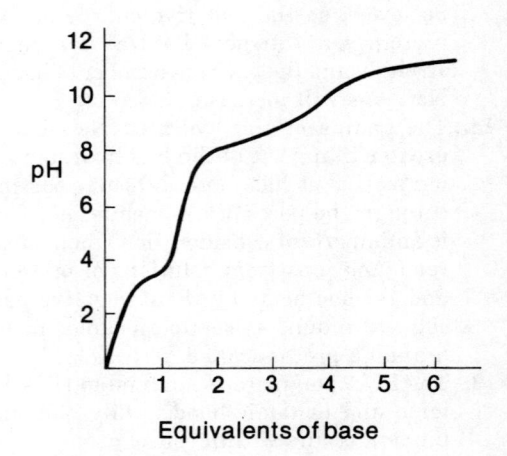

ANSWERS TO PROBLEMS

28.

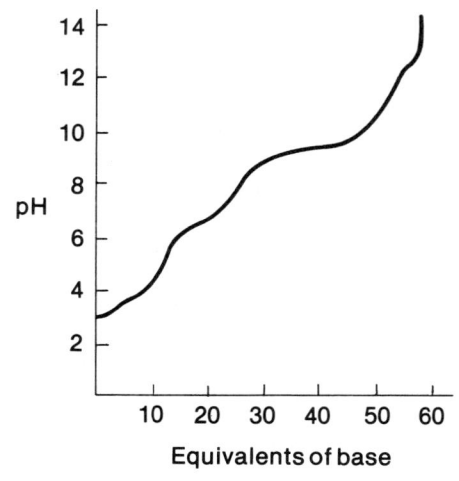

[Graph: pH vs Equivalents of base, x-axis from 10 to 60, y-axis from 2 to 14, showing a titration curve rising from about pH 3 to pH 13]

29. Four histidines are not titrated and are probably buried within the protein molecule.
30. Only four lysines are seen, suggesting that the other four are buried. The addition of HCHO eliminates all amino acids with $pK = 10$, confirming that four of the lysines are external.
31. $[CaEDTA] \approx 0.003$ M and $[EDTA] = 0.01 - 0.003 = 0.007$ M. Therefore, as in Example 9–J, $[Ca^{2+}] = 0.003/(0.007 \times 10^{16}) = 4 \times 10^{-17}$ M.
32. EDTA binds lead ions more tightly than Ca^{2+} ions. Thus the Ca^{2+} ions are displaced and PbEDTA is excreted.
33. No, it is monodentate because only a single oxygen atom can coordinate. A chelating agent, by definition, is at least bidentate.
34. Yes. When ionized, two of the oxygens have free electron pairs. Thus it can form a ring compound, as shown below (M = metal ion).

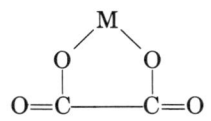

[Structure: M bonded to two O atoms, each connected to C=O groups forming a ring]

35. Metal ions are probably required to stabilize the structure.
36. DNA has many oxygen atoms in the sugar-phosphate backbone that can chelate metal ions. These ions will certainly affect the density, since CsDNA and MgDNA will have different densities. In the presence of EDTA all contaminating ions will be eliminated and all of the DNA will be CsDNA.

CHAPTER 10

1. (a) $1.180 - 0.13 = 1.05$ V. (b) $-0.799 - (0.337) = -0.462$ V. (c) $0.763 - 2.375 = -1.612$ V. (d) $0.763 - (-0.799) = 1.562$ V.
2. (a) $\mathscr{E}° = 0.44$ V. (b) $\mathscr{E}° = 0.64$ V. (c) $\mathscr{E}° = 1.58$ V. Each reaction is spontaneous.
3. MnO_4^- is a powerful oxidizing agent. The SO_4^{2-} ion cannot be oxidized, so that the only possible oxidation is $Cu^+ \rightarrow Cu^{2+} + e^-$, for which $\mathscr{E}° = -0.153$ V. The other half-reaction is $Mn^{2+} + 4H_2O \rightarrow MnO_4^- + 8H^+ + 5e^-$, for which $\mathscr{E}° = -1.51$ V. The oxidation takes place at the anode and the reduction at the cathode; the standard cell potential is $-0.153 - (-1.51) = 1.357$ V. The electrons flow through the connecting wire from the Pt electrode in the $CuSO_4$ solution to the Pt electrode in the $KMnO_4$ solution. The cell reaction is $5Cu^+ + MnO_4^- + 8H^+ \rightarrow 5Cu^{2+} + Mn^{2+} + 4H_2O$.
4. $Ag \rightarrow Ag^+ + e^-$, with $\mathscr{E}° = -0.799$, and $Cu^{2+} + 2e^- \rightarrow Cu$, with $\mathscr{E}° = 0.337$. The standard potential for the reaction, $2Ag + Cu^{2+} \rightarrow 2Ag^+ + Cu$, is $\mathscr{E} = -0.462$ V. Two electrons are transferred. To determine the voltage at the given concentration, we use the Nernst equation,

$$\mathscr{E} = \mathscr{E}° - \frac{0.0256}{2} \ln \frac{[Ag^{2+}]^2[Cu]}{[Ag]^2[Cu^{2+}]}$$

$$= -0.462 - 0.0128 \ln \frac{(0.2)^2(1)}{1^2(10^{-3})}$$

$$= -0.509 \text{ V} .$$

This is less than zero, so that the reaction is not spontaneous.

5. When written as oxidation reactions, the potentials are 0.440 and -1.360 V. The total reaction will be $Fe + Cl_2 \rightarrow Fe^{2+} + 2Cl^-$. The cell will have a voltage of 1.800 V, and the electrons flow from the Fe anode to the $Pt|Cl_2$ cathode.

6. (a) $Cr \rightarrow Cr^{3+} + 3e^-$ and $Cu^{2+} + 2e \rightarrow Cu$.
   (b) $0.74 + 0.34 = 1.08$ V.
   (c) $K = [Cr^{3+}]^2/[Cu^{2+}]^3$. Note that Cr and Cu do not appear in the expressions because they are solid.
   (d) The relevant equation is $\ln K = n\mathscr{F}\mathscr{E}°/RT$. To determine n, the number of electrons involved in the reaction, we must rewrite the half-reactions so that they add to give the balanced equation: that is, $2Cr \rightarrow 2Cr^{3+} + 6e^-$ and $3Cu^{2+} + 6e^- \rightarrow 3Cu$. This indicates that $n = 6$. Therefore $\ln K = 6(96500)(1.08)/(8.314)298 = 252.4$. Therefore $K = e^{252.4}$.

7. (a) The cell is a concentration cell, for which $n = 2$, since the half-reactions involve two electrons. The voltage $\mathscr{E}$ is $-\frac{1}{2}(0.0257)\ln(0.01/0.5) = 0.05$ V.
   (b) The electrode in the 0.01-M solution will dissolve and more Zn will be deposited on the electrode in the 0.5-M solution.
   (c) The voltage is zero and the concentrations in each cell are the same, namely 0.255 M.

8. (a)
$$Cd^{2+} + 2e^- \rightarrow Cd,$$
$$\mathscr{E}° = -0.403$$
$$Ag + Cl^- \rightarrow AgCl + e^-,$$
$$\mathscr{E}° = -0.222$$

$$Cd^{2+} + 2e^- + 2Ag + 2Cl^- \rightarrow$$
$$Cd + 2AgCl + 2e^-$$

   (b) $-0.403 + (-0.222) = -0.625$ V.
   (c) 2.
   (d) $\mathscr{E} = \mathscr{E}° - (RT/2\mathscr{F})\ln\{[Cd][AgCl]^2/[Cd^{2+}][Ag]^2[Cl^-]^2\} = -0.802$ V.
   (e) $Cd + 2AgCl \rightarrow Cd^{2+} + 2Ag + 2Cl^-$.

9. (a)
$$H_2 \rightarrow 2H^+ + 2e^-$$
$$AgCl + e^- \rightarrow Ag + Cl^-$$

   (b) Adding the half-reactions to generate a balanced equation yields $2AgCl + H_2 + 2e^- \rightarrow 2Ag + 2H^+ + 2Cl^- + 2e^-$, indicating that *two* electrons participate.
   (c) Neither $\mathscr{E}°$ nor $\Delta G°$ are affected by the electrode material if it is inert, since $\mathscr{E}°$ and $\Delta G°$ are functions of the chemical reaction that occurs in the cell.

10. +2.12 V.

11. The half-reactions are $Zn \rightarrow Zn^{2+} + 2e^-$ and $2MnO_2 + NH_4^+ + 2e^- \rightarrow Mn_2O_3 + NH_3 + OH^-$. Thus one electron is transferred *per molecule* of MnO_2 consumed or one faraday per mole. Two gram of $MnO_2 = 0.023$ mole. Thus, $(0.23)(96500) = 2218$ coulomb are transferred. The current, 5 milliampere, equals 5×10^{-3} coulomb per second. Thus the light would be on for $2218/(5 \times 10^{-3}) = 4.44 \times 10^5$ sec $= 5.1$ day.

12. We are asked for the composition at equilibrium, so we must calculate the equilibrium constant. From $\ln K = n\mathscr{F}\mathscr{E}°/RT$, we obtain $K = 1.6 \times 10^{37}$. Since the cell reaction is $Zn + Cu^{2+} \rightarrow Zn^{2+} + Cu$, then $1.6 \times 10^{37} = [Zn^{2+}]/[Cu^{2+}]$. If we have a standard cell, the large value of K shows that $[Zn_2^+] \approx 2$ M at equilibrium. Therefore $[Cu^{2+}] = 2/(1.6 \times 10^{37}) = 1.3 \times 10^{-37}$ $M \approx 0$.

13.

ΔG	$\mathscr{E}$	Reaction
> 0	< 0	Spontaneous
$= 0$	$= 0$	Equilibrium
< 0	> 0	Not spontaneous

14. $\Delta G° = n\mathscr{F}\mathscr{E}° = 69.48$ kJ mol^{-1}; $\Delta H° = -n\mathscr{F}[\mathscr{E}° - T\,d\mathscr{E}/dT] = 62.12$ kJ mol^{-1}; and $\Delta S° = n\mathscr{F}\,d\mathscr{E}°/dT = 24.70$ J K^{-1} mol^{-1}.

15. The half-reactions and the potentials are

$$Zn \rightarrow Zn^{2+} + 2e^-, \quad (\mathscr{E}° = 0.763 \text{ V})$$

and

$$2Cu^+ + 2e^- \rightarrow Cu, \quad (\mathscr{E}° = -(-0.521) = 0.521 \text{ V});$$

the concentrations are 1 M and the temperature is 25°C; we have standard conditions, so that 1.284 V is the potential of the cell. Two electrons are transferred, so that $n = 2$ and $\Delta G = -n\mathscr{F}\mathscr{E} = -2(96500)(1.284) = -247.8$ kJ mol^{-1}.

16. Use the equation $\mathscr{E}°'(pH = x) = -0.06x + \mathscr{E}°$. Of course, for a reaction in which the H$^+$ ion does not participate, $\mathscr{E}° = \mathscr{E}°'$(pH 7). Thus the values of (a) and (b) are unchanged. For (c), (d), and (e), the values are $0.318 - 0.420 = -0.102$ V, $0.224 -$

0.420 = −0.196 V, and 0.433 − 0.420 = 0.013 V, respectively.

17. The half-reactions are

$$Fe(CN)_6^{4-} \rightarrow Fe(CN)_6^{3-} + e^-,$$
$$\mathscr{E}° = -0.356 \text{ V}.$$
$$2H_2O + O_2 + 4e^- \rightarrow 4OH^-,$$
$$\mathscr{E}° = 0.401 \text{ V}.$$

Thus, $\mathscr{E}° = 0.045$ V. Inasmuch as $\mathscr{E}° = (0.0257/n)\ln K$, we have $K = 1.1. \times 10^3$.

18. The reaction is $AgCl \rightarrow Ag^+ + Cl^-$, which is the sum of the half-reactions $AgCl + e^- \rightarrow Ag + Cl^-$, with $\mathscr{E}° = 0.222$ V and $Ag \rightarrow Ag^+ + e^-$, with $\mathscr{E}° = -0.799$ V. For the total reaction, $\mathscr{E}° = 0.222 - 0.799 = -0.577$ V. One electron is transferred so that from $\ln K = \mathscr{F}(-0.577)/RT$, $K = 1.74 \times 10^{-10}$.

19. First we must calculate $\mathscr{E}°_{cell}$. Write the half-reactions as $Pb \rightarrow Pb^{2+} + 2e^-$, with $\mathscr{E}° = -0.126$, and $2AgCl + 2e^- \rightarrow 2Ag + 2Cl^-$, with $\mathscr{E}° = 0.222$. To get the cell reaction, we must reverse the Pb half-reaction, so that $\mathscr{E}° = +0.126$. Thus $\mathscr{E}°_{cell} = 0.222 + 0.126 = 0.348$. Now $a_{PbCl_2} = a_\pm^3$, since there are three ions in the molecule; $a_\pm^3 = \gamma_\pm^3 (c_{Pb})(2c_{Cl^-})^2 = 4\gamma_\pm^3 c^3$, in which $c = 0.05$ M. Since two electrons are transferred,

$$\mathscr{E}_{cell} = \mathscr{E}°_{cell} - \frac{RT}{2\mathscr{F}} \ln a_{PbCl_2}$$

$$= \mathscr{E}°_{cell} - \frac{RT}{2\mathscr{F}} \ln 4\gamma_\pm^3 c^3,$$

or

$$0.522 = 0.348 - \frac{RT}{2\mathscr{F}} \ln 4\gamma_\pm^3 c^3,$$

so that $\gamma_\pm = 0.14$.

20. The anode reaction has $\mathscr{E}° = -0.377$ V. The cathode (calomel) has $\mathscr{E}° = 0.280$ V. $\mathscr{E} = 0.020$ V, then $\mathscr{E}_{cathode} = 0.280 - 0.020 = 0.260$ V. Thus $\mathscr{E}_{anode} = -0.260$ V. According to the Nernst equation $-0.260 = -0.337 - (0.0257/2)\ln[Cu^{2+}/[Cu]$ or $[Cu^{2+}] = 0.0025$ M.

21. The voltage would be calculated as $(RT/\mathscr{F})\ln(2.63/0.263)$, if the ion were Hg^+, and would be $(RT/2\mathscr{F})\ln(2.63/0.263)$, if it were Hg^{2+}. These values are 0.0577 V and 0.0289 V, respectively, so that the ion is Hg_2^{2+}.

22. The difference $0.3243 - 0.3111 = 0.0132 = (RT/2\mathscr{F})\ln(0.01)/[Ca^{2+}]$, so that $[Ca^{2+}] = 0.00348$ M.

23. The cation is M^{3+}, so that $n = 3$. The atomic weight of M is $3(2.05)(96500)/2400 = 247.3$.

24. The half-reactions are $2H_2O \rightarrow O_2 + 4H^+ + 4e^-$ and $Fe^{2+} + 2e^- \rightarrow Fe$. Since four and two electrons are used in producing O_2 and Fe, respectively, the same charge (or current per unit time) releases two moles of Fe per mole of released O_2. Thus the time required to release one half mole of Fe is the time required to release one-fourth mole or eight grams of O_2. That is, $t = (8/2.1)(10) = 38.1$ min.

25. The glass electrode is sensitive to Na^+ ions as well as to H^+ ions and indicates an excessive value of $[H^+]$ when Na^+ ions are present.

26. The pH is 1.03.

CHAPTER 11

1. Zeroth order with respect to D, first order with respect to A and C, and second order in B.
2. (b), (c), and (d) are probably not elementary, since reactions having molecularity greater than two are not common.
3. (a) and (b) are bimolecular; (c), (d), and (e) are trimolecular. Note that the number of products is irrelevant.
4. (a) $d[A_2B]/dt = k[A_2][B]$.
   (b) $d[A_2B_2]/dt = k[A_2][B]^2$.
   (c) $d[ABC]/dt = k[A][B][C]$.
   (d) $d[AB]/dt = k[A_2][B_2]$.
5. (a) $d[A_2B_2]/dt = -d[A_2]/dt$.
   (b) $d[AB]/dt = -2d[A_2]/dt$.
   (c) $d[A]/dt = -d[AB]/dt$.
   (d) $d[B]/dt = -2d[A_2B_2]/dt$.
6. k is always greater than zero, by

definition.
7. $d[C]/dt = k[A][B]^2$.
8. The velocity is proportional to a and independent of b. Without considering complexities, the rate equals $k[A]$.
9. The predicted rate law is $d[C]/dt = [A][B]^2$. The data show, however, that the rate is proportional to a^2 and to b, suggesting a rate law $d[C]/dt = [A]^2[B]$. Thus the hypothesis is incorrect.
10. The molecular weight of NH_2NO_2 is 62.032, so that initially 0.806 mmol were present. The number of moles of N_2O formed, assuming that N_2O is an ideal gas, is $PV/RT = 0.261$ mmol. Thus, $0.806 - 0.261 = 0.545$ mmol NH_2NO_2 remain. For a first-order reaction, the number of moles equals e^{-kt} times the initial number of moles. Therefore $k = (1/t)\ln(n_0/n) = (1/70)\ln(0.806/0.545) = 0.00559$. Also $t_{1/2} = \ln 2/k = 0.693/0.00559 = 124.0$ minutes.
11. (a) The rate equation is $d[A]/dt = -k[A][B] = -k[A]^2$, because [A] always equals [B]. If a is the initial concentration of A, the equation is $1/[A] - 1/a = kt$. For $t = 50$ sec, $[A] = 0.9a$, so that $ka = 2.22 \times 10^{-3}$ sec. To reach half the initial value, $[A] = 0.5a$ and the time takes a value of $t = (1/2.22 \times 10^{-3})(1/0.5 - 1) = 450$ sec.
(b) Since $t_{1/2} = 1/ka$, doubling a will halve $t_{1/2}$, so that it would take 225 seconds to halve the initial value if the concentration were doubled.
12. If this is an elementary reaction and the equation is balanced, the reaction should be second order. Thus $k = x/ta(a - x)$, in which $x = [C]$ at time t and a is the initial concentration of A and of B. Plotting x versus $ta(a - x)$ yields $k = 0.0125$ l mol^{-1} min^{-1}.
13. For a reaction of order n and a single reactant, $-d[A]/dt = k[A]^n$. This can be written as: $\log(-d[A]/dt) = \log k + n \log[A]$. The rate of increase of pressure dP/dt is proportional to $-d[A]/dt$ or $-cd[A]/dt$, in which c is a constant. In a plot of dP/dt versus [Q], in which P is the pressure, the slope of the line is n; the value is $n = 0.92 \approx 1$.

14. (a) A_2, zero; B, first; H^+, first.
(b) $d[AB]/dt = k[B][H^+]$.
(c) About 300 M^{-1} sec^{-1}.
15. k is equal to $(1/t)\ln(a/a - x)$, which equals $(1/20)\ln\{100/(100 - 54.8)\} = 3.97 \times 10^{-2}$ min^{-1}.
16. No, both are third order. This can be shown in the following way: for reaction 1, $d[ABC]/dt = 2k[ABC_2][AB] = 2kK[AB]^2[C_2]$; for reaction 2, $d[ABC]/dt = 2k'[A_2B_2][C_2] = 2k'K'[AB]^2[C_2]$.
17. Since the reaction is second order in A, a simple mechanism would include the reaction $A + A \rightarrow A_2$; this is rate-determining so it must be slow. The net reaction must regenerate an A and be fast—this can be accomplished with the reaction $A_2 + B \rightarrow C + D + A$.
18. The two rates are equal when $k[A] = k'[A][H^+]$ so that $[H^+] = k/k'$ or pH $= -\log(k/k')$. Thus pH $= 1.26$.
19. (a) 25 min;
(b) 8×10^8 cells per ml;
(c) 225 min;
(d) $\ln 2/25$ min $= 0.028$ min^{-1}.
20. First order; $k = 0.01444$.
21. Since $[A] = [B]$, the rate law is $d[A]/dt = -k[A]^2$, which has the solution $1/[A] - 1/a = kt$, in which a is the initial concentration of A. At $t = 2$ min, $[A] = 0.9a$, so that $k = 1.23 \times 10^{-2}$ mol^{-1} sec^{-1}. At 50 percent completion $[A] = 0.5a$, so that $t = 810$ sec.
22. Zeroth order.
23. That $k/k' = [H^+] = 10^{-2}$.
24. (a) (i) An essential component of the reaction has been removed or reduced in concentration. (ii) There are two parallel reactions and the activity with the higher rate has been lost.
(b) There are two parallel reactions. The faster reaction is second order and the slower one is first order. The second order reaction has been lost during the purification.
25. $d[X]/dt = k'[A][B] + 2k''[A_2][B]^2$.
26. (a) The concentration decrease is linear with time, so the reaction is zeroth order.
(b) The reaction probably consists of two steps. In the first step A absorbs light and

is activated to form A*; in the second step A* + B → AB. If the first reaction is slow and its rate is determined entirely by the intensity of the light, and the second reaction is fast, then $d[A]/dt = -d[AB]/dt$ and the reaction will be zeroth order.

27. (a) Step 1.
    (b) Step 1.
    (c) Two.
    (d) $d[B]/dt = -k[A][B]$.
    (e) $d[B]/dt = +k[A][B]$.
28. $k_{-1} \gg k_2$, so that B returns to A more rapidly than B goes to C; $k_3 \gg k_2$ and $k_3 > k_{-2}$, so that C goes to D more rapidly than C returns to B or than C is made.
29. For A it is (ii) because A first appears in this reaction; for AD it is (iii), because this is the only slow step.
30. (a) (i), $d[A_2]/dt = 2(k_2 k_1/k_{-1})([A_3]^2/[A_2])$; (ii), $d[A_2]/dt = 3k[A_3]^2$.
    (b) Measure the initial rate of production of A_2 as a function of the initial concentration of A_2. For mechanism (i), the initial velocity decreases with increasing [A_2]; for mechanism (ii), the initial velocity is independent of [A_2].
31. (a) Since $k_3 \gg k_2$, $d[\phi N_2^+]/dt$ equals $d[ONBr]/dt$ in step 2, which is $k_2[H_2NO_2^+][Br^-]$. Inasmuch as the first step is an equilibrium step, $k_1/k_{-1} = [H_2NO_2^+]/([H^+][HNO_2])$, which, when combined with the preceding expression, yields $d[\phi N_2^+]/dt = (k_1 k_2/k_{-1})[H^+][HNO_2][Br^-]$.
    (b) $d[H_2NO_2^+]/dt = 0 = k_1[H^+][HNO_2] - k_{-1}[HNO_2^+] - k_2[H_2NO_2^+][Br^-]$, which, when combined with $d[\phi N_2^+]/dt = k_2[H_2NO_2^+][Br^-]$ from part (a) yields $d[\phi N_2^+]/dt = (k_1 k_2[H^+][HNO_2][Br^-])/(k_{-1} + k_2[Br^-])$.
    (c) $k_{-1} \gg k_2[Br^-]$.
32. At equilibrium $k_{foreward}/k_{back} = K$, in which k_{back} is the rate constant for the back reaction and K is the equilibrium constant. Therefore $r_{back} = k_{back} K[A][B][C]$. $K = [D][E][F]/[A][B]^3[C]^2$, so that $r_{back} = k_{back}[D][E][F]/[B]^2[C]$.
33. Since $\ln r = -kt$, $\ln 2$ would equal $5670k$, so that k would equal 1.22×10^{-4}. Thus the age of the log is $t = (1/k)\ln(15.3/3.2) = 12{,}825$ years.

34. The boiling point of the water is lower, so that the egg must cook at a lower temperature.
35. 84.64 kJ mol^{-1}.
36. 8.8 kJ mol^{-1}.
37. Since an enzyme is a catalytic protein, one must consider the effect of [NaCl] on the catalytic activity. Frequently at high ionic strength the protein molecule undergoes a shape change that markedly reduces its catalytic activity.
38. Use Equation 70 to calculate the slope of $\log(k'/k'')/(\sqrt{I'} - \sqrt{I''}) = 1.018$ (H)(z). Remember to use $\log_{10}$ and not ln. It will be found that $z = -2$ so that C^{2-} is probably the second reactant.
39. Comparing Equations 62 and 66 one can see that $A = (\mathbf{k}T/h)e^{\Delta S^*/R}$ so that $\Delta S^* = R\ln(hA/\mathbf{k}T) = -30.42$ J K^{-1} mol^{-1}. Since $\Delta S > 0$, the transition state is probably more complex than the component molecules and its formation requires considerable rearrangement.
40. 1, the products, AB and C; 2, the activation complex (ABC)*; 3, the reactants A and BC; 4, the separated atoms, A, B, and C.
41. (a) 1 is the activation energy for the forward reaction or, in terms of collision theory, the collision energy necessary to initiate the reaction.
    (b) 2 is the activation energy for the reverse reaction.
    (c) Endothermic, because a net energy input is required.
42. Clearly the figure shows two consecutive reactions. A represents the initial reactants, B is the transition state, and C is both the product of the first reaction and the reactant for the second reaction. Thus, C is an intermediate, D is the transition state for the second reaction, and E is the product.
43. A = E + S; C = ES; E = EP; and G = E + P.
44. From Equation 67 it can be seen that the term $(r_A + r_B)^2/r_A r_B = 4$ when the two reacting molecules have the same radius. Thus, in this case the rate constant is independent of molecular size. The ratio of the rate constants is then just the recipro-

cal of the ratio of the viscosities. Thus, the reaction in hexane occurs three times as fast as the reaction in water.

45. No, because η depends on temperature.
46. Energy is required to displace the solvent molecules from one another in order that the reactants reach one another. The energy is the cohesive energy between solvent and solute molecules and between solvent and solvent molecules.

CHAPTER 12

1. According to Equation 64 of Chapter 11, the rate of the reaction increases by a factor of $\exp[(75 - 30)/2.477] = 7.76 \times 10^7$.
2. Only the surface of a heterogeneous catalyst is active and this vastly increases the surface-to-mass ratio.
3. The enzyme begins to lose its catalytic activity above 40°C.
4. The Lineweaver-Burk or Eadie-Hofstee plot yields a reliable extrapolated value of V_m by accurate measurements of the initial velocity at low substrate concentrations.
5. The same as [S], or moles per liter.
6. Use Equation 22 and plot $1/V$ versus $1/[S]$. The slope is $0.0081 = K_m/V_m$. The y intercept is 1.61, so that V_m is 0.62. Therefore, $K_m = 0.005$.
7. To A.
8. The value of $V_m = 1.79 \times 10^{-6}$ moles of Q per second, or 6.2 moles of Q per second per mole of enzyme; $K_m = 0.0158\ M$.
9. (a) The velocity is $0.0312/30 = 0.00104$ mol^{-1} sec^{-1}, so that the turnover number is $0.00104/10^{-5} = 104$ sec^{-1}.
   (b) $0.00104 \times 6 \times 10^{23} = 6.24 \times 10^{20}$ molecules converted per second per liter.
10. 5×10^{-6} mol of substrate react per minute when there are $2 \times 10^{-6}/27,000 = 7.4 \times 10^{-11}$ mol of enzyme. The turnover number is $5 \times 10^{-6}/7.4 \times 10^{-11} = 6.76 \times 10^4$ min^{-1} = 1.3×10^3 sec^{-1}.
11. $6.4 \times 10^{-6}/336 = 1.9 \times 10^{-8}$ mol of nucleotides are released per minute by $5 \times 10^{-6}/23,000 = 2.2 \times 10^{-10}$ mol of enzyme. Thus, the turnover number is $1.9 \times 10^{-8}/2.2 \times 10^{-10} = 87.4$ min^{-1} = 1.46 sec^{-1}.
12. (a) The catalytic activity is reduced but the binding of the substrate is the same. Presumably the substrate has greater difficulty reaching the transition state. This is very likely due to a subtle change in shape of the catalytic region of the enzyme. This is an example of *noncompetitive inhibition*.
    (b) If K_m is decreased, binding must be tighter. Thus, R probably causes a change in shape in the region to which the substrate binds. For example, binding often is accompanied by a change in shape of the substrate-binding site; if the change is induced by R, binding of the substrate will be facilitated. Since V_m is unaffected, the rate-limiting step in the reaction, which may be the formation of the transition state or formation of the product from the transition state, is unaffected.
13. The substrate is an inhibitor of the enzyme at high [S]. This is usually because the enzyme has a binding site for the substrate. When the substrate is bound, the enzyme is modified, usually by a small change in shape, so that the catalytic site of the enzyme is less effective.

CHAPTER 13

1. 5.95 cm.
2. $\gamma = \frac{1}{2}\rho ghr$, so that $h = 2\gamma/\rho gr$; $h_1 - h_2 = \Delta h = (2\gamma/\rho g)\,[(r_2 - r_1)/r_1 r_2]$, so that $\gamma = \frac{1}{2}\Delta h \rho g\,[r_1 r_2/(r_2 - r_1)]$.
3. 54.5 dyn cm^{-1}.
4. A substance that raises the surface tension of a liquid must raise the free energy of the surface. Therefore it will not spontaneously concentrate at the surface because ΔG should be less than 0. This fact is expressed quantitatively by the Gibbs isotherm.
5. The one containing the surfactant will spread out and form a much flatter drop-

let.
6. From Equation 3, $2.5 = [2\gamma \cos 10°]/[(0.998 - 0.825)(980)(0.03)]$. Thus, $\gamma = 6.46$ dyn cm^{-1}.
7. There are roughly 10^3 droplets per cm^3 or 10^{18} droplets per km^3. The surface area of a sphere is $4\pi r^2$ or 1.26×10^{-5} cm^2 per droplet, or 1.26×10^{13} cm^2 for the fog. If all of the water had been contained in a single large sphere, the radius of this sphere would be 10^3 cm and its surface area would be 1.26×10^7. Thus, the difference in the area of the large sphere and the surface area of the fog is $1.26 \times 10^{13} - 1.26 \times 10^7 \approx 1.26 \times 10^{13}$ cm^2. The free energy change is $(1.26 \times 10^{13})(72.8) = 9.17 \times 10^{14}$ erg $= 9.17 \times 10^7$ joule. This is a lot of energy but on a molar basis, it is small—4.2×10^9 cm^3 of water is $(1/18)(4.2 \times 10^9) = 2.3 \times 10^8$ mol, so that the free energy change per mole is 9.17×10^7 J/2.3×10^8 mol or 0.40 J mol^{-1}.
8. At an air-water interface, the amino group is in the water and the butyl group will point toward the air. At an air-benzene interface the direction is reversed.
9. (a) The *n*-octyl portion.
   (b) The ring.
   (c) The C-C chain will lie horizontally on the air side.
   (d) The C-C chain.
   (e) The butane portion.
10. The number of molecules in the film are $(8 \times 10^{-6})(6 \times 10^{23})/854 = 5.6 \times 10^{15}$. The cross-sectional area of a square containing one molecule is $18.5/(5.6 \times 10^{15}) = 3.30 \times 10^{-15}$ cm^2 = 33.0 Å^2. If it is cylindrical, the radius is $\frac{1}{2}\sqrt{33.0} = 2.87$ Å and the area is $\pi(2.87)^2 = 25.88$ Å^2.
11. Chemisorption has occurred because there has been a chemical change. Presumably the diatomic molecule is converted to the single atoms when adsorbed. These must migrate on the surface so that the arrangement of the single atoms is randomized. During desorption the single atoms recombine to form diatomic molecules.
12. Desorption of $C^{18}O^{16}O$.
13. A, because the binding is clearly much weaker than to B.
14. (a) Plastic.
    (b) Leucine, because it has a large hydrophobic portion.
15. Use Equation 31 and plot c/x versus c. The values of b and m are 5430 and 5×10^{-5}, respectively. By inserting these in the equation, one obtains 49.9 micromoles.
16. Yes, because in the Langmuir model adsorption and desorption are in equilibrium.
17. (a) $V_m = 1.81$ cm^3. (b) The volume of gas adsorbed per gram of ZnO is $1.81/5 = 0.362$ cm^3. This is $0.362/22,400 = 1.62 \times 10^{-5}$ mol or 9.7×10^{18} molecules. The area taken up by each close-packed molecule is $(7$ Å$)^2 = 4.9 \times 10^{-15}$, so the area per gm of ZnO is 4.8×10^4 cm^2.
18. Follow example 13-E.
    (a) 6.06 cm^3.
    (b) 16.1 kJ mol^{-1}.
    (c) 3.55×10^4 cm^2.
19. Often there are significant changes in shape during the adsorption process. These changes may be irreversible and destroy the activity of the enzyme.

CHAPTER 14

1. The first compartment contains 0.52 M Na$^+$ and 0.52 M Cl$^-$. The other compartment contains 0.58 M Na$^+$, 0.48 M Cl$^-$, and 0.1 M P$^-$.
2. If the subscript I denotes the compartment containing the protein and x is the concentration moving from compartment II to I, then $[A^+]_I = 6 \times 10^{-4} + x$, $[X^-]_I = x$, $[A^+]_{II} = 10^{-2} - x$ and $[X^-]_{II} = 10^{-2} - x$. By Equation 4, $x = 4.85 \times 10^{-3}$ M. Thus $[A^+]_I = 5.45 \times 10^{-3}$ M and $[A^+]_{II} = 5.15 \times 10^{-3}$ M, and by Equation 6, $\Delta\phi = 1.54$ mV.
3. First, calculate the concentration. The molarity of the DNA, expressed as moles of nucleotide per liter, is $1/336 = 2.98 \times 10^{-3}$ M. If x is the decrease in concentration of NaCl in the DNA solution at equilibrium, the concentrations are $[Na^+]_l = 2.98 \times 10^{-3} + 0.01 - x$, $[Cl^-]_l = 0.01 -$

x, $[Na^+]_r = x$, and $[Cl^-]_r = x$. From Equation 4, $x = 5.65 \times 10^{-3}$ M. Thus, the concentrations at equilibrium are $[Na^+]_l = 7.33 \times 10^{-3}$ M, $[Cl^-]_l = 4.35 \times 10^{-3}$ M, $[Na^+]_r = 5.65 \times 10^{-3}$ M, and $[Cl^-]_r = 5.65 \times 10^{-3}$ M; and $[DNA] = 10^{-7}$ M. The osmotic pressure, Π, is $RT\Delta c$ in which Δc is the difference in molar concentrations of all solutes across the membrane, or, $7.33 \times 10^{-3} + 4.35 \times 10^{-3} + 10^{-7} - 5.65 \times 10^{-3} - 5.65 \times 10^{-3} = 3.80 \times 10^{-4}$ M, and $\Pi = (0.082)(298)(3.80 \times 10^{-4}) = 0.0093$ atm.

4. $\Delta\phi = (RT/z\mathscr{F})\ln(0.015/3 \times 10^{-4}) = 49.4$ mV. The potential is lower on the high-concentration side because positive ions have been transferred to the low-concentration side.
5. $\Delta\phi = (RT/z\mathscr{F})\ln(0.1/0.0035) = 42.32$ mV. The potential is higher on the $ZnSO_4$ side because SO_4^{2-} ions are transferred to the other side.
6. (a) The concentration ratio expected for 89 mV is defined by $\Delta\phi = (RT/z\mathscr{F}) \ln (c_{inside}/c_{outside})$ or $c_{inside}/c_{outside} = 28$, as observed. Thus, transport is not against an electrochemical gradient, and there is no need to hypothesize active transport.
(b) This is greater than 28, so there may well be a pump for concentrating K^+ ions.
(c) This is less than 28, so there is probably a pump that *removes* K^+ ions from inside the cell.

CHAPTER 15

1. The molecular weight of glycine is 75 but, in a protein, one oxygen and two hydrogens have been removed in forming each peptide bond. Thus, the molecular weight of the amino acid residue is 57. Therefore, $10,600/57 \approx 186$, which is the degree of polymerization.
2. It would probably dimerize with the two hydrophobic patches in contact.
3. They create a basic shape so that weak bonds can form; in other words, they bring together groups that can form weak bonds but that would not otherwise do so.
4. $M_n = 36,835$; $M_w = 40,283$.
5. Since $M_n = M_w$, the sample is probably homogeneous.
6. $M_w = 15.7 \times 10^6$; $M_n = 21.6 \times 10^6$.
7. One cannot answer this question until it is known whether there is a high- or low-molecular-weight contaminant.
8. The sample is a mixture of monomers and dimers with twice as many dimers as monomers. The monomer molecular weight is 15,000. Thus, $M_n = 25,000$ and $M_w = 27,000$. Boiling converts all of the molecules to monomers.
9. $R_{G, sphere} = 14.8$ Å. For a random coil we can calculate R_G from h; the calculated value is $(36 \times 10^{-8})/\sqrt{6} = 14.7$ Å. These values agree; so the molecule is probably spherical.
10. $R_{G, sphere} = 16.95$ Å; $R_{G, molecule}$ is 3.81 times as large, so that the molecule is somewhat extended, perhaps ellipsoidal. It is not rodlike, because $R_{G, obs}/R_{G, sphere}$ would be 10 or more for a rod.
11. The center of mass is at the center of the square. Each sphere is $\sqrt{50}$ cm from the center of mass. Thus

$$R_G^2 = \frac{2(5)(\sqrt{50})^2 + 2(1)(\sqrt{50})^2}{1 + 5 + 1 + 5} = 50.$$

Thus, $R_G = \sqrt{50} = 7.07$ cm.
12. No, because the molecule would behave like a somewhat rigid, long loop, and the end-to-end distance would be the distance between the ends of the loop.
13. (a) For a random coil, it would increase, because the increased number of collisions with the solvent molecules would increase the average size of the molecule.
(b) If the rod were rigid, there would be no change.
14. In 1 M NaCl the molecule becomes flexible, so that h and R_G decrease. In 0.01 M NaCl the flexibility is nearly the same, so that the decrease in h and R_G is very slight.
15. It is determined by the thermal stability of the more stable of the two interactions.
16. A, because the two hydrogen bonds are adjacent to one another and therefore

ANSWERS TO PROBLEMS

each enhances the stability of the other.

17. No, because the transition is only half complete.
18. Helicity is lost, so the optical activity should decrease substantially.
19. It should be a denaturant by reacting with amino groups on the bases in transiently broken base pairs, thereby eliminating the possibility of reforming hydrogen bonds. Thus, the DNA should slowly denature without heating.
20. $(1/3) [2(1.3) + 1.7] = 1.433$ g cm^{-3}.
21. (a) The final density is $\frac{1}{2}(1.3 + 1.7) = 1.5$ g cm^{-3}. The volume is equal to $(5 \times 10^7)/(6 \times 10^{23} \times 1.5) = 5.5 \times 10^{-17}$ cm^3, so that the radius is 2.36×10^{-6} cm; $D = kT/6\pi a\eta = 9.08 \times 10^{-8}$.
    (b) Smaller because the asymmetry of the particle will increase friction.
22. Using the method of Example 15-F, the value of D for a sphere would be 1.26×10^{-7}. The actual value is about six times too small. This suggests that the molecule is rather long and thin.
23. The molecule has six subunits. The values of D for the protein and one subunit, if each were spherical, would be 4.45×10^{-7} and 8.1×10^{-7}, respectively. The values of f/f_0 for the subunit and the protein are 1.30 and 1.03, respectively. Thus, the axial ratio of the subunit is 6 and that of the protein is 2, and six subunits could not be arranged end-to-end but could be in a hexagonal array.
24. The velocity $v = s\omega^2 r$. Expressed in radians per second, $\omega = 4189$. Thus $v = (22 \times 10^{-13})(4189)^2(6.0) = 2.32 \times 10^{-4}$ cm sec^{-1}. Thus the distance moved in twenty min or 1200 sec is $(2.32 \times 10^{-4})(1200) = 0.278$ cm.
25. The sphere, because it encounters less friction when moving.
26. The circle, because it will encounter less friction than the linear molecule.
27. (a) No, because two molecules with different molecular weights and different shapes may have the same sedimentation coefficient.
    (b) No, because two molecules having the same molecular weight may have different shapes as, for example, in the case of a linear and a circular DNA molecule.
28. The frictional coefficient f is $m(1 - \bar{v}\rho)/s$, in which m is expressed in grams (that is, $50,000 \times 1.6 \times 10^{-24} = 8 \times 10^{-20}$ g), and $\bar{v} \approx 1/1.35 = 0.74$. Then $f = 9.9 \times 10^{-8}$. If the particle were spherical, $f_0 = 6\pi\eta a$, in which $\eta = 0.01$ poise and a is the radius of the sphere. The volume of the sphere is $8 \times 10^{-20}/1.35 = 5.93 \times 10^{-20}$ cm^3, so $a = 2.42 \times 10^{-7}$ cm. Thus, $f_0 = 4.56 \times 10^{-8}$. Since $f/f_0 = 2.17$, the particle is not spherical and is probably rodlike.
29. $M = 3.6 \times 10^5$.
30. $M = 1.37 \times 10^5$.
31. $M = 2.06 \times 10^4$.
32. Since $\bar{v} \approx 1/$density, then the DNA and lipoproteins are more and less dense, respectively, than the solvent. Thus the DNA molecule will be at the bottom of the tube and the lipoprotein will form a layer floating on the surface of the solvent.
33. The width is proportional to $M^{-1/2}$ and to $\rho^{1/2}$. Therefore, the band width of the tRNA would be $(25 \times 10^6/25 \times 10^3)^{1/2} (1.8/1.7)^{1/2} (1.3) = 42.3$ mm.
34. From the chemical formula for DNA, one can determine that the average base whose molecular weight is 336 has 9.5 carbon atoms, so that its molecular weight increases from 336 to 345. Thus the density becomes $(345.5/336)(1.7) = 1.748$. The separation is $0.048/0.1070 = 0.45$ cm.
35. 3.2×10^6 is the molecular weight for Cs DNA. The DNA used in the light scattering experiment is K DNA whose molecular weight will be equivalent to $[(39 + 330)/(137 + 330)](3.2 \times 10^6) = 2.53 \times 10^6$.
36. There are molecules whose molecular weight is less than that of the bulk of the material, although their density is the same. Thus the band observed is the sum of small but broader band superimposed on the narrower major band.
37. No, because they may have different shapes.
38. From Equation 51, $\log(M_1/M_2)$ is equal to $-a(x_1 - x_2)$, in which x is the distance moved. From the values given, $a = -0.1$. Thus $\log(M/35,400) = 0.1(2.8 - 4.6)$, so

that $M = 23,390$.

39. The 10^{-1} microcuries corresponds to a quantity of protein = $0.1/1500 = 6.67 \times 10^{-5}$ μmol. Since there is only one NH_2 terminus, 6.67×10^{-5} μmol per 12 μg means that the molecular weight is 12 μg/6.67×10^{-5} μmol = 180,000.
40. Lower, because the molecule is less extended after joining the ends.
41. No. The viscosity of compact molecules will increase very slightly, whereas the viscosity of an extended molecule will decrease. The magnitude of the viscosity changes depends on the degree of compaction or extension.
42. The conversion to linearity is a first-order reaction. Since the fraction of fast-moving material represents the remaining circular molecules, a plot of ln(fraction fast) against dose is linear, and the dose yielding the fraction $1/e$ is the dose producing, on the average, one double-strand break per molecule. This is 2000 rad.
43. At the isoelectric point the molecule is uncharged. At other values the OH^- group becomes charged. At any pH all tyrosines have the same charge. The repulsion causes the polypeptide to expand, thus increasing friction and lowering the mobility. In 0.5 M NaCl the charges are neutralized by counterions and there is no repulsion.

CHAPTER 16

1. Equation 7 gives points that are more equally spaced.
2. The enzyme may undergo a change in shape such that its binding site has greater affinity for the substrate. Alternatively, the binding may be inhibited by charge repulsion between two like charges, one near the binding and one on the substrate. At high ionic strength these are shielded from one another.
3. The protein may undergo a shape change that increases the affinity of the binding site for A. Alternatively, the binding may be a result of electrostatic attraction; decreasing the dielectric constant by the addition of methanol will decrease the shielding and increase the attraction.
4. (a) Since the concentration is constant after fifteen hours, the system has come to equilibrium. There is binding because the concentration on side I is greater than that on side II. According to Equation 27, $K_d = (10^{-3})(6.12 \times 10^{-5})$ divided by $(6.54 \times 10^{-5} - 6.12 \times 10^{-5}) = 1.46 \times 10^{-2}$.
(b) X and Y bind at the same site. Y binds much more tightly than X. Since no X is bound when [Y] = [protein], it suggests that there is only one binding site for X. There is not enough data to say whether there are additional binding sites for Y to which X cannot bind.
(c) Z does not bind to the protein because [Z] is the same on both sides of the membrane. The binding of X is markedly decreased. Z certainly inhibits the binding of X. Since it does not bind to the protein, Z must act directly on X. One possible explanation is that Z binds to X and alters X, so that X binds less strongly. Another possibility is an electrostatic effect on the system. For instance, if the protein-X binding is the result of an attraction between positive and negative charges and Z is strongly dipolar, it would produce Coulombic shielding, as would high ionic strength.
5. Cytochrome c binds to DNA in 0.01 M NaCl but not in 0.3 M NaCl.
6. Two.
7. Binding is irreversible.
8. The x intercept would be constant but the slope would change.
9. (a) The x intercept is four and the curve is linear. Thus there are four independent binding sites. This suggests (but, of course, does not prove) that the enzyme contains four identical subunits. Other subunits that do not bind S might also be part of the enzyme.
(b) The y intercept is $10^{-7} = n/K_d$, so $K_d = 4 \times 10^7$.
10. For a monoprotic acid a straight line will result whose slope is $-K$. For a triprotic

acid the x intercept will be three; since the equilibrium constants will rarely be the same, the plot will be curved.
11. If the binding is cooperative, the protein molecules will be clustered as shown in panel (b).
12. The binding is cooperative, with five binding sites; $K_d = 10^{-7}$.
13. (a) Each subunit binds one molecule of X so that the tetramer binds four times as much per molecule as the monomer does.
(b) The sigmoidal shape of the H curve suggests cooperative binding.
(c) A Scatchard plot would show curvature. However a Hill plot would be more informative. If binding is cooperative, a slope of four will be found in the Hill plot.
14. K_d for region I is K_T; K_d for region II is K_R.

CHAPTER 17

1. (a) The frequency is $2989 \times c = (2989)(3 \times 10^{10}) = 8.97 \times 10^{13}$ sec^{-1}. The wavelength is $1/2989 = 33{,}456$ Å.
(b) In the infrared.
2.

ΔE	λ
$E_1 - E_0$	497
$E_2 - E_1$	662
$E_3 - E_2$	994
$E_2 - E_0$	284
$E_3 - E_1$	398

All values are in nanometers.
3. At very high temperature, the NaCl is converted to atomic sodium and atomic chlorine by intermolecular collisions. Many of the Na atoms receive sufficient energy in the collisions that they are raised to an excited state. The atoms return to the ground state by emission of light. The flame appears yellow because one of the most intense lines of the Na spectrum is in the range the eye perceives as yellow.
4. No, it is too small. The mass of a two-ton car is 1.816×10^6 g. The velocity is 1341 cm sec^{-1}. Thus $\lambda = h/mv = 2.7 \times 10^{-36}$ cm.
5. The energy of the 660-nm photon is hc/λ $= 3.01 \times 10^{-12}$ erg. Since this is the minimal energy that can cause photoelectron production, this must be the value of the work function. Thus, the energy of the electron produced by a 450-nm photon is $hc/\lambda - 3.01 \times 10^{-12} = 1.4 \times 10^{-12}$ erg.
6. Infinitely far from the nucleus.
7. The average kinetic energy is kT or 4×10^{-14} erg at room temperature. A typical electronic transition in the red region of the spectrum requires 3×10^{-12} erg per atom. Even though the kinetic energy is Boltzmann-distributed, it would be very rare that 3×10^{-12} erg could be transferred in a collision between two atoms.
8. $\Delta E = -\mathcal{R}[(1/n_1^2) - (1/n_2^2)] = -\mathcal{R}[(1/25) - (1/81)] = 6.028 \times 10^{-13}$ erg, so that $\lambda = hc/\Delta E = 3297$ nm.
9. Since Δn can have any integral value, there are fifteen possible transitions.
10. After one electron is removed from a helium atom, the ionized He$^+$ is just a one-electron atom whose nuclear charge is $+2$ rather than the $+1$ charge of the the hydrogen nucleus. Referring to the Bohr theory or the quantum mechanical analysis of the hydrogen atom, one finds that the energy associated with electronic transitions is proportional to the square of the nuclear charge. Thus the ionization energy for the second electron of helium is $2^2 = $ four times that of hydrogen, or 54.32 eV.
11. As the temperature increases, the kinetic energy of the atom increases. Thus, during interatomic collisions, energy is transferred. At any particular temperature, the Boltzmann distribution tells us the fraction of the population of atoms that possess various amounts of energy. As the temperature increases, more and more atoms are in the electronic states having greater energy—that is, the average value of the quantum number n increases. Thus, absorption occurs with greater frequency in states for which $n > 1$, and the higher-order series spectra become more intense.
12. Atoms and atomic spacing are generally smaller than 5 Å. Thus the wavelength of

the radiation should not be greater than that value and, thus, must be in the X-ray region. This is the reason for the development of the powerful technique of X-ray diffraction.

13. The Raman spectrum of the amide group suggests that the overall three-dimensional configuration of the main polypeptide chain determined by X-ray diffraction is that which exists in solution. However, one or more tyrosines must be in a slightly different arrangement when the protein is in solution than when it is dry, as the X-ray diffraction of the dry material shows.

14. The absorption-maximum of tyrosine will be at a longer wavelength than the maximum for isoleucine. Also, the absorption per mole will be greater, but this is not something you would know from the information given in this chapter.

15. The transition will be for $\Delta J = +1$ from $J = 0, 1, 2, 3$. Thus we must calculate the energies of the first four levels by using Equation 38. These are 0, 1.112×10^{-15}, 3.335×10^{-15}, and 6.671×10^{-15} erg, respectively. The values of ΔE are 1.112×10^{-15}, 2.223×10^{-15}, and 3.336×10^{-15} erg respectively, so that the frequencies are 1.678×10^{11}, 3.356×10^{11}, and 5.034×10^{11} sec^{-1}, respectively.

16. Use Equations 41 and 42. The value of I is 1.24×10^{-38} g cm^2 and $r = 2.36 \times 10^{-8}$ cm.

17. No, because changes in rotational levels are not accompanied by a change in dipole moment, which is zero for helium.

18. The frequency bands that correspond to the rotational spectra of N_2 and O_2 are fairly narrow, so you need only select a range of frequencies for your generator that would not overlap these absorption spectra. The noble gases (argon, krypton, and so forth) are of no concern, because they do not produce rotational spectra.

19. The number of moments of inertia determines the number of rotational bands. A linear polyatomic molecule has two modes of rotation, but both modes have the same amount of inertia, and there is only one rotational band. A nonlinear molecule has three modes of rotation, each with its own value of I (some of which might be equal), so there are at most three rotational bands. One can learn quite a lot about the three-dimensional structure of a molecule from determination of the values of I when there is more than one value.

20.

All modes are active because in each there is a change in dipole moment.

21. For HCl, $\bar{\nu} = 2886$ cm^{-1} corresponds to $\nu = 8.66 \times 10^{13}$ sec^{-1}. The reduced mass is 1.62×10^{-24} g. Thus, $K = 4\pi^2\nu^2\mu = 4.78 \times 10^5$ dyn cm^{-1}. For CO, $\nu = 6.51 \times 10^{13}$ sec^{-1}. The reduced mass is 1.14×10^{-23} g. Thus $K = 1.90 \times 10^6$ dyn cm^{-1}. Hence the atoms of CO are bonded more tightly.

22. Yes, because the reduced mass would change. Since the force constant would be unchanged, the frequency will change.

23. According to the harmonic oscillator model, which is a good model for vibrational spectroscopy, the frequency of the absorbed radiation is proportional to $(K/m)^{1/2}$, in which K is the force constant and m is the mass of the vibrating atom. Thus, changing m without changing K alters the frequency. Since deuterium has the same chemical properties as hydrogen, the values of K for D and H are the same. The values of the reduced mass μ of O and H and of O and D differ and, in fact, $(\mu_{O-D}/\mu_{O-H})^{1/2} = 1.38$, so that $\nu_{O-D}/\nu_{O-H} = 1.38$. Thus the water absorption band shifts from 2.94×10^4 to 4.06×10^4 Å, or, using the more conventional notation for infrared spectroscopy, from 3400 cm^{-1} to 2464 cm^{-1}.

24. II extends to higher wavelengths and must be the fluorescent spectrum.

25. (a) A-(proline)$_8$-B, because A and B are

nearer.

(b) A-(alanine)$_7$-B, because this molecule is more flexible than A-(proline)$_7$-B, and A and B will be nearer.

26. (a) The tryptophan could be internal, in a crevice too narrow for the iodide to enter, or adjacent to negatively charged amino acids.

(b) All tryptophans need not have the same fluorescent intensity because of different environments. Some tryptophans may be accessible to the solvent but near a cluster of negative charges that repel the iodide. Some tryptophans may be in a narrow crevice in the molecule.

27. The tryptophans may already be quenched. The ligand may remove one quenching factor in the protein yet introduce another. Binding of the ligand may produce a conformational change in the entire molecule, so that quenching of the fluorescence of the tryptophans in the binding site is counteracted by enhancement of the fluorescence of tryptophans that are elsewhere in the molecule and were partially quenched before ligand binding.

28. No. A quencher removes excitation energy before fluorescence can occur. It does not alter the energy of an emitted photon.

CHAPTER 18

1. 2537 Å = 2537 × 10^{-8} cm. The frequency = c/λ = $(3 \times 10^{10})/(2.537 \times 10^{-5})$ = 1.18×10^{15} cycles per second.

2. 543.8 nm.

3. 2500 nm = 2.5×10^{-4} cm. Thus ν equals $(3 \times 10^{10})/(2.5 \times 10^{-4})$ = 1.2×10^{14} cycles per second, and $\bar{\nu}$ = $1/(2.5 \times 10^{-4})$ = 4000 cm^{-1}.

4. The one with the shorter wavelength always has the greater energy.

5. Far UV, 196 nm; near UV, 3550 Å; visible, 540 nm; infrared, 28,000 Å.

6. OD = 1.8.

7. $\log(I/I_0)$ = -0.5. Thus I/I_0 = 0.32; 32 percent is transmitted.

8. (a) A_{260} = 0.62/5 = 0.124.

(b) ϵ = A/0.5 = 0.248.

9. (0.25/1)(0.52) = 0.13.

10. Absolutely nothing.

11. The Beer-Lambert Law is not obeyed above a molarity of 0.5.

12. (1) The molecules aggregate at 1 M in water but do not aggregate in alcohol. (2) As the molecules in aqueous solution come very close to one another, there is a spectral shift, so that the molar absorption coefficient is reduced. This does not occur in alcohol.

13. The final concentration of X and Y are 0.4 M and 0.15 M, respectively. The A_{260} value at these concentrations are (0.4)(0.82) = 0.328 and (0.3)(1.3) = 0.39. Therefore, the total A_{260} is 0.328 + 0.39 = 0.718.

14. Use Equation 11: [X] = $2.47 \times 10^{-4} M$; [Y] = $8.04 \times 10^{-5} M$.

15. Referring to Equation 11, divide the equation for [X] = DNA by that for [Y] = [phenol], and set [X]/[Y] = 100. Designate by r the ratio A_{260}/A_{280} and substitute A_{260} = $r(A_{280})$. This yields r = 1.72. When $r >$ 1.72, [phenol]/[DNA] is less than 1 percent.

16. The slope of the part of the curve due to scattering is $(0.085 - 0.062)$ divided by $(550^4 - 600^4)$ = -6.037×10^{-13}, which also equals $(x - 0.062)/(350^4 - 600^4)$, so that x = 0.13. Therefore the true A_{350} = $0.64 - 0.13$ = 0.51, and the concentration is $0.51/4520$ = $1.13 \times 10^{-4} M$.

17. The molecule must be able to absorb the shorter wavelength.

18. (a) The sample is thicker, so that the deeper regions of the volume are receiving less light.

(b) Irradiate longer or increase the volume without changing the solution depth (that is, increase the surface area).

19. The rates for each reaction are: (a) $d[HX]/dt = I_{abs}$, in which I_{abs} is the intensity of the absorbed light. (b) $d[HX]/dt = k_2[H][HX]$. (c) $d[X_2]/dt = I_{abs} + k_2[H][HX]$. Since H is made in the first step and consumed in the second step, the steady-state equation yields $d[H]/dt = 0 = I_{abs} - k_2[H][HX]$. Combining the equa-

tions for $-d[HX]/dt$ with the steady-state equation yields $-d[HX]/dt = 2I_{abs}$. Since by definition $Q = -d[HX]/dt/I_{abs}$, then $Q = 2$.

20. $E = N_0 hc/\lambda = 4.59 \times 10^{12}$ erg = 459 kJ.
21. One einstein of light of 2537 Å is 469 kJ. Therefore H_2 but not O_2 can be dissociated in this way.
22. Ninety percent of the light is absorbed, or, the energy absorbed is $(0.9)(5850)(60) = 3.16 \times 10^5$ J. The energy of one einstein of this light is 4.59×10^5 J, so that the number of einsteins absorbed is equal to $(3.16 \times 10^5)/(4.59 \times 10^5) = 0.688$. The quantum yield is then $0.05/0.688 = 0.073$.
23. (a) For ninety minutes N/N_0 is equal to $\exp[-(.693)(24)/1.5] = 1.5 \times 10^{-5}$;
    (b) For 11 days $N/N_0 = \exp[-(0.693)/(11)] = 0.94$;
    (c) For five years, the value of N/N_0 is $\exp[-(.693)(1)/(5)(365)] = 0.9996$.
24. The half-life of ^{32}P is 14.2 days or 20,448 minutes. Thus

    $$N/N_0 = \exp[-(0.693/\ 20{,}448)]$$
    $$= 0.99996611,$$

    so that 0.00003389 of the material decays in the first minute. One microgram of ^{32}P is $10^{-6}/32 = 3.125 \times 10^{-8}$ moles = 1.875×10^{16} atoms, so that $(1.875 \times 10^{16})(0.00003389) = 6.3544 \times 10^{11}$ decays per minute or $(6.3544 \times 10^{11})/(3.7 \times 10^{10})(60) = 0.286$ curies.
25. (a) ^{14}N;
    (b) ^{12}C;
    (c) ^{1}H;
    (d) ^{234}Th;
    (e) ^{35}Cl.
26. (a) Age in years = $(5730/\ln 2)\ln(^{14}C$ content of live object/present ^{14}C content) = $8267 \ln(14.5/\text{present }^{14}C \text{ content})$.
    (b) $8267 \ln(14.5/5.2) = 8478$ years.
27. (a) We use the equation $N = N_0 e^{-\lambda t}$. We substitute $N = I$. Since in general some D is present before the measurement begins, $N_0 = I + D$. Thus $I = (I + D)e^{-\lambda t}$, which is rearranged to $t = -(1/\lambda)\ln(1 + D/I)$.
    (b) The assumption is made that all D is derived from I.
    (c) $t = 1.14 \times 10^9$ years.

28. (a) Since the energy of the beam is an average, then according to the Maxwell-Boltzmann distribution, some electrons will have a much lower energy and $-dE/dx$ will be much higher for these electrons. Furthermore, the energy loss is also Boltzmann-distributed, and occasionally a sufficiently large amount of energy will be given up that bond breakage can occur. This is the reason for the ion clusters described in the text.
    (b) In solution, indirect effects, such as formation of H_2O_2 can occur, and such substances can attack the protein.
29. You do not know what to do without additional measurement. If the energy is too high, the beam can traverse the tumor without depositing energy. On the other hand, if the energy is too low, it may be totally absorbed by healthy tissue before reaching the tumor.
30. (a) 12,500 rad/min.
    (b) At the concentration of $FeSO_4$ used, 120,000 rad convert *all* of the $FeSO_4$ to $Fe_2(SO_4)_3$.
31. 3.26×10^{16} quanta per second.
32. 0.005 mole represents $(0.005)(6 \times 10^{23}) = 3 \times 10^{21}$ molecules. The sample absorbs $(0.75)(3.15 \times 10^{16})(60)(60)(10) = 8.5 \times 10^{20}$ quanta in ten hours. Thus Q equals $(8.25 \times 10^{20})/(3 \times 10^{21}) = 0.28$.
33. Assuming that the Beer-Lambert Law is obeyed, one can calculate the percent transmission after one hour. However, one must know the rate at which the percent transmission changes to determine the total amount of energy absorbed in one hour.
34. The number of joule per photon for the wavelength is $hc/\lambda = 5.6 \times 10^{-19}$, so that the number of photons absorbed is $8.25/5.6 \times 10^{-19} = 1.47 \times 10^{19}$. The number of molecules decomposed equals $(6.2 \times 10^{-5})(6 \times 10^{23}) = 3.72 \times 10^{19}$. The quantum yield is $(1.47 \times 10^{19})/(3.72 \times 10^{19}) = 0.40$.
35. 155 disintegrations per second.
36. No, because the mass *must* decrease.
37. 500 rad.
38. 3.14×10^5 J mol^{-1} = $\mathcal{N}hc/\lambda$. Thus $\lambda = 378$

nm.
39. Adenine absorbs ultraviolet light and was simply serving as a screen.
40. No, because diffusion will be slower than in aqueous solution, and many of the reactive chemicals will lose activity before finding a DNA molecule.
41. The cells have clumped during the centrifugation. A clump can have many dead cells and still yield a colony. These clumps comprise the resistant fraction. The clumps gradually dissociate; the next morning all clumps will have dissociated.
42. (b)
43. (a) 60 percent. (b) 36×10^6.

CHAPTER 19

1. $(\frac{1}{2}, \frac{1}{2}, 0), (\frac{1}{2}, \frac{1}{2}, 1), (1, \frac{1}{2}, \frac{1}{2}), (0, \frac{1}{2}, \frac{1}{2}), (\frac{1}{2}, 1, \frac{1}{2})$, and $(\frac{1}{2}, 0, \frac{1}{2})$.
2. (a) 111.
   (b) 326.
   (c) 122.
   (d) $3\bar{2}\bar{2}$.
3. (a) 100.
   (b) 200.
   (c) 110.
   (d) 220.
4. Use Equation 2, substituting $h = 3$, $k = 2$, and $l = 2$. The spacing $d = 1.61$ Å.
5. Using the Bragg equation, d_{min} will occur at $(\sin \theta)_{max}$, which theoretically is 1. Thus, $d_{min} = 0.708/2 = 0.354$ Å. Actually, the lines corresponding to this limit are not usually observable, because the planes contain relatively few scattering centers and hence the lines are very faint.
6. Direct the heterochromatic beam against any face of the NaCl crystal and use radiation reflected at any angle. This radiation will be monochromatic, for it will have arisen by reflection of (usually) a unique wavelength from a unique plane. It is possible, of course, that two different wavelengths reflecting from different planes might be reflected at the same angle.

7. The molecular weight of KF is 58.1. From the crystal density the molar volume is $58.1/2.48 = 23.4$. The molecular volume is 23.4 divided by Avogadro's number, or 3.89×10^{-23} cm^3. The NaCl structure is shown in the text (analysis of Table 19-3) to contain four molecules per unit cell. Therefore, the volume of the unit cell (which is a cube of side a) is $4 \times 3.89 \times 10^{-23} = 1.55 \times 10^{-22}$ cm^3, so $a = 5.38$ Å.
8. The volume of the unit cell is $(4.59)^2(2.96) = 62.36$ Å^3 $= 62.36 \times 10^{-24}$ cm^3. The molecular weight is 79.9, so the mass per cell is $(2 \times 79.9)/(6.025 \times 10^{23}) = 2.65 \times 10^{-22}$ g. Thus, the density is $(2.65 \times 10^{-22})/(62.36 \times 10^{-24}) = 4.25$ g cm^{-3}.
9. There are, on the average, eight carbon atoms per unit cell or a mass of 1.59×10^{-22} g. From the density the length of the side of the cubic unit cell is 3.57 Å. Refer to the figure and draw a triangle that consists of three atoms—one at a corner, one in the center of a face, and the third at the center of a tetrahedron. The length of the side of the triangle on the face of the unit cell is one-half the length of the diagonal, or 2.52 Å. This side of the triangle is opposite the tetrahedral angle of 109°. The other angles each have a value of $(\frac{1}{2})(180 - 109)$ degrees. Two C—C bonds form the other sides of the triangle. Use the law of sines to determine the length of one of the sides. The value is 1.54 Å.
10. (a) The molecular weight of iron is 55.85. The average number of atoms per unit cell is 2 for the body-centered cubic (bcc) cell and 4 for the face-centered cubic (fcc) cell; thus, the values of the mass per unit cell are 1.85×10^{-22} and 3.7×10^{-22} g, the volumes of the unit cells are 24.39×10^{-24} and 49.83×10^{-24} cm^3, and thus, the densities are 7.59 and 7.42 g cm^{-3} for bcc and fcc, respectively.

    (b) One would expect pressure to induce a transition from the less dense to the more dense form—that is, fcc to bcc.
11. (a) For the 100 plane, $d = 4.20$ Å. Using the Bragg equation with $n = 1, 2, 3, 4,$

and 5 yields 10.56°, 21.50°, 33.36°, 47.16°, and 66.44°, respectively.

(b) For the 110 plane, $d = 2^{1/2} \times 4.20 = 5.94$ Å. Thus, the angles are 7.45°, 15.03°, 22.89°, 31.23°, and 40.40°, respectively.

12. The volume of the unit cell is 2.371×10^{-19} cm^3. Since there are, on the average, 8 molecules per unit cell, the volume per molecule is 2.964×10^{-20} cm^3. The density of the crystal is 1.242 but, correcting for the water, yields a "protein density" in the crystal of $0.644 \times 1.242 = 0.800$ g cm^{-3}. Thus, the mass per molecule is the volume per molecule times the protein density times Avogadro's number or 14,279—2.7 percent higher than the true value. This agreement is fairly good, but one might worry that the dimensions of the unit cell are slightly in error. However, usually these values are quite accurate and the small difference in the observed and expected values is an error in the determination of the fraction of the weight of the crystal that is protein. It is quite difficult to remove all of the bound water in a protein crystal, so, in fact, the fraction of the weight of the crystal that is protein is probably slightly less than 64.4 percent.

13. Body-centered cubic.

14. At the higher temperature the crystal probably expanded slightly, as is the case for most solids. This would cause the spacing between the component molecules to increase. With a different value of d, the diffraction angles would have to change in accord with the Bragg equation.

INDEX

Absolute entropy, 248–250
Absolute zero, definition, 9
Absorbance, 667
  scattering correction for, 676–677
Absorption of light, 665–673
Absorption spectra, 608–609
Acetone, molecular rotation of, 630–631
Acid, strong
  strong, 277
  weak, 277–284
Acridine orange, 659–660, 687–689
Actinometry, 679–680
Activation energy, 184–186, 380
Active transport, 477, 482–485
Activity, 164, 221, 229–232, 763–764
  calculation of, 231
  definition, 229–230
  equilibrium constant and, 232
Activity coefficient, 230–235, 253–256, 258–259
  concentration and, 259
  electrolytes of, 258

Activity coefficient *(Continued)*
  ionic strength dependence of, 258
  measurement of, 231–232, 253–256, 340–341
  units, interconversion of, 234–235
Adenosine triphosphate, 186–188, 485
Adiabatic, definition, 90
Adiabatic expansion, 121–122
Adsorption, 400, 423, 439–456, 463, 474
  from solution, 479–482
  of gas to solid, 443–446, 448
  of H_2 to platinum, 445
  of polymers, 453
Adsorption chromatography, 455
Adsorption coefficient, 424
Adsorption isotherm, 400, 443–451
Ag-AgCl electrode, 334, 342, 345
Agarose, 542
Alanine, synthesis of, 189
Allosteric activation, 588–589
Allosteric inhibition, 401

Allosteric protein, 584
Allostery, 584–594
α decay, 690–691
α helix, 501–504
  infra-red spectra, 655
  ultraviolet spectra, 649–650
α particle, 690–691, 699, 703
Alternating copolymer, 495
Amino acids, 75, 301–304
  formulas of, 496–499
  titration curve of, 302–304
  ultraviolet spectra, 647–653
Ammonium sulfate
  fractionation, 273–274
Amphipathic molecules, 426
Anesthesia, 221
anion, definition, 246
Apoenzyme, 582
Apparent dissociation constant, 279
Area of molecules, measurement of, 430–431
Arrhenius equation, 185
Atomic spectra, 605–611
ATP, 186–188
Attractive forces, 41, 44, 50, 52
Avogadro's Law, 8
Axial ratio, 535–536

801

INDEX

Bacteria, 487, 669–670
  determination of concentration of, 669–670
Bacteriophage
  inactivation by X rays, 713–714
  osmotic rupture of, 216
Barometric formula, 31
Base stacking, 73, 510
Beer-Lambert Law, 665–673
Bending, 669
BET isotherm, 446–448
β decay, 690–691
β spectra, 691
β structure, 503–504, 650, 655
  infrared spectra of, 655
  ultraviolet spectra of, 650
Bicarbonate buffer, 298–299
Binding
  cooperative, 578–600
  free energy of, 173, 414–415
  measurement of, 595–600
Binding energy, 172, 414–415
Binding sites, 571–601
  multiple, 574–606
  single, 573–574
  tight, 574–575
  two different, 581–582
  weak, 575–578
Biochemical standard state, 175–176
Blood, gases in, 215–216
Bohr effect, 593–595
Bohr radius, 608
Bohr theory of the atom, 605–611
Boiling point, 62–64, 170, 202–203
  elevation, 202–203
  hydrogen bonds and, 64
  van der Waals force and, 62–63
Boltzmann distribution, 20, 473–474
Bond length, 109–110
Boundary conditions, 616
Boyle's Law, 6–8, 19
Bragg equation, 734, 738
Branched polymer, 501
Bravais lattices, 725–726

Brønsted-Lowry theory, 278
Brownian motion, 36–38
Buffering capacity, 288–292
Buffers, 263, 287–292, 297–301
  capacity of, 278–292
  ionic strength, effect of, 260, 290–291
  temperature, effect of, 300–301

Calomel electrode, 335–336, 343, 475
Calorie, 95
Calorimetry, 111–114
Capillary rise, 421–422
Carbon dating, 393
Carbon dioxide, vibration of, 23, 24, 632
Carbonyl bend, 628
Carbonyl bond, vibration of, 630–631
Carnot cycle, 125–135
Carriers, 482, 486–489
Catalysis, 393–415
  covalent, 404
  electrostatic, 403
  enzymatic, entropic advantage of, 410–415
  heterogeneous, 398–402
  homogeneous, 402–415
  platinum, 398
  poisoning of, 401
Catenane, 502
Cation, definition of, 244
Chain-breaking step, 378
Channels, 482, 487, 491–494
Charged membranes, 488–494
Charles's Law, 10
Chelating agent, 316
Chemical bonds
  breakage by light, 678
  breakage by ionizing radiation, 705
  breakage by transmutation, 695–697
  types, 66–70
Chemical potential
  definition, 194

Chemical potential (Continued)
  electrolyte of, 252
  standard, 195
Chemisorption, 398, 441–443, 446
Chlorophyll, 685
Cholesteric, 743–745
Chromatography, 455–456
Circular polymer, 501–502
Cofactor, 401, 570, 581
Cohesive force, 199, 228, 419, 439, 441
Coil—helix transition, 522–530
Collagen, 501, 509, 511
Colligative properties, 201–210, 246–248, 505, 549
  ionic solutions, 246–248
  measurement of molecular weight by, 211–212
Collisions, 25–28
Competing reactions, 368–369
Competitive inhibition, 401, 410–412
Compton recoil, 703–704
Concentration cell, 341–345
Condensation, 42
Conductivity, 457–458
Consecutive reactions, 370
Constants, fundamental, 5
Contact potential, 474–476
Contour length, 507
Conversion factors, 5
Cooperative binding, 578–600
  negative, 583
  positive, 583
Cooperative transitions, 513–517
Cooperativity, significance of, 582–583
Coordination compounds, 265, 312–317
Coordination number, 313
Copolymer, 495
$c_o t$ analysis, 524–530
Coulomb's Law, 245
Counterions, 177–179, 471
Coupled reactions, 177–179
Coupled transport, 483
Covalent catalysis, 404

INDEX

Covalent circle, 502
Critical micelle concentration, 458, 670
Critical pressure, 42, 46
Critical temperature, 42, 46
Critical volume, 42, 46
Crystallography, X-ray, 733–741
Crystals, 723–741
  basis, 724
  covalent, 730, 732
  ionic, 730–731
  lattice of, 724–727
  molecular, 732–733
  unit cells of, 725–727
CsCl, uses of, in centrifugation, 553
Curie, 694
Cyclic processes, 123
Cytochromes, 351

D_2O, conversion to H_2O, 36
Dalton's Law, 12
  kinetic theory of, 19
Debye-Hückel limiting law, 257, 769
Debye-Hückel theory, 256–265, 766–769
Debye length, 261
Definite integrals, 754
Degree of dissociation, 243–248, 283
Degree of freedom, 225
Denaturation
  of DNA, 271, 513–517
  of protein, 308–312
  surface, 520–521
Density gradient, 552–556
Derivatives, 750–753
Detergents, 460–462
Diamond, 732
Diffuse layer, 472
Diffusion, 40, 477, 531–537, 706–709, 717
  facilitated, 384
Diffusion coefficient, 385–386, 532–537, 549–550

Diffusion-controlled reactions, 384–387
Dipole moment, 48–52, 410, 557, 674
  oscillation of, 632
Direct effect, 706
Dissociation constant
  acid and base, 283
  apparent, 279
  K_d, 572
DNA
  adsorption of, 454
  base stacking, 73–74, 515–517
  breathing of, 517–518
  damage by light, 685–686, 688
  damage by ^{32}P-decay, 696–697
  damage by ^{3}H-decay, 697
  denaturation of, 271, 516, 652
  denatured, structure of, 522–524
  effect of ionic strength on, 266–267
  effect of X rays on, 713–717
  electron microscopy of, 438–440
  hydration, 271
  hydrogen bonds in, 655–657
  intercalating agents, effect of, 687–689
  isolation, 221
  melting curves, 654
  radius of gyration, 508
  relation between structure and function, 509–510
  renaturation of, 517–530
  repair of, 687–688
  sodium trifluoroacetate and, 518–519
  ultraviolet spectrum of, 651–656
Domains, protein, 523
Donnan effect, 478–482, 596
Dose-response curves, 707
Dosimetry, 705
Double-reciprocal plot, 570, 578, 580

Eadie-Hofstee plot, 408–409
EDTA, 315–316
Efficiency of an engine, 127
Effusion, 24–27
Einstein, definition, 677–678
Einstein equation, 37
Einstein-Smoluchowski equation, 37–38
Einstein-Stark Law, 684–686
Electrical conductivity, 248
Electrical double layer, 471–474
Electrical potential, 479, 485
  across surface, 469
Electrochemical cell, 321–340
  in series, 327
Electrochemical potential, 479, 485
Electrochemical reactions, 321–340
Electrodes, 323–325
  inert, 325
Electrolytes
  activity of, 250
  activity coefficient of, 256–257
  chemical potential of, 252
  definition, 243
  strong, 248
  weak, 248
Electromotive force, 323
Electron beam, 692, 698
Electron microscopy, 28, 564–565, 611
Electronegativity, 65–69
Electrophoresis, 540–546
Electrostatic catalysis, 403
Elementary reaction, 355
Emission spectra, 608–609
End-group labeling, 561–562
End-to-end distance, 506–508
Endothermic, 97
Energy barrier, 185
Energy loss by radiation, 700–704
Energy transfer, fluorescent, 644–646
Engine, 123
Enthalpy
  definition, 91–92

Enthalpy *(Continued)*
  standard, 97
  temperature dependence of, 111
Enthalpy of activation, 383
Enthalpy of formation, 104
Enthalpy of fusion, 98
Enthalpy of reaction, 103, 108–109
Enthalpy of transition, 99
Enthalpy of vaporization, 98
Entropy, 129–147, 258–260, 339–340, 384
  absolute, 144–146
  irreversibility and, 133
  measurement with electrochemical cells, 339–340
  probability and, 135
  residual, 146–147
Entropy change
  Carnot cycle, 133
  evaluation of, for various systems, 140–144
  isothermal expansion, 132
  standard, 147–148
  universe, of, 137–140
Entropy-controlled processes, 157
Entropy of activation, 384
Entropy of formation, 138
Entropy of solution, 149
Entropy of vaporization, 66, 147
Equation of state, 83
Equilibrium, conditions for, 159, 163, 170, 397
Equilibrium constant activity and, 232
  Debye-Hückel theory and, 262–263
  ΔG and, 170–174
Equilibrium dialysis, 595–597
Equipartition of energy, 21–24
Ethidium bromide, 659–660, 688–689
Evaporation retardation, 439
Excluded volume, 58, 227, 231–232, 505

Exothermic, 96
Eyring equation, 383

Facilitated diffusion, 384
Fast reactions, 372
Fatty acids, 425–426
Feedback inhibition, 401
Fick's Laws of diffusion, 532–533
Film, 432–440, 470, 487
  charged, 470
  determination of molecular weight with, 432–433
  effect of ionic environment, 435–436
  protein, 437–439
  surface, 73, 486
Film balance, 429
First Law of thermodynamics, 82–91, 156, 363–367
  reversible reactions and, 366–367
  statement of, 81, 90
First-order reactions, 361–362
Flash photolysis, 682
Flotation, 462–464
Flow system, 373–374
Fluid mosaic model, 461
Fluor, 641
Fluorescence, 639–646, 657–660, 677
  energy transfer and, 644–646
  ethidium bromide of, 659–660
  spectroscopy, 657–660
  tryptophan, of, 657–658
Fluorescent lifetime, 642
Force-area curve, 431, 434, 436
Force constant, 629
Formation constant, 313
Franck-Condon principle, 636–637
Free energy, 157–190
  additivity of, 162
  concentration, 164–167
  definition, 157

Free energy *(Continued)*
  standard, 162
Free energy of activation, 397
Free energy of biological process, 177–190
Free energy of dilution, 167
Free energy of hydrophobic interactions, 183–184
Free energy of solution, 180–181
Free expansion, entropy of, 136–137
Free radicals, 379
Freezing point depression, 204–206, 211–212, 246
  determination of molecular weight from, 211–212
Frequency, 663
Freundlich isotherm, 449–451
Frictional force, 534–537
Frictional ratio, 535–536
Fugacity, 236–239

G, definition, 157
ΔG
  concentration and, 164–167
  pressure and, 169–170
  temperature and, 167–169
$\Delta G°$, standard electrode potential and, 335–337
Galvanic cell, 323–330
  temperature dependence of emf, 328
γ ray, 691, 703
Gas, ideal, *See* Ideal gas
Gas, real. *See* Real gas
Gas, solubility of, 216–218
Gas chromatography, 455–456
Gas constant, value of, 5, 11
Gay-Lussac's Law, 9
Gel electrophoresis, 542, 545–546
Gibbs absorption isotherm, 425
Gibbs free energy, 156–190
Glass electrode, 342–344
Glycine, synthesis of, 107

INDEX

Glycogen, 501
Good buffers, 299
Gouy-Chapman theory, 472–474
Gouy layer, 472, 488, 490
Graham's Law of diffusion, 25
Graphite, 732
Ground state, 608
Group transfer potential, 186–188
G value for radiation damage, 705

H^+ ion, hydration of, 272
Half-expansion temperature, 435–436
Half life, 693
Half reaction, 321–325
   convention in writing, 322
   definition, 321
Harmonic oscillator, 619–622
HCl, vibrational spectrum of, 633–634
Heat, 87–88
   sign convention for, 88
Heat and work, interchange of, 122
Heat capacity
   definition, 92–93
   measurement of, 94
   temperature dependence of, 111
Heat of combustion, 105–107
Heat of dilution, 101–102
Heat of formation, 105–107
Heat of solution, 101–102
Heat transfer, mechanism of, 120
Helix-coil transition, 513–522, 651–653
Helix destabilizing protein, 518
Helmholtz double layer, 472
Hemoglobin, 182, 593–595
Hemolysis, 215
Henderson-Hasselbach equation, 286–287

Henry's Law, 202–203, 216–217, 229
Hertz, definition, 663
Hess's Law, 99, 103–104, 106, 162
Heterogeneous catalysis, 398–402
High energy compounds, 186–190
High energy phosphate bond, 190
Hill equation, 580
Hill plot, 580–581
Hit theory, 707–715
Holoenzyme, 582
Homogeneous catalysis, 402–415
Homopolymer, 495
Hydration, 243, 246, 268–271, 312
Hydrogen atom, quantum theory of, 622
Hydrogen bond, 63–66, 655–657
   boiling point, relation to, 64
   infrared spectra of, 655–657
Hydrogen-bonded liquid, 66
Hydrogen ion, hydration of, 272
Hydrophobic interaction, 71–74, 150, 183–184, 305, 456–462, 510
   entropy and, 150
Hypertonic, 214–215
Hyperchromicity, 652
Hypochromicity, 652
Hypotonic, 214–215

i factor, 247–248
Ideal gas, 10–11, 18
   molecular definition of, 18
   work on, 128–129, 152–153
Ideal gas law, failure of, 41
Ideal solution, 197, 225
Independent surface action, principle of, 427
Induced polarization, 51

Infrared spectra, 653–657
Inhibition of enzymes, 401–413
Initial rates, 366
Initiation step, 378
Intensity, 664
Interaction energy, 53–55
Intercalating agent, 687–689
Interface, 423
Internal energy, definition, 190
Ion, definition of, 242–243
Ion cloud, 473
Ion cluster, 64, 260–261
Ion concentration, measurement, 341–343
Ion exchange chromatography, 268–269
Ion exchanger, 268
Ion pump, 483
Ionic size effect, 258
Ionic solution, structure of, 260–261
Ionic strength
   buffers and, 290–291
   calculation of, 253
   definition, 252
   effect on activity coefficient, 258
Ionization energy, 607
Ionizing radiation, 689–707
   interaction with matter, 689–697, 699–707
Ionophore, 486–487
Ions
   conductivity and, 457
   hydration of, 268–271
   reactions between, 387–388
Ion-selective electrode, 344–345
Irreversibility, 129–133
Irreversible chemical reactions, 358
Isoelectric point, 302, 541, 671
Isoionic point, 302
Isosbestic point, 668–669, 672
Isotherm, definition, 8
Isothermal expansion, reversible, 123–126, 132–134, 152–153

806 INDEX

Isothermal process, definition, 91
Isotonic, 214–215
Isotopes, 691, 693–694

k_{cat}, 406–410
K_m, 405–406
Kinetic energy, 80, 82
  of a gas, 20
Kinetic independence of particles, 249–250
Kleinschmidt method, 438–440

"L," mnemonic rule of, 331
Lac system, 386–387
Langmuir adsorption isotherm, 400, 444–447, 572
Langmuir film balance, 429
Latent heat, 96
Lecithin, 426
Lewis theory, 312–313
Ligand, 313, 570–596
Light scattering, 556–561, 676–677
Linear polymers, 500–505
Lineweaver-Burk plot, 408–411
Lipid bilayer, 460
Liposome, 460–461
Liquid crystals, 742–748
Liquefaction, 42
Liquid junction potential, 474–476
London forces, 55–57, 61–62, 70
Luciferin, 684
Lungs, partial pressure of gases in, 12
Lyotropic series, 271

Macromolecules, 233–234, 266, 495–565
  circular, 501–502
  diffusion of, 532–537
  dimensions, 506–509
  effect of ionic strength on shape, 266–267

Macromolecules *(Continued)*
  excluded volume, 505
  linear, 500–505
  molecular weight of, 546–552
  native structure, 513
  subunits, 510–512
  tertiary structure of, 502
  volume of, 233–234
Maxwell-Boltzmann distribution, 32–35
Mean free path, 29
Mean ionic activity, 251
Melting, 160, 513
Melting curve, 653–654
Melting point, sharpness of, 226
Melting temperature, 516
Membrane, 226–227, 456–461, 468, 476–493
  charged, 488–493
  conductivity of, 488–493
  definition, 476
Membrane filtration, 588
Membrane potential, 477–481
Metal-ion complexes, 265, 312–318
Micelles, 456, 670–671
Michaelis constant, 397–398
Michaelis-Menten kinetics, 404–413
Miller indices, 727–729, 736–740
Mixing, final temperature of, 93
Mobility, 269–270, 540–541, 562
Molal volume, 196
Molality, definition, 193
Molar absorption coefficient, 667
Molarity, definition, 193
Mole fraction, definition, 193
Molecular length, 432
Molecular rotation, 629–637
Molecular vibration, 637
Molecular weight
  average, 547–549
  determination of, 31–32, 553–562

Molecular weight *(Continued)*
  effect of impurities, 212–214
  end-group labeling, by, 562
  light scattering, by, 556–560
  macromolecules, of, 546–552
  number average, 213, 548–549
  weight average, 213
Molecularity of reaction, 356–360
Moment of inertia, 631–632
Monodispersity, 505
Monolayer, 426–439
  types, 434–436
Multihit curves, 709–710

Native structure, 513
Nematic, 743–745
Nernst equation, 337
Nerve membrane, 492–493
Nicked circle, 502
Noncompetitive inhibition, 412–413
Nonideality, experimental correction for, 232
Nucleic acids
  fluorescent spectra of, 659–660
  infrared spectra of, 655–657
  ultraviolet spectra of, 646–653
  See also DNA
Nucleotide, 72, 653
  absorbance of, 653
Number average molecular weight, 213, 548–549

Oligomer, 495
Operator, in quantum mechanics, 614
Order, reaction, 356–360
Osmotic pressure, 200–216, 457–458
  tonicity and, 214–216
Oxidation, 321
Oxidation-reduction, 332
Oxidative phosphorylation, 484
Ozone layer, 683–684

INDEX

^{32}P-decay, 696–697
Parallel reaction, 368–369
Partial molal volume, 197
Partial pressure, 12, 19–20, 230
Particle in a box, 617–619
Partitioning, 221–224
Passive transport, 482
Path
 definition, 83
 example, 85
Peptide bond, 190, 499, 503
Peptide group, 499, 503, 648–649
 ultraviolet absorbance of, 648–649
Permeability, selective, 477
pH
 definition, 281
 dissociation constant and, 281–282
 measurement of, 341–343, 475–476
 pK and, 286, 295
pH meter, 475–476
Phages, 454, 464, 555
 adsorption of, 454
 density of, 555
Phase diagram, 223
Phase rule, 224–227
Phases, definition of, 224
Phosphate transfer potential, 187
Phosphorescence, 642–643
Phosphoric acid, ionization of, 292–294
Photochemical damage, 685–689
Photochemical mechanisms, 677–685
Photochemical reactions, rates of, 684–685
Photochemistry, 677–688
Photodecomposition, 680, 682
Photodynamic action, 687
Photoelectric absorption, 703
Photoelectric effect, 585, 691–692
Photon, 664
Photoreactivation, 686

Photosensitization, 680–687
Photosynthesis, 682–684
Physisorption, 398, 441–449
pK, definition, 286
Planck's constant, 604–605
Platinum catalyst, 399
Platinum-H$_2$ electrode, 326
Pleated sheet, 494
Pockels point, 430–431, 435
Poisoning of catalysts, 401
Poisson distribution, 757–759
Polar pockets, 403
Polarizability, 50–55, 403, 557, 674
Polonium electrode, 471
Polydispersity, 505
Polymers
 adsorption of, 453
 branched, 501
 circular, 501–502
 losses by adsorption, 453–454
Polynucleotide, 501
 See also DNA; Nucleic acids
Polypeptide, 499
 See also Proteins
Polyprotic acids, 278, 292–297
Potential, electrochemical, 337–341
 concentration dependence of, 338
 temperature dependence of, 339–340
Potential energy, 80
Potential energy diagram, 635–639
Powder diagram, 736–737
Pressure, kinetic theory of, 18–19
Primary kinetic salt effect, 388, 393
Probability, 119–121, 135–139
Propagating step, 378
Protein
 α helices in, 501–504
 β structure in, 503–504
 denaturation of, 521
 density, 308–310
 diffusion coefficient, 535–539
 domains in, 522

Protein (*Continued*)
 effect of detergents on, 462
 fibrous, 502
 fluorescent spectra, 657–658
 globular, 502
 hydrophobic patches in, 512
 hydrophobic pockets in, 305
 infrared spectra of, 653–655
 molecular weight by radiosensitivity, 714–715
 radiation damage in, 700–704
 sedimentation analysis of, 538–539
 separation of, 268
 shape, 74, 535–539
 side-chain interactions in, 502
 solubility of, 273
 structure of newly synthesized, 516
 titration, 304–311
 ultraviolet spectra of, 646–653
 unfolding, 437–438
Protein films, 437–442
 in electron microscopy, 438–440
Pseudo-order reactions, 365
P-V work, 84

Quantum, 604, 664
Quantum mechanics, 612–627
Quantum theory, history of, 604–612
Quantum yield, 642, 678

"R," mnemonic rule of, 323
R, value of, 5, 11
Rad, definition, 705
Radiation
 after-effect, 716–717
 damage to biological systems, 707–715
 measurement of, 704–705
 therapy, 702
Radioactive decay, 293, 692–694

Radiocarbon dating, 393
Radius of gyration, 506–509, 559
Raman spectra, 637–639
Random coil, 502, 508, 650, 655
  infrared spectra of, 655
  ultraviolet spectra of, 650
Raoult's Law, 200–202, 220, 228–233, 505, 761–762
  activity formulation of, 230
  deviations from, 201, 202, 220, 228–229
Rate constant, 357–372
  sign convention of, 357
Rate law, 356–372
Rate-determining step, 358
Reaction coordinate, 394–395, 402–403
Reaction intermediate, 176
Reaction, ionic, 387–388
Reaction mechanism, 355–372
Reaction, order of, 356–360
Reaction rates, 356–386, 678–685
  photochemical reactions, of, 678–685
  temperature dependence of, 379–381
  viscosity and, 385
Reactions, enzymatic, 404–415
Reactive intermediate, 370–372
Real gas, 14, 42–45
Real solution, 225–232
Recoil, 696–697
Red blood cell, 213
Reduced mass, 624, 629
Reduction, definition, 321
Reduction potential of biochemical reactions, 347–349
Reference electrode, 327–335
Relaxation methods, 374–378
Relaxation time, 375–378
Renaturation, 522–530
Rep, definition, 705
Repair, 711
Repressor, 597
Residual entropy, 146–147
Retinal, 684
Reversibility, 131, 143–144
Rhodopsin, 684

Rise of sap in trees, 422
Roentgen, definition, 705
Rotation, molecular, 629–637
Rotational energy levels, 631
Rydberg constant, 603

Salting in, 264
Salting out, 215, 264, 270, 272–274
Salts of weak acids or bases, 284–285
Saturated solution, 180–181, 219–220, 255
Scatchard equation, 577
Scattering, light, 665–670, 673–677
Schrödinger equation, 612–627
Screening, 245
Second Law of thermodynamics, 117–136, 155, 397
  statements of, 121, 127–128, 132, 134, 136
Second-order reactions, 362–364, 367–368
Sedimentation, 537–540, 599–600
  coefficient, 538, 562
  equilibrium, 550–552
Sedimentation-diffusion method, 549–550
Selection rules, 626
Semipermeable membranes, 205
Shielding, 245
SI units, 2–5, 469
Side-chain interactions, 502
Silent carrier, 486
Silver–silver chloride electrode, 334, 342, 345
Single-hit curves, 708
Smectic, 743–745
Smog, 684
Soap bubbles, 460–461
Solubility
  activity and, 254–255
  definition, 273
  factors determining, 219–221
  of a gas, 216–219
  ionic strength and, 263–264
  nonideal, 227–234

Solute
  definition, 194
  distribution between solvents, 221–224
  volatile, 202
Solutions of gases, 216–222
  removal of gases from, 202
Solvation, 246, 510
Solvent, definition, 194
Solvent perturbation method, 651
Specific heat, 95
Spectra, 608
  molecular, 627–629
Spectroscopic ruler, 644
Spontaneity, 119–122, 158
Spreading of films, 428–430
Stacking of bases, 73
Standard electrode potential, 325–328, 335–337, 347
  biochemical, 347
  ΔG and, 335–337
Standard enthalpy change, 175
Standard entropy change, 147–148, 177
Standard free energy change, 175–176
Standard reduction potential, 329–342
  table of, 330
Standard state, 97–98, 161, 179–180
  biochemical, 179–180
State function, 83, 130
  definition, 83
Statvolts, 4
Steady state, definition, 371
Steady-state approximation, 371, 407
Stern layer, 473–474
Stirling approximation, 758
Stirred-flow reactor, 373
Stretching, 629
Strong acid, 277
Subunits, 536–537, 584
Sublimation, 35
Sunset, colors at, 675
Supercoil, 501–502
Supercooled liquid, 220
Superhelix, 501
Surface charge, 478

INDEX

Surface denaturation, 520–521
Surface energy, 420, 427
Surface film, 426–439
Surface potential, 469–471, 480–482
   measurement of, 471
Surface tension, 418–426, 431, 457–458
Surface, work in forming, 87
Surfactant, 425–426
Survival curves, 707–715

Target volume, 713
Temperature jump, 374–377
Terminal oxidation chain, 350–351
Tertiary structure, 502
Thermionic effect, 692, 697
Thermopile, 679
Third Law of thermodynamics, 139–140
Thymine dimer, 686, 688
Titration
   amino acids, of, 302–304
   proteins, of, 304–311
   with formaldehyde, 311
Titration curve, 286–288
T_m, 516, 518–520, 524, 654
   base composition and, 519–520
Total differential, definition, 752
Transition state, 381–384, 388, 397
Transmutation, 694–695
Transport, 482–483
Traube's rule, 452–453
Trigonometric relations, 755–756
Trees, rise of sap in, 422
Tritium decay, 697

Trouton's rule, 98, 147
Tryptophan, 657–658
Tunnel effect, 627
Tyrosine, ultraviolet spectrum of, 648

Ultraviolet spectra, 646–653
   of α helix, 649
   of β structure, 650
   effect of solvent on, 647–648
   random coil, of, 650
Uncompetitive inhibition, 411–412
Unit cell, 725–727, 729, 732
Units, 1, 468–469
Urea, synthesis of, 187
Ussing equation, 483–485

V_{max}, 405
Vacuum evaporation, 28–29
Valinomycin, 486–487
Van der Waals constants, 44–45, 57, 59
Van der Waals forces, 60–63, 68–69, 78, 398, 459, 511, 512
   boiling point and, 62–63
Van der Waals law, 44–48, 331
Van der Waals radius, 47, 58, 60, 70
Van't Hoff equation, 174, 210
Vapor pressure, 35, 200–206, 208–210
   Boltzmann distribution and, 35
Vapor pressure lowering, 203–204
Velocity distribution, 32–35

Vibration
   effect on fluorescence, 645
   energy of, 22
   modes of, 21–22, 621
   molecular, 629–637
Vibrational bands, 633
Vibrational spectra, 633, 636–637
   of oxygen, 657
Viruses, 512
Viscosity, 40
   of cytoplasm, 38
Volume, excluded, 58

Water, structure of, 71, 149
Wave function, 613
Wave number, 664
Wave theory of electron, 610–611
Wavelength, 663
Weak acid, 277–284
Weak bonds, 70
Weight-average molecular weight, 548–549
Weight percent, 195
Work, 126–128, 152–153, 158
   electrical 86, 336
   moving an ion, 257
   relation to heat, 88
   sign convention, 87

X rays, 697–704, 708
   diffraction of, 734
   line spectra of, 698–699

Zero-point energy, 618
Zeroth-order reaction, 360
Zimm plot, 559
Zwitterion, 301–302

BARD COLLEGE LIBRARY
Main
QD 453.2 .F73 1985
Principles of physical chemistry

WITHDRAWN

BARD COLLEGE LIBRARY
Annandale-on-Hudson, N. Y. 12504